W0230419

Advancing Culture of Living with Landslides

Kyoji Sassa · Matjaž Mikoš
Yueping Yin
Editors

Advancing Culture of Living with Landslides

Volume 1 ISDR-ICL Sendai Partnerships 2015–2025

Editors
Kyoji Sassa
International Consortium on Landslides (ICL)
Kyoto
Japan

Matjaž Mikoš
Faculty of Civil and Geodetic Engineering
University of Ljubljana
Ljubljana
Slovenia

Yueping Yin
China Institute of Geo-Environment Monitoring
China Geological Survey
Beijing
China

Associate editors
Mauri McSaveney
GNS Science
Lower Hutt
New Zealand

Eileen McSaveney
GNS Science
Lower Hutt
New Zealand

Khang Dang
International Consortium on Landslides (ICL)
Kyoto
Japan

ISBN 978-3-319-53500-5 ISBN 978-3-319-59469-9 (eBook)
DOI 10.1007/978-3-319-59469-9

Library of Congress Control Number: 2017939909

Hiroshima landslide disasters in August 2014, Hiroshima, Japan (PASCO Corporation—Kokusai Kogyo Co., Ltd. All Rights Reserved)

Printed on acid-free paper

This Springer imprint is published by Springer Nature
The registered company is Springer International Publishing AG
The registered company address is: Gewerbestrasse 11, 6330 Cham, Switzerland

Foreword By Irina Bokova

Every year, disasters induced by natural hazards affect millions of people across the world. The loss of life is tragic, impacting on communities for the long term.

The costs are also economic, as disasters are responsible for estimated annual economic losses of around USD 300 billion. With the rising pressures of climate change, overpopulation, and urbanization, we can expect costs to increase ever more.

We cannot prevent disasters, but we can prepare for them better. This is the importance of the *International Consortium on Landslides*, supported actively by UNESCO, to advance research and build capacities for mitigating the risks of landslides. Led by Prof. Kyoji Sassa, the Consortium has become a success story of international scientific cooperation at a time when this has never been so vital.

This is especially important as the world implements the *2030 Agenda for Sustainable Development* and the Paris Agreement on Climate Change, as well as the *Sendai Framework for Disaster Risk Reduction 2015–2030*—adopted in Sendai, Japan, to assess global progress on disaster risk reduction and set the priority actions.

The International Strategy for *Disaster Risk Reduction—International Consortium on Landslides Sendai Partnerships 2015–2025* is the key outcome relating to landslides from the 3rd World Conference on Disaster Risk Reduction, held in Sendai. On this basis, every member of the *International Consortium of Landslides* is redoubling efforts to understand, foresee, and reduce landslide disaster risk across the world.

Led by the Consortium, the Landslide Forum is a triennial milestone event that brings together scientists, engineers, practitioners, and policy makers from across the world—all working in the area of landslide technology, landslide disaster investigation, and landslide remediation. Meeting in Slovenia, the 4th Landslide Forum will explore the theme, "Landslide Research and Risk Reduction for Advancing Culture of Living with Natural Hazards," focusing on the multidisciplinary implementation of the Sendai Framework to build a global culture of resilient communities.

Against this backdrop, this report includes state-of-the-art research on landslides, integrating knowledge on multiple aspects of such hazards and highlighting good practices and recommendations on reducing risks. Today, more than ever, we need sharper research and stronger scientific cooperation. In this spirit, I thank all of the contributors to this publication and I pledge UNESCO's continuing support to deepening partnerships for innovation and resilience in societies across the world.

January 2017

Irina Bokova
Director-General of UNESCO

Foreword By Robert Glasser

Landslides are a serious geological hazard. Among the host of natural triggers are intense rainfall, flooding, earthquakes or volcanic eruption, and coastal erosion caused by storms that are all too often tied to the El Niño phenomenon. Human triggers including deforestation, irrigation or pipe leakage, and mining spoil piles, or stream and ocean current alteration can also spark landslides.

Landslides occur worldwide but certain regions are particularly susceptible. The UN's Food and Agriculture Organization underlines that steep terrain, vulnerable soils, heavy rainfall, and earthquake activity make large parts of Asia highly susceptible to landslides. Other hotspots include Central, South, and Northwestern America.

Landslides have devastating impact. They can generate tsunamis, for example. They can bring high economic costs, although estimating losses is difficult, particularly so when it comes to indirect losses. The latter are often confused with losses due to earthquakes or flooding.

Globally, landslides cause hundreds of billions of dollars in damages and hundreds of thousands of deaths and injuries each year. In the US alone, it has been estimated that landslides cause in excess of US$1 billion in damages on average per year, though that is considered a conservative figure and the real level could be at least double.

Given this, it is important to understand the science of landslides: why they occur, what factors trigger them, the geology associated with them, and where they are likely to happen.

Geological investigations, good engineering practices, and effective enforcement of land use management regulations can reduce landslide hazards. Early warning systems can also be very effective, with the integration between ground-based and satellite data in landslide mapping essential to identify landslide-prone areas.

Given that human activities can be a contributing factor in causing landslides, there are a host of measures that can help to reduce risks, and losses if they do occur. Methods to avoid or mitigate landslides range from better building codes and standards in engineering of new construction and infrastructure, to better land use and proper planned alteration of drainage patterns, as well as tackling lingering risks on old landslide sites.

Understanding the interrelationships between earth surface processes, ecological systems, and human activities is the key to reducing landslides disaster risks.

The Sendai Framework for Disaster Risk Reduction, a 15-year international agreement adopted in March 2015, calls for more dedicated action on tackling underlying disaster risk drivers. It points to factors such as the consequences of poverty and inequality, climate change and variability, unplanned and rapid urbanization, poor land management, and compounding factors such as demographic change, weak institutional arrangements, and non-risk-informed policies. It also flags a lack of regulation and incentives for private disaster risk reduction investment, complex supply chains, limited availability of technology, and unsustainable uses of natural resources, declining ecosystems, pandemics and epidemics.

The Sendai Framework also calls for better risk-informed sectoral laws and regulations, including those addressing land use and urban planning, building codes, environmental and resource management and health and safety standards, and underlines that they should be updated, where needed, to ensure an adequate focus on disaster risk management.

The UN Office for Disaster Risk Reduction (UNISDR) has an important role in reinforcing a culture of prevention and preparedness in relevant stakeholders. This is done by supporting the development of standards by experts and technical organizations, advocacy initiatives, and the dissemination of disaster risk information, policies, and practices. UNISDR also provides education and training on disaster risk reduction through affiliated organizations, and supports countries, including through national platforms for disaster risk reduction or their equivalent, in the development of national plans and monitoring trends and patterns in disaster risk, loss, and impacts.

The International Consortium on Landslides (ICL) hosts the Sendai Partnerships 2015–2025 for the global promotion of understanding and reducing landslide disaster risk. This is part of 2015–2025, a voluntary commitment made at the Third UN World Conference on Disaster Risk Reduction, held in 2015 in Sendai, Japan, where the international community adopted the Sendai Framework.

The Sendai Partnerships will help to provide practical solutions and tools, education and capacity building, and communication and public outreach to reduce landslides risks. As such, they will contribute to the implementation of the goals and targets of the Sendai Framework, particularly on understanding disaster risks including vulnerability and exposure to integrated landslide-tsunami risk.

The work done by the Sendai Partnerships can be of value to many stakeholders including civil protection, planning, development and transportation authorities, utility managers, agricultural and forest agencies, and the scientific community.

UNISDR fully support the work of the Sendai Partnerships and the community of practice on landslides risks, and welcomes the 4th World Landslide Forum to be held in 2017 in Slovenia, which aims to strengthen intergovernmental networks and the international programme on landslides.

Robert Glasser
Special Representative of the Secretary-General
for Disaster Risk Reduction and head of UNISDR

Preface

The International Consortium on Landslides (ICL) organized the ICL-IPL Conference in Kyoto, Japan in 2013, and discussed and prepared the 2014 Beijing Declaration to be adopted at the World Landslide Forum 3 in Beijing, China in June 2014. ICL wrote the draft of ICL-IPL Sendai Partnerships 2015–2025—Landslide disaster risk reduction for a safer geo-environment to be examined in Sendai, Japan, in March 2015. **The 2004 Beijing Declaration**—Landslide mitigation toward a safer Geo-environment was examined at a high-level panel discussion with the participation of the Director-General of UNESCO, Ms, Irina Bokova and was adopted at the end of WLF3 in Beijing, China, which was held on June 2–6, 2014 (Sassa et al. 2015).

ICL organized the Steering Committee meeting in Kyoto on October 7–9 , 2014, together with the International Forum "Urbanization and Landslide Disaster"—Hiroshima landslide disaster, in August, 2014 and Japan's contribution to the post-2015 framework for Disaster Risk Reduction. This forum was planned as a preparatory meeting of the ICL-IPL Sendai Partnerships Conference on March 11–15, 2015. Key members of ICL, UNESCO, UNISDR, MEXT, and the Cabinet Office and the Ministry of Land, Infrastructure, Transport and Tourism (MLIT), Government of Japan attended and discussed the global collaborative framework contributing to the Third World Conference on Disaster Risk Reduction.

Establishment of the ISDR-ICL Sendai Partnerships 2015–2025

ICL initially proposed a thematic session "Urbanization and Geodisasters" to be considered as part of the Third UN World Conference on Disaster Risk Reduction (WCDRR). This topic was not retained among the topics of the Conference. Thereafter, ICL became a co-organizer of Working Session No. 4 (WS 4) "Underlying Risk Factors" (Priority No. 4 of the Hyogo Framework for Action), together with MLIT, UNESCO and other organizations under the initiative of ISDR. ICL proposed a Sendai Partnership on Landslides to the session. It was changed from the initial proposal of "ICL-IPL Sendai Partnerships 2015–2024—Landslide disaster risk reduction for a safer geo-environment" to the "Sendai Partnerships for the Global Promotion of Understanding Disaster Risk" (Priority 1 of the Sendai Framework for Disaster Risk Reduction 2015–2030) to widen the scope beyond just landslides. However, an opinion was expressed that it was too broad, and the session should focus on specific disasters within the interest of organizers of the Working Session No. 4. It was then changed to the "ISDR-ICL: Sendai Partnerships 2015–2024 for Global Promotion of Understanding and Reducing Landslide, Flood and Tsunami Disaster Risk—Tools for Implementing and Monitoring the Post-2015 Framework for Disaster Risk Reduction and the Sustainable Development Goals." This version was circulated to the expected intergovernmental, international and national organizations on 21 January 2015. However, it was suggested that because this partnership is under the initiative of the International Consortium on Landslides, it is better to focus on landslides. As a result, it was finally returned to only landslides (Sassa 2015; Wahlström 2015)

The revised title of the finally agreed Sendai Partnerships was:

Header: Voluntary commitment to the World Conference on Disaster Risk Reduction, Sendai, Japan, 2015

Title: ISDR-ICL Sendai Partnerships 2015–2025 for global promotion of understanding and reducing landslide disaster risk

Subtitle: *Tools for Implementing and Monitoring the Post-2015 Framework for Disaster Risk Reduction and the Sustainable Development Goals*

Based on this frame for the Sendai Partnerships, ICL members and ICL advisory members discussed its main contents before and during the ICL-IPL Sendai Partnerships Conference on March 11–15, 2015 in Sendai.

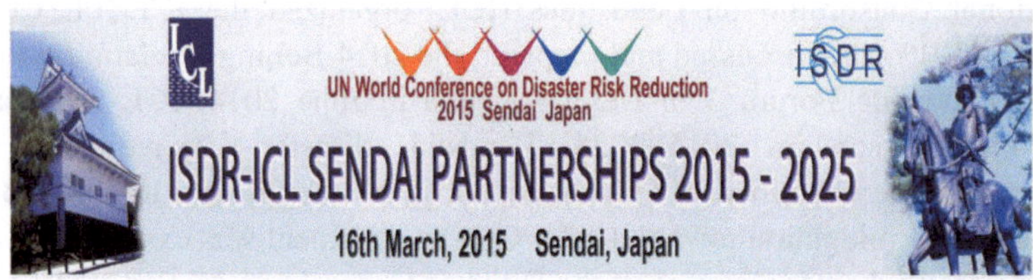

ISDR-ICL Sendai Partnerships 2015–2025 was adopted at a session "Underlying risk factors" of 3rd WCDRR in the morning of 16 March 2015 and it was signed by 16 signatory organizations in the afternoon of the same day in Sendai, Japan. The WCDRR Conference hall was constructed in front of Sendai Castle (*left*) build by Mr. Masamune Date (*right*) in 1601. He sent a mission of 180 people lead by Mr. Tsunenaga Hasekura on a mission to Spain and Rome for international trade and cooperation from 1613 to 1620.

The above is the poster displayed at the preparatory meeting and also the signing ceremony. It has the logos of ICL and ISDR, as well as the logo of the Third UN Conference on Disaster Risk Reduction. The agreed major content is presented below.

We acknowledgee that:

- Landslide disasters are caused by exposure to hazardous motions of soil and rock that threaten vulnerable human settlements in mountains, cities, coasts, and islands.
- Climate change will intensify the risk of landslides in some landslide-prone areas through an increase in the frequency and/or magnitude of heavy rainfall, and shifts in the location and periodicity of heavy rainfall.
- Developments in mountains and coastal areas, including construction of roads and railways and expansion of urban areas due to population shifts, increase exposure to hazards of landslides.
- Although they are not frequent, strong earthquakes have potential to trigger rapid and long-runout landslides and liquefaction. Earthquake-induced coastal or submarine large-scale landslides or megaslides (with depths on the order of hundreds of meters to one thousand meters) in the ocean floor can trigger large tsunami waves. These hazardous motions of soil and water impacting on exposed and vulnerable population can result into very damaging effects.
- The combined effects of triggering factors, including rainfall, earthquakes, and volcanic eruptions, can lead to greater impacts through disastrous landslides such as lahars, debris flows, rock falls, and megaslides.
- Understanding landslide disaster risk requires a multi-hazard approach and a focus on social and institutional vulnerability. The study of social and institutional as well as physical vulnerability is needed to assess the extent and magnitude of landslide disasters and to guide formulation of effective policy responses.
- Human intervention can make a greater impact on exposure and vulnerability through, among other factors, land use and urban planning, building codes, risk assessments, early

warning systems, legal and policy development, integrated research, insurance, and, above all, substantive educational and awareness-raising efforts by relevant stakeholders.
- The understanding of landslide disaster risk, including risk identification, vulnerability assessment, time prediction, and disaster assessment, using the most up-to-date and advanced knowledge, is a challenging task. The effectiveness of landslide disaster risk reduction measures depends on scientific and technological developments for understanding disaster risk (natural hazards or events and social vulnerability), political "buy-in", and on increased public awareness and education.
- At a higher level, social and financial investment is vital for understanding and reducing landslide disaster risk, in particular social and institutional vulnerability through coordination of policies, planning, research, capacity development, and the production of publications and tools that are accessible, available free of charge and are easy to use for everyone in both developing and developed countries.

We agree on the following initial fields of cooperation in research and capacity building, coupled with social and financial investment:

- Development of people-centered early warning technology for landslides with increased precision and reliable prediction both in time and location, especially in a changing climate context.
- Development of hazard and vulnerability mapping, vulnerability and risk assessment with increased precision and reliability, as part of multi-hazard risk identification and management.
- Development of improved technologies for monitoring, testing, analyzing, simulating, and effective early warning for landslides.
- Development of international teaching tools that are always updated and may be used free of charge by national and local leaders and practitioners, in developed and developing countries through the Sendai Partnerships 2015–2025.
- Open communication with society through integrated research, capacity building, knowledge transfer, awareness-raising, training, and educational activities to enable societies to develop effective policies and strategies for reducing landslide disaster risk, to strengthen their capacities for preventing hazards from developing into major disasters, and to enhance the effectiveness and efficiency of relief programs.
- Development of new initiatives to study research frontiers in understanding landslide disaster risk, such as the effect of climate change on large-scale landslides and debris flows, the effective prediction of localized rainfall to provide earlier warning and evacuation, especially in developing countries, the mechanism and dynamics of submarine landslides during earthquakes that may cause or enhance tsunamis, and geotechnical studies of catastrophic megaslides for prediction and hazard assessment.

All of the items above came from the discussions of ICL and its partners. Within those items, one of the most discussed parts are the effects of climate change on landslides. It was mentioned in two places as a high priority. Climate changes are studied by meteorologists, but not studied by landslide scientists and engineers. It is not easy to prove the effects of climate change on landslides in a decisive and a quantitative way. However, all agreed, in the following sentence, to add "in some landslide prone areas."

Climate change will intensify the risk of landslides in some landslide-prone areas through an increase in the frequency and/or magnitude of heavy rainfall, and shifts in the location and periodicity of heavy rainfall.

Figure 1 presents a comparison of extreme rainfalls in Japan of over 100 mm/day and extreme rainfalls in Vietnam of over 51 mm/day. One is the number of days and another is the number of events, and the monitoring period is different. However, both present an increasing

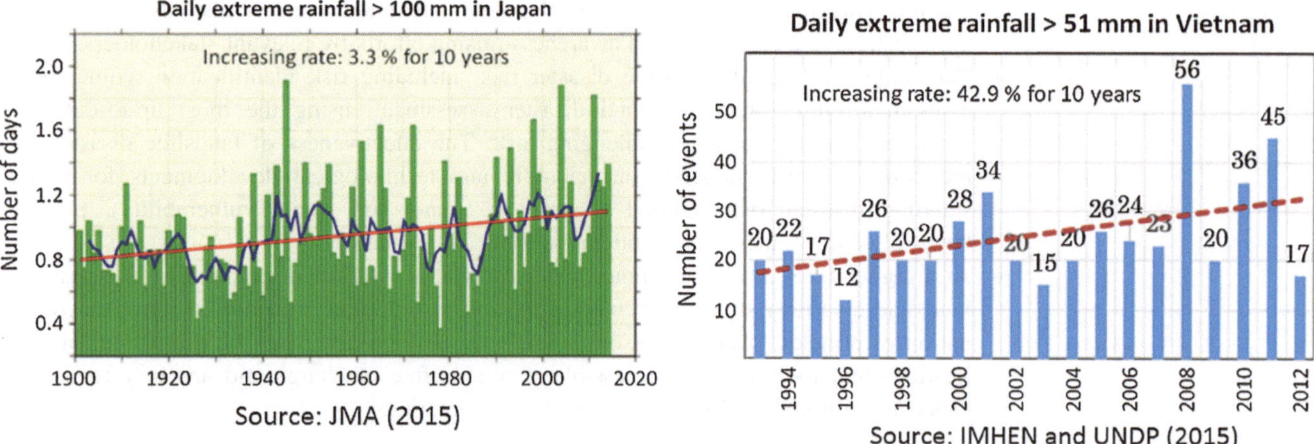

Fig. 1 Comparison of extreme rainfalls in Japan of over 100 mm/day and extreme rainfalls in Vietnam of over 51 mm/day. *Source 1* Japanese Meteorological Agency (JMA) (2015) Climate Change Monitoring Report 2014. *Source 2* IMHEN (Institute of Meteorology, Hydrology and Climate Change, Vietnam) and UNDP (United Nations Development Program) (2015) Viet Nam Special Report on Managing the Risks of Extreme Events and Disasters to Advance Climate Change Adaptation

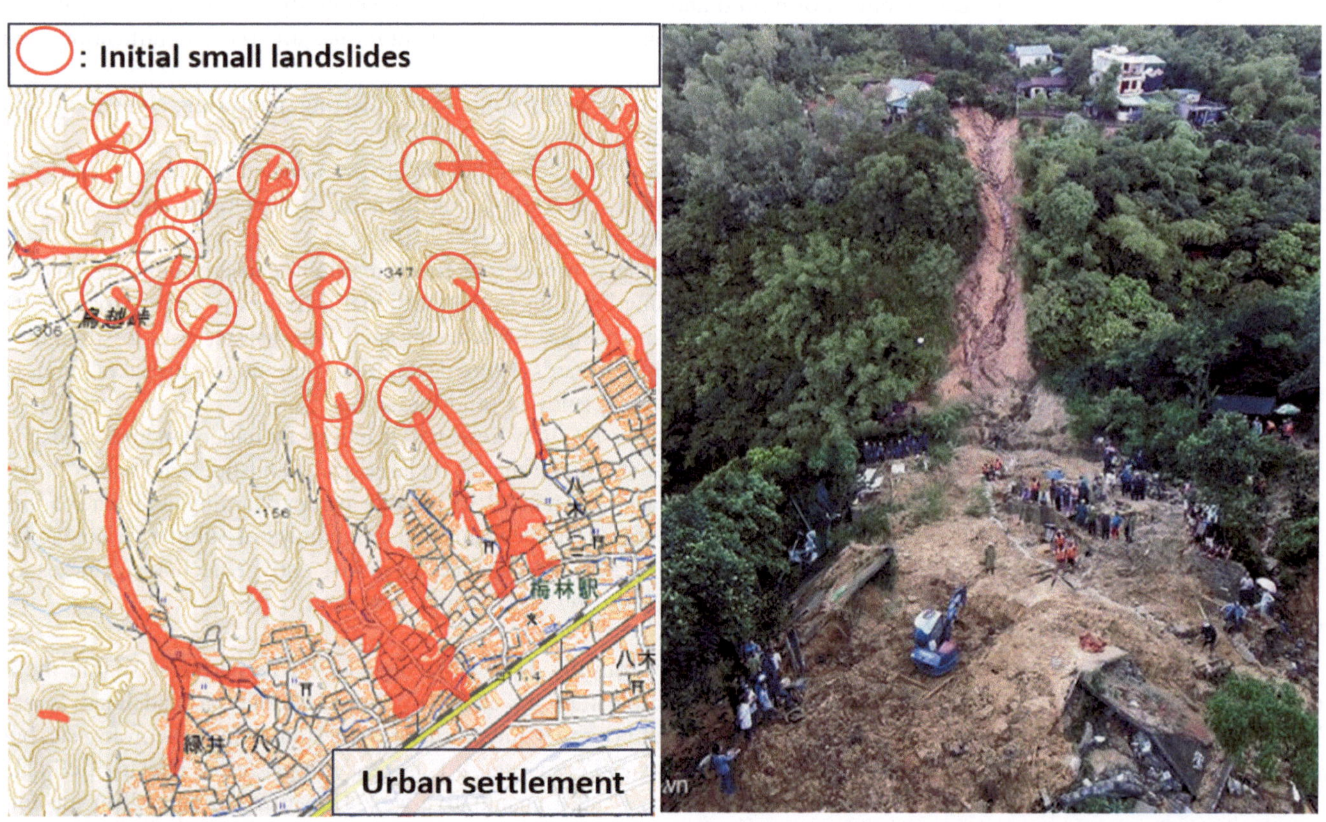

Fig. 2 Two examples of small-scale landslides due to extreme rainfall that destroyed human settlements at the toe of slopes in Japan and in Vietnam. Source of Landslide distribution: Geospatial Information Authority of Japan (GSI) Red circles showing the locations of the initial small landslides are added by K. Sassa. Source of UAV photo of Ha Long landslide: Vietnamese news company "Zing. vn"

trend of frequency of extreme rainfall, and the rate in Vietnam is one order higher than that in Japan. Climate change effects will be different in countries or regions.

Figure 2 presents recent typical landslide disasters in Japan and Vietnam in 2014 and in 2015. Local heavy rainfall (the maximum rainfall was 121 mm/h, and 217 mm for 3 h) deluged urban areas of Hiroshima city. Many small-scale shallow landslides occurred in the mountains (shown by red circles in the left figure). As they flowed down they increased in volume, and destroyed urban settlements and killed 66 persons in this area. A shallow landslide triggered by heavy rainfall (the maximum rainfall was 87 mm/h, 277 mm for 5 h) in Ha Long city and the landslide debris destroyed three houses and killed eight people as shown in the right figure.

Global Promotion of Understanding and Reducing Landslide Disaster Risk

The ICL and signatory organizations of the Sendai Partnerships 2015–2025 wish to voluntarily commit to the Sendai framework for disaster risk reduction 2015–2030 and the United Nations Sustainable Development Goals No.11 *"Make cities and human settlements inclusive, safe, resilient and sustainable."*

The successful result of each item proposed in the Sendai Partnership cannot be achieved without close cooperation with the signatory organizations and other related organizations. In order to plan the milestones of the Sendai Partnership, ICL and Sendai Partnership groups are organizing a High-Level Panel Discussion on May 30, 2017 as a Plenary session of the Fourth World Landslide Forum in Ljubljana, Slovenia:

High-Level Panel Discussion "Strengthening Intergovernmental Network and the International Programme on Landslides (IPL) for *"ISDR-ICL SENDAI PARTNERSHIPS 2015–2025 for global promotion of understanding and reducing landslide disaster risk*"

Objectives: The International Programme on Landslides (IPL) is a programme of the International Consortium on Landslides (ICL). ICL proposed the IPL in a thematic session of the Second World Conference on Disaster Reduction (WCDR) in Kobe, 2005. The activities of IPL were defined in the 2006 Tokyo Action Plan "Strengthening research and learning on landslides and related earth system disasters for global risk preparedness" at the Round Table Discussion held in Tokyo, 2006. IPL was supported by seven global stakeholders—UNESCO, WMO, FAO, UNISDR, UNU, ICSU, and WFEO, and ICL exchanged Memorandums of Understanding to promote the Tokyo Action Plan with each of them in 2006.

The activities of IPL include the triennial organization of the World Landslide Forum, the implementation of various IPL Projects, identification of World Centres of Excellence on Landslide Risk Reduction (WCoE), and the publication of the ICL bimonthly journal Landslides. Based on this background, ICL proposed the ISDR-ICL Sendai Partnerships 2015–2025 for global promotion of understanding and reducing landslide disaster risk at the 3rd WCDRR in Sendai, 2015 which was accepted and signed by 17 global and national stakeholders, including the governments of Croatia, Italy and Japan. This high-level panel discussion aims to strengthen networking with governments in landslide-prone countries and governments supporting landslide disaster risk reduction efforts in developing countries. The close cooperation within governments, United Nation Organizations and International NGOs is necessary and effective to implement the ISDR-ICL Sendai Partnerships 2015–2025 and the International Programme on Landslides (IPL) in its infrastructure.

Following the discussion result in the high-level panel discussion, a **Round Table Discussion on the follow-up of the high-level panel discussion and implementation planning** will be held at 13:30–17:30 on May 31, 2017 at Club CD as a parallel session. Representatives of 17 signatory organizations, ICL-IPL members, and potential new members of the Sendai Partnerships are invited. All participants will examine an action plan/road map/Addendum to the Partnerships to implement and further develop the Sendai Partnerships, effectively contributing to the SENDAI Framework for Disaster Risk Reduction. At the end of the session, signing to the Sendai Partnerships by new members may be organized. To strengthen the Sendai Partnerships cooperation network, we wish to invite new signatory organizations and also the new members of ICL. Those organizations are invited to the high-level panel discussion and also the round-table discussion.

The WLF4 will publish five volumes of books. Volumes 2–5 are the proceedings of technical papers presented at the forum. Volume 1 includes **Part 1 ISDR-ICL Sendai Partnerships**: (1) three forum lectures (Rupestrian world heritage sites at landslide risk, Subaerial landslide-generated (tsunami) waves, Rock fall occurrence and fragmentation) to present leading landslide issues, (2) Contribution of signatory organizations to provide basic information for the high-level panel discussion, (3) a planning initiative from ICL to Sendai Partnerships to create "Landslide Dynamics-ISDR-ICL Landslide Interactive Teaching Tools (LITT)" to broaden the availability of landslide technologies for landslide risk reduction for capacity development, and to examine the initial stage of "ICL World Report on Landslides" to share landslide information and technologies within WRL contributors and users. **Part 2 International Programme on Landslides (IPL)** is a programme of ICL contributing to ISDR with support from seven global stakeholders. IPL consists of IPL projects proposed and implemented by ICL member organizations, and the activities of World Centres of Excellence on Landslide Risk Reduction (WCoEs), which are updated at the triannual world landslide forum. **Part 3** includes papers from the Session 3 Landslides and Society.

Vol. 1 Sendai Partnerships 2015–2025 will be published initially as a free online access book and then as a printed book distributed to all participants in the Fourth World Landslide Forum. We would like to ask the readers of this volume to join this voluntary commitment to the Sendai Framework for Disaster Risk Reduction. Such support shall promote and enable the realization of many of the difficult tasks proposed in the Sendai Partnerships.

Call for Cooperation

ICL acknowledges the dedicated support from ICL's many supporting organizations and cooperating individuals for the International Programme on Landslides (IPL) including the editing and publication of an international journal "Landslides: Journal of International Consortium on Landslides," and we request further support for ICL and IPL and the activities of the Sendai Partnerships. Those organizations and individuals are invited to the Fourth World Landslide Forum and the high-level panel discussion on May 30, 2017 and the round-table discussion to follow on May 31, 2017, to join this global initiative 2015–2025. Information on the Fourth World Landslide Forum is uploaded at the WLF4 website: https://www.wlf4.org/. Inquiries and cooperation for the Sendai Partnerships 2015–2025 should be addressed to the ICL Secretariat secretariat@iclhq.org.

References

Sassa K (2015) ISDR-ICL Sendai Partnerships 2015–2025 for global promotion of understanding and reducing landslide disaster risk. Landslides, Vol 12 (4), pp 631–640

Sassa K, Yin Y and Canuti P (2015) The third world landslide forum, Beijing, China. Landslides, Vol 12 (1), pp 177–192

Wahlström M (2015) Preface. Landslides, Vol 12 (4), pp 629

Kyoji Sassa
Executive Director of ICL
Kyoto, Japan

Matjaz Mikos
A Vice President of ICL
Ljubljana, Slovenia

Yueping Yin
President of ICL
Beijing, China

Organizers

International Consortium on Landslides (ICL)

International Programme on Landslides (IPL)

University of Ljubljana

Geological Survey of Slovenia (GeoZS)

Co-organizers

REPUBLIC OF SLOVENIA
MINISTRY OF THE ENVIRONMENT AND SPATIAL PLANNING

Republic of Slovenia Ministry of the Environment and Spatial Planning

REPUBLIC OF SLOVENIA
MINISTRY OF INFRASTRUCTURE

Republic of Slovenia Ministry of Infrastructure

Slovenian National Platform for Disaster Risk Reduction

Slovenian Chamber of Engineers (IZS)

- Društvo Slovenski komite mednarodnega združenja hidrogeologov (SKIAH)—International Association of Hydrogeologists Slovene Committee (SKIAH)
- Društvo vodarjev Slovenije (DVS)—Water Management Society of Slovenia (DVS)
- Geomorfološko društvo Slovenije (GDS)—Geomorphological Association of Slovenia (GDS)
- Inštitut za vode Republike Slovenije (IzVRS)—Institute of Water of the Republic of Slovenia (IzVRS)
- Slovensko geološko društvo (SGD)—Slovenian Geological Society (SGD)
- Slovensko geotehniško društvo (SloGeD)—Slovenian Geotechnical Society (SloGeD)
- Slovenski nacionalni odbor programa IHP UNESCO (SNC IHP)—Slovenian National Committee for IHP (SNC IHP)
- Slovensko združenje za geodezijo in geofiziko (SZGG)—Slovenian Association of Geodesy and Geophysics (SZGG)

Organizing Committee

Honorary Chairpersons

Borut Pahor, President of the Republic of Slovenia*
Irina Bokova, Director General of UNESCO
Robert Glasser, Special Representative of the United Nations Secretary-General for Disaster Risk Reduction*
José Graziano Da Silva, Director General of FAO*
Petteri Talaas, Secretary General of WMO
David Malone, Rector of UNU
Gordon McBean, President of ICSU
Toshimitsu Komatsu, Vice President of WFEO
Roland Oberhaensli, President of IUGS
Alik Ismail-Zadeh, Secretary General of IUGG
Hisayoshi Kato, Director General for Disaster Management, Cabinet Office, Government of Japan
Kanji Matsumuro, Director, Office for Disaster Reduction Research, Ministry of Education, Culture, Sports, Science and Technology, Government of Japan
Fabrizio Curcio, Head, National Civil Protection Department, Italian Presidency of the Council of Ministers, Government of Italy
Jadran Perinic, Director General, National Protection and Research Directorate, Republic of Croatia
Takashi Onishi, President of Science Council of Japan
Juichi Yamagiwa, President of Kyoto University
Ivan Svetlik, Rector of University of Ljubljana, Slovenia
Walter Ammann, President/CEO, Global Risk Forum Davos
*Note: Honorary chairpersons are Leaders of signatory organizations of the ISDR-ICL Sendai Partnerships. * to be confirmed.*

Chairpersons

Matjaž Mikoš, Chairman, Slovenian National Platform for Disaster Risk Reduction
Yueping Yin, President, International Consortium on Landslides
Kyoji Sassa, Executive Director, International Consortium on Landslides

International Scientific Committee

Che Hassandi Abdulah, Public Works Department of Malaysia, Malaysia
Biljana Abolmasov, University of Belgrade, Serbia
Basanta Raj Adhikari, Tribhuvan University, Nepal
Beena Ajmera, California State University, Fullerton, USA
Irasema Alcántara Ayala, Universidad Nacional Autonoma de Mexico, Mexico
Guillermo Avila Alvarez, Universidad Nacional de Colombia, Colombia
Željko Arbanas, University of Rijeka, Croatia
Behzad Ataie-Ashtiani Sharif, University of Technology, Iran
Mateja Jemec Auflič, Geological Survey of Slovenia, Slovenia
Yong Baek, Korea Institute of Civil Engineering and Building Technology, Korea
Lidia Elizabeth Torres Bernhard, Universidad Nacional Autónoma de Honduras, Honduras
Matteo Berti, University of Bologna, Italy

Netra Prakash Bhandary, Ehime University, Japan
He Bin, Chinese Academy of Sciences, China
Peter Bobrowsky, Geological Survey of Canada, Canada
Giovanna Capparelli, University of Calabria, Italy
Raul Carreno, Grudec Ayar, Peru
Nicola Casagli, University of Florence, Italy
Filippo Catani, University of Florence, Italy
Byung-Gon Chae, Korea Institute of Geoscience and Mineral Resources, Korea
Buhm-Soo Chang, Korea Infrastructure Safety and Technology Corporation, Korea
Giovanni Battista Crosta, University of Milano Bicocca, Italy
Sabatino Cuomo, University of Salerno, Italy
A.A. Virajh Dias, Central Engineering Consultancy Bureau, Sri Lanka
Tom Dijkstra, British Geological Survey, UK
Francisco Dourado, University of Rio de Janeiro State, Brasil
Erik Eberhardt, University of British Columbia, Canada
Luis Eveline, Universidad Politécnica de Ingeniería, Honduras
Teuku Faisal Fathani, University of Gadjah Mada, Indonesia
Paolo Frattini, University of Milano Bicocca, Italy
Hiroshi Fukuoka, Niigata University, Japan
Rok Gašparič, Ecetera, Slovenia
Ying Guo, Northeast Forestry University, China
Fausto Guzzetti, National Research Council, Italy
Javier Hervas, ISPRA, Italy/EU
Daisuke Higaki, Japan Landslide Society, Japan
Arne Hodalič, National Geographic Slovenija, Slovenia
Jan Hradecký, University of Ostrava, Czech Republic
Johannes Hübl, University of Natural Resources and Life Sciences, Austria
Oldrich Hungr, University of British Columbia, Canada
Sangjun Im, Korean Society of Forest Engineering, Korea
Michael Jaboyedoff, University of Lausanne, Switzerland
Jernej Jež, Geological Survey of Slovenia, Slovenia
Pavle Kalinić, City of Zagreb, Croatia
Bjørn Kalsnes, Norwegian Geotechnical Institute, Norway
Dwikorita Karnawati, University of Gadjah Mada, Indonesia
Asiri Karunawardana, National Building Research Organization, Sri Lanka
Ralf Katzenbach, Technische Universitaet Darmstadt, Germany
Nguyen Xuan Khang, Institute of Transport Science and Technology, Vietnam
Kyongha Kim, National Institute of Forest Science, Korea
Dalia Kirschbaum, NASA Goddard Space Flight Center, USA
Jan Klimeš, Academy of Sciences of the Czech Republic, Czech Republic
Marko Komac, University of Ljubljana, Slovenia
Kazuo Konagai, University of Tokyo, Japan
Hasan Kulic, Albanian Geological Survey, Albania
Santosh Kumar, National Institute of Disaster Management, India
Simon Loew, ETH Zürich, Switzerland
Jean-Philippe Malet, Université de Strasbourg, France
Claudio Margottini, ISPRA, Italy
Snježana Mihalić Arbanas, University of Zagreb, Croatia
Gabriele Scarascia Mugnozza, University of Rome "La Sapienza", Italy
Chyi-Tyi Lee, National Central University, Chinese Taipei
Liang-Jeng Leu, National Taiwan University, Chinese Taipei
Ko-Fei Liu, National Taiwan University, Chinese Taipei
Janko Logar, University of Ljubljana, Slovenia

Ping Lu, Tongji University, China
Juan Carlos Loaiza, Colombia
Mauri McSaveney, GNS Science, New Zealand
Matjaž Mikoš, University of Ljubljana, Slovenia
Ashaari Mohamad, Public Works Department of Malaysia, Malaysia
Hirotaka Ochiai, Forest and Forest Product Research Institute, Japan
Igwe Ogbonnaya, University of Nigeria, Nigeria
Tomáš Pánek, University of Ostrava, Czech Republic
Mario Parise, National Research Council, Italy
Hyuck-Jin Park, Sejong University, Korea
Cui Peng, Chinese Academy of Sciences, China
Luciano Picarelli, Second University of Naples, Italy
Tomislav Popit, University of Ljubljana, Slovenia
Saowanee Prachansri, Ministry of Agriculture and Cooperatives, Thailand
Boštjan Pulko, University of Ljubljana, Slovenia
Paulus P. Rahardjo Parahyangan Catholic University, Indonesia
Bichit Rattakul Asian Disaster Preparedness Center, Thailand
K.L.S. Sahabandu, Central Engineering Consultancy Bureau, Sri Lanka
Kyoji Sassa, International Consortium on Landslides, Japan
Wei Shan, Northeast Forestry University, China
Z. Shoaei, Soil Conservation and Watershed Management Research Institute, Iran
Mandira Shrestha, International Centre for Integrated Mountain Development, Nepal
Paolo Simonini, University of Padua, Italy
Josef Stemberk, Academy of Sciences of the Czech Republic, Czech Republic
Alexander Strom, JSC "Hydroproject Institute", Russian Federation
S.H. Tabatabaei, Building & Housing Research Center, Iran
Kaoru Takara, Kyoto University, Japan
Dangsheng Tian, Bureau of Land and Resources of Xi'an, China
Binod Tiwari, California State University, Fullerton & Tribhuvan University, USA
Veronica Tofani, University of Florence, Italy
Adrin Tohari, Indonesian Institute of Sciences, Indonesia
Oleksandr M. Trofymchuk, Institute of Telecommunication and Global Information Space, Ukraine
Emil Tsereteli, National Environmental Agency of Georgia, Georgia
Taro Uchimura, University of Tokyo, Japan
Tran Tan Van, Vietnam Institute of Geosciences and Mineral Resources, Vietnam
Timotej Verbovšek, University of Ljubljana, Slovenia
Pasquale Versace, University of Calabria, Italy
Vít Vilímek, Charles University, Czech Republic
Ján Vlčko, Comenius University, Slovak Republic
Kaixi Xue, East China University of Technology, China
Yueping Yin, China Geological Survey, China
Akihiko Wakai, Japan Landslide Society, Japan
Fawu Wang, Shimane University, Japan
Gonghui Wang, Kyoto University, Japan
Huabin Wang, Huazhong University of Science and Technology, China
Janusz Wasowski, National Research Council, Italy
Patrick Wassmer, Université Paris 1, France
Mike Winter, Transport Research Laboratory, UK
Sabid Zekan, University of Tuzla, Bosnia and Herzegovina
Oleg Zerkal, Moscow State University, Russian Federation
Ye-Ming Zhang, China Geological Survey, China

Local Organizing Committee

Biljana Abolmasov, Faculty of Mining and Geology, University of Belgrade, Serbia
Željko Arbanas, Faculty of Civil Engineering, University of Rijeka, Croatia
Miloš Bavec, Geological Survey of Slovenia
Nejc Bezak, Faculty of Civil and Geodetic Engineering, University of Ljubljana
Mitja Brilly, Slovenian National Committee for IHP
Darko But, Administration for Civil Protection and Disaster Relief, Ministry of Defence of the Republic of Slovenia
Lidija Globevnik, Water Management Society of Slovenia
Arne Hodalič, National Geographic Slovenia
Mateja Jemec Auflič, Geological Survey of Slovenia
Jernej Jež, Geological Survey of Slovenia
Vojkan Jovičić, Slovenian Geotechnical Society
Robert Klinc, Faculty of Civil and Geodetic Engineering, University of Ljubljana
Janko Logar, Faculty of Civil and Geodetic Engineering, University of Ljubljana
Matej Maček, Faculty of Civil and Geodetic Engineering, University of Ljubljana
Snježana Mihalić Arbanas, Faculty of Mining, Geology and Petroleum Engineering, University of Zagreb, Croatia
Matjaž Mikoš, Faculty of Civil and Geodetic Engineering, University of Ljubljana
Zlatko Mikulič, International Association of Hydrogeologists Slovene Committee
Gašper Mrak, Faculty of Civil and Geodetic Engineering, University of Ljubljana
Mario Panizza, University of Modena and Reggio Emilia, Italy
Alessandro Pasuto, National Research Council, Padua, Italy
Ana Petkovšek, Faculty of Civil and Geodetic Engineering, University of Ljubljana
Tomislav Popit, Faculty of Natural Sciences and Engineering, University of Ljubljana
Boštjan Pulko, Faculty of Civil and Geodetic Engineering, University of Ljubljana
Jože Rakovec, Slovenian Association of Geodesy and Geophysics
Črtomir Remec, Slovenian Chamber of Engineers
Mauro Soldati, University of Modena and Reggio Emilia, Italy
Timotej Verbovšek, Faculty of Natural Sciences and Engineering, University of Ljubljana
Sabid Zekan, Faculty of Mining, Geology and Civil Engineering, University of Tuzla, Bosnia and Herzegovina

Contents

Part I

ISDR-ICL Sendai Partnerships 2015–2025

The ISDR-ICL Sendai Partnerships 2015–2025: Background and Content

Kyoji Sassa

Abstract

The International Consortium on Landslides proposed the ISDR-ICL Sendai Partnerships 2015–2025 for global promotion of understanding and reducing landslide disaster risk at the session "Underlying risk factors" of the Third United Nations World Conference on Disaster Risk Reduction (WCDRR) in the morning of 16 March 2015, in Sendai, Japan. The proposal was accepted and signed by 16 United Nations, international and national organizations in the afternoon of the same day in a Japanese restaurant "Junsei", Sendai, Japan. This article describes the background and content of the Partnerships including example of major landslide disaster in the world with the full text of the partnerships and the list of signatory organizations.

Keywords

Landslides • International Consortium on Landslides (ICL) • International Strategy for Disaster Risk Reduction (ISDR) • World Conference on Disaster Risk Reduction (WCDRR)

Introduction

Part 1 ISDR-ICL Sendai Partnerships 2015–2025 describes: 1.1 The ISDR-ICL Sendai Partnerships 2015–2025: Background and Content, 1.2 Three selected forum lectures as examples of recent landslide research as the scientific base of the partnerships, 1.3 Contributions from signatory organizations of the Sendai Partnerships as basic information for the high-level panel discussion for Strengthening Intergovernmental Networks and the International Programme on Landslides (IPL) for "ISDR-ICL SENDAI PARTNERSHIPS 2015–2025 for global promotion of understanding and reducing landslide disaster risk", 1.4 One of the contributions from ICL to the Partnership "Landslide Dynamics-ISDR-ICL Landslide Interactive Teaching Tools (LITT)", 1.5 The planned common platform for landslide case reports for the promotion of cooperation.

This chapter presents a visual overview of some landslide disasters around the world to the wider communities that are partly involved in landslide disaster risk reduction, showing first the significance of the Partnerships, then the background of the Partnerships and the full content of the partnerships.

Examples of Landslide Disasters Around the World

"Landslide disasters are caused by exposure to hazardous motions of soil and rock that threaten vulnerable human settlements in mountains, cities, coasts, and islands" (from the Sendai Partnership Resolution). When large-scale landslides have occurred and caused major disasters, they are reported. When small scale-landslides have occurred and caused disasters in urban areas in National capitals or Provincial capitals such as Hiroshima city in Japan and Ha Long city in Vietnam (introduced in the Preface), those are reported. However, small-scale landslides that killed people living in a few houses in rural areas are not always recorded

K. Sassa (✉)
International Consortium on Landslides, Kyoto, Japan
e-mail: sassa@iclhq.org

K. Sassa et al. (eds.), *Advancing Culture of Living with Landslides*,
DOI 10.1007/978-3-319-59469-9_1

in many countries. Both the number and frequency of small-scale landslides are some order of magnitude higher than that of large-scale landslides. To achieve the UN Sustainable Development Goals No. 11 "Make cities and human settlements inclusive, safe, resilient and sustainable", disaster reduction should be fostered by "the development of people-centered early warning technology for landslides with increased precision and reliable prediction both in time and location, especially in a changing climate context" (from the Sendai Partnership Resolution) and by applying it to rural areas as well as urban areas.

Unfortunately, small-scale landslides occur in many places and so frequently that they are neither remarked nor recorded, in contrast with the cases of earthquakes, volcanic eruptions, and typhoons/hurricanes. However, big landslide disasters are reported and may be found in Wikipedia or other sources on the internet.

Definition of landslides have varied around the world. As a voluntary commitment to the International Decade of Natural Disaster Reduction (1990–2000), the landslide-related communities in the International Geotechnical Societies and UNESCO established a working party for the World Landslide Inventory to establish a definition of landslides. The discussed result was published in "Landslide Types and Processes" by David Cruden and David Varnes in Landslides—Investigation and Mitigation, Transportation Research Board, US National Research Council in 1996. In order to disseminate this new definition and classification of landslides, including debris flows, earth flows, rock falls, rock toppling and other types of very slow to very rapid movements of rock, debris or soils, the Landslide Handbook—A Guide to Understanding Landslides was edited as an International Programme on Landslides IPL 106 Best Practice handbook for landslide hazard mitigation (2002–2007), and it was published by U. S. Geological Survey in 2008. This handbook with many illustration and photographs, has been translated and published in Portuguese and Spanish, Japanese, and Chinese. This project received the IPL Award for Success at the 2nd World Landslide Forum at the Food and Agriculture Organization of the United Nations (FAO) headquarters in Rome, Italy.

As a contribution to the Sendai Partnerships, ICL are editing the Landslide Dynamics: ISDR-ICL Landslide Interactive Teaching Tools (LITT) (two volumes of around 1600 pages) for capacity development necessary as a key component of Sendai Partnerships. The revised landslide handbook "Landslide types: Description, illustration and photos" including more illustrations and photos and "Landslide Dynamics for risk reduction" for the assessment of landslide initiation and motion are written and included as the fundamental part of the Landslide Interactive Teaching Tools (LITT), which is introduced in this volume.

Landslide researchers know major landslide disasters, and showing some examples to scientists, engineers, and policy makers who are partly involved to landslide risk reduction efforts is useful. Table 1 presents an outline of major landslide disasters in the world.

Table 1 A list of major landslide disasters around the world

No	Date	Place	Casualties
1	21 May 1792	Nagasaki, Japan	16,000
2	19 May 1919	Kelud, Indonesia	5110
3	16 December 1920	Ningxia, China	>100,000
4	25 August 1933	Sichuan, China	−3100
5	5 July 1938	Kwansai, Japan	−1000
6	13 December 1941	Ancash, Peru	4000–6000
7	10 July 1949	Oblast,Tajikistan	800–4000
8	18 July 1953	Wakayama, Japan	1046
9	26 September 1958	Shizuoka. Japan	1094
10	10 January 1962	Ranrahirca, Peru	4000–5000
11	09 October 1963	Longarone, Italy	= 2000
12	31 May 1970	Yungay, Peru	18,000
13	18 March 1971	Chungar, Peru	400–600
14	13 November 1985	Tolima.Colombia	23,000
15	30 October 1998	Mt. Casita, Nicaragua	2000
16	16 December 1999	Vargas,Venezuela	30,000
17	17 January 2001	El Salvador	500–1700
18	17 February 2006	Leyte, Philippines	1144

(continued)

Table 1 (continued)

No	Date	Place	Casualties
19	9 August 2009	Kaohsiung,Taiwan	500–600
20	8 August 2010	Gansu, China	1287
21	11 January 2011	Rio de Janeiro. Brazil	>1.000
22	16 June 2013	Uttarakhand, India	5700
23	02 May 2014	Badakhshan, Afghanistan	500
24	1 October 2015	El Cambray Dos. Guatemala	220

Examples of Large-Scale Landslides and Their Disasters Around the World

Photos and summary information is presented on several large-scale landslides in which the depth of the sliding surfaces are the order of 10 to 100 meters. Those differ from small-scale shallow landslides, in which the depth of sliding surfaces of the initial landslides are a few meters (as presented in the Preface) (Fig. 1).

The author investigated (1) Mayuyama landslide-tsunami disaster, (7) Las Colinas earthquake-induced landslide, (8) Leyte rainfall + earthquake induced landslide, and (12) Potential landslides in Machu Picchu. The author visited (2) Vajont landslide, (5) Salerno landslides-debris flows, (11) Usoy earthquake-induced landslide and the landslide-dammed Lake Sarez. The author did not visit (3) Huascaran debris avalanche, (4) Nevado del Ruiz debris flow, (6) Vargas debris flow, (9) Uttarakhand landslide-

Fig. 1 **a** Mayuyama Landslide was triggered on the Unzen volcano by a nearby earthquake on 21 May 1792. The landslide mass moved into the Ariake Sea and triggered a tsunami wave. The landslide and the landslide-induced tsunami killed 15,153 people. It was the largest landslide disaster and the largest volcanic disaster in Japan. *Left top* is a Google photo of the landslide. *Left bottom* is the landslide cross-section. *Right bottom* shows the reproduction of the landslide and the resulting tsunami wave, using computer simulations of the landslide and the tsunami (*Sources* Landslides Vol. 11(5) 2014 Vol. 13 (6), 2016) **b** Vajont landslide was triggered by water-level changes in a dam reservoir on 9 October 1963. A large-scale rapid landslide mass entered into the reservoir of the Vajoint dam. The water in the reservoir overflowed over the dam and wiped out a community along the river. *Left photo* shows the landslide mass fill in the dam reservoir. *Right two photos* show the community of Longarone village, Italy before and after the landslide and flood. The village disappeared and around 2000 people were killed. **c** Nevados Huascaran debris avalanche was triggered by an earthquake on 31 May 1970 in Peru. The rapidly moving large-scale landslide mass destroyed the town of Yungay and killed around 18,000 people. *Left photo* is an air photo showing the source of landslide and the debris covering the Yungay town. *Right photos* show that nothing remained of Yungay town. **d** A large-scale landslide-debris flow was triggered by the eruption of Nevado del Ruiz volcano on 13 November 1985 in Columbia. The resulting volcanic debris flow destroyed the town of Armero in Tolima, Columbia, killing 20,000–23,000 people. The map on the *top left* shows the debris flow path and volcanic hazard zones of nearby areas. All *three photos* show the town after the disaster. **e** A group of many small landslides were triggered by heavy rainfall in Salerno, Italy on 5 May 1998. This is a similar type of disaster to the 2014 Hiroshima landslide disaster introduced in the Preface. Initial small and shallow landslides moved debris down to the lower slope and torrents increased the flow in volume, overwhelming the urban settlement. 280 persons were killed in Salerno town. **f** A storm on December 14–16, 1999 struck the State of Vargas along the Caribbean Sea in Venezuela. It triggered thousands of landslides and large-scale debris flows that killed 10,000–30,000 persons. **g** An earthquake triggered a rapid landslide on January 13, 2001 in Las Colinas, El Salvador. The landslide mass travelled through a densely populated urban area and killed 500–1700 people **h** Accumulated rainfall of 674 mm from 8 to 17 February 2006 hit Guinsaugon, Leyte, Philippine. A very small earthquake (Ms 2.6) then occurred on 17 February 2006. This small earthquake after the long rainfall triggered a large-scale rapid landslide in volcano-clastic debris that killed more than 1144 people. *Source* Landslides Vol. 7(3), 2010) **i** Heavy rainfall from 14 to 17 June 2013 struck the Indian state of Uttarakhand. This heavy rainfall caused snow melting of a glacier, triggered landslides and led to floods. The death toll was 5700 people. **j** Mudslides occurred on 2 May 2014 in Badakhshan, Afghanistan. A week before the mudslides, there had been torrential rain. The sliding mass flowed over a settlement and killed around 500 people. *Source* Wikipedia **k** The Usoy landslide is a very large landslide, with a depth of 700 m, which was triggered by an earthquake in 1911. The landslide mass blocked the river, forming a landslide dam lake called Lake Sarez. The water level of this lake has continued to increase and is currently near the top of the landslide dam (height is 567 m). *Top-left* is an air photo of the Usoy landslide dam and Lake Sarez, and *bottom-left* shows a wide area satellite photo. *Lower right* is a ground photo of the landslide dam and the lake. *Top right* is a record of the increasing water level since 1940. This landslide dam is threatened by further gradual increases in water level and also landslide-induced tsunami, which may be triggered by a landslide from the slopes along the shore of the dam lake (*Source* Science Vol. 326, 2009). **l** The Inca World Heritage site at Machu Picchu shows signs of potential landslides. *Left-top* shows the Machu Picchu citadel constructed on the sliding surface of a big landslide between two peaks of Huayna Picchu and Machu Picchu. The sliding surface and another potential sliding surface (*yellow*) are along gently dipping shear bands. Close up photos of the shear bands are in (*b*) and (*c*). *Left-bottom* is the movement record of an extensometer installed in the lower slope of Machu Picchu citadel. *Right figure* shows ground radar investigation along a part of the *red dotted line*. It suggested the Plaza (*flat area*) was formed by filling a crack (it might be a head scarp of the potential retrogressive landslide). *Sources* Proc. 1st World Landslide Forum ▶

(a)

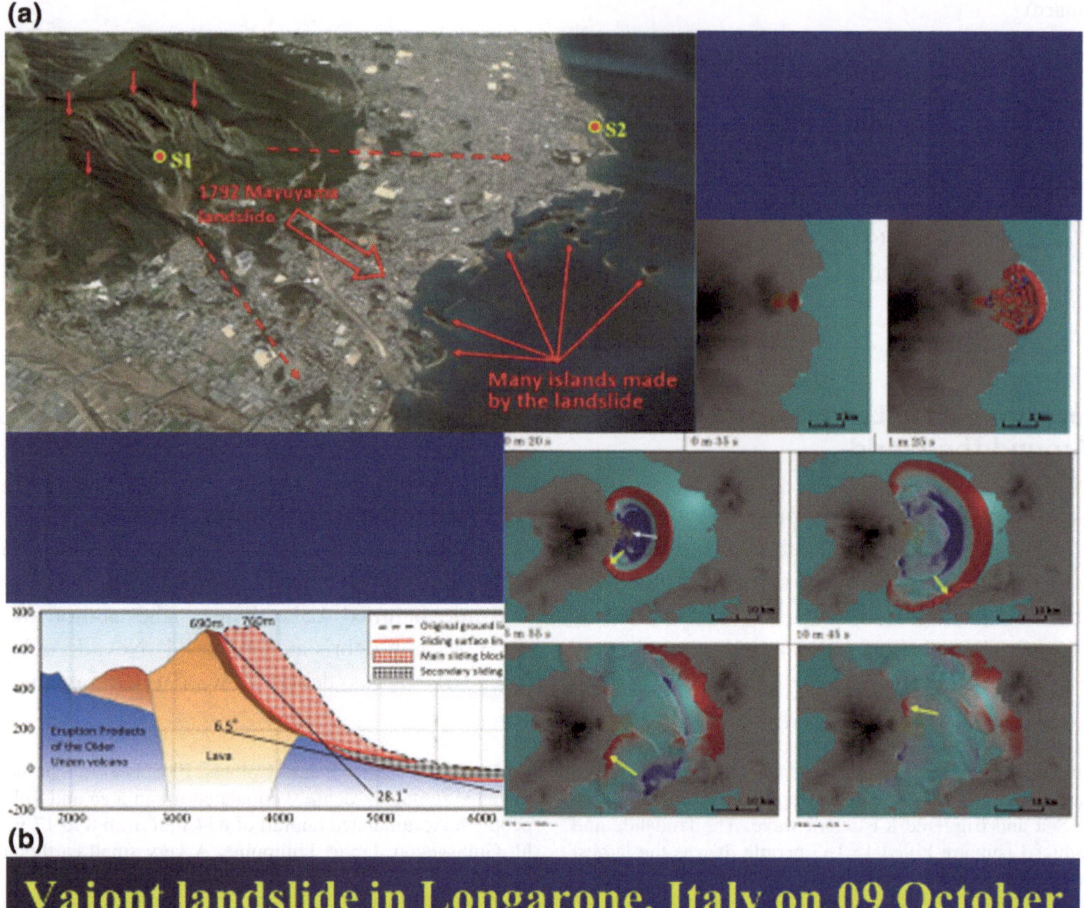
S1
S2
1792 Mayuyama landslide
Many islands made by the landslide
690m
700m
6.5°
28.1°
Eruption Products of the Older Unzen volcano
Lava

(b)

Vajont landslide in Longarone, Italy on 09 October 1963 (Death toll: 2,000)
Before landslide
After landslide
http://landslides.usgs.gov

(c)

(d)

Fig. 1 (continued)

(e)

(f)

Fig. 1 (continued)

(g)

Las Colinas landslide, El Salvador on 17 January 2001
(Death toll: 500 - 1,700)

(h)

Leyte Landslide, Philippines on 17 February 2006
(Death toll: 1,144)

Fig. 1 (continued)

(i)

(j)

Fig. 1 (continued)

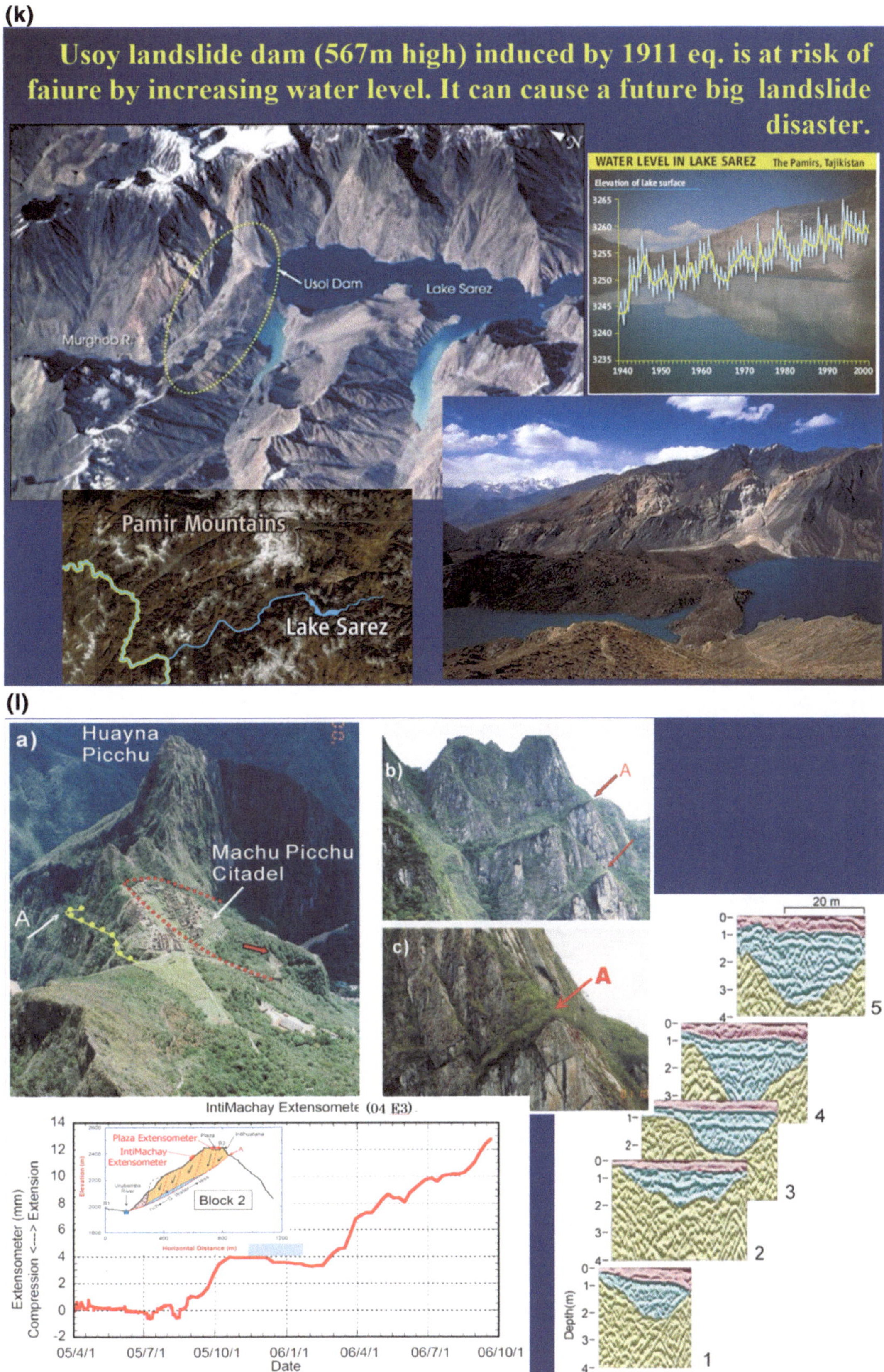

Fig. 1 (continued)

Fig. 2 Photo of the high-level panel who discussed "Initiative to create a safer geoenvironment toward WCDRR 2015 and forward. The high-level panel was chaired by Hans van Ginkel (Former Rector of UNU). UNESCO (Director-General Irina Bokova), UNISDR, WMO, ICSU/IRDR, China Geological Survey, and ICL together discussed aspects from the floor. The 2014 Beijing Declaration "Landslide Risk Mitigation: Toward a Safer Geoenvironment" was adopted on 6 June 2014 following this panel discussion, which was the preparation for the ISDR-ICL Sendai Partnerships 2015–2025. 531 people, 211 national and international organizations from 40 countries and 5 organizations of United Nations System participated in WLF3

Fig. 3 A group photo after the signing ceremony of ISDR-ICL Sendai Partnerships 2015–2025 on 16 March 2015 in Sendai, Japan

debris flows and floods, or (10) Badakhshan mudslides. The cases of (3), (4), (6), (9) are indirect information from Wikipedia (2016.10.3) below and other websites.

https://en.wikipedia.org/wiki/List_of_landslides#1976.E2.80.932000.

Background of the ISDR-ICL Sendai Partnerships 2015 -2025

The very beginning of ISDR-ICL Sendai Partnerships came from the foundation of ICL in January 2002 which UNISDR supported and sent its delegate. The concept of an ICL contribution to the post-2015 Framework for Disaster Risk Reduction started from the 10th anniversary Conference on 17–20 January 2012 in Kyoto, Japan, with financial support from the Japan Science and Technology Agency (JST). Participants reviewed the first decade of ICL and IPL activities and examined the second decade of ICL-IPL activities. As a result, **ICL Strategic plan 2012–2021—To create a safer geoenvironment** was adopted. This conference approved the establishment of four regional networks and five thematic networks of ICL to expand the activities of ICL members and cooperation with non-ICL members in the specific region and themes. ICL organized the ICL-IPL Conference in Kyoto, Japan in 2013 with financial support from JST. At this conference, ICL discussed and prepared the 20014 Beijing Declaration to be adopted in the World Landslide Forum 3 in Beijing, China on 2–6 June 2014. Furthermore ICL examined and drew up the draft of ICL-IPL Sendai Partnerships 2015–2025—Landslide disaster risk reduction for a safer geo-environment to be examined in Sendai, Japan, in March 2015. **The 2004 Beijing Declaration—Landslide mitigation toward a safer Geoenvironment** was examined in the high-level panel discussion with the participation of the Director-General of UNESCO Ms Irina Bokova and was adopted at the end of WLF3 in Beijing, China, which was held on 2–6 June 2014 (Fig. 2).

ICL organized the Steering Committee meeting in Kyoto on 7–9 October 2014, together with the **International Forum "Urbanization and Landslide Disaster"—Hiroshima landslide disaster** in August, 2014 and Japan's contribution to the Post-2015 framework for Disaster Risk Reduction. This forum, together with the ICL Steering Committee meeting, was planned as a preparatory meeting for the ICL-IPL Sendai Partnerships Conference on 11–15 March 2015. Key members of ICL, UNESCO, UNISDR, MEXT, and the Cabinet Office and the Ministry of Land, Infrastructure, Transport and Tourism (MLIT), Government of Japan attended and discussed the global collaborative framework contributing to the Third World Conference on Disaster Risk Reduction.

The ISDR-ICL Sendai Partnerships 2015–2025

The signing ceremony of the ISDR-ICL Sendai Partnerships was organized in a Japanese Restaurant "Junsei" in Sendai, Japan from 12:00–13:30 on 16 March 2015. 16 intergovernmental, international and national organizations signed the Sendai Partnerships. The heads of some organizations attended and signed there, some organizations nominated an officer in-charge of disaster reduction to sign the documents, while some organizations signed it in advance and sent a representative to bring the signed partnerships to this signing ceremony. Following are the organizations which agreed and signed the Sendai partnerships on 16 March 2015. The World Meteorological Organization (WMO) signed the Sendai Partnerships on 15 April 2016. ICL took 6–8 months to obtain the signatures from seven global stakeholders (UNESCO, WMO, FAO, UNISDR, UNU, ICSU and WFEO) for the Letter of Intent aiming to provide a platform for a holistic approach in research and learning on 'Integrated Earth System Risk Analysis and Sustainable Disaster Management' after WCDR in 2005 and also exchange MoUs with each of the same global stakeholders to promote the 2006 Tokyo Action Plan—Strengthening Research and Learning on Landslides and Related Earth System Disasters for Global Risk Preparedness. ICL planned to establish the Sendai Partnerships during the 3rd WCDRR in Sendai, Japan. Figure 3 shows the memorial photo after signing the Sendai Partnerships in a Japanese restaurant "Junsei" in Japan.

The full text of the ISDR-ICL Sendai Partnerships is show below as an appendix of this article. The partnerships are updated due to a change of ICL members and also signatory organizations. The objectives of partnerships will be better realized by new signatory organizations and new ICL members.

Call for Cooperation

ICL wishes to implement and develop this Sendai Partnerships 2015–2025 together with all organizations and individuals. We wish to invite those organization and individuals to join the high-level panel discussion and the round-table discussion on 30–31 May 2017 during the Fourth World Landslide Forum in Ljubljana, Slovenia. The information of the WLF4 is uploaded at https://www.wlf4.org/.

The information of the International Consortium on Landslides can be obtained from http://icl.iplhq.org/category/home-icl/.

Information on the International Programme on Landslides (IPL, a programme of the ICL for ISDR) can be obtained from http://iplhq.org/. All inquiries on ICL, IPL and the Sendai Partnerships should be addressed to ICL Secretariat secretariat@iclhq.org.

Voluntary commitment to the World Conference on Disaster Risk Reduction
Sendai, Japan, 2015

ISDR-ICL SENDAI PARTNERSHIPS 2015-2025 FOR GLOBAL PROMOTION OF UNDERSTANDING AND REDUCING LANDSLIDE DISASTER RISK

***Tools for Implementing and Monitoring** the Post-2015 Framework for Disaster Risk Reduction and the Sustainable Development Goals*

At the 2nd United Nations World Conference on Disaster Reduction, which was held in Kobe, Japan, on 18-22 January 2005, the International Consortium on Landslides (ICL) co-organized a session which resulted in a global partnership and platform taking a holistic approach to research and learning on 'Integrated Earth system risk analysis and sustainable disaster management'. This partnership was forged through a "Letter of Intent", that was signed by UNESCO, UNISDR, WMO, FAO, UNU, ICSU, and WFEO. It further led to the adoption and implementation of the 2006 Tokyo Action Plan, thus creating a global partnership on Landslides, i.e., the current International Programme on Landslides (IPL) of ICL.

At the 3rd World Conference on Disaster Risk Reduction (WCDRR), which was convened by the United Nations and hosted by Japan in Sendai from 14 to 18 March 2015, the ICL and its IPL contributed further to the UN International Strategy for Disaster Reduction (ISDR) and co-organized the Working Session "Underlying Risk Factors" together with UNESCO, the Japanese Ministry of Land, Infrastructure, Transport and Tourism (MLIT) and other pertinent organizations.

At the Working Session, the causes that create risk and their cumulative effects, as well as the relevant achievements of the Hyogo Framework for Action 2005-2015, were reviewed. Steps to address the principal drivers of vulnerability and exposure and to support hazard and risk assessment were suggested. In addition, the participating scientific and academic institutions and governmental and non-governmental organizations proposed that the ***Sendai Partnerships 2015-2025 for Global Promotion of Understanding and Reducing Landslide Disaster Risk*** be established. This sound global platform will be mobilized in the coming decade to pursue prevention, to provide practical solutions, education, communication, and public outreach to reduce landslide disaster risk. These Partnerships will engage all significant stakeholders concerned with the challenge of understanding and reducing disaster risk, including relevant international, national, local, governmental, and non-governmental institutions, programmes and initiatives. The Partnerships will focus on delivering tangible and practical results that are directly related to the implementation of the goals and targets of the post-2015 Framework for Disaster Risk Reduction.

The ***Sendai Partnerships 2015-2025 for Global Promotion of Understanding and Reducing Landslide Disaster Risk*** are hereby established. They represent **Tools for Implementing and Monitoring** the Post-2015 Framework for Disaster Risk Reduction and the Sustainable Development Goals.

Partners in the "Partnerships" adopt the following Resolution:

We acknowledge that:

- ✓ Landslide disasters are caused by exposure to hazardous motions of soil and rock that threaten vulnerable human settlements in mountains, cities, coasts, and islands.

Voluntary commitment to the World Conference on Disaster Risk Reduction Sendai, Japan, 2015

- ✓ Climate change will intensify the risk of landslides in some landslide prone areas through an increase in the frequency and/or magnitude of heavy rainfall, and shifts in the location and periodicity of heavy rainfall.
- ✓ Developments in mountains and coastal areas, including construction of roads and railways and expansion of urban areas due to population shifts, increase exposure to hazards of landslides.
- ✓ Although they are not frequent, strong earthquakes have potential to trigger rapid and long runout landslides and liquefaction. Earthquake-induced coastal or submarine large-scale landslides or megaslides (with depths on the order of hundreds of meters to one thousand meters) in the ocean floor can trigger large tsunami waves. These hazardous motions of soil and water impacting on exposed and vulnerable population can result into very damaging effects.
- ✓ The combined effects of triggering factors, including rainfall, earthquakes, and volcanic eruptions, can lead to greater impacts through disastrous landslides such as lahars, debris flows, rock falls, and megaslides.
- ✓ Understanding landslide disaster risk requires a multi-hazard approach and a focus on social and institutional vulnerability. The study of social and institutional as well as physical vulnerability is needed to assess the extent and magnitude of landslide disasters and to guide formulation of effective policy responses.
- ✓ Human intervention can make a greater impact on exposure and vulnerability through, among other factors, land use and urban planning, building codes, risk assessments, early warning systems, legal and policy development, integrated research, insurance, and, above all, substantive educational and awareness-raising efforts by relevant stakeholders.
- ✓ The understanding of landslide disaster risk, including risk identification, vulnerability assessment, time prediction, and disaster assessment, using the most up-to-date and advanced knowledge, is a challenging task. The effectiveness of landslide disaster risk reduction measures depends on scientific and technological developments for understanding disaster risk (natural hazards or events and social vulnerability), political "buy-in", and on increased public awareness and education.
- ✓ At a higher level, social and financial investment is vital for understanding and reducing landslide disaster risk, in particular social and institutional vulnerability through coordination of policies, planning, research, capacity development, and the production of publications and tools that are accessible, available free of charge and are easy to use for everyone in both developing and developed countries.

We agree on the following initial fields of cooperation in research and capacity building, coupled with social and financial investment:

- ✓ Development of people-centered early warning technology for landslides with increased precision and reliable prediction both in time and location, especially in a changing climate context.
- ✓ Development of hazard and vulnerability mapping, vulnerability and risk assessment with increased precision, and reliability as part of multi-hazard risk identification and management.
- ✓ Development of improved technologies for monitoring, testing, analyzing, simulating, and effective early warning for landslides.
- ✓ Development of international teaching tools that are always updated and may be used free of charge by national and local leaders and practitioners, in developed and developing countries through the Sendai Partnerships 2015-2025.
- ✓ Open communication with society through integrated research, capacity building, knowledge transfer, awareness-raising, training, and educational activities to enable societies to develop effective policies and strategies for reducing landslide disaster risk, to strengthen their capacities for preventing hazards to develop into major disasters,

Voluntary commitment to the World Conference on Disaster Risk Reduction
Sendai, Japan, 2015

and to enhance the effectiveness and efficiency of relief programs.

✓ Development of new initiatives to study research frontiers in understanding landslide disaster risk, such as the effect of climate change on large-scale landslides and debris flows, the effective prediction of localized rainfall to provide earlier warning and evacuation especially in developing countries, the mechanism and dynamics of submarine landslides during earthquakes that may cause or enhance tsunamis, and geotechnical studies of catastrophic megaslides for prediction and hazard assessment.

We further agree to advocate that activities should be balanced at regional, national, and community levels in order to empower and engage more professionals, practitioners and decision-makers in formulating policies and establishing programmes for the benefit of disaster risk reduction efforts.

We further agree that progress made in the contribution of the ***Sendai Partnerships 2015-2025 for Global Promotion of Understanding and Reducing Landslide Disaster Risk*** toward the implementation of the Post-2015 Framework for Disaster Risk Reduction will be reported and emerging challenges will be discussed every two years at the Global Platform for Disaster Risk Reduction in Geneva.

A Call for joining the Partnerships

Competent global, regional, national, and local institutions participating in the 3rd WCDRR and in the implementation of the Post-2015 Framework for Disaster Risk Reduction are invited to support this initiative by joining and signing these Partnerships through participation in clearly defined projects related to the issues and objectives of these Partnerships. The potential partners are requested to be in contact with the secretariat of the host organization.

Host Organization and Secretariat

The International Consortium on Landslides (ICL) hosts the Sendai Partnerships 2015-2025 as a voluntary commitment to the United Nations World Conference on Disaster Risk Reduction, Sendai, Japan. The ICL Secretariat in Kyoto, Japan, serves as the Secretariat of the Sendai Partnerships.

Signatories:

Kyoji Sassa	Margareta Wahlström
Mr. Kyoji Sassa **Executive Director** **International Consortium on Landslides** **Host organization of the Partnerships**	**Ms. Margareta Wahlström** **Special Representative of the UN Secretary-General for Disaster Risk Reduction** **Chief of UNISDR**
16/03/15	16 March 2015 in Sendai
Date	Date

Voluntary commitment to the World Conference on Disaster Risk Reduction
Sendai, Japan, 2015

Mr. Qunli Han Director Division of Ecological and Earth Sciences United Nations Educational, Scientific and Cultural Organization	Mr. Dominique Burgeon Resilience Coordinator, Director Emergency and Rehabilitation Division Food and Agriculture Organization of the United Nations
16 March 2015 Date	16 March 2015 Date
Mr. Kazuhiko Takeuchi Senior Vice-Rector United Nations University	Mr. Petteri Taalas Secretary-General World Meteorological Organization
16 March 2015 Date	15.4.16 Date
Mr. Gordon McBean President International Council for Science	Mr. Toshimitsu Komatsu Vice President World Federation of Engineering Organizations
16/03/2015 Date	March 16, 2015 Date
Mr. Roland Oberhänsli President International Union of Geological Sciences	Mr. Alik Ismail-Zadeh Secretary-General International Union of Geodesy and Geophysics
16/03/2015 Date	16 MARCH 2015, SENDAI, JAPAN Date

Voluntary commitment to the World Conference on Disaster Risk Reduction
Sendai, Japan, 2015

Mr. Kaoru Saito
Director
Disaster Preparedness and
International Cooperation Division
Disaster Management Bureau
Cabinet Office, Government of Japan

16/03/2015
Date

Mr. Hideaki Maruyama
Director
Office for Disaster Reduction Research
Minstry of Education, Culture, Sports,
Science and Technology, Japan

16. 03. 2015.
Date

Mr. Takashi Onishi
President
Science Council of Japan

March 16, 2015
Date

Ms.Kayo Inaba
Executive Vice President for Gender
Equality, International Affairs, and
Public Relations
Kyoto University

16. 03. 15
Date

Mr. Prefetto Franco Gabrielli
Head
National Civil Protection Department
Italian Presidency of the Council of
Ministers
Government of Italy

16. 03. 2015
Date

Mr. Jadran Perinic
Director General
National Protection and Rescue Directorate
Republic of Croatia

16.03.2015.
Date

Mr. Walter Ammann
President/CEO
Global Risk Forum GRF Davos

16 March 2015
Date

Voluntary commitment to the World Conference on Disaster Risk Reduction
Sendai, Japan, 2015

ANNEX to the ISDR-ICL SENDAI PARTNERSHIPS 2015-2025

ICL member organizations (registered on 1 April 2015)

1. Albanian Geological Survey, ALBANIA
2. The Geotechnical Society of Bosnia and Herzegovina, BOSNIA AND HERZEGOVINA
3. Geological Survey of Canada, CANADA
4. China Geological Survey, CHINA P.R.
5. Institute of Cold Regions Science and Engineering, Northeast Forestry University, CHINA P.R.
6. Institute of Mountain Hazards and Environment, Chinese Academy of Sciences, CHINA P.R.
7. Bureau of Land and Resources of Xi'an, China P.R.
8. Nanjing Institute of Geography and Limnology, Chinese Academy of Sciences, CHINA P.R.
9. Tongji University, College of Surveying and Geo-Informatics, CHINA P.R.
10. Universidad Nacional de Colombia, Colombia
11. Croatian Landslide Group from University of Rijeka and University of Zagreb, CROATIA
12. City of Zagreb, Emergency Management Office, CROATIA
13. Charles University, Faculty of Science, CZECH REPUBLIC
14. Institute of Rock Structure and Mechanics, Czech Academy of Sciences, Department of Engineering Geology, CZECH REPUBLIC
15. Joint Research Centre (JRC), EUROPEAN COMMISSION
16. Cairo University, EGYPT
17. Technische Universitat Darmstadt, Institute and Laboratory of Geotechnics, GERMANY
18. Department of Geology of National Environmental Agency of Georgia, GEORGIA
19. Universidad Politécnica de Ingeniería, UPI, HONDURAS
20. Instituto Hondureño de Ciencias de la Tierra, IHCIT /Universidad Nacional Autónoma de Honduras UNAH, HONDURAS
21. National Institute of Disaster Management, New Delhi, INDIA
22. Amrita Vishwa Vidyapeetham, Amrita University, INDIA
23. Gadjah Mada University, INDONESIA
24. Parahyangan Catholic University, INDONESIA
25. Research Center for Geotechnology-Indonesian Institute of Sciences, INDONESIA
26. Building & Housing Research Center, IRAN
27. Soil Conservation and Watershed Management Research Institute, IRAN
28. University of Firenze, Earth Sciences Department, ITALY
29. ISPRA-Italian Institute form Environmental Protection and Research, ITALY

Voluntary commitment to the World Conference on Disaster Risk Reduction
Sendai, Japan, 2015

30. University of Calabria, Laboratory of Environmental Cartography and Hydraulic and Geological Modeling, ITALY
31. Istituto di Ricerca per la Protezione Idrogeologica (IRPI), of the Italian National Research Council (CNR), ITALY
32. Kyoto University, Disaster Prevention Research Institute, JAPAN
33. University of Tokyo, Geotechnical Engineering Group, JAPAN
34. Niigata University, Research Institute for Natural Hazards and Disaster Recovery, JAPAN
35. Forestry and Forest Product Research Institute, JAPAN
36. Japan Landslide Society, JAPAN
37. Korea Institute of Geoscience and Mineral Resources (KIGAM), REPUBLIC OF KOREA
38. Korea Forest Research Institute, REPUBLIC OF KOREA
39. Korea Infrastructure Safety & Technology Corporation, REPUBLIC OF KOREA
40. Korea Institute of Construction Technology, REPUBLIC OF KOREA
41. Korean Society of Forest Engineering, REPUBLIC OF KOREA
42. Slope Engineering Branch, Public Works Department of Malaysia, MALAYSIA
43. Institute of Geography, UNAM, MEXICO
44. International Centre for Integrated Mountain Development (ICIMOD), NEPAL
45. Department of Geology, University of Nigeria, Nsukka, NIGERIA
46. Norwegian Geotechnical Institute (NGI), Oslo, NORWAY
47. Grudec Ayar, PERU
48. Department of Engineering and Ecological Geology, Moscow State University, RUSSIA
49. JSC "Hydroproject Institute", RUSSIA
50. University of Belgrade, Faculty of Mining and Geology, SERBIA
51. Comenius University, Faculty of Natural Sciences, Department of Engineering Geology, SLOVAKIA
52. University of Ljubljana, Faculty of Civil and Geodetic Engineering (ULFGG), SLOVENIA
53. Geological Survey of Slovenia, SLOVENIA
54. University of Ljubljana, Faculty of Natural Sciences and Engineering (UL NTF) , SLOVENIA
55. Central Engineering Consultancy Bureau (CECB), SRI LANKA
56. National Building Research Organization, SRI LANKA
57. Landslide group in National Central University from Graduate Institute of Applied Geology, Department of Civil Engineering, Center for Environmental Studies, CHINESE TAIPEI
58. National Taiwan University, Department of Civil Engineering, CHINESE TAIPEI
59. Ministry of Agriculture and Cooperatives, Land Development Department, THAILAND
60. Asian Disaster Preparedness Center, THAILAND
61. Institute of Telecommunication and Global Information Space, UKRAINE
62. California State University, Fullerton, USA & Tribhuvan University, Institute of Engineering, Nepal, USA/NEPAL
63. Institute of Transport Science and Technology, Ministry of Transport, VIET NAM
64. Vietnam Institute of Geosciences and Mineral Resources, Ministry of Natural Resources and Environment, VIET NAM

Rupestrian World Heritage Sites: Instability Investigation and Sustainable Mitigation

Claudio Margottini, Peter Bobrowsky, Giovanni Gigli, Heinz Ruther, Daniele Spizzichino, and Jan Vlcko

Abstract

Rupestrian settlements were among the first man-made works in the history of humanity. The most relevant masterpieces of such human history have been included in the UNESCO World Heritage List. These sites and their associated remains are not always in equilibrium with the environment. They are continuously impacted and weathered by a variety of internal and external factors, both natural and human-induced, with rapid and/or slow onset. These include major sudden natural hazards, such as earthquakes or extreme meteorological events, but also slow cumulative processes such as the erosion of rocks, compounded by the effects of climate change, as well as the role of humans, especially in conflict situations. Many rupestrian sites have been carved into soft rock, generally with UCS < 25 MPa (ISRM in Int J Rock Mech Min Sci Geomech Abs 18:85–110, 1981), in vertical cliffs, and show major conservation issues in the domain of rock slope stability and rock weathering. This paper reports the experience of rock fall mitigation in rupestrian sites, mainly from the UNESCO World Heritage List (Bamiyan in Afghanistan; Lalibela in Ethiopia; Petra in Jordan, Vardzia in Georgia and others). The general approach, implemented in the activities, includes a very detailed interdisciplinary study, with the objective to understand degradation processes and causative factors, followed, as a subsequent step, by proper field conservation work. The latter is mainly related to re-discovering and potential application of traditional knowledge and sustainable practices, and is primarily based on local conservation techniques.

Keywords

Rupestrian site • Geotechnical threats • Investigations • Monitoring • Sustainable conservation policies

C. Margottini (✉) · D. Spizzichino
Department of Geological Survey of Italy, ISPRA, via Brancati, 60, 00143 Rome, Italy
e-mail: claudio.margottini@isprambiente.it

D. Spizzichino
e-mail: daniele.spizzichino@isprambiente.it

P. Bobrowsky
Geological Survey of Canada, Natural Resources Canada, 9860 West Saanich Road, 6000Sidney, BC V8L 4B2, Canada
e-mail: Peter.Bobrowsky@canada.ca

G. Gigli
University of Florence, Department of Earth Sciences, Largo E.Fermi, 2, 50125 Florence, Italy
e-mail: giovanni.gigli@unifi.it

H. Ruther
Principal Investigator "African Cultural Heritage Sites and Landscapes" Project Zamani Project Division of Geomatics (APG), University of Cape Town, 7701 Private Bag, Rondebosch, South Africa
e-mail: heinz.ruther@zamaniproject.org

J. Vlcko
Department of Engineering Geology, Faculty of Natural Sciences, Comenius University, Bratislava Ilkovicova 6, Mlynska Dolina, Slovakia
e-mail: vlcko@fns.uniba.sk

K. Sassa et al. (eds.), *Advancing Culture of Living with Landslides*,
DOI 10.1007/978-3-319-59469-9_2

Introduction

Rupestrian settlements were among the first man-made works in the history of humanity. During some 3 million years, humankind survival relied on two basic activities: hunting (or fishing) and gathering edible items (from plants to insects). A radical change came roughly 10,000 years ago, after the last glacial age, when people first learned to cultivate crops and to domesticate animals, in what can certainly be considered one of the most significant developments in human history (Margottini and Spizzichino 2014).

A subsequent significant change concerned the process of planning, designing and building structures for human settlements. Architectural works, in the material form of buildings and/or rupestrian settlements, are often perceived as cultural symbols and as works of art, and civilizations are often identified by their architectural achievements. However, architectural works are not merely entities of physical construction. They represent the synthesis of a complex system that, throughout time, has been guided by human genius, and depends on the availability and types of construction materials (natural geological resources). Their form is determined by social and economic conditions, by local morphological situations (e.g. defensive settlements on top of cliff), and is influenced by local meteo-climatic conditions.

The most relevant masterpieces of such human history have been included in the UNESCO World Heritage List.

Major Threats Affecting Rupestrian Sites

The sites and remains discussed herein have not always been in equilibrium with the environment. They are continuously impacted and degraded by several internal and external factors, both natural and human-induced, with rapid and/or slow onset. These include major sudden natural hazards, such as earthquakes or extreme meteorological events, but also slow, cumulative processes such the erosion of rocks, compounded by the effects of climate change, including the role of humans, especially in conflict situations.

Many rupestrian sites have been carved into soft rock, generally with UCS < 25 MPa (ISRM, 1981), in vertical cliffs and show major conservation issues in the domain of rock slope stability and rock weathering. The low strength range of rock might be influenced by physical characteristics, such as size, saturation degree, weathering and mineral content. Finally, the investigations generally show that the strength reduces significantly with saturation (Agustawijaya 2007).

The low strength of the rock, together with the discontinuity pattern in steep slopes and the weakening of the cliff produced by the man-made settlements, pose a serious concern for the long term stability of the sites.

As a confirmation, the following Fig. 1 shows the relationship between UCS and porosity for some rupestrian sites discussed in this paper. It is evident that the low UCS value is generally coupled with high porosity, especially in volcanic materials (Lalibela and Vardzia). On the other hand, the continental/sedimentary geological formations of Bamiyan and Petra exhibit a relevant vertical heterogeneity, so values reported here are rough estimates.

Here we report on some advances developed during rock fall and weathering analysis, monitoring and mitigation in rupestrian sites, mainly belonging to the UNESCO world heritage list (Bamiyan in Afghanistan; Lalibela in Ethiopia; Petra in Jordan and Vardzia in Georgia) and under UNESCO coordination. The general approach, implemented in the conservation activities, include a very detailed interdisciplinary study, to understand rock degradation processes and causative factors, followed by field conservation work. The latter is mainly related to re-discovering traditional knowledge and sustainable practices and is based on the application of local conservation techniques (Margottini et al. 2016a, b).

The described methodology is then primarily aimed at the protection of the heritage sites but, at the same time and when possible, at the empowerment of local communities to independently manage the site. Protecting heritage from natural threats is, in fact, not a luxury but a fundamental step to be given priority together with other humanitarian concerns. This is especially relevant at a time when traditional knowledge and sustainable practices, which ensured a certain level of protection from the worst effects of natural hazards or human-made disasters, are being progressively abandoned and/or forgotten (Margottini et al. 2016a, b).

Following is a brief description of some advanced techniques employed by the authors and sustainable conservation practices in selected case histories.

Investigation Techniques

Laser Scanner

As a result of the outstanding development of terrestrial laser scanner (TLS) in recent years, rock slopes and rupestrian settlements can now be investigated and mapped through high resolution point clouds.

Laser scanning for the generation of surface models acquires 3D surface information by determining the xyz co-ordinates of large numbers of surface points, referred to as a point cloud. The scanner (Fig. 2) can only measure and record surfaces in the field-of-view of the instrument and it is therefore necessary to move the instrument to multiple positions to cover an entire object. The scans obtained in this way, must overlap each other sequentially to allow the

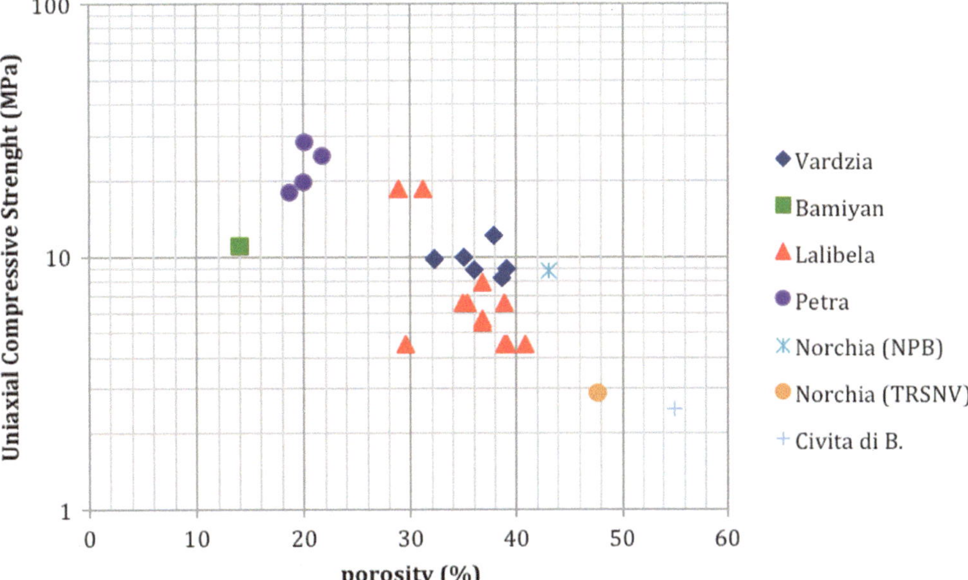

Fig. 1 The relationship between Uniaxial compressive strength and porosity for some of the case studies discussed in this paper

combination of all acquired scans into a single point cloud. In subsequent processing the individual points are connected to form a triangulated mesh which can be further processed to create a complete, full surface. Finally, photos can be draped over the surface to create a photo-realistic appearance. The interval between points of the point cloud, referred to as resolution, can be chosen by the operator of the laser scanner depending on the required detail and the complexity of the surface.

Finally, TLS can also be utilised to identify the different members of a stratigraphic sequence. In fact, the instrument records the RGB values and measures the reflected ray energy, providing the reflectivity index I of the micro-portion surface (Ercoli et al. 2013). Therefore, the reflectivity index associated with each TLS point can provide an indirect estimation of the lithothypes (Fig. 3). Intensity variations as a result of different angles of incident proved insignificant in this context.

Fig. 2 Laser scanner (Z&F) above Al Khazneh (Petra, Jordan)

The Siq of Petra

The Siq, the main entrance to the Nabatean archaeological site of Petra (Jordan), is a 1.2 km long, naturally formed gorge, with an irregular horizontal shape and a complex vertical slope.

For the rock-mechanics analysis of the Siq, a 3D computer model of the rock walls on both sides of the Siq was created (Ruther 2015). Ideally this model should have covered the full height of both surfaces over the 1.2 km length of the Siq which would have resulted in a model of 2.4 km length with heights varying from about 70 to 120 m (Fig. 4).

Two methods were available for the development of the 3D surface model, laser scanning and photogrammetry. In lengthy investigations and discussions, the accuracy potential, the practicality and the cost of the two techniques were explored including considerations regarding the use of Unmanned Aerial Vehicles (UAVs), scaffolds and poles. It became obvious that neither method was capable in a practical sense of providing surface data beyond the areas which were visible from the ground, unless unrealistically high expenses were accepted. The final decision favoured laser scanning as the most feasible method and an extensive laser scan survey, limited to the rock surfaces visible from ground level, was subsequently completed by the Zamani research team, from the University of Cape Town. A number of individual boulders and sections of rock walls were modelled by PNP surveyors using Structure-from-Motion-Photogrammetry.

For the Siq survey, the scanner was positioned on the Siq floor at intervals of not more than 10 meters (Ruther et al. 2014). The lower wall areas over the entire length of the Siq were thus captured up to a height varying from 20 to 80 m, depending on visibility (Fig. 5a). In addition to this, a

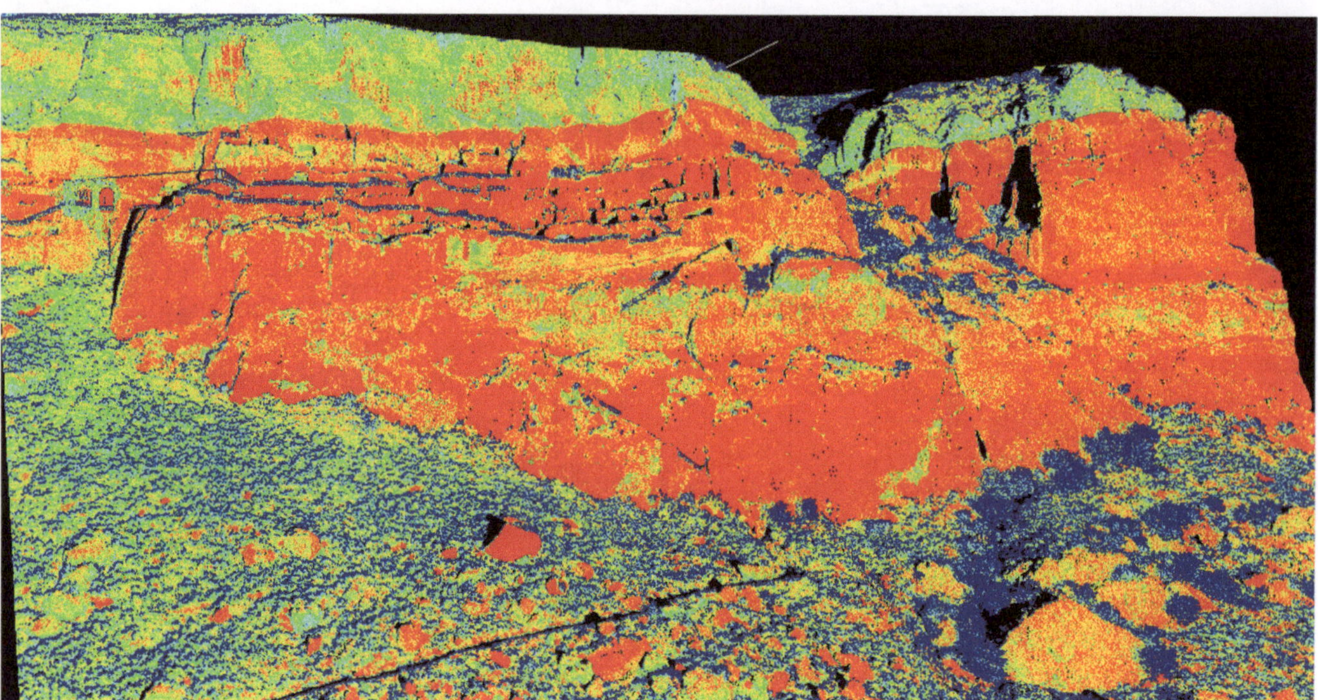

Fig. 3 Textured 3D model and reflectivity map of the cave monastery of Vardzia (Georgia) from TLS techniques

number of scans could be taken from points at the upper edge of the Siq (Fig. 5b). The scans at high elevations provided, in some cases, surface data from Siq floor to the top of the rock walls. The total number of scans in the Siq was 220 with an average point interval of approximately 3 cm, resulting in a point cloud of five billion points. This is only a part of the complete point cloud which the Zamani team acquired for the Siq, Wadi Musa and Wadi Farasa. This collective point cloud comprised 25 billion points. The relative accuracy of points, i.e. the accuracy of neighbouring points is estimated to be in the order of a cm or better, whereas the absolute accuracy, i.e. the accuracy of points over the entire length of the Siq is in the one or two decimeter range. The final model (example in Fig. 6) can be viewed in 3D viewing and processing software. Coordinate and dimension measurements can be taken in applications such as the open source system Meshlab (meshlab.sourceforge.net). The 3D model and the GIS are referenced to UTM coordinates on WGS 84 (ITRF8) determined via GNSS survey by the Zamani team.

Cross-Sections

The 3D model of the Siq walls also had to be available in 2D formats for processing and display in a variety of CAD and analysis software applications (Ruther 2015). For this purpose sections and ortho-images were generated.

One hundred-and-twenty-one cross-sections at approximately right angle to the central axis of the Siq were generated from the 3D model at intervals of 10 m (Fig. 7).

Ortho Images

Sixty nine ortho-images (Fig. 8) of the Siq wall were created covering the full extent of the Siq rock walls (Ruther 2015). It is noteworthy that, due to the non-planar nature of the rock surface, it is not possible to create ortho-images which can be used for accurate measurements, as would normally be possible for ortho-images (Fig. 9). Thus these images are primarily useful for inspection of wall sections and annotations, whereas measurements must be done on the 3D model in 3D viewing and measurement software (such as Meshlab or Cloud Compare). Measurements on the ortho image are only correct for surfaces which are parallel to the projection plane of the ortho image. Deviations of the rock surface from this condition result in errors.

Panoramas

One hundred and two full dome photographic panoramas were acquired for the Siq of Petra, at an average interval between panorama stations of about 10 m (Ruther 2015). Each panorama was generated from 21 to 49 images

SIQ Sector - Elevations overview (with Satellite image)

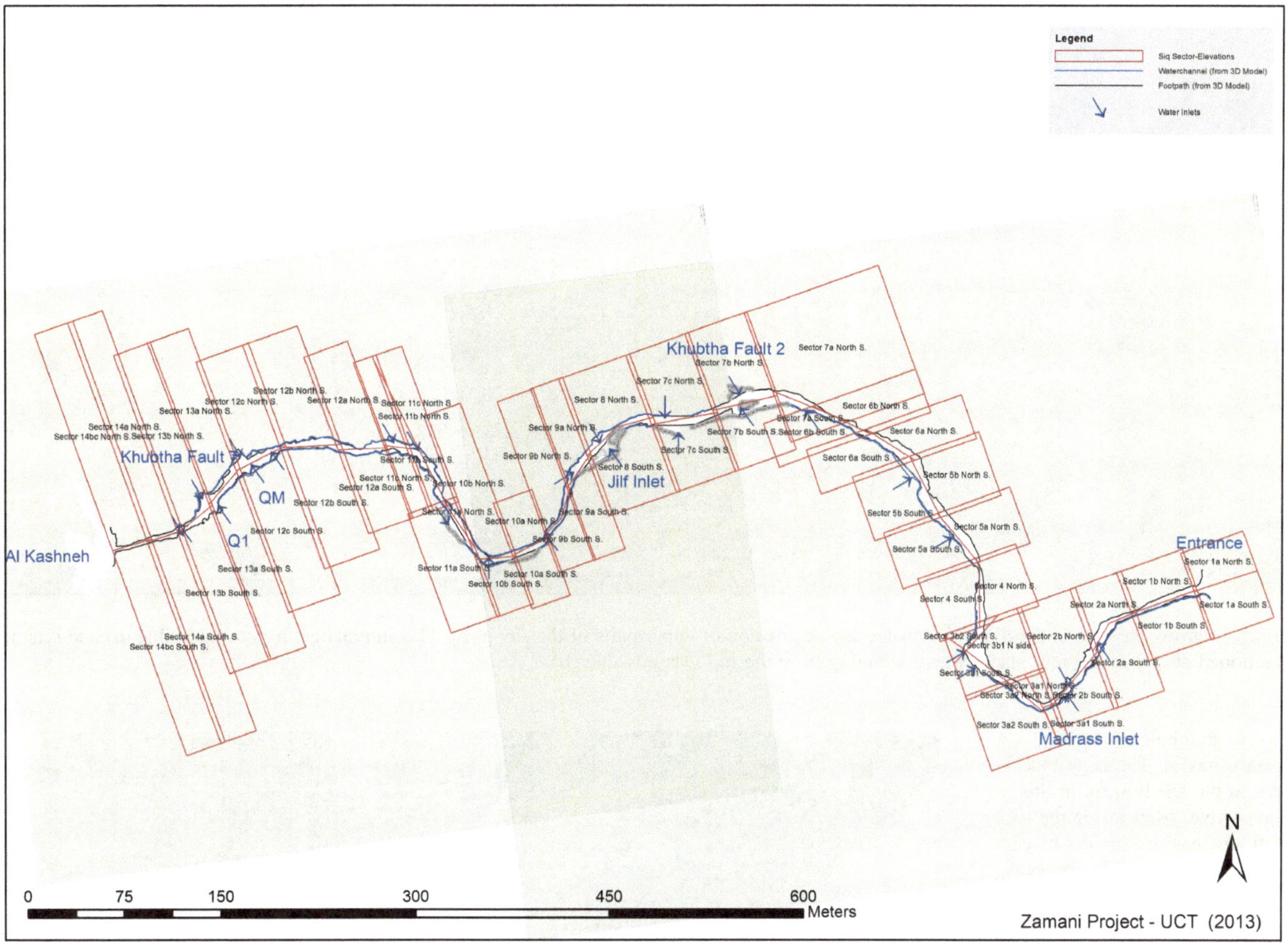

Fig. 4 Top view of 1.2 km Siq point cloud (Courtesy of Zamani Project)

captured with a Nikon D 200 camera equipped with a 10.5 mm lens. In each case the camera was pointed into six horizontal and one zenith direction. Images were then taken with a view to High Dynamic Range (HDR) processing and depending on the light conditions 3, 5 or 7 bracketed images were acquired in each of the seven directions.

The panoramas were combined into a panorama tour which makes it possible to view the entire length of the Siq from the position of a person walking from the Siq entrance to Al Khazneh.

Such materials provided a very useful description and verification tool in the elaboration of the project, especially when not on site.

Mineralogy and Petrography

Weathering of rock can have a wide range of causes, all of which must be considered in an integrated approach before embarking on conservation activities. An example for such an investigation is the conservation of the rock hewn churches of Lalibela in Ethiopia.

The rock-hewn churches of Lalibela, included since 1978 in the UNESCO's World Heritage List, attracted the attention of the conservation science community because of their severe chemical weathering and physical decay (Fig. 10). Thin section study, X-ray diffraction (XRD) and SEM-EDS mineral composition of samples from seven rock-hewn churches (Biet Medhane-Alem, Biet Mariam, Trinity Church, Biet Giyorgis, Biet Emanuel, Biet Abba-Lebanos and Biet Gabriel/Rufael) were carried out. These investigations showed that the churches were carved in hydrothermally-altered and nearly aphyric vesicular basalts with incipient lateritization (Ruther et al. 2014). This is in contrast to the earlier literature on Lalibela which often reports the rock-hewn churches as being carved into "weathered basic tuffs". In reality, the geological materials are hydrothermally altered scoriaceous unit/s of massive to

Fig. 5 Narrow passage in the Siq which makes the acquisition of upper parts of the slopes by TLS impractical if not impossible (a) and scanner positioned at one of the few places where scanning from the top was possible (b)

Fig. 6 Example of the Siq surface model. The large black area to the left is seen, in this perspective, from inside the rock wall whereas the smaller black patches inside the Siq are areas invisible to the scanner

slightly fractured basaltic lava still present as bedrock, and not a pyroclastic level sensu stricto. In fact, late-stage and post-magmatic phases (smectites, zeolites and calcite) scattered in the groundmass and filling the large subspherical vesicles represent a typical hydrothermal facies of continental flood basalts. Appropriate modal mineralogy and petrography of the Lalibela churches provided useful insights to unravel causes of deterioration of these World Heritage monuments.

Considering the heavy deterioration affecting the investigated rocks, the correct petrographic (basalt) and lithological (partly lateritized vesicular lava) definitions, coupled with the comprehensive hydrothermal mineralogy are of paramount importance to better understand the causes of

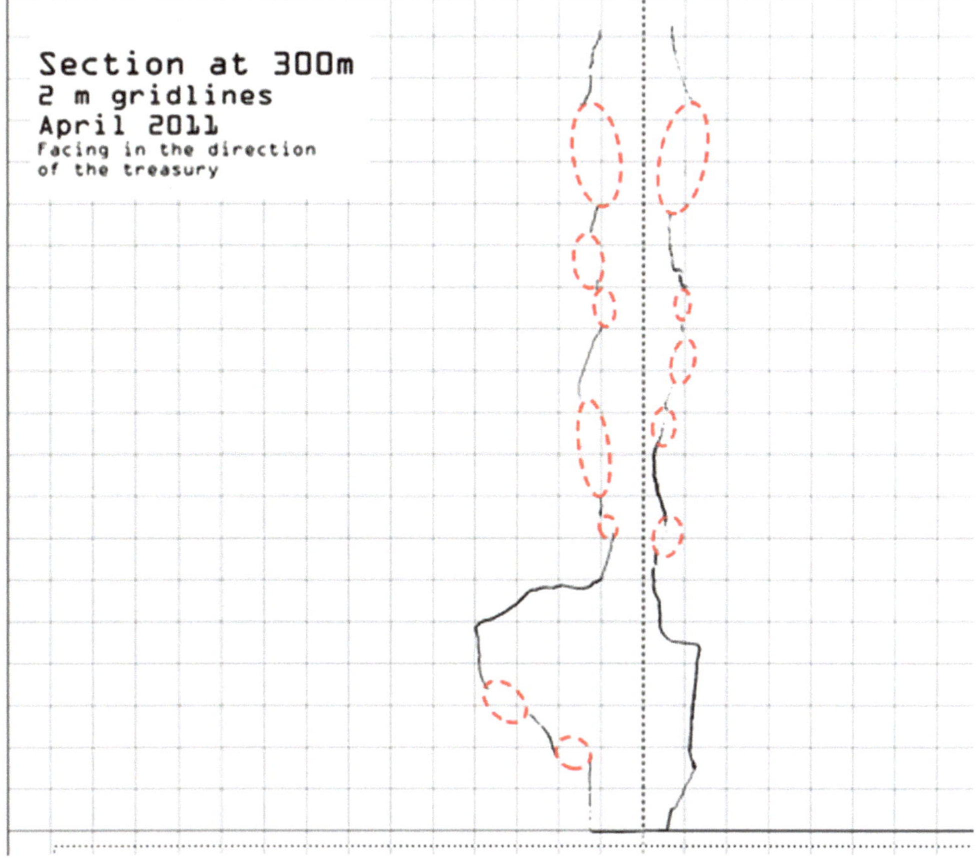

Fig. 7 Cross-section through the Siq. Invisible areas are indicated in *red*

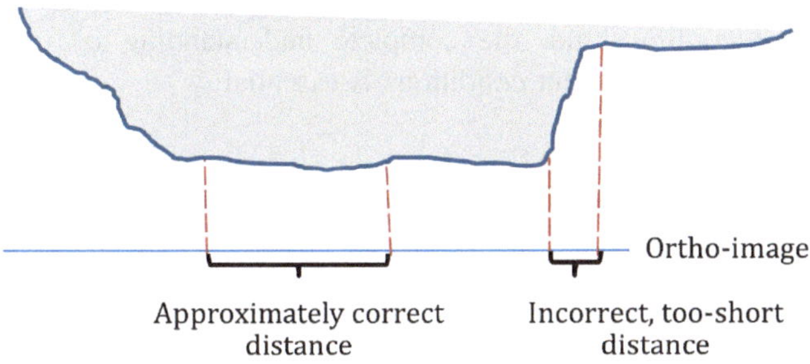

Fig. 8 Distortions in ortho image due to irregular surface shapes

their chemical weathering and physical decay and to plan conservation actions. The presence of swelling clay minerals (due to hydrolytic weathering of silicates), minor dissolution-crystallization processes of soluble salts and the whole inexorable lateritization processes, typical of regions with dry and rainy seasons, all concur in the deterioration of the rocks. Additionally, a recent study (Renzulli et al. 2011) emphasises how the high microporosity of the abundant zeolites, coupled with the wide range of apparent density and porosity to water saturation should now be considered and further investigated in the framework of future conservation studies of the Lalibela rock-hewn churches.

In conclusion, very often weathering processes depend on a variety of factors, to be jointly investigated in a comprehensive manner. Such weathering, even if dependent on micro processes may play a large role in the entire site conservation, as in the case of Lalibela reported in this paper.

Rock Mechanic Parameters

Rock mechanic parameters are of paramount relevance for the conservation of rupestrian sites. Generally, since such sites are primarily carved into soft rocks, the strength and deformability parameters of rock material refer mainly to the rock mass, which is composed of the intact rock and the joint family, depending on the scale of analysis. The intact rock is well described and characterized by UCS values whereas the rock mass is described and characterized by indices (e.g. RMR, Q_system, GSI) as a result of UCS, joints and water condition combination. The following example is taken from the investigations for the consolidation of the remains of the destroyed Buddha Statues in Bamiyan (Afghanistan).

The rock mass of Bamiyan (Afghanistan) is composed of continental lithotypes, likely coming from the dismounting of surrounding morphological peaks and deposition in a

Fig. 9 Ortho image of rock wall showing surface details

flood plain and small lagoon. The subsequent river erosion produced the present morphology for the part where the rupestrian settlement is located (Delmonaco and Margottini 2014).

The cliff and niches are composed of alternating conglomerate and siltstone (yellow at the bottom and red in the middle of the cliff); the conglomerate, with different sized pebbles (coarse, mid and fine-grained), is the predominant material in the cliff and exhibits a moderate cohesion. The siltstone is interlayered with the conglomerate strata. The mean value for the uniaxial compressive strength of the conglomerate is 2.99 MPa, whereas the value for the uniaxial compressive strength of the siltstone is 6.91 MPa (Feker 2006). Point load data10 show that the average UCS values are 11.0 MPa for siltstone and 5.6 MPa for conglomerate.

Accordingly, the siltstone is apparently stronger than siltstone. A simple test with immersion in water, clearly shows a potential and relevant slaking attitude of the material, clearly not visible on site because of the low rainfall rate in the area (annual average rainfall equal to 162.56 mm) (Margottini 2014a, b).

The mechanism for the swelling is likely dependent on the presence of a cement carbonate in the conglomerate and the absence of such permanent cement into the sandstone, as revealed by microscopic thin section and x-ray diffraction (Fig. 11b, c).

As a general conclusion we note that relevant geomechanical parameters also depend on the local environment and a global and detailed investigation must be performed to avoid unexpected problems. In this case, considering the importance of water with respect to the adherence of potential anchors, as well as the use of water or air as drilling head cooling fluid, the complete understanding of rock behavior in different conditions is essential.

Semi-automatic Discontinuity Detection and Analysis

In order to implement a kinematic analysis and further hazard analysis in rock slopes, discontinuities must be identified and mapped. Availability of TLS data make this step more advanced, allowing the structural analysis to be implemented through a semi-automatic method. As an example, the case study of the rock-cut city of Vardzia is reported. Vardzia is a cave monastery located in south-western Georgia, excavated from the slopes of the Erusheti mountains on the left bank of the Mtkvari River. The main period of construction was the second half of the twelfth century. The caves stretch along the cliff for some eight hundred meters and range in height over fifty meters within the rocky wall. The monastery consists of more than six hundred hidden rooms spread over thirteen floors, which protected the monastery from the Mongol domination (Fig. 12).

Discontinuities data were obtained, in Vardzia (Georgia) both during field survey and through the elaboration of TLS

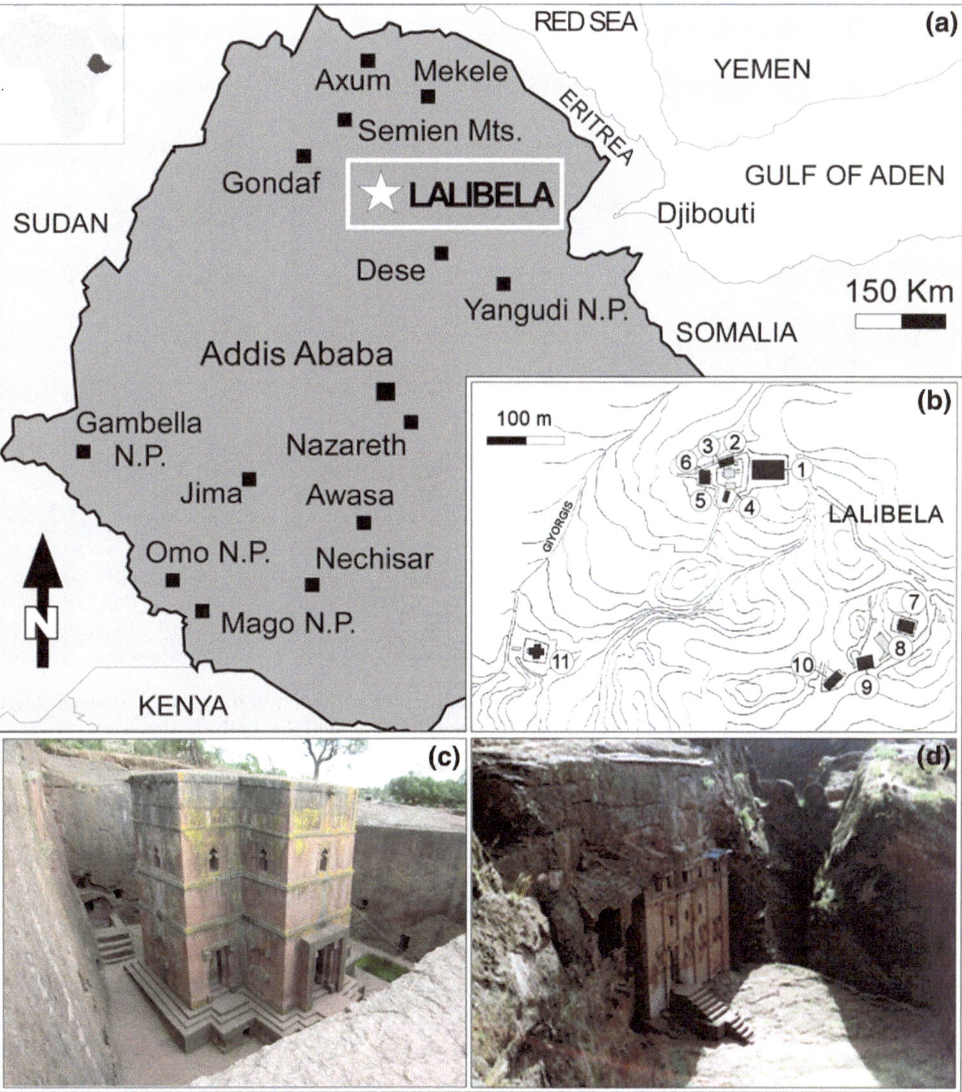

Fig. 10 **a** Location of the Lalibela site; **b** distribution of the 11 rock-hewn churches:*1*: Biet Medhane-Alem (BA); *2*: Biet Maryam (BM); *3*: Biet Masqal; *4*: Biet Danagel; *5* and *6*: Trinity Church (BT, i.e., Biet Debre-Sina and Biet Mikael-Golgota); *7*: Biet Amanuel (BE); *8*: Biet Maqorewos; *9*: Biet Abba-Libanos (BL); *10*: Biet GabrielRufael (BR); *11*: Biet Giyorgis (BG); **c** view of Biet Giyorgis church; **d** view of Biet Abba-Lebanos church (Renzulli et al. 2011)

with a specific software, Coltopo3D®, suitable to detect slope and slope direction of the many outcropping surfaces11. Coltop3D software performs structural analysis by using digital elevation model (DEM) and 3D point clouds acquired through terrestrial laser scanners. A colour representation merging slope aspect and slope angle is used in order to obtain a unique code of color for each orientation of a local slope (Fig. 13). Thus a continuous planar structure appears in a unique colour (Metzger et al. 2009).

Kinematic Analysis

Most classifications of mass movements and hazard assessment in rock slopes use relatively simple, idealized geometries for the basal sliding surface (BSS), like planar sliding, wedge sliding, toppling or columnar failures. For small volumes, the real BSS can often be well described by such simple geometries (Oppikofer et al. 2011). Extended and complex rock surfaces, however, can exhibit a large number of mass movements, also showing various kinds of kinematics. As a consequence, the real situation in large rock surfaces with a complicated geometry is generally very complex and a site dependent analysis, such as fieldwork and compass, cannot comprehensively reveal the real situation.

The availability of digital slope surface data can offer a unique chance to determine potential kinematics in a wide area for all the investigated geomorphological processes. In more detail, the proposed method is based on least-squares-fitting of planes to point clusters extracted by moving a sampling cube over the point cloud. If the associated standard deviation is below a defined threshold, the cluster is considered acceptable. By applying geometric criteria it is possible to join all clusters lying on the same surface; in this way discontinuity planes can be

Fig. 11 Stability of conglomerate (*left* in **a**) and degradation of siltstone (on the *right* in **a**) when saturated. Thin section of conglomerate with white carbonate cement (**b**) and thin section of siltstone (**c**) (Margottini 2014a, b)

Fig. 12 General view of rock-cut city of Vardzia (Georgia)

reconstructed, rock mass geometrical properties can be calculated and, finally, potential kinematics established.

As previously mentioned, the Siq of Petra (Jordan), is a 1.2 km naturally formed gorge, with an irregular horizontal shape and a complex vertical slope that represents the main entrance to Nabatean archaeological site. In the Siq, discontinuities of various type (bedding, joints, faults), mainly related to geomorphological evolution of the slope, lateral stress released, stratigraphic setting and tectonic activity can be recognized. Rock-falls have been occurring, even recently, with unstable rock mass volumes ranging from 0.1 m^3 up to over some hundreds m^3. Slope instability, acceleration of crack deformation and consequent increasing of rock-fall hazard conditions, could threaten the safety of tourists as well as the integrity of the heritage site.

Thus, for the identification of the main rockfall source areas, a spatial kinematic analysis for the whole Siq was performed, by using discontinuity orientation data extracted from the point cloud by means of the software Diana (Casagli and Gigli 2011). Orientation, number of sets, spacing/frequency, persistence, block size and scale dependent roughness were obtained combining fieldwork and automatic analysis. This kind of analysis is able to establish where a particular instability mechanism is kinematically feasible, given the geometry of the slope, the orientation of discontinuities and shear strength of the rock.

Fig. 13 The exposed rock faces of Vardzia cliff and detected slope aspect and slope angle, also on a Schmidt-Lambert stereonet (Margottini et al. 2016a, b, c, d)

The final outcome of this project was a detailed landslide kinematic index map, reporting main potential instability mechanisms for a given area. The kinematic index was finally calibrated for each instability mechanism (plane failure; wedge failure; block toppling; flexural toppling) surveyed in the site.

A synthetic view was also provided, integrating all the different probability of occurrence for the different kinematic conditions, into a single evaluation. This elaboration, as in Fig. 14, may help in the general management of the site, identifying the most endangered places, however a detailed case by case investigation is still required.

Infrared Thermographic Analysis (IRT)

InfRared (IRT) Termography is a non invasive technique that has also recently been applied to the rapid detection of slope portions, mainly investigating thermal anomalies in the rock mass and along the discontinuities.

Thermal imaging is based on the detection of electromagnetic waves in the thermal infrared spectrum, which are converted to electrical signals. All objects with temperature above 0 K (−273.15 °C) emit characteristic IR radiation, which can then be displayed in the visible spectrum by thermal imaging cameras.

The IRT detection method relies on air circulation and ventilation through open joints and cracks when, in winter, the warmer subsurface communicates with the (colder) ground surface; cracks can thereby be identified as relatively warm areas using IRT. The contrast between temperatures deep in the rock, which at depths of several meters generally correspond to the local mean annual temperature, and the actual temperature at the ground surface is accentuated by airflow in the open joints, cracks, and cavities. In winter, air circulating through open joints absorbs heat from deep within the rock mass, becomes warmer and lighter than the ambient air, rises, and is expelled through joints and caverns at the ground surface (Baron et al. 2014).

Within the obtained surficial temperature maps (thermograms) shown in Fig. 15, the temperature is represented by means of a colour scale, in which the higher temperatures are displayed by the lighter colours, whereas the colder temperatures by the darker ones. In order to obtain a comparison with TLS data, IRT data can be referenced to the laser scan data (Gigli et al. 2013).

The higher the temperature of an object of interest, the greater will be the intensity of emitted radiation. This means that larger temperature differences in an assessed area present brighter contrast levels in the image.

Water leaks and moisture zones are also potentially visible and detectable by using IR images, due to the lowering

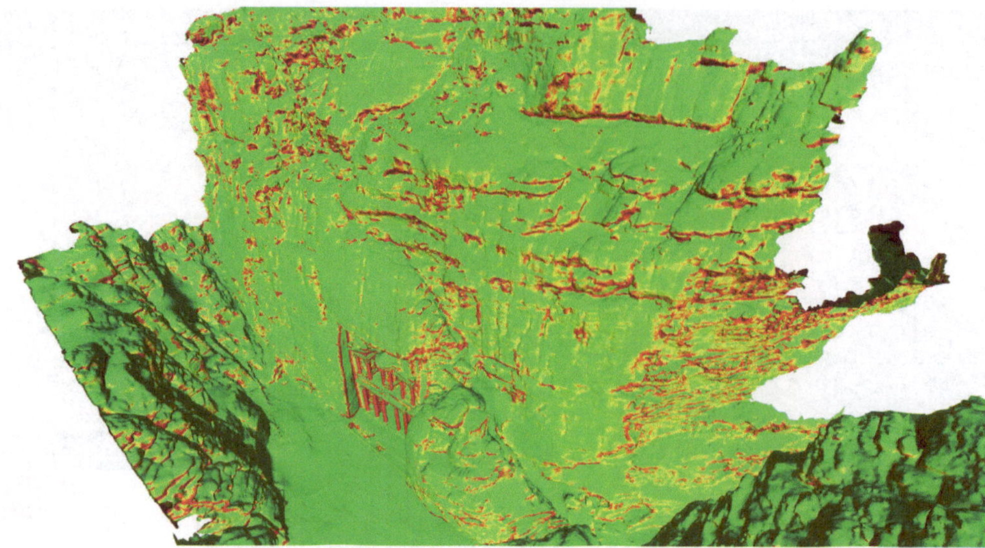

Fig. 14 Integrated result from the 3D individual kinematic analysis for the different analyzed instability mechanisms (block toppling, flexural toppling, wedge failure, plane failure and free fall) in the Treasury area (Petra, Jordan). The probability of occurrence (hazard) is expressed following a common colour scale (0–30% of probability to be involved in a kinematic process)

Fig. 15 Thermal image of the Palace Tomb in Petra, compared with the image as seen by a normal camera

effect of evaporation which cools moist areas (Clark et al. 2003).

Thermal Influence

The changing climatic conditions apparently accelerate the deterioration processes of rocks as the host environment of several rupestrian sites or as the building material, especially when the historic monuments are in sight of the study. As can be confirmed by many researchers moderate and cyclic nature of temperature changes may facilitate thermally driven processes even in mid-latitudes (Gomez-Heras et al. 2008a).

The heat flow penetrating into the rock mass produces a temperature field which changes with heating and cooling phases as a function of depth. Temperature oscillations produce cyclic thermal strains in rock masses which result in deterioration processes associated with crumbling, cracking and jointing, opening/closing of joints/fractures, and in jointed rock masses may initiate thermally driven displacements ranging from slow creep (block ratcheting in terms of Pasten et al. (2015), further toppling, and tilting to rapid movement in the form of rock cliffs overturning to rock falls. There are several slope failures, where no significant trigger was proposed and the rock failure mechanism remained unclear were recently recognized as hazards triggered by meteorological factors, including temperature variations (Dehn et al. 2000; Vargas et al. 2004; Krähenbühl et al. 2004; Gunzburger et al. 2011; Hales and Roering 2007; Vlcko et al. 2009; Bakun-Mazor et al. 2011, 2013; Gischig et al. 2011a, b; Grøneng et al. 2011; Hatzor 2004).

As an example we may present our experience with Spis Castle (UNESCO site) from eastern Slovakia where since 1980 and 1992 three TM-71 type crack-gauges (TM-71) installed at the castle subgrade operate. Within this period mean annual displacement recorded at this site reached the value of 0.3–0.9 mm (mean value 0.33 mm/yr). According to the rate of displacements the slope movements maybe assigned as extremely slow (Varnes 1978). Significant for the monitoring records is that regardless of rainfall data as the most frequent slope movements trigger, the increment in the rate of displacement is quasi "stable" and as can be seen corresponds to the cyclic temperature variations (Fig. 16).

In order to get better support of this idea, thermomechanical behavior of the travertines were tested. Cylindrical samples of 34 mm diameter and 50 mm height, cut in three axis were prepared, surface faces of specimens remained unpolished in order to avoid sealing of their pores and cavities. The samples were subject to three temperature variations simulating real seasonal insolation cycles within the range of −20 to 50 °C (measured directly in the centre the rock sample) using a thermodilatometer VLAP (Vlčko et al. 2007), Fig. 17. Thermal expansion was measured with quartz rods attached to linear variable differential transformers (LVDT) in six consecutive thermal cycles with temperature rate 0.3 °C/min. The number of thermal cycles corresponds to the amount of mean annual rock face temperature reaching the temperature difference of 35 °C.

The highest value of linear thermal residual strain or plastic deformations (in some papers assigned as residual dilation) was determined with the max value $\Delta Lr = 3.9 \times 10^{-6}$ m, which means that 10% of recorded rate of displacements are triggered by the temperature fluctuations. If we calculate that in the natural conditions volumetric (three dimensional) deformation is the usual case, the percentage is even higher.

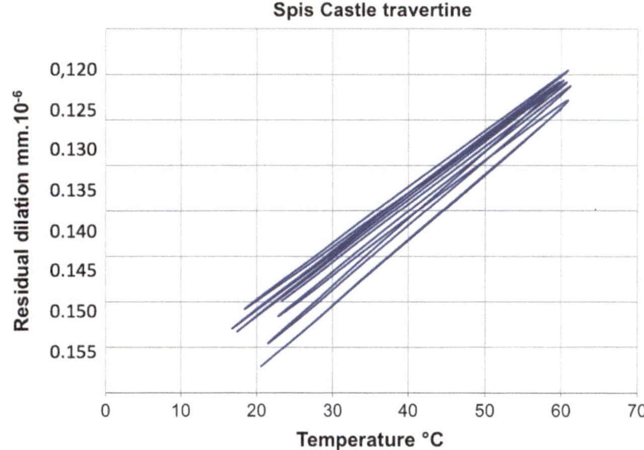

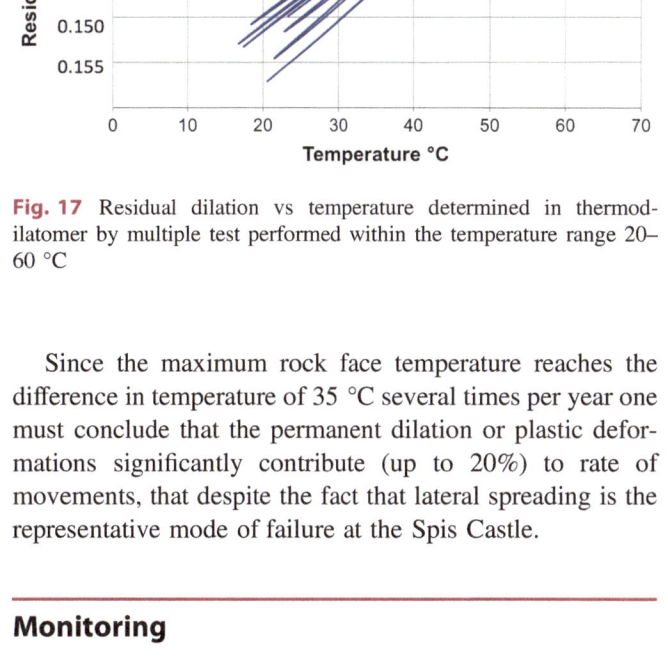

Fig. 17 Residual dilation vs temperature determined in thermodilatomer by multiple test performed within the temperature range 20–60 °C

Since the maximum rock face temperature reaches the difference in temperature of 35 °C several times per year one must conclude that the permanent dilation or plastic deformations significantly contribute (up to 20%) to rate of movements, that despite the fact that lateral spreading is the representative mode of failure at the Spis Castle.

Monitoring

Landslide monitoring is required for a wide variety of reasons. These may include: the determination of the extent, magnitude and style of landslide movement, risk and even emergency risk management assessments and/or assistance with the design and implementation of site remedial and/or mitigation works.

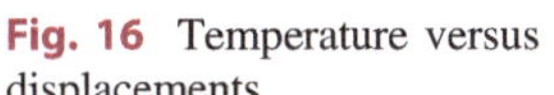

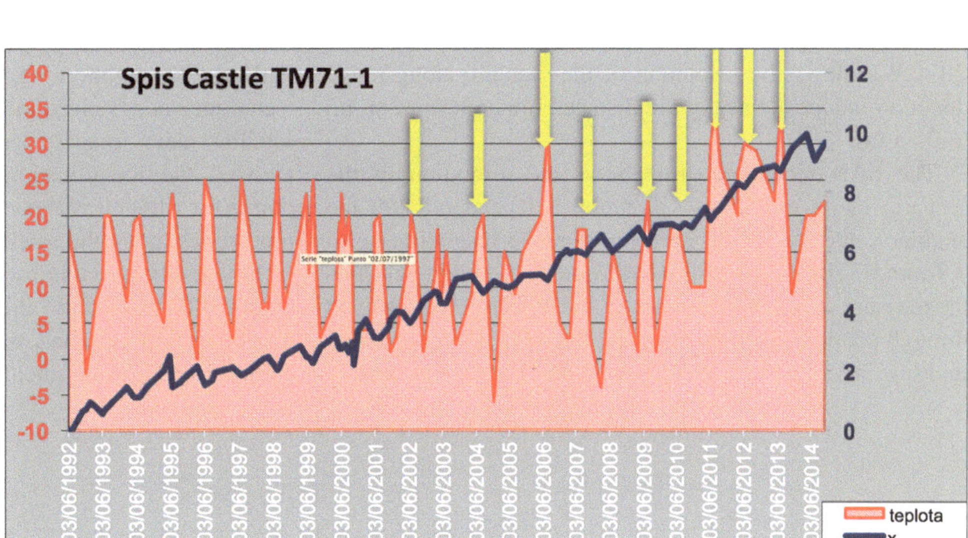

Fig. 16 Temperature versus displacements

Since each monitoring project has specific requirements, the measuring device used for deformation monitoring depends on the application, the chosen method and the required regularity and accuracy. Therefore, monitoring of slopes or landslide areas can only be defined, designed and realized in an interdisciplinary approach (Wunderlich 2006). A close cooperation with experts from geology, geophysics and hydrology together with experts from any measurement discipline such as geodesy and remote sensing and other academic fields is an indispensable requirement.

Following are some examples of advanced monitoring systems for the control of important archaeological sites.

Ground Based Radar Interferometry

Large vertical rupestrian sites, in open space conditions, can be effectively monitored, with high resolution and accuracy, through a new ground based remote sensing investigation: radar interferometry. An application is presented in the case study of Vardzia (Margottini et al. 2015).

Considering the morphological settings of Vardzia, (slope extent ca. 105 m^2) and slope instability processes (different typologies in size, magnitude and probability of occurrence), a new advanced simple and flexible monitoring system has been implemented in order to obtain measurements, processing and remote control in real time, and to transform in future, the monitoring system into a warning system. The system adopted for the monitoring of the entire cliff is based on ground based interferometric radar. This equipment allows the monitoring of ground surface displacement, along the line of sight, with a resolution of mm.

The radar system is the Stepped-Frequency Continuous Wave (SF-CW) coherent radar with SAR and interferometric capabilities. The acquisition station has been realized with the valuable support of the NACHPG and the pre-acquisition and start up activities were finalised and calibrated during the last field mission. The above mentioned technique (SF-CW) allows the resolution of the scenario along range direction independently of the distance (range resolution up to 0.75 m).

The SAR technique also allows the resolution of the scenario along cross-range direction independently (in the angular value) of the distance (cross-range resolution up to 4.3 mrad). The differential interferometry technique enables the measure of the displacement of the objects to be resolved through coupling SF-CW analysis. The system was installed in May 2012. As of this paper it has been continuously operating. The radar configuration adopted is reported in the following table, the "selection mask" contains about 50,000 points (Table 1).

The TLS derived DTM was used as 3D model for the visualization of the main monitored quantities (displacement and velocity) as collected and stored in real time by the monitoring system. The monitoring system is actually close to the end of the first 6 months of acquisition and the preliminary results are quite stable and promising.

With the exception of some ostensible changes in individual control points (mainly due to noise factors related to vegetation) the investigated area is stable and under control (Fig. 18). In the next 6 months on site verification of the main critical outcomes of the monitoring systems will be carried out along the cliff in order to calibrate and correct the results and define the most active zones in which downscaling of the landslide hazard and risk assessment is recommended.

Recent elaboration from Ilia University in Tbilisi, covering the time period 2013–2014, clearly identified 5 major areas of deformation (Fig. 16), one of which was stabilized in summer 2015.

Satellite Radar Interferometry

Petra is a famous archaeological city carved out of stone, hidden by towering sandstone mountains in Jordan. Although uninhabited today, during ancient times Petra was a wealthy trading town, capital of the Nabataean kingdom. The archaeological site is composed by many historical monuments and the Siq, a 1.2 km naturally formed gorge in the Sandstone Mountains that served, during the Nabataean times, and still today, as the main entrance to Petra. Slopes are continuously affected by rock falls and local sliding events, involving volumes from less than 1 m^3 to a few hundreds m^3 (Alberti et al. in printing).

An area of about 50 km^2, including Petra Archaeological Park and its neighbouring, was analysed with the SqueeSAR technique, an advanced InSAR algorithm. The analysis of 38 satellite radar images, acquired between 2003 and 2010, allowed the identification of about 62,000 measurement points for which it was possible to estimate the displacement time series along the satellite line of sight.

Potentially unstable rock blocks with medium-large size (> 50–100 m3) and already occurred rock falls, also involving heritage sites, were investigated, also through the comparison between satellite and ground based geotechnical

Table 1 Main parameters of radar configuration in Vardzia (Georgia)

Topic	Dimension	Value
Distance from the slope	(m)	350–500
Antenna beam width	(deg)	>70
Number of points	–	50,000
Range resolution	(m)	0.5
Cross range resolution	(mrad)	4.3
Scanning time	(min)	5

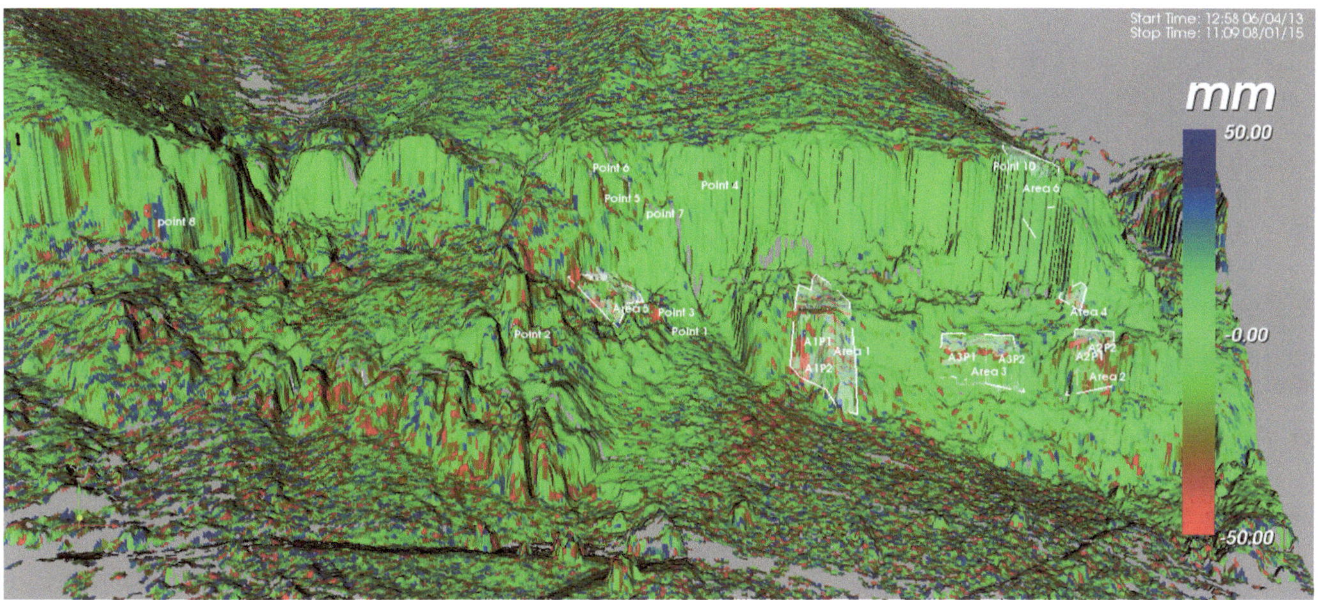

Fig. 18 Displacement detected from ground based radar interferometry on the rock cliff of Vardzia in the period 6 April 2013–8 January 2015. Displacements are plotted against distance from the Radar installation

Fig. 19 Satellite radar interferometry investigating the long term stability of El Dir (Monastery) in Petra

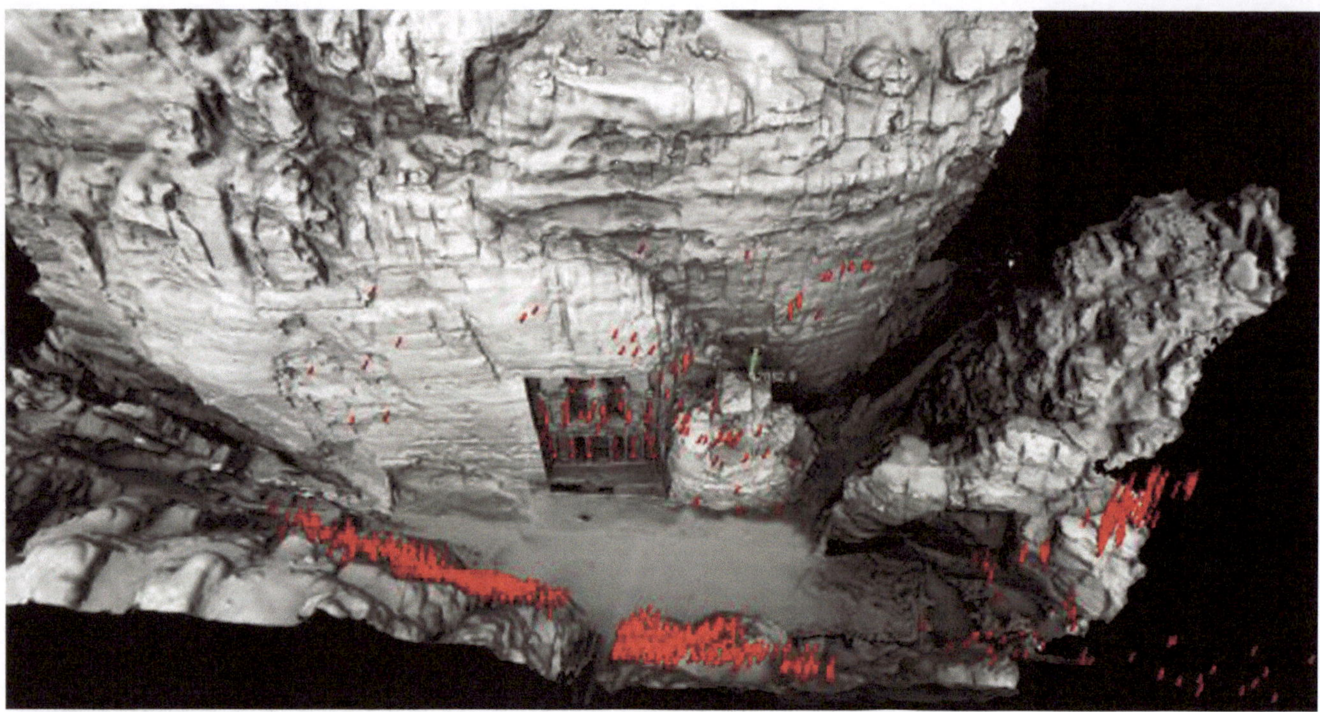

Fig. 20 Reflector-less surveyed points in the Treasury area (Petra, Jordan), superimposed on the terrestrial laser scanner

monitoring. Results of the analysis show that the area of Petra Archaeological Park is relatively safe from medium-large size rock fall (Fig. 19).

A possible seasonal effect on rock slope and monuments is evident, without a major permanent ground deformation, potentially leading to incipient collapse.. Minor rock falls (<50m^3) are clearly not detectable with medium-resolution satellite imagery, such as the one used in this study. In any case, further analyses may consider the use of the new satellite radar sensors (e.g. COSMO-SkyMed), characterized by a significant increase in both, spatial resolution of the data, and temporal sampling of the acquisitions, compared to those available until 2007.

Reflector-Less Robotic Surveying

For monitoring rock slope instability processes, geomatics methods are largely applied. Landslides are monitored by using Total Stations (Stiros et al. 2004; Tsai et al. 2012) and laser scanners, both conventional (Kasperski e al. 2010; Lichun et al. 2008) and recent models capable of full-waveform analysis (Mallet and Bretar 2009; Pirotti et al. 2013).

All geomatics-related activities should be integrated with geomorphological and geotechnical surveys, in order to better understand the potential failure mode of the landslide and set up a model; this analysis is also useful for early warning.

In the Siq of Petra, a robotic total station (LEICA TM 30), in reflector-less modality, was selected, to get the benefit of a high accuracy method as well as the largest capability to detect an analyze individual observation points selected by the rock mechanic specialists.

From a practical point of view, in any selected area one or more stations can be established. Such stations do not require precise permanent positions or ground markers, instead, a new highly precise position is determined for every new survey epoch, using a method referred to as Free-Station-Adjustment. This approach can be adopted if at least three permanently fixed prisms are visible from the Total Station set up point.

At present, more than 1600 individual points have been selected in potentially unstable areas of the Siq for reflector-less surveying. Figure 20 shows the point distribution in the treasury area.

Traditional Geotechnical Monitoring in Wireless Environment

The low environmental impact, even from a monitoring system, is a pre-requisite for any intervention in a high value cultural heritage site.

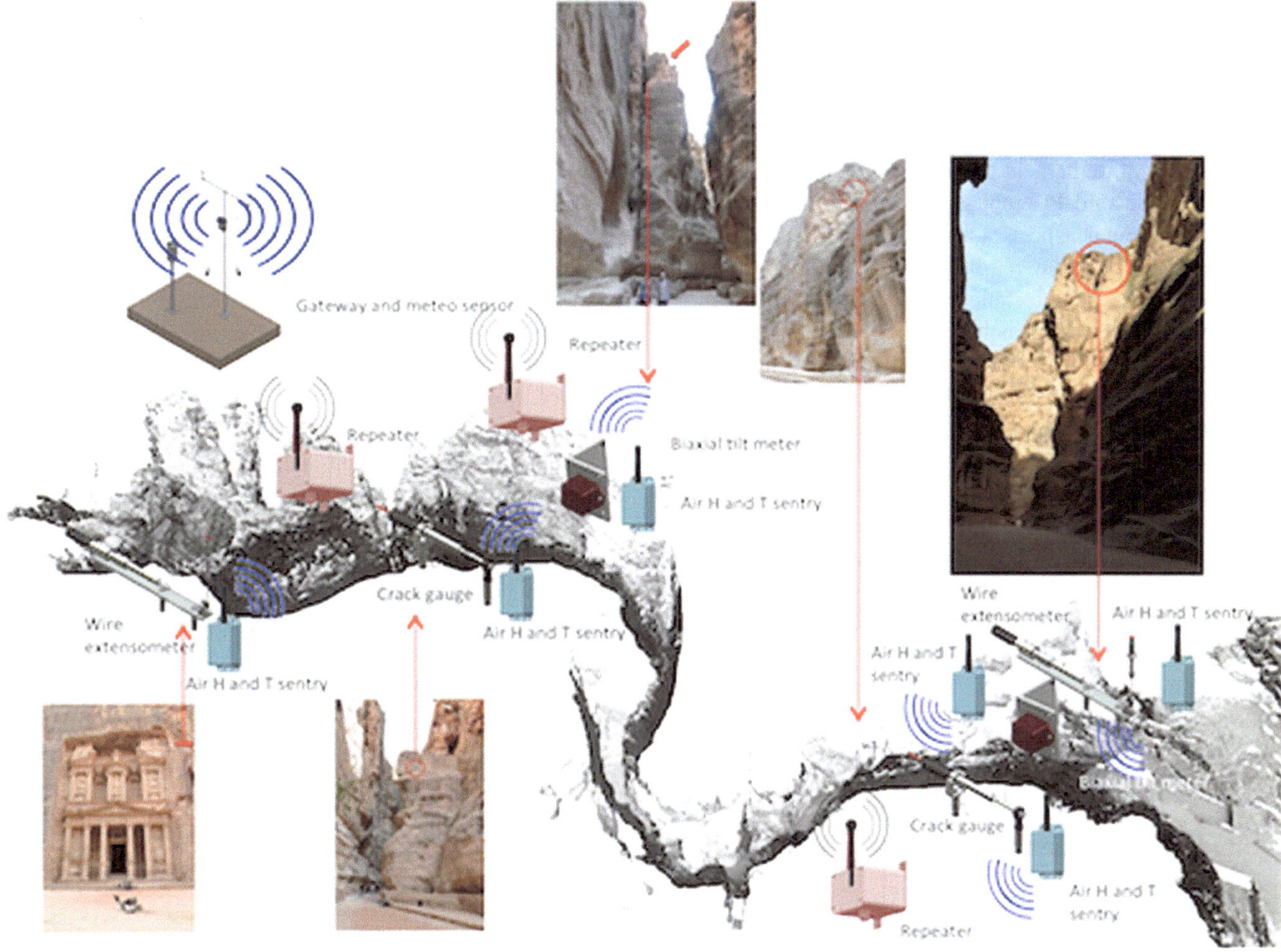

Fig. 21 Details of the geotechnical wireless monitoring network in the Siq of Petra (Jordan), superimposed on the 3D laser scanner of Siq

In order to define possible deformation paths and with the final aim to measure active displacement upon selected medium-large rock blocks affected by potential instability process, a proper permanent monitoring network was designed, implemented and installed along the entire Siq of Petra. The monitoring network (Fig. 21) consists of:

- n.2 tiltmeter sentries;
- n.2 wire deformometer sentries;
- n.2 crack meter/gauges sentries;
- n.6 air humidity and temperature sentries;
- n.1 meteorological station.

Geomechanical sentries are manufactured by Sisgeo and re-assembled by Minteos to provide longer duration measurement periods (about five years with four measurements per hour).

In order to solve the problem of the high visual impact of the monitoring network and related energy supplies and cables in the archaeological site of Petra, a wireless network equipped with long lasting batteries was designed, customized, implemented and installed. The present network ensures flexibility, remote control and future possibility to shift the system from monitoring to warning. The potentially unstable rock blocks were selected from the landslide inventory map, due to their potential hazard, size and typologies.

Conservation Policies and Management

Conservation of rupestrian sites is a complex endeavour, requiring expertise from the sciences of conservation, geotechnical engineering and earth science. On the other hand there is a high need for innovation addressing the central and exclusive role of conservators in the past and looking forward toward a truly holistic and interdisciplinary approach. In fact, measures to be adopted need to be as much as possible:

(i) effective
(ii) non-invasive and
(iii) feasible for the employment of local materials and manpower.

The first requirement is obviously aimed at solving the problem; the second to emphasize the maximum preservation of the original aspect of the site, whereas the third is meant to maximize the reproducibility, both in time and space, of the adopted techniques in case of further interventions. The latter also includes the involvement of local community in protection and development of the site. This is not a minor point, since local expertise and traditional knowledge may enrich the sustainability of the conservation policies, ensuring local management and the long term maintenance.

Following are some examples of rupestrian sites where the above approach was pursued. Clearly, each case is unique, showing peculiar natural processes and social elements. As a consequence each site must be investigated with high attention to avoid the imposition of techniques and solutions that are not balanced in the specific cultural, natural and human environment, then balancing mitigation strategies with local expertise and traditional knowledge.

Bamiyan (Afghanistan)

The historical site of Bamiyan is affected by geomorphological deformation processes which were worsened by the explosion of the Buddhas in March 2001, which destroyed the statues that date back to the 6th century AD (Fig. 22). Not only was invaluable cultural heritage irremediably lost, but the consequences of the explosions as well as the collapse of the giant statues also added greatly to the geological instability of the area. Traces of rocks which recently slid and fell are relevant proofs of the deterioration of its stability conditions and most parts of the site now appear prone to collapse in the near future.

Under the coordination of UNESCO, a global project to assess the feasibility conditions for the site's restoration was developed (Margottini 2014a, b); field data were collected and a mechanism for the potential cliff and niches' evolution was provided. In the meantime some consolidation works were carried out in the most critical rock fall-prone areas, to avoid any further collapse in the coming winter season, but also to enable archaeologists to safely catalogue and recover the Buddha statues' remains, still lying on the floor of the niches. The emergency activities started in October 2003 and finished on 2012, and included:

- the installation of a monitoring system, to evaluate in real time any possible deformation of the cliff. Sensors were designed to monitor the entire working area, connected with an alarm system, to guarantee the safety of those working in the site;
- the realization of temporary protection with steel ropes, and two iron beams suitable to avoid lateral deformation, inside the niche, from blocks destabilized by the explosion. Among the temporary work, just after the consolidation of the niche's wall, a wire net was installed over the back wall of both niches to allow archaeologists to work on the ground floor in safe conditions;
- the final stabilization of the East niche. In this area anchors, nails and grouting were introduced (Fig. 20), in order to reduce the risk of rock fall and collapse; particular care was given to the problem of grouting material because of the very high slaking capability of siltstone. The anchors placed in 2003 were pre-grouted to avoid any oxidation and then percolated inside the niche. In 2004 it was decide to use only stainless steel materials, even if not pre-grouted.
- minimization of intervention (anchor/nail head finishing) complete the execution of work. Anchor and nail heads were designed to be placed slightly inside the rock and then covered by a mortar allowing a total camouflage of the work. A number of tests for the best mixture of cement, local clay/silt and water, used to cover the anchor/bolt heads, were also developed in 2003, in cooperation with ICOMOS experts. The results provided the best chromatic, stability and robustness of the mixture.

Some minor consolidation and impermeabilisation works were implemented in the Western Buddha niche in 2009.

Consolidation works were mainly implemented by professional climbers, directly operating on the cliff as well as by means of inner niche scaffolding (Fig. 23).

Finally, after the stabilization of the external part of the Eastern Buddha niche, in 2009-2012 the inner back wall (shear zone of explosion) was stabilized by means of small anchors and limited grouting also aimed at fixing the still existing original part of the Statue plaster, jointly executed by Engineering Geologists and Conservators (Fig. 24).

Lalibela (Ethiopia)

Lalibela is located in the northern-central part of Ethiopia. The town, which has a population of about twelve thousand, is situated at an altitude of 2600 m in the Lasta province of the Amhara region.

The construction of the eleven rock-hewn churches is attributed to King Lalibela (1167–1207) of the Zagwe dynasty. They are still in daily use for religious practices and ceremonies, and on important religious occasions large

Fig. 22 Explosion and destruction of the Western (*left*) and Eastern (*right*) Giant Buddha Statues

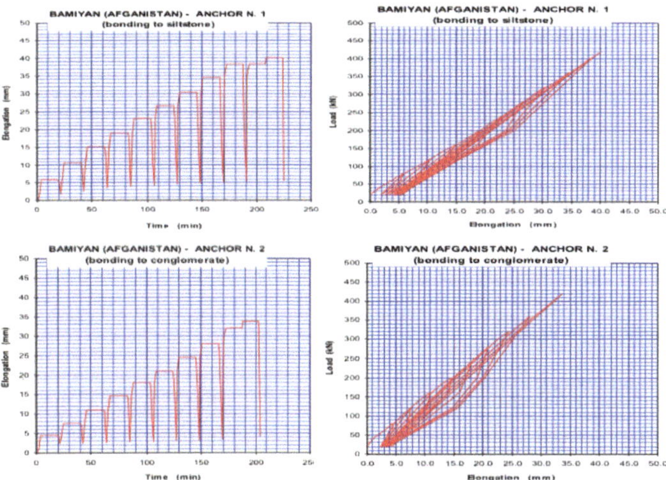

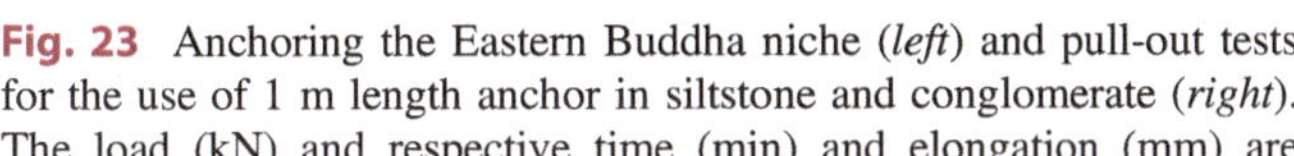

Fig. 23 Anchoring the Eastern Buddha niche (*left*) and pull-out tests for the use of 1 m length anchor in siltstone and conglomerate (*right*). The load (kN) and respective time (min) and elongation (mm) are reported showing, till 40 ton, the uphold of elastic domain and still the missing of any permanent deformation for the tested anchors

crowds of believers and pilgrims gather at the site. The eleven churches and their surrounding area form a complex that is unique in the world. In 1978 the churches of Lalibela were included in UNESCO's World Heritage List.

The churches have been exposed to physical erosion for approximately eight hundred years. As a result, their condition has worsened over the years, and has now become critical. To protect the churches from direct exposure to the rain, five churches, namely Bete Medhane-Alem, Bete Maryam, Bete Masqal, Bete Amanuel and Bete Abba-Libanos, have been covered by temporary shelters.

The shelter design had to respect the following requirements:

- complete reversibility;
- perspiration;
- non alteration of the aesthetic qualities and absolute respect of the harmonic shape of the complex and of the texture and colours of the materials;
- to be implemented in a way to allow local management and maintenance.

As of this paper no proper investigation or restoration has been implemented. Only new shelters were introduced, but with an impact that was altering the aesthetic value of the site. Thus, the shelters are protected from water infiltration and can be considered as a temporary reliable

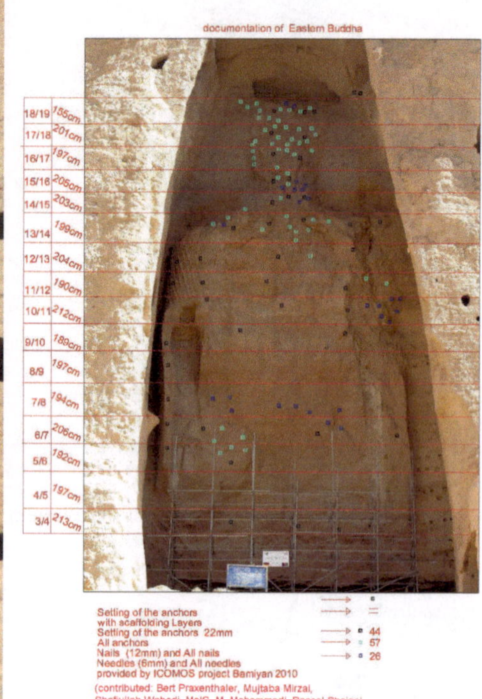

Fig. 24 The scaffolding for the stabilisation of back wall shear zone and the position and typology of installed anchors (diam. 22 mm, 12 mm, 6 mm)

measurement before a correct conservation plan can be realized (Fig. 25).

Petra (Jordan)

The Siq of Petra is entirely formed of fractured rock slopes and potential detachment of rock materials represents possible hazard to people. Slope stability mitigation techniques are briefly described according to the materials involved, as well as typology and magnitude of potential failures in the Siq slopes, focusing on the most common and feasible typologies that can be successfully applied to the mitigation of landslides.

The selection of a specific mitigation typology for the stabilization of blocks/slopes in the Siq has to be done according to several basic conditions that take into account the following: volume of the unstable block; height of the block above the ground; potential impact on archaeological remains; local technical feasibility; and cost/benefit analysis.

Field geological and geo-structural investigation of potential rock slope failures conducted in the Siq have determined that failure modes affecting the rock masses of the Siq slopes can be classified into the following categories, according to the type and degree of structural control (kinematic movement): (i) Planar failures; (ii) Wedge failures; (iii) Toppling failures; (iv) Free fall; (v) Unstable loose blocks and debris.

According to the inventory map produced for the Siq Stability project (Margottini 2015), the volumes of potentially unstable rock blocks have been differentiated into 3 classes: (i) small blocks with volume <5 m^3; (ii) medium blocks with 5–15 m^3 volume; and (iii) large blocks with volume >15 m^3.

Depending on typology and volume, several kinds of mitigation works have been suggested. In any case, such techniques cannot be applied without a local investigation on the site. In detail, the selection of potential working typologies has to take into account their impact on the peculiarity of the site geo-cultural environment (geomorphology, landslide types, presence of archaeological remains) so that only some specific typologies of consolidation interventions are recommended. Clearly, any intervention has to take into consideration the minimization of the environmental/visual impact of works, the local sustainability and feasibility, and the transfer of know-how to the local system. All these details are included in specific Guidelines (Margottini 2015) for the benefit of Jordanian authorities. Particular attention is given to the traditional techniques and the proper maintenance of such practices (Fig. 26).

Norchia (Italy)

The ancient Etruscan town of Norchia (Central Italy, 80 km North of Rome) is situated on a long volcanic plateau

Fig. 25 The modern shelters covering the rock-hewn churches of Lalibela (Ethiopia) (source www.flickr.com)

surrounded by steep slopes, at the confluence of rivers Pile and Acqua Alta into the river Biedano. It has been constructed along the ancient Via Clodia, a short-range route intended for commercial traffic between Rome and the colonies in Etruscan lands.

The flourishing of the town, evidenced by the beautiful necropolis, is placed between the end of the fourth and half of the second century BC. With its necropolis Norchia is the most significant example of funerary architecture rock Hellenistic period (IV-II century BC.). Its rock-cut tombs are among the most important archaeological sites of Etruscan civilization. They are an important and rare example of rock architecture and one of the few preserved in Italy. Also, the necropolis, with an extension of more than 100 hectares, is composed of rock-cut tombs of various types (façade, half-cube, false-cube and temple type) and dimensions (4–10 m in height), exhibiting a remarkable similarity with Asian tombs. From a geological point of view, the area is exhibiting the overly rigid volcanic products from both Vico and Volsini volcanic apparatus; as bedrock, a plastic clay formation is positioned (Fig. 27).

The rock-cut tombs were excavated on two main volcanic levels, following the natural profile of tuff outcrops. The tombs located in the upper part of the necropolis have been excavated in a Red Tuff from Vico volcanic district, whereas those in lower level are dug into a grey tuff (Nenfro) from Vulsini volcanic apparatus. Recent investigations revealed the presence of many threats affecting the conservation of the site that include: surface rock weathering, water percolation and infiltration, surface vegetation and biological colonization, instability and collapse of the cliff. The purpose of this study was mainly focused to verify whether the geological, geomorphological and geomechanical processes that have allowed the creation of a typical "butte" landscape, later inhabited by Etruscans, are still active. Field survey and historical data collection revealed the presence of many rock slope instabilities that have affected the site. Particularly meaningful is the presence of a large debris fan, just at the toe

Fig. 26 Manual and proper cleaning of stone dams, to protect the Siq of Petra from flash flood

of the most relevant archaeological feature, where the half-cube rock-cut tombs are positioned, testifying to the importance of rock-falls after the excavation of the necropolis.

In recent periods such instabilities have been magnified by the weathering of the tuff and the growth of vegetation that is disconnecting original blocks.

The Necropolis is currently exhibiting a lack of maintenance. The preliminary investigation (Margottini et al. 2016a, b, c, d) revealed the need of stabilization of exposed rock cliff by means of anchor and dowels, as well as the reduction of water infiltration in existing joints and cracks. Such discontinuities have been probably generated by the stress release resulting in the formation of the valley and now widening as a consequence of vegetation. Eliminating vegetation in a very large area (>10 Ha) is not an easy task. Indeed, in recent years the presence of farm animal (cows and sheep) highly reduced the growing of vegetation. In a situation of low budget availability, the implementation of a natural control system for vegetation through sheep and cow-farming was proposed by the local community. The latter can help to make the site more accessible to visitors and reduce the growth of vegetation affecting the slope stability. This maintenance solution, even if very trivial, can be implemented by means a public-private partnership, for the benefit a unique cultural heritage site, presently abandoned.

Vardzia (Georgia)

Vardzia represents an excellent example of a rock-cut city, which unites architectural monuments with an outstanding natural-geological environment. Such monuments are particularly vulnerable and their restoration and conservation requires a complex approach. The site is carved in various layers of volcanic tuffs and covers several hectares, with chronologically different segments of construction. This monument, as many similar monuments worldwide, is subjected to a slow but permanent process of destruction, through the following factors: surface weathering of rock, active tectonics (seismic displacement along the active faults and earthquakes), interaction between lithologically different rock layers, existence of major cracks and associated complex block structure, surface rainwater runoff and infiltrated

Fig. 27 The temple tombs of Norchia, as reconstructed form Canina (1851) and after the cleaning of vegetation in Spring 2016

ground water, temperature variations, etc. During its lifetime, Vardzia was heavily damaged by historical earthquakes, such as in 1283, and only partly restored afterwards.

Currently there is major threat related to rock fall and rock slide (Margottini et al. 2015).

During the summer of 2015, after a joint collaboration with local Universities and Research Agencies, a first consolidation intervention was established. The mitigation project was based on an advanced monitoring system (Margottini et al. 2015) and the following field survey:

- Geomechanical rock mass classification through scan lines, in order to derive the main geomechanical characteristics and indexes (e.g. RMR, GSI);
- Tilt test;
- Schmidt-hammer test on joint surfaces and intact rock block for in situ analysis of UCS (unconfined compressive strength);
- Point load test to provide UCS data from sampled blocks (ISRM 1981, 1978);
- Strength and deformation parameters from scientific and technical literature (ISRM 1978; Hoeck 2007; Barton 1973) as well as from local technical reports;
- laboratory tests on tuff rock blocks and cores (Uniaxial and tensile strength parameters in dry and saturated conditions).

The main results are summarized in Table 2.

All the performed investigations allowed the definition of the most unstable areas. The map is reported in Fig. 29.

Fig. 28 Map of a large debris fan, just at the toe of the most relevant archaeological place, where the half-cube rock-cut tombs are positioned, testifying important rock-falls after the excavation of the necropolis (Colonna Di Paolo and Colonna 1978)

Table 2 Main geomechanical parameters of Vardzia rocks

Lithology	Unit weight γ (KN/m3)	Porosity (%)	σc dry MPa	σc sat MPa	Basic friction angle (ϕ°)	GSI	σt dry MPa	σt sat MPa
Grey tuff	16.3	37.2	10.3	3.6	22°–32°	70	0.8	0.3
White tuff	15.9	38.8	8.7	2.8	22°–32°	65	0.9	0.3

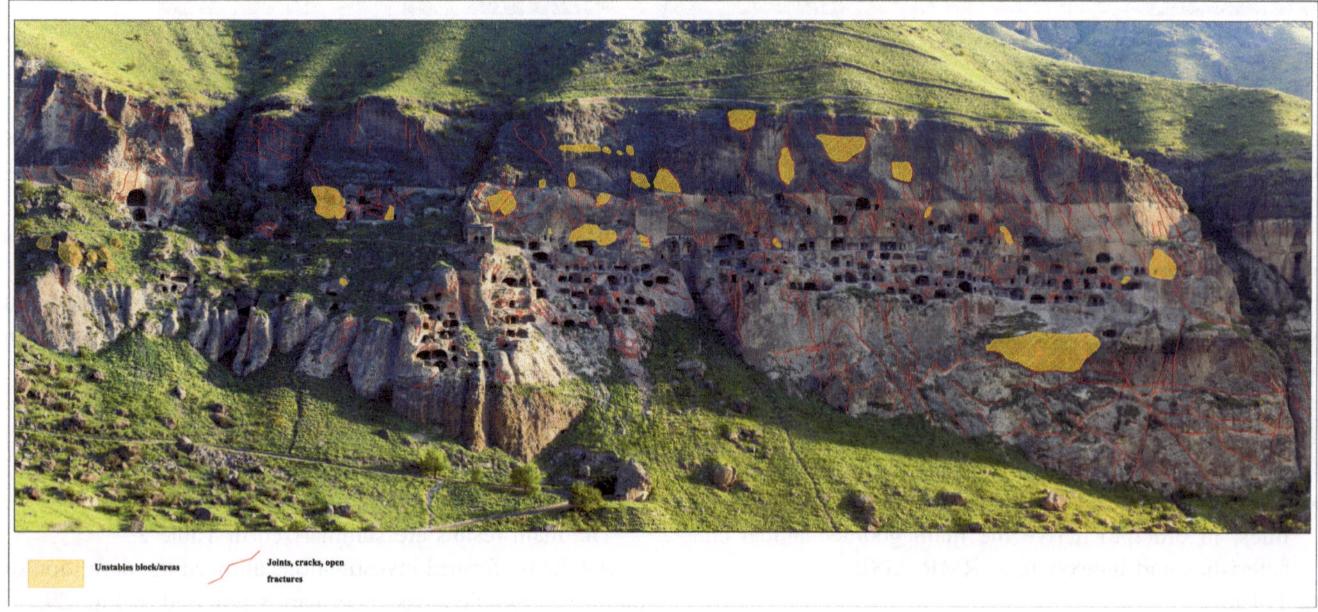

Fig. 29 Joints, cracks and unstable blocks in Vardzia cliff

It was then decided to stabilize one of the potentially unstable blocks, located in the lower-right portion of Fig. 28, the largest among the selected ones. Such a block is also interesting because of the presence of an archaeological tunnel, connecting the flood plain with the rock-cut city.

The block, partially overhanging, is delimited on its back by a joint characterized by a high dip (on average 80°) and dip direction between 170° and 178°. Its length is about 11.5 m along this section, whereas the area of the block is about 20.8 m^2. The more likely failure mechanism, based on numerous in situ observations conducted along the cliff, is that of sliding along the joint previously described. This discontinuity is also characterized by a significant opening in some of its portions, highlighting the precarious stability conditions of the rock-mass under investigation.

The stability of the block was then evaluated adopting the limit equilibrium approach, with reference to possible sliding along the joint above described, together with the effectiveness of the proposed mitigation measurements consisting of passive rock dowels. The safety factor FS for the potential sliding can be expressed as:

$$FS = \frac{C_{rockbridges} + N\ \tan\left(\varphi_r + JRC\log_{10}\frac{JCS}{\sigma_n}\right)}{T}$$

where the symbol $C_{\text{rock bridges}}$ indicates the strength contribution of the rock bridges potentially acting along the joint.

The final calculation was then elaborated, also taking into consideration the block reinforcement constituted by the installation of 32 mm diameter steel dowels, characterized by improved adherence, with reference to a Feb44 k concrete. The dowels should show a minimum characteristic yielding stress equal to 430 MPa, for a corresponding yielding load of 346 kN.

Unfortunately no experienced drilling company was available in Georgia, nor was a proper scaffolding for the site. It was then decided to initiate a large capacity building exercise of a local company contracted by the Agency for

Fig. 30 The wood scaffolding realised in Vardzia

Fig. 31 Positioning of the the 48 installed dowels

Cultural Heritage Preservation of Georgia. An Italian engineer with considerable field work experience then accompanied the purchasing of equipment and the field work, including the construction of the scaffolding. The final successful result is shown in the following figure showing the realized wood scaffolding and the 48 installed dowels (Figs. 30 and 31).

Conclusions

The protection of rupestrian sites, from geotechnical and geological hazards is an interdisciplinary effort, involving, minimally, the Science of Conservation of Cultural Heritages and Earth Science. The conservator has to develop the proper restoration project, taking into consideration and having understood geological processes acting on the site and the monument; in the mean time, the engineering geologist has to implement a mitigation plan and monitoring system which fulfill the request for low impact and perfect integration of solutions into the archaeological contest. A typical example of connection points between these two major branches of science in heritage conservation, is the usage of solutions with low environmental impact, that cannot damage the site or the cultural landscape, while clearly reducing the natural processes acting on the site; similarly it is required to use materials that, over time, cannot loose original properties, generating salts, oxides, etc., that may affect the integrity and conservation of the heritage site. In this context also the proposed monitoring systems need to fulfill the requirement of low environmental impact and minimum interference with the archaeological remains.

The above reflections, without obviously being exhaustive, clearly underline the impact that the Earth Sciences have had in understanding and monitoring threats, as well as in the conservation of the cultural properties; it is self evident that the same disciplines have to assume, today and in the future, a fundamental role in all the policies that are necessary for the protection and conservation of the heritage.

This aspect of conservation has never been very clear in the past, since the archaeology and the conservation aspects had a strong centrality and autonomy. This point of view is now less evident, with more attention to the integration of different sciences. Indeed it is possible to affirm that the protection of cultural heritage represents an interdisciplinary process (and not multi-disciplinary) at the border of art, history, science, policies for management and exploitation.

The present paper demonstrates how the conservation of rupestrian sites requires an interdisciplinary approach, developing comprehensive field investigations and monitoring processes but, finally, implementing conservation practices and management of the sites that are based on local expertise and traditional knowledge. Clearly this paper does not report the comprehensive methodology, generally well known, but rather highlights the contribution of emerging technologies and interdisciplinary approaches to some specific issues. In this light, major recent achievements for application in cultural heritage, without being exhaustive, involve the following. In the field of basic data collection: laser scanning, new panorama photo utilization, mineralogy and petrography, rock mechanic parameters, semi-automatic discontinuities detection and analysis; kinematic analysis, infrared thermographic analysis. In the field of monitoring: ground based radar interferometry; reflector-less robotic surveying, laser scanning and traditional geotechnical monitoring in a wireless environment.

The collected data, jointly with traditional information mainly from rock mechanic and rock fall/slide investigation, allow the deep understanding of processes affecting a given site. The consequent mitigation strategy can then be prepared in order to ensure effectiveness of adopted solutions, non-invasive infrastructures and hopefully, feasible employment of local materials and manpower. Moreover, a proper understanding of local expertise joined with traditional and indigenous knowledge should address the proposed solutions in order to facilitate their execution and maintenance in time, and then, in other words, to enhance sustainability.

Acknowledgements The present paper was based on many interdisciplinary projects, mainly coordinated by UNESCO. The authors are very grateful to UNESCO colleagues and friends in Paris (WHC), Kabul and Amman, as well as to colleagues and friends of the National Agency for *Cultural Heritage Preservation* of *Georgia*.

References

Agustawijaya DS (2007) The uniaxial compressive strength of soft rock. Civ Eng Dimension 9(1):1, 9–14

Alberti S, Ferretti A, Margottini C, Spizzichino D (in printing) Surface deformation data in the archaeological site of Petra from medium-resolution satellite radar images. Submitted to J Cult Heritage

Bakun-Mazor D, Hatzor YH, Glaser SD, Santamarina JC (2011) Climatic effects on key-block motion: evidence from the rock slopes of Masada world heritage site. 45th US symposium in rock mechanics, San Francisco, California, pp 480–487

Bakun-Mazor D, Hatzor YH, Glaser SD, Santamarina JC (2013) Thermally versus seismically induced block displacements in Masada rock slopes. Intl J of Rock Mech and Ming Sci 61:196–211

Baroň I, Bečkovský D, Míča L (2014) Application of infrared thermography for mapping open fractures in deep-seated rockslides and unstable cliffs. Landslides 11:15–27

Barton NR (1973) Review of a new shear-strength criterion for rock joints. Eng Geol 7:287–332

Canina L (1851) Canina 1851: L'antica Etruria marittima, vol II Tavole, Roma

Casagli N, Gigli G (2011) Semi-automatic extraction of rock mass structural data from high resolution LIDAR point clouds. Int J Rock Mech Min Sci 48:187–198

Clark MR, McCann DM, Forde MC (2003) Application of infrared thermography to the non-destructive testing of concrete and masonry bridges. UK, NDT&E International 36:265–275

Colonna Di Paolo E, Colonna G (1978) Norchia—Le necropoli rupestri dell'Etruria meridionale, vol 1 and 2, Roma

Dehn M, Bürger G, Buma J, Gasparetto P (2000) Impact of climate change on slope stability using expanded downscaling. Eng Geol 55:193–204

Delmonaco G, Margottini C (2014) General environmental conditions. In: Margottini C (ed) After the destruction of Giant Buddha statues in Bamiyan (Afghanistan) in 2001. A UNESCO's emergency activity for recovering and rehabilitation of cliff and niches. Springer-Verlag, Berlin Heidelberg, pp 69–100

Ercoli L, Megna B, Nocilla A, Zimbardo M (2013) Measure of a limestone weathering degree using laser scanner. Int J Architect Heritage 7:591–607

Fecker E (2006) Report on rock mechanical aspects concerning the Eastern Buddha Niche, ICOMOS Internal Report to UNESCO

Gigli G, Frodella W, Garfagnoli F, Morelli S, Mugnai F, Menna F, Casagli N (2013) 3-D geomechanical rock mass characterization for the evaluation of rockslide susceptibility scenarios. Landslides 1: 1–14

Gischig V, Moore J, Evans K, Amann F, Loew S (2011a) Thermo-mechanical forcing of deep rock slope deformation. 1. Conceptual study of a simplified slope. J of Geoph Res: Earth Surface, 116, F04010

Gischig V, Moore J, Evans K, Amann F, Loew S (2011b) Thermo-mechanical forcing of deep rock slope deformation. 2. The Randa rock slope instability. J of Geoph Res: Earth Surface, 116, F04011

Gomez-Heras M, Smith BJ, Fort R (2008) Influence of surface heterogeneities of building granite on its thermal microenvironment and its potential for the generation of thermal weathering. Environ Geol 56:547–560

Grøneng G, Christiansen H, Nilsen B, Blikra LH (2011) Meteorological effects on seasonal displacements of the Aknes rockslide, western Norway. Landslides 8:1–15

Gunzburger Y, Merrien-Soukatchoff V (2011) Near-surface temperatures and heat balance of bare outcrops exposed to solar radiation. Earth Surf Process and Land 36:1577–1589

Hales TC, Roering JJ (2007) Climate controls on frost cracking and implications for the evolution of bedrock landscapes. J. Geoph. Res-Earth Surf 112:1–14

Hatzor YH (2004) Keyblock stability in seismically active rock slopes – the Snake Path cliff—Masada. J of Geotech and Geoenv Eng 129:697–710

Hoek E (2007) Practical rock engineering. http://www.rocscience.com. Access Sept 2015

Kasperski J, Delacourt C, Allemand P, Potherat P, Jaud M, Varrel E (2010) Application of a Terrestrial Laser Scanner (TLS) to the Study of the Séchilienne Landslide (Isère, France). Remote Sens 2:2785–2802

Krähenbühl R (2004) Temperatur und Kluftwasser als Ursachen von Felssturz. Bulletin fur angewandte Geologie 9:19–35

ISRM—International Society of Rock Mechanics—Commission On Standardization of Laboratory And Field Tests (1978) Suggested methods for the quantitative description of discontinuities in rock masses. Int J Rock Mech Min Sci Geomech Abs 15(6):319–368

ISRM—International Society of Rock Mechanics—Commission on the Classification of Rocks and Rock Masses (1981) Basic geotechnical description of rock masses. Int J Rock Mech Min Sci Geomech Abs 18:85–110

Lichun S, Wang X, Zhao D, Qu J (2008) Application of 3D laser scanner for monitoring of landslide hazards. Int Arch Photogrammetry, Remote Sens Spat Inf Sci 37:277–281

Mallet C, Bretar F (2009) Full-waveform topographic Lidar: State-of-the-art. ISPRS J Photogrammetry and Remote Sens 64:1–16

Margottini C (ed) (2014) After the destruction of Giant Buddha statues in Bamiyan (AFghanistan) in 2001. A UNESCO's emergency activity for the recovering and rehabilitation of cliff and niches. Springer-Verlag, Berlin-Heidelberg

Margottini C (2014) Properties of local materials. In: Margottini C (ed) After the destruction of Giant Buddha statues in Bamiyan (Afghanistan) in 2001. A UNESCO's emergency activity for recovering and rehabilitation of cliff and niches. Springer-Verlag, Berlin Heidelberg, pp 153–171

Margottini C (ed) (2015) Guidelines for the sustainable mitigation and management of landslides at the Siq—Petra World Heritage Site. ISPRA Internal Report to UNESCO-Draft

Margottini C, Antidze N, Corominas J, Crosta GB, Frattini P, Gigli G, Giordan D, Iwasaky I, Lollino G, Manconi A, Marinos P, Scavia C, Sonnessa A, Spizzichino D, Vacheisvili N (2015) Landslide hazard assessment, monitoring and conservation of Vardzia Byzantine monastery complex, Georgia. Landslides 12:193–204

Margottini C, Gigli G, Ruther H, Spizzichino D (2016a) Advances in geotechnical investigations and monitoring in rupestrian settlements inscribed in the UNESCO's World Heritage List. In: Proceedings of the fourth Italian workshop on landslides, Procedia, Elsevier

Margottini C, Gigli G, Ruther H, Spizzichino D (2016b) Advances in sustainable conservation practices in rupestrian settlements inscribed in the UNESCO's World Heritage List. In: Proceedings of the fourth Italian workshop on landslides, Procedia, Elsevier

Margottini C, Spizzichino D (2014) How geology shapes human settlements. In: Bandarin F, van Oers R (eds) (2014) Reconnecting the city. The historic urban landscape approach and the future of Urban Heritage. Wiley Blackwell, Chichester, pp 47–84

Margottini C, Spizzichino D, Argento A, Russo A (2016) Rock-fall hazard in the Etruscan archaeological site of Norchia (Central Italy). Geophys Res Abstracts 18:14434 (EGU General Assembly 2016)

Margottini C, Spizzichino D, Crosta GB, Frattini P, Mazzanti P, Scarascia Mugnozza G, Beninati L (2016d) Rock fall instabilities and safety of visitors in the historic rock cut monastery of Vardzia (Georgia). In: Rotonda T, Cecconi M, Silvestri F, Tommasi P (eds) Volcanic rocks and soils. CRC Press Balkema, Taylor and Francis Group, pp 371–378

Metzger R, Jaboyedoff M, Oppikofer T, Viero A, Galgaro A (2009) Coltop3D: a new software for structural analysis with high resolution 3D point clouds and DEM. AAPG Search and Discovery Article #90171 CSPG/CSEG/CWLS GeoConvention 2009, Calgary, Alberta, Canada, 4–8 May 2009

Oppikofer T, Jaboyedoff M, Pedrazzini A, Derron MH, Blikra LH (2011) Detailed DEM analysis of a rockslide scar to characterize the basal sliding surface of active rockslides. J Geophys Res 116:1–22

Pirotti F, Guarnieri A, Vettore A (2013) Ground filtering and vegetation mapping using multi-return terrestrial laser scanning. ISPRS J Photogrammetry and Remote Sens 76:56–63

Pasten C, García M, Cortes DD (2015) Physical and numerical modelling of the thermally induced wedging mechanism. Geotechnique Lett 5(3):186–190

Renzulli A, Antonelli F, Margottini C, Santi P, Fratini F (2011) What kind of volcanite the rock-hewn churches of the Lalibela UNESCO's world heritage site are made of? J Cult Heritage 12:227–235

Rüther H (2015) Surface model of the Siq. In: Margottini C (ed) Guidelines for the sustainable mitigation and management of landslides at the Siq—Petra World Heritage Site. ISPRA Internal Report to UNESCO-Draft, pp 7–11

Rüther H, Bhurtha R, Wessels S (2014) Spatial documentation of the Petra World Heritage Site . Africa Geo 1

Stiros SC, Vichas C, Skourtis C (2004) Landslide monitoring based on geodetically derived distance changes, J Surv Eng-ASCE, 130 (4):156–162

Tsai Z, You GJY, Lee HY, Chiu YJ (2012) Use of a total station to monitor post-failure sediment yields in landslide sites of the Shihmen reservoir watershed. Geomorphology 139–140:438–451

Vargas JR, Castro JT, Amaral C, Figueiredo RP (2004) On mechanism for failure of some rock slopes in Rio de Janeiro, Brasil: thermal fatigue? In: Lacerda WA, Ehrlich M, Fontoura SAB, Sayao ASF (eds) Proc of 9th Int Symp on Landslides. Rio de Janeiro, Brazil, pp 1007–1013

Varnes DJ (1978) Slope movement types and processes. In: Schuster RL, Krizek RJ (eds) Landslides, analysis and control. Transportation research board Sp. Rep. No. 176, Nat Acad of Sciences, pp 11–33

Vlcko J, Greif V, Grof V, Jezny M, Petro L, Brcek M (2009) Rock displacement and thermal expansion study at historic heritage sites in Slovakia. Environ Geol 58:1727–1740

Wunderlich T (2006) Geodätisches Monitoring—Ein fruchtbares Feld für interdisziplinäre Zusammen-arbeit. Vermessung & Geoinformation 1 + 2:50–62

Vlcko J, Jezny M, Pagacova Z (2005) Influence of thermal expansion on slope displacements. In: Sassa K, Fukuoka H, Wang F, Wang G (eds) Proceedings of the first general assembly of the international consortium on landslides. Landslides: Risk Analysis and Sustainable Disaster Management. Springer Verlag, Washington

Subaerial Landslide-Generated Waves: Numerical and Laboratory Simulations

Saeedeh Yavari-Ramshe and Behzad Ataie-Ashtiani

Abstract

Subaerial landslide-generated waves (SALGWs) are among destructive hazards which have been not often studied in comparison with earthquake-generated tsunamis and submarine landslide-generated waves. This paper represents a brief review of the physical and numerical studies on SALGWs. Samples of the laboratory experiments are provided and it is highlighted that all the available data should be combined and studied collectively to overcome the discrepancies and improve our understandings of SALGWs. Commonly applied numerical approaches to simulate SALGWs are discussed. A Boussinesq-type model (LS3D) considering landslide as a rigid body, and a two-layer shallow-water type model (2LCMFlow) considering landslide as a layer of a Coulomb mixture are utilized to investigate the effects of landslide deformations on the characteristics of the landslide-generated waves (LGWs) based on a set of available experimental data. With a rigid landslide assumption, the maximum height of LGW is about 16% overestimated. Dense material deformes into a thick front—thin tail profile and induce a LGW consists of a larger wave crest than the wave trough while loose material shows a dam-break type behaviour with a LGW having a larger wave trough. A real case of SALGW is simulated by both models. The maximum LGW height predicted by the 2LCMFlow model which is closer to the physics is about 14% less than the equivalent value predicted by the LS3D model. On the other hand, the LS3D model, with the 4th order of accuracy of wave dispersion, simulates the LGW propagation stage more efficiently and with around 30% less runtime. Assessing the effects of the landslide initial submergence on the LGW characteristics shows that a semi-submerged, a submarine, and a subaerial landslides induce the largest wave crest, wave trough, and landslide runout distance, respectively. Combining different conceptual and mathematical models at the various stages of SALGWs initiation, propagation, transformations and runup can advance the current numerical practice, in this field, both from accuracy and computational efficiency point of views.

Keywords

Subaerial landslide • Tsunami • Impulsive wave • Numerical modelling • Physical model • Case study

---

S. Yavari-Ramshe (✉) · B. Ataie-Ashtiani
Department of Civil Engineering, Sharif University of Technology, Tehran, Iran
e-mail: yavari@mehr.sharif.edu

B. Ataie-Ashtiani
e-mail: ataie@sharif.edu

K. Sassa et al. (eds.), *Advancing Culture of Living with Landslides*,
DOI 10.1007/978-3-319-59469-9_3

Nomenclature

$\mathbf{A}$	Constant $\nabla(\nabla \cdot \mathbf{u}_0)$
a	Wave amplitude
a_0	Characteristic wave amplitude
a_{nm}	Maximum negative wave amplitude
a_{pm}	Maximum positive wave amplitude
B	Parameter $\nabla \cdot (h\mathbf{u}_0) + h_t/\varepsilon$
b	Bottom topography
C	Parameter $f(\mathbf{A}, \mathrm{B})$
g	Gravitational acceleration
H	Wave height
h	Water depth
h_0	Still water depth
h_1	Water layer depth
h_2	Landslide layer depth
K	Earth pressure coefficient
L	Wave length
l_0	Characteristic length
l_S	Landslide length
P	Pressure
P_{zz}	Normal pressure
q	Discharge hu
r	Relative density ρ_2/ρ_1
S	Slide initial distance from water surface
T	Wave period
T_S	Landslide thickness
t	Time
U_S	Rigid landslide velocity vector
$\mathbf{u}$	Horizontal velocity vector (u, v)
u	Velocity component in x direction
$\mathbf{u}_0$	Constant
$\mathbf{u}_1$	Constant
$\mathbf{u}_2$	Constant
V_S	Landslide volume
v	Velocity component in y direction
w	Velocity component in z direction
w_S	Landslide width
x	Cartesian coordinate component
y	Cartesian coordinate component
z	Cartesian coordinate component
z_a	A distinct water depth
z_b	A distinct water depth
z_S	Slide initial height due to water surface
$\tilde{z}$	Characteristic depth
β	Weighting parameter
δ	Basal friction angle
δ_0	Angle of repose
δ_{mod}	Modified δ
ε	Wave nonlinearity index a_0/h_0
ζ	Water surface fluctuations
θ	Slope angle
Λ_1	Parameter $\lambda_1 + K(1 - \lambda_1)$

Λ_2	Parameter $r\lambda_2 + K(1 - r\lambda_2)$
λ'	Constant parameter
λ_1	Constitutive coefficient
λ_2	Constitutive coefficient
μ	Wave dispersion index h_0/l_0
ρ	Density
σ_c	Critical basal stress
ϕ	Internal friction angle
$\Im$	Coulomb friction term
∇	Gradient vector $(\partial/\partial x, \partial/\partial y)$
Δt	Time step
Δx	Grid size in x direction
Δy	Grid size in y direction

Introduction

Impacts of earthquakes, landslides, volcanic eruptions and meteorites, into a water body such as lakes, reservoirs, and oceans generate impulsive gravity water waves. The most common cause of such impulsive wave events is seismic displacements of the seafloor which initiate earthquake-generated waves (EGWs) or tsunamis. Tsunamis are devastating disasters with fatal consequences due to overtopping dams, shorelines, and their defensive structures, and subsequent flooding. Landslide-generated waves (LGWs) are gravity waves generated by the impulsive impacts of a submarine or subaerial landslide into a water body. In some cases, a combination of seismic displacements and mass failures are required to explain the large values of the observed induced wave heights and run-ups (Watts and Borrero 2001). Large-scale LGWs are extreme natural hazards with destructive and fatal consequences on infrastructures and communities.

Landslides and volcanos are the most common sources of tsunamis after earthquakes (Yavari-Ramshe and Ataie-Ashtiani 2016). LGWs are the sources of the biggest tsunami hazards to the US Atlantic coast even though EGWs are more often (ten Brink et al. 2014). The term "landslide" includes all types of natural gravity mass movements mobilizing soil, rock, lava, ice, pyroclastic material, and snow (Hungr et al. 2001). When a landslide is initially located beneath the water surface it is called submarine landslide (SML) and when a landslide starts its motion outside the water, it is a subaerial landslide (SAL).

For the past decades, engineers and researchers have investigated various aspects of LGW phenomena including the LGW generation, propagation, and other related consequences. Two recent samples of the endeavors to describe the state of the art regarding the tsunami waves in general and LGWs are pointed out in the following. The October 2015 Theme issue of *Philosophical Transactions of Royal Society A* was devoted to the "Tsunamis: bridging science, engineering and society" (Kanoglu et al. 2015). In a decade after 2004 Boxing Day tsunami, the issue presented the advances in tsunami science, including methodologies, standards for warnings, and challenges. A recent issue of *Landslides* was published as a thematic issue "Landslide-generated tsunami waves" (Ataie-Ashtiani 2016). The issue consisted of papers regarding LGWs and the interactions between landslide and water body. The issue presented the laboratory experiment investigations, review of the numerical and analytical aspects of the LGWs modelling, and application of numerical simulation in association with field measurements and examinations of LGW events and case studies.

The 2004 Indian Ocean earthquake and tsunami with 280,000 casualties is the deadliest natural disaster since 2000. Due to the catastrophic impacts of tsunami events in the coastal area, a major part of the available technical literature emphasis on the submarine landslide generated waves (SMLGWs). The study of LGW processes, e.g. slides initiation (or triggering), motion, interaction with water and air, impulsive wave generation, propagation, run-up, overtopping, and inundation, is a multidisciplinary and challenging task and is a vital venture for the informed assessment and management of the LGW risks. The objective of this work is to provide a brief review of the recent developments regarding subaerial landslide generated tsunami waves (SALGWs). This work does not aim to provide a comprehensive review, though to present some samples from applications of different tools and methods that have been applied so far, and a particular emphasize is given to the authors' previous experiences and works and their research prospects.

Subaerial Landslide-Generated Waves

When a SAL moves into a water body, it generates SALGWs. The wave characteristics (e.g. amplitudes,

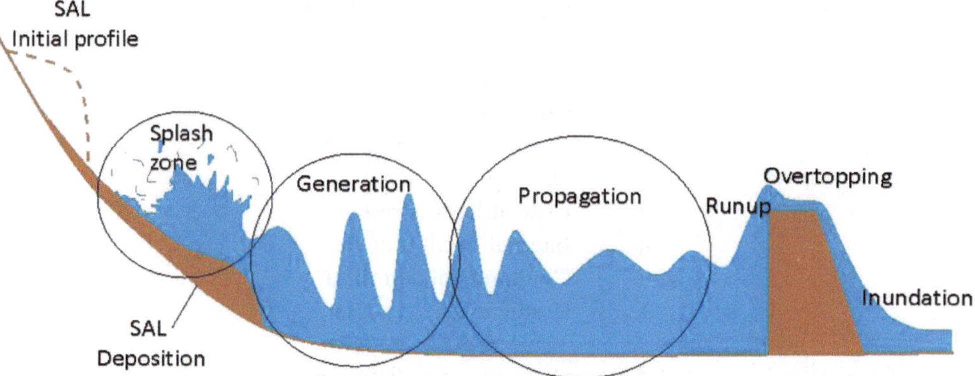

Fig. 1 A schematic of different stages involved in a SALGW event

velocity, wavelength and period) are governed by the water body geometry, depth, volume, and dimensions as well as the slide characteristics, and the slope angle (Fritz et al. 2003; Panizzo et al. 2005a; Zweifel et al. 2006; Najafi-Jilani and Ataie-Ashtiani 2008). A schematic of a typical SALGW development at different stages of impulsive wave generation, propagation, runup, overtopping, and inundation is shown in Fig. 1. The impact of SAL initiates a violent flow mixed of slide, water, and air which is shown as splash zone in Fig. 1. The generated waves, due to the impact, propagate into the water body (e.g., ocean, dam reservoir) and are subjected to wave transformations. Finally, LGW run-up and the following inundation may cause destructive damages in the downstream.

Some examples of the most catastrophic historical SALGW events are mentioned here. An earthquake with a moment magnitude of M_w 7.8 triggered a rockslide of 30 million cubic Meters to fall from several hundred Meters into the narrow inlet of Lituya Bay, Alaska in 1958. The rockslide initiates the largest recorded tsunami in the world with the maximum runup of about 520 m (Miller 1960). The largest SALGW event occurred in 1963 in Italy as 260 million cubic Meters of rock fell into the reservoir of the Vajont Dam, built in 1959, producing an enormous flooding due to at least 50 million cubic Meters of water. The dam did not suffer any serious damage, but flooding due to the overtopping height of about 245 m above the dam crest destroyed several villages in the valley and killed almost 2000 people. (Fritz et al. 2001). The 1792 Unzen–Mayuyama megaslide (3.4×10^8 m^3) was the greatest LGW disaster in the history of Japan with 15,000 fatalities (Unzen Restoration Office of the Ministry of Land, Infrastructure and Transport of Japan 2002). Roberts et al. (2014) compiled a global catalogue of LGWs caused by SALs including 254 events from the fourteenth century AD to 2012. Norway, one of the most hazardous area in the world regarding SAL events, has endured three major rockslide-generated tsunamis of Leon, at 1905 and 1936, and Tafjord, at 1934, with total 174 fatalities (Harbitz et al. 2014).

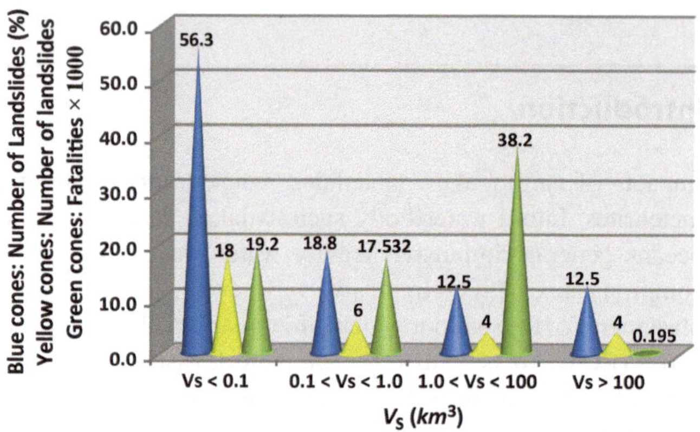

Fig. 2 Landslide volume distributions (%) for historical LGW events with the recorded number of fatalities for each category

Yavari-Ramshe and Ataie-Ashtiani (2016) presented a list of 34 cases of the historical LGW hazards and their properties. Based on their work, the slide volume varies from small amounts of about 105 m^3 to large values of more than 100 km^3. Figure 2 shows that more than 55% (18 events) of these cases have a slide volume less than 0.1 km^3. More than 25% of death toll due to LGW hazards is due to the landslide events with a volume less than 0.1 km^3 and landslides with the volume of 1.0–100 km^3 are caused 38,200 fatalities, about 51% of the total loss of life due to LGW events. 75% of these events are caused by SALs.

Laboratory Studies

Physical models are generally used to gain a better intuition of LGW behaviour and to identify the influential parameters on LGW characteristics such as the slide geometrical, geomechanical, and dynamic parameters, the slope angle, and water body depth and geometry and to develop predictive equations as a function of such factors. Noda (1970) was among the first researchers who experimentally studied SALGWs caused by solid blocks. He categorized SALGWs

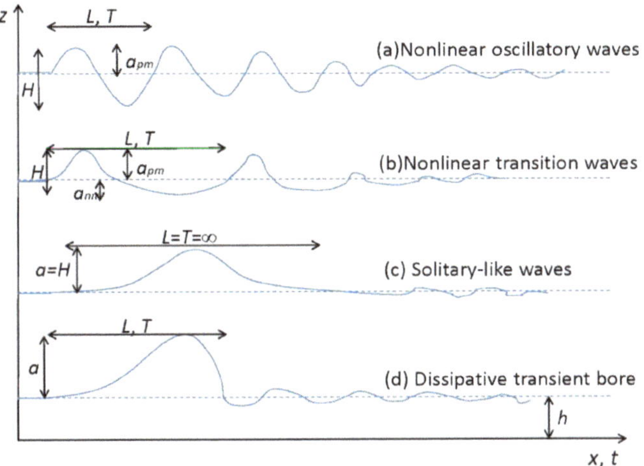

Fig. 3 A schematic of the four different patterns of SALGWs proposed by Noda (1970)

into four different patterns as nonlinear oscillatory, nonlinear transition, solitary-like, and dissipative transient bore which are schematically shown in Fig. 3. These four LGW types consist of several waves with respectively a symmetrical wave profile, a longer wave through than the wave crest, one dominant wave crest, and one dominant wave containing a large amount of air in the wave front (Heller and Hager 2011). They also hold the smallest to the largest amounts of mass transports, respectively (Di Risio et al. 2011).

The laboratory investigations of Kamphuis and Bowering (1970) on SALGWs showed that the LGW characteristics were mainly a function of the slide volume and the Froude number of the slide upon impact with the water. Moreover, the wave propagation speed can be approximated based on the solitary wave theory. Huber and Hager (1997) carried out a set of three-dimensional experiments using granular landslide falling into a water tank; they suggested the dependence of wave height to the values of the non-dimensional landslide volume. Walder et al. (2003) studied the near-field characteristics of SALGWs by solid blocks in a two-dimensional physical model. It was displayed that the quantities controlling near-field wave properties were non-dimensional landslide volume per unit width, non-dimensional submerged time of motion and non-dimensional vertical impact speed. Fritz et al. (2004) performed elegant two-dimensional laboratory experiments with granular materials. They presented empirical equations for predicting pattern of LGWs based on the slide properties and also presented empirical equations for predicting energy conservation from the slide to water. Panizzo et al. (2005b) carried out three-dimensional experiments and concluded that the maximum generated wave height was influenced predominantly by a non-dimensional time of underwater landslide motion and the surface of the landslide front. A dependence on the inclination angle of the landslide movement was also observed.

The effects of bed slope angle, water depth, slide impact velocity, geometry, shape and deformation on impulse wave characteristics were inspected by Ataie-Ashtiani and Nik-Khah (2008). The experimental work of Ataie-Ashtiani and Nik-Khah (2008) is explained briefly as a sample of the experimental set-ups here. They performed 120 laboratory tests using rigid, confined and deformable slide masses. Their experimental set-up is shown in Fig. 4. The impulse wave features such as amplitude, period and also energy conversion were studied. Recorded data at near-field and far-field showed a general pattern of SALGW consists of a wave train with positive leading wave amplitude. The second wave crest of this train has the maximum amplitude which is followed by smaller oscillatory waves.

Figure 5 demonstrates a sample of the SALGW patterns features based on the experimental data. The maximum wave crest amplitude, a_{pm} was strongly affected by bed slope angle, landslide impact velocity, thickness, kinematics and deformation and weakly by landslide shape. Slide deformation made a maximum reduction in wave crest amplitude down to 35% and a maximum increasing up to 30% on period. Slide shape was not strongly affecting the general feature of impulse wave and at most the amplitude. The energy transfer from landslide into the wave was generally increased where the slide Froude number of landslide decreases. Increasing of slide rigidity caused similar effects. The experimental data of Ataie-Ashtiani and Nik-Khah (2008) has been used for the benchmarking SALGW simulations by numerical models such as LS3D (Ataie-Ashtiani and Najafi-Jilani 2007) and 2LCMFlow (Yavari-Ramshe and Ataie-Ashtiani 2015), as will be described in the following section.

Mohammed and Fritz (2012) physically modelled SALGWs using deformable granular landslides in a three-dimensional wave basin. It was shown that the wave characteristics are dominantly controlled by the landslide Froude number. In their study, 1–15% of the landslide kinetic energy at impact was converted into the wave energy. Heller and Spinneken (2013) conducted tests to study the influences of the slide Froude number, the relative slide thickness, and the relative slide mass on SALGW characteristics. Their experiments showed that block slides do not necessarily generate larger waves than granular slides, in disagreement with Ataie-Ashtiani and Nik-Khah (2008). However, they found both the wave amplitude and period in agreement with the findings of Ataie-Ashtiani and Nik-Khah (2008). Heller and Spinneken (2015) investigated the effects of the water body geometry on the wave characteristics in the near- and far-field, using both two- and three-dimensional physical models. It was observed that for

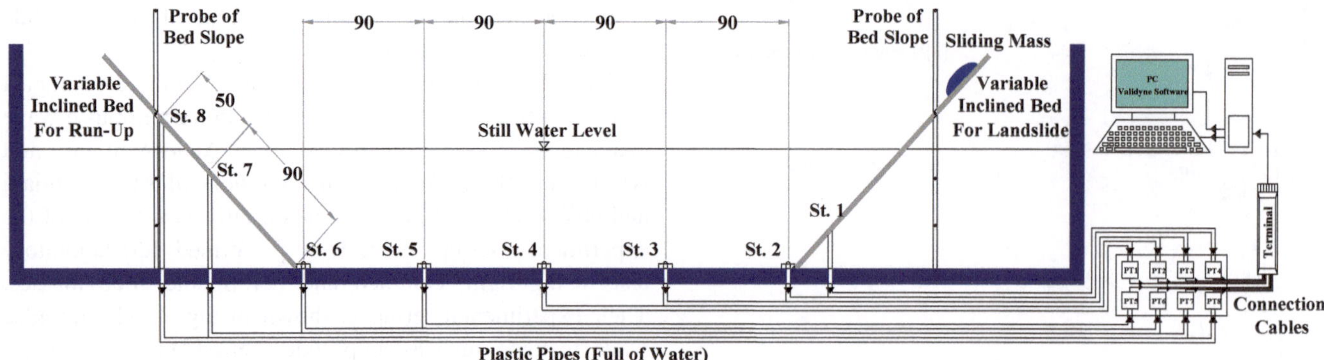

Fig. 4 Schematic of the experimental set up of Ataie-Ashtiani and Nik-Khah (2008) for SALGWs; all dimensions are in centimeter

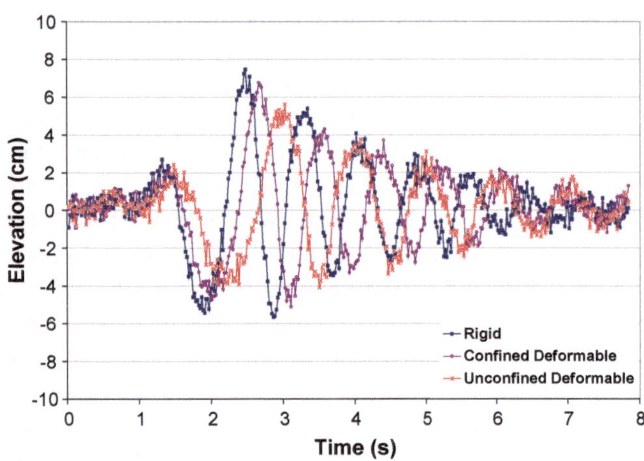

Fig. 5 A sample of SALGWs formation pattern for different landslide rigidity in laboratory tests of Ataie-Ashtiani and Nik-Khah (2008)

a small slide Froude number, relative slide thickness and relative slide mass, the three-dimensional wave heights were considerably smaller both in the near- and far-field, compared to those for two-dimensional. Recently, Lindstrøm (2016) investigated SALGWs experimentally in a two-dimensional wave tank. Five different slide materials including block slide and four granular slides with grain diameter ranging from 3 to 25 mm were employed. It was perceived that block slide generated waves showed amplitudes which were considerably larger than LGW amplitudes caused by granular slides, similar to Ataie-Ashtiani and Nik-Khah (2008) observations. Table 1 summarizes some of the most important previous laboratory studies that have performed for SALGWs.

Although, there is still a necessity for further detailed experimental investigations in well-controlled laboratory-scale (Yavari-Rameshee and Ataie-Ashtiani 2016), the first step is to make the best use of the available data. All the available experimental data should be collected and compiled together to provide a more consistent and improved understanding of the physics of LGWs based on the collective analysis of all these data in order to advance our capability of hazard assessment for potential landslides. This is a research front that the authors are currently working on.

Numerical Modelling

Numerical models are essential tools to simulate and predict the LGW initiation, propagation, transformations, and to manage LGW consequences and hazards. Recently, Yavari-Ramshe and Ataie-Ashtiani (2016) presented a comprehensive review of numerical modelling of subaerial and submarine landslide generated tsunami waves and scrutinized the recent advances and future challenges. They discussed conceptual, mathematical, and numerical structures of LGW numerical modelling.

Yavari-Ramshe and Ataie-Ashtiani (2016) categorized the numerical simulation of LGWs into three general approaches. In the first approach, models impose the effects of landslide motion as an equivalent water surface initial condition or an ad hoc boundary condition and then simulate the water wave propagation in the considered domain. In other words, this approach solely simulates water wave propagation. The second approach applies a single LGW numerical model which simulates the combination of landslide and water. These models can conceptualize landslide as a rigid or a deformable moving slide. In the third approach, more than one model is applied and each one simulates a part of the involved processes. In this approach, wave generation is simulated using a more accurate model and beyond the wave generation area where the LGWs enter the long wave interval, a simpler model is applied to simulate far-field wave propagation rather than a full three-dimensional model. The results of each model are applied as the initial conditions of the following model.

LGW phenomena are involved with the flow of a three-phase (water-grain-air) mixture, large deformations of moving free surfaces, and water splashes. Accordingly,

Table 1 Specification of some of the previous laboratory investigations on SALGWs

Reference	Tank dimension			Bed slope (°)	Slide mass specifications	Model dim.	Wave stage
	L (m)	W (m)	H (m)				
Noda (1970)	Shallow water tank			–	Solid rectangular box	–	G
Kamphuis and Bowering (1970)	45	1	0.23–0.46	45	Steel box	2VD	G
Huber and Hager (1997)	30.33	0.5	0.5	28–60	Granular material	3D	G, P
Walder et al. (2003)	3.0	0.285	1.0	10–20	Hollow rectangular Nylon Box	2VD	G
Fritz et al. (2004)	11	0.5	1.0	45	Granular material and PLG	2VD	G, P
Panizzo et al. (2005b)	11.5	6	0.8	16–36	Solid rectangular box	3D	G, P
Ataie-Ashtiani and Nik-Khah (2008)	3.6	2.5	0.8–0.5	15–60	Solid, confined, and granular	2VD	G, P
Mohammed and Fritz (2012)	48.8	26.5	0.3–1.2	27.1	Granular	3D	G, P
Heller and Spinneken (2013)	24.5	0.6	0.3, 0.6	45	Solid	2D	G, P
Heller and Spinneken (2015)	21,20	0.6,7.4	0.24, 0.48	45	Solid	2D, 3D	G, P
Lindstrøm (2016)	7.3	0.2	0.1	35	Solid, and granular	2D	G, P

2VD two-vertical dimensional; *3D* three dimensional; *PLG* pneumatic landslide generator; *G* generation of impulse wave; *P* propagation of impulse wave; *Dim* dimensions

Lagrangian mesh-less approaches such as smooth particles hydrodynamics (SPH) seems to be more adequate than Eulerian's for numerical modelling of such hazards (e.g. Ataie-Ashtiani and Shobeyri 2008; Ataie-Ashtiani and Mansour-Rezaei 2009; Pastor et al. 2009; Capone et al. 2010; Gómez-Gesteira et al. 2012a, b; Heller et al. 2016). However, particle-based methods are computationally cumbersome due to the large number of domain particles. This makes SPH type approaches inefficient for a real-scale problem (Xie et al. 2013). Among Eulerian approaches, finite difference method (FDM) and finite volume method (FVM) are the most frequent and the most recently applied methods, respectively (Yavari-Ramshe and Ataie-Ashtiani 2016). Mesh-based Eulerian methods improve the shortcomings regarding moving free surfaces and nearshore hydrodynamics processes using non-hydrostatic pressure assumption (Young and Wu 2009; Zijlema and Stelling 2008) and interface tracking techniques (e.g. Harlow and Welch 1965; Hirt and Nichols 1981; Osher and Fedkiw 2003).

Regarding the mathematical formulations, based on the Yavari-Ramshe and Ataie-Ashtiani (2016) categorization, researchers either apply the direct solution of Navier-Stocks equations (NSEs) or most commonly the depth-averaged approximation of NSEs including Boussinesq-type wave equations (BWEs) and Shallow water type equations (SWEs). NSEs have the drawback of extensive memory usage and computational costs, especially for real-scale cases. A number of models such as NHWAVE (Ma et al. 2012), TSUNAMI 3D (Horrillo et al. 2013), and Dual-SPHysics (Crespo 2015) apply advanced techniques such as parallel processing and GPU accelerator to provide the prospects for real-scale and real-time simulation of LGWs and related hazards.

The numerical models developed based on the DAEs are the most commonly applied models to simulate LGW hazards. Table 2 provides a categorized list of numerical models for the simulation of SALGWs considering rigid and deformable landslides. The table is an extract of Yavari-Ramshe and Ataie-Ashtiani (2016) review article. As it can be observed in Table 2, BWEs have been only applied in SALGW numerical models which consider a rigid landslide. The numerical models with a deformable landslide consideration solve either NSEs for a multi-phase flow or SWEs for a two-layer flow where each layer may contain a single-, two-, or three-phase fluid. In the following sections, two examples of the depth-averaged models, based on the authors' previous works, are presented. The first model describes landslide as a rigid body and the second one consider the effects of landslide deformations on LGW characteristics.

Depth-Averaged Models

Boussinesq-Type Model with Solid Landslide

Boussinesq-type formulations for water waves are developed by considering a polynomial approximation of the vertical profile of horizontal flow velocities. The classic Boussinesq equations are derived assuming a second-order variation of velocity in vertical direction (Peregrine 1967). Classic Boussinesq-type formulations are usually restricted to weak dispersion and nonlinearity (Ataie-Ashtiani and Najafi-Jilani 2007).

The extended Boussinesq formulations have developed to address the shortcomings of classic Boussinesq formulations with higher accuracy of wave nonlinearity and dispersion. Ataie-Ashtiani and Najafi-Jilani (2007) developed a fourth-order Boussinesq approximation for the simulation of SMLGW. Higher-order perturbation analyses were applied to derive the mathematical formulations. The final system of model equations was derived as follow.

which respectively represent the vector of the horizontal velocity components and the vertical velocity component in the z direction, are expanded into

$$\mathbf{u} = \mathbf{u}_0 + \mu^2 \cdot \mathbf{u}_1 + \mu^4 \cdot \mathbf{u}_2 \tag{3}$$

$$w = \mu^2 \cdot w_1 + \mu^4 \cdot w_2 \tag{4}$$

in perturbation analysis with μ^2 as the basic small parameter. $\tilde{z}$ is a characteristic variable depth defined as a weighted average of two distinct water depths z_a and z_b based on $\tilde{z} = [\beta \cdot z_a + (1-\beta) \cdot z_b]$. β is an optimized weighting parameter. Moreover, $\mathbf{A} = \nabla(\nabla \cdot \mathbf{u}_0)$, $\mathrm{B} = \nabla \cdot (h\mathbf{u}_0) + \frac{1}{\varepsilon} h_t$, and $\mathrm{C} = f(\mathbf{A}, \mathrm{B})$.

$$\begin{aligned}
&\frac{1}{\varepsilon} h_t + \zeta_t + \nabla \cdot \{ (\varepsilon\zeta + h)\mathbf{u_0} \\
&\quad + \mu^2 \left[\frac{-1}{6}(\varepsilon^3\zeta^3 + h^3)\mathbf{A} + \frac{1}{2}\tilde{z}^2(\varepsilon\zeta + h)\mathbf{A} - \frac{1}{2}(\varepsilon^2\zeta^2 - h^2)(\nabla \cdot B) + \tilde{z}^2(\varepsilon\zeta + h)(\nabla \cdot B) \right] \\
&\quad + \mu^4 \left[\frac{1}{120}(\varepsilon^5\zeta^5 + h^5)\nabla(\nabla \cdot \mathbf{A}) - \frac{1}{24}(\varepsilon\zeta + h)\tilde{z}^4\nabla(\nabla \cdot \mathbf{A}) - \frac{1}{12}(\varepsilon^3\zeta^3 + h^3)\nabla(\nabla \cdot (\tilde{z}^2 \cdot \mathbf{A})) \right. \\
&\quad + \frac{1}{4}(\varepsilon\zeta + h)\tilde{z}^2\nabla(\nabla \cdot (\tilde{z}^2\mathbf{A})) + \frac{1}{24}(\varepsilon^4\zeta^4 - h^4)\nabla(\nabla \cdot (\nabla B)) - \frac{1}{6}(\varepsilon\zeta + h)\tilde{z}^3\nabla(\nabla \cdot (\nabla B)) \\
&\quad - \frac{1}{6}(\varepsilon^3\zeta^3 + h^3)\nabla(\nabla \cdot (\tilde{z}\nabla B)) + \frac{1}{2}(\varepsilon\zeta + h)\tilde{z}\nabla(\nabla \cdot (\tilde{z}\nabla B)) \\
&\quad \left. + \frac{1}{2}(\varepsilon^2\zeta^2 - h^2)\nabla C - (\varepsilon\zeta + h)\tilde{z}\nabla C \right] \Big\} = O(\varepsilon^6, \mu^6)
\end{aligned} \tag{1}$$

$$\begin{aligned}
&\mathbf{u_{0t}} + \varepsilon(\nabla \cdot \mathbf{u_0})\mathbf{u_0} + \varepsilon\left(w_1|_{z=0}\right)\mathbf{u_{0z}} \\
&\quad + \mu^2 \left[\mathbf{u_{1t}}|_{z=0} + \varepsilon(\nabla \cdot (\mathbf{u_1}|_{z=0}))\mathbf{u_0} + \varepsilon(\nabla \cdot \mathbf{u_0})\left(\mathbf{u_1}|_{z=0}\right) + \varepsilon\left(w_2|_{z=0}\right)\mathbf{u_{0z}} + \left(w_1|_{z=0}\right)\left(\mathbf{u_{1z}}|_{z=0}\right) \right] \\
&\quad + \mu^4 \left[\mathbf{u_{2t}}|_{z=0} + \varepsilon(\nabla(\mathbf{u_2}|_{z=0}))\mathbf{u_0} + \varepsilon(\nabla \cdot (\mathbf{u_1}|_{z=0}))\left(\mathbf{u_1}|_{z=0}\right) + \varepsilon(\nabla \cdot \mathbf{u_0})\left(\mathbf{u_2}|_{z=0}\right) \right. \\
&\quad \left. + \varepsilon\left(w_2|_{z=0}\right)\left(\mathbf{u_{1z}}|_{z=0}\right) + \left(w_1|_{z=0}\right)\left(\mathbf{u_{2z}}|_{z=0}\right) \right] \\
&\quad + \nabla\left(P|_{z=0}\right) = O(\varepsilon^6, \mu^6)
\end{aligned} \tag{2}$$

Equation (1) represents the continuity equation and Eq. (2), which is a two-component vector equation, represents the depth-averaged momentum equations in the two horizontal x and y directions. $\varepsilon = \frac{a_0}{h_0}$ and $\mu = \frac{h_0}{l_0}$ are two indexes indicating wave nonlinearity and dispersive behaviour. a_0, l_0, and h_0 stand for a characteristic wave amplitude, wave length and water depth, respectively. The subscripts represent the partial derivative (e.g. $h_t = \frac{\partial h}{\partial t}$). t is time, h water depth, ζ the water surface fluctuations, p the water pressure, and $\nabla = \left(\frac{\partial}{\partial x}, \frac{\partial}{\partial y}\right)$ the horizontal gradient vector. The velocity domain components $\mathbf{u} = (u, v)$ and w

A sixth-order multi-step finite difference method was utilized for spatial discretization and a sixth-order Runge–Kutta method was implemented for temporal discretization of the higher-order depth-integrated governing equations and boundary conditions. This model (LS3D) describes landslide as a rigid body with a hyperbolic-shape geometry which its effects is inserted in the model equations as a time variable bottom boundary condition (Ataie-Ashtiani and Najafi-Jilani 2007). The LS3D model parameters are schematically illustrated in Fig. 6.

Ataie-Ashtiani and Yavari-Ramshe (2011) extended the LS3D model to handle SALGW cases and successfully

Table 2 An overview of the numerical models of SALGWs

Slide type	No	Developer	Model name	Governing equations	Num. method	Model dim.	Wave simulation stage	Real case study
Rigid landslide	1	Monaghan and Kos (2000)	–	SWE	SPH	DA	G, P	–
	2	Fine et al. (2003)	–	SWE	FDM	DI	G, P, R	–
	3	Lynett and Liu (2005)	COULWAVE	BWE	FDM	DA	G, P, R	–
	4	Yuk et al. (2006)	COBRAS	RANS	FDM	1D	G, P	–
	5	Koo and Kim (2008)	–	PFE	BEM	2D	G, P, R	–
	6	Ataie-Ashtiani and Yavari-Ramshe (2011)	LS3D	BWE	FDM	DI	G, P	Maku dam reservoir
	7	Bosa and Petti (2011)	–	SWE	FVM	DA	G, P, R	Vajont reservoir
	8	Gómez-Gesteira et al. (2012a, b)	SPHysics	NSE	SPH	1D	G, P, R	–
	9	Huang et al. (2013)	–	PFE	FDM/Lg	WA	G, P, R	–
Deformable landslide	10	Gittings (1992)	SAGE	NSE	FVM	3D	G, P	Cumbre Vieja Volcano
	11	Imran et al. (2001)	BING-1D	SWE	FDM	1D	G	–
	12	Thomson et al. (2001)	–	SWE	FDM	3D	G, P	Skagway, Alaska
	13	Fine et al. (2003, 2005)	–	SWE	FDM	DA	G, P, R	Grand Banks, Canada
	14	Quecedo et al. (2004)	–	NSE	FEM	3D	G, P, R	Lituya Bay
	15	Shigihara et al. (2006)	–	SWE	FDM	DA	G, P	–
	16	Pasenow et al. (2008)	–	NSE	FEM	3D	G, P	–
	17	Abadie et al. (2008, 2010)	THETIS	NSE	FVM/Lg	WA	G	Cumbre Vieja Volcano
	18	Fernández-Nieto et al. (2008)	–	SWE	FVM	1D	G, P, R	–
	19	Biscarini (2010)	FLUENT	NSE	FVM	3D	G, P	Lituya Bay
	20	Kelfoun et al. (2010)	VolcFlow	SWE	FDM	DA	G, P	Réunion Islands, Gulmar flank, Canary Islands
	21	Davies et al. (2011)	Fluidity	NSE	FEM	3D	G, P	Storegga slide
	22	Pudasaini and Miller (2012a, b)	–	SWE	FVM	1D	G, P, R	–
	23	Horrillo et al. (2013)	TSUNAMI3D	NSE	FDM	3D	G, P, R	Gulf of Mexico
	24	Zhao et al. (2015)	–	NSE	FEM	3D	G, P, R	Lituya Bay
	25	Yavari-Ramshe and Ataie-Ashtiani (2015)	2LCMFlow	SWE	FVM	1D	G, P, R	–

(continued)

Table 2 (continued)

Slide type	No	Developer	Model name	Governing equations	Num. method	Model dim.	Wave simulation stage	Real case study
	26	Macías et al. (2015)	HySEA	SWE	FVM	2D	G, P, R	Al-Borani
	27	Fu and Jin (2015)	–	NSE	Lg/MPS	3D	G, P, R	–
	28	Ma et al. (2015)	–	SWE	FVM	DI	G, P, R	–
			NHWAVE	NSE		3D		

SWE shallow water equations, *SPH* smoothed particle hydrodynamics, *MPS* moving particle semi-implicit, *G* generation stage
BWE Boussinesq wave equations, *FDM* finite difference method, *Lg* Lagrangian, *P* propagation stage
RANS Reynolds-Averaged NSE, BEM boundary element method, *DA* depth-averaged, *R* runup stage
PFE potential flow equations, *FVM* finite volume method, *DI* depth-integrated, *Num* numerical
NSE Navier-Stokes equations, *FEM* finite element method, 1D one-dimensional, *Dim* dimensions

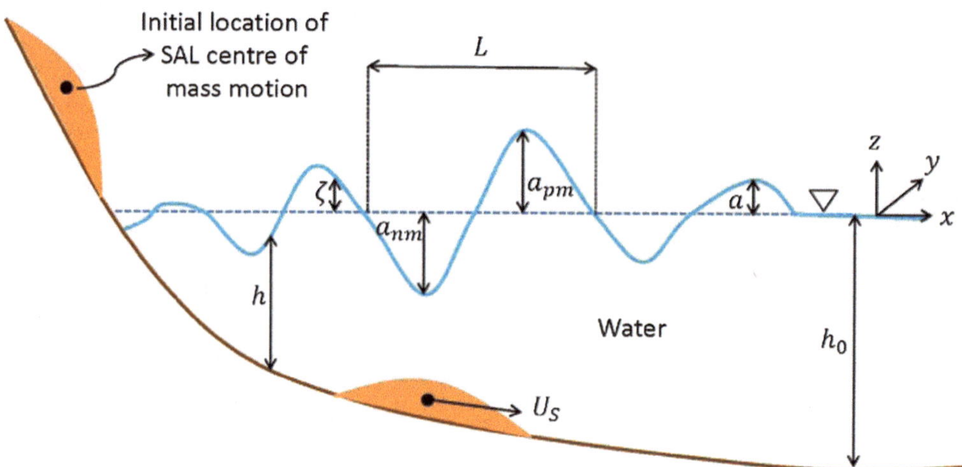

Fig. 6 Schematic definition of the LS3D model parameters and assumptions

validated the model versus experimental data of Ataie-Ashtiani and Nik-Khah (2008). They applied the new LS3D model to a real case to study SALGWs in the Maku dam reservoir located in the north of Iran. Further explanations are provided in the "Sensitivity analysis" and "Case study" sections.

Shallow-Water Type Model with Deformable Landslide

Depth-averaged equations (DAEs) have been frequently applied to simulate SMLs (Yavari-Ramshe and Ataie-Ashtiani 2016) with considering a solid landslide. To consider the slide deformations, especially for SALs which are more complicated in the wave generation zone, solving NSEs is a typical option (Biscarini 2010). An alternative which is also frequently applied by researchers is solving SWEs for a two-layer flow including a layer of granular material moving beneath a layer of water (e.g. Fernández-Nieto et al. 2008; Kelfoun et al. 2010; Yavari-Ramshe and Ataie-Ashtiani 2015; Macías et al. 2015; Ma et al. 2015).

The two-layer flow model of Yavari-Ramshe and Ataie-Ashtiani (2015) solves the incompressible Euler equations based on a state of the art Roe-type finite volume method introduced by Yavari-Ramshe et al. (2015). The computational domain considered in the 2LCMFlow model is illustrated in Fig. 7 including the schematic definition of the model parameters. The sliding mass is described as a Coulomb mixture; a two-phase mixture of water and solid grains where its interaction with the bottom follows a Coulomb-type friction law and the normal and longitudinal stresses of the solid phase are related with a coefficient K. The final system of mathematical equations for this model which is called the two-layer Coulomb mixture flow (2LCMFlow) model is:

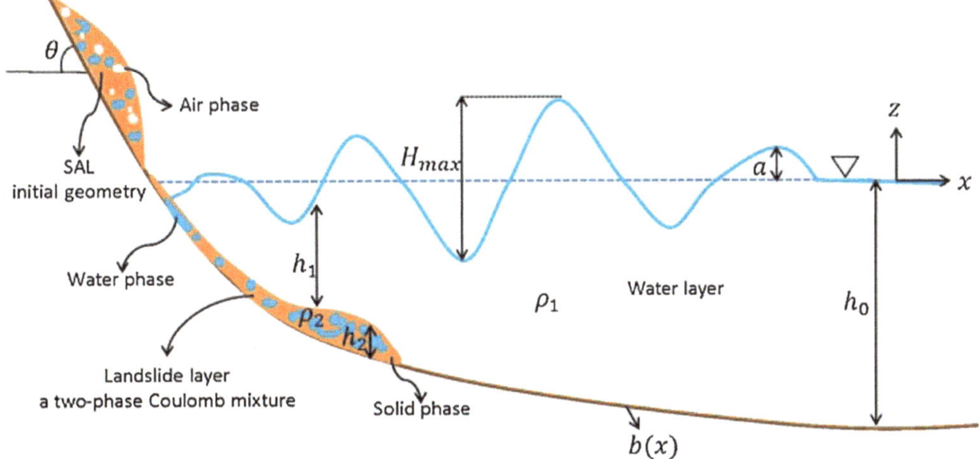

Fig. 7 Schematic definition of the 2LCM Flow model parameters and assumptions

$$\begin{cases} h_{1t} + (q_1\cos\theta)_x = 0 \\ q_{1t} + \left(h_1 u_1^2\cos\theta + g\frac{h_1^2}{2}\cos^3\theta\right)_x = -gh_1\cos\theta b_x + g\theta_x\frac{h_1^2}{2}\sin\theta\cos^2\theta - gh_1\cos\theta(h_2\cos^2\theta)_x \\ h_{2t} + (q_2\cos\theta)_x = 0 \\ q_{2t} + \left(h_2 u_2^2\cos\theta + g\Lambda_2\frac{h_2^2}{2}\cos^3\theta\right)_x = -gh_2\cos\theta b_x + g\theta_x\frac{h_2^2}{2}\sin\theta\cos^2\theta - rgh_2\Lambda_1\cos\theta(h_1\cos^2\theta)_x + \frac{\Im}{\cos\theta} \end{cases} \quad (5)$$

where subscript 1 and 2 stands for the water and the granular layers, respectively. θ is the local bed slope and $b(x)$ represents the bottom topography. Moreover, $q = hu$, $\Lambda_1 = \lambda_1 + K(1-\lambda_1)$, and $\Lambda_2 = r\lambda_2 + K(1 - r\lambda_2)$ where $K = 2\left(1 - \text{sgn}(u_{2x})\sqrt{1 - \left(\frac{\cos\phi}{\cos\delta_{mod}}\right)^2}\right)/\cos^2\phi - 1$ is the coefficient of the earth pressure and $\tan\delta_{mod} = \tan\delta - \lambda' K h_{2x}$. $r = \frac{\rho_2}{\rho_1}$ is the relative density of the landslide.

ϕ and δ represent the internal and the basal friction angles of the sliding mass, respectively. δ_{mod} is the dynamically modified basal friction angle. ρ_1 and ρ_2 are the water and the landslide densities, respectively and λ' is a constant. Finally, $\Im$ stands for the Coulomb friction term defined as:

$$\begin{cases} \Im = -(g(1-r)h_2\cos^2\theta + h_2 u_2^2\cos\theta \times (\sin\theta)_x)\frac{q_2}{|q_2|}\tan\delta_{mod} & |\Im| \geq \sigma_c \\ q_2 = 0 & |\Im| < \sigma_c \end{cases} \quad (6)$$

where $\sigma_c = g(1-r)h_2\cos\theta\tan\delta_0$ is a basal critical stress which is defined based on δ_0, the angle of repose of the granular material. Accordingly, the sliding mass stops moving when its angle is less than the angle of repose. The constitutive structure of the sliding material is defined based on Eq. (7) with the two parameters λ_1 and λ_2 which distribute the water layer pressure between the solid and the fluid phases of the second layer on the interface and along the second layer, respectively (Yavari-Ramshe and Ataie-Ashtiani 2015).

$$\begin{cases} P_{2zz}^f = \lambda_1\rho_1 h_1\cos\theta + \lambda_2\rho_1(h_2 - z)\cos\theta \\ P_{2zz}^s = (1-\lambda_1)\rho_1 h_1\cos\theta + (\rho_2 - \lambda_2\rho_1)(h_2 - z)\cos\theta \end{cases} \quad (7)$$

In this equation, P_{zz} is the normal stress and the superscripts f and s stands for the fluid and the solid phases of the second layer (the sliding mass), respectively. The 2LCMFlow is able to capture the simultaneous appearance of the static/flowing regions within the landslide path. The model is also capable of simulating the interactions between water and a variety of granular material with different water content from rockslide and dry cohesion-less material to loose and muddy flows and even sediment transport based on the considered rheological structure for landslide. The finite volume structure of the model applies a well-balanced second-order Roe-type scheme to solve the model equations in a conservative form.

Sensitivity Analysis

Model Validation

The experimental measurements of Ataie-Ashtiani and Nik-Khah (2008) are applied to assess the ability of both

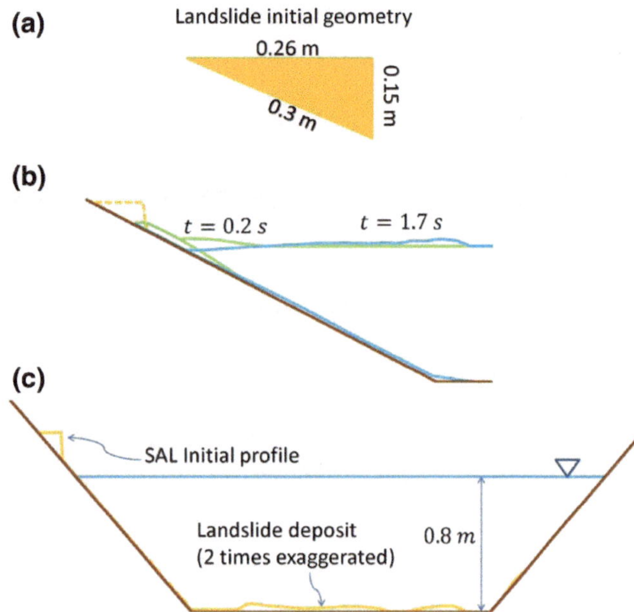

Fig. 8 **a** Landslide initial geometry, **b** landslide profiles and water surface elevations at $t = 0.4$ s and $t = 1.7$ s, and **c** the initial conditions and the final landslide deposit

LS3D and 2LCMFlow models in simulating SALGWs. Two groups of solid and deformable masses are considered in these experiments. The deformable landslides include two kinds of material; confined and unconfined masses of sand with the density of 1900 kg/m^3 representing cohesive and non-cohesive material, respectively (Yavari-Ramshe and Ataie-Ashtiani 2015). All the deformable masses have an initial triangular geometry which is schematically shown in Fig. 8a. The experiments 5, 104, and 113 of Ataie-Ashtiani and Nik-Khah (2008) are selected to be simulated by both LS3D and 2LCMFlow models. These experiments have the same initial and boundary conditions, which is shown in Fig. 8c, but different landslide rigidity including a rigid, an unconfined, and a confined landslide, respectively. The wedge-shaped landslide is released from the initial vertical distance of 0.15 m due to the water level surface. The sliding slope and the water depth are 30° and 0.8 m, respectively.

The time history of water surface fluctuations at the wave generation stage recorded by the first gauge (ST. 1) in Fig. 4 are shown in Fig. 9. This graph includes the recorded water surface fluctuations at ST. 1 for tests no. 5, 104, and 113 and the equivalent values estimated by LS3D and 2LCMFlow models. As it can be observed in this figure, the estimated LGWs by LS3D model are closer to the experimental measurements of test 5 with a rigid landslide. The computational error is less than 8%. The rigid landslide in the LS3D model is considered to have a hyperbolic-shaped geometry while in experiment 5 the sliding mass has a wedge-shaped form. This fact results in a higher difference between the experimental and the numerical LGWs. On the other hand, the numerical results of 2LCMFlow model are in a better agreement with the experimental measurements of test 104 with an unconfined mass of sand. The computational error for this case is less than 5%. Comparison between the numerical results of LS3D and 2LCMFlow models in Fig. 9 demonstrates that with considering a rigid landslide the maximum generated wave is commonly overestimated. The maximum wave height of test 104, $H_{max} = a_{pm} + a_{nm} = 0.11m$, is overestimated by the LS3D model with a relative error of about 16% while the 2LCMFlow model predicted H_{max} as 0.1094 m which has a relative error of around 0.51% with the equivalent experimental measurment.

With comparing the numerical results of LS3D and 2LCMFlow models with the equivalent experimental measurements at higher times, the ability of each model in simulating the wave dispersion will be evaluated. Figure 8b reveals up to 8% time phase difference between the numerical results of LS3D and the experimental data of test 5. This time difference which is probably caused by the combination effects of numerical dispersion and depth-averaged modelling of real three-dimensional experiments gradually makes the numerical LGWs far from the experimental LGWs. The LGWs predicted by LS3D model disperse faster than the experimental LGWs. 2LCMFlow solves the non-dispersive Euler equations. As a result, it can be observed in Fig. 9 that the LGWs estimated by 2LCMFlow moves ahead the experimental LGWs of test 104. It means that the numerical LGWs are dispersing and descending slower than the equivalent experimental LGWs.

Predicting the topographic changes of the bottom and observing the simultaneous interactions between water surface fluctuations and landslide deformations are among the advantages of considering the rheological behaviour of landslide in LGW modelling. Figure 8b shows the landslide profiles and the water surface elevations predicted by 2LCMFlow model 0.4 and 1.7 s after releasing the sliding mass. Landslide deposit is also illustrated in Fig. 8a. As it can be observed in Fig. 8b at $t = 0.4$ s, the impacts of the sliding mass within its intrusion into the water generate a series of wave crests which move towards the runup surface. Landslide moves and elongates along the sliding surface until it is completely beneath the water surface. Afterwards, as it is shown in Fig. 8b at $t = 1.7$ s, a wave trough starts to shape due to the motion of the submerged landslide. These interactions between the water body and the landslide stops when the sliding mass comes to rest and forms its final deposit as shown in Fig. 8c.

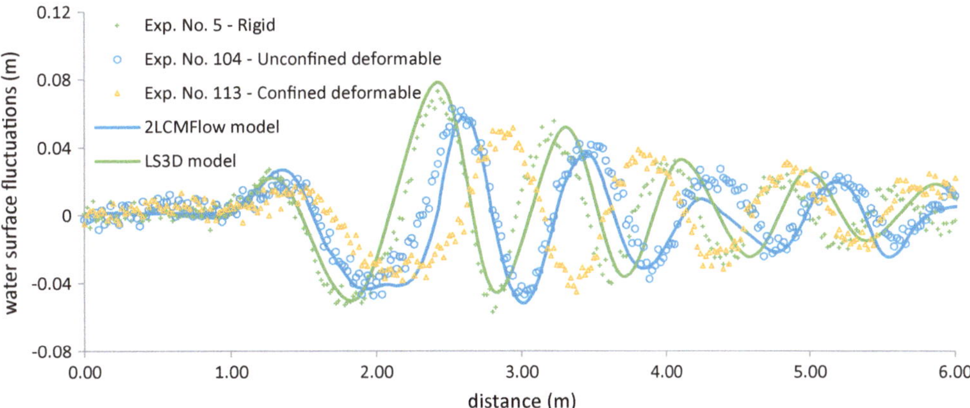

Fig. 9 Time histories of the water surface fluctuations at the wave generation stage recorded at ST.1 for tests no. 5, 104, and 113 of Ataie-Ashtiani and Nik-Khah (2008) accompanied with the equivalent values predicted by LS3D and 2LCMFlow models

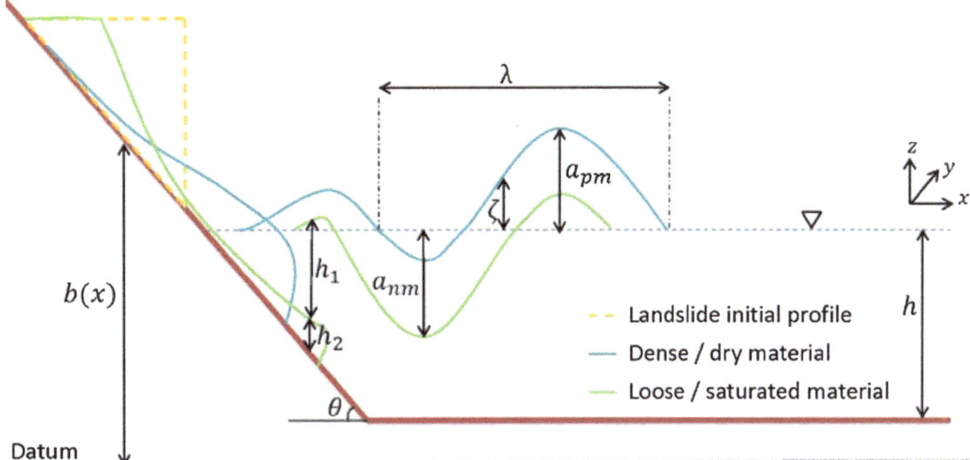

Fig. 10 A schematic of the landslide deformation and the SALGW formation patterns for different landslide material

Sensitivity Analysis

Yavari-Ramshe and Ataie-Ashtiani (2015) performed a series of sensitivity analysis on the effects of landslide constitutive structure and landslide rigidity on both the sliding mass deformations and the water surface fluctuations. The authors demonstrated the importance of landslide rheological behaviour, constitutive structure and deformations on LGW characteristics based on the comparison of 2LCMFlow numerical results with two sets of experimental data on SMLs (Ataie-Ashtiani and Najafi-Jilani 2008) and SALs (Ataie-Ashtiani and Nik-Khah 2008). The key results of their simulations are schematically illustrated in Figs. 10 and 12.

The 2LCMFlow model is able to simulate the interactions between water and a variety of the sliding material from dense and dry material to loose and saturated masses (Yavari-Ramshe and Ataie-Ashtiani 2015). This ability is based on applying different values of landslide constitutive parameters, λ_1 and λ_2, in Eq. (7). As it can be observed Fig. 10, dense material deforms into a thick front and thin tail profile and initiates a LGW consist of a larger wave crest than the wave trough. Dense material which are defined by low values of λ_1 and λ_2 implies a stronger impact to the water surface which results in a larger wave crest. On the other hand, loose material has a dam break behaviour; a sudden withdrawal which forms a thin front and thick tail profile. Due to the sudden reduction of the landslide layer thickness beneath the water layer, the resulting LGW has a large wave trough accompanied with a smaller wave crest. Moreover, landslides with loose material elongate faster than dense masses.

The LS3D model considers a hyperbolic-shaped geometry for the rigid landslide (Ataie-Ashtiani and Najafi-Jilani 2008). The same assumption has been also considered by some other researchers such as Grilli et al. (2002) and Lynett and Liu (2004). Our sensitivity analysis propose that landslide gradually deforms to a hyperbolic-shaped geometry along its runout path if it moves on a smooth surface with averaged values of λ_1 and λ_2. For example, Fig. 11 illustrates the deformed shape of a landslide for different values of λ_1 and λ_2, 0.8 s after its motion. As it can be observed, the initial wedge-shaped geometry of the landslide has gradually deformed to a hyperbolic-shaped geometry for $\lambda_1 = \lambda_2 = 0.5$.

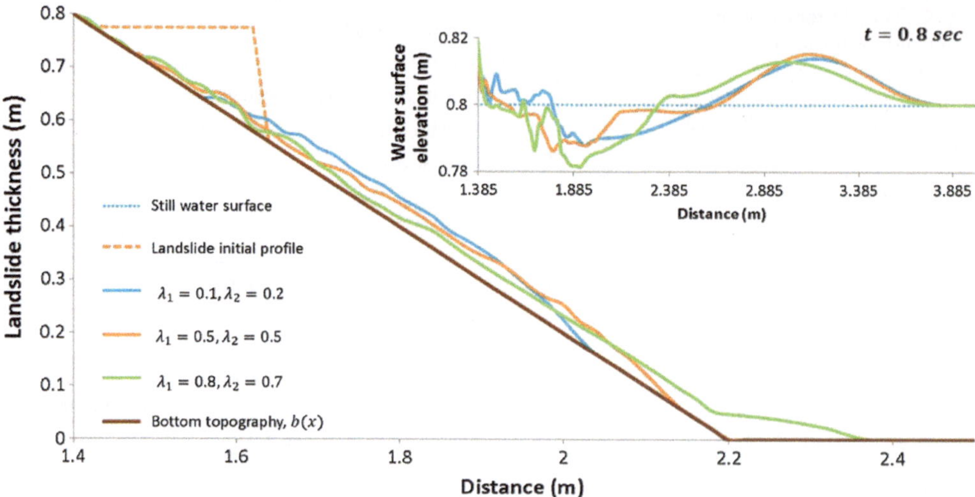

Fig. 11 Landslide profiles and water surface elevations for different values of λ_1 and λ_2

However, for $\lambda_1 = 0.1$ and $\lambda_2 = 0.2$ the sliding mass has a thick front and thin tail and for $\lambda_1 = 0.8$ and $\lambda_2 = 0.7$ the sliding mass tail is thicker than its front. Moreover, the assumption of a hyperbolic-shaped landslide is not applicable for a general topography having a rough surface. The water surface fluctuations are also shown in Fig. 11. When $\lambda_1 = 0.1$ and $\lambda_2 = 0.2$, the generated LGW has a larger wave crest than wave trough and moves faster due to the more strenght of the landslide impact. With increasing λ_1 and λ_2, the generated wave consists of a wave trough larger than the wave crest which moves slower.

Figure 12 illustrates a schematic of the key numerical results regarding the effects of landslide rigidity on the LGW characteristics based on the comparison between the experimental measurements (Ataie-Ashtiani and Nik-Khah 2008) and the 2LCMFLow (Yavari-Ramshe and Ataie-Ashtiani 2015) and the LS3D (Ataie-Ashtiani and Yavari-Ramshe 2011) simulations. A more rigid landslide initiates a larger LGW which moves faster. Confined material represents cohesive masses such as saturated muds and clays. On the other hand, unconfined material exemplifies the sliding mass with negligible cohesion such as rockfalls and loose sands. Cohesion-less material moves faster than cohesive masses and imposes a stronger impact to the water body which results in a larger LGW. Cohesion may play an important role in the landslide triggering mechanism. Although, the cohesive interconnects between the slide grains gradually fail as it starts its movement such that cohesion almost disappears within the collapsed parts of the sliding mass, even for clay-rich soils (Skempton 1964, 1985). Therefore, as it is also demonstrated by Iverson and Vallance (2001), the cohesion-less Coulomb law applied in 2LCMFlow model is an optimal choice for describing the dynamic behaviour of the granular type flows.

In the next section, the application of LS3D and 2LCMFlow models for a real case study will be explained

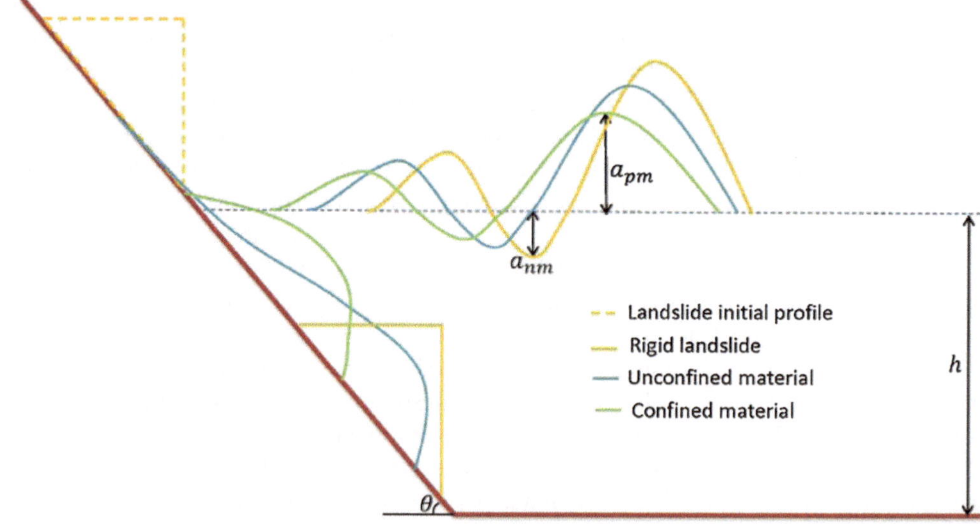

Fig. 12 A schematic of the SALGW formation patterns for different landslide rigidity

and the advantages and disadvantages of each one are examined.

Case Study

Maku Dam Reservoir

The Maku dam is located in the southern part of the Maku town, West Azarbaijan province, Iran, on the Zangmar River. The Zangmar River originates in the mountains above the Maku town, along the Turkish–Iranian border, not far from Mount Ararat and flows south and east into the Araxes River at the town of Pol-Dasht (Fig. 13). The Maku dam is a 75 m high earth dam with a reservoir capacity of 135 M m^3 having the average length, width, and water depth of about 510, 185, and 50 m respectively (Ataie-Ashtiani and Yavari-Ramshe 2011). The length and width of the dam are 350 and 10 m, respectively. The dam crest level is 1690 m from the sea level (Mahab Ghods 1999).

Geological investigations (Mahab Ghods 1999) have identified a large number of tensile cracks along the reservoir borders forming some areas of instability. One of the most hazardous SAL scenarios is selected for the following numerical simulation; a circular-shape instability located on the Eastern beach with the horizontal distance of 230 m from the dam axis (Ataie-Ashtiani and Yavari-Ramshe 2011). The potential SAL characteristics are summarized in Table 3.

Numerical Simulations

Ataie-Ashtiani and Yavari-Ramshe (2011) applied the LS3D model (Ataie-Ashtiani and Najafi-Jilani 2008), to simulate the potential SALGWs in the Maku dam reservoir. The model parameters are available in Table 3. Figure 14 illustrates the first generated SALGW and its characteristics estimated by the LS3D model for this scenario. Based on this figure, a 18 m LGW, with a maximum positive wave amplitude, a_{pm}, of about 10.1 m and a maximum negative

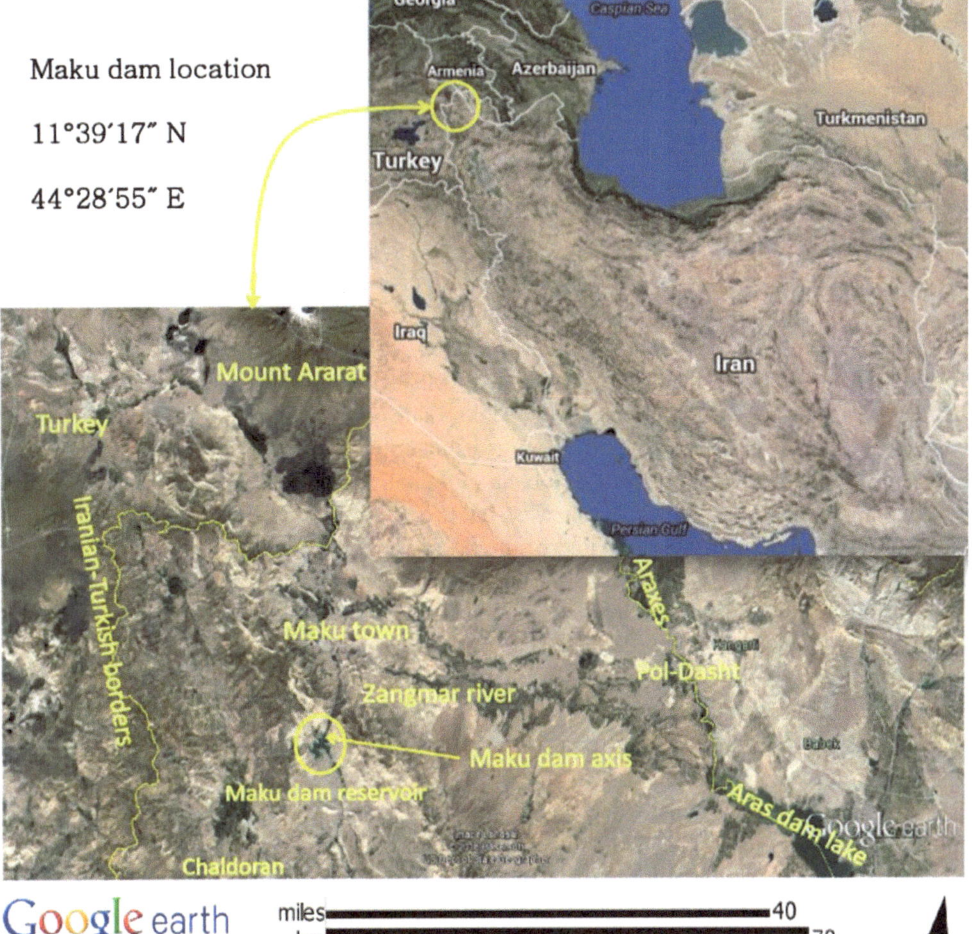

Fig. 13 Geographical location of the Maku dam reservoir (Google earth map)

Table 3 The input data of the LS3D model for simulating the potential SALGW of the Maku dam reservoir

SAL characteristics	Length	l_s (m)	60
	Thickness	T_s (m)	10
	Width	w_s (m)	250
	Volume	V_s (M m^3)	0.15
	Density	ρ_s (gr/cm^3)	1.9
	Initial height	z_s (m)	5
	Sliding slope	$\theta(\circ)$	32
Model parameters	Grid size in x direction	Δx (m)	5
	Grid size in y direction	Δy (m)	5
	Time step	Δt (s)	0.05

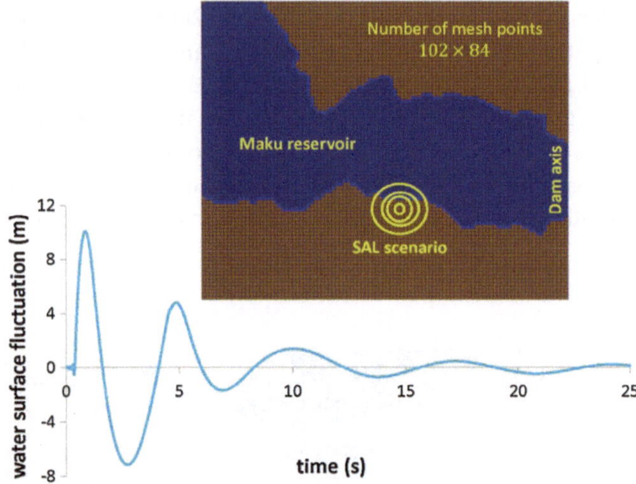

Fig. 14 The first LGW caused by the SAL scenario of the Maku dam reservoir and its characteristics, predicted by the LS3D model with a solid landslide consideration

wave amplitude, a_{nm}, of about 7.2 m, will be triggered and propagate towards the reservoir shorelines and dam axis.

A cross section of the Maku dam reservoir (Fig. 15a) is considered to simulate the same potential SALGW with 2LCMFlow. The topographic map of the Maku dam reservoir and the location of the potential SAL are illustrated in Fig. 15a. As it can be observed, the east bank of the lake has a mild slope of about 30° while the west bank is steeper (about 45°–60°) and the average width of the lake is around 185 m. Accordingly, a basin with a 30° slipping surface and a 45° runup surface is considered as the computational domain for the following simulations. The still water depth is considered to be the same as the average water depth of the Maku dam lake of about 50 m. The resulting computational domain is illustrated in Fig. 15b.

The sliding material consists of cohesion-less debris deposits and rock fragments with the volume of several cubic meters on an impermeable bed of marl (Mahab Ghods 1999). As a result, water infiltration is probably the main factor in triggering the potential landslides. The permeated water traps and flows along the impermeable interface of debris deposits and marl bed and initiates a landslide with decreasing the frictional resistance of the contact surface. Therefore, the basal friction angle is considered to have the small value of about $\delta = 10°$. The internal friction angle is also supposed to be $\phi = 38°$ (Denlinger 2007). The geological investigations have estimated the porosity of landslide material to be about 0.4% (Mahab Ghods 1999). Accordingly, both constitutive coefficients of λ_1 and λ_2 are considered to have the same value of 0.4. As it is illustrated in Fig. 15b, the sliding mass is supposed to have an initial wedge-shaped geometry with the density of $\rho_2 = 1900\,\mathrm{kg/m^3}$, the length of $l_s = 26\,\mathrm{m}$, and the maximum thickness of $T_S = 15\,\mathrm{m}$.

We have applied this cross section of the Maku dam reservoir to simulate the generation stage of its potential SAL with both the LS3D and the 2LCMFlow models. The grid size and the time step are supposed to be $\Delta x = 1.0\,\mathrm{m}$ and $\Delta t = 0.1\,\mathrm{s}$ in both models. The time history of water surface fluctuations close to the impact area is shown in Fig. 16. As it is expected, the first LGW estimated by the LS3D is about 14% larger than the first LGW predicted by the 2LCMFlow due to considering a rigid landslide. The LS3D model simulates wave dispersion more accurately because of solving the 4th order Boussinesq-type equations. This explains the up to 5% phase difference between the numerical results of LS3D and 2LCMFlow which steadily makes the predicted LGWs by LS3D far from the estimated LGWs by 2LCMFlow. Moreover, higher accuracy of wave dispersion in the LS3D model makes the wave amplitudes to be diminished faster in comparison with the wave amplitudes in 2LCMFlow such that after about 15 s the maximum positive LGW amplitude, a_{pm}, has around 25 and 50% reduction based on the 2LCMFlow and the LS3D results, respectively. This demonstrates that considering a rigid landslide is not always a conservative choice. The maximum LGW may be estimated conservatively, but the wave dispersion may also be overestimated numerically (Ataie-Ashtiani and Yavari-Ramshe 2011) which result in

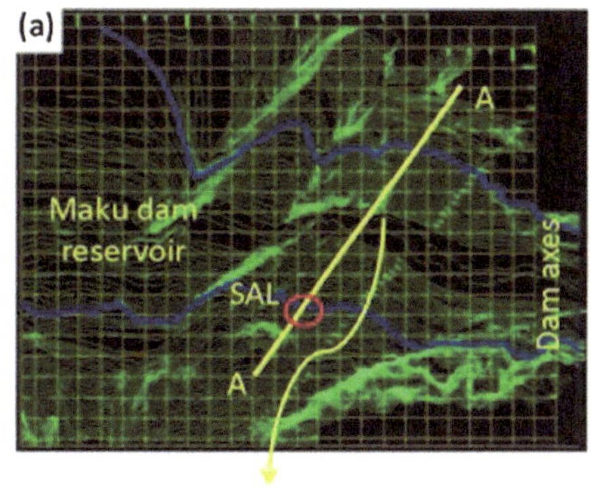

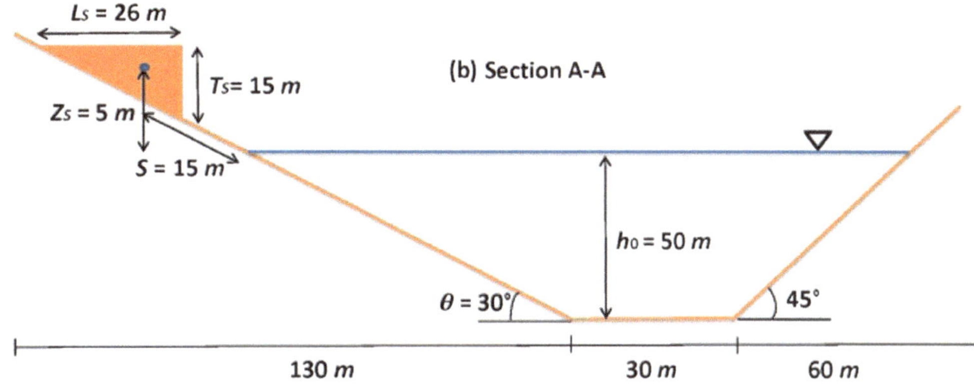

Fig. 15 **a** The topographic map of the Maku dam reservoir, and **b** a cross section of the Maku dam reservoir as the considered computational domain in 2LCMFlow and LS3D models including the initial geometry and location of the Potential SAL

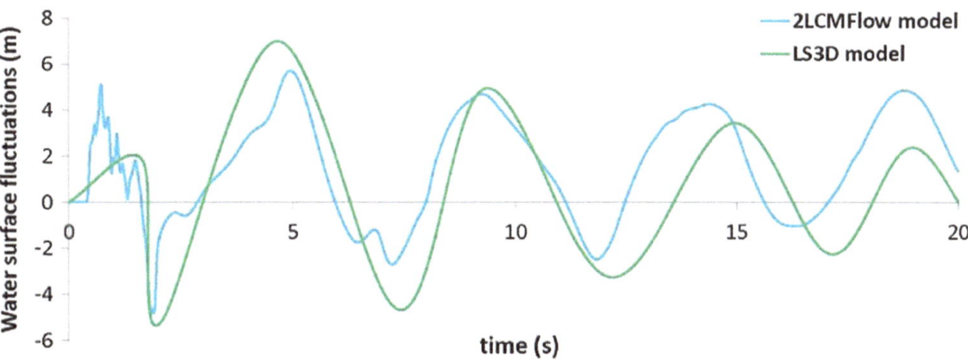

Fig. 16 Time history of water surface fluctuations close to the impact area predicted by LS3D and 2LCMFlow models for the potential SAL of the Maku dam reservoir

the underestimation of propagated wave amplitudes, wave runups, overtopping heights, and flood volumes.

The wave formation pattern predicted by the 2LCMFlow model is closer to the physics of the phenomena due to considering the effects of landslide deformations (Yavari-Ramshe and Ataie-Ashtiani 2015). As it can be observed in Fig. 16, a rigid landslide, considered by LS3D model, imposes a single impact at the moment of its intrusion into the water and consequently generates a single wave crest. On the other hand, with considering a deformable landslide in the 2LCMFlow model, a set of consecutive wave crests are originated due to the successive impacts of the landslide trailing with different speeds.

Figure 17 illustrates the simultaneous interactions between the water surface fluctuations and the landslide deformations predicted by the 2LCMFlow model at several times after releasing the SAL. The sliding mass continuously elongates towards the reservoir bottom. As it can be observed in Fig. 17 for example at $t = 2.2$ s, the 2LCMFlow model has properly captured the sudden appearance of the flowing/static regions along the landslide path (Yavari-Ramshe and Ataie-Ashtiani 2015). The generated waves have grown close to the opposite side of the lake due to the topographic changes of the seafloor and the shallowness of the coastline (See Fig. 17 at $t = 2.2$ s). Then, the enhanced wave has run up with a vertical runup height of about 10 m along the water side (Fig. 17 at $t = 3$ s).

The final landslide deposition can be observed in Fig. 17 at $t = 20$ s. After deposition, landslide profile gradually deforms until it forms a stable geometry based on the angle of repose of the sliding material (Yavari-Ramshe and Ataie-Ashtiani 2015). Predicting the topographic changes of

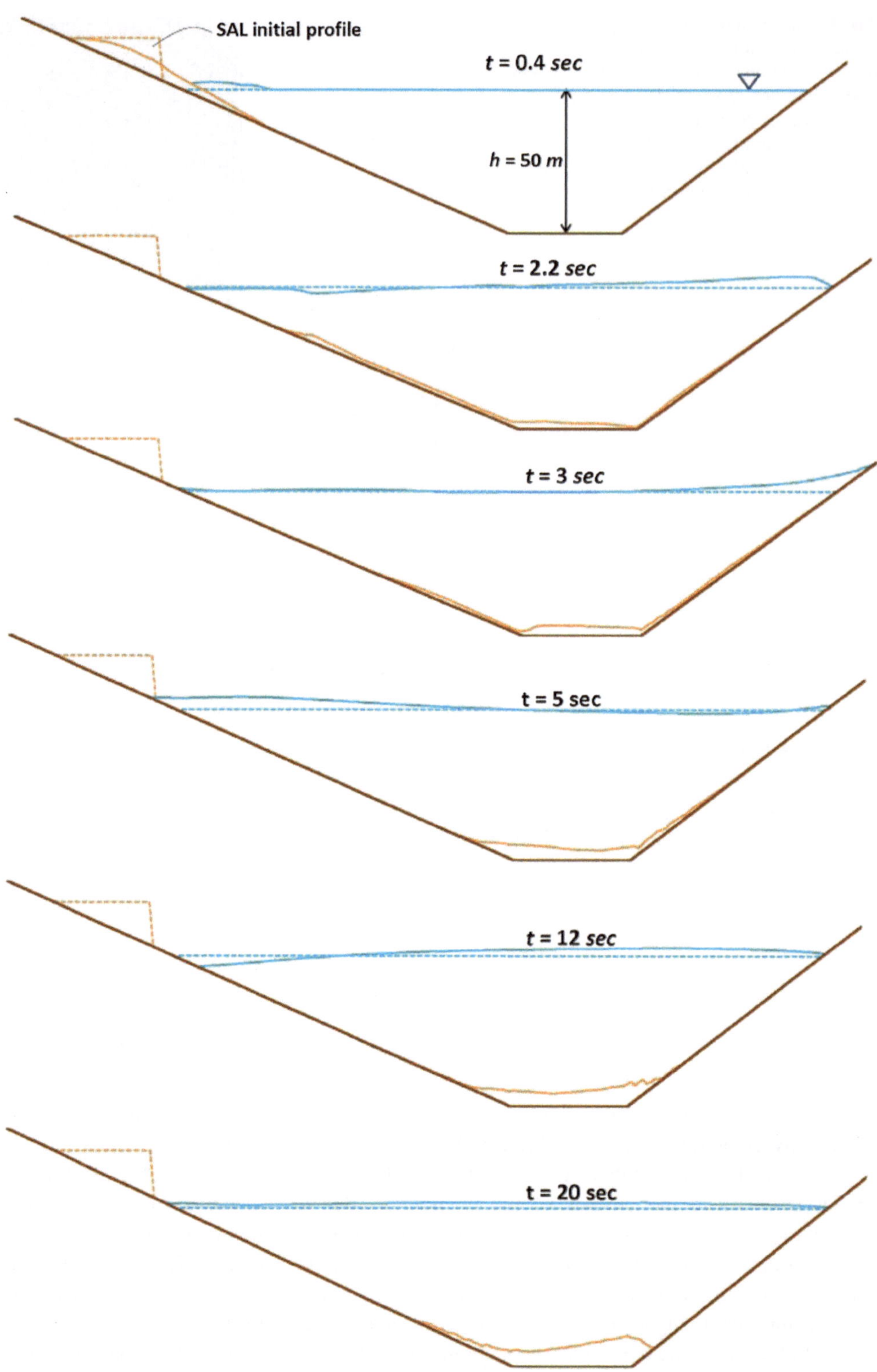

Fig. 17 Landslide profiles and water level elevation predicted by 2LCMFlow model at different times for the potential SAL of the Maku dam reservoir

the seafloor after a landslide is one of the advantages of the numerical models such as 2LCMFlow which consider a deformable landslide.

Based on the comparison of the numerical results of 2LCMFlow and LS3D models with the laboratory measurements of Ataie-Ashtiani and Nik-Khah (2008) for SALGWs, Yavari-Ramshe and Ataie-Ashtiani (2015) demonstrated that the 2LCMFlow is more accurate in estimating the characteristics of the first generated LGW and the LS3D model better estimates the wave propagation stage.

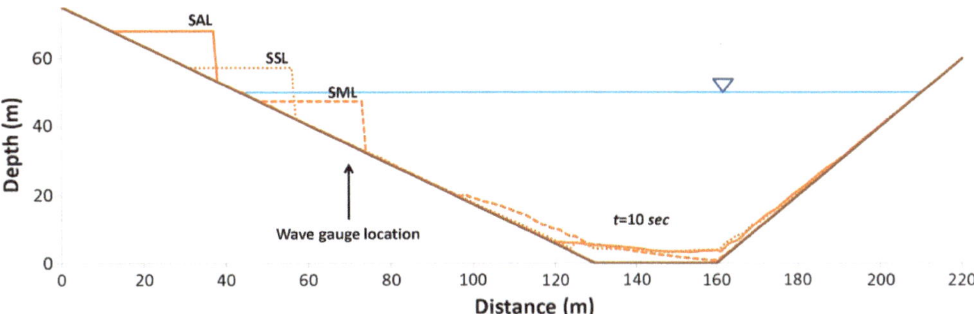

Fig. 18 The initial situations of three landslide scenarios, the location of the considered wave gauge, and landslide profiles after 10 s

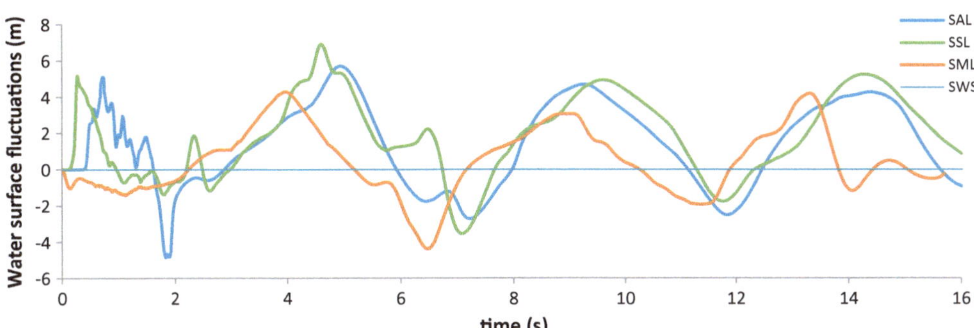

Fig. 19 Time histories of water surface fluctuations predicted by 2LCMFlow model for three SAL, SSL, and SML scenarios of Fig. 18 at $x = 70\,\text{m}$

These facts are also confirmed based on the present simulations. Both models are easy to apply and need reasonable input data for running. The LS3D takes around 30% less computational time than the 2LCMFlow. This is because the later model solves a coupled system of model equations in a finite volume framework while the former one applies a finite different method to solve a system of BWEs.

Research Prospect

Authors are extending the results of these preliminary comparisons to a comprehensive sensitivity analyses on the influences of the physical parameters and numerical approaches in their future research efforts. In this section, a sample case of these sensitivity analysis is represented which describes the effects of the landslide initial submergence on the characteristics of the generated landslide tsunamis.

The sensitivity analysis are conducted on the same computational domain described in Fig. 15b. The sliding mass has the same geometrical and geotechnical parameters as the potential SAL of the Maku dam reservoir but three different initial locations including $S_1 = 15\,\text{m}$, $S_2 = -5\,\text{m}$, and $S_3 = -25\,\text{m}$ which indicate a SAL, a semi-submerged landslide (SSL), and a SML, respectively. The initial situations of these landslide scenarios are shown in Fig. 18. Parameter S which is schematically defined in Fig. 15b represents the bottom-ward distance of the landslide due to the still water surface. The model parameters are chosen to have the same values of $\Delta x = 1.0\,\text{m}$ and $\Delta t = 0.1\,\text{s}$.

Figure 19 illustrates the time histories of the LGWs in the near-field (wave generation stage) for all three cases estimated by the 2LCMFlow model. This figure illustrates the predicted values of the water surface fluctuations at $x = 70\,\text{m}$ shown as the wave gauge location in Fig. 18. As it can be observed in this figure, the largest wave crest is about 6.9 m and is generated by the SSL. a_{pm} for SSL is less than 21% larger than the equivalent value generated by the motion of SAL. As it is illustrated in Fig. 17 for $t = 0.4\,\text{s}$, the SAL spreads and deforms before reaching to the water surface. As a result, the impacts induced by the SSL to the water is probably stronger than the impacts of the SAL and subsequently induce a larger wave crest. The deepest wave trough is on the other hand induced by the SML due to the sudden loss of the landslide layer thickness beneath the water layer which is caused by the instantaneous discharge of the SML material. This maximum a_{nm} is about 6.03 m and happens at $x = 57\,\text{m}$ which is closer to the beach compared to the location of the considered wave gauge in Fig. 18. The maximum a_{nm} is about 24.5% larger than the a_{nm} of the SAL.

The landslide profile of each scenario at $t = 10\,\text{s}$ is illustrated in Fig. 18. Based on this figure, the SAL and the SSL have a longer runout distance. They both move up the 45°

runup surface and reach to the vertical elevations of about 21 and 14 m, respectively. However, the SML reaches a small vertical distance of about 2 m along the runup surface.

Conclusions

An overview of the laboratory and numerical investigations of the SALGWs are provided with an explicit consideration to the previous works of the authors. Although a wide ranges of physical factors and methods have been applied in related experimental studies so far, still there is a need for further laboratory works on this subject and more immortally, the available laboratory measurements on SALGWs should be unified to achieve a standard catalogue on the LGW characteristics based on the effective parameters such as the landslide geometrical, geological, and geotechnical properties and water body geometry.

SALGWs are multi-phase in nature and are nonlinear and dispersive waves. As a result, the computational costs and the accuracy of wave nonlinearity and dispersion are among important factors in numerical modelling of such events. NSEs are computationally expensive and require advanced techniques such as parallel processing and GPU accelerator to practically be applied for real cases. DAEs are commonly applied to simulate LGWs. Two numerical models based on the general approaches of Boussinesq-type equations with considering a rigid landslide (LS3D model) and two-layer shallow-water type equations with assuming landslide as a layer of granular flow (2LCMFlow model) are considered and utilized in this study.

These two models are validated based on the experimental data of Ataie-Ashtiani and Nik-Khah (2008) on SALGWs. Based on the comparisons between the experimental measurments and the numerical results, the LS3D model overstimate the characteristics of LGWs due to considering a rigid landslide. Besides, because of considering a solid hyperbolic shape for the rigid slide, LS3D results in a more than 7% computational error in estimating the characteristics of LGWs caused by a rigid box with different geometry. On the other hand, 2LCMFlow is able to estimate the characteristics of both rigid and deformable landslides appropriately with proper definition of λ_1 and λ_2. Considering the deformability of a landslide has also the advantage of predicting the topographic changes of the sea bottom.

These two approaches are compared based on simulating a potential SAL in the Maku dam reservoir located in the West Azerbaijan province, Iran. The 2LCMFlow model is more realistic in estimating the formation pattern and the characteristics of the first LGW. The maximum LGW height approximated by 2LCMFlow model is about 14% less than the equivalent value estimated by LS3D. However, LS3D is more accurate in modeling wave dispersion in the LGW propagation stage. During the 15–20 s after triggering the potential SAL, the LGW height reduced as much as the 50–60% of the maximum generated LGW height by LS3D while this value is about 25–30% for 2LCMFlow. Besides, an up to 5% time phase difference gradually appears between numerical LGWs of LS3D and 2LCMFlow. LS3D is more efficient than 2LCMFlow computationally with around 30% less runtime.

Authors are also working on the effects of landslide characteristics including the initial geometry, initial submergence, and its geotechnical and constitutive parameters on the LGW characteristics. An example of the effects of landslide initial submergence is included in the last section of this paper. The SSL induces the largest wave crest which is about 21% larger than the a_{pm} caused by the impacts of the SAL. The largest wave trough, which is about 24.5% larger than the a_{nm} of SAL, is induced by the sudden motion of the SML. The run out distances of the SAL and the SSL scenarios are longer than the SML. The longest runout distance is travelled by the SAL. The SML almost stop its motion on the flat part of the flume with a vertical motion of around 2 m along the 45° runup surface.

Based on the comparisons between the numerical results of both models, the most efficient tactic is applying the models coupling which is one of the author's research fronts. The LGW generation stage can be estimated using the 2LCMFlow model and then the results can be applied as the input data for LS3D model to predict the LGW propagation. Moreover, the available laboratory measurements need detailed processing and analyses to provide a standard and worldwide database for SALGWs.

Acknowledgements The authors wish to thank continuous support of Civil Engineering department of Sharif University of Technology, Tehran, Iran, for this research topic. The second author appreciates the contributions of his former graduate students including: Dr. A. Najafi-Jilani, Dr. G.R. Shobayri, Eng. A. Nik-Khah, Dr. L. Farhadi, Dr. S. Malek-Mohammadi, Dr. S. Mansour-Rezaei, and Dr. R. Jalali-Farahani, on this research topic.

References

Abadie S, Morichon D, Grilli S, Glockner S (2008) VOF/Navier-Stokes numerical modelling of surface waves generated by subaerial landslides. La Houille Blanche 1:21–26. doi:10.1051/lhb:2008001

Abadie S, Morichon D, Grilli S, Glockner S (2010) Numerical simulation of waves generated by landslide using a multiple-fluid

Navier-Stokes model. Coast Eng 57:779–794. doi:10.1016/j.coastaleng.2010.03.003
Ataie-Ashtiani B, Najafi-Jilani A (2007) A higher-order Boussinesq-type model with moving bottom boundary: applications to submarine landslide tsunami waves. Int J Numer Meth Fluid 53 (6):1019–1048. doi:10.1002/fld.1354
Ataie-Ashtiani B, Najafi-Jilani A (2008) Laboratory investigations on impulsive waves caused by underwater landslide. Coastal Engineering 55(12):989–1004. doi:1016/j.coastaleng.2008.03.003
Ataie-Ashtiani B, Nik-khah A (2008) Impulsive waves caused by subaerial landslides. Environ Fluid Mech 8(3):263–280. doi:10.1007/s10652-008-9074-7
Ataie-Ashtiani B, Shobeyri G (2008) Numerical simulation of landslide impulsive waves by incompressible smoothed particle hydrodynamics. Int J Numer Methods Fluids 56(2):20932. doi:10.1002/fld.1526
Ataie-Ashtiani B, Mansour-Rezaei S (2009) Modification of weakly Compressible Smoothed Particle Hydrodynamics for preservation of angular momentum in simulation of impulsive wave problems. Coastal Eng J 51(4):363–386. doi:10.1142/S0578563409002077
Ataie-Ashtiani B, Yavari-Ramshe S (2011) Numerical simulation of wave generated by landslide incidents in dam reservoirs. Landslides 8:417–432. doi:10.1007/s10346-011-0258-8
Ataie-Ashtiani B (2016) Preface: thematic issue "landslide-generated tsunamis waves". Landslides 13(6):1321–1321. doi:10.1007/s10346-016-0732-4
Biscarini Ch (2010) Computational fluid dynamics modeling of landslide generated water waves. Landslides 7:117–124. doi:10.1007/s10346-009-0194-z
Bosa S, Petti M (2011) Shallow water numerical model of the wave generated by the Vajont landslide. Environ Model Softw 26:406–418. doi:10.1016/j.envsoft.2010.10.001
Capone T, Panizzo A, Monaghan JJ (2010) SPH modeling of water waves generated by submarine landslides. J Hydraul Res 48:80–84. doi:10.3826/jhr.2010.0006
Crespo AJC, Domínguez JM, Rogers BD, Gómez-Gesteira M, Longshaw S, Canelas R, Vacondio R, Barreiro A, García-Feal O (2015) DualSPHysics: open-source parallel CFD solver based on Smoothed Particle Hydrodynamics (SPH). Comput Phys Commun 187(2):204–216. doi:10.1016/j.cpc.2014.10.004
Davies DR, Wilson CR, Kramer SC (2011) Fluidity: a fully unstructured anisotropic adaptive mesh computational modeling framework for geodynamics. Geochem, Geophys, Geosyst, AGU Geomech Soc 12(6):20 pp. doi:10.1029/2011GC003551
Denlinger RP (2007) Simulations of potential runout and deposition of the Ferguson rockslide, Merced River Canyon, California. US Geological Survey Open File Report 2007-1275. http://pubs.usgs.gov/of/2007/1275
Di Risio M, De Girolamo P, Beltrami G (2011) Forecasting landslide generated tsunamis: a review. In: Mörner NA (ed) The Tsunami threat—research and technology. ISBN 978-953-307-552-5/ISSN, 26 pp
Fernández-Nieto ED, Bouchut F, Bresch B, Castro Díaz MJ, Mangeney A (2008) A new Savage-Hutter type model for submarine avalanches and generated tsunami. J Comput Phys 227:7720–7754. doi:10.1016/j.jcp.2008.04.039
Fine IV, Rabinovich AB, Thomson RE, Kulikov EA (2003) Numerical modeling of tsunami generation by submarine and subaerial landslides. Submarine landslides and tsunamis. In: Yalçıner AC, Pelinovsky E, Okal E, Synolakis CE (eds) Kluwer, Dordrecht/Boston, vol 21, pp 69–88. doi:10.1007/978-94-010-0205-9_9
Fine IV, Rabinovich AB, Bornhold BD, Thomson RE, Kulikov EA (2005) The Grand banks landslide generated tsunami of November 18, 1929: preliminary analysis and numerical modeling. Mar Geol 215(1–2):45–57. doi:10.1016/j.margeo.2004.11.007
Fritz HM, Hager WH, Minor HE (2001) Lituya Bay case: rockslide impact and wave run-up. Sci Tsunami Hazards 19(1):3–22
Fritz HM, Hager WH, Minor HE (2003) Landslide generated impulse waves. 2. Hydrodynamic impact craters. Exp Fluids 35(6):520–532. doi:10.1007/s00348-003-0660-7
Fritz HM, Hager WH, Minor HE (2004) Near field characteristics of landslide generated impulse waves. J Waterway Port Coastal and Ocean Eng 130(6):287–302. doi:10.1061/(ASCE)0733-950X(2004)130:6(287)
Fu L, Jin YC (2015) Investigation of non-deformable and deformable landslides using meshfree method. Ocean Eng 109:192–206. doi:10.1016/j.oceaneng.2015.08.051
Gittings ML (1992) 1992 SAIC's Adaptive Grid Eulerian code. Defense Nuclear Agency Numerical Methods Symposium, pp 28–30
Gómez-Gesteira M, Rogers BD, Crespo AJC, Dalrymple RA, Narayanaswamy M, Dominguez JM (2012a) SPHysics—development of a free-surface fluid solver—Part 1: theory and formulations. Comput Geosci 48:289–299. doi:10.1016/j.cageo.2012.02.029
Gómez-Gesteira M, Crespo AJC, Rogers BD, Dalrymple RA, Dominguez JM, Barreiro A (2012b) SPHysics—development of a free-surface fluid solver—Part 2: efficiency and test cases. Comput Geosci 48:300–307. doi:10.1016/j.cageo.2012.02.028
Grilli ST, Vogelmann S, Watts P (2002) Development of a 3D numerical wave tank for modeling tsunami generation by underwater landslides. Eng Anal Bound Elem 26:301–313. doi:10.1016/S0955-7997(01)00113-8
Harbitz CB, Glimsdal S, Løvholt F, Kveldsvik V, Pedersen GK, Jensen A (2014) Rockslide tsunamis in complex fjords: from an unstable rock slope at Åkerneset to tsunami risk in western Norway. Coastal Eng 88:101–122. doi:10.1016/j.coastaleng.2014.02.003
Harlow FH, Welch JE (1965) Numerical calculation of time-dependent viscous incompressible flow of fluid with free surface. Phys Fluids 8:2182–2189. doi:10.1063/1.1761178
Heller V, Hager WH (2011) Wave types of landslide generated impulse waves. Ocean Eng 38(4):630–640. doi:10.1016/j.oceaneng.2010.12.010
Heller V, Spinneken J (2013) Improved landslide-tsunami prediction: effects of block model parameters and slide model. J Geophys Res Oceans 118:1489–1507. doi:10.1002/jgrc.20099
Heller V, Spinneken J (2015) On the effect of the water body geometry on landslide–tsunamis: physical insight from laboratory tests and 2D to 3D wave parameter transformation. Coastal Eng 104:113–134. doi:10.1016/j.coastaleng.2015.06.006
Heller V, Bruggemann M, Spinneken J, Rogers BD (2016) Composite modelling of subaerial landslide–tsunamis in different water body geometries and novel insight into slide and wave kinematics. Coast Eng 109:20–41. doi:10.1016/j.coastaleng.2015.12.004
Hirt CW, Nichols BD (1981) Volume of fluid (VOF) method for the dynamics of free boundaries. J Comput Phys 39:201–225. doi:10.1016/0021-9991(81)90145-5
Horrillo J, Wood A, Kim GB, Parambath A (2013) A simplified 3-D Navier-Stokes numerical model for landslide-tsunami: application to the Gulf of Mexico. J Geophys Res: Oceans 118:6934–6950. doi:10.1002/2012JC008689
Huang TY, Hsiao SC, Wu NJ (2013) Nonlinear wave propagation and run-up generated by subaerial landslides modeled using meshless method. Comput Mech 53(2):203–214. doi:10.1007/s00466-013-0902-3
Huber A, Hager WH (1997) Forecasting impulse waves in reservoirs. Proc 19th Congrès des Grands Barrages, Florence, ICOLD, Paris, pp 993–1005
Hungr O, Evans SG, Bovis MJ, Hutchinson JN (2001) A review of the classification of landslides of the flow type. Environ Eng Geosci VII (3):221–238. doi:10.2113/gseegeosci.7.3.221

Imran J, Parker G, Locat J, Lee H (2001) 1D numerical model of muddy subaqueous and subaerial debris flows. J Hydraul Eng 127:959–968. doi:10.1061/(ASCE)0733-9429(2001)127:11(959)

Iverson RM, Vallance JW (2001) New views of granular mass flows. Geology 29(2):115–118. doi:10.1130/0091-7613(2001)029<0115:NVOGMF>2.0.CO;2

Kamphuis JW, Bowering RJ (1970) Impulse waves generated by landslides. Proc 12th Coastal Eng 1:575–588, ASCE, Washington DC, USA. doi:10.9753/icce.v12.%25p

Kanoglu U, Titov V, Bernard E, Synolakis C (2015) Tsunamis: bridging science, engineering and society. Phil Trans R Soc A 373 (2053 20140369). doi:10.1098/rsta.2014.0369

Kelfoun K, Giachetti T, Labazuy P (2010) Landslide-generated tsunamis at Réunion Island. J Geophys Res 115(F04012), 17 pp. doi:10.1029/2009JF001381

Koo WC, Kim MH (2008) Numerical modeling and analysis of waves induced by submerged and aerial/sub-aerial landslides. KSCE J Civil Eng 12(2):77–83. doi:10.1007/s12205-0077-1

Lindstrøm EK (2016) Waves generated by subaerial slides with various porosities. Coast Eng 116:170–179. doi:10.1016/j.coastaleng.2016.07.001

Lynett P, Liu PLF (2005) A numerical study of the run-up generated by three-dimensional landslides. J Geophys Res 110(C03006). doi:10.1029/2004JC002443

Lynett P, Liu PLF (2004) A two-layer approach to wave modeling. Proc Roy Soc Lond 460:2637–2669. doi:10.1098/rspa.2004.1305

Ma G, Shi F, Kirby JT (2012) Shock-capturing non-hydrostatic model for fully dispersive surface wave processes. Ocean Model 43–44:22–35. doi:10.1016/j.ocemod.2011.12.002

Ma G, Kirby JT, Hsu TJ, Shi F (2015) A two-layer granular landslide model for tsunami wave generation: theory and computation. Ocean Model 93:40–55. doi:10.1016/j.ocemod.2015.07.012

Macías J, Vázquez JT, Fernández-Salas LM, González-Vida JM, Bárcenas P, Castro MJ, Díaz-del-Río V, Alonso B (2015) The Al-Borani submarine landslide and associated tsunami, A modelling approach. Mar Geol 361:79–95. doi:10.1016/j.margeo.2014.12.0066

Mahab Ghods Inc (1999) Final report of the Maku dam reservoir project (in Persian)

Miller DJ (1960) Giant waves in Lituya Bay Alaska. USGS Professional Paper 354-C, Shorter Contributions to General Geology

Mohammed F, Fritz HM (2012) Physical modeling of tsunamis generated by three-dimensional deformable granular landslides. J Geophys Res 117(C11015). doi:10.1029/2011JC007850

Monaghan JJ, Kos A (2000) Scott Russell's wave generator. Phys Fluids 12(3):622–630

Najafi-Jilani A, Ataie-Ashtiani B (2008) Estimation of near-field characteristics of tsunami generation by submarine landslide. Ocean Eng 35(5–6):545–557. doi:10.1016/j.oceaneng.2007.11.006

Noda E (1970) Water waves generated by landslides. J Waterw Port Coastal Ocean Div, Am Soc Civ Eng 96(4):835–855

Osher S, Fedkiw R (2003) Level set methods and dynamic implicit surfaces. Springer-Verlag

Panizzo A, De Girolamo P, Di Risio M, Maistri A, Petaccia A (2005a) Great landslide events in Italian artificial reservoirs. Nat Hazards Earth Syst Sci 5:733–740

Panizzo A, De Girolamo P, Petaccia A (2005b) Forecasting impulse waves generated by subaerial landslides. J Geophys Res 110 (C12025). doi:10.1029/2004JC002778

Pasenow F, Zilian A, Dinkler D (2008) Numerical model for tsunami generation by subaerial landslides. PAMM Proc Appl Math Mech 8:10519–10520. doi:10.1002/pamm.200810519

Pastor M, Haddad B, Sorbino G, Cuomo S, Drempetic V (2009) A depth-integrated, coupled SPH model for flow-like landslides and related phenomena. Int J Numer Anal Methods Geomech 33:143–172. doi:10.1002/nag.705

Peregrine DH (1967) Long waves on a beach. J Fluid Mech 27:815–827. doi:10.1017/S0022112067002605

Pudasaini SP, Miller SA (2012a) Buoyancy induced mobility in two-phase debris flow. Am Inst Phys Proc 1479:149–152. doi:10.1063/1.4756084

Pudasaini SP, Miller SA (2012b) A real two-phase submarine debris flow and tsunami. Am Inst Phys Proc 1479:197–200. doi:10.1063/1.4756096

Quecedo M, Pastor M, Herreros MI (2004) Numerical modelling of impulse wave generated by fast landslides. Intl J Numer Meth Engng 59:1633–1656. doi:10.1002/nme.934

Roberts NJ, McKillop R, Hermanns RL, Clague JJ, Oppikofer T (2014) Displacemnet waves from subaerial landslides. Landslide science for a Safer Geoenvironment, pp 687–692. doi:10.1007/978-3-319-04996-0_104

Shigihara Y, Goto D, Imamura F, Kitamura Y, Matsubara T, Takaoka K, Ban K (2006) Hydraulic and numerical study on the generation of a subaqueous landslide-induced tsunami along the coast. Nat Hazards 39:159–177. doi:10.1007/s11069-006-0021-y

Skempton AW (1964) Long-term stability of clay slopes. Geotechnique 14:75–101. doi:10.1680/geot.1964.14.2.77

Skempton AW (1985) Residual strength of clays in landslides, folded strata and the laboratory. Geotechnique 35:3–18. doi:10.1680/geot.1985.35.1.3

ten Brink US, Chaytor JD, Geist EL, Brothers DS, Andrews BD (2014) Assessment of tsunami hazard to the U.S. Atlantic margin. Marine Geol 353:31–54. doi:10.1016/j.margeo.2014.02.011

Thomson RE, Rabinovich AB, Kulikov EA, Fine IV, Bornhold BD (2001) On numerical simulation of the landslide-generated Tsunami of November 3, 1994 in Skagway Harbor, Alaska. In: Hebenstreit GT (ed) Tsunami research at the end of a critical decade. Kluwer Academic, Dordrecht, pp 243–282

Unzen Restoration Office of the Ministry of Land, Infrastructure and Transport of Japan (2002) The Catastrophe in Shimabara—1791–92 eruption of Unzen-Fugendake and the sector collapse of Mayu-Yama. An English leaflet (23 p)

Walder JS, Watts P, Sorensen OE, Janssen K (2003) Tsunamis generated by subaerial mass flows. J Geophys Res 108(B5):2236. doi:10.1029/2001JB000707

Watts P, Borrero JC (2001) Probability distributions of landslide tsunamis. ITS 2001 Proc 6(6–8):697–710

Xie J, Tai YC, Jin YC (2013) Study of the free surface flow of water–kaolinite mixture by moving particle semi-implicit (MPS) method. Int J Numer Anal Meth Geomech 38:811–827. doi:10.1002/nag.2234

Yavari-Ramshe S, Ataie-Ashtiani B, Sanders BF (2015) A robust finite volume model to simulate granular flows. Comput Geotech 66:96–112. doi:10.1016/j.compgeo.2015.01.015

Yavari-Ramshe S, Ataie-Ashtiani B (2015) A rigorous finite volume model to simulate subaerial and submarine landslide generated waves. Landslides 14(1):203–221. doi:10.1007/s10346-015-0662-6

Yavari-Ramshe S, Ataie-Ashtiani B (2016) Numerical simulation of subaerial and submarine landslide generated tsunami waves—recent advances and future challenges. Landslides 13(6):1325–1368. doi:10.1007/s10346-016-0734-2

Young CC, Wu CH (2009) An efficient and accurate non-hydrostatic model with embedded Boussinesq-type like equations for surface wave modeling. Int J Numer Meth Fluids 60:27–53. doi:10.1002/fld.1876

Yuk D, Yim SC, Liu PLF (2006) Numerical modeling of submarine mass-movement generated waves using RANS model. Comput Geosci 32(7):927–935. doi:10.1016/j.cageo.2005.10.028

Zhao L, Mao J, Bai X, Liu X, Li T, Williams JJR (2015) Finite element simulation of impulse wave generated by landslides using a three-phase model and the conservative level set method. Landslides. doi:10.1007/s10346-014-0552-3

Zijlema M, Stelling GS (2008) Efficient computation of surf zone waves using the nonlinear shallow water equations with non-hydrostatic pressure. Coastal Eng 55:780–790. doi:10.1016/j.coastaleng.2008.02.020

Zweifel A, Hager WH, Minor HE (2006) Plane impulse waves in reservoirs. J Waterway Port Coastal Ocean Eng ASCE 132(5):358–368. doi:10.1061/(ASCE)0733-950X(2006)132:5(358)

Rockfall Occurrence and Fragmentation

Jordi Corominas, Olga Mavrouli, and Roger Ruiz-Carulla

Abstract

Rockfalls are very rapid and damaging slope instability processes that affect mountainous regions, coastal cliffs and slope cuts. This contribution focuses on fragmental rockfalls in which the moving particles, particularly the largest ones, propagate following independent paths with little interaction among them. The prediction of the occurrence and frequency of the rockfalls has benefited by the rapid development of the techniques for the detection and the remote acquisition of the rock mass surface features such as the 3D laser scanner and the digital photogrammetry. These techniques are also used to monitor the deformation experienced by the rock mass before failure. The quantitative analysis of the fragmental rockfalls is a useful approach to assess risk and for the design of both stabilization and protection measures. The analysis of rockfalls must consider not only the frequency and magnitude of the potential events but also the fragmentation of the detached rock mass. The latter is a crucial issue as it affects the number, size and the velocity of the individual rock blocks. Several case studies of the application of the remote acquisition techniques for determining the size and frequency of rockfall events and their fragmentation are presented. The extrapolation of the magnitude-frequency relationships is discussed as well as the role of the geological factors for constraining the size of the largest detachable mass from a cliff. Finally, the performance of a fractal fragmentation model for rockfalls is also discussed.

Keywords

Rockfalls • Magnitude-frequency relations • Fragmentation • Hazard analysis

Introduction

Rockfalls are widespread phenomena in mountain ranges, coastal cliffs, volcanos, river banks, and slope cuts. Although most of them take place in remote places, they also threaten residential areas and transport corridors (Hungr et al. 1999; Chau et al. 2003; Corominas et al. 2005). The unpredictable nature often attributed to the rockfalls events is cause of concern of the authorities and decision makers. Although rockfalls have a limited size, they are extremely rapid processes that exhibit high kinetic energies and damaging capability. Turner and Jayaprakash (2012) prepared an exhaustive list of rockfall events which demonstrates that even relatively small volumes of rocks may cause significant damage and traffic disruption, particularly in railroads. Recent studies (Petley 2012) have shown that losses due to landslides and rockfalls are concentrated in less developed countries in which deficit of research exists, and often lack of the appropriate resources.

This type of events may be mitigated with stabilization and protection works but often engineers have to make

J. Corominas (✉) · O. Mavrouli · R. Ruiz-Carulla
Department of Civil and Environmental Engineering, Universitat Politècnica de Catalunya, Jordi Girona 1-3, 08034 Barcelona, Spain
e-mail: Jordi.corominas@upc.edu

O. Mavrouli
e-mail: olga.mavrouli@upc.edu

R. Ruiz-Carulla
e-mail: roger.ruiz@upc.edu

K. Sassa et al. (eds.), *Advancing Culture of Living with Landslides*,
DOI 10.1007/978-3-319-59469-9_4

difficult judgements due to the uncertainties associated to the prediction of the size and frequency of the potential events.

Rockfalls are defined as the detachment of a rock from a steep slope along a surface on which little or no shear displacement takes place (Cruden and Varnes 1996). The main feature is that the mass descends very rapidly through the air by falling, bouncing, and rolling. Furthermore, almost no interaction takes place between the most mobile moving fragments, which interact mainly with the substrate (Hungr et al. 2014). Rockfalls are considered relatively small mass movements confined to the detachment of an individual rock or a relatively small rock mass (Selby 1982). Detachments of large-scale rock masses are defined as rockslides and rock avalanches (Cruden and Varnes 1996).

It is widely accepted that rockfalls and rock avalanches exhibit different propagation mechanisms. However, there is still a debate on the characterization of rockfalls, particularly on their dimensions. Some researchers attempted to restrict the term rockfall based on a maximum kinetic energy (Spang and Rautenstrauch 1988) or on volumetric terms such as debris falls (<10 m^3), boulder falls (10–100 m^3), block falls (>100 m^3), cliff falls (10^4–10^6 m^3) and Bergsturz (>10^6 m^3) (Whalley 1984). The current practice indicates that terms such as rockfalls, rockslides and rock avalanches are often used in a non-coincident way (i.e. Hungr et al. 1999; Chau et al. 2003; Dussauge-Peisser et al. 2002; Guzzetti et al. 2003) and that the agreement on the terms has not yet reached. Turner and Jayaprakash (2012) found preferable to maintain the definition of rockfall as one involving significant velocities and some measure of free flight without an upper limit on the volume or kinetic energy of individual blocks. An interesting consideration was made by Rochet (1987), who distinguished between stone fall (chute de pierre) up to few hundred of cubic meters, in which no interaction exists between the rock fragments which follow independent trajectories; rock mass fall (éboulement en masse) up to few hundreds of thousands of cubic meters in which the interaction between particles is weak as they follow independent trajectories or soon they become independent; very large rock mass fall (>10^5–10^6 m^3) showing strong interaction of particles within the moving mass with the development of internal pressures (possible fluidification) and low energy dissipation; mass propagation (déplacement en masse) (>10^6 m^3) that progresses mostly by a translational displacement. Evans and Hungr (1993) introduced the term fragmental rockfall to describe the events in which the individual fragments move as independent rigid bodies interacting with the substrate by means of episodic impacts. They usually involve volumes smaller than 10^5 m^3. For larger volumes, the blocks propagate as granular flows and are considered rock avalanches (Hungr et al. 2014). Distinguishing between these terms is relevant because fragmental rockfalls are modelled by means of ballistic trajectories while rock avalanches are simulated as granular flows (Bourrier et al. 2013). The passage from a falling of independent particles to a granular flow is gradual and both mechanisms can coexist in some events. The transition may take place at volumes as small as 5×10^4 m^3 (Davies and McSaveney 2002) although other authors raise it up to 10^7 m^3 (Hsü 1978). In light of these considerations, in the authors' opinion it is preferable not to propose a specific volume threshold between them as suggested by Turner and Jayaprakash (2012). This communication will focus on fragmental rockfalls only.

Large rockfalls, rockslides and rock avalanches are sometimes difficult to distinguish one from the other (Chau et al. 2003). Rockfall affected areas are characterized by the presence of scars at the rock face and by the presence of rock blocks scattered over the slope, which are less evident if the vegetation cover is present. In case of very active rock walls, the repeated fall of rock fragments generates a talus deposit at the foot of the slope, in which the segregation of blocks by volume takes place. Fine material is found near the apex while the mean block size increases downslope (Evans and Hungr 1993). This sorting is consequence of the lesser kinetic energy of the small rock blocks, which can be easily stopped by obstacles and trapped in depressions between larger rocks of the talus slopes (Statham 1976; Dorren 2003). As the size of the falling mass increases, blocks located underneath the mass during the impact may become crushed, generating a large amount of minor particles.

Rock avalanches are expected to travel far beyond the distal edge of the talus slopes (Wieczorek 2002). The generated debris sheet often spreads in the direction transverse to the flow and has a chaotic arrangement in which huge blocks define an extremely irregular, rough morphology (Soeters and Van Westen 1996). The drainage pattern may be seriously disturbed and the accumulated mass may be large enough to dam the stream and generate a lake. Lobate shapes and either longitudinal or transverse ridges are often found, sometimes close to the front of the deposit (Hewitt 1999; Hewitt et al. 2008). The debris is composed of crushed and pulverized rock and contains the same lithology from de largest blocks up to the smallest ones. In case several lithotypes are involved they maintain their identity and form bands that respect the original stratigraphic order (Davies et al. 2009; Hewitt et al. 2008).

Rockfall Source Characterization

The failure of a rock slope is controlled by the lithology, strength and structure of the rock mass. The existing discontinuities, their orientation, spacing and persistence determine the failure mechanism (i.e. planar, wedge, toppling), which can be assessed by stability analyses (Hoek and Bray 1981). The failure of the rock mass and the onset of a rockfall depend on additional factors such as rock strength, degree of weathering, cleft water pressures, and erosion (Budetta 2004). To account for the interaction between the discontinuities and rock shearing, the conventional kinematic and limit equilibrium techniques to simulate simple failures are replaced by numerical continuum–discontinuum codes with fracture simulation capabilities, well suited to complex instabilities (Eberhardt et al. 2004). The latter requires effective data collection in the field and data interpretation for potential failure modes.

The characterization of the rock mass fracture pattern has traditionally been carried out in the field, with systematic sampling of discontinuities. The most rigorous way is by scanlines (Priest and Hudson 1981; Priest 1993), for which the distribution function of the different joint sets and their spacing is obtained. This procedure, performed in situ has obvious limitations due to the inaccessibility of the outcrops (especially in steep cliffs), high time consumption for data collection and the possibility of measurement errors. Modern techniques allow us to characterize the structure visible from a remote and safe position. The LiDAR technology has experienced rapid expansion and growing with multiple applications in architecture, science and engineering (Heritage and Large 2009). Equipment mobility, accuracy and data acquisition rate, compared to conventional surveying and photogrammetric methods, permits an unprecedented level of detail and is useful for geotechnical purposes. The ability to remotely capture the position of the points on the exposed rock and its processing has significant advantages ranging from safety to efficiency, resulting in a high resolution record.

The advances in techniques of data capture have contributed of the development of methodologies that exploit these high-resolution data in the domain of the structural geology and rock mechanics. These include the generation of high resolution Digital Surface Models (Slob et al. 2002; Kemeny et al. 2006; Jaboyedoff et al. 2007), the identification and characterization of discontinuity sets (Kemeny and Post 2003; Sturzenegger and Stead 2009a, b; Slob et al. 2004; Jaboyedoff et al. 2009; Riquelme et al. 2014), the measurement of the spacing of the joint sets (Slob 2010; Oppikofer et al. 2011; Riquelme et al. 2015) or the definition of potentially movable rock masses (Lato et al. 2009). In the following sections some applications of the use of Lidar, also in combination with digital photogrammetry for analysis of the rockfalls, will be presented.

The laser scanner equipment can be terrestrial (TLS) or mounted on an aircraft (airborne-based or ALS). The operation of the LiDAR equipment for data acquisition and software packages for processing are becoming more affordable. It generates a large database of points defined by the coordinates x, y, z (3D point cloud). The capture rate ranges from 2500 to 500,000 points per second, depending on the type of scanner used and allows creating scenes. To build the point cloud it is required several scans from different points of view, which will be aligned with each other. This is done to avoid occlusion of the points and to increase the density of the point cloud (Jaboyedoff et al. 2012).

The ability to convert a set of points of known positions, as the compiled by the LiDAR, in a database of structural measures, requires several phases (see details in Jaboyedoff et al. 2012). Points must be collected and processed, then manipulated and analyzed for characterization. Data processing should be done with knowledge of the structure to be analyzed. To identify a particular surface in the point cloud, the number of captured points must be sufficient for it to be displayed. This value depends on the extent of the visible surface, and the density of points that can be acquired, which in turn depends on the distance from the recognized surface and the beam divergence of specific equipment. The more variable orientation of the surface or larger the roughness, the more the data density are required.

The TLS still has limitations in the viewing distance and the emergence of areas of occlusion when the escarpment has large vertical development or in the case of rock protrusions (Sturzenegger and Stead 2009b). Some of these restrictions may be overcome by combining TLS/ALS with digital photogrammetry. This technique allows the production of 3D point clouds from a set of photographs taken of the object of interest. It also allows the generation of meshes that are texturized with the same images, obtaining 3D models for the characterization of geological 3D structure (Lim et al. 2005; Sturzenegger and Stead 2009b). The photographs necessary for the use of digital photogrammetry must have enough quality and can be taken from the ground or from the air. The digital photogrammetry reaches its highest potential when the camera is mounted on an Unmaned Aerial Vehicle (UAV or drone). It carries navigation and inertial systems (gyroscopes, accelerometers, altimeters and GPS) and may perform flights of predefined trajectories of a rock face. The use of UAVs offset the lower image resolution with a closer approach to the rock face safely and different angles of observation. The reader will find more details about the performance, applications and limitations of these techniques in several review papers and in the references therein (Wehr and Lohr 1999; Haneberg 2008; Jaboyedoff et al. 2012).

Rockfall Prediction

The temporal prediction of the slope failure has been traditionally the weakest point in the rockfall hazard analysis and management. For long time, rockfalls were assumed as the sudden detachment of a rock mass from a cliff but they are not. Successive Lidar data captures and DinSAR images have demonstrated that the progressive deformation of the slope takes place before failure.

Failure of rock masses is frequently preceded by creep, progressive deformation, and extensive internal disruption of the slope mass (Stead et al. 2006). The prediction of time to failure can be based on measurements of either surface or subsurface displacements, repeated over time (Hungr et al. 2005). This approach does not take necessarily into account the underlying mechanism of failure and concentrates on the evolution of the slope face. It was first proposed by Saito (1965) and later developed by Voight (1989) and Fukuzono (1990). Plotting the inverse displacement rate versus time generates trend-lines that are projected considering different rheological creep models to the zero value on the abscissa (time axis) and calculate the failure time. The practical application of these methods to failure prediction requires that velocity threshold values be set. It is worth noting that the interpretation of these trends is sometimes difficult due to the fluctuation of the external slope conditions (i.e. rainfall), that may induce deviations of the rate of displacement. Comprehensive discussion and reviews of these methods are provided by Petley et al. (2002), Crosta and Agliardi (2003), Rose and Hungr (2007).

Nowadays, the monitoring of slope movements has become standard practice in most mining and geotechnical projects. Slope displacements a well as other visual precursors such as cracking, localized falls, audible noise are recorded to predict the time of failure. Techniques such as the LiDAR equipment, digital photogrammetry and radar interferometry provide not only the analysis temporal patterns of the precursors before a large slope failure but also the spatial one, thus allowing the definition of the size of the unstable volumes (Oppikofer et al. 2008; Ferrero et al. 2011; Stock et al. 2012; Royán et al. 2014, 2015). Periodic surveys using TSL have shown an increase of the rate of small-size rockfall events prior to the failure of large masses, which are mostly concentrated in the detachment zone (Rosser et al. 2007; Royán et al. 2014; Kromer et al. 2015). The above observations are accompanied with an increase in the displacements of the moving mass away from a background level. The larger the volume of the unstable mass the sooner and greater the number of these features noticed (Rosser et al. 2007).

Although several uncertainties remain associated on whether every large rockfall has precursors and to the determination of the exact time of failure using the creep curves, the measurement of the slope displacements which may be analyzed in combination with other indicators (Amitrano et al. 2005), enables the identification of potential catastrophic failures, so that future events can be recognized beforehand and the society can adapt to the hazard (Gischig et al. 2009; Hermanns et al. 2013).

Rockfall Risk Assessment

The land-use planning and the safety of the transportation corridors require the appropriate analysis of the rockfall hazard. This task is facilitated by the Quantitative Risk Analysis (QRA) approach. It is a powerful management tool in which assumptions and uncertainties are explicitly expressed and considered (Fell et al. 2005). QRA facilitates the objective decision making as it eliminates the use of ambiguous terms, the results are repeatable and consistent, and provides the ingredients for cost-benefit analysis for different scenarios (Corominas and Mavrouli 2011; Corominas et al. 2014). QRA in rockfalls, requires the determination of the probability of the slope failure for a range of volumes; the expected trajectories and the fragmentation of the detached mass, and the evaluation of the probability of impact and damage to the exposed elements. As the available information is often limited, Lee and Jones (2004) warned that the probability of the slope failure and the value of adverse consequences are only estimates.

The risk for a given rockfall location may be expressed analytically as follows (modified from Corominas et al. 2014):

$$R = \sum_{M_i} P(M_i)P(X_j|M_i)P(T|X_j)V_{ij}C \qquad (1)$$

Where

R	Risk due to the occurrence of a rockfall of magnitude M_i on an element at risk located at a distance X from the landslide source
$P(M_i)$	Probability of occurrence of a rockfall of magnitude M_i
$P(X_j \mid M_i)$	Probability of the rockfall reaching a point located at a distance X from the landslide source with an intensity j
$P(T \mid X_j)$	Probability of the element being at the point X at the time of the rockfall occurrence
V_{ij}	Vulnerability of the element being impacted by a rockfall of magnitude i and intensity j
C	Value of the element at risk

The rupture of the rock wall and its probability $P(M_i)$ define the rockfall initiation and is the most challenging part of the rockfall hazard assessment. The potential for failure can be approached from either rational (geomechanical approach) or empirical methods.

Rational Approach

In the geomechanical approach, the stability of the slope can be evaluated using analytical tools (Hoek and Bray 1981) or numerical calculation considering the strength of the rock mass (Eberhardt 2008; Stead et al. 2006). The geomechanical analysis is the appropriate tool for the understanding of the underlying mechanisms driving the instability. The assignment of the properties of the rock mass (rock strength, water pressure, joint orientation and persistence, among others) is however subjected to a high degree of uncertainty.

The classical result of the slope stability analysis is the Factor of Safety, which is only a qualitative expression of the probability of failure. The latter may be addressed using probabilistic slope stability analyses which incorporate and quantify the uncertainties associated to the variability of the input parameters by means of their statistical distributions (El-Ramly et al. 2002). It must be noted that the statistical analyses should be based on sufficient amount of data, otherwise they can be misleading. If data are sparse, a simplified analysis using both the most likely parameters' values and the most unfavorable ones is preferable (Hungr 2016).

Even though the outputs of these methods can be implemented on GIS platforms and may be used to prepare maps showing the potential for rockfall failure from a source area, time is not explicitly taken into account and because of this, they are not currently used in hazard analyses and zoning. Furthermore, procedures for the estimation of the model parameters values at regional scale have yet to be developed.

Frequency Analysis

The empirical methods calculate the probability of failure by statistical analysis using inventories of past events, which also allow preparing magnitude-frequency (MF) relations of the events (Hungr et al. 1999; Dussauge-Peisser et al. 2002; Guzzetti et al. 2003).

It must be taken into account that landslides do not repeat themselves. Despite of this, rockfalls as well as debris flows are landslide types that are usually treated as repetitive events (Corominas and Moya 2008). The temporal occurrence of landslides may be expressed in terms of frequency, return period, or exceedance probability. The frequency expresses the number of events in a certain time interval (e.g. annual frequency) and it can be assessed from the rockfall inventories.

The statistical analyses applied to landslides have found that magnitude versus the cumulative number or frequency of landslides is scale invariant and follows a power law distribution (Hovius et al. 1997; Pelletier et al. 1997; Guzzetti et al. 2002b). The size distribution is similar to the observed by Gutenberg and Richter for earthquakes, and may be expressed in the following way:

$$Log_{10}N(>M) = a - bM \quad (2)$$

Where N is the cumulative number of landslides greater than magnitude M;

While "a" and "b" are coefficients. "a" is a measure of the level of landslide activity; and "b" is the gradient of the relation, where higher b-value indicates a larger proportion of small landslides, and lower b-value a smaller proportion of small landslides. As M is measured on a logarithmic scale, then a linear relationship is obtained in the log-log scale.

This type of relations have been obtained for debris flows (Guthrie and Evans 2004; Hungr et al. 2008) and rockfalls (Hungr et al. 1999; Dussauge-Peisser et al. 2002; Chau et al. 2003; Guzzetti et al. 2003). In order to perform this analysis it is required an inventory as complete as possible (Hungr et al. 1999; Dussauge-Peisser et al. 2002).

The TLS has been used for the identification and calculation of volumes of rock blocks that have fallen, from the images of sequential scans (Abellan et al. 2006; Royán et al. 2015). The subtraction of the Digital Surface Models (DSM) obtained with the respective point clouds can accurately calculate the missing volumes. If the geometry prior to the failure is unknown, the volumes disappeared can be calculated by the reconstruction of the original topography on a DSM generated based on earlier pictures of the event. Oppikofer et al. (2009) detected scars on a rocky wall surfaces formed by the failure and reconstructed the original relief by fitting discontinuity planes to the scars, using specific programs and limited number of scars. Unfortunately, rockfall inventories are not always available or are incomplete. Most historical records cover a limited period while the TLS equipment has been available for only the last few years. This may lead to obviate the occurrence of large infrequent failures.

For landslide risk management purposes, one of the most important questions to answer is what the likelihood of infrequent events is. In that respect several questions arise about the validity of the obtained power laws in both space and time domains and whether they can be extrapolated or not. Dussauge-Peisser et al. (2002) discussed the interpretation of the fitting parameters a and b-values. They consider that if the data do not fit any law, the inventory can be used to estimate an overall frequency in the range of volumes covered by the data. However, in case a power law can be fitted and statistical tests certify the completeness, then it is possible to extrapolate the frequency to larger volumes. This point of view is shared by other researchers (Guzzetti et al. 2003; Picarelli et al. 2005). In any case, as noted by Corominas and Moya (2008) the frequency-magnitude relation for landslides is not purely linear in a log-log scale. A rollover effect (a flattening of the curve) is observed at small landslide magnitudes. This implies that the number of observed small size landslides, is lower than the one expected from the above relationship. It has been suggested that rollover might be due to the incompleteness (censoring) of landslide records because small landslides are not detected in aerial photographs (Hungr et al. 2008; Stark and Hovius 2001) or that some kind of physical constraint must exist that justifies this rollover at small sizes (Guthrie and Evans 2004). However, the lack of fit to the power law may also take place at large landslide magnitudes (Guzzetti et al. 2002b; Malamud et al. 2004) and, consequently other causes should be explored in order to explain such behavior. These types of constraints are not well understood yet. This issue becomes critical when dealing with rockfall risk management strategies.

Obtaining the Distribution of Missing Volumes in Cliffs

Obtaining long records of rockfalls is a challenging task because in road and railways cuts, the rock blocks are immediately removed after their occurrence in order to resume the traffic conditions. In some exceptional cases, the maintenance teams have collected the necessary information for the preparation of MF relations (Hungr et al. 1999; Ferlisi et al. 2012). Unfortunately, the situation is worse in natural slopes. A few number of historical rockfall inventories are available that allow the analysis to the scale of up to a hundred years (Wieczorek et al. 1998; Guzzetti et al. 2003; Hantz et al. 2003). On the other hand, the rock blocks deposited at the slope below the rockfall sources cannot be usually used to prepare data sets because it is not possible to discriminate which of the blocks belong to the different rockfall events that have generated the deposit.

Instead, the cliff face where the rockfalls have been generated may keep the record of the events occurred during the last hundreds or thousands of years in the form of rockfall scars. The scar is an area of rupture on the rock wall or cliff, generated by the separation of a rock mass in a single or multiple events (Fig. 1). The distribution of volume or density of scars, are indicative of the occurrence of rockfalls and can be used as an indirect measure of the frequency of the events.

We have developed a supervised and stepwise methodology for calculating the volume distribution of rockfall scars using the TLS. This section summarizes the work carried out in the steep and intensely fractured granodiorite rock wall of Forat Negre in the Solà de Santa Coloma, Principality of Andorra (Santana et al. 2012). In the last decades buildings were built in the Solà de Santa Coloma and are threatened by rockfalls. The talus deposit at the bottom of the slope contains blocks with volumes between 0.5 and 270 m^3. The detailed description of the area and the actions undertaken protection are found at Copons (2007), Copons et al. (2004), or Corominas et al. (2005).

The input data are the point clouds from which the planes present in the slope surface are extracted, the main discontinuity sets are identified, the area of the exposed surfaces of each of the scars are calculated, their height as well and finally the distribution of rock volumes that have disappeared from the wall is calculated. The analysis is based on several assumptions (Santana et al. 2012): (i) the detachment of the rock blocks at a particular point is due to the presence of unfavorable dipping discontinuities and/or its intersection with other discontinuities; (ii) the volume of the scars is approached by a prism formed by the intersection of discontinuities in the rock wall, as shown in Fig. 2; (iii) each scar basal plane corresponds to at least one rockfall event; and (iv) the detachment over adjacent sliding surfaces with the same orientation, but separated by a step of at least 20 cm jump are considered independent events. According to these assumptions, step-path failures are taken into consideration but only for steps smaller than 0.2 m.

The analysis of the point cloud allowed to identify 7 main discontinuity sets (Table 1). Kinematic analysis indicates that two sets, F3 and F5, generate potential planar failures, which is predominant failure mode for rockfalls in the study area as confirmed by the inventoried cases (Copons 2007). F1 and F7 sets bound the unstable rock blocks and act as tension crack and lateral release plane, respectively. These four sets account for 78% of the discontinuities present at the rock mass (Table 1). To obtain the volume distribution of the scars, the points of the point cloud belonging to each one of the four sets F1, F3 and F5, and F7 were filtered and planes were adjusted to them. These planes correspond to the scar edges. Their areas, the maximum width (along the strike) and length (along the dip direction) were measured. The heights of the scars are defined by the intersections of the planes of joints F1 and F7, and their distribution is defined by the common length of these planes. Eventually, the size distribution of the scars was calculated by means of a Monte Carlo simulation, multiplying the scar basal areas with the heights. The scar volumes were assumed to be prismatic as the angles between the height and the basal area are greater than 60º and the inaccuracy imposed by considering angles of 90° is limited (Palmstrom 2005).

For a sample of 5000 scars, which is of the same order of magnitude as the identified scars on the point cloud, the maximum calculated scar volumes are of the order of few thousands of m^3 ($\sim$3000 m^3). This is the maximum rock mass size that has been detached from the slope face leaving

Fig. 1 Example of rockfall scars (*blue polygons*) observable at the Degotalls wall, Montserrat massif, Catalonian Coastal Ranges, Spain

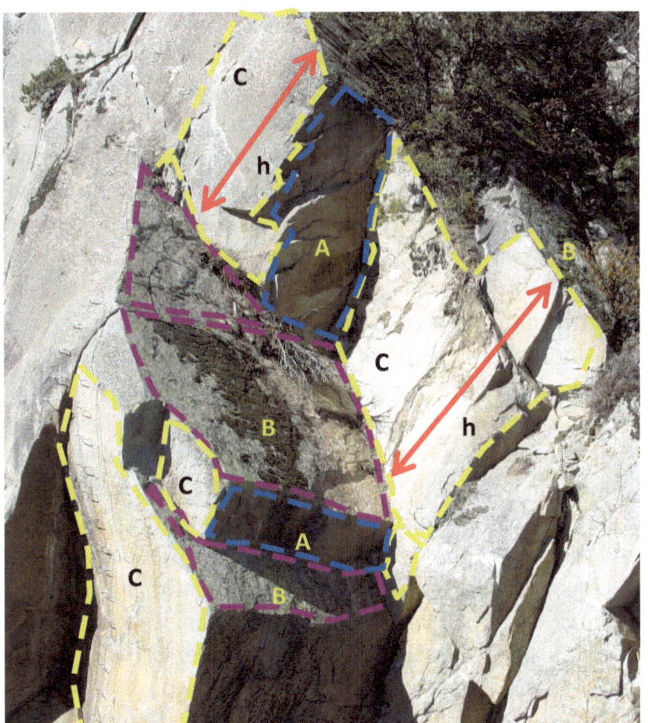

Fig. 2 Rockfall scar defined by three intersecting joint sets. The detached block was resting on a basal plane (*B*) which is bounded by planes (*A*) and (*C*). The *height* of the scar (*h*) may involve several spacings

a scar on it. The largest observed scar base was 213 m^2, indicating a respective scar height of the order of 15 m. The resolution of the technique allows the detection of volumes as small as 0.02 m^3.

For scar volumes larger than 0.75 m^3, the distribution is well fitted by the following inverse power law (Fig. 3):

$$N(v > V) = 1919v^{-0.9} \tag{3}$$

Where N is the accumulated number of rockfall scars larger than a volume V, in cubic meters

It is worth noticing that in the Solà de Santa Coloma, the occurrence of rockfalls (larger than 1 m^3) during the last 50 years is one every two years as an average (Moya et al. 2010). The obtained distribution may thus represent the result of the rockfall activity in the slope for the last few thousands of years. Due to the assumption adopted, this volume distribution should be considered an upper envelop for the rockfall volumes because the scars may be the result of one or several failure events. However, this distribution might underestimate the occurrence of the largest failures because only the step-path surfaces with steps smaller than 0.2 m have been measured in the followed approach. Despite of this, neither the presence of large scarps suggesting the occurrence of any large failure nor the presence of a massive rockfall or avalanche deposits have been identified in the slope and valley bottom respectively.

Table 1 Discontinuity sets of the Forat Negre slope, relative frequency and maximum and minimum areas obtained with TLS (modified from Santana et al. 2012)

Joint set	Dip direction/angle (°)	Relative frequency (%)	Maximum area (m^2)	Minimum area (m^2)
F1	056/63	13.5	121.6	0.3
F2[a]	320/60			
F3	155/57	25.7	236.3	0.7
F4A	247/45	1.4	11.6	0.4
F4B	266/64	2.7	20.3	0.4
F5	187/54	17.7	144.4	0.7
F6	187/75	9.1	23.8	0.5
F7	155/87	21.5	213.7	0.5
F8	092/57	4.2	34.6	0.4

[a]Merged with F7 in the LiDAR analysis

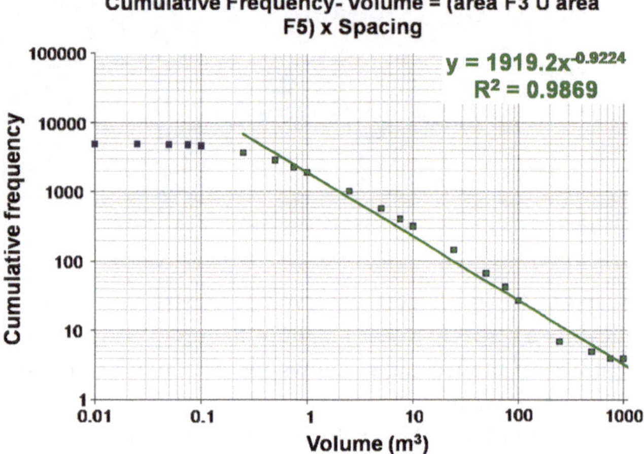

Fig. 3 Magnitude (volume in m^3)—Cumulative frequency of rockfall scars, calculated from the point cloud (Santana et al. 2012)

Are There Geologic Controls of the Rockfall Volume Distribution?

The a-value in equation 2 represents the rate of rockfall activity, which unless normalized, it is also function of the size of the study area. The higher the b-value, the less important the contribution of the larger failures is, and vice versa. Compared to other published works (Dussauge et al. 2003; Guzzetti et al. 2003), the absolute b-value in Forat Negre is relatively high (Table 2). Dussauge et al. (2003), argue that rockfalls for sub-vertical cliff and for a wide range of volumes (10^2–10^{10} m^3) have b-values of 0.5 ± 0.2. The high b-value in the Andorran case could be of course attributed to the exclusion of staggered failures. However, as mentioned above no geomorphological evidences of large slope failures are found in the valley. The maximum volume with this procedure is few thousand cubic meters.

Several authors relate the b-value with the lithology and the level of fracturing of the slope (Dussauge-Peisser et al. 2002). Brunetti et al. (2009) after analyzing 19 data sets (including several rockfall data sets) concluded that variation of the scaling behavior of the non-cumulative distributions is independent of the lithological characteristics, morphological settings, triggering mechanisms, length of period and extent of the area covered by the datasets. They argue that the statistics of landslide volume is conditioned primarily by the geometrical properties of the slope or rock mass and that difference between the scaling exponents of rockfalls and landslides is consequence of the disparity of mechanisms. On the other hand, the fact that rockfalls and rock slides they studied exhibit the same frequency-volume relationship made Guzzetti et al. (2003) and Dussauge-Peisser et al. (2002) to conclude that there is no statistical difference between these types of landslides. By comparing the results of four data sets, Hungr et al. (1999) suggested that the lower exponents could correspond to areas of massive rocks with the possibility to produce a relatively greater proportion of large-magnitude structurally controlled failures.

In the case of the Forat Negre in Andorra, the structural analysis of the joint sets suggests that a geological control on the size of rockfalls may exist that could justify the greater b-value of the scar volume distribution. A field survey has been carried out of the granodiorite massif of Forat Negre aiming at determining the relative chronology of the structural features in the rock mass (Fig. 4). It was performed at key outcrops where discontinuities are better exposed. The outcrops were studied combining scanlines and detailed structural observations. It is found that set F6 was formed first as it is affected by other sets that interrupt and displace its planes. A second phase is characterized by sets F2 poorly identified with LiDAR and merged with F7. They include both very persistent conjugate faults and joints. F3 is a joint set that could be associated to this phase. It shows high scattering and undulation with amplitude up to 20 cm. The last phase is characterized by the occurrence of F1 and F4. They should be interpreted as conjugate faults that are

Table 2 Exponents of the power law fitted distributions obtained for different rockfall inventories

Reference	Location	Length of the record (year)	Range of volumes fitted (m^3)	Number of events (N)	Scaling parameter (b)
Hungr et al. (1999)	Highway 99 British Columbia	40	10^1–8×10^8	390	−0.43
	BCR line	12	10^0–10^4	403	−0.4
	Highway 1		10^0–10^4	226	−0.7
	CP Line	22	10^0–10^4	918	−0.65
Gardner (1970)[a]	Lake Louis	Two summers	10^{-1}–10^3	409	−0.72
Chau et al. (2003)	Hong Kong, China			201	−0.87
Dussauge-Peisser et al. (2002)	Upper Arly, gorge French Alps		10^0–10^4	59	−0.45
	Grenoble, French Alps	60	10^{-2}–10^6	87	−0.41
	Yosemite, USA	77	10^0–10^5	101	−0.46
Royán et al. (2015)	Puigcercós	6.87	10^{-2}–10^2	3096	−0.72

[a]Cited in Hungr et al. (1999)

Fig. 4 (*Left*) Outcrop of conjugated faults F4 and F1; (*right*) intersection of planes of sets F1, F3 and F2

superposed and interrupt the rest of sets. F3 and F5 are joint sets usually limited and confined by fault planes of the two main deformation phases.

This survey has highlighted the frequent interruption of the basal planes (discontinuities F3 and F5) at their intersection with the tension crack and lateral release planes F7 and F1, respectively (F7 that usually plays the role of tension crack). This interruption may prevent the formation of large failures along the basal sliding surface. The analysis of the scar planes of Santana et al. (2012) has also permitted measuring the spacing of the involved discontinuity sets, as the perpendicular distances between successive planes (Mavrouli and Corominas 2017). Using the same data, the visible length (along the dip) of the scar edges was also calculated, as the maximum edge distance along the dip of a plane (Table 3). In these table, the average length of F3 and F5 (the basal instability planes) are very similar to the average spacing of the planes of F7, although obtained by independent procedures. However, this does not always happen, the maximum measured length of F3 and F5 (27.08 and 14.65 m respectively) is much longer than the maximum spacing of F7 and one order of magnitude longer than its average spacing. This suggests that in the Forat Negre slope, the failure surface may generate by coalescence of several (although few) unfavorable dipping F3/F5 planes and/or by brittle failure of minor rock bridges. The maximum volume will then depend on the length of the basal plane and on the resistance of the rock bridges, if any.

Assessment of the Largest Credible Volume

The assessment of the largest credible volume is crucial for the management of rockfall risk. Residential areas located below rockfall susceptible slopes have often developed strategies of risk mitigation using a combination of land use planning, stabilization and protective works (protection fences, embankments). The design of these measures and the delimitation of the hazardous areas are based on analyses for a range of expected potential rockfall volumes (Corominas et al. 2005; Abbruzzese et al. 2009; Agliardi et al. 2009; Li et al. 2009). Although some risks analyses (i.e. Hungr et al. 1999) have shown that the highest risk is associated to mid-size rockfall events ($1–10^3$ m^3), the occurrence of rockfall events larger than the used for the design of the protective works would not be manageable and the population might be exposed at an unacceptable risk level. The question posed is what the largest credible rockfall volume can be. It is usually characterized by volumes of rock masses of several orders of magnitude greater than the events commonly observed in the study area. It must be kept in mind that the largest credible rockfall event is a reasonable largest event, not the largest conceivable event.

The analysis of the failure of a rocky slope is intrinsically linked to the knowledge of the fracture pattern of the massif which, on one side determines the volume of kinematically unstable rock mass and on the other side determines the mechanism of rupture. The instability mechanism may involve displacement along existing discontinuities either fully persistent or not and brittle failure of intact rock. It may involve single large blocks bounded by discontinuities or rock masses composed of smaller blocks. Figure 5 shows some conceptual schemes of the fracture patterns associated to the failure mechanisms. Figure 5a: simple planar failure is a rock mass affected by a fully persistent joint set. The volume mobilized is directly determined by the orientation and dip of the discontinuity and the surface topography. This setting may generate the largest failure in the slope; Fig. 5b: if several joint sets interrupt and/or displace others, a stepped failure may develop by the coalescence of the discontinuities, which may also mobilize large rock mass volumes; Fig. 5c: In this case, a stepped failure may also develop. The failure surface of the rock mass develops across the existing joints and rock bridges. These ruptures are more difficult to define and predict due to the uncertainty associated with the persistence of the discontinuities in the rock mass. The potential for failure and the moveable volume is determined by the structure of the rock mass and the strength of the rock bridges. Volumes can be large but generally smaller than in the previous cases; Fig. 5d: failure exclusively defined by the intersection of planes of different joint sets. The interruption of the discontinuities by others results in relatively small moveable volumes.

The volumes calculated by the statistical distribution of the rockfall events are the result of the observations. The simplest way to estimate the maximal rockfall size is considering the greatest inventoried one, independently of the rock mass properties. This is feasible if the inventory used covers a long span of time (hundreds to thousands of years). Instead, Malamud et al. (2004), Picarelli et al. (2005), Brunetti et al. (2009) suggested the extrapolation of the power-law magnitude-frequency relations, for a preliminary assessment of the largest events. Malamud et al. (2004) found that the scaling factor (exponent) held for a wide range of sizes (from 10^{-3} to 10^6 m^3). The question that arises is whether this can lead to an overestimation of volumes and an oversizing of the solutions. For the detachment of large rock masses the local characteristics of the jointed rock mass such as continuity lengths of the discontinuity sets and spacings are relevant. These characteristics might be a possible reason for the truncation of the afore-mentioned distributions, as the found in some rockfall records. Brunetti et al. (2009) found that some landslide data sets exhibit a deviation of the power-law fitted tail, for large landslides. They suggested that this deviation may be the result of undersampling or due to geometrical constraints such as that a landslide cannot be larger than the slope where the failure occurs (Guzzetti et al. 2002b). In fact, deviations are observed at both high and low magnitudes (Stark and

Table 3 Measured areas, lengths and spacings of the discontinuity sets (Mavrouli and Corominas 2017)

Discontinuity	F1	F3	F5	F7
Max spacing (m)	8.09	6.35	4.11	5.60
Average spacing (m)	2.11	1.84	0.76	1.22
Median spacing (m)	1.63	1.22	0.53	1.00
Max area (m^2)	121	236	144	213
Max length (m)	19.85	27.08	14.65	19.14
Average length (m)	0.96	1.19	0.99	1.45
Median length (m)	0.76	1.04	0.73	0.89

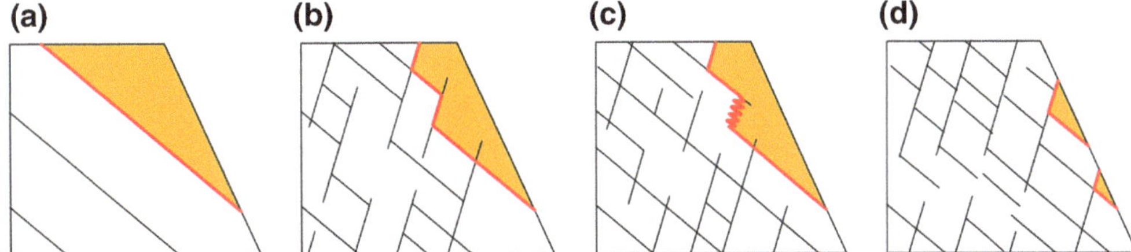

Fig. 5 Conceptual scheme of fracture pattern and development of a rock mass failure. Notice that the size of the moveable mass diminishes from **a** to **d**

Hovius 2001; Brardinoni and Church 2004; Guthrie and Evans 2004).

The assessment of the maximum possible rockfall volume at a rock slope presents several uncertainties. The unstable volumes are not always visible on the surface and additional information is required about both the orientation, spacing and persistence of discontinuities in the rock mass, which is rarely available (Palmström 2001; Elmouttie and Poropat 2012; Lambert et al. 2012; Wyllie and Mah 2004). These are critical factors for the determination of failure mechanisms too. Instability is much more likely to occur if spacings are dense and joints are fully persistent, given the usually much higher resistance of intact rock compared to that of joints (Einstein et al. 1983). The above procedures are usually applied in the detection of sizes commonly observed detachment, but do not include extreme events that could be produced in particularly adverse conditions. Setting a realistic maximum volume of detachment is still a challenge.

The role of the cliff morphology and structure on the rockfall detachment has been highlighted by Frayssines and Hantz (2006). Rosser et al. (2007) suggested that small rockfall detachments may accentuate unstable morphological features such as rock spurs and overhangs. These features may become the source of larger failures in the area and the small rockfall events could appear in this locations as precursors of the failure. In fact sustained rockfall activity for tens of years was observed before the occurrence of rock avalanches in Mount Fletcher in New Zealand (McSaveney 2002). The assessment of the maximum credible rockfall volume in the Forat Negre, that could not be identified by the analysis of the rockfal scars, is presented here summarizing the work of Mavrouli et al. (2015) and Mavrouli and Corominas (2017). The cinematically detachable rock masses are identified using a digital elevation model (DEM) and applying the Markland criterion (Hoek and Bray 1981) in each cell in the DEM where the outcrop of unfavorable discontinuities is assumed. The size of the unstable areas are defined by the addition of all the adjacent cells that meet the criteria of Markland. The distribution of potentially unstable volumes are calculated based on the areas of unstable cells in the DEM which are subsequently transformed into equivalent volumes (Fig. 6).

This analysis describes a conservative scenario with the following assumptions: (i) the whole of the unstable mass is separated at a time, regardless the possible occurrence of successive failures; (ii) all representative joint sets of the rock mass are present in each cell of the slope and (iii) for calculating the volume is considered that discontinuities outcrop in the lowest part of the unstable cells. No restriction by lateral confinement is imposed for the detachment of the rock mass. The procedure has been applied in the sector Forat Negre at the Solà de Santa Coloma, Andorra and implemented in a GIS.

The sectors containing rock volumes kinematically movable were overlapped to orthophotos in order to verify instability and to delineate smaller masses within the larger ones.

The areas of these sectors are indicative of the size of the rockfalls that can occur in this slope. In order to convert the area A, in volume V, two simple shapes and alternative rock masses mobilized are assumed: cubic or prismatic. For both, the base corresponds to the area A. The height depends on the persistence of basal plane within the rock mass (see joint length in Fig. 6). In the cubic form this length is taken as $L = \sqrt{A}$ and prismatic as $L = 0.5\sqrt{A}$. The cubic and prismatic volumes are calculated respectively by Eqs. (4) and (5).

$$V = A^{3/2} \tag{4}$$

$$V = 0.5A^{3/2} \tag{5}$$

The maximum volumes obtained by this analysis are 50,000–25,000 m^3 for cubic and prismatic volumes respectively (Fig. 7). The largest basal area obtained is 1361 m^2. The tails of the volume distributions obtained are fitted to negative power laws whose b-values are −0.57 and −0.55, respectively. As the concurrence of all the above mentioned hypotheses (i) to (iii) is highly unlikely and the assumptions conservative, these volumes set an upper limit for rockfalls in the study area.

The size distribution of scars observed (see Fig. 3) is an empirical evidence of rockfalls that occurred in the past. Instead, the kinematically movable rock masses indicate

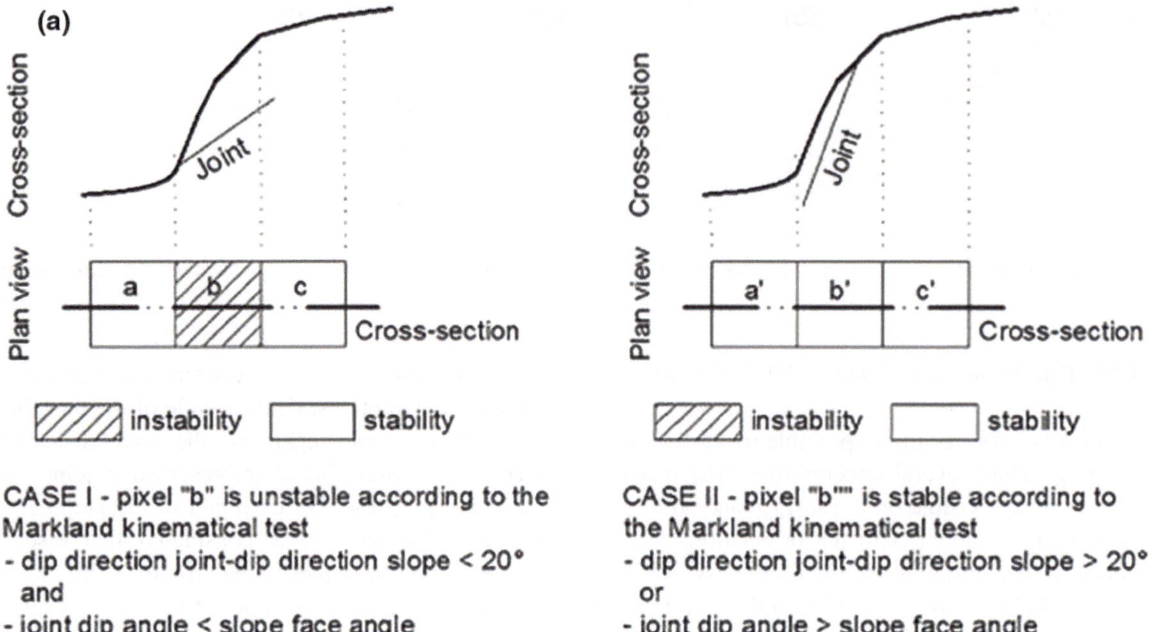

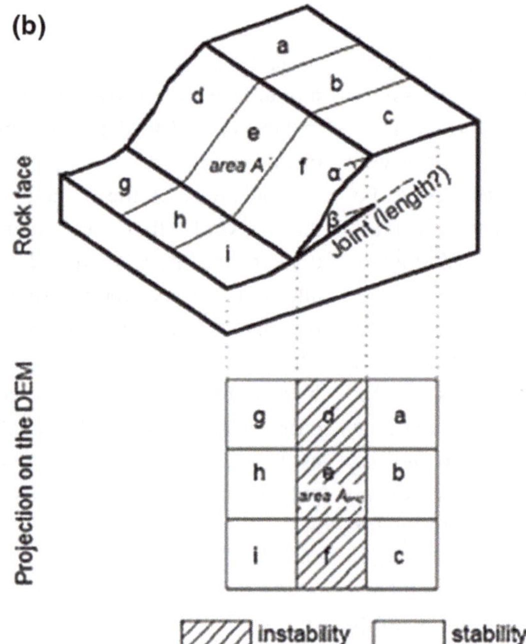

Fig. 6 Detection cinematically movable rock volumes in a DEM. *Left* **a** stability test for the cell and **b** formation of unstable volumes by aggregation kinematically unstable adjacent cells (modified from Mavrouli et al. 2015)

hypothetical rockfalls that could occur in the future. Comparing the two distributions, the difference between the maximum volumes of scars (about 3000 m^3) and the kinematically movable masses of rock (25,000 or 50,000 m^3) is one order of magnitude. This difference is mainly attributable to the divergence of the assumptions made. Calculating volumes scars it is based on the dimensions of the basal planes of unstable joints, which is the part of the discontinuity in the slope that remains exposed after rupture. Conversely, on detecting the mass of kinematically movable rock, discontinuities are considered persistent and it is assumed that F3 and F5 joint sets outcrop in each cell of the DEM. Therefore, the extent of the surface of discontinuity which is involved in the detachment is greater. It is worth noticing that the kinematically movable rock masses generate continuous basal planes up to 1361 m^2. However, these

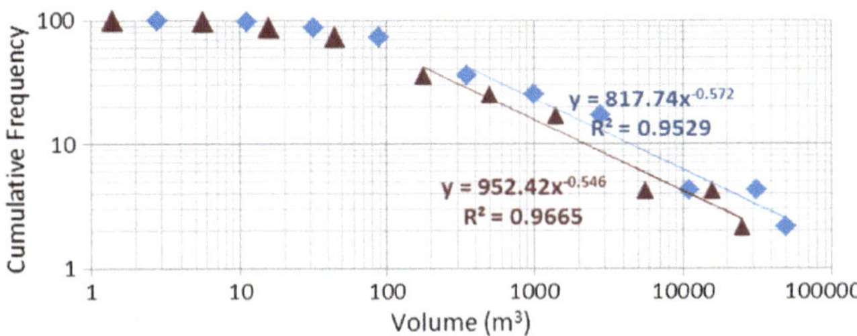

Fig. 7 Distribution (cumulative frequency) of largest potential rockfall volumes for cubic forms (*diamond*) and prismatic (*triangles*) shapes (Mavrouli et al. 2015)

planes are not observed on the actual slope, where the maximum detected area is 213 m^2 (Mavrouli and Corominas 2017) and in any case, the largest potential kinematically detachable volume calculated is less than 10^5 m^3.

Defining Credible Risk Scenarios

In terms of rockfall hazard assessment, answering to the question on whether the MF relations can be used to predict the probability of occurrence of large events (rock slides, rock avalanches), larger than those inventoried, is not a trivial issue. As mentioned before, the occurrence of events larger than the observed in historical and/or prehistorical records can be unmanageable and may pose a risk considered socially unacceptable. Powerful analytical and numerical tools are nowadays available and despite the uncertainties in relation to the fracture pattern and the properties of the joints and rock mass several, they can be used to analyze critical slopes. However, their use for regional (or spatially distributed) analysis is still in its infancy.

Various reasons may be accounted for checking the validity of the extrapolation of the MF for the prediction of large slope failures in contexts such as in the case of Andorra. Large massive rock slope failures (large rock mass falls, large rockslides or rock avalanches) have not been identified in the area. The occurrence of large rockslides and rock avalanches has geomorphic consequences which can be deciphered by means of the analysis of the landscape (Fig. 8). The detachment of large rock volumes from the source area can be often identified by the presence of large prominent scarps or arcuate depressions on the rock wall as well as by the long-runout accumulation of debris (Soeters and Van Westen 1996; Hewitt 2002; Ballantyne and Stone 2004).

On the other hand, the distribution of large slope failures in mountain ranges does not look like random. Large landslides are found in sectors of mountain fronts representing distinct topographic, geomorphic, and geologic characteristics that either individually or in combination favors mountain front collapse (Hermanns and Strecker 1999; Jarman et al. 2014). Large size rockslides and rock avalanches are often associated to unfavorable geostructural settings and a variety of geometries have been identified as having a potential for catastrophic failure (Hutchinson 1988). The role of both the strength and structure of the rock mass has been intensively discussed by Fell et al. (2007) and Glastonbury and Fell (2010) and found that single or two persistent intersecting discontinuities (bedding planes, schistosity, faults, stress release joints) are systematically involved in occurrence of rapid rockslides. Similar structural controls have been found elsewhere (i.e. Hermanns and Strecker 1999; Hungr and Evans 2004; Brideau et al. 2009).

A recent study of large-scale rock slope failures in the Eastern Pyrenees (including the Principality of Andorra) using imaginery and field surveys (Jarman et al. 2014), has

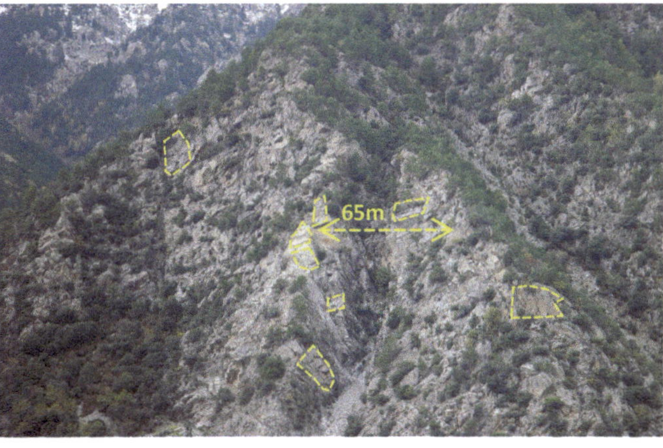

Fig. 8 Two granodiorite rock mass outcrops in the Pyrenees, showing different pattern of instability. *Yellow dashed lines* define large sliding surfaces. (*Left*) Pala de Morrano, Aigüestortes-Sant Maurici National Park, Central Pyrenees. Exposed basal sliding planes (030°/52°) either single or step-path may generate surfaces over 4000 m^2; (*Right*) Forat Negre-Borrassica in the Solà de Santa Coloma, Principality of Andorra, Eastern Pyrenees. The largest basal sliding plane (155°/57°) measured has an area of 213 m^2

identified 30 main large slope failures and further 20 smaller or uncertain cases. This inventory shows no obvious regional pattern or clustering and a surprisingly sparse population that affects 45–60 km^2 or 1.5–2.0% of the 3000 km^2 glaciated core of the mountain range, with others in fluvial valleys just beyond. From them, only 27% can be considered as large catastrophic events (rock or debris avalanches) and none of them in the Valira river valley where the slope of Forat Negre is located. For comparison, in the Alps, 5.6% of the entire 6200 km^2 montane area is affected by deep-seated gravitational slope deformations alone (Crosta et al. 2013) and up to 11% in the Upper Rhone basin (Pedrazzini et al. 2016). This sparsity has been interpreted by a low-intensity glaciation and less subsequent debuttressing, relative tectonic stability and small fluvial incision (Jarman et al. 2014). In the case of Forat Negre, this type of large failure has not occurred during the last thousands of years and should not be considered as a credible scenario.

Fragmentation in Rockfalls

It is assumed that the detached mass may break up on impact (Cruden and Varnes 1996; Hungr et al. 2014), however little attention has been paid to rockfall fragmentation. Rockfalls may involve a rock mass including discontinuities, which usually disintegrates along the path. The rockfall fragmentation is the process by which the detached mass loses its integrity while falling from a steep slope and breaks up into smaller pieces. Normally, this occurs during the first impacts on the ground (Wang and Tonon 2010). The fragmentation (Figs. 9 and 10) may consist of the separation of the rock blocks existing in the detached rock mass bounded by discontinuities (disaggregation), the breakage of the rock blocks during the impacts, or both (Ruiz-Carulla et al. 2015b).

Fragmentation invariably leads to a reduction of the particle size. The importance of the rockfall fragmentation in risk analysis has been discussed by Corominas et al. (2012). The definition of the initial volume of the rockfalls is a basic input parameter for trajectographic analysis. Rock breakage reduce the kinetic energy of the individual particles. Analyses performed with the volume of a non-fragmented rock mass produce results significantly different from the obtained if the resultant rock fragments are used instead (Okura et al. 2000; Dorren 2003). Working with the initial volume of a non-fragmented rock mass leads to the overestimation of the kinetic energy and the reach. Large blocks follow straight paths and display farther stopping points than the small ones. These effects may change significantly the way rockfalls interact with the terrain and affect the probability of impact on exposed elements, their vulnerability and the design of the protective elements (Volkwein et al. 2011).

Fig. 9 Block fragmentation by breakage in a real scale test in Vallirana (Barcelona, Spain)

Fig. 10 Rock fall fragmentation by the disaggregation of the original rock mass in Estany Gento, Central Pyrenees, Spain. Most of the block faces are joint planes present in the in situ rock mass. Notice that crushing of the fragments is virtually absent

Furthermore, if fragmentation is not accounted, the frequency and probability of impact is underestimated. The original rock mass is divided into a large number of fragments, which leads to multiply the probability of impact by a factor "n" equal to the number of new blocks generated (Corominas et al. 2012).

Currently available simulation programs for modelling the trajectory of the rockfalls (i.e. Jones et al. 2000; Dorren et al. 2006; Bourrier et al. 2009) allow calculating the distance travelled, height of jump, the kinetic energy at different points of the path and make a zonation of the exposed area. The major limitation of most of these programs is that they assume that any rock mass detached from a wall or cliff, regardless of the volume, arrives intact at the arresting point, which is not real. Some codes incorporate a fragmentation module for propagation analysis such as HY-STONE (Guzzetti et al. 2002a; Agliardi and Crosta 2003) which includes a trained neural network. The model is efficient for predicting whether a block of rock breaks or not but it may have difficulties in defining the number and size of fragments observed in reality. Salciarini et al. (2009) used a model of discrete elements to simulate the effects of fragmentation using software UDEC, and simulation results indicate that both the position of the blocks and the extent of the accumulation zone are strongly affected by fragmentation process of the rock mass.

Fragmentation in rockfalls is a complex physical mechanism, still little known and difficult to simulate (Chau et al. 2002; Zhang et al. 2000). The analysis of fragmentation is performed by measuring the size of the resultant fragments. This can be done either manually or by means of image analysis (Crosta et al. 2007; Locat et al. 2006). The degree of fragmentation may be calculated by the comparison of parameters representing the size distribution of the fragments before and after the fragmentation, such as the d50 diameter or the mean size. This approach is used in mining industry to assess the efficiency of blasting and it can be associated to the explosive energy and powder factors (Kuznetsov 1973). Similarly, Locat et al. (2006) determined the degree of fragmentation for rock avalanches, by comparing the mean diameters of the blocks within the intact rock mass and the deposited fragments.

If data are available, both the block size distribution of the in situ rock mass (IBSD) and that of the entire fragmented deposit can be used (Latham and Lu 1999). The block size distribution is typically represented as a grain size curve, in terms of percentage of material passing a certain size, typically with 3 or 4 orders of magnitude. Several researchers have found that the BSD of the fragments follows a power law (Turcotte 1986; Poulton et al. 1990; Locat et al. 2006; Crosta et al. 2007) with a negative exponent, whose value increases with the violence of the fragmentation process (Hartmann 1969). Fragmental rockfalls also show a similar fractal pattern. Ruiz-Carulla et al. (2016a) found that the deposits of six rockfall events yield volume distributions of the fragments that can be fitted to power laws (Fig. 11). The rock fragments were measured directly in the field one by one with a tape. In case the deposit formed a continuous young debris cover with a high number of blocks to be measured, the methodology proposed by Ruiz-Carulla et al. (2015a) was followed. The rockfall volume involved in these events ranges from 2.6 m^3 to 10,000 m^3.

Despite the apparent similarity of the distributions shown in Fig. 11, they contain significant differences. The first one is the scaling factors of the tails whose values range between 0.5 and 1.3 (Table 4). The scaling factors are an expression of the intensity of the fragmentation process. This can be observed in Fig. 12, where the number of fragments generated by the breakage of each individual block is plotted against the exponents of the fitted power laws of their volume distributions. There exists a positive correlation between the number of blocks generated and the exponents of the power law. The meaning of the exponents in the case of rockfall deposits is however, less evident as it is generated from an in situ block size distribution (IBSD) of the detached mass. The fragmentation in rockfalls is a function of several variables (Dussauge et al. 2003; Wang and Tonon 2010; Hantz et al. 2014): the presence of discontinuities in the initial rock mass and their persistence, their orientation at the time of impact, the energy and angle of impact and the ground stiffness. The highest exponent of the rockfall inventoried corresponds to the case of Vilanova de Banat with a value of 1.27 and 60,000 blocks generated from a volume of 10,000 m^3. It has both the largest volume and highest height of fall (Table 4). As it will be shown latter, the

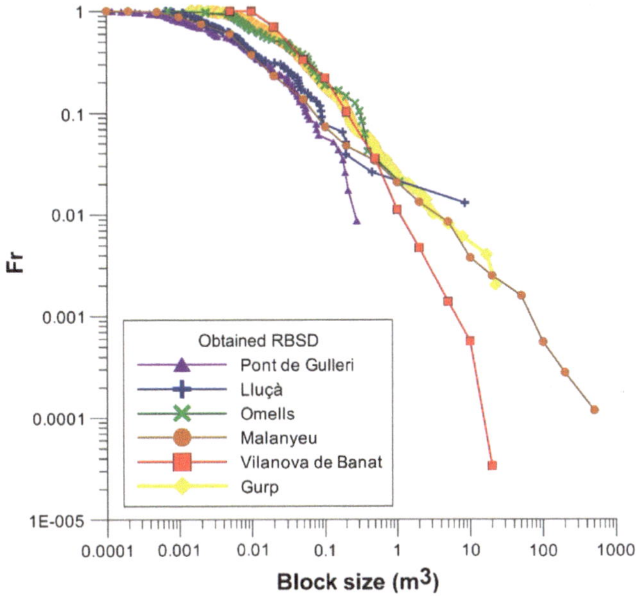

Fig. 11 Rockfall Block Size Distribution (RBSD) from 6 fragmental rockfalls events inventoried in Catalonia (from Ruiz-Carulla et al. 2017)

number of fragments generated is an order of magnitude bigger than the original IBSD. Most of the new fragments are small and appear concentrated forming a young debris cover at the base of the cliff (Ruiz-Carulla et al. 2015a). The number of fragments smaller than 0.01 m^3 represents more than 60% of the total. The cases Lluçà and Omells are the opposite situation. The exponents of the fitted distribution are small (0.51–0.53) which is consistent with the small height of fall (0.6–0.8). The amount of fragments generated is partly due to the low strength of the involved rocks. The cases of Malanyeu, Gurp and Pont de Gullerí are intermediate situations with a height of fall of 10 or more meters. The higher value of the exponent in the Pont de Gullerí case is basically due to the IBSD as most of the fragments are bounded by preexisting joint faces. The rockfall block size distribution (RBSD) in this case is best fitted to an exponential law and can be considered as case of pure disaggregation.

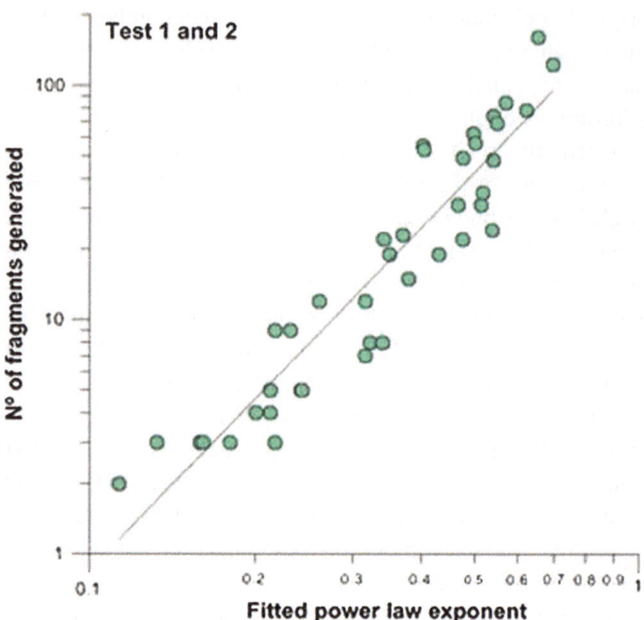

Fig. 12 Exponents of the fitted power laws of the volume distribution of the fragments generated by breakage of single blocks in real scale tests carried out in the Vallirana quarry, Spain (Ruiz-Carulla et al. 2016b)

Fractal Fragmentation Model

Most of current approaches used to obtain the RBSD consider the energy required to convert the IBSD into a new fragment-size distribution (Latham and Lu 1999; Lu and Latham 1999) and only few of them are applied to the analysis of the fragmentation of rock avalanches (Locat et al. 2006; Bowman et al. 2012; Charrière et al. 2015).

We have developed a fractal fragmentation model to characterize fragmental rockfalls (Ruiz-Carulla et al. 2015a, 2016b). The procedure aims at obtaining the rockfall block size distribution (RBSD) from the volume of the initial detached mass and its fracture pattern (IBSD) (Ruiz-Carulla et al. 2015b, 2016b). This model is based on a generic fractal fragmentation model (Perfect 1997) that considers a cubic block of unit length which is broken into small pieces according to a power law. The size distribution of the elements in a fractal system is given by (Eq. 6):

$$N(1/b^i) = k\left[1/b^i\right]^{-D_f}; \quad i = 0, 1, 2, \ldots\infty \tag{6}$$

Where N $(1/b^i)$ is the number of elements at the level "i" of the hierarchy; "k" is the number of initiators of unit length; "b" is a scaling factor >1; and D_f is the fractal dimension of fragmentation, which can be defined as:

$$D_f = 3 + \frac{log[P(1/b^i)]}{log[b]} \tag{7}$$

Where P $(1/b^i)$ or P_f: It is the probability of fracture that determines the proportion of the original block that breaks and generates new fragments. P_f is physically related to the interfaces of the subunits and maximum (limit) strength. The

Table 4 Characteristics of the inventoried rockfalls (modified from Ruiz-Carulla et al. 2016b)

Joint set	Pont de Gulleri	Lluçà	Omells	Malanyeu	Vilanova de Banat	Gurp
Lithology	Schist	Sandstone	Sandstone	Limestone	Limestone	Conglomerate
Total rockfall volume (m^3)	2.6	10.7	4.2	5000	10000	100
Free fall height (m)	12	0.6	0.8	10	40	80
Max block (m^3)	0.28	8.5	1.1	445	31	22
Min. volume measured (m^3)	0.0001	0.0007	0.0007	0.0001	0.01	0.01
# of blocks measured	116	78	48	2721	1524	500
Total # of blocks calculated	116	78	48	25,500	60,000	500
Exponent of the fitted power law	0.92	0.51	0.53	0.72	1.27	0.74
Min. block size fitted (m^3)	0.001	0.001	0.01	0.001	0.01	0.01
R^2 of the fitted power law	0.94	0.95	0.89	0.98	0.95	0.98

interfaces may correspond to the surfaces of existing joints, to rock anisotropy, or non-persistent joints (Perfect 1997). The range of the probability of failure is $b^{-3} < P(1/b^i) < 1$. When $P(1/b^i) = 1$ and $D_f = 3$ the whole block breaks, while for $P(1/b^i) \leq b^{-3}$ the block remains intact. The model performance is summarized in Fig. 13

The fractal fragmentation model has been adapted for the case of the rockfall. First, instead of k initial volumes of unit length, the IBSD is used as input, classifying it in bins. Second, not all the blocks of the IBSD break upon impact on the ground. To consider this, a survival rate, S_r, representing the proportion of blocks that do not break is defined.

The FFM has been applied to several cases inventoried in the Spanish Pyrenees. Here, the case of Vilanova de Banat is presented. This rockfall took place in November 2011, affecting a volume of about 10,000 m^3 of limestone (Ruiz-Carulla et al. 2015a). The model uses as input data the size of the rockfall (the unstable volume) and the discontinuity pattern of the detached rock mass (joint set orientations and spacing) to obtain the ISBD. A Nikon D90 digital camera with a focal length of 60 mm and 12Mp resolution was used to generate the digital surface model (DSM) of the rockfall scar. The following step was to reconstruct the volume of the detached rock mass by subtracting the DSM of the scar from the available topographic map at 1:5000 scale (before the failure). Then, the joint sets and their spacing were identified using the DSM texturized with the images and matched to the detached rock mass. Given that neither high quality photos of the source area nor a detailed digital surface model (DSM) prior to the occurrence of the rockfall were available, the volume and the IBSD obtained are subjected to a high degree of uncertainty. There is a difference between the total volume measured in the detachment zone ($\sim$10,000 m^3) and the measured in the deposit (8000 m^3). However, the difference could be explained by the proportion of smaller blocks of 0.015 m^3, which were not measured in the field.

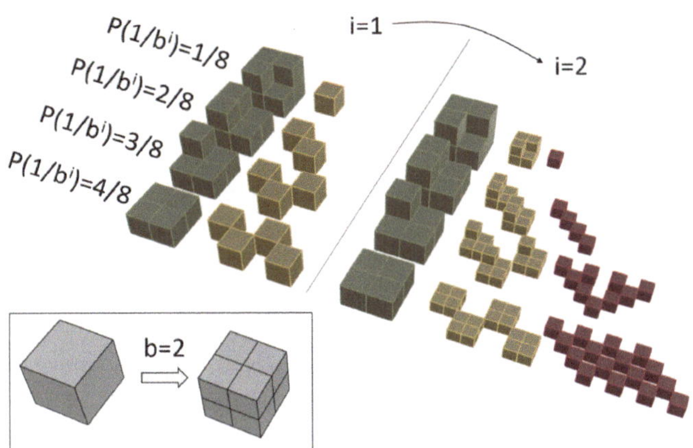

Fig. 13 Rockfall fragmentation model (Ruiz-Carulla et al. 2017)

Five joint sets using both semiautomatic and manual techniques were identified. The fracture pattern was applied to the missing rock mass volume, assuming joints of infinite persistence. Finally the IBSD is generated. We considered two different shapes for the detached mass: (a) the irregular volume reconstructed directly from the scar ($\sim$10,000 m^3) and (b) a prismatic shape with the same volume, to simplify the cutting tasks. In order to account for the uncertainties associated to the reconstruction of the rock mass, two volumes were used and 4 or 5 fully persistent joint sets. Figure 14 shows the pattern of both the prismatic and irregular reconstructed volumes. The IBSDs have been generated with the mentioned assumptions: prismatic and irregular; 10,000 m^3 volume and 5000 m^3; and 4 and 5 joint sets. The IBSDs obtained are shown on the right side of the figure. All tails are fitted to exponential laws with coefficients of determination close to 1.

We calibrated the model parameters in the Vilanova de Banat case study in order to obtain a RBSD that fits that observed in the field. Several combinations of the model parameters produce well fitted block volume distributions. Their values range between 0.05 and 0.34 for S_r, between 0.73 and 0.80 for P_f and between 1.6 and 3.4 for "b". The Xi^2 was used to optimize P_f, S_r and to check the goodness of results, obtaining a range of values between 0.02 and 0.11 for the four IBSD. Figure 15 shows the IBSD obtained from the irregular reconstructed volume, the RBSD measured in the field and RBSD obtained from the calibration of fractal fragmentation model (FFM). The results show that the RBSD can be successfully generated from the ISBDs.

It is worth noticing that fragmentation has a significant effect in the size-range between 1 and 10 m^3. The number of blocks within this range has been reduced up to one order of magnitude. Conversely, the number of blocks smaller than 1 m^3 has increased more than one order of magnitude. This effect has direct consequences on the kinematics of the fragments with a reduction of the distances travelled, velocities. Despite of this, the survival rates (S_r) obtained indicate that up to one third of the original blocks have remained unbroken as illustrated by the 272 large scattered blocks that were measured in the field and travelled far away from the young debris cover (Ruiz-Carulla et al. 2015a).

Furthermore, the accumulated volume of the smallest blocks (<0.01 m^3) has risen from an insignificant value up to more than 15% of the total. This represents an expenditure of the breakage energy that cannot be obviated.

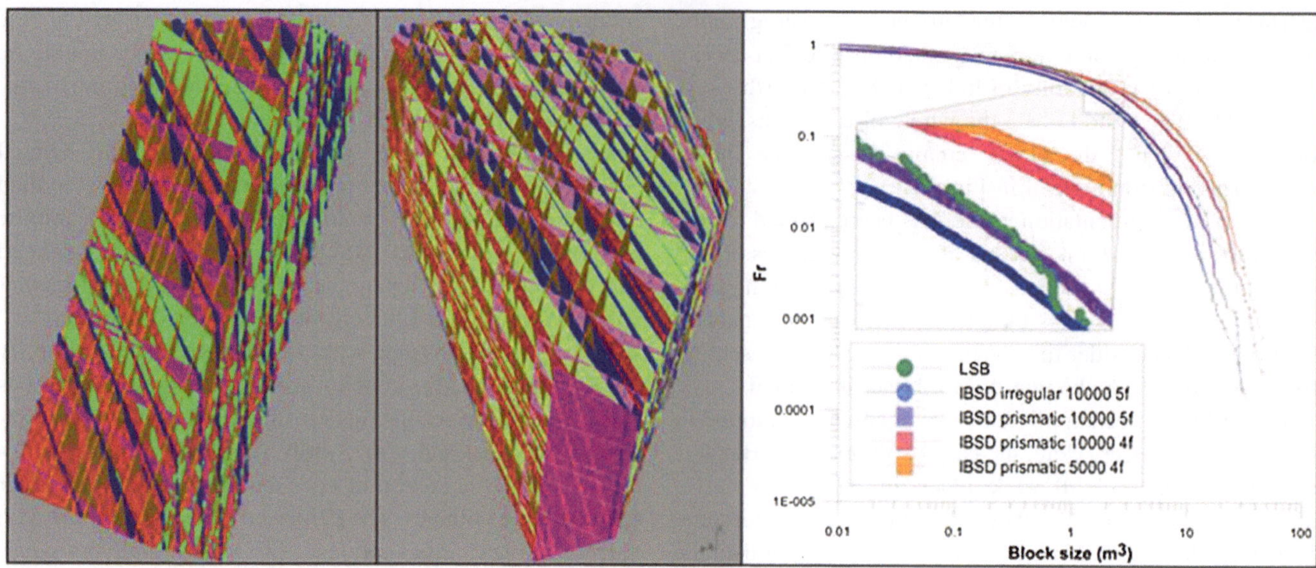

Fig. 14 Fracture model generated from the detached rock mass volume cut by the discontinuities with their actual spacings using a prismatic shape (*left*) and the reconstructed irregular shape volume of the rockfall source (*center*). Plot of the resultant IBSD distributions considering 4 or 5 sets of fully persistent joints (*right*)

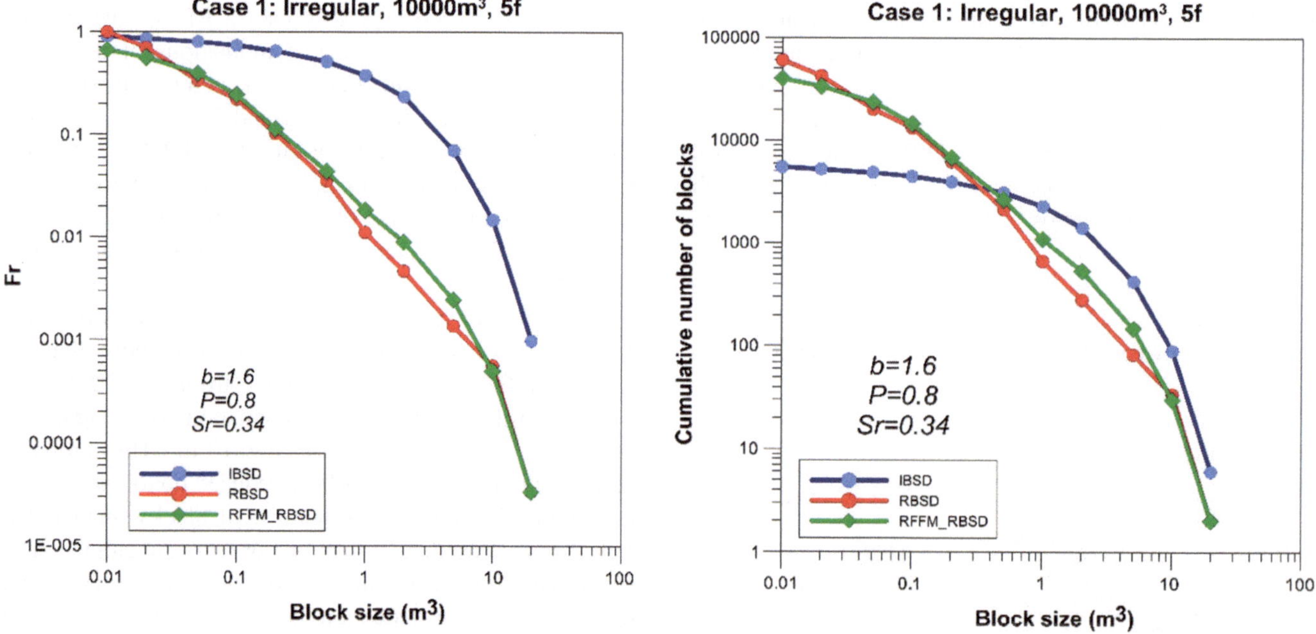

Fig. 15 The IBSD of the reconstructed volume, the observed RBD and the RBSD generated with fractal fragmentation model (RBSD—FFM) in terms of relative frequency (*left*) and cumulative number of blocks (*right*) versus block size

Further work is needed for the performance of the FFM. Several parameters are involved in the fragmentation process and the resultant RBSD such as the IBSD, the total volume detached, the impact energies and the morphology and the rigidity of the ground. The procedure in the example of Vilanova de Banat is iterative until the fit between the observed and modeled RBSD is achieved. In order to use the FFM as predictive tool it is required the analysis of more cases to relate P_f, S_r and the scaling factor to the local geological conditions as well as to the geomechanical, and morphological characteristics of the detached rock mass and of the slope.

Final Remarks

FM relations are fundamental for performing the QRA. Most of the rockfall volume distributions are characterized by a negative power law. It has been argued that, if the statistical tests are fulfilled, they can be used to calculate the frequency of large volumes and that both rockfalls and rockslides display the same b-value. However, a critical issue is the definition of the maximum credible volume to be used in the hazard analyses and for implementing risk mitigation measures. The case of Andorra provides empirical evidence that rockfalls could be size-constraint due to the geological context as no rock avalanche deposits are found in the Valira river basin. Two analyses on the rockfall size distribution have been carried out at slope of Forat Negre. The first analysis corresponds to the observed size distribution of the rockfall scars, and it is an empirical evidence of past rockfalls. The second one calculates the kinematically detachable rock masses, indicating hypothetical rockfalls that might occur in the future. These two independent approaches differ on an order of magnitude only with a maximum credible volume between 25,000 and 50,000 m^3. The volume distribution of the rockfall scars is well fitted by a power law with b-value of −0.92, and suggests that large rockfall events are less abundant than in other mountainous regions.

The clue for such a behavior could be in the persistence of the discontinuity sets. In the slope of Forat Negre, F1 and F7 are fault sets that intersect and displace the rest of discontinuity sets as they have probably been generated during the last deformation phases in this sector of the range. They exert a control over the length of the planes of the discontinuities F3 and F5 and imply a limitation on their persistence. This volume restriction can be overcome to some extent either by coalescence of basal planes or through step-path failures involving the breakage of rock bridges. This situation however, will necessarily involve smaller volumes than in the case of fully persistent basal joints. In any case, the worst case scenarios that may be foreseen can be nowadays faced with better tools thanks to the development of the remote data collection equipment, particularly the LiDAR, digital photogrammetry and interferometric techniques. As several experiences have already shown, these techniques provide the deformation pattern and rate of movement over large terrain areas thus allowing the identification and delineation of potentially unstable masses and the implementation of EWS and evacuation strategies (i.e. Froese et al. 2009; Hermanns et al. 2013; Michoud et al. 2013).

Despite the difficulties and uncertainties associated to the generation of the IBSD and the RBSD, the results of the six rockfall events inventoried indicate that breakage of the particles is a fundamental mechanism in all-size fragmental rockfalls. The number of new generated particles increases with the size of the rockfall and is the result of the interplay of other factors such as the strength of the rock, the height of fall and stiffness of the impact ground surface.

The FFM has successfully generated a RBSD that fits well to the observed in the field. It is able to consider both the disaggregation and breakage mechanisms of rockfalls while considering successive iterations will increase the number of small-size fragments generated. It is simple enough to be incorporated into rockfall trajectory analyses. However, its application at present is not straightforward. The model runs using the IBSD as input data and requires the definition of three additional parameters: the survival rate of the blocks present in the detached rock mass, the probability of failure that determines the proportion of each initiator block that breaks, and the scaling factor or ratio of sizes between the initial block and the fragments. In order to use this model as a predictive tool more case histories are needed to calibrate it. Finally, we argue that fragmentation cannot be obviated and must be incorporated in the rockfall hazard analyses.

Acknowledgements The authors acknowledge the support of the Spanish Economy and Competitiveness Ministry to the Rockrisk research project (BIA2013-42582-P) and of the Government of Andorra (Edicte de 10/04/2013, BOPA nº18 17/04/2014).

References

Abbruzzese JM, Sauthier C, Labiouse V (2009) Considerations on Swiss methodologies for rock fall hazard mapping based on trajectory modelling. Nat Hazards Earth Syst Sci 9(4):1095–1109

Abellan A, Vilaplana JM, Martinez J (2006) Application of a long-range terrestrial laser scanner to a detailed rockfall study at Vall de Nuria (Eastern Pyrenees, Spain). Eng Geol 88:136–148

Agliardi F, Crosta GB (2003) High resolution three-dimensional numerical modelling of rockfalls. Int J Rock Mech Min Sci 40:455–471

Agliardi F, Crosta GB, Frattini P (2009) Integrating rockfall risk assessment and countermeasure design by 3D modelling techniques. Nat Hazards Earth Syst Sci 9:1059–1073

Amitrano D, Grasso JR, Senfaute G (2005) Seismic precursory patterns before a cliff collapse and critical point phenomena. Geophys Res Lett 32: L08314

Ballantyne CK, Stone JO (2004) The Beinn Alligin rock avalanche, NW Scotland: cosmogenic 10Be dating, interpretation and significance. Holocene 14:448–453

Bourrier F, Dorren L, Nicot F, Berger F, Darve F (2009) Toward objective rockfall trajectory simulation using a stochastic impact model. Geomorphology 110(3):68–79

Bourrier F, Dorren L, Hungr O (2013) The use of ballistic trajectory and granular flow models in predicting rockfall propagation. Earth Surf Proc Land 38:435–440

Bowman ET, Take AW, Rait KL, Hann C (2012) Physical models of rock avalanche spreading behaviour with dynamic fragmentation. Can Geotech J 49:460–476

Brardinoni F, Church M (2004) Representing the landslide magnitude-frequency relation: Capilano river basin, British Columbia. Earth Surf Proc Land 29:115–124

Brideau MA, Yan M, Stead D (2009) The role of tectonic damage and brittle rock fracture in the development of large rock slope failures. Geomorphology 103:30–49

Brunetti MT, Guzzetti F, Rossi M (2009) Probability distributions of landslide volumes. Nonlin Processes Geophys 16:179–188

Budetta P (2004) Assessment of rockfall risk along roads. Nat Hazards Earth Syst Sci 4:71–81

Charrière M, Humair F, Froese C, Jaboyedoff M, Pedrazzini A, Longchamp C (2015) From the source area to the deposit: collapse, fragmentation, and propagation of the Frank Slide. Geol Soc Am Bull 128:332–351

Chau KT, Wong RHC, Wub JJ (2002) Coefficient of restitution and rotational motions of rockfall impacts. Int J Rock Mech Min Sci 39:69–77

Chau KT, Wong RCH, Liu J, Lee CF (2003) Rockfall hazard analysis for Hong Kong based on rockfall inventory. Rock Mech Rock Eng 36:383–408

Copons R (2007) Avaluació de la perillositat de caigudes de blocs rocosos al Solà d'Andorra la Vella. Monografies del CENMA

Copons R, Vilaplana JM, Corominas J, Altimir J, Amigó J (2004) Rockfall risk management in high-density urban areas. The Andorran experience. In: Glade T, Anderson M, Crozier MJ (eds) Landslide hazard and risk. Wiley, Chichester, pp 675–698

Corominas J, Mavrouli O (2011) Rockfall quantitative risk assessment. In: Lambert S, Nicot F (eds) Rockfall engineering. ISTE Ltd. & Wiley, pp 255–301

Corominas J, Moya J (2008) A review of assessing landslide frequency for hazard zoning purposes. Eng Geol 102:193–213

Corominas J, Copons R, Moya J, Vilaplana JM, Altimir J, Amigó J (2005) Quantitative assessment of the residual risk in a rock fall protected area. Landslides 2:343–357

Corominas J, Mavrouli O, Santana D, Moya J (2012) Simplified approach for obtaining the block volume distribution of fragmental rockfalls. In: Eberhardt E et al (eds) Proceedings of the 11 International Symposium on Landslides, Banff, Canada, vol 2. CRC Press/Balkema, Leiden, pp 1159–1164

Corominas J, van Westen C, Frattini P, Cascini L, Malet JP, Fotopoulou S, Catani F, Van Den Eeckhaut M, Mavrouli O, Agliardi F, Pitilakis K, Winter MG, Pastor M, Ferlisi S, Tofani V, Hervás J, Smith JT (2014) Recommendations for the quantitative analysis of landslide risk. Bull Eng Geol Environ 73:209–263

Crosta GB, Agliardi F (2003) Failure forecast for large rock slides by surface displacement measurements. Can Geotech J 40:176–191

Crosta GB, Frattini P, Fusi F (2007) Fragmentation in the Val Pola rock avalanche, Italian Alps. J Geophys Res 112:F01006

Crosta GB, Frattini P, Agliardi F (2013) Deep seated gravitational slope deformations in the European Alps. Tectonophysics 605:13–33

Cruden DM, Varnes DJ (1996) Landslide types and processes. In: Turner AK, Schuster RL (eds) Landslides investigation and mitigation. National Research Council, Transportation Research Board, Special Report 247, pp 36–75

Davies TR, McSaveney MJ (2002) Dynamic simulation of the motion of fragmenting rock avalanches. Can Geotech J 39:789–798

Davies TR, McSaveney MJ, Hodgson KA (2009) A fragmentation–spreading model for long-runout rock avalanches. Can J Geotech 36:1096–1110

Dorren LKA (2003) A review of rockfall mechanics and modeling approaches. Prog Phys Geogr 27(1):69–87

Dorren L, Berger F, Putters US (2006) Real size experiments and 3D simulation of rockfall on forested and nonforested slopes. Nat Hazards Earth Syst Sci 6:145–153

Dussauge C, Grasso JR, Helmstetter A (2003) Statistical analysis of rockfall volume distributions: implications for rockfall dynamics. J Geophys Res B6(108):2286

Dussauge-Peisser A, Helmstetter A, Grasso JR, Hanz D, Desvarreux P, Jeannin M, Giraud A (2002) Probabilistic approach to rockfall hazard assessment: potential of historical data analysis. Nat Hazards Earth Syst Sci 2:15–26

Eberhardt E (2008) The role of advanced numerical methods and geotechnical field measurements in understanding complex deep-seated rock slope failure mechanisms. Can Geotech J 45:484–510

Eberhardt E, Stead D, Coggan JS (2004) Numerical analysis of initiation and progressive failure in natural rock slopes—the 1991 Randa rockslide. Int J Rock Mech Min Sci 41:69–87

Einstein HH, Veneziano D, Baecher GB, O'Reilly KJ (1983) The effect of discontinuity persistence on rock slope stability. Int J Rock Mech Min Sci Geomech Abstracts 20:227–236

Elmouttie MK, Poropat GV (2012) A method to estimate in situ block size distribution. Rock Mech Rock Eng 45(3):401–407

El-Ramly H, Morgenstern NR, Cruden DM (2002) Probabilistic slope stability analysis for practice. Can Geotech J 39:665–683

Evans S, Hungr O (1993) The assessment of rockfall hazard at the base of talus slopes. Can Geotech J 30:620–636

Fell R, Ho KKS, Lacasse S, Leroi E (2005) A framework for landslide risk assessment and management. In: Hungr O, Fell R, Couture R, Eberthardt E (eds) Landslide risk management. Taylor and Francis, London, pp 3–25

Fell R, Glastonbury J, Hunter G (2007) Rapid landslides: the importance of understanding mechanisms and rupture surface mechanisms. Q J Eng Geol 40:9–27

Ferlisi S, Cascini L, Corominas J, Fabio M (2012) Rockfall risk assessment to persons travelling in vehicles along a road: the case study of the Amalfi coastal road (southern Italy). Nat Hazards 62:691–721

Ferrero AM, Migliazza M, Roncella R, Rabbi E (2011) Rock slopes risk assessment based on advanced geostructural survey techniques. Landslides 8(2):221–231

Frayssines M, Hantz D (2006) Failure mechanisms and triggering factors in calcareous cliffs of the Subalpine Ranges (French Alps). Eng Geol 86(4):256–270

Froese CR, Moreno F, Jaboyedoff M, Cruden DM (2009) 25 years of movement monitoring of South Peak, Turtle Mountain: understanding the hazard. Can Geotech J 46:256–269

Fukuzono T (1990) Recent studies on time prediction of slope failure. Landslide News 4:9–12

Gischig V, Loew S, Kos A, Raetzo H, Lemy F (2009) Identification of active release planes using ground-based differential InSAR at the Randa rock slope instability, Switzerland. Nat Hazards Earth Syst Sci 9:2027–2038

Glastonbury J, Fell R (2010) Geotechnical characteristics of large rapid rock slides. Can Geotech J 47:116–132

Guthrie RH, Evans SG (2004) Analaysis of landslide frequencies and characteristics in a natural system, coastal British Columbia. Earth Surf Proc Land 29:1321–1339

Guzzetti F, Crosta G, Detti R, Agliardi F (2002a) STONE: a computer program for the three-dimensional simulation of rock-falls. Comput Geosci 28:1079–1093

Guzzetti F, Malamud BD, Turcotte DL, Reichenbach P (2002b) Power-law correlations of landslide areas in Central Italy. Earth Planet Sci Lett 195:169–183

Guzzetti F, Reichenbach P, Wieczorek GF (2003) Rockfall hazard and risk assessment in the Yosemite Valley, California, USA. Nat Hazards Earth Syst Sci 3:491–503

Haneberg WC (2008) Using close range terrestrial digital photogrammetry for 3-D rock slope modeling and discontinuity mapping in the United States. Bull Eng Geol Environ 67:457–469

Hantz D, Vengeon JM, Dussauge-Peisser C (2003) An historical, geomechanical and probabilistic approach to rock-fall hazard assessment. Nat Hazards Earth Syst Sci 3(6):693–701

Hantz D, Rossetti JP, Servant F, D'Amato J (2014) Etude de la distribution des blocs dansun éboulement pour l'évaluation de l'aléa. Proceedings of Rock Slope Stability 2014, Marrakesh, p 10

Hartmann WK (1969) Terrestrial, lunar and interplanetary rock fragmentation. Icarus 10:201–213

Heritage GL, Large ARG (2009) Laser scanning for the environmental sciences. Wiley-Blackwell, London, 288 p. ISBN 978-1-4051-5717-9

Hermanns R, Strecker MR (1999) Structural and lithological controls on large Quaternary rock avalanches (sturzstroms) in arid northwestern Argentina. Geol Soc Am Bull 111:934–948

Hermanns R, Blikra LH, Anda E, Saintot A, Dahle H, Oppikofer T, Fischer L, Bunkholt H, Böhme M, Dels JF, Lauknes TR, Redfiled TF, Osmundse T, Eiken T (2013) Systematic mapping of large unstable rock slopes in Norway. In: Margottini C et al (eds) Landslide science and practice, vol 1. Springer, Berlin, pp 29–34. doi:10.1007/978-3-642-31325-7_3

Hewitt K (1999) Quaternary moraines vs. catastrophic rock avalanches in the Karakoram Himalaya, Northern Pakistan. Quat Res 51:220–237

Hewitt K (2002) Styles of rock avalanche depositional complexes conditioned by very rugged terrain, Karakoram Himalaya, Pakistan. In Evans SG and DeGraff IV Catastrophic landslides: effects, occurrence, and mechanisms, Reviews in Engineering Geology 15, 345–377

Hewitt K, Clague JJ, Orwin JF (2008) Legacies of catastrophic rock slope failures in mountain landscapes. Earth Sci Rev 87:1–38

Hoek E, Bray JW (1981) Rock slope engineering, Institution of Mining and Metallurgy. London

Hovius N, Stark CP, Allen PA (1997) Sediment flux from a mountain belt derived by landslide mapping. Geology 25:231–234

Hsü KJ (1978) Albert Heim: observations of landslides and relevance to modern interpretations. In: Voight B (ed) Rockslides and avalanches; 1, Natural phenomena. Elsevier, Amsterdam, pp 70–93

Hungr O (2016) A review of landslide hazard and risk methodology. In: Aversa S, Cascini L, Picarelli L, Scavia C (eds) Landslides and engineered slopes: experience, theory and practice, vol 1. CRC press, pp 3–27. ISBN 978-1-138-02989-7

Hungr O, Evans SG (2004) The occurrence and classification of massive rock slope failure. Felsbau 22:1–12

Hungr O, Evans SG, Hazzard J (1999) Magnitude and frequency of rock falls and rock slides along the main transportation corridors of southwestern British Columbia. Can Geotech J 36:224–238

Hungr O, Corominas J, Eberhardt E (2005) Estimating landslide motion mechanisms, travel distance and velocity. In: Hungr O, Fell R, Couture R, Eberthardt E (eds) Landslide risk management. Taylor and Francis, London, pp 99–128. ISBN 041538043X

Hungr O, McDougall S, Wise M, Cullen M (2008) Magnitude–frequency relationships of debris flows and debris avalanches in relation to slope relief. Geomorphology 96:355–365

Hungr O, Leroueil S, Picarelli L (2014) The Varnes classification of landslides types, an update. Landslides 11:167–194

Hutchinson JN (1988) Morphological and geotechnical parameters of landslides in relation to geology and hydrogeology. In: Bonnard C (ed) Landslides. Proceedings of the 5th International Conference on Landslides, vol 1. Lausanne, pp 3–35

Jaboyedoff M, Metzger R, Oppikofer T, Couture R, Derron MH, Locat J, Turmel D (2007) New insight techniques to analyze rock-slope relief using DEM and 3D-imaging cloud points: COLTOP-3D software. In: Eberhardt E, Stead D, Morrison T (eds) Rock mechanics: meeting society's challenges and demands, vol 1. Taylor & Francis, pp 61–68

Jaboyedoff M, Couture R, Locat P (2009) Structural analysis of Turtle Mountain (Alberta) using digital elevation model: toward a progressive failure. Geomorphology 103:5–16

Jaboyedoff M, Oppikofer T, Abellán A, Derron MH, Loye A, Metzger R, Pedrazzini A (2012) Use of LIDAR in landslide investigations: a review. Nat Hazards 61:5–28

Jarman D, Calvet M, Corominas J, Delmas M, Gunnell Y (2014) Large-scale rock slope failures in the eastern Pyrenees: identifying a sparse but significant population in paraglacial and parafluvial contexts. Geogr Ann A 96(3):357–391

Jones CL, Higgins JD, Andrew RD (2000) Colorado rockfall simulation program, version 4.0. Colorado Geological Survey. http://www.geosurvey.state.co.us

Kemeny J, Post R (2003) Estimating three-dimensional rock discontinuity orientation from digital images of fracture traces. Comput Geosci 29(1):65–77

Kemeny J, Norton B, Turner K (2006) Rock slope stability analysis utilizing ground-based LiDAR and digital image processing. Felsbau 24:8–16

Kromer RA, Hutchinson DJ, Latto MJ, Gauthier D, Edwards T (2015) Identifying rock slope failure precursors using LiDAR for transportation corridor hazard management. Eng Geol 195:93–103

Kuznetsov VM (1973) The mean diameter of fragments formed by blasting rock. J Min Sci 9:144–148

Lambert C, Thoeni K, Giacomini A, Casagrande D, Sloan S (2012) Rockfall hazard analysis from discrete fracture network modelling with finite persistence discontinuities. Rock Mech Rock Eng 45 (5):871–884

Latham J, Lu P (1999) Development of an assessment system for the blastability of rock masses. Int J Rock Mech Min Sci 36:41–55

Lato M, Diederichs MS, Hutchinson J, Harrap R (2009) Optimization of LiDAR scanning and processing for automated structural evaluation of discontinuities in rock masses. Int J Rock Mech Min Sci 46:194–199

Lee EM, Jones DKC (2004) Landslide risk assessment. Thomas Telford, London, p 454. ISBN 9780727731715

Li ZH, Huang HW, Xue YD, Yin J (2009) Risk assessment of rockfall hazards on highways. Georisk 3:147–154

Lim M, Petley DN, Rosser NJ, Allison RJ, Long AJ (2005) Digital photogrammetry and time-of-fight laser scanning as integrated approach to monitoring cliff evolution. Photogramm Rec 20 (110):109–129

Locat P, Couture R, Leroueil S, Locat S (2006) Fragmentation energy in rock avalanches. Can Geotech J 43:830–851

Lu P, Latham JP (1999) Developments in the assessment of in-situ block size distributions in rock masses. Rock Mech Rock Eng 32:29

Malamud B, Turcotte L, Guzzetti F, Reichenbach P (2004) Landslide inventories and their statistical properties. Earth Surf Proc Land 29:687–711

Mavrouli O, Corominas J (2017) Comparing rockfall scar volumes and kinematically detachable rock masses. Eng Geol 219:63–74

Mavrouli O, Corominas J, Jaboyedoff M (2015) Size distribution for potentially unstable rock masses and in situ rock blocks using LiDAR-generated digital elevation models. Rock Mech Rock Eng 48(4):1589–1604

McSaveney MJ (2002) Recent rockfalls and rock avalanches in Mount Cook National Park, New Zealand. In: Evans SG, DeGraff JV (eds) Catastrophic landslides: effects, occurrence, and mechanisms, vol 15 (Rev Eng Geol). Geological Society of America, pp 35–70

Michoud C, Bazin S, Blikra LH, Derron MH, Jaboyedoff M (2013) Experiences from site-specific landslide early warning systems. Nat Hazards Earth Syst Sci 13:2659–2673

Moya J, Corominas J, Perez-Arcas J, Baeza C (2010) Tree-ring based assessment of rockfall frequency on talus slopes at Solà d'Andorra, Eastern Pyrenees. Geomorphology 118:393–408

Okura Y, Kitahara H, Sammori T, Kawanami A (2000) The effects of rockfall volume on runout distance. Eng Geol 58(2):109–124

Oppikofer T, Jaboyedoff M, Keusen HR (2008) Collapse of the eastern Eiger flank in the Swiss Alps. Nat Geosci 1:531–535

Oppikofer T, Jaboyedoff M, Blikra L, Derron MH (2009) Characterization and monitoring of the Aknes rockslide using terrestrial laser scanning. Nat Hazards Earth Syst Sci 9:1643–1653

Oppikofer T, Jaboyedoff M, Pedrazzini A, Derron MH, Blikra LH (2011) Detailed DEM analysis of a rockslide scar to characterize the basal sliding surface of active rockslides. J Geophys Res Earth Surf 116

Palmström A (2001) Measurement and characterization of rock mass jointing. In: Sharma VI, Saxena K (eds) In-situ characterization of rocks. A. A. Balkema publishers

Palmstrom A (2005) Measurement of and correlations between block size and rock quality designation (RQD). Tunn Undergr Space Technol 20:362–377

Pedrazzini A, Humair F, Jaboyedoff M, Tonini M (2016) Characterisation and spatial distribution of gravitational slope deformation in the Upper Rhone catchment (Western Swiss Alps). Landslides 13:259–277

Pelletier JD, Malamud BD, Blodgett TA, Turcotte DL (1997) Scale-invariance of soil moisture variability and its implications for the frequency-size distribution of landslides. Eng Geol 48:254–268

Perfect E (1997) Fractal models for the fragmentation of rocks and soils: a review. Eng Geol 48:185–198

Petley DN (2012) Landslides and engineered slopes: protecting society through improved understanding. In: Eberhardt E, Froese C, Turner AK, Leroueil S (eds) Landslides and engineered slopes, vol 1. CRC Press, London, pp 3–13

Petley DN, Bulmer MHK, Murphy W (2002) Patterns of movement in rotational and translational landslides. Geology 30:719–722

Picarelli L, Oboni F, Evans SG, Mostyn G, Fell R (2005) Hazard characterization and quantification. In: Hungr O, Fell R, Couture R, Eberthardt E (eds) Landslide risk management. Taylor and Francis, London, pp 27–62 ISBN 041538043X

Poulton MM, Mojtabai N, Farmer IW (1990) Scale invariant behaviour of massive and fragments rock. Int J Rock Mech Min Sci Geomech Abstracts 27:219–221

Priest SD (1993) Discontinuity analysis for rock engineering. Chapman and Hall, London, p 473

Priest SD, Hudson JA (1981) Estimation of discontinuity spacing and trace length using scanline surveys. Int J Rock Mech Min Sci Geomech Abstracts 18:183–197

Riquelme AJ, Abellán A, Tomás R, Jaboyedoff M (2014) A new approach for semi-automatic rock mass joints recognition from 3D point clouds. Comput Geosci 68:38–52

Riquelme AJ, Abellán A, Tomás R (2015) Discontinuity spacing analysis in rock masses using 3D point clouds. Eng Geol 195:185–195

Rochet L (1987) Application des modeles numeriques de propagation a l'etude des eboulements rocheux. Bull Lab Ponts et Chaussées 150 (151):84–95

Rose ND, Hungr O (2007) Forecasting potential rock slope failure in open pit mines using the inverse-velocity method. Int J Rock Mech Min Sci 44:308–320

Rosser N, Lim M, Petley D, Dumming S, Allison R (2007) Patterns of precursory rockfall prior to slope failure. J Geophys Res 112: F04014

Roýan MJ, Abellán A, Jaboyedoff M, Vilaplana JM, Calvet J (2014) Spatio-temporal analysis of rockfall pre-failure deformation using Terrestrial LiDAR. Landslides 11:697–709

Roýan MJ, Abellán A, Vilaplana JM (2015) Progressive failure leading to the 3 December 2013 rockfall at Puigcercós scarp (Catalonia, Spain). Landslides 12:585–595

Ruiz-Carulla R, Corominas J, Mavrouli O (2015a) A methodology to obtain the block size distribution of fragmental rockfall deposits. Landslides 12(4):815–825

Ruiz-Carulla R, Corominas J, Mavrouli O (2015b) An empirical approach to rockfall fragmentation. In: Eurock 2015—ISRM European Regional Symposium—the 64th Geomechanics Colloquium, 7–10 October 2015, Salzburg, Austria

Ruiz-Carulla R, Corominas J, Mavrouli O (2016a) Comparison of block size distribution in rockfalls. In: Aversa S, Cascini L, Picarelli L, Scavia C (eds) Landslides and engineered slopes: experience, theory and practice, vol 3. CRC press, pp 1767–1774. ISBN 978-1-138-02991-0

Ruiz-Carulla R, Matas G, Prades A, Gili JA, Corominas J, Lantada N, Buill F, Mavrouli O, Núñez-Andrés A, Moya J (2016b) Analysis of rock block fragmentation by means of real-scale tests. In: 3rd RSS Rock Slope Stability conference, Lyon 2016, p 2

Ruiz-Carulla R, Corominas J, Mavrouli O (2017, available online) A fractal fragmentation model for rockfalls. Landslides. doi: 10.1007/s10346-016-0773-8

Saito M (1965) Forecasting the time of occurrence of a slope failure. Proceedings of the 6th International Conference on Soil Mechanics and Foundation Engineering. Pergamon Press, Montréal, Oxford, pp 537–541

Salciarini D, Tamagnini C, Conversini P (2009) Numerical approaches for rockfall analysis: a comparison. In: Proceedings 18th World IMACS/MODSIM Congress, Cairns, Australia

Santana D, Corominas J, Mavrouli O, Garcia-Sellés D (2012) Magnitude-frequency relation for rockfall scars using a Terrestrial Laser Scanner. Eng Geol 145–146:50–64

Selby MJ (1982) Hillslope materials and processes. Oxford University Press, New York

Slob S (2010) Automated rock mass characterisation using 3-D terrestrial laser scanning. Ph.D. thesis, TU Delft. URL: http://www.narcis.nl/publication/RecordID/oai:tudelft.nl:uuid:c1481b1d-9b33-42e4-885a-53a6677843f6

Slob S, Hack R, Turner A K (2002) An approach to automate discontinuity measurements of rock faces using laser scanning techniques. In: Da Gama CD, Sousa LRE (eds) ISRM EUROCK 2002, Lisbon, Portugal. ISBN: 972-98781-2-9, pp. 87-94

Slob S, Hack R, van Knapen B, Kemeny J (2004) Automated identification and characterization of discontinuity sets in outcropping rock masses using 3D terrestrial laser scan survey techniques. In: Proceedings of the ISRM Regional Symposium EUROCK 2004 & 53rd Geomechanics Colloquy, Salzburg, pp 439–443

Soeters R, Van Westen CJ (1996) Slope instability, recognition, analysis and zonation. In: Turner AT, Schuster RL (eds) Landslides—investigation and mitigation, Transportation Research Board Special Report No 247. National Academy Press, Washington DC, pp 129–177

Spang RM, Rautenstrauch RW (1988) Empirical and mathematical approaches to rock fall protection and their practical application. In: Bonnard C (ed) 5th International Congress on Landslides, vol 2. Lausanne. Balkema, Rotterdam, pp 1237–1243. ISBN 90-6191-838-3

Stark CP, Hovius N (2001) The characterization of the landslide size distributions. Geophys Res Lett 28:1091–1094

Statham I (1976) A scree slope rockfall model. Earth Surf Proc Land 1:43–62

Stead D, Eberhardt E, Coggan JS (2006) Developments in the characterization of complex rock slope deformation and failure using numerical modelling techniques. Eng Geol 83:217–235
Stock GM, Martel SJ, Collins BD, Harp EL (2012) Progressive failure of sheeted rock slopes: the 2009–2010 Rhombus Wall rock falls in Yosemite Valley, California, USA. Earth Surf Proc Land 37:546–561
Sturzenegger M, Stead D (2009a) Quantifying discontinuity orientation and persistence on high mountain rock slopes and large landslides using terrestrial remote sensing techniques. Nat Hazards Earth Syst Sci 9(2):267–287
Sturzenegger M, Stead D (2009b) Close-range terrestrial digital photogrammetry and terrestrial laser scanning for discontinuity characterization on rock cuts. Eng Geol 106:163–182
Turcotte D (1986) Fractals and fragmentation. J Geophys Res 91 (B2):1921–1926
Turner AK, Jayaprakash GP (2012) Introduction. In: Turner AK, Schuster RL (eds) Rockfall characterization and control. Transportation Research Board, National Academy of Sciences, Whasington D.C., pp 3–20
Voight B (1989) A relation to describe rate-dependent material failure. Science 243:200–203
Volkwein A, Schellenberg K, Labiouse V, Agliardi F, Berger F, Bourrier F, Dorren LKA, Gerber W, Jaboyedoff M (2011) Rockfall characterisation and structural protection—a review. Nat Hazards Earth Syst Sci 11:2617–2651
Wang Y, Tonon F (2010) Discrete element modeling of rock fragmentation upon impact in rock fall analysis. Rock Mech Rock Eng 44:23–35
Wehr A, Lohr U (1999) Airborne laser scanning–an introduction and overview. ISPRS J Photogramm Remote Sens 54:68–82
Whalley WB (1984) Rockfalls. In: Brunsden D, Prior DB (eds) Slope instability. Wiley, New York, pp 217–256
Wieczorek GF (2002) Catastrophic rockfalls and rockslides in the Sierra Nevada, USA. In: Evans SG, DeGraff J (eds) Catastrophic landslides: effects, occurrence and mechanisms. Rev Eng Geol, vol 15. Geological Society of America, pp 165–190
Wieczorek GF, Morrissey MM, Iovine G, Godt J (1998) Rock fall Hazards in the Yosemite Valley. US Geological Survey, Open File Report. 98–467
Wyllie DC, Mah C (2004) Rock slope engineering; civil and mining, 4th edn. SponPress, New York, p 431
Zhang ZX, Kou SQ, Jiang LG, Lindqvist PA (2000) Effects of loading rate on rock fracture: fracture characteristics and energy partitioning. Int J Rock Mech Min Sci 37:745–762

International Consortium on Landslides (ICL)— The Proposing Organization of the ISDR-ICL Sendai Partnerships 2015–2025

Kyoji Sassa, Yueping Yin, and Paolo Canuti

Abstract

The International Consortium on Landslides (ICL) was founded in January 2002 during the UNESCO-Kyoto University Joint IGCP symposium "Landslide Risk Mitigation and Protection of Cultural and Natural Heritage". It proposed and adopted the Letter of Intent in 2005 during the 2nd UN World Conference on Disaster Reduction, Kobe, Japan, adopted the Tokyo Action Plan in 2006, and the ISDR-ICL Sendai Partnerships 2015–2025 in 2015. This paper describes the history of ICL from preparation to present in a table of the chronology of events since 1987-present including the organization of ICL until 2020 when the Fifth World Landslide Forum will be held in Kyoto, Japan.

Keywords

Landslides • International consortium on landslides (ICL) • International strategy for disaster risk reduction (ISDR) • World conference on disaster risk reduction (WCDRR) • Varnes medal

Introduction

The International Consortium on Landslides (ICL) is an international non-governmental and non-profit scientific organization promoting landslide research and capacity building for the benefit of society and the environment. The ICL was founded in January 2002 in Kyoto, Japan. It was registered as a legal body (No. 1300-05-005237) under Japanese law in Kyoto Prefectural government in August 2002. In March 2007 the ICL was approved as a scientific research organization (No. 94307) which can receive the scientific grants of the Ministry of Education, Culture, Sports, Science and Technology (MEXT), Japan.

ICL, Kyoto University and UNESCO established the UNITWIN (University Twining and Networking) Cooperation Programme on Landslide risk mitigation for society and the environment in March 2003 and expanded its activity area to Landslide and water-related disaster risk management for society and the environment in November 2010. The ICL has been approved as an NGO having operational relations with UNESCO in April 2007, and reclassified as an NGO with a consultative partnership with UNESCO in March 2012.

ICL exchanged Memorandums of Understanding (MOU) with each of five UN organizations (UNESCO, WMO, FAO, UNISDR, UNU) and two global stakeholders in Science and Technology (ICSU and WFEO) to promote the 2006 Tokyo Action Plan "Strengthening Research and Learning on Landslides and Related Earth System Disasters for Global Risk Preparedness". It is the follow-up to the "2005 Letter of Intent", an outcome of the session titled "New international Initiatives for Research and Risk Mitigation of Foods (IFI) and Landslides" (IPL: International

K. Sassa (✉)
International Consortium on Landslides (ICL), 138-1 Tanaka-Asukai cho, Sakyo-ku, Kyoto, 606-8226, Japan
e-mail: sassa@iclhq.org

Y. Yin
China Institute of Geo-Environment Monitoring, CGS, 100081 Beijing, China
e-mail: yyueping@mail.cgs.gov.cn

P. Canuti
University of Firenze, Via La Pira, 4, 50121 Florence, Italy
e-mail: canuti37@gmail.com

K. Sassa et al. (eds.), *Advancing Culture of Living with Landslides*,
DOI 10.1007/978-3-319-59469-9_5

Programme on Landslides) organized in the 2005 United Nations World Conference on Disaster Reduction in Kobe, Japan, 2005.

Landslides are studied in many fields of science, technology and disaster reduction, and understandings of landslides differ in various fields. To create a common understanding of landslides as a distinct science, an international journal is absolutely necessary. The first project of ICL was to found an international journal "Landslides": IPL C100. There are many different terminologies used in landslides in different fields. However, one information source these fields have in common are full color photos of landslides. The most important condition was that the new journal should be printed in full color. The first full color scientific journal "Landslides: Journal of International Consortium on Landslides" was inaugurated in 2004. It is not easy to launch a new journal and it was developed based on the past 15 years of publication of the three-color printed International Newsletter "Landslide News" from 1987 to 2003 by the Japan Landslide Society. The journal "Landslides" has a latest impact factor 3.049 and is published bimonthly, with 300 pages per issue from 2017.

ICL has organized World Landslide Forums together with IPL Partners in 2008, 2011, 2014, and will organize it in 2017 in Ljubljana, Slovenia, and in 2020 in Kyoto, Japan. The current greatest mission is to implement "ISDR-ICL Sendai Partnerships 2015–2025 for global promotion of understanding and reducing landslide disaster risk", proposed, accepted and signed during the Third United Nations World Conference on Disaster Risk Reduction in Sendai, Japan, 2015. The history of ICL was compiled in this article as the basic information for the High-Level Panel Discussion "Strengthening Intergovernmental Network and the International Programme on Landslides (IPL) for "ISDR-ICL Sendai Partnerships 2015–2025 for global promotion of understanding and reducing landslide disaster risk".

Objectives of ICL

The objectives of ICL are written in its statutes. The principal objectives are to:

(a) promote landslide research for the benefit of society and the environment, and capacity building, including education, notably in developing countries;
(b) integrate geosciences and technology within the appropriate cultural and social contexts in order to evaluate landslide risk in urban, rural and developing areas including cultural and natural heritage sites, as well as to contribute to the protection of the natural environment and sites of high societal value;
(c) combine and coordinate international expertise in landslide risk assessment and mitigation studies, thereby resulting in an effective international organization which will act as a partner in various international and national projects; and
(d) promote a global, multidisciplinary Programme on landslides, the International Programme on Landslides.

ICL Members

Members are those organizations that support the objectives of ICL intellectually, practically and financially. Membership is for a minimum period of two years. Members will come from one of four categories:

(a) Intergovernmental organizations
(b) Non-governmental organizations
(c) Governmental organizations and public organizations
(d) Other organizations and entities.

Management of ICL

All matters of management are examined and decided by the Board of Representatives (BOR). Members of BOR are the representatives of each ICL member organization.

- Full power for the management of the affairs of the Consortium is vested in the Board of Representatives, which will meet at least annually. The quorum and internal regulations are defined by the bylaws.
- The Board of Representatives shall be composed of representatives of the Member organizations. Each Member organization shall designate one Representative and one Alternative Representative.
- In the absence of a Member's Representative from any meeting of the Board of Representatives, the alternative representative may attend the meeting and exercise all the rights, powers and privileges of the absent Representative. Alternatively, the Representative may delegate his rights, powers and privileges to another Member of ICL for that particular meeting, or authorize him/her to act and vote on his behalf.
- The Board of Representatives shall:
  (a) determine general policy;
  (b) initiate scientific programmes and decide on future priorities for the activities of ICL;
  (c) approve or change, if necessary, the budget and accounts;

(d) examine and decide on each application for Member, Associate or Supporter status;
(e) elect the Officers of ICL in accordance with the Bylaws;
(f) terminate the status of any Member, Supporter of ICL which has failed to fulfill any of its obligations or when the association is no longer considered appropriate, in accordance with the Bylaws
(g) change the Statutes and Bylaws;
(h) deal with other items which may be referred to it.

- Voting will be decided on a simple majority. Each Member shall have one vote. Normally the President of ICL will not vote but in the event of a tie, the President may have the casting vote.

One of the characteristic of ICL management is the direct democracy. All matters are decided by the BOR. The number of ICL members increased from 33 to 64 in 2002–2016. The number is small enough to discuss in one room, in a round-table discussion style.

History of ICL to Propose the Partnerships

The history of ICL was compiled and chronologically presented from the preparatory stage (from June 1987, the publication of the first issue of the international Newsletter "Landslide News", to January 2001). The 2001 Tokyo Declaration "Geoscientists tame landslides" proposed an International Consortium on Landslides, so the interim stage was from the foundation of ICL in January 2002 by adoption of Statutes and selection of an interim President to the organization of the first session of the Board of Representatives at UNESCO Headquarters, in Paris. The first period was from 1st January 2003 to 31 December 2005, the second period from 1st January 2006 to 31 December 2008, the third period from 1 January 2009 to 31 December 2011, the fourth period from 1 January 2012 to 31 December 2014, and the fifth period from 1 January 2015 to 31 December 2017.

ICL HISTORY—from preparation to present

Date	Chronology of events
Preparatory stage	
June 1987	The First issue of the International Newsletter "Landslide News" was published by the Japan Landslide Society. It was a newsletter with three-color (red, blue and black) printing and a three-column layout. The newsletter was planned based on discussions at the 4th International Conference and Field Workshop on Landslides held in 1985. No.1-No.15 were published from 1987 to 2003. 5000 copies were printed and 2000 copies were distributed to the world free of charge, funded by Japanese companies and agencies supporting this initiative. UNESCO, U.S. National Council, United Nations Disaster Relief Coordinator (UNDRO) contributed Prefaces. It was a timely contribution to the International Decade for Natural Disaster Reduction (IDNDR). The International Editor was Robert Schuster. Chair of the publication committee was Toshio Taniguchi, Masami Fukuoka, then Kyoji Sassa. This newsletter was moved to "Recent Landslides" within the full-color quarterly (bimonthly from Vol. 10, 2013) Journal "Landslides: Journal of the International Consortium on Landslides"
1991–1998	Japan-China Joint Research on Assessment of Landslide Hazard in Lishan, Xian, China was proposed by Kyoji Sassa and funded by Japanese Ministry (MEXT) as a part of IDNDR from 1991 for 8 years. Lishan is a steep mountain slope (fault scarp) behind the Royal Resort Palace of the Tang Dynasty (618–907), China. It is at risk of a possible landslide disaster, though no landslide has occurred at the royal resort palace
July 1997	The Japan-China Joint Project group organized the International Symposium on Landslide Hazard Assessment, Xian, China, in July 1997 as a concluding meeting of the joint research. It was organized as the Committee for Prediction of Rapid Landslide Motion of the IUGS Working Group on Landslides (WGL/RLM). A representative from UNESCO-Cultural Heritage, the leader of IGCP programme and others attended and were impressed by the research results, which assessed the landslide hazard with convincingly detailed monitoring of slope deformation and undrained ring shear testing results for the site, which has lasted more than 1000 years. The group was invited to apply for an IGCP project because the project is very much suitable for the new policy of IGCP "Geoscience for the Society"
February 1998	An IGCP project proposed by Kyoji Sassa was approved by the IGCP Board in February 1998. The UNESCO-IUGS joint project, International Geological Correlation Programme (IGCP) No. 425, Landslide Hazard Assessment and Mitigation for Cultural Heritage Sites and Other Locations of High Societal Value began. IGCP is currently renamed as the Geoscience Programme

(continued)

ICL HISTORY—from preparation to present	
Date	Chronology of events
November–December 1998	The UNESCO-IUGS-IGCP Joint Symposium on Natural Hazards and Cultural Heritage was held at the Canadian Embassy in Tokyo from 31 November to 1 December 1998
September 1999	IGCP-425 group and UNESCO organized The International Conference "Cultural Heritage at Risk" at UNESCO Headquarters, Paris, France. Leader: Kyoji Sassa, Japan, Deputy leaders: Paolo Canuti, Italy and Raul Carreno, Peru
November–December 1999	Following the IGCP-425 activities, the Memorandum of Understanding between UNESCO and DPRI, Kyoto University, Japan concerning "Cooperation in Research for Landslide Risk Mitigation and Protection of the Cultural and Natural Heritage as a Key Contribution to Environmental Protection and Sustainable Development in the First Quarter of the Twenty-First Century" was signed by Koichiro Matsuura on 26 November and by Shuichi Ikebuchi on 3 December 1999
Mach 2000	Sassa et al. investigated the Machu Picchu Citadel site from the ground and also from a chartered helicopter with special permission to fly over Machu Picchu from the Instituto Nacional de Cultura (INC) as a part of IGCP subproject "Protection of Cultural Heritage on Landslide Zones at Cusco. They deduced possible geomorphic processes from observation and made a hypothesis that the Inca people constructed the citadel on the sliding surface after the mass of a large-scale landslide slid down, with another landslide in a deeper shear band parallel to the current ground surface/previous sliding surface
January 2001	The UNESCO/IGCP Symposium on Landslide Risk Mitigation and Protection of Cultural and Natural Heritage was organized at the headquarters of the Science Council of Japan, Tokyo. The initial monitoring result (November to December 2000) of extensometers installed in Machu Picchu site were reported. The result was widely reported by Japanese newspapers and the British magazine "New Scientists" and others
January 2001	2001 Tokyo Declaration "Geoscientists tame landslides" was adopted by the participants in the UNESCO/IGCP Symposium on Landslide Risk Mitigation and Protection of Cultural and Natural Heritage. The Tokyo Declaration proposed the establishment of the International Consortium on Landslides (ICL)
Foundation of the International Consortium on Landslide and its Interim Stage	
Interim President: Kyoji Sassa	
January 2002	The UNESCO-Kyoto University joint symposium "Landslide Risk Mitigation and Protection of Cultural and Natural Heritage" was organized in Kyoto, Japan. It was attended by six directors from Earth Science, Water Sciences, Cultural Heritage, Engineering Science of UNESCO, and deputy Secretary General of WMO, a representative from UNISDR, and also from the Ministry of Foreign Affairs and the Ministry of Education, Culture, Sports, Science and Technology (MEXT), as well as IGCP-425 members from around the world
January 2002	ICL was established by adopting the 2002 Kyoto Appeal "Establishment of a New International Consortium on Landslides" and the statutes of ICL and deciding the Interim President on 21 January 2002 during the above symposium. The main objectives of ICL foundation is to establish a new International Programme on Landslides (IPL) after the IGCP-425. Most of 31 subproject leaders of IGCP-425 received promotional benefits of their landslide research from authorization as the subproject leaders of the UNESCO/IUGS joint programme and wished to establish a new International Programme on Landslides (IPL)
August 2002	ICL was registered as a legal body under Japanese law in the Kyoto Prefectural Government, Japan in August 2002
November 2002	The First Session of the Board of Representatives (BOR) of ICL was organized at UNESCO Headquarters on 19–21 November 2002. The initial 33 ICL member organizations, mostly from IGCP-425 subproject leaders, attended, as well as the officers for the first three years. They decided to found an international journal "***Landslides***" as the initial IPL project Coordinating Project C100 as a core activity of ICL. ICL is basically a bottom up self-supporting organization with supports from UNESCO, WMO, FAO, UNISDR, Government of Japan and others
The First Period: 1 January 2003–31 December 2005	
President: Kyoji Sassa	
Vice Presidents: Peter Bobrowsky, Paolo Canuti, Romulo Mucho, Peter Lyttle	
Executive Director: Kaoru Takara Treasurer: Claudio Margottini	
Number of ICL Member organizations: 33	
March 2003	ICL, Kyoto University and UNESCO established the UNITWIN (University Twining and Networking) Cooperation Programme on Landslide Risk Mitigation for Society and the Environment in March 2003. The establishment of UNITWIN Cooperation Programme was initially suggested by participants from UNESCO to authorize the International Programme on Landslides (IPL) as a UNESCO network activity

(continued)

ICL HISTORY—from preparation to present	
Date	Chronology of events
May 2003	The First ICL Steering Committee Meeting was organized at the headquarters of the Food and Agriculture Organization of the United Nations (FAO) in Rome, Italy. The steering committee meetings were organized annually or at necessary times later on
October 2003	The 2nd Session of the Board of Representatives (BOR) of ICL was organized at Simon Fraser University Harbour Centre, Vancouver, Canada on 28 October–1 November 2003
September 2004	UNITWIN Headquarters Building was constructed by ICL and Kyoto University at the Kyoto University Uji Campus in September 2004
October 2004	The 3rd Session of the Board of Representatives (BOR) of ICL was organized at Druzba Hotel, Bratislava, Slovakia on 19–22 October 2004
April 2004	ICL founded a new full color quarterly Journal "*Landslides:* Journal of the International Consortium on Landslides" in 2004, in cooperation with Springer after negotiating with several international publishers
January 2005	The World Conference on Disaster Reduction (WCDR) was held on 18–22 January 2005 in Kobe, Japan
January–June, 2005	ICL proposed and organized a session titled "New international Initiatives for Research and Risk Mitigation of Foods (IFI) and Landslides (IPL)" in cooperation with Flood group. ICL was well prepared for the session and proposed a ***"Letter of Intent*** aiming to provide a platform for a holistic approach in research and learning on "Integrated Earth System Risk Analysis and Sustainable Disaster Management". A Letter of Intent is much lighter than agreements and Memorandums of Understanding, and was suggested by Hans van Ginkel (Rector of United Nations University). It was agreed and signed by heads of seven global stakeholders of UNESCO, WMO, FAO, UNISDR, UNU, ICSU and WFEO from January to June 2005
October 2005	The First General Assembly of ICL, together with the 4th Session of the Board of Representatives (BOR) of ICL, was organized at Keck Center of the National Academy of Sciences, Washington D.D., USA on 12–14 October 2005

(continued)

ICL HISTORY—from preparation to present	
Date	Chronology of events
The Second Period: 1 January 2006–31 December 2008	
President: Kyoji Sassa	
Vice Presidents: Peter Bobrowsky, Paolo Canuti, Oddvar Kjekstad, Peter Lyttle	
Executive Director: Kaoru Takara Treasurer: Hiroshi Fukuoka	
Number of ICL Member organizations: 48	
January 2006	The Round Table Discussion "Strengthening Research and Learning on Earth System Risk Analysis and Sustainable Disaster Management within UN-ISDR as Regards "Landslides" was co-organized by ICL, UNESCO, WMO, FAO, UNISDR, UNEP, UNU and Kyoto University at United Nations University, Tokyo, Japan on 18–20 January 2006. This discussion aimed to implement the 2005 Letter of Intent
January 2006	The 2006 Tokyo Action Plan Strengthening Research and Learning on Landslides and Related Earth System Disasters for Global Risk Preparedness was adopted by the participants of the Round Table Discussion on 20 January 2006. The Tokyo Action Plan prosed a new stage of ***International Programme on Landslides (IPL)***, a programme of ICL for ISDR, which is managed by the IPL Global Promotion Committee consisting of all ICL member organizations and 7 global stakeholders (UNESCO, WMO, FAO, UNISDR, UNU, ICSU, and WFEO). New IPL activities include IPL projects, the World Landslide Forum (WLF) every three years, and the World Centres of Excellence on Landslide Risk Reduction (WCoEs) to be identified at each WLF. The First WLF succeeded the First General Assembly of ICL in 2005
April–December 2006	ICL exchanged Memorandums of Understanding to promote IPL with UNESCO, WMO, FAO, UNISDR, UNU, ICSU, WFEO within 2006
November 2006	The 5th Session of the Board of Representatives (BOR) of ICL was organized at UNESCO Headquarters on 23–24 November 2006
March 2007	ICL was approved as a scientific research organization (No. 94307) which can receive the scientific grant of the Ministry of Education, Culture, Sports, Science and Technology (MEXT), Japan in March 2007. Thereafter, ICL can apply for Scientific Grants and other Scientific programmes

(continued)

ICL HISTORY—from preparation to present	
Date	Chronology of events
April 2007	ICL was approved to be an NGO having operational relations with UNESCO in April 2007
November 2007	The 6th Session of the Board of Representatives (BOR) of ICL was organized at UNESCO Headquarters on 14–16 November 2007
May 2008	ICL was registered in the cross-ministerial research and development management system (e-Rad) of all ministries of Japan in May 2008
November 2008	The 7th Session of the Board of Representatives (BOR) of ICL was organized at the United Nations University, Tokyo, Japan on 17 November 2008
November 2008	The First World Landslide Forum (WLF1) was organized by ICL, IPL and Partners at the United Nations University, Tokyo, 18–21 November 2008. (The Second General Assembly of ICL to disseminate ICL-IPL activities was organized as the First World Landslide Forum from 2008.)
The Third Period: 1 January 2009–31 December 2011	
President: Paolo Canuti	
Vice Presidents: Oddvar Kjekstad, Peter Lyttle, Kaoru Takara	
Executive Director: Kyoji Sassa Treasurer: Hiroshi Fukuoka	
Number of ICL Member organizations: 48	
November 2009	The 8th Session of the Board of Representatives (BOR) of ICL was organized at UNESCO Headquarters on 17–19 November 2009
November 2010	The UNITWIN programme was updated and developed to ***Landslide and water-related disaster risk management for society and the environment*** in a wider scope in November 2010
November 2010	The 9th Session of the Board of Representatives (BOR) of ICL was organized at UNESCO Headquarters on 16–19 November 2010
October 2011	The 10th Session of the Board of Representatives (BOR) of ICL was organized at FAO Headquarters, Rome, Italy on 3–5 October 2011
October 2011	The Second World Landslide Forum (WLF2) was organized by ICL, IPL and partners at the Food and Agriculture Organization Headquarters of the United Nations, Rome, 3–9 October 2011

(continued)

ICL HISTORY—from preparation to present	
Date	Chronology of events
The Fourth Period: 1 January 2012–31 December 2014	
President: Paolo Canuti	
Vice Presidents: Kaoru Takara, Yueping Yin, Claudio Margottini, Irasema Alcantara-Ayara	
Executive Director: Kyoji Sassa Treasurer: Hirotaka Ochiai	
Number of ICL Member organizations: 53	
January 2012	ICL organized 10th anniversary meeting, Kyoto, Japan in January 2012 and adopted the ICL Strategic Organization of Plan 2012–2021
March 2012	ICL was reclassified as an NGO with a consultative partnership with UNESCO in March 2012
November 2012	The 11th Session of the Board of Representatives (BOR) of ICL was organized at UNESCO Headquarters, Paris, France on 20–23 November 2012
2013	"Landslides" Journal became a bimonthly journal from Vol. 10 in 2013
November 2013	The 12th Session of the Board of Representatives (BOR) of ICL was organized at Yamanouchi Hall, Shiran Kaikan, Kyoto University, Kyoto, Japan on 19–22 November 2013
June 2014	The 13th Session of the Board of Representatives (BOR) of ICL was organized at China National Convention Center, Beijing, China on 2 June 2014
June 2014	The Third World Landslide Forum (WLF3) was organized by ICL, IPL and partners at the National Convention Center of China, Beijing, in June 2014. ICL adopted the 2014 Beijing Declaration "Landslide Risk Mitigation: Toward a Safer Geoenvironment" at the WLF3 in Beijing
The Fifth Period: 1 January 2015–31 December 2017	
President: Yueping Yin	
Vice Presidents: Claudio Margottini, Irasema Alcantara-Ayara, Matjaz Mikos, Dwikorita Karnawati	
Executive Director: Kyoji Sassa Treasurer: Kaoru Takara	
Number of ICL Member organizations: 57 (64 in 2016.10)	
March 2015	The 14th Session of the Board of Representatives (BOR) of ICL was organized at Tohoku Gakuin University, Sendai, Japan on 11–15 March 2015
March 2015	ICL proposed the "***ISDR-ICL Sendai Partnerships 2015-2025 for global promotion***

(continued)

ICL HISTORY—from preparation to present	
Date	Chronology of events
	of understanding and reducing landslide disaster risk" in the Third United Nations World Conference on Disaster Risk Reduction in Sendai, Japan, 2015
June 2015	Landslides: Journal of International Consortium on Landslides reached a 2015 Impact Factor of 3.049
March 2016	The 15th Session of the Board of Representatives (BOR) of ICL was organized at Kyoto University Uji Campus, Kyoto, Japan on 7–11 March 2016
January 2017	The number of pages for one issue of Landslides: Journal of the International Consortium on Landslides will be increased from around 200 pages/issue in 2016 to around 300 pages/issue in 2017
May 2017	The Fourth World Landslide Forum (WLF4) will be organized by ICL, IPL and partners at the Cultural and Congress Centre in Ljubljana, Slovenia, from 29 May to 2 June 2017
The Sixth Period: 1 January 2018–31 December 2020	
President: Peter Bobrowsky	
Vice Presidents: under nomination	
Executive Director: Kaoru Takara Treasurer: Kyoji Sassa	
November 2020	The Fifth World Landslide Forum (WLF5) will be organized in 2–6 November 2020 by ICL, IPL and the ISDR-ICL Sendai Partnerships partners at the Kyoto International Conference Center, Kyoto, Japan where "the Kyoto Protocol" to the United Nations Framework Convention on Climate Change" was adopted at COP3 in 1979. The Forum is organized in the mid-term of the Sendai Partnerships 2015-2025. The mid-term review and planning for the latter half will be examined. The Kyoto 2020 Commitment will be examined during the forum to establish a stable global framework to implement further the ISDR-ICL Sendai Partnerships 2015–2025 and pursue and enhance thereafter efforts towards landslide disaster risk reduction

ICL Award "Varnes Medal"

ICL was founded to create a new International Programme on Landslides (IPL) to promote science and technology for landslide risk reduction by mobilizing funds and the efforts of active landslide researchers and their organizations. Dr. David Varnes (1919–2002) of U.S. Geological Survey studied many types of landslides and proposed his landslide classification. His classification was the base of the internationally agreed upon definition of landslides during the International Decade for Natural Disaster Reduction (IDNDR) from 1990 to 2000 (Landslide Dynamics: ISDR-ICL Landslide Interactive Teaching Tools (LITT) by Sassa et al. in this volume). It is the base of the international journal "Landslides: Journal of the International Consortium on Landslides". David Varnes is the most respected researcher in the landslide community. ICL created the "Varnes Medal" as the highest award of the consortium. The current criteria of the Varnes Medal and the past recipients are introduced below.

Criteria of the Varnes Medal

The Varnes Medal is the highest award provided by the International Consortium on Landslides; it recognizes professional excellence in landslide research. Nominees for the Varnes Medal must meet at least two of the following criteria:

- Professional excellence in landslide research
- Professional excellence in landslide disaster risk reduction (new criteria)
- Significant contribution to the development of ICL and IPL (new criteria)
- Significant contribution to public education regarding landslide hazards
- International recognition for a professional career involving landslides
- Influential landslide research or development of methods or techniques
- Teacher of students who work on landslide issues.

Note: Two new criteria were added from 2016 to extend the scope to persons other than landslide researchers. From 2003 to 2016, recipients of the Varnes Medal were landslide researchers. The Varnes Medal for 2017 will be awarded to Badaoui Rouhban of UNESCO, as the first recipient other than landslide researchers, at Ljubljana, Slovenia on 30 May 2017.

Past Recipients of the Varnes Medal

- Robert Schuster (US Geological Survey, USA), 2nd Session of BOR/ICL at Simon Fraser University, Vancouver, Canada in October 2003
- John Hutchinson (Imperial Colleague, UK), 3rd Session of BOR/ICL at Druzba Hotel, Bratislava, Slovakia in October 2004
- Masami Fukuoka (University of Tokyo, Japan), 4th Session of BOR/ICL at Keck Center of the National Academy of Science, Washington D.C., USA in October 2005

- Norbert R. Morgenstern (University of Alberta, Canada), 5th Session of BOR/ICL at Bonvin Building, UNESCO Headquarters in Paris, France in November 2006
- Edward Derbyshire (University of Leicester, UK), 6th Session of BOR/ICL at Bonvin Building, UNESCO Headquarters in Paris, France in November 2007
- David Cruden (University of Alberta, Canada), 7th Session of BOR/ICL at the United Nations University, Tokyo, Japan in November 2008
- Zaiguan Lin (Leader of China-Japan Joint research on Landslide hazard assessment at Cultural heritage sites in Xi'an. It developed into IGCP-425 "Landslide Hazard Assessment and Cultural Heritage" which was further developed at ICL and IPL), 9th Session of BOR/ICL at Bonvin Building, UNESCO Headquarters, Paris, France in November 2010
- Rajendra Kumar Bhandari (Central Building Research Institute, India), 11th Session of BOR/ICL at Bonvin Building, UNESCO Headquarters, Paris, France in November 2012
- Luciano Picarelli (Seconda Università di Napoli), 13th Session of BOR/ICL at National Convention Center, Beijing, China, in June 2014
- Oldrich Hungr (University of British Columbia, Vancouver, Canada), 14th Session of BOR/ICL at Tohoku Gakuin University, Sendai, Japan, in March 2015
- Jordi Corominas (Technical University of Catalonia-UPC, Spain), 15th Session of BOR/ICL at Disaster Prevention Research Institute, Kyoto University, Kyoto, Japan, in March 2016
- Badaoui Rouhban (Special Advisor to the Assistant Director-General for Sciences, UNESCO as well as the Advisor to the International Programme on Landslides), decided at 16th Session of BOR/ICL, at UNESCO Headquarters, Paris, France, in November 2016 for 2017.

Until 2014, the Varnes medal was originally produced in Egypt and is a pure silver medal (200 g), in which the recipient's name is engraved on the back side. A set of new medals were ordered from a company in Kyoto, Japan in 2015. It is a 95% silver medal (200 g) for better maintenance. The recipient's name and the award place and date are printed in a plate in the left side of the award holder, and the medal is placed in the right side when it is open, as shown below.

The 2015, 2016 and 2017 Varnes medals will be awarded to three recipients on 30 May 2017 during the Fourth World Landslide Forum in Ljubljana, Slovenia.

The new version of the Varnes Medal for Jordi Corominas of Spain.

A group photo of participants at the ICL foundation meeting is presented at the end of this article.

Group photo commemorating the establishment of the International Consortium on Landslides on 23 January 2005 during the UNESCO-Kyoto University joint symposium "Landslide Risk Mitigation and Protection of Cultural and

Natural Heritage" at the Kyoto Campus Plaza in Kyoto, Japan. Participants included those from ICL supporting organizations, Andras Szollosi-Nagy (Director of Water Sciences, UNESCO), Wolfgang Eder (Director of Earth Sciences, UNESCO), Badaoui Rouhban (Chief of Engineering Sciences and Technology, UNESCO), Michel Jarraud (Deputy Secretary-General of WMO), Pedro Basabe (UN/ISDR), three members of the Division of Cultural Heritage of UNESCO (Laurent Levi-Strauss, Galia Saouma-Forero and Christian Manhart), three representatives of the Government of Japan from the Ministry of Foreign Affairs (Multilateral Cultural Cooperation Division) and MEXT (Offices of the Disaster Prevention Research and the Director-General for International Affairs). Robert Schuster of the U.S. Geological Survey, Paolo Canuti and ICL founding members, and other landslide colleagues from the world also participated.

Acknowledgements and Call for Cooperation

ICL was established in 2002 based on the membership fund from ICL members. ICL also has been supported by various organizations and leaders from outside of the landslide community. UNESCO has supported ICL from IGCP-425 (1998–2003) from its foundation in 2002 and during its development until today, intellectually, practically and financially. The strong support from UNESCO is very much appreciated.

ICL developed its global supporting network through the "Letter of Intent" aiming to provide a platform for a holistic approach in research and learning on "Integrated Earth System Risk Analysis and Sustainable Disaster Management", which was adopted at a thematic session of the World Conference on Disaster Reduction (WCDR in Kobe, Japan in 2005) and signed by UNESCO, WMO, FAO, UNISDR, UNU, ICSU, WFEO. The letter was inspired by Han van Ginkel (Rector of the United Nations University). UNESCO-DG (Koïchiro Matsuura) and UNISDR-Director (Salvano Briceno) and WMO Secretary General (Michel Jarraud) presented the opening remarks to this session, and signed the Letter of Intent soon after the session. Other invited organizations also agreed and signed within a few months. Hans van Ginkel and Salvano Briceno greatly contributed to the 2006 Tokyo Action Plan to found a new stage of IPL (International Programme on Landslides). Salvano Briceno served as the chair of the IPL-Global Promotion Committee (2006–2014). Both are very much appreciated for their contribution to ICL and IPL development.

The foundation of ICL is based on the international activities of the Japan Landslide Society. A major activity of ICL has been the publication over 15 years (1987–2003) of a three-color printed international newsletter: "Landslide News" by the Japan Landslide Society. 5000 copies were annually printed and 2000 copies were distributed to landslide researchers abroad free of charge. Robert Schuster of USGS worked as an international chief editor of this newsletter and supported the Landslide News Publication Committee (Kyoji Sassa: Chair). Schuster's contribution was highly valued, and he was then nominated to be the first recipient of the Varnes Medal.

ICL applied for and obtained major research funding from the Japan Science and Technology Agency (JST) and the Japan Society for the Promotion of Science (JSPS), and the Ministry of Education, Culture, Sports, Science and Technology, Japan (MEXT) and from the IDNDR special budget of the Japan-China Joint Research on the Assessment of Landslide Hazards in Lishan, Xian (1991–1998). ICL obtained considerable funding through UNISDR from the Ministry of Foreign Affairs (MOFA) of Japan in 2007 to prepare and organize the first World Landslide Forum in 2008 and also the second World Landslide Forum in 2011. Those budgets played a critical role in the success of the foundation of the series of World Landslide Forums. At this opportunity, 15 years after its foundation, ICL extends sincere thanks to ICL supporting organizations (UNESCO, WMO, FAO, UNISDR, UNU, ICSU, WFEO and IUGS), the government of Japan, and also private companies in Japan which have supported ICL, most of them from the publication of Landslide News in 1987.

The basic funding of ICL comes from the ICL membership fee. However, in-kind service, cooperation and periodical financial support for activities have contributed to ICL management and development. Without cooperation from ICL's supporting organizations, ICL could not have developed through this 15 years. We proposed the ISDR-ICL Sendai Partnerships 2015–2025 which was agreed on and signed by 17 global stakeholders. ICL has made its greatest efforts since March 2015, and strongly wishes to further develop this Partnerships toward its midterm 2020 conference. ICL calls for cooperation and participation in this initiative as a voluntary commitment to the Sendai Framework for Disaster Risk Reduction 2015–2030.

United Nations Office for Disaster Risk Reduction (UNISDR)—UNISDR's Contribution to Science and Technology for Disaster Risk Reduction and the Role of the International Consortium on Landslides (ICL)

Chadia Wannous and German Velasquez

Abstract

The Sendai Framework for Disaster Risk Reduction 2015–2030 was agreed at the Third UN World Conference on Disaster Risk Reduction in Sendai, Japan in March 2015 and endorsed by the UN General Assembly in June 2015. The goal of the Sendai Framework is to prevent new and reduce existing disaster risk. UNISDR coordinates and ensures synergies among the disaster reduction activities of the United Nations system and regional organizations and stakeholders The role of science and technology in providing the evidence and knowledge on risk features prominently in the Sendai Framework. Expanding the interface between science, technology and policy is therefore essential for effective disaster risk reduction. In January 2016, UNISDR hosted the Science and Technology Conference on the Implementation of the Sendai Framework. The main outcome of the conference was the launching of the Science and Technology Partnership and the endorsement of the science and technology roadmap that outlines expected outcomes, actions, and deliverables under each of the four priority actions of the Sendai Framework. Over the last twenty years, the majority of disasters have been caused by floods, storms, heatwaves and other weather-related events. Most of these disasters can cause landslides, which in turn cause hundreds of billions of dollars in damage and hundreds of thousands of deaths and injuries each year The International Consortium on Landslides (ICL) 2015–2025 and The Sendai Partnerships promotes global understanding and reduction of landslide disaster risk. They will contribute significantly to the implementation of the science and technology roadmap by providing practical solutions and tools, education and capacity building, and communication and public outreach to reduce landslides risks. UNISDR fully supports the work of the Sendai Partnerships and the community of practice on landslides risks

Keywords

UNISDR • Sendai framework • Disaster risk reduction • Science and technology

The Sendai Framework for Disaster Risk Reduction (2015–2030)

The Sendai Framework for Disaster Risk Reduction 2015–2030 was agreed at the Third UN World Conference on Disaster Risk Reduction in Sendai, Japan in March 2015 and endorsed by the UN General Assembly in June 2015 (UNISDR 2015a).

The goal of the Sendai Framework is to prevent new and reduce existing disaster risk through the implementation of integrated and inclusive economic, structural, legal, social, health, cultural, educational, environmental, technological, political and institutional measures that prevent and reduce hazard exposure and vulnerability to disasters, increase preparedness for response and recovery, and thus strengthen resilience.

C. Wannous (✉) · G. Velasquez
UNISDR, 9-11 Rue de Varembe, 12020 Geneva, Switzerland
e-mail: wannous@un.org

K. Sassa et al. (eds.), *Advancing Culture of Living with Landslides*,
DOI 10.1007/978-3-319-59469-9_6

The expected outcome up to 2030 is to achieve a substantial reduction in disaster risk and losses in lives, livelihoods and health in the economic, physical, social, cultural and environmental aspects of people, the private sector, communities and countries. There are four priorities, seven targets, thirteen principles and suggested actions for stakeholders at global, regional, national and local levels.

The Sendai Framework sets out a path to ensure that disaster risk is factored into planning and development at all levels across all sectors, as well as in disaster preparedness, recovery and reconstruction. The Sendai Framework is wide in scope. It applies to the risk of small-scale and large-scale, frequent and infrequent, sudden and slow-onset disasters, caused by natural or man-made hazards, as well as related environmental, technological and biological hazards and risks.

The Sendai Framework identifies four priorities for action, which call for focused action within and across sectors by States at local, national, regional and global levels in the following four priority areas:

1. **Understanding disaster risk**

Policies and practices for disaster risk management should be based on an understanding of disaster risk in all its dimensions of vulnerability, capacity, exposure of persons and assets, hazard characteristics and the environment. Such knowledge can be leveraged for the purpose of pre-disaster risk assessment, for prevention and mitigation, and for the development and implementation of appropriate preparedness and effective response to disasters.

2. **Strengthening disaster risk governance to manage disaster risk**

Disaster risk governance at the national, regional and global levels is of great importance for an effective and efficient management of disaster risk. Clear vision, plans, competence, guidance and coordination within and across sectors, as well as participation of relevant stakeholders, are needed. Strengthening disaster risk governance for prevention, mitigation, preparedness, response, recovery and rehabilitation is therefore necessary and fosters collaboration and partnership across mechanisms and institutions for the implementation of instruments relevant to disaster risk reduction and sustainable development.

3. **Investing in disaster risk reduction for resilience**

Public and private investment in disaster risk prevention and reduction through structural and non-structural measures are essential to enhance the economic, social, health and cultural resilience of persons, communities, countries and their assets, as well as the environment. These can be drivers of innovation, growth and job creation. Such measures are cost-effective and instrumental in saving lives, preventing and reducing losses and ensuring effective recovery and rehabilitation.

4. **Enhancing disaster preparedness for effective response, and to "Build Back Better" in recovery, rehabilitation and reconstruction**

Disasters have demonstrated that the recovery, rehabilitation and reconstruction phase, which needs to be prepared for ahead of a disaster, is a critical opportunity to "Build Back Better", including through integrating disaster risk reduction into development measures, making nations and communities resilient to disasters.

To support the assessment of global progress in achieving the outcome and goal of this framework, seven global targets have been agreed. These targets will be measured at the global level. National targets and indicators will contribute to the achievement of the outcome and goal of this framework and an intergovernmental expert working group is developing indicators to measure global progress on the Framework's seven agreed targets.

The seven global targets are:

1. Substantially reduce global disaster mortality by 2030, aiming to lower average per 100,000 global mortality between 2020–2030 compared to 2005–2015.
2. Substantially reduce the number of affected people globally by 2030, aiming to lower the average global number per 100,000 between 2020–2030 compared to 2005–2015.
3. Reduce direct disaster economic loss in relation to global gross domestic product (GDP) by 2030.
4. Substantially reduce disaster damage to critical infrastructure and disruption of basic services, among them health and educational facilities, including through developing their resilience by 2030.
5. Substantially increase the number of countries with national and local disaster risk reduction strategies by 2020.
6. Substantially enhance international cooperation to developing countries through adequate and sustainable support to complement their national actions for implementation of this framework by 2030.
7. Substantially increase the availability of, and access to, multi-hazard early warning systems and disaster risk information and assessments to the people by 2030.

The UN Office for Disaster Risk Reduction (UNISDR)

The mandate of the UN Office for Disaster Risk Reduction (UNISDR) has been defined by a number of United Nations General Assembly Resolutions "to serve as the focal point in the United Nations system for the coordination of disaster reduction and to ensure synergies among the disaster reduction activities of the United Nations system and regional organizations and activities in socio-economic and humanitarian fields" (UN General Assembly Resolution 56/195).

UNISDR focuses on the following key activities:

- Coordinates international efforts for disaster risk reduction.
- Monitors progress in the implementation the Sendai Framework for Disaster Risk Reduction (2015–2030).
- Works closely with countries for building capacities for creating resilient societies.
- Provides tools and guidelines on DRR.
- Organizes regional and global platforms for DRR to report on progress and foster collaboration between countries and stakeholders.

Summary of Implementation of the Sendai framework 2016 as outlined in the UN Secretary General report (UNISDR 2016a).

During the first year of implementation, plans and approaches at all levels have been reviewed and revised by many Member States to align them with the Sendai Framework. In addition, many new partnerships have been established.

Effective integration of disaster risk management into sustainable development policies, practices and investments at all levels is critical. Therefore, concerted efforts have been made to establish coherence with other internationally agreed agendas and frameworks, including the Addis Ababa Action Agenda of the Third International Conference on Financing for Development, the 2030 Agenda for Sustainable Development and the Paris Agreement on climate change.

Progress of the implementation of the Sendai Framework will be reviewed regularly at global and regional platforms for disaster risk reduction. To facilitate this, extensive work has been undertaken on the global monitoring system for the Sendai Framework, including on the indicators for the global targets and updating the DRR terminology. This work will be concluded by end of 2016.

The Sendai Framework target on achieving a substantial increase in the number of countries with national and local disaster risk reduction strategies by 2020 is the first milestone. To achieve this target, over the next four years concerted efforts will be required to better understand existing levels of disaster risk and trends, to develop strategies based on a sound knowledge of current challenges and to set clear priorities and targets. This will include the establishment or enhancement of data systems to record disaster losses and to establish hazard and vulnerability assessments and disaster risk estimates.

Continued collaboration and commitment by the public and private sectors to integrate disaster risk reduction into their respective policies, practices and investments is emphasized in the Sendai Framework and a call for a substantial increase in investment in disaster risk reduction is required. While some progress has been made in this regard, still more resources are needed to ensure substantial reduction of disaster risk and losses in lives and livelihoods, and assets of persons, businesses, communities and countries.

Science and technology is expected to provide valuable contributions to measuring progress through the Sendai Framework monitor, and by providing the evidence-base for the development and implementation of the national and local disaster risk reduction strategies and in public and private investment.

The Strong Emphasis on Science and Technology in Sendai Framework

A main feature of the Sendai Framework, in comparison to its predecessor (the Hyogo Framework of Action), is the shift of focus from managing 'disasters' to managing 'risks'. Such a shift requires a better understanding of risk in all its dimensions of hazards, exposure and vulnerability. Therefore, the role of science and technology in providing the evidence and knowledge of risk features heavily in the Sendai Framework.

There are a number of references to science and technology in the Sendai Framework. Paragraph 36 (b) for example, requests: "Academia, scientific and research entities and networks to: focus on the disaster risk factors and scenarios, including emerging disaster risks, in the medium and long term; increase research for regional, national and local application; support action by local communities and authorities; and support the interface between policy and science for decision-making".

More specifically, paragraph 25 (g) states: "Enhance the scientific and technical work on disaster risk reduction and its mobilization through the coordination of existing networks and scientific research institutions at all levels and all regions with the support of the UNISDR Scientific and Technical Advisory Group in order to: strengthen the evidence-base in support of the implementation of this framework; promote scientific research of disaster risk patterns, causes and effects; disseminate risk information with the best use of geospatial information technology; provide guidance on methodologies and standards for risk assessments, disaster risk modelling and the use of data; identify research and technology gaps and set recommendations for research priority areas in disaster risk reduction; promote and support the availability and application of science and technology to decision-making; contribute to the update of the 2009 UNISDR Terminology on Disaster Risk Reduction; use post-disaster reviews as opportunities to enhance learning and public policy; and disseminate studies".

Science and technology have a fundamental role to play in forecasting disasters, building resilient infrastructure and accurately calculating disaster losses. Expanding the interface between science, technology and policy is therefore essential for effective disaster risk reduction. Science and technology stakeholders participated actively in the World Conference on Disaster Risk Reduction, including announcing the launch of a number of science and technology initiatives and commitments to support the implementation of the Sendai Framework.

Given the call in the Sendai Framework for science and technology-based DRR, in addition to a number of commitments made by the scientific and technology community at the Third UN World Conference on Disaster Risk Reduction, a need for a stronger partnership has been identified with a clearer direction and strategy for implementation as a mechanism to 'foster collaboration across global and regional mechanisms and institutions for the implementation and coherence of instruments and tools relevant to disaster risk reduction" around common goals and actions identified in the road map.

The International Conference on Science and Technology for DRR

In January 2016, UNISDR hosted the Science and Technology Conference on the Implementation of the Sendai Framework for Disaster Risk Reduction 2015–2030 (UNISDR 2016b, c). The main outcome of the conference, which was attended by more than 700 science and technology experts representing numerous scientific institutions and societies, as well as young scientists, was the launch of the Science and Technology Partnership, which consists of major scientific and technical institutes and organizations, research centers, networks and platforms, and UN science-based organizations working on the different disciplines of advancing science and technology for disaster risk reduction (UNISDR 2015b).

The principal goal of the Scientific and Technical Partnership and its Advisory Group is to provide scientific and technical expertise and guidance to strengthen the evidence-base in support for the implementation of Sendai Framework.

The conference also endorsed the development of a comprehensive 15-year road map to define the expected outcomes of the science and technology work under the four Sendai Framework priorities for action and ways to monitor progress and needs.

The 'Science and Technology Roadmap to Support the Implementation of the Sendai Framework'

The 'Science and Technology Roadmap to Support the Implementation of the Sendai Framework for Disaster Risk Reduction 2015–2030' includes expected outcomes, actions, and deliverables under each of the four priority of actions of the Sendai Framework (UNISDR 2016d). The science and technology community can then link to and plan around the implementation of the Roadmap (Table 1).

Work plans for several of the deliverables (with responsibilities, outputs and a timeline) in the Roadmap can then be developed as appropriate. These can be developed on a needs basis with identified partners with the support of the UNISDR Science and Technology Advisory Group.

The partnerships that have been developed both for the Third UN World Conference on Disaster Risk Reduction in March 2015 and the UNISDR Science and Technology Conference in January 2016 are a core part of implementation of the Roadmap. The science and technology partnerships and initiatives help to complement and strengthen collaboration among the partners, within their respective mandates and expertise.

There are also a number of cross-cutting actions such as capacity development, gender equity, citizen engagement, public-private sector partnership, and coherence or alignment with other post-2015 global agendas such as Sustainable Development Goals and climate change conventions, which will need to be linked with other stakeholders actions in the implementation of the Sendai Framework.

Table 1 Summary of the expected outcomes of the Science and technology road map

Sendai framework priority for action	Science and technology expected outcomes
1. Understanding disaster risk	1.1 Assess and update the current state of data, scientific and local and indigenous knowledge and technical expertise availability on disaster risks reduction and fill the gaps with new knowledge 1.2 Synthesize, produce and disseminate scientific evidence in a timely and accessible manner that responds to the knowledge needs of policy-makers and practitioners 1.3 Ensure that scientific data and information support are used in monitoring and reviewing progress towards disaster risk reduction and resilience building 1.4 Build capacity to ensure that all sectors and countries have access to, understand and can use scientific information for better informed decision-making
2. Strengthening disaster risk governance to manage disaster risk	2.1 Support a stronger involvement and use of science to inform policy- and decision-making within and across all sectors at all levels
3. Investing in disaster risk reduction for resilience	3.1 Provide scientific evidence to enable decision-making of policy options for investment and development planning
4. Enhancing disaster preparedness for effective response, and to "Build Back Better" in recovery, rehabilitation and reconstruction	4.1 Identify and respond to the needs of policy- and decision-makers at all levels for scientific data and information to strengthen preparedness, response and to "Build Back Better" in Recovery, Rehabilitation and Reconstruction to reduce losses and impact on the most vulnerable communities and locations

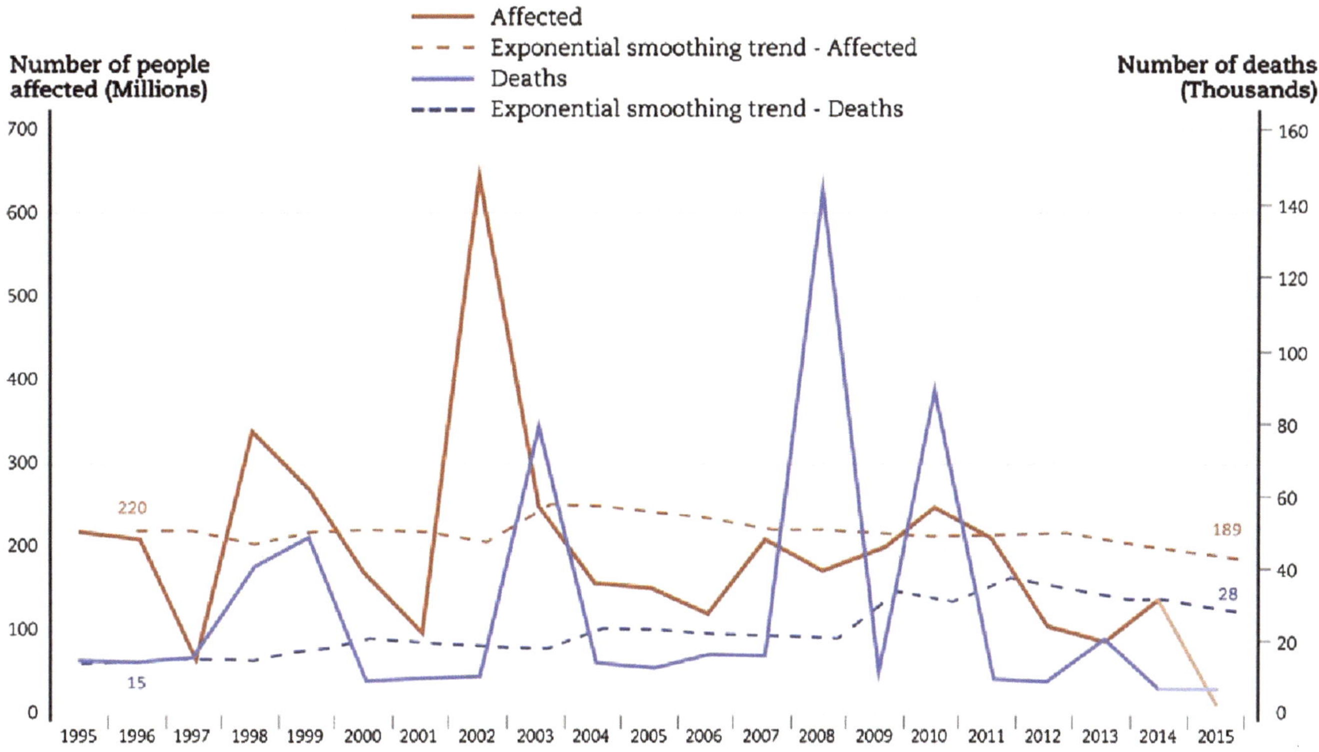

Fig. 1 Trends in the numbers of people affected and killed annually by weather-related disasters worldwide (1995–2015) (CRED 2016a)

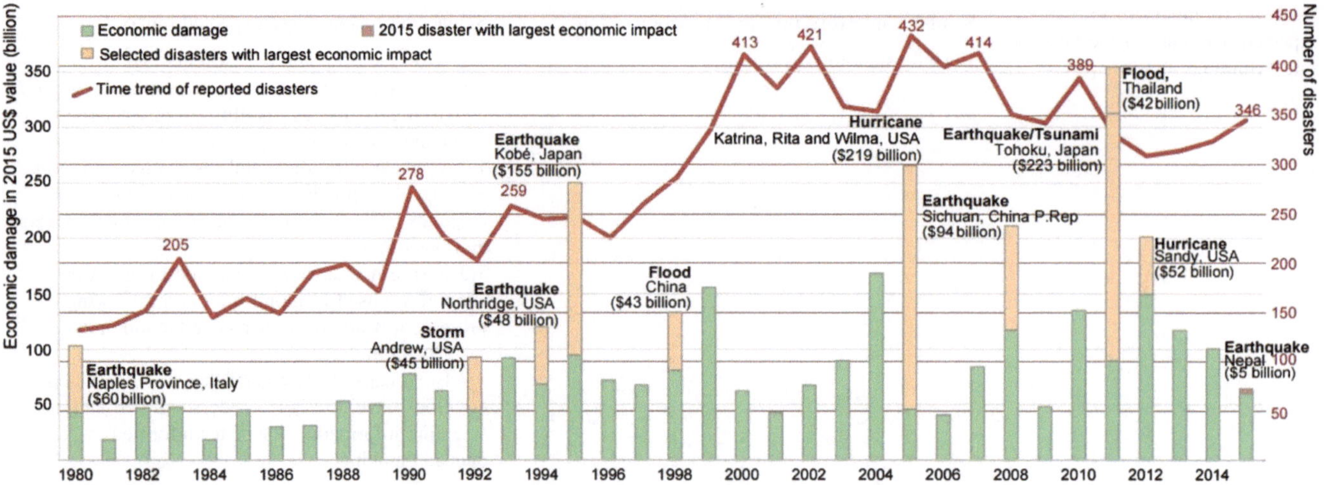

Fig. 2 Annual occurrence and reported economic damages from disasters 1980–2015 (CRED 2016b)

The Role of ICL Sendai Partnership

Enhancing international cooperation and the provision of means of implementation to support least developed countries and small islands in developing States in the implementation of the Sendai Framework and, in that context, making bilateral and multilateral development assistance programs risk-informed is essential.

Over the last twenty years, the majority of disasters (90%) have been caused by floods, storms, heatwaves and other weather-related events as outlined in Fig. 1 (CRED 2016a).

Most of these disasters can cause landslides, which in turn cause hundreds of billions of dollars in damage (Fig. 2) and hundreds of thousands of deaths and injuries each year (CRED 2016b). In the US alone, it has been estimated that landslides cause in excess of US$1 billion in damages on average per year, though that is considered a conservative figure and the real level could be at least double. Given this, it is important to understand the science of landslides: why they occur, what factors trigger them, the geology associated with them, and where they are likely to happen.

The International Consortium on Landslides (ICL) 2015–2025 promotes global understanding and reduction of landslide disaster risk. The Sendai Partnerships will contribute significantly to the implementation of the science and technology roadmap by helping to provide practical solutions and tools, education and capacity building, and communication and public outreach to reduce landslides risks. As such, they will contribute to the implementation of the goals and targets of the Sendai Framework, particularly on understanding disaster risks, including vulnerability and exposure to integrated landslide-tsunami risk.

UNISDR fully support the work of the Sendai Partnerships and the community of practice on landslides risks, and welcomes the 4th World Landslide Forum to be held in 2017 in Slovenia, which aims to strengthen intergovernmental networks and the international programme on landslides.

References

CRED Crunch (2016a) Disaster Data: A Balanced Perspective. Issue No. 41, February. Available at http://reliefweb.int/sites/reliefweb.int/files/resources/CredCrunch41.pdf. Accessed 19 Oct 2016

CRED Crunch (2016b) What is the human cost of weather-related disasters (1995–2015) Issue No. 42, April. Available at http://reliefweb.int/sites/reliefweb.int/files/resources/CredCrunch42.pdf. Accessed 19 Oct 2016

UNISDR (2015) Sendai Framework for Disaster Risk Reduction 2015–2030. In: UN world conference on disaster risk reduction, 2015 March 14–18, Sendai, Japan. Geneva: United Nations Office for Disaster Risk Reduction. Available at http://www.unisdr.org/files/43291_sendaiframeworkfordrren.pdf. Accessed 19 Oct 2016

UNISDR (2015b) Terms of Reference of the Scientific and Technical Partnership for the implementation of the Sendai Framework for Disaster Risk Reduction 2015–2030, 13 July 2015. Available at http://www.preventionweb.net/files/45270_torofunisdrstpartnership.pdf. Accessed 19 Oct 2016

UNISDR (2016a) Implementation of the Sendai Framework for Disaster Risk Reduction 2015–2030. Report of the Secretary-General. Available at https://unisdr.org/files/resolutions/N1624116.pdf. Accessed 19 Oct 2016

UNISDR (2016b) Launching UNISDR Science and Technology Partnership and The Science and Technology Road Map to 2030. Concept Note. Geneva. United Nations office for Disaster Risk Reduction. Available at http://www.preventionweb.net/files/45270_conceptnoteunisdrstconference2729ja.pdf. Accessed 19 Oct 2016

UNISDR (2016c) UNISDR Science and Technology Conference on the implementation of the Sendai Framework for Disaster Risk

Reduction 27–29 January 2016, Geneva International Conference Centre. Summary Outcomes. Available at http://www.preventionweb.net/files/45270_unisdrscienceandtechnologyconferenc[2].pdf. Accessed 19 Oct 2016

UNISDR (2016d) The Science and Technology Roadmap to Support the Implementation of the Sendai Framework for Disaster Risk Reduction 2015–2030. Available at http://www.preventionweb.net/files/45270_unisdrscienceandtechnologyroadmap.pdf. Accessed 19 Oct 2016

United Nations Educational, Scientific and Cultural Organization (UNESCO)—UNESCO's Contribution to the Implementation of UNISDR's Global Initiative and ICL

Giuseppe Arduino, Rouhban Badaoui, Soichiro Yasukawa, Alexandros Makarigakis, Irina Pavlova, Hiroaki Shirai, and Qunli Han

Abstract

UNESCO operates at the interface between natural and social sciences, education, culture and communication, playing a vital role in constructing a global culture of resilient communities. UNESCO assists countries to build their capacities in managing disaster and climate risk and with their ability to cope with disasters. The Organization provides a forum for governments to work together and it provides essential scientific and practical advice in disaster risk reduction. UNESCO's programmes in relation to the International Strategy for Disaster Reduction (ISDR) cut across all of its areas of competence (education, natural and social sciences, culture and communication). Working alone or in collaboration with both UN Agencies and other scientific entities, UNESCO has been a catalyst for international, inter-disciplinary cooperation in many aspects of disaster risk reduction and mitigation. Since the establishment of ICL in 2002, UNESCO has continuously supported ICL's activities as a part of its contributions to ISDR, namely the Hyogo and now Sendai Frameworks for action.

Keywords

UNESCO • Hyogo framework • Sendai framework

UNESCO Disaster Risk Reduction Activities

UNESCO has been strongly involved in disaster risk reduction (DRR) since the 1960s, with studies on earthquakes and oceanography. Its programme has since expanded into other categories of hazards and many areas, as it pursues multidisciplinary actions to study natural hazards and mitigate their effect.

In the 1990–2000s, UNESCO kept supporting natural hazard-related studies and mitigation activities during the International Decade for Natural Disaster Reduction (IDNDR) proclaimed by the United Nations (Clayson 1991).

UNESCO promotes scientific exchange and collaborative efforts in order to establish effective early warning systems for different hazards such as tsunamis, landslides, volcanoes, earthquakes, floods and droughts. UNESCO helps Member States to collectively achieve effective early warning and hazard-monitoring, helps coordination between existing research centers and educates communities at risk about preparedness measures, including setting up warning and emergency response Standard Operating Procedures and community drill exercises. UNESCO promotes community-based approaches in the development of response plans and awareness campaigns, which strongly involve educational institutions and local community actors.

G. Arduino (✉) · A. Makarigakis · H. Shirai
Division of Water Sciences, Natural Sciences Sector, UNESCO, 7, Place de Fontenoy, 75352 Paris 07 SP, France
e-mail: g.arduino@unesco.org

R. Badaoui
Natural Sciences Sector, UNESCO, 7, Place de Fontenoy, 75352 Paris 07 SP, France

S. Yasukawa · I. Pavlova · Q. Han
Division of Ecological and Earth Sciences, Natural Sciences Sector, UNESCO, 7, Place de Fontenoy, 75352 Paris 07 SP, France
e-mail: s.yasukawa@unesco.org

K. Sassa et al. (eds.), *Advancing Culture of Living with Landslides*,
DOI 10.1007/978-3-319-59469-9_7

The scientific and technical work in disaster reduction is essentially promoted by four of the Organization's International and Intergovernmental Science Programmes, namely the International Geoscience and Geoparks Programme (IGGP) (UNESCO 2016d), the Man and Biosphere (MAB) Programme (UNESCO 2016e), the programmes of the Intergovernmental Oceanographic Commission (IOC) (IOC-UNESCO 2013) and the International Hydrological Programme (IHP) (UNESCO 2016c). A cross-sectoral group working on Disaster Risk Reduction serves to coordinate the organization's work on DRR (UNESCO 2016b).

Each Programme focuses on particular hazards. IGGP deals mostly, but not exclusively, with geohazards (including earthquakes, volcanic eruptions, landslides), and contributes to multihazard approaches, the IHP and MAB specialize in hydro-meteorological hazards, such as floods, drought and desertification, IOC's activities concentrate on tsunamis and storm surges. UNESCO promotes multi-hazard assessments of disaster risks to identify hotspots and aid decision-making for policy and planning, and recognizes the need to understand exposure and vulnerability as key components of risk assessments. In addition, UNESCO actively contributes to understanding and measuring natural hazard resilience, in an effort to build safer and more resilient communities.

Through these undertakings, UNESCO also contributes to the three global observing systems, the Global Ocean Observing System (GOOS) hosted in IOC (GOOS 2016), the Global Climate Observing System (GCOS) (GCOS 2016) and the Global Terrestrial Observing System (GTOS) (GTOS 2011). These are joint initiatives of UNESCO, IOC, Food and Agriculture Organization (FAO), World Meteorological Organization (WMO), United Nations Environment Programme (UNEP) and the International Council for Science (ICSU).

UNESCO advocates for accessible, safe and inclusive educational facilities by promoting school safety assessments and providing Member States with practical information on the risks their schools are exposed to (using a multi-hazard approach), as well as on priority areas for intervention and an estimation of the investments needed for upgrading school facilities. (UNESCO 2016g). Furthermore, UNESCO's Education Sector promotes a variety of materials on information, education and public awareness and the inclusion of disaster risk reduction in formal and informal education.

Through its Culture Sector, UNESCO participates in the operations undertaken to safeguard the cultural heritage, monuments and other works of art which are at risk. Finally, in the framework of the Management of Social Transformations (MOST) Programme and other major activities falling under Social and Human Sciences Sector, approaches to the social aspects and perception of risks and risk prevention and preparedness are facilitated and exchanges encouraged between scientists, community leaders and policy makers (UNESCO 2016f).

UNESCO's Contribution to the Hyogo Framework for Action and Sendai Framework for Disaster Risk Reduction

From 18 to 22 January 2005 in Kobe (Hyogo, Japan), the World Conference on Disaster Reduction took stock of progress in disaster risk reduction accomplished since the Yokohama Conference of 1994 and made plans for the next ten years, encapsulated in the Hyogo Framework of Action 2005–2015 (UNISDR 2005a) and the Hyogo Declaration (UNISDR 2005b).

The Hyogo Declaration and Framework for Action include many of UNESCO's concerns in the field of disaster reduction: capacity-building, research on natural hazards, interdisciplinary approaches, and integration of disaster reduction into developmental plans. In cooperation with the Inter-Agency Secretariat of the International Strategy for Disaster Reduction (UNISDR), and other members of the Inter-Agency Task force on Disaster Reduction, UNESCO committed to playing an active part in the follow-up to the World Conference on Disaster Reduction held in Kobe.

In its contribution to the implementation of the Hyogo Framework for Action, UNESCO sought to promote a better understanding of the distribution in time and space of natural hazards such as earthquakes, landslides, volcanic eruptions, floods, tsunamis and of their intensity, to help set up reliable early warning systems, to encourage rational land-use plans, to secure the adoption of suitable building design, to protect educational buildings and cultural monuments, to strengthen environmental protection for the prevention of natural disasters, to enhance preparedness and public awareness through education and training, and to foster post-disaster investigation, recovery and rehabilitation, notably for educational buildings and cultural sites (UNESCO 2016b).

UNESCO's work is based on a multidisciplinary approach. Nearly half of all DRR-related activities were focused on multihazards. UNESCO has established tsunami, flood and drought early warning systems and set up earthquake and landslide monitoring systems. It has provided technical assistance and trainings in all hazards worldwide (Fig. 1).

According to multiple country indicators, such as disaster risk and vulnerability indexes (WorldRisk Index), populations in Asia and Pacific and Latin America and the Caribbean regions are the most exposed to natural hazards. Due to the high number of disaster occurrences and their severe impact in these regions, much of UNESCO's risk reduction efforts were carried out in Asia and Latin America.

Almost half of the activities implemented during the 2005–2015 Hyogo Framework for Action included a raising

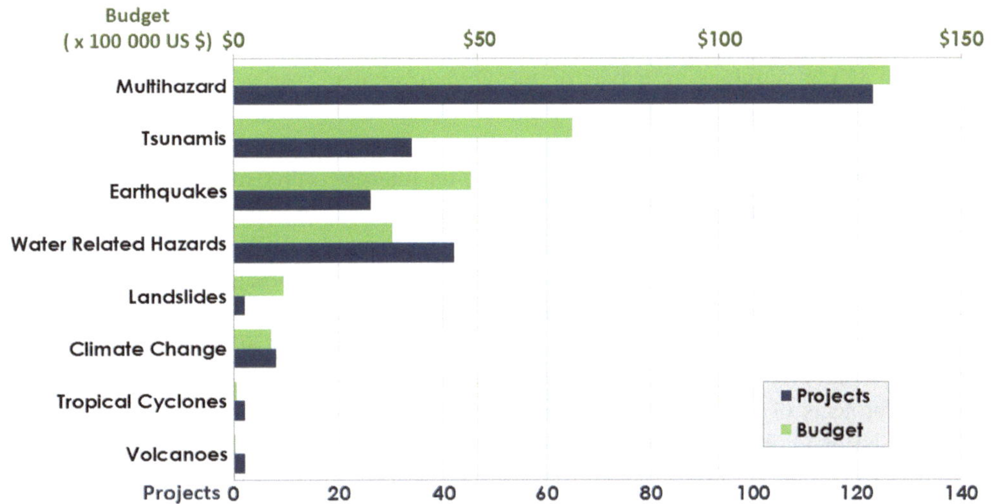

Fig. 1 Distribution of DRR projects by hazards with their budgets for the 2005–2015 period, based on UNESCO reporting system

awareness component. Networking was the focus of at least a quarter of UNESCO's work, whereas research, technical support and education activities were presented in more than 20% of the projects in DRR. Supporting the establishment of a DRR-related policies component was included in one out of 6 activities implemented, while technical training was provided in 15% of DRR-related projects worldwide.

The main beneficiary of UNESCO's projects was the general public, as they were involved in more than 90% of DRR projects. Policy-makers were the primary beneficiaries of at least 80% of the activities implemented. The scientific community was integrated in nearly half of the activities, while 35% of all DRR projects included collaboration with international, regional, and local NGOs. UNESCO cooperated with other UN agencies in one out of three projects implemented. The private sector was involved in less than 10% of DRR projects.

At the 3rd World Conference on Disaster Risk Reduction (WCDRR), which was convened by the UN ISDR and hosted by Japan in Sendai from 14 to 18 March 2015, UNESCO committed to operating in line with the "Sendai Framework for Disaster Risk Reduction 2015–2030" (UN 2015a) and in accordance with its four Priorities for Action:

Priority 1: Understanding disaster risk
Priority 2: Strengthening disaster risk governance to manage disaster risk
Priority 3: Investing in disaster risk reduction for resilience
Priority 4: Enhancing disaster preparedness for effective response and to Build Back Better in recovery, rehabilitation and reconstruction.

As outcomes of numerous WCDRR working sessions in which UNESCO was actively participating, the Organization committed to a number of initiatives. UNESCO supports the establishment of the international partnership of Science and Technology to support the implementation of the Sendai Framework for Disaster Risk Reduction 2015–2030 by mobilising relevant institutions, networks and initiatives. UNESCO, on behalf of the Global Alliance for Disaster Risk Reduction and Resilience in the Education Sector, committed to providing tools and technical support to interested Governments for school safety implementation according to the three pillars of the Worldwide Initiative for Safe Schools. Among various actions on Education and Knowledge in Building a Culture of Resilience, UNESCO will work further to strengthen the link between Disaster Risk Reduction Education with Education for Sustainable Development and Climate Change Education. Regarding the Integrated Water Resource Management, UNESCO committed to strengthening education and capacity building in order to help Member States better cope with the hydrological extremes of floods and droughts. UNESCO agreed to highlight the need to further enhance our understanding of climate-related mega disasters, their underlying factors and impact on livelihoods, and to identify ways to adapt and respond to these disasters in the future. UNESCO actively participated in the working session on Resilient Cultural Heritage, which provided a vision for the protection and safeguarding of cultural heritage in disasters and conflicts and promoted its recognition as an important element of community resilience and local development. UNESCO is among other UN agencies which continue and enhance the collaborative initiative between member states and UN agencies to develop a strategy for effective measures for building resilience (UN 2015b).

UNESCO's Partnership with ICL

ICL was founded in 2002 during the UNESCO-Kyoto University Joint Symposium on "Landslide Risk Mitigation and Protection of Cultural and Natural Heritage" as an activity of UNESCO project "IGCP-425 Landslide hazard assessment and mitigation for cultural heritage sites and other locations of high societal value". Over the past fifteen years since its establishment, UNESCO has continuously supported ICL's activities as a part of the Organization's strategic contribution to ISDR, namely to the Hyogo and now Sendai frameworks for action.

The Kyoto Declaration on "Establishment of an International Consortium on Landslides (ICL)" was signed on 21 January 2002. The First Session of the Board of Representatives (BOR) of ICL was organized at UNESCO Headquarters on 19–21 November 2002. Initial members of ICL agreed to launch the International Programme on Landslides (IPL) and adopted eight coordinating projects and 14 member projects of IPL. Since the establishment of ICL in 2002, UNESCO has continuously supported ICL/IPL activities (Sassa et al. 2005).

The UNITWIN (University Twinning and Networking) Cooperation Programme on Landslide Risk Mitigation for Society and the Environment was established by ICL in cooperation with UNESCO and Kyoto University in March 2003. The programme was extended to the UNESCO/KU/ICL Landslide and Water-related Disaster Risk Management for Society and the Environment Cooperation Programme to include water-related disaster and also disaster risk management in 2010 (ICL 2015).

In 2005, a Letter of Intent to promote further joint global activities on disaster reduction and risk prevention was signed by heads of seven global stakeholders, namely UNESCO, WMO, FAO, UNISDR, UNU, ICSU, and WFEO. Based on this Letter of Intent, ICL, UNESCO, WMO, FAO, UNISDR, UNEP, UNU, and Kyoto University jointly organized the Round Table Discussion on 18–20 January 2006 in Tokyo, Japan. The 2006 Tokyo Action Plan, adopted as a result of the Round Table Discussion, aimed at promoting further joint global activities in disaster risk reduction and risk prevention through "Strengthening research and learning on landslides and related earth system disasters for global risk preparedness". This Action Plan has developed IPL into a new global International Programme on Landslides.

Further cooperation between UNESCO and ICL on the implementation of the 2006 Tokyo Action Plan on Landslides was formalized in August 2006 by a Memorandum of Understanding. Furthermore, ICL was approved to be a NGO having operational relations with UNESCO in April 2007. It was reclassified as an Organization having a consultative status and partnership with UNESCO in March 2012.

UNESCO actively participated in, and supported the organization of, the First (Tokyo in 2008), Second (Rome in 2011) and Third (Beijing in 2014) World Landslide Forums. Experts, scholars and officials from across the world attended Forums.

The initiative of identifying World Centres of Excellence (WCoE) on Landslide Risk Reduction was introduced in the 2006 Tokyo Action Plan by the IPL Global Promotion Committee. WCoEs are identified at the World Landslide Forum organized every 3 years within eligible organizations, such as universities, institutions, NGOs, government ministries and local governments, contributing to "Risk Reduction for Landslides and Related Earth System Disasters" (ICL 2012). Fifteen World Centres of Excellence (WCoEs) for 2014–2017 were identified at the Third World Landslide Forum in June 2014 at Beijing, China. Certificates of WCoEs were awarded from Ms. Irina Bokova, the Director-General of UNESCO to each leader of WCoEs in Beijing (Fig. 2) (Sassa 2015a).

UNESCO supports the publication of Landslides: Journal of the International Consortium on Landslides. The Journal is the core project of the IPL, managed by the IPL Global Promotion Committee (ICL 2012). UNESCO contributes to the Editorial and Advisory Boards. Through its various communication networks, UNESCO ensures wide dissemination of this journal. UNESCO circulates journal issues via IHP national committees, in particular in the Asia Pacific regions.

Among other activities, UNESCO is involved in the evaluation of the IPL Awards for Success. This award is given at each World Landslide Forum for up to three successful projects implemented within IPL, based on an evaluation of the previous three years' activities. UNESCO has chaired the evaluation committee of the IPL Awards for Success since 2011 (ICL 2012).

Way Forward with ICL

During the Third UN World Conference on Disaster Risk Reduction (WCDRR) in March 2015, ICL took the initiative of organizing, together with IPL, the Japanese Ministry of Land, Infrastructure, Transport and Tourism (MLIT), UNESCO and others, the Working Session "Underlying Risk Factors". As an outcome of this Working session, the "ISDR-ICL Sendai Partnerships 2015–2025 for global promotion of understanding and reducing landslide disaster risk" was signed by a number of key international organizations, including UNESCO (Fig. 3) (Sassa 2015b).

These Partnerships, a sound global platform, will be mobilized in the coming decade to pursue prevention, and to provide practical solutions, education, communication, and public outreach to reduce landslide disaster risk. Partners have agreed on a number of initial fields of cooperation in research and capacity building, coupled with social and financial investment. These common interests will focus on new initiatives to study research frontiers in understanding landslide disaster risk, on development of people-centered early warning technologies for landslides, hazard and vulnerability mapping, and international teaching tools, as well as open communication of these results with society.

UNESCO is committed to the promotion and implementation of the "Sendai Partnerships 2015–2025 for Global Promotion of Understanding and Reducing Landslide

Fig. 2 (*Top*) A certificate of the World Centre of Excellence on Landslide Risk Reduction (WCoE) 2014–2017 was conferred by Director-General of UNESCO, Irina Bokova on Dwikirita Karnawati, Gadjah Mada University, Indonesia, leader of one of 15 identified WCoEs for 2014–2017. (*Bottom*) An IPL Award for Success was given to the IPL project leader (Ogbonnaya Igwe) from the Department of Geology, University of Nigeria. S. Diop received the US$3000 award from Director-General of UNESCO on behalf of O. Igwe (Sassa 2015a)

Fig. 3 The signers from the first 16 signatory organizations with ICL officers after the signing of the ISDR-ICL Sendai Partnerships 2015–2025 document (Sassa 2015b)

Disaster Risk". In this context and according to its mandate in DRR, UNESCO will continue to support the development of global, regional and national multi-hazard early warning systems to natural hazards, including landslides; improve the scientific basis to develop technologies and tools to cope with landslides as part of multi-risk identification and management; strengthen capacity for floods and landslide monitoring and forecasting; enhance the capacity of schools and local communities to prepare for and respond to environmental and other threats and disasters; provide policy support and technical assistance for capacity enforcement in landslide disaster risk reduction; enhance research, partnerships and international scientific cooperation on landslide disaster risk reduction; and collaborate with international partners and cross sectors, UNESCO field Offices, UNESCO Category 2 Centres and UNESCO Chairs on the topic of landslide disaster risk reduction.

An action plan of the Sendai Partnerships will be discussed during the 16th Board of Representatives of ICL and 12th IPL Global Promotion Committee, which will be held at the UNESCO Headquarters in Paris on 15–18 November 2016. The action plan would be adopted at the forth World Landslide Forum to be held in 2017 in Slovenia with continuous support of UNESCO.

References

Clayson A (1991) Standing up to natural disasters. UNESCO contributions to the International Decade for Natural Disaster Reduction 1990–2000. UNESCO, 51 pp

GCOS (2016) GCOS-Global Climate Observing System. Available at: http://www.wmo.int/pages/prog/gcos/. Accessed 3 Aug 2016

GOOS (2016) The Global Ocean Observing System. Available at: http://www.ioc-goos.org/. Accessed 3 Aug 2016

GTOS (2011) GTOS-Global Terrestrial Observing System. Available at: http://www.fao.org/gtos/. Accessed 3 Aug 2016

ICL (2012) ICL Leaflet 2012. Available at: http://icl.iplhq.org/category/icl/leaflet-and-publications/. Accessed 3 Aug 2016

ICL (2015) UNITWIN Programme. Available at: http://icl.iplhq.org/category/icl/unitwin-programme-icl/. Accessed 3 Aug 2016

IOC-UNESCO (2013) Intergovernmental Oceanogaphic Commission. Available at: http://www.ioc-unesco.org/. Accessed 8 Aug 2016

Sassa K (2015a) The Third World Landslide Forum, Beijing 2014. Landslides 12(1):177–192

Sassa K (2015b) ISDR-ICL Sendai Partnership 2015–2025 for global promotion of understanding and reducing landslide disaster risk. Landslides 12(4):631–640

Sassa K, Fukuoka H, Wang F, Wang G (eds) (2005) Landslides—risk analysis and sustainable disaster management. In: Proceedings of the 1st General Assembly of the International Consortium on Landslides. Springer, 385 pp

UN (2015a) Sendai framework for disaster risk reduction 2015–2030. Available at: http://www.unisdr.org/we/inform/publications/43291. Accessed 8 Aug 2016

UN (2015b) Proceedings of the 3rd UN World conference on Disaster Risk Reduction. Available at: http://www.unisdr.org/we/inform/publications/45069. Accessed 8 Aug 2016

UNESCO (2016a) Contributing to the implementation of the hyogo framework for action 2005–2015. Available at: http://www.unesco.org/new/en/natural-sciences/priority-areas/sids/disaster-preparedness/. Accessed 8 Aug 2016

UNESCO (2016b) Disaster risk reduction. Available at: http://www.unesco.org/new/en/natural-sciences/special-themes/disaster-risk-reduction/. Accessed 8 Aug 2016

UNESCO (2016c) Hydrology (IHP). Available at: http://en.unesco.org/themes/water-security/hydrology. Accessed 8 Aug 2016

UNESCO (2016d) International Geoscience and Geoparks Programme (IGGP). Available at: http://www.unesco.org/new/en/natural-sciences/environment/earth-sciences/international-geoscience-and-geoparks-programme/. Accessed 8 Aug 2016

UNESCO (2016e) Man and the Biosphere Programme. Available at: http://www.unesco.org/new/en/natural-sciences/environment/ecological-sciences/man-and-biosphere-programme/. Accessed 8 Aug 2016

UNESCO (2016f) Management of Social Transformations (MOST) Programme. Available at: http://www.unesco.org/new/en/social-and-human-sciences/themes/most-programme/. Accessed 8 Aug 2016

UNESCO (2016g) School safety. Available at: http://www.unesco.org/new/en/natural-sciences/special-themes/disaster-risk-reduction/school-safety/. Accessed 8 Aug 2016

UNISDR (2005a) Hyogo framework for action 2005–2015. Building the resilience of nations and communities to disasters. Available at: http://www.unisdr.org/2005/wcdr/intergover/official-doc/L-docs/Hyogo-framework-for-action-english.pdf. Accessed 8 Aug 2016

UNISDR (2005b) Hyogo declaration. Available at: http://www.unisdr.org/2005/wcdr/intergover/official-doc/L-docs/Hyogo-declaration-english.pdf. Accessed 8 Aug 2016

United Nations University (UNU)—The United Nations University: Research and Policy Support for Environmental Risk Reduction

Jakob Rhyner

Abstract

The United Nations University (UNU) was established in 1973 in Tokyo, Japan, as the academic arm of the United Nations. In its role as a think tank for the UN system it engages in policy-relevant research to generate science-based knowledge and solutions to urgent global challenges across a variety of comprehensive themes. UNU's research focuses on three broad thematic clusters: Peace and Governance, Global Development and Inclusion as well as Environment, Climate and Energy. In addition, research is complemented by important themes in science, technology and innovation. Research is carried out by a global network of institutes and programmes (see Fig. 1), each with a specific thematic focus. This paper provides an overview of United Nations University, with particular attention to one of its institutes, namely the Institute for Environment and Human Security (UNU-EHS) in Bonn, Germany.

Keywords

United Nations University • Global think tank • Research and education • Policy • Global development • UNU-EHS • Environment • Human security • Risk

The UNU: Founding and Conception

The United Nations University (UNU) functions as a global think tank engaged in research, postgraduate teaching and the dissemination of knowledge that is aimed at furthering the purposes and principles of the United Nations Charter. Following the proposition by United Nations Secretary-General U Thant in his 1969 Annual Report that a "United Nations University, truly international in character and devoted to the Charter objectives of peace and progress" be created, the UNU was founded as the academic arm of the United Nations in 1973.

Headquartered in Tokyo, Japan, the mission of the UNU is centrally defined by its effort to "contribute, through collaborative research and education, dissemination, and advisory services, to efforts to resolve the pressing global problems of human survival, development and welfare that are the concern of the United Nations, its Peoples and Member States" (cited in UNU Strategic Plan 2015–2019). In carrying out this mission, the UNU works with leading universities and research institutes in UN Member States, functioning as a bridge between the international academic community and the United Nations system.

UNU Research and Objectives

Overview

In its role as a think tank for the United Nations system, the UNU conducts focused, policy-relevant research that offers objective, science-based perspectives in the service of policy debate and development. UNU researchers aim to develop innovative approaches to today's pressing problems, to

J. Rhyner (✉)
United Nations University Institute for Environment and Human Security, UN Campus, Platz der Vereinten Nationen 1, 53113 Bonn, Germany
e-mail: rhyner@ehs.unu.edu

K. Sassa et al. (eds.), *Advancing Culture of Living with Landslides*,
DOI 10.1007/978-3-319-59469-9_8

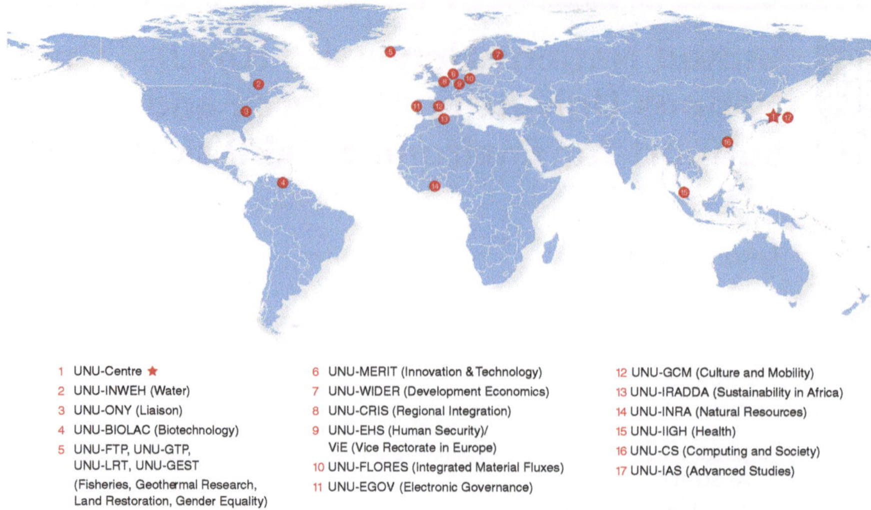

Fig. 1 The Global UNU Campus

provide proactive analyses of emergent challenges, and to propose feasible coping mechanisms and solutions. Guided by the UNU Strategic Plan, which prescribes the basic direction, agenda, and priorities of this research, UNU produces evidence-based, policy-relevant research outcomes that seek to contribute to the UN policymakers' toolkit and inform deliberations on key policy issues.

Research Agenda and Strategy

Following its Charter mandate, the UNU investigates various aspects of global sustainability and human well-being, with a particular emphasis on the challenges and needs of developing countries. This encompasses issues of peace and human security, governance and human rights, human health, gender equality, development and poverty reduction, environmental protection and natural resources management, climate change, energy, the impact of science and technology, and other important contemporary concerns.

The focus of UNU's research has continually evolved in response to the changing state of global ecosystems, the social impact of scientific discoveries, and the evolution of the international system. UNU aims to maintain the high quality of its outputs by focusing its research activities into specified thematic areas. The three overarching thematic clusters prescribed by the *UNU Strategic Plan 2015–2019* are:

- Peace and Governance;
- Global Development and Inclusion;
- Environment, Climate and Energy.

These clusters incorporate gender dimensions, and are complemented by relevant, cross-cutting research themes in the areas of science, technology, and innovation.

A 'Truly International University'

The work and research of UNU is carried out by a decentralized network of institutes and programmes spread across a range of international sites. The researchers across UNU's varied units collaborate with each other and diverse external partners such as leading universities and research centers, think tanks and other UN entities, to more effectively respond to the range of today's multifaceted global challenges.

As Fig. 1 illustrates, UNU's campus spans a range of countries and world regions. With its headquarters in Tokyo and its Vice-Rectorate for Europe located in Bonn, Germany (the only establishment of this kind outside Japan), the UNU continues to pursue partnerships and the establishment of UNU sites in developing countries.

Comprising a total of ca. 650 staff and personnel members in 2015, the UNU is characterised by a diversity of institutional approaches across its global sites. Thus, while some institutes strive to build capacity from within, and therefore emphasize internally conceived projects, others place a stronger focus on external proposals. Although the majority of UNU projects comprise a small team of researchers (who collaborate across UNU's diverse sites), externally proposed and financed projects occasionally involve multiple institutions and large networks of researchers.

Dissemination

UNU disseminates the findings of its research through a variety of channels: these include books, reports, policy briefs, journal articles, and the internet as well as conferences, seminars, lectures, and workshops. The goal is to provide practical knowledge, in a timely manner and convenient form, to those who can make best use of it. Target audiences include UN officials, government leaders, scientists, scholars, public and private sector decision-makers, and on-the-ground practitioners.

UNU's Educational Focus

Beyond its research and investigative work, educational programmes pose a centrepiece of UNU's work as the academic arm of the United Nations. UNU has established and institutionalised a number of master's programmes and doctoral fellowships across a range of international sites. In the Netherlands, for example, the double M.Sc. programme between UNU-MERIT and Maastricht University in Public Policy and Human Development was ranked number one in the category of Political Science and Public Administration, adding to the success of that Institute's internationally renowned joint doctoral programme. In Japan, UNU has partnered with the University of Tokyo to develop a joint diploma programme M.Sc. degree aimed at training future leaders in the area of sustainability science. In Germany, the Technical University of Dresden, recently designated by the German federal and state governments as a university of excellence, is partnering with UNU-FLORES to offer an innovative joint doctoral programme on the integrated management of water, soil and waste. In Bonn, UNU has developed an entirely new, fully accredited joint M.Sc. degree programme with the University of Bonn, combining the areas of geography, environmental change, and human security. The UNU programmes in Iceland offer education and training in the areas of fisheries (UNU-FTP), geothermal energy (UNU-GTP), and soil erosion (UNU-LRT), the latter being of particular relevance to landslide problems.

Setting of UNU-EHS in UNU

Research on environmental risks is carried out at several UNU institutes, e.g., at the Institute for Advanced Sustainability (UNU-IAS). In the following we concentrate on the Institute for Environment and Human Security (UNU-EHS), which has environmental risks as its core mandate. UNU-EHS was established in Bonn in 2003. The institute shares its administration, information technology and communication units with the UNU Vice-Rectorate in Europe (UNU-ViE), which also hosts the programme UNU-SCYCLE, focused on the development of sustainable production, consumption/usage, and disposal of ubiquitous goods, with a special focus on electrical and electronic equipment.

UNU-EHS is an integral part of the UNU system. It receives its core funding from the German Ministry for Education and Research and the State of Northrhine Westphalia.

Mission and Vision of UNU-EHS

The mission of UNU-EHS is to act as a think tank for the advancement of human security and resilience through knowledge-based approaches to handle environmental risks. UNU-EHS spearheads UNU's policy-oriented research and capacity building activities in the interdisciplinary field of risk and vulnerability, including knowledge-based adaptation and resilience strategies. The institute's mission is based on the observation that worldwide:

- the occurrence of extreme sudden-onset and slow-onset anthropogenic and natural hazards is on the rise;
- due to global environmental, social, economic and political changes, an increasing number of people are exposed and becoming more vulnerable to the impacts of these hazards;
- these impacts increasingly influence development processes in a negative way.

This development is caused by a multitude of global, regional and local drivers. To mention just a few: global warming, urbanization and other migration processes, conflicts, and economic inequalities.

Two basic guiding principles form the frame of the approach UNU takes when discussing environmental risks:

- When talking about "environmental" risk, the notion "environmental" does not claim that the risks are "caused" by the (biophysical) environment, or that the environment is even the main driver. Risks always occur from an interaction of the biophysical environment and the socio-economic, cultural and political settings. It is only with this comprehensive view that risks can be discussed with the goal of formulating appropriate comprehensive solutions.
- A medium to long-term development perspective should provide the frame for the identification of risk reduction solutions and the evaluation of, possibly competing, options. Without such an underlying perspective, risk mitigation solutions are prone to be of a short term, non-sustainable nature, and may prove to become inappropriate and obsolete in the longer run.

More specifically UNU-EHS focuses on four Programmes:

- Vulnerability Assessment, Risk Management and Adaptive Planning (*VARMAP*): Methods and concepts for risk and vulnerability assessment; institutional and planning aspects of environmental risk, with particular focus on risks in rural-urban interfaces;
- Environmental Vulnerability and Ecosystem Services (*EVES*): Ecosystem aspects of environmental risks;
- Environmental Migration, Social Vulnerability and Adaptation (*EMSVA*): Impact of climate change on environmental risks;
- Education programmes (*EduSphere*).

These programmes are selective and there are several areas of environmental risks that are not covered by any of the four programmes, e.g., health aspects, which are the focus of UNU-IIGH (International Institute for Global Health) in Kuala Lumpur, Malaysia.

Positioning in View of Global Policy Processes

The global policy agenda is characterized by at least five UN-led processes, which have all culminated in major conferences in 2015/16. These include (in chronological order):

1. Sendai Framework for Disaster Risk Reduction 2015–2030 (SFDRR)
2. Addis Ababa Action Agenda (AAAA)
3. The Agenda 2030 (Sustainable Development Goals, SDGs)
4. Paris Agreement (UNFCCC 21st Conference of the parties)
5. Outcomes of the World Humanitarian Summit.

The Sendai process and the climate negotiations are of direct thematic relevance to UNU-EHS, and the institute is involved in multiple ways. For the Sendai Framework, besides projects relevant to the process, UNU-EHS is represented in UN interagency teams for the implementation as well as in the organizing committee for scientific support processes, e.g., in the organizing committee of the *UNISDR Science and Technology Conference on the implementation of the Sendai Framework for Disaster Risk Reduction 2015–2030*.

With regards to climate negotiations, the EMSVA programme in particular draws on a long-term collaboration with the Munich Climate Insurance Initiative (MCII), and involvements in Loss and Damage projects, science support seminars with negotiators, etc.

The Sustainable Development Goals process requires scientific support, mainly in the development of indicators and assessment and evaluation processes. A successful step has been made, e.g., by the conference "Measuring Sustainable Development", co-organized by UNU-ViE and the German Research Foundation (DFG) in New York in April 2015.

UNU-EHS Programme Overview

The *VARMAP* programme develops and applies conceptual frameworks, theoretical approaches and scientific methods to assess, monitor and understand risk and vulnerability in the context of natural hazards, environmental change and societal transition. The programme explores new opportunities for risk reduction and evaluates competing adaptation options against a multi-dimensional set of criteria. In doing so, VARMAP examines ways to strengthen adaptive governance, i.e., the continuous adaptation of institutions which is necessary for any successful adaptation process and goes beyond the adjustment of physical infrastructure and economic practices.

The programme is particularly interested in understanding the synergies but also rifts between state and non-state action in adaptation processes and promotes integrated governance solutions. Drawing on scenario techniques and dynamic assessment tools, it examines the trends in vulnerability and adaptive capacity as societies undergo wider cultural, economic, demographic and political transformation. Analyzing the interplay between urbanization pathways and shifts in risk profiles is one of the programme's specific areas of research.

One of the most important projects of the programme is the World Risk Index, which has been developed in cooperation with the Alliance Development Works and is published annually in the World Risk Report since 2011. The World Risk Index emphasises the view that environmental risks are not simply determined by the exposure of people and assets to hazardous environmental processes but also societal factors such as the lack of coping and adaptation capacities.

The *EVES* programme conducts research addressing the vulnerability and resilience of social-ecological systems, where the concept of ecosystem services is of central importance. Specifically, EVES aims to understand how social-ecological systems are affected by diverse environmental hazards such as water and land degradation, floods, and droughts, as well as how ecosystems can support disaster risk reduction and climate change adaptation. Besides original research, the programme advocates for ecosystem-based solutions, for example, through engagement in the Partnership for Environment and Disaster Risk Reduction (PEDRR). EVES collaborates with other UN entities including UNEP, UN-WATER, UNESCO, and the UN Country Team in Indonesia and develops research-based policy recommendations to inform governments around the world.

The *EMSVA* programme undertakes research on how society manages climate-related stresses such as floods, droughts, storms, sea level rise, and climatic shifts. The programme works with governments, the private sector, and civil society to identify and design innovative climate risk management solutions to build resilience. Specific themes of adaptation-related research include (1) human mobility, (2) livelihood resilience, (3) loss and damage, and legal dimensions of global environmental change, and (4) comprehensive climate risk management including insurance approaches, the latter within the Munich Climate Insurance Initiative (MCII).

The *EduSphere* programme addresses the activities of UNU-EHS with respect to enhancing the educational capabilities of its partner institutions in developing and transitional countries. The programme has developed a range of modular education activities, always with a strong focus on creating new partnerships. The flagship among these programmes is the Joint Master of Science "The Geography of Environmental Risks and Human Security", together with the University of Bonn.

After its successful start in October 2013, the programme developed an 'Integrative Modular Education Framework' consisting of the following different educational components offered at UNU-EHS: M.Sc. programme, collaborative Ph.D. programme, affiliated degree programme, non-degree courses and Technology enhanced learning (e-Learning).

The two-year master's programme educates students in an interdisciplinary and transdisciplinary manner on how to investigate and manage various resources related to environmental hazards by implementing science-based principles and methodologies to disaster risk management.

Finally, the research in many of UNU-EHS' projects is carried out within Ph.D. theses work. While EHS secures the funding, defines the research plan, and provides the scientific supervision, it does not grant the Ph.D. degree itself. All Ph.D. theses are carried out in partnership with a degree-granting partner university.

Concluding Remarks

Environmental risks will be an area of increasing concern in the future. The main drivers are not only environmental processes per se, but also the socio-economic systems which interact with environmental processes. A successful management of environmental risks needs to be based on integral development planning. This intimate connection is taken into account in all research projects of UNU and also by a comprehensive approach to education and training.

References

Munich Climate Insurance Initiative (MCII). URL: http://www.climate-insurance.org. Last accessed 18 Aug 2016

Partnership for Environment and Disaster Risk Reduction (PEDRR): http://pedrr.org/

Thant U (1969) Annual Report of the Secretary-General on the Work of the Organization: 16 June 1968–15 June 1969. United Nations General Assembly, Twenty-Fourth Session Supplement No. 1 (A/7601). New York, USA

United Nations University (2014) United Nations Strategic Plan 2015–2019. Tokyo, Japan

United Nations University Institute for Environment and Human Security Office of Communication (2016) About EHS. URL: http://ehs.unu.edu/about/about-ehs. Last accessed 22 June 2016

United Nations University Office of Communication (2016) UNU Research. URL: http://unu.edu/research. Last accessed 26 July 2016

World Risk Report and World Risk Index. URL: http://ehs.unu.edu/research/world-riskreport.html#outline. Last accessed 18 Aug 2016

World Meteorological Organization (WMO)—Concerted International Efforts for Advancing Multi-hazard Early Warning Systems

Jochen Luther, Alasdair Hainsworth, Xu Tang, John Harding, Jair Torres, and Margherita Fanchiotti

Abstract

Recent international agreements such as the Sendai Framework for Disaster Risk Reduction 2015–2030, the 2030 Agenda for Sustainable Development and the Paris Agreement have all recognized the importance of developing and operationalising multi-hazard early warning systems that integrate the specificities of single-hazard early warning systems in a holistic, systematic and coordinated manner to promote synergies and maximize efficiency. While much progress has been made in recent years towards the advancement of knowledge and practice related to early warning systems worldwide, the lack of multi-disciplinary and transboundary cooperation among and across communities of scientists, decision-makers and practitioners continues to be a key challenge for the successful establishment and operation of these systems. To address this gap, major international and national organizations have collaborated to establish the International Network for Multi-Hazard Early Warning Systems (IN-MHEWS), with the aim of facilitating knowledge sharing and capacity development for multi-hazard early warning systems around the globe. This paper presents an overview of advances and challenges in promoting a multi-hazard and systematic approach to early warning, as well as the aim, objectives and expected contributions of this newly established Network.

Keywords

Multi-hazard early warning systems • Sendai framework for disaster risk reduction 2015–2030 • Multi-stakeholder partnerships • Knowledge sharing • Disaster risk management

J. Luther · A. Hainsworth · X. Tang (✉)
Disaster Risk Reduction Services Division, World Meteorological Organization (WMO) Secretariat, 7 bis, Avenue de la Paix, P.O. Box 2300, 1211Geneva 2, Switzerland
e-mail: xtang@wmo.int

J. Luther
e-mail: jluther@wmo.int

A. Hainsworth
e-mail: ahainsworth@wmo.int

J. Harding
The United Nations Office for Disaster Risk Reduction (UNISDR), 9-11, Rue de Varembé, 1202 Geneva, Switzerland
e-mail: harding@un.org

J. Torres · M. Fanchiotti
Natural Science Sector, Earth Sciences and Geo-Hazards Risk Reduction Section, United Nations Educational, Scientific and Cultural Organization (UNESCO), 7, Place de Fontenoy, 75352 Paris 07 SP, France
e-mail: j.torres@unesco.org

M. Fanchiotti
e-mail: m.fanchiotti@unesco.org

K. Sassa et al. (eds.), *Advancing Culture of Living with Landslides*,
DOI 10.1007/978-3-319-59469-9_9

Introduction

Over the past two decades, early warning systems (EWS) have received increasing local, national, regional and international attention and are now recognised as a critical and solid component of national disaster risk management (DRM) arrangements aimed at reducing risk, saving lives and minimising losses from hazard events such as floods, storm surges, earthquakes, tsunamis and epidemics (UNISDR 2015a, 2006a, b; UNEP 2012; Basher 2006; Chang Seng 2012). The importance of EWS for disaster risk reduction (DRR) has been repeatedly highlighted in major international agendas, including the Yokohama Strategy and Plan of Action for a Safer World 1994 (UN 1994), the Hyogo Framework for Action 2005–2015: Building the Resilience of Nations and Communities (HFA) (UNISDR 2005), major multilateral environmental agreements and action plans such as under the United Nations Framework Convention on Climate Change (UNFCCC), and the United Nations Convention to Combat Desertification (UNCCD) as well as, most recently, in the Sendai Framework for Disaster Risk Reduction 2015–2030 (UNISDR 2015b), the 2030 Agenda for Sustainable Development (UN 2015) and the Paris Agreement on Climate Change 2015 (UNFCCC 2015).

While EWS have often been developed to target specific hazards and consequences, the need to adopt a people-centred, all-hazard approach to early warning has been acknowledged based on the understanding that citizens, decision makers and practitioners are more concerned with integrated DRM strategies rather than the specificities of each individual hazard, especially when dealing with their concatenated or cascading impacts (Basher 2006). In some cases, EWS are already operated for multiple hazards, particularly in the context of hydrometeorological phenomena and increasingly in disease outbreaks and humanitarian emergencies, and to a lesser extent for geophysical hazards. However, more efforts are needed to develop and operationalise MHEWS that integrate the characteristics of single EWS through "a coordinated 'system of systems'" (Basher 2006: 2171) to build synergies and promote multi-disciplinary collaboration (UNISDR 2006a, b; UNEP 2012). Through the Sendai Framework adopted at the Third United Nations (UN) World Conference for Disaster Risk Reduction (WCDRR) in Sendai, Japan, in March 2015, UN Member States agreed on the necessity of investing in, developing, maintaining and strengthening people-centred multi-hazard early warning systems (MHEWS), including telecommunication systems for hazard monitoring and emergencies, simple and low-cost early warning equipment and facilities, and broadened release channels for warning information that is tailored to different user needs and sectors.

One of the seven global DRR targets of the Sendai Framework (target g) aims to "*substantially increase the availability of and access to "Multi-Hazard Early Warning Systems (MHEWS)" and disaster risk information and assessments to the people by 2030*". In this document States also called for the "*further development of and investment in effective, nationally compatible, regional multi-hazard early warning mechanisms, where relevant, also contributing to the Global Framework for Climate Services (GFCS)*", and the facilitation of information sharing and exchange across all countries. To a large extent, achieving this target will depend on strengthened regional and international cooperation and on developing and applying science- and community-based methodologies and tools for MHEWS.

Against this background the article presents the recently established International Network for Multi-Hazard Early Warning Systems (IN-MHEWS), a major multi-stakeholder initiative proposed at WCDRR and launched in 2016. To provide essential context, the following section looks at the evolution of the conceptual understandings of EWS and MHEWS and their key components but also at recent advances and remaining gaps and challenges with them.

Context—from Early Warning Systems to Multi-hazard Early Warning Systems

Every societal entity has its own set of threats to be concerned about and therefore an interest in early warning. Governments and many non-governmental organizations (NGOs) have a legal and ethical obligation to protect their citizens and economies by issuing early warnings. While the risks from some hazards can be reduced to a tolerable level and their impacts can be well prepared for, other risks cannot be eradicated—although one can raise the awareness of these hazards, their likelihood and the severity of their impacts. For most of these (perceived or real) threats a mix of informal or formal warning systems exists—often focussing on the same or similar hazards, but operated by several players at the same time, in any given area, and at the levels of individuals, communities, businesses, governments or international organisations. Collectively, these systems provide a first defence against a variety of hazards.

Understanding the concept and components of (MH)EWS is a key requirement to develop and strengthen such systems, to prioritize investment and international cooperation and to measure effectiveness of and progress with them. This quickly proves to be complex, and a universally accepted definition of an (MH)EWS does not—and may never—exist.

What Is a (Multi-hazard) Early Warning System?

Based on the practice across warning systems for different hazard types there is a general agreement among practitioners and decision-makers that an EWS is made up of several components rather than being the issuance of warnings alone (Fig. 1). This is reflected in the well-established definition of EWS contained in the 2009 Terminology on DRR which is based on the outcomes of three International Conferences on Early Warning in 1998, 2003 and 2006 (EWC I-III): "*The set of capacities needed to generate and disseminate timely and meaningful warning information to enable individuals, communities and organizations threatened by a hazard to prepare and to act appropriately and in sufficient time to reduce the possibility of harm or loss*" (UNISDR 2009: 12).

However, through the Sendai Framework, States tasked the United Nations Office for Disaster Risk Reduction (UNISDR) to facilitate an intergovernmental process to update this terminology, including the term EWS and the new term MHEWS. While these definitions will only become available in early 2017, EWS are being described as an interrelated set of hazard monitoring and prediction, risk assessment, communication and preparedness sub-systems and processes/activities that enable individuals, communities, governments, businesses and others to take timely action to reduce their risks in advance of hazardous events. The definition will be complemented by a annotation that essentially states that effective "end-to-end" and "people-centred" EWS are still comprised of four interrelated key elements: (1) risk knowledge based on the systematic collection of data and risk assessments; (2) detection, monitoring, analysis and forecasting of the hazards and possible consequences; (3) dissemination and communication of timely, accurate and actionable warnings and associated information on likelihood and impact; and (4) preparedness and response capabilities at different levels

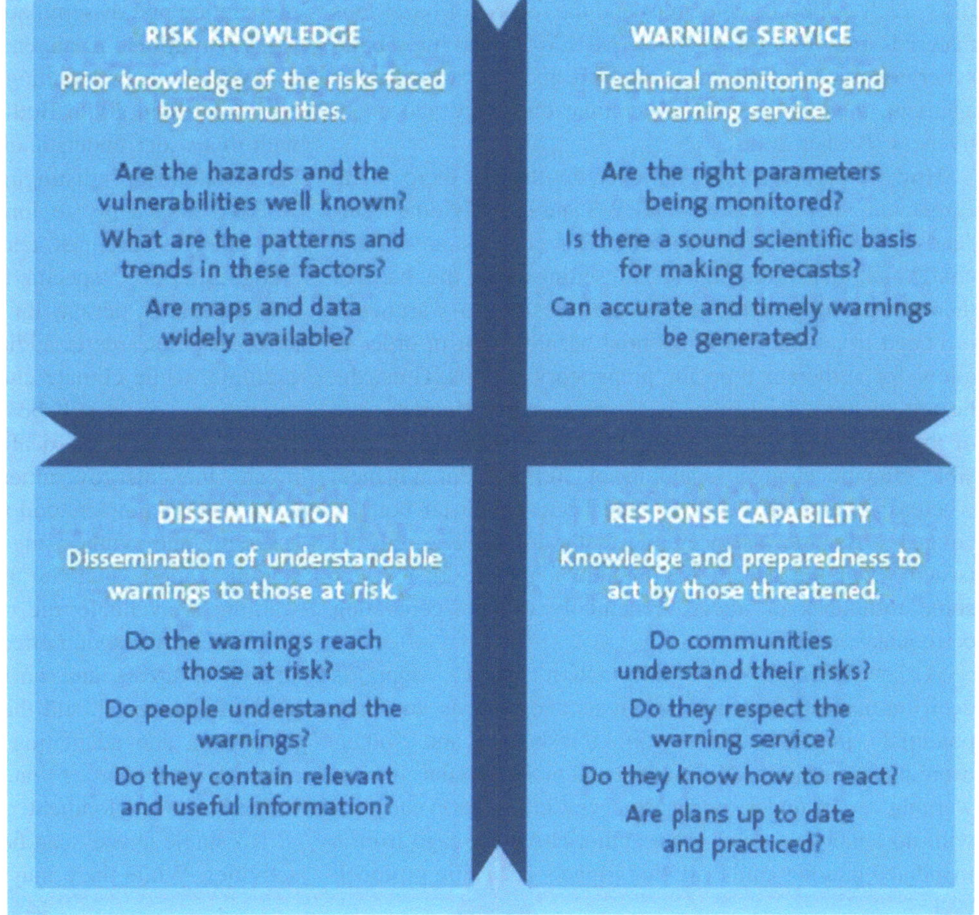

Fig. 1 The four elements of effective early warning systems [*Source* Information brochure of the International Early Warning Programme (IEWP) (http://www.unisdr.org/2006/ppew/info-resources/docs/IEWP.pdf [Last accessed: 4 November 2016])], (UNISDR 2006b)

to respond to the warnings received (UNISDR 2016b). These four interrelated elements need to be coordinated within and across sectors and multiple levels for the system to work effectively and effective feedback mechanisms need to be in place for continuous improvement. Failure in one component, or lack of coordination across them, could lead to the failure of the whole system—a critical issue since an EWS that does not warn effectively will not be trusted (Golnaraghi 2012).

EWS can be developed for specific hazards and specific consequences or for multiple hazards and a range of impacts. The latter are termed MHEWS—a concept that is relatively new—and currently defined (UNISDR 2016b) as EWS which are designed to be used in multi-hazard contexts where hazardous events may occur alone, simultaneously or cumulatively over time, and taking into account the potential interrelated effects. A MHEWS with the ability to warn of one or more hazards increases the efficiency and consistency of warnings through coordinated and compatible mechanisms and capacities, involving multiple disciplines for updated and accurate hazards identification, mapping and monitoring and dissemination of warnings from various sources with a "single authoritative voice" and through standardized formats, codes and definitions. They may also involve international cooperation to address transboundary risks, such as floods, epidemics and the release of hazardous materials into the air or water. MHEWS, therefore, build synergies for data collection, analysis and operational management, thus enhancing cooperation, efficiency and effectiveness (Basher 2006).

However, in practice and even in theory, there is no agreement what is meant by an EWS and its individual terms "early", "warning" or "system". The answers are highly contextual, depending on social settings and the hazards themselves. In general, "early" indicates that a warning is provided in advance of a potential hazard event in order to allow for sufficient time for preparatory actions Thus, the hazard-specific "lead time" becomes a crucial characteristic of the EWS. However, "early" in terms of warning of possible climatic changes could mean 100 years in advance, whereas for a flash flood it could be less than an hour in advance. In some instances, the first warnings can only be issued when the hazard has already materialised, e.g. for some tornadoes or after a potentially tsunami-generating earthquake.

A "warning" is information that supports decision-making dedicated to threats and hazards and their potential impacts based on thresholds and other, user-specific criteria. However, caution is advised since a warning can be understood as a general information category with no legal obligation or an authoritative message from a mandated national authority that triggers a specific protocol. It may also include specific actions which people or organizations should take.

A "system", from a systems thinking perspective, is made up of entities that are linked through flows and interactions, with inputs and outputs. For EWS as a "system of systems" this includes for example monitoring, forecasting, warning generation, communication, emergency response and feed-back systems as well as their legal and institutional basis, and the people involved.

EWS need to focus on vulnerabilities as part of the day-to-day lives of the people they serve who display different forms and degrees of vulnerabilities and capacities. Kelman and Glantz (2014) argue that EWS are a social process that involves technical components embedded in their social context, which contrasts with technical views that an EWS comprises only the technical equipment detecting or forecasting a hazard event and sending its parameters to a decision-making authority. This process is on-going and rooted in day-to-day and decade-to-decade functioning of society and is not only triggered when a hazard is about to strike. It is used to educate people, train them about response (e.g. through drills), gather baseline data and map risks. In fact it can be observed that there is a narrower understanding of EWS (i.e. detection, monitoring, modelling, and forecasting of the hazards as well as warning preparation and dissemination, based on risk knowledge and legal/institutional arrangements), and a wider understanding of EWS (i.e. corresponding to the 4 components) as adopted by EWC III in 2006. Both are widely in use. The problem with this wider understanding of EWS is that an EWS then becomes almost indistinguishable from DRM.

EWS also apply to long-term, "creeping" changes that can change baselines and indicate trends – often only recognised once a specific threshold is crossed and often not being hazards per se but influencing other hazards and slow-onset hazards in addition to the quick-onset ones. For example, while climate change may not be a hazard itself, the process could still be warned of, partly to address the causes and partly to deal with the consequences. In this regard the Intergovernmental Panel on Climate Change (IPCC) can be understood as an EWS for climate change by assessing and synthesising climate change science and presenting actions that are needed. In theory, the earlier a warning is available, the more time there is to prepare for and hopefully respond effectively to the potential impacts of stressors, threats and hazards. An EWS for slower and gradual changes should therefore give more time to design prevention and adaptation to new hazard regimes, plan a response and integrate that response into day-to-day life and longer-term development.

No single agency can be responsible for all EWS-related activities. While there may be an officially designated EWS for specific hazard types with authorities having a mandate for these hazard types or the warnings of them, there are many other channels through which people receive or look for warning information and advice.

Recent Advances in Early Warning Systems

In line with Priority for Action 2 of the HFA, regions and countries across the world have made significant progress over the past 10 years in strengthening end-to-end, people-centred EWS—often for multiple hazards. Progress has been evident in the development of observation and monitoring systems and the strengthening of communication and information on risks, as part of the overall efforts to strengthen disaster resilience. Today, EWS are established and operational in many countries of the world, focusing on a variety of natural and human-induced hazards and utilizing available scientific knowledge and latest technologies.

The recent decade has witnessed a significant evolution in information and communication technologies (ICT). Access to space-based data is now more open, made possible by the changes in data policies. The development of personal mobile devices, such as smart phones and tablets, and the use of geographic information systems and geo-viewers has advanced quickly. This has built the foundations for a new generation of highly sophisticated EWS making use of high-accuracy information and advanced processing techniques to provide warnings in near real-time conditions, as is the case with earthquake EWS (Meissen and Voisard 2010). In this regard, there is a need to determine and share the best ways of applying these advances in ICT to EWS operations worldwide.

With regards to EWS for hydrometeorological hazards, which remain the trigger of most disaster events, significant advances have been made in predicting weather, water, and climate extremes. On average, a five-day weather forecast of today is more reliable than a two-day weather forecast of two decades ago. There has also been consistent progress in risk assessment and hazard mapping, and the recognition of indigenous warning knowledge in enhancing the operations of such EWS.

With regards to non-hydrometeorological and non-geophysical hazards (such as biological and technological hazards, famine and other societal hazards), technical advances have been made in individual fields that have, for example, improved the detection and monitoring of epidemic-prone diseases; enhanced preparedness for potential humanitarian emergencies from various causes (e.g. lack of food availability and access can lead to food insecurity); increased safety of air, food and water quality; and reduced the risk of chemical and radiological exposures. Considerable progress has also been made on EWS for volcanic eruptions, landslides, avalanches, tsunamis, and—more recently—for earthquakes. Geophysical hazards require rigorous hazard, vulnerability and overall risk assessments in order to develop effective long-term plans and some of these systems are still relatively novel and present many theoretical and operational challenges. In general, there is a strong need to enhance the scientific, technological and operational capacities of countries, both in those that are already operating such EWS and in those that could greatly benefit from their future implementation. In this regard, international initiatives such as the establishment of the "International Platform on Earthquake Early Warning Systems" (IP-EEWS) under the United Nations Educational, Scientific and Cultural Organization's (UNESCO) lead contribute to fostering scientific knowledge on least developed EWS and enhancing collaboration between scientists and practitioners for capacity development, while at the same time promoting the integration of these systems into MHEWS. Furthermore, the increase of extreme geo-hazard events due to climate change has been subject of a large body of literature in recent years, which has investigated how global warming may trigger a broad range of geo-hazards including earthquakes and volcanic hazards (cf. McGuire 2013; Liggins et al. 2010; Deeming et al. 2010). While further research is needed to explore these interactions, their potential implications for policy and practice reinforce the need for integrated, multi-hazard systems.

EWS, together with other DRR measures, have in many regions led to a substantial reduction in the number of lives lost due to natural hazards. Moreover, in the 2011 Global Assessment Report on DRR (GAR 2011), UNISDR reported that in most parts of the world, the risk of being killed by a tropical cyclone or a major river flood is lower today than it was in 1990, also thanks to EWS. However, the economic losses from disasters are now reaching an average of US$250 billion to US$300 billion each year (UNISDR 2015a). More critically, the mortality and economic loss associated with extensive risks, i.e. the risk of low-severity, high-frequency disasters mainly associated with localised hazard events, in low and middle-income countries are trending up (UNISDR 2015a).

Gaps and Challenges Related to Early Warning Systems

Notwithstanding these advances in EWS in the past decade, many countries, in particular least developed countries (LDCs), small island developing states (SIDS), and landlocked developing countries (LLDCs), still have not benefited from them as much as they could have. Developed countries operate more EWS than developing countries, for which the sustainability of EWS is a major challenge

(UNISDR 2006a, b; UNEP 2012). Despite tremendous efforts and progress with EWS, the world has seen increasing economic losses and continued high tolls of death, injuries and illnesses from the impact of natural hazards such as storms in the USA, the Philippines, and Myanmar; floods in European countries, Thailand, India and Pakistan; the outbreak of diseases such as Ebola in West Africa; droughts in Africa; heat waves in Europe and Asia; tsunamis in the Indian Ocean and the Northwest Pacific Ocean and disastrous earthquakes in Haiti, Ecuador and China.

What has been defined as the "last mile" of EWS (Shah 2006)—i.e. reaching the most remote and vulnerable population with timely, meaningful, and actionable warning information and integrating a gender-sensitive and inclusive perspective into EWS—has been difficult. Most EWS focus on the hazard component of risk. But the necessity to promote an understanding of social vulnerability as a fundamental element of risk information and early warning to address the needs of the most vulnerable and account for the differential disaster impacts has been internationally recognised, and so has the need to tailor DRM strategies and practices to local capacities and needs (cf. UNISDR 2015b; Hewitt 1983). Thus, operational challenges exist for technical agencies, such as national geological, meteorological and hydrological services, to work together with DRM/civil protection agencies, statistical offices, and other relevant stakeholders such as NGOs, to incorporate in their warning activities relevant vulnerability and impact information (including hazard and risk maps) that is requisite to forewarn, empower and guide at-risk individuals, groups and communities to assume a more proactive role in the delivery of services for DRR.

Several gaps persist due to weak multi-disciplinary coordination and collaboration among the actors and agencies concerned, lack of standard operating procedures (SOP) and interoperable information systems, limited public awareness and participation in risk management, insufficient political commitment, and limited public/private financial support for the implementation of these systems (UNISDR 2006a, b; UNEP 2012; Clinton et al. 2016). Additional efforts are needed to institutionalize and strengthen MHEWS that provide multi-hazard risk communication messages tailored to the needs of specific individuals and communities, while following common standards such as the Common Alerting Protocol (CAP) and colour-coded warning messages based on risk matrixes. All these factors add to the scientific and technical challenges of developing some of these systems and are critical prerequisites for their operationalization (cf. e.g. Clinton et al. 2016).

Global societal changes such as rapid urbanization, increased mobility of populations and the growing exposure of people and assets to hazards are resulting in a highly dynamic, complex and globalised state of disaster risk. This is further exacerbated by global threats such as climate change, antimicrobial resistance, and other emerging and re-emerging disease threats. As proven by recent calamities, the impact of natural hazards can cascade into more serious consequences. For example, the 2010 eruptions of Iceland's Eyjafjallajökull volcano created havoc in the airline industry, triggering many cancellations and delays in flights, and affecting economic, political and cultural activities in Europe and across the world. This and similar cascading disaster events such as the 2006 Philippines landslide, 2011 Thailand floods, and the 2013 Typhoon Haiyan in the Philippines and its storm surge manifest the need to broaden the scope of EWS to address and cope with multiple hazards and risks, at multiple spatial (including regional and even global levels) and temporal scales.

At the same time, significant gaps remain with advancing the development of EWS for specific hazards, particularly for fastest onset hazards such as earthquakes and slowest onset hazards such as droughts and eventually climate change impacts. Most EWS exist for hydrometeorological hazards while they are generally less developed for geophysical hazards (UNISDR 2006b; UNEP 2012) such as landslides which are often triggered by hydrometeorological events. This adds to the challenges of linking these systems together to establish MHEWS. In this regard, international and regional collaboration of multiple stakeholders such as internationally-agreed strategies for data sharing, together with sustainable funding for MHEWS is critically necessary, given the transboundary nature of most natural hazards.

In particular, findings from surveys (UNISDR 2006a, b; WMO 2006, 2014; UNISDR 2015a, b) highlighted further gaps and challenges in implementing EWS worldwide (Table 1). Overall, there is a present need for a prominent "voice" for early warning at the international level that could advance a coordinated agenda on MHEWS worldwide, raise their visibility and advocate the strengthening of MHEWS in global and regional platforms and among key stakeholders, such as donors, private sector partners and academia.

Tracking Progress

In 2005, at the request of the UN Secretary-General, a global survey of EWS was undertaken with a view to advancing the development of EWS for all natural hazards (UNISDR 2006b). Already this survey report concluded that while some EWS are well advanced, there are numerous gaps and shortcomings, especially in developing countries and in terms of effectively reaching and serving the needs of those at risk. After the global survey and the adoption of the HFA, UNISDR facilitated biennial government reviews of

Table 1 Main gaps and challenges with regards to early warning systems

Risk knowledge
Need for quality-controlled historical time series of extreme hazard event and disaster occurrences in terms of intensity or magnitude, location, duration, timing, impacts
Risk and impact information often not or insufficiently integrated into EWS due to e.g. lack of cooperation between technical agencies responsible for collecting hazard data and stakeholders collecting vulnerability and exposure data as well as a lack of availability or access to (reliable) loss and impact information
Even if risk knowledge is incorporated, often still an inadequate representation of all dimensions of vulnerability (e.g. in urban areas, future dynamics)
Monitoring and warning service
Many regions lack modern monitoring and communication systems such as Doppler radars
More research and development is needed to improve observations, monitoring, data processing, modelling, forecasting and prediction and related applications
Lack of policy and legal frameworks to ascertain authority and accountability
Lack of resources for sustainable operations of agencies
Insufficient transboundary information sharing
Warning dissemination and communication
Insufficient focus on the uptake and use of warning messages, including capacity to use the information for longer term interventions
Proliferation of information and communication technologies leading to loss of single authoritative voice and to warning messages from unofficial sources (also due to ineffective engagement with the media and private sector)
Warning messages sometimes not clear and incomplete, e.g. due to a lack of standardized nomenclature and non-technical, actionable language and because uncertainties are often not well specified and explained
Communication networks break down during disasters (lack of back-up systems)
New technologies and support for dissemination and communication of warnings are often not available in least developed countries
Challenges in promoting public/private partnerships, market access, and incorporation of indigenous knowledge
Response capability
Response capabilities vary between countries and depending on the hazard
Lack of education and training for response
Role of non-governmental responders not well reflected in policies and legislation, missing out on opportunities for partnerships that do not rely on the central government
Lack of participation of the public and local emergency management agencies in the development of response plans

progress in implementing the HFA at the local, national, regional and international levels, using the HFA monitor.[1]

While the surveys conducted between 2005 and 2012, including the HFA reports, provide an overview of EWS operated in many countries and regions of the world (WMO 2006), this was largely restricted to disasters triggered by natural hazards only and there has not yet been a new comprehensive, global inventory of EWS for individual hazards, let alone of MHEWS that could serve as a baseline for monitoring progress in the coming decades. In addition, at this moment in 2016 there is no official mechanism yet for detailed reporting on national, regional or global efforts on early warning, especially on advances since the 2004 Indian Ocean Tsunami.

However, as recommended in paragraph 50 of the Sendai Framework, the UN General Assembly established via resolution 69/284 (adopted on 3 June 2015) an open-ended intergovernmental expert working group (OIEWG DRR) in 2015, comprised of experts nominated by States, and supported by UNISDR and with involvement of relevant stakeholders, for the development of a set of possible indicators and an update of the 2009 UNISDR Terminology on DRR by December 2016. This is carried out in coherence with the work of the inter-agency and expert group on sustainable development indicators (IAEG-SDG) and will allow for the measurement of global progress in the

[1]http://www.preventionweb.net/english/hyogo/hfa-monitoring [Last accessed: 4 November 2016].

implementation of the Sendai Framework,˙ including with target g), using the Sendai Monitor.

The currently proposed indicators (UNISDR 2016a) include the number of countries that have multi-hazard monitoring and forecasting systems, the number of people who are covered by and have access to MHEWS (per 100,000 persons), the number of local governments having a preparedness and/or evacuation plan with SOP, the number of countries that have multi-hazard national risk assessments/information available in an accessible, understandable and usable format and the number of local governments that have multi-hazard risk assessment/risk information available in an accessible, understandable and usable format for stakeholders and people. All these indicators combined yield the compound indicator of the number of countries that have MHEWS.

There is thus a need both for a new baseline as well as for a periodic account and review of (MH)EWS implemented and operated worldwide. Governments and stakeholders will need, however, guidance, support and good practice examples on how to strengthen MHEWS and on how to report against the respective indicators.

An International Network for Multi-hazard Early Warning Systems (IN-MHEWS)

To address these issues, key international and national agencies and organizations announced the establishment of an International Network for Multi-Hazard Early Warning Systems (IN-MHEWS) at the WCDRR Working Session on Early Warning as one of its major outcomes. This partnership under the Sendai Framework will foster coordination, cooperation, collaboration, and networking with the aim of strengthening MHEWS. Building on their respective programmes and activities and institutional mechanisms for cooperation, the IN-MHEWS partners will work together to promote a holistic and integrated multi-hazard (natural, including biological, and human-induced hazards), multi-stakeholder and multi-level approach to early warning with a common priority agenda and plan of action.

With this IN-MHEWS builds on earlier international efforts on early warning. The three International Early Warning Conferences (EWCs I-III) hosted by the Government of Germany were specifically dedicated to the scientific, operational and coordination aspects of EWS around the world, presenting innovative projects that would minimise the impact of natural hazards through the implementation of people-centred EWS. In response to the call for establishing a suitable framework for advancing early warning as an essential risk management tool, EWC II in 2003 proposed the International Early Warning Programme (IEWP). In line with the international efforts to promote early warning at the time, the World Conference for Disaster Reduction (WCDR) in 2005 Kobe, Japan, which adopted the HFA, also launched the IEWP. As an implementation mechanism for the IEWP, the Platform for the Promotion of Early Warning (PPEW) was created in 2004 in Bonn, Germany and remained operational until 2008. IN-MHEWS wants to give new momentum to these efforts under the Sendai Framework, without being a new institution nor a global operational EWS.

Objectives

As a broad-based networking initiative on early warning, IN-MHEWS will exemplify the importance of multi-stakeholder cooperation in MHEWS as a way to guide and advocate their implementation and improvement, share lessons learnt and increase the efficiency of investments in MHEWS for enhanced societal resilience. It aims at serving as the preferred source of information on MHEWS and related efforts worldwide. Responding to the calls by States in the Sendai Framework, the key objectives of IN-MHEWS are to:

(a) Promote synergies and partnerships between and among stakeholders and those directly involved in (in charge of) MHEWS at national, regional and international levels and to strengthen respective user-interface platforms;
(b) Identify effective strategies and actions to promote and strengthen MHEWS in support of the implementation of the Sendai Framework and other international agreements;
(c) Facilitate the sharing of good practice and making available to governments and key stakeholders expertise and policy-relevant guidance to enhance and sustain MHEWS and related services as an integral component of their national strategies for DRR and climate change adaptation (CCA) in their strive for a resilient and sustainable development;
(d) Provide a sound conceptual and scientific understanding of MHEWS and advocate the usefulness of a multi-hazard and systems approach to early warning in regional and international platforms and processes and among key stakeholders, including donors, from all sectors;
(e) Assess the progress made by individual EWS for specific hazards or hazard clusters, the existing relations within and between them and the potential synergies facilitating their integration into an effective, people-centred MHEWS; and,
(f) Identify new areas of, and promote further, scientific research on and technological development of EWS for

single hazards (hazard clusters) while advocating for their integration into MHEWS as well as the application of these latest scientific and technical advances.

Expected Outcomes and Initial Activities

Derived from the objectives, a first expected key outcome of IN-MHEWS is a viable and active community of institutions and practitioners working on EWS and MHEWS, including the communication tools required for such a network (a website, calendar of events and internal meetings, mailing lists, online work space, etc.), an appropriate governance mechanism, a common priority agenda on MHEWS throughout the lifetime of the Sendai Framework with a work plan for the near term, expert teams, etc. This requires federating a broad range of key international stakeholders in MHEWS, including their organizations' policy stances and attitudes on all aspects relating to MHEWS, and promote IN-MHEWS in their respective constituencies as well as multi-stakeholder fora. Activities include supporting and expanding this network of individuals, organizations and programmes working to improve early warning and related efficient response, stimulating dialogue and collaboration through further networking and partnerships (e.g. among major UN agencies concerned with early warning, such as UNESCO, WMO, the Food and Agricultural Organization (FAO), the World Food Programme (WFP), the United Nations Environment Programme (UNEP), and the UNFCCC Secretariat, amongst others).

A second key outcome is a baseline and inventory of EWS/MHEWS on different levels, based on a global review (e.g. in the form of a survey among countries and organizations). Such a survey could be conducted on a regular basis (e.g. every four years) as a benchmark that identifies gaps and emerging issues related to early warning from an international or global perspective (e.g. cascading effects of natural hazard impacts, climate change impacts, urbanization, amplification of risks at the global scale, ethical, accountability and liability issues in EWS operation, etc.). This may yield an improved conceptual understanding of MHEWS over time. It may also allow for better coordination of major programmes, funding initiatives and related projects on MHEWS that all strive to enhance the capacities of countries to generate and disseminate early warnings for multiple and/or cascading hazards and to respond to them effectively.

A third key outcome is providing policy-relevant guidance with evidence that is actually used and helpful. This requires e.g. facilitating the development of guidelines for countries to review and measure the effectiveness of and progress with EWS for single hazards/ hazard clusters and MHEWS in line with the priorities, targets and the monitoring mechanism of the Sendai Framework ("Words into Action" Sendai Implementation Guide), including how to measure the number of people with access to early warning and risk information and assessments against the number of people exposed to related risks; the potential impact of disasters, (including disaster-related mortality and morbidity) and the extent to which gender, youth and vulnerable groups' perspectives are reflected in these systems. Another set of guidelines will address multi-stakeholder partnerships for MHEWS and the mainstreaming of related goals and strategies into development processes, including legislation, policy development, institutional frameworks required and planning of development programmes and investments at international, regional, national, and community levels. These need to be based on identified user requirements, good practices and existing guidance material.

Lastly, an important expected outcome is a mechanism that is in place for sharing of good practices and expertise in relation to MHEWS across regions, countries, cities, and local communities (e.g. through lessons learnt from the use of indigenous knowledge in early warning, regional demonstration projects, etc.) and in a manner that enables countries and key stakeholders to use this information effectively. The Network will facilitate dialogue between and among stakeholders on the scientific, technological, and social issues concerning individual hazard EWS and MHEWS, and regular publications and their open access dissemination (e.g. via web portals such as PreventionWeb, media exchange, etc.) promulgating the case studies, lessons learnt and emerging issues on MHEWS, as well as related policy developments in countries, will be made available. The Network will support the conduct of regular forums, seminars and conferences to discuss current and emerging issues and to share information and knowledge, including the application of advances in science and technology to MHEWS and to provide visibility to MHEWS in established international and national discussions and platforms for DRR.

As a first major activity, encouraged by the 17th World Meteorological Congress in 2015 and as a commitment under the UN Plan of Action on DRR for Resilience, WMO and UNISDR have taken the initiative to organize an international conference on MHEWS (the earlier ones were on EWS) on 22 and 23 May 2017 in collaboration with international and national partners. Ten years after EWC-III, this conference will be the next global opportunity to merge knowledge on how to design and implement MHEWS, following the UNISDR Science & Technology Conference[2] in January 2016 that addressed early warning in one of its four main work streams. Its outcomes will directly feed into the 2017 Global Platform for DRR (in particular its Special Session on Early Warning), also in May 2017, the first one since the adoption of the Sendai Framework.

Partners, Governance and Structure

As a networking partnership IN-MHEWS is open to all stakeholders committed to sustaining the achievements of countries in implementing especially HFA Priorities 2 and 5, and to promoting a holistic, integrated, and multi-hazard approach to early warning in accordance with the Sendai Framework. Network partners can be grouped in six categories operating at three levels (global, regional, and national):

(1) National governments;
(2) UN entities and their regional bodies;
(3) Other intergovernmental organizations and their regional/national counterparts;
(4) Non-governmental organizations and civil society in the broadest sense;
(5) Academia;
(6) Private sector (including media); and,
(7) Financial institutions.

IN-MHEWS is governed by an international Steering Committee, comprised of representatives from a number global, regional and national partners. As of August 2016 these are the following 11 organizations: International Telecommunication Union (ITU), UNESCO, Intergovernmental Oceanographic Commission of UNESCO (IOC-UNESCO), United Nations Economic Commission for Asia and the Pacific (UNESCAP), United Nations Development Programme (UNDP), UNISDR, United Nations Office for Outer Space Affairs (UNOOSA)/United Nations Platform for Space-based Information for Disaster Management and Emergency Response (UN-SPIDER), WMO, World Health Organization (WHO), as well as the International Federation of Red Cross and Red Crescent Societies (IFRC), the Helmholtz-Centre Potsdam—GFZ German Research Centre for Geosciences (GFZ), and the Deutsche Gesellschaft für Internationale Zusammenarbeit (GIZ). Its co-chairmanship rotates on a yearly basis (July to June) basis, providing coordinating staff resources. For the first year, the IN-MHEWS Steering Committee is co-chaired by UNISDR and WMO.

Apart from these core partners predominantly from the global/international level represented in the Steering Committee, IN-MHEWS is made up of an open network of "associated partners" from around the world. IN-MHEWS will furthermore constitute multi-disciplinary Expert Groups to support the collaborative activities of IN-MHEWS in response to specific requirements/requests of countries and which would identify relevant stakeholders groups at different levels.

It is suggested that IN-MHEWS be structured in clusters, corresponding to:

(a) Interrelated hazard types which address also cascading impacts;
(b) The four functional components of MHEWS; and,
(c) Regional components of IN-MHEWS.

The most appropriate core partner could take on the lead for coordinating the members and activities related to the individual hazard and functional clusters, as well as for the regional components, of IN-MHEWS. Table 2 shows a suggestion of possible clusters with examples of IN-MHEWS partners with relevant mandates and activities in the respective field.

Implementation Approach

The underpinning strategy for IN-MHEWS is to utilize existing frameworks, partnerships and fora to complement current and emerging strategies for DRR, CCA, and sustainable development with enhanced early warning efforts. IN-MHEWS will also build on the experience, good practice, and achievements of States and the international community in this field (incl. IEWP and PPEW).

The Network will provide policy-relevant advice to countries to strengthen the linkages between national technical agencies, providing data on hydrometeorological, geophysical, and other hazards, and national DRM agencies, statistical offices and other relevant institutions providing data on vulnerability, losses and damages. This would include, for example, greater engagement of sectoral/technical agencies, such as National Meteorological and Hydrological Services (NMHSs) or National Ministries of Health, to reinforce the paradigm shift underway from current providers of hazard forecasts and early warnings to providers of impact-based forecasts and risk-informed warnings (WMO 2015) and as more proactive players in DRM.

In some regions, countries are already collaborating on early warning issues. Such Regional Networks for MHEWS

[2]http://www.preventionweb.net/events/view/45270?id=45270 [Last accessed: 4 November 2016].

Table 2 Suggested hazard-, function- and region-specific components of IN-MHEWS

Hazard (and consequence) clusters
Hydrometeorological hazards (WMO, UNESCO, etc.)
Geophysical hazards (UNESCO, IOC-UNESCO, ICL, etc.)
Technological hazards (IAEA, etc.)
Biological/infectious disease hazards (WHO, IAEA, UNEP, etc.)
Food security (FAO, WFP, etc.)
Others
Functional clusters
Risk knowledge (UNISDR, etc.)
Detection, monitoring, analysis, forecasting of the hazards and respective risk assessment and generation of warnings (WMO, UNOOSA, etc.)
Dissemination and communication of timely, accurate and actionable warnings and associated likelihood and impact information (ITU, WMO, etc.)
Preparedness and response capabilities at different levels to respond to the warnings received (IFRC, UNOCHA, etc.)
Others
Regional components
Asia (UNESCAP, RIMES, etc.)
Europe (United Nations Economic Commission for Europe (UNECE), European Commission Joint Research Centre (EC JRC) DRM Knowledge Centre (DRMKC), MeteoAlarm (developed for EUMETNET, the Network of European Meteorological Services), etc.)
Others

(RN-MHEWS) are developing in Southeast Asia [e.g. linked to the Regional Integrated Multi-Hazard Early Warning System for Africa and Asia (RIMES)] and Europe [e.g. linked to Meteoalarm[3] and the European Commission Disaster Risk Management Knowledge Centre (DRMKC)[4]].

The ultimate beneficiary of IN-MHEWS is the population of the UN Member States, where a country's EWS/MHEWS with its respective national stakeholders, legal and institutional frameworks and fora for exchange can be considered as the national counterpart/component of IN-MHEWS. Operating these systems and issuing official warnings remains a national responsibility.

Strategic Linkages

In addition to directly contributing to achieving the seventh global target of the Sendai Framework and shaping the "Words into Action" Implementation Guide for this target g), IN-MHEWS will inform the revised United Nations Plan of Action on Disaster Risk Reduction for Resilience: Towards a Risk informed and Integrated Approach to Sustainable Development (UNISDR 2016c) and contribute to achieving the Sustainable Development Goals 3 and 13 which recommend to "*strengthen the capacity of all countries, in particular developing countries, for early warning, risk reduction and management of national and global health risks*" and "*improve education, awareness-raising and human and institutional capacity on climate change mitigation, adaptation, impact reduction and early warning*" respectively (UN 2015), as well as Articles 7-8 of the Paris Agreement (UNFCCC 2015) and the DRR Priority Area of the GFCS.[5]

IN-MHEWS will also contribute to strengthening capacity for the implementation of the International Health Regulations (IHR) (WHO 2016). The IHR are an international legal instrument that is binding on 196 countries across the globe, to help the international community prevent and respond to acute public health risks that have the potential to cross borders and threaten people worldwide.

IN-MHEWS will also be closely linked to the Climate Risk and Early Warning Systems (CREWS) initiative which was initiated by France and launched at the Paris Climate Change Conference in December 2015.[6] CREWS as a

[3]http://www.meteoalarm.eu [Last accessed: 4 November 2016].

[4]http://drmkc.jrc.ec.europa.eu [Last accessed: 4 November 2016].

[5]http://www.wmo.int/gfcs [Last accessed: 4 November 2016].

[6]http://www.cop21.gouv.fr/en/launch-of-crews-climate-risk-early-warning-systems [Last accessed: 4 November 2016].

project implementation initiative leverages IN-MHEWS as a corresponding and complementing expert forum and platform for knowledge sharing and dissemination. To this end CREWS also supports the international conference on MHEWS in 2017.

Conclusion

In conclusion, the need to promote the development and operationalization of people-centred, MHEWS has been recognized as a global target (UNISDR 2015b). While several countries have made significant progress in the establishment of EWS worldwide, in some cases even for multiple hazards, much remains to be done to successfully support the establishment of integrated and coordinated systems that work across hazards to maximize synergies and efficiencies. Moreover, the lack of multi-disciplinary and international cooperation between different players in the field of (MH)EWS has been identified as a key challenge for their implementation (UNISDR 2006a, b; UNEP 2012). The sharing of expertise and good practice on (MH)EWS will help to strengthen them around the world and make them an integral component of national strategies for DRR, CCA, and resilience building. To this end, major international and national organizations have joint efforts and established IN-MHEWS as a commitment made at WCDRR to support the implementation of the Sendai Framework. Especially achieving its global target related to MHEWS requires extensive yet coordinated action on all levels. Sound definitions of the terms, clear indicators for measuring effectiveness and progress with (MH)EWS are requirements to develop and strengthen such systems and to prioritize international cooperation. IN-MHEWS aims to provide this support. As such, IN-MHEWS welcomes the participation of the International Consortium on Landslides (ICL), while many of its core partners can also contribute to the work of the Sendai Partnerships within the framework of this partnership, with the aim of facilitating knowledge sharing and capacity development for MHEWS worldwide. Guidelines, good practices and other outcomes of the work of the Network will be shared with the international community to advocate for MHEWS and contribute to knowledge and practice advancement in this field.

Acknowledgements The article at hand is largely based on the IN-MHEWS concept paper (available from the WMO DRR Programme extranet website[7]) that was produced by the IN-MHEWS Steering Committee in a consultative process over several months in 2015 and 2016.

[7]http://www.wmo.int/pages/prog/drr/documents/IN-MHEWS/IN-MHEWS.html [Last accessed: 4 November 2016].

References

Basher R (2006) Global early warning systems for natural hazards: systematic and people-centred. Philos Trans R Soc A 364 (1845):2167–2182

Chang Seng D (2012) Improving the governance context and framework conditions of natural hazard early warning systems. J Integr Disaster Risk Manage 2(1):1–25

Clinton J, Zollo A, Marmureano A, Zulficar C, Parolai S (2016) State-of-the art and future of earthquake early warning in the European region. Bull Earthq Eng 14:2441–2458

Deeming K, McGuire WJ, Harrop P (2010) Climate forcing of volcano lateral collapse: evidence from Mount Etna, Sicily. Philos Trans R Soc A 368:2559–2577

Golnaraghi M (ed) (2012) Institutional partnerships in multi-hazard early warning systems: a compilation of seven national good practices and guiding principles. Springer, Berlin

Hewitt K (1983) Interpretations of calamity from the viewpoint of human ecology. The risks and hazards series 1. Allen and Unwin, London

Kelman I, Glantz M (2014) Early warning systems defined. In: Singh A, Zommers Z (eds) Reducing disaster: early warning systems for climate change. Springer, Dordrecht

Liggins F, Betts RA, McGuire WJ (2010) Projected future climate changes in the context of geological and geomorphological hazards. Philos Trans R Soc A 368:2347–2367

McGuire WJ (2013) Walking the giant: how a changing climate triggers earthquakes, tsunamis, and volcanoes. Oxford University Press, Oxford

Meissen U, Voisard A (2010) Towards a reference architecture for early warning systems. In: Proceedings of the 2010 International Conference on Intelligent Networking and Collaborative Systems

Shah HC (2006) The last mile: earthquake risk mitigation assistance in developing countries. Philos Trans R Soc A 364:2183–2189

United Nations (UN) (1994) Yokohama strategy and plan of action for a safer world. International Decade for Natural Disaster Reduction (IDNDR). UN, New York

United Nations (UN) (2015) Transforming our world: the 2030 Agenda for Sustainable Development. UN, New York

United Nations Environment Programme (UNEP) (2012) Early warning systems: state-of-art analysis and future directions. UNEP, Nairobi

United Nations Framework Convention on Climate Change (UNFCCC) (2015) Adoption of the Paris Agreement. Conference of the Parties. Twenty-first session. 30 November–11 December 2015. Paris, France

United Nations Office for Disaster Risk Reduction (UNISDR) (2005) Hyogo framework for action 2005–2015: building the resilience of nations and communities (HFA). UNISDR, Geneva

United Nations Office for Disaster Risk Reduction (UNISDR) (2006a) Platform for the promotion of early warning. Basics of early warning. URL: http://www.unisdr.org/2006/ppew/whats-ew/basics-ew.htm. Last accessed 1 Sep 2016

United Nations Office for Disaster Risk Reduction (UNISDR) (2006b). Global survey on early warning systems. An assessment of capacities, gaps and opportunities towards building a comprehensive global early warning system for all natural hazards. UNISDR, Geneva

United Nations Office for Disaster Risk Reduction (UNISDR) (2009) 2009 UNISDR terminology on disaster risk reduction. UNISDR, Geneva

United Nations Office for Disaster Risk Reduction (UNISDR) (2015a) Making development sustainable: the future of disaster risk management. Global assessment report on disaster risk reduction. UNISDR, Geneva

United Nations Office for Disaster Risk Reduction (UNISDR) (2015b) Sendai framework for disaster risk reduction 2015–2030. UNISDR, Geneva

United Nations Office for Disaster Risk Reduction (UNISDR) (2016a). Updated technical non-paper on indicators for global targets A, B, C, D, E and G of the Sendai Framework for Disaster Risk Reduction (30 Sep 2016). Sessional and Inter-sessional documentation and information of the Open-ended Intergovernmental Expert Working Group on Disaster Risk Reduction and Terminology (OIEWG). URL: http://www.preventionweb.net/documents/oiewg/Updated%20technical%20non-paper%20on%20indicators%20for%20global%20targets%20ABCDEG%20of%20the%20Sendai%20Framework%20for%20DRR.pdf. Last accessed 28 Oct 2016

United Nations Office for Disaster Risk Reduction (UNISDR) (2016b). Terminology related to Disaster Risk Reduction—updated technical non-paper (30 Sep 2016). Sessional and Inter-sessional documentation and information of the Open-ended Intergovernmental Expert Working Group on Disaster Risk Reduction and Terminology (OIEWG). URL: http://www.preventionweb.net/documents/oiewg/Terminology%20related%20to%20Disaster%20Risk%20Reduction%20-%20updated%20technical%20non-paper%2030%20September%202016.pdf. Last accessed 28 Oct 2016

United Nations Office for Disaster Risk Reduction (UNISDR) (2016c) United Nations plan of action on disaster risk reduction for resilience. Towards a risk-informed and integrated approach to sustainable development. UNISDR, Geneva

World Health Organization (WHO) (2016) International Health Regulations (2005), 3rd edn. WHO, Geneva

World Meteorological Organization (WMO) (2006) First coordinated capacity assessment of National Meteorological and Hydrological Services (NMHSs) in support of disaster risk reduction. WMO, Geneva

World Meteorological Organization (WMO) (2014) Synthesis of status and trends with the development of early warning systems. Background Paper prepared for the Global Assessment Report 2015, Geneva

World Meteorological Organization (WMO) (2015) Valuing weather and climate: economic assessment of meteorological and hydrological services. WMO, no. 1153, Geneva

International Council for Science (ICSU)—On the Future Challenges for the Integration of Science into International Policy Development for Landslide Disaster Risk Reduction

Irasema Alcántara-Ayala, Virginia Murray, Philip Daniels, and Gordon McBean

Abstract

In 2015 four UN landmark agreements were developed: the Sendai Framework for Disaster Risk Reduction 2015–2030 (hereafter referred to as the Sendai Framework); the agenda related to Financing for Development; the Sustainable Development Goals and the Paris Agreement on Climate Change. These can be regarded as the main guiding documents to galvanise action to address the new or emerging global challenges. The Science and Technology community are asked to support the implementation of the Sendai Framework, in order to 'prevent new and reduce existing disaster risk' by 'enhancing the scientific and technical work on disaster risk reduction and its mobilization through the coordination of existing networks and scientific research institutions at all levels and all regions with the support of the UNISDR Scientific and Technical Advisory Group (STAG)' (UNISDR in Sendai Framework for Disaster Risk Reduction 2015–2030, 2015a, Paragraph 25g). Within the Sendai Framework agenda, the commitment of STAG and the Integrated Research on Disaster Risk Program (IRDR) is focusing the integration and collaboration between science, policy and practice. IRDR is a multi-disciplinary, all-hazards approach, supported by the International Council for Science (ICSU), the International Social Science Council (ISSC) and the United Nations Office for Disaster Risk Reduction (UNISDR), to strengthen capacity at global, regional and local levels to address hazards and generate science-based decisions on actions to reduce their impact (IRDR in Integrated research on disaster risk strategic plan 2013–2017, 2013). Along the line of critical actions identified by STAG and IRDR, particular efforts are being undertaken by the International Consortium on Landslides (ICL) to understand the configuration of landslide disaster risk and reduce its impacts. ICSU, via IRDR, as one of the voluntary signatories of the International Strategy

I. Alcántara-Ayala (✉)
Institute of Geography, National Autonomous University of Mexico (UNAM), Circuito Exterior, Ciudad Universitaria, Coyoacán, 04510 Mexico City, Mexico
e-mail: irasema@igg.unam.mx

V. Murray · P. Daniels
Public Health England, London, SE1 8UG, UK
e-mail: Virginia.Murray@phe.gov.uk

P. Daniels
e-mail: Philip.Daniels@phe.gov.uk

G. McBean
Department of Geography, Social Science Centre Rm 2322, The University of Western Ontario, London, ON N6A 5C2, Canada
e-mail: gmcbean@uwo.ca

G. McBean
International Council for Science, 75116 Paris, France

K. Sassa et al. (eds.), *Advancing Culture of Living with Landslides*,
DOI 10.1007/978-3-319-59469-9_10

for Disaster Reduction–International Consortium on Landslides (ISDR-ICL) Sendai Partnerships 2015–2025 for Global Promotion of understanding and reducing landslide disaster risk, is committed to enhance such endeavours. In this paper, attention is drawn to identifying some of the main future challenges for the integration of science into local, national, regional and international policy development for Landslide Disaster Risk Reduction within the Sendai Framework.

Keywords
Landslides • Science • Policy making • Practice • Disaster risk reduction • Integrated research • Challenges

Introduction—Landslides as a Major Hazard

According to the World Bank, some 5% of the total population on the earth, circa 300 million inhabitants live in an area of 3.7 million km^2 susceptible to landslides (Dilley et al. 2005). It is considered that the impact of landslide disasters will continue to mount dramatically as a result of climate change (IPCC 2012, p. 15), population growth, and urbanization processes, leading to the establishment of regular and irregular human settlements on unstable slopes in both developed and developing countries.

Landslides are the concrete expression of slope instability, resulting from natural processes, human-induced factors and, quite commonly, the interactions of both. Deforestation processes, mining, excavations, loading of slopes, inappropriate terracing, artificial induced vibrations, deficient drainage and sewage systems are among the main anthropogenic determining factors of landslides (Alcántara-Ayala 2016).

A recent systematic review of landslides (Kennedy et al. 2015) showed that health impacts of landslides are poorly recorded. Estimates of mortality are frequently inaccurate (both over- and underestimated) and tend to emphasise the short-term impacts. Factors leading to high morbidity and mortality include the volume of material displaced, speed of displacement and vulnerability of the population.

Short-term physical health impacts include crushing, asphyxiation and inhalation of debris. Longer term physical health impacts include illness resulting from disruption and destruction of sanitation systems, as well as mental ill health and trauma resulting from direct involvement and/or from experience of loss and or trauma to loved ones. Economic losses can be significant, although are frequently greatest in high income countries, whereas injury and loss of life are lower than in developing countries due to improved mitigation, response and recovery.

Reasons for poor recording have been cited as:

- The location of many landslides in remote, often highly mountainous, areas in poor countries in which levels of reporting of lower impact events is likely to be variable;
- Misidentification of landslides, for example, a debris flow may be described as a flood, leading to them being incorrectly recorded;
- The co-occurrence of landslides with another (triggering) process, such as an earthquake, leading to the cause of death being incorrectly recorded;
- The difficulty of tracking down the occurrence of post-event mortality (Petley 2012)

Disaster risk is understood as the probability of loss of life, injury or damage/destruction of assets of a system, society or a community, determined by the interaction among hazard, vulnerability and exposure (ICSU-LAC 2010; UNISDR 2015b (see Fig. 1). While hazards can be of a natural, socio-natural or technological nature (Lavell 1996, 2003a; ICSU-LAC 2010; Oliver-Smith et al. 2016), vulnerability involves historically rooted economic, social, political, cultural and institutional multi-dimensions (Blaikie et al. 1994; Wisner et al. 2004; Oliver-Smith 2013; Oliver-Smith et al. 2016). Exposure refers to the physical location of individuals and societies, assets (including environmental), livelihoods and infrastructure in areas susceptible to hazards that are likely to be affected by potential losses or damage (UNISDR 2009, b).

Social processes shaped by the structure of development models and economic patterns of transformation have determined the social construct of disaster risk (Lavell 2003b).

The UN Landmark agreement in 2015, the Sendai Framework for Disaster Risk Reduction 2015–2030, was adopted by 187 countries in March 2015 and endorsed by the UN General Assembly in June 2015. It states that there is a need for focused action within and across sectors by states at local, national, regional and global levels in the four priority areas, with the understanding of risk being the first priority. For landslides, understanding risk therefore requires a full comprehension of causality, in other words, identifying, and assessing root causes of disaster risk, in addition to dynamic pressures or drivers of risk in order to be properly addressed (Burton 2010, 2015).

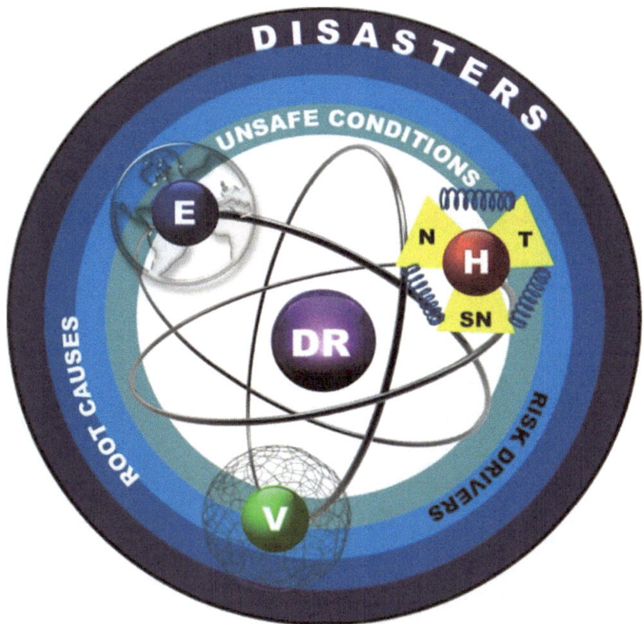

Fig. 1 The key relationships and processes in the social construction of risk (*E* exposure; *V* vulnerability; *H* hazard, *N* natural, *T* technological and *SN* socio-natural; *DR* disaster risk). *Source* Oliver-Smith et al. (2016)

Degradation of the environment, land-use changes, urbanization, migration, climate change and deficient governance are among drivers associated with exposure and vulnerability increasing risk, as they amplify unsafe conditions that potentially lead to landslides. These patterns of risk influence the creation of new risks, including landslides, at global, regional, national and local scales (Oliver-Smith et al. 2016) (Fig. 1).

When destabilized by physical characteristics and processes, coupled with the potential to affect societies, slopes can be transformed into hazards. Likewise, when human intervention on the environment takes place in areas susceptible to landslides, enhancing their occurrence and potential impact on societies, they are considered as socio-natural hazards. Excluding large mass movement processes resulting from volcanic or seismic activity and/or tsunami, most rainfall triggered landslides that are transformed into disasters are of socio-natural type (Alcántar-a-Ayala 2016).

Global Frameworks to Reduce Disaster-Risks, Hazards and Vulnerability

Reflecting the complex causality of disasters outlined above and following several decades of international negotiations, strategies and initiatives aimed at preventing and lessening the impact of disasters, 2015 became a watershed year for disaster risk reduction, science and policy making, with four United Nations landmark international agreements being agreed (Fig. 2). These four agreements are listed below:

Fig. 2 Intertwined convergence of the key four 2015 landmarks to achieve DRR and DRM. SDGs are sustainable development goals

- Third United Nations World Conference on Disaster Risk Reduction in Sendai, Japan (Sendai Framework for Disaster Risk Reduction 2015–2030
- Third United Nations International Conference on Financing for Development in Addis Ababa, Ethiopia
- United Nations Sustainable Development Goals in New York, USA (SDG's); and
- The United Nations Framework Convention on Climate Change in Paris, France

The Sendai Framework for Disaster Risk Reduction 2015–2030

Derived from its predecessor Hyogo Framework for Action 2005–2015 (HFA), the Sendai Framework for Disaster Risk Reduction consists of seven targets, four priorities for action and thirteen guiding principles which are summarised in Fig. 3.

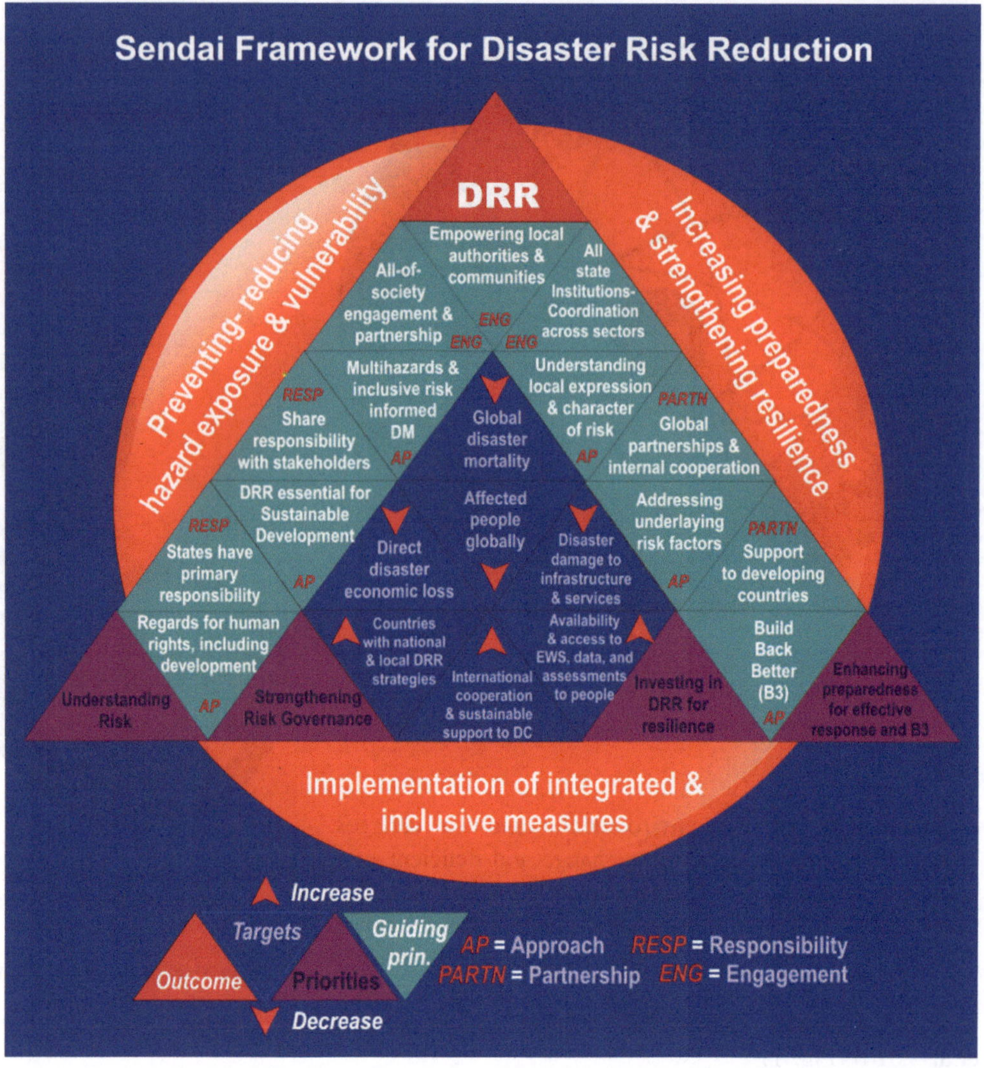

Fig. 3 A diagrammatic interpretation of the structure of the Sendai framework for disaster risk reduction (2015–2030) reflecting its seven global targets, its four priorities for action and its thirteen guiding principles. *Source* adapted from UNISDR (2015a)

Its scope and purpose is directed towards guiding '*the multi-hazard management of disaster risk in development at all levels as well as within and across all sectors*' by taking into account '*the risk of small-scale and large-scale, frequent and infrequent, sudden and slow-onset disasters, caused by natural or manmade hazards as well as related environmental, technological and biological hazards and risks*' (UNISDR 2015a, Paragraph 15).

The Sendai Framework aims to achieve '*the substantial reduction of disaster risk and losses in lives, livelihoods and health and in the economic, physical, social, cultural and environmental assets of persons, businesses, communities and countries*' (UNISDR 2015a, Paragraph 16). To achieve this outcome, the following goal must be pursued '*Prevent new and reduce existing disaster risk through the implementation of integrated and inclusive economic, structural, legal, social, health, cultural, educational, environmental, technological, political and institutional measures that prevent and reduce hazard exposure and vulnerability to disaster, increase preparedness for response and recovery, and thus strengthen resilience*' (UNISDR 2015a, Paragraph 17).

Third United Nations International Conference on Financing for Development

This agreement, confirmed in Addis Ababa in July 2015 contains three main tasks (AAAA 2015):

- *to follow-up on commitments and assess the progress made in the implementation of the Monterrey Consensus and the Doha Declaration;*
- *to further strengthen the framework to finance sustainable development and the means of implementation for the universal post-2015 development agenda; and*
- *to reinvigorate and strengthen the financing for development follow-up process to ensure that the actions to*

which we commit are implemented and reviewed in an appropriate, inclusive, timely and transparent manner.

The agreement aims to end poverty and hunger, protect the environment, and promote inclusive economic growth and social inclusion (Paragraph 1). The Agenda also recognizes the importance of aligning climate, humanitarian and development finance (Paragraphs 62–66), as well as developing data, disaggregated by age, sex and socioeconomic group, in order to support research and programming (Paragraph 126) (AAAA 2015).

Crucially, in the context of this paper, the agenda pledged that, by 2020, the United Nations (AAAA 2015):

- "*will increase the number of cities and human settlements adopting and implementing integrated policies and plans towards inclusion, resource efficiency, mitigation and adaptation to climate change and resilience to disasters. We will develop and implement holistic disaster risk management at all levels in line with the Sendai Framework. In this regard, we will support national and local capacity for prevention, adaptation and mitigation of external shocks and risk management*" (34);
- *"Encourage consideration of climate and disaster resilience in development financing to ensure the sustainability of development results"* (64);
- *"Enable countries to prevent or combat situations of chronic crisis related to conflicts or natural disasters ... [recognising] the need for the coherence of developmental and humanitarian finance to ensure more timely, comprehensive, appropriate and cost-effective approaches to the management and mitigation of natural disasters and complex emergencies. We commit to promoting innovative financing mechanisms to allow countries to better prevent and manage risks and develop mitigation plans. We will invest in efforts to strengthen the capacity of national and local actors to manage and finance disaster risk reduction and to enable countries to draw efficiently and effectively on international assistance when needed" (66).*

The agreement further encouraged consideration of further debt relief steps, where appropriate, and/or other measures for countries affected by disasters.

The 2030 Agenda for Sustainable Development—the Sustainable Development Goals

Unlike the Millennium Goals, their successors, the Sustainable Development Goals (SDGs), known officially as the 2030 Agenda for Sustainable Development, are characterized by being 'Universal', that is to say, they should be applicable to all countries. They are also intended to be an integrated and transformative vision for a better world, as a function of five-sided baselines: People, Planet, Prosperity, Peace and Partnership. The SDGs aim at ending poverty and hunger; protecting the planet from degradation; ensuring that all human beings can enjoy prosperous and fulfilling lives at the same time that development is in equilibrium with nature; fostering peaceful, just and inclusive societies; and enhancing Global Partnerships for the Implementation of the Sustainable Development Agenda (United Nations 2015).

The urgent need to move towards disaster risk reduction (DRR) and management (DRM) is clearly embedded into the 2030 Agenda for Sustainable Development, which comprises 17 Sustainable Development Goals (SDGs) and 169 associated global targets.

Disaster risk reduction is a cross-cutting issue in different aspects and sectors of development and reflecting this, there are 25 targets, in 10 out of the 17 SDGs, that are strongly related to DRR (Table 1). They mainly address poverty, ending hunger, ensuring healthy lives, education, sustainable management of water, building resilient infrastructure, resilient cities, climate change and marine and terrestrial ecosystems goals.

The 2015 Paris Agreement on Climate Change

After more than two decades of intense negotiations, some commitments and fewer influential actions, the 2015 Paris Agreement reached an international consensus to address climate change by strengthening emission reduction targets to keep global temperature rise well below 2.0 °C and engaging into efforts to limit the increase to 1.5 °C under a stronger system of transparency and accountability for measuring progress. The 'all countries' policy includes the support actions of non-party stakeholders, such as businesses, investors, states, provinces, cities, authorities at sub-national and local level, financial institutions, and civil societies, among others, to reduce the emissions and building resilient societies in the light of a sustainable future (UNFCCC 2015).

The agreement comprises twenty-nine articles, dealing with mitigation, adaptation, loss and damage, finance, technology development and transfer, capacity-building, and transparency of action, global stock-take and support. Accordingly, support for climate action to reduce emissions and building resilience in developing countries will continue to be provided by developed nations, although also other countries are persuaded to offer or continue the provision of such support on a voluntarily basis. Likewise, from a local or national scale, all signing Parties have a legally binding obligation to prepare, communicate and contribute to

Table 1 Disaster risk reduction identified as cross-cutting issues in 10 of the 17 SDGs (adapted from UNISDR 2015c)

Sustainable development goals related to disaster risk reduction	
G1. End poverty in all its forms everywhere	Target 1.5.—Building the resilience of the poor
G2. End hunger, achieve food security and improved nutrition and promote sustainable agriculture	Target 2.4.—Advancing actions in mainstreaming disaster risk reduction and climate adaptation into agriculture sector planning and investments in order to promote resilient livelihoods, food production and ecosystems
G3. Ensure healthy lives and promote well-being for all at all ages	Target 3.d.—Strengthening early warning and risk reduction of national and global health risks (resilient health systems)
G4. Ensure inclusive and equitable quality education and promote lifelong learning opportunities for all	Target actions 4.7 and 4.a.—Building and upgrading education facilities and promoting education for sustainable development, contribute significantly to resilience-building in the education sector
G6. Ensure availability and sustainable management of water and sanitation for all	Target 6.6.—Protecting and restoring water-related ecosystems, will significantly contribute to strengthening the resilience of communities to water-related hazards and mainstreaming ecosystem-based approaches
G9. Build resilient infrastructure, promote inclusive and sustainable Industrialization and foster innovation	Targets 9.1 and 9.a.—Developing sustainable and resilient infrastructure development to protect existing and future infrastructure investments
G11. Make cities and human settlements inclusive, safe, resilient and sustainable	Action targets 11.1, 11.3, 11.4, 11.5, 11.b and 11. c.—Upgrading urban slums, integrated urban planning, reducing social and economic impacts of disaster risk, building the resilience of the urban poor, adopting and implementing urban policies
G13. Take urgent action to combat climate change and its impacts	Targets 13.1–13.3 and 13.a–13.b.—Strengthening the integration between disaster and climate resilience to protect broader development paths at all levels and influencing the provision of long-term financing for addressing disaster and climate risk, aiming at a transformative change
G14. Conserve and sustainably use the oceans, seas and marine resources for sustainable development	Target action 14.2.—Reducing disaster risk and increase in demand for healthy marine and coastal ecosystems
G15. Protect, restore and promote sustainable use of terrestrial ecosystems, sustainably manage forests, combat desertification, and halt and reverse land degradation and halt biodiversity loss	Target actions 15.1–15.4 and 15.9.—Contributing to resilience building by managing and restoring forests, combating land degradation and desertification, conserving mountain ecosystems and their biodiversity and integrating ecosystem and biodiversity values into national and local planning, development processes, poverty reduction strategies

mitigation, but Parties are not legally bound at the international level to achieve their targets (UNFCCC 2015).

Cross Cutting Themes and Issues Between Declarations

In many respects it is possible to see evident linkages between climate change, disaster risk reduction and sustainable development as well as in the financing for development, that have been identified in the 2015 UN Landmark agreements. By its very nature, the interaction of weather-related and climate hazards with exposed and vulnerable individuals and communities around the globe should not be neglected. For example, from the year 2005 to 2014, although this data is not complete, it was determined that approximately 700,000 people died, 1.7 billion inhabitants were affected, and economic losses mounted $1.4 trillion in disasters worldwide (UNFAO 2015).

As cited by the Intergovernmental Panel on Climate Change (IPCC 2012), a changing climate modifies the frequency, intensity, spatial extent, duration, speed of onset and timing of weather related and climate hazards. Moreover, the diversity, complexity and severity of the impacts of both, extreme and non-extreme weather or climate events on vulnerability to future extreme events by influencing resilience, coping capacity, and adaptive capacity has been widely recognized (IPCC 2012). A further problem is that the severity of extreme events is strongly associated with the degree of vulnerability and exposure and will be a major driver for future temporary and permanent population displacement which, in addition to migration due to other causes, will also be a challenging issue for all governments in the years to come.

In the absence of equality and social wealth, coping and adaptive capacity at local and sub-national levels, lives and livelihoods are threatened and generating challenges for disaster risk management and adaptation.

On the Missing Link Between Science and Technology, Policy Making and Practice

Preliminary Considerations

Within the last few years there has been a growing interest in linking science and technology, policy-making, and practice for disaster risk reduction and management (UNISDR STAG 2013 and 2015 reports). This concern has been indirectly pointed out since the Rio Declaration in 1992, when the issue of the precautionary principle as an approach to disaster risk management was introduced. Accordingly, it was stated that 'in order to protect the environment, the precautionary approach shall be widely applied by States according to their capabilities. Where there are threats of serious or irreversible damage, lack of full scientific certainty shall not be used as a reason for postponing cost-effective measures to prevent environmental degradation' (Rio Declaration 1992).

The unification of science and policy making around practice is crucial for tailoring efficient and effective implementation strategies for DRR and DRM. Uhlenbrock et al. (2014) argue that scientists have a key role in informing public policy and science should serve as a foundation to the decision-making process. A central feature of this positioning indeed unveils a delicate situation since 'science is such a powerful and omnipresent way to frame policy questions. But, on the other hand, this very framing means that the way to attack an opponent's position is to attack the science' (Boehlert 2007).

The work by Kahneman (2011) suggests that the majority of decisions are taken fast, as a function of feelings, previous experiences, associations, habits, trivial consequences, or evident preferences. Along this line of thought, scientific knowledge should be influential in decision-making processes based on beliefs and values, so that a solid bond to decision making can be constructed (von Winterfeldt 2013).

The focus of science should not be restricted to provide 'the right information, at the right time, for the right people', but scientific research should also be 'useful, usable and used' (Boaz and Haydn 2002). Nonetheless, there is not a quick route to build a sound interface between science and technology, policy making and practice, unless integrated research into disaster risk is carried out (Alcántara-Ayala et al. 2015; Cutter et al. 2015).

Integrated Research on Disaster Risk

Integrated Research on Disaster Risk (IRDR) is a multi-disciplinary approach to strengthen capacity at global, regional and local levels to address hazards and generating scientific-based decisions on actions to reduce their impacts. Aiming at developing trans-disciplinary, multi-sectorial alliances for in-depth, practical disaster risk reduction research studies and the implementation of effective evidence-based disaster risk policies and practices, the Integrated Research on Disaster Risk program was set up under the umbrella of the International Council for Science (ICSU), the International Social Science Council (ISSC) and the United Nations Office for Disaster Risk Reduction (UNISDR). The main purposes of IRDR involve:

Fig. 4 Goals of the integrated research on disaster risk (IRDR) programme

(a) "Characterizing hazards, vulnerability, and risk by identifying hazards and vulnerability leading to risks, and forecasting, assessing, and dynamic modeling";
(b) "Understanding decision-making in complex and changing risk contexts by identifying decision-making systems, their contexts, and their interactions, and improving the quality of decision-making practice"; and
(c) "Reducing risk and curbing losses through knowledge-based actions through vulnerability assessments, and the analysis of effective approaches to risk reduction" (IRDR 2013)

In order to summarise these purposes Fig. 4 provides a visual summary.

Other Science and Technology Mechanisms for Implementation of the UN Frameworks for Informing Policy and Practice

In seeking to advance the negotiations and discussions process of the post-2015 Framework for Disaster Risk Reduction, three major priority areas for action and scientific engagement in DRR and DRM were recognized by the UNISDR Scientific and Technical Advisory Group (STAG) and the Major Group on Science and Technology, in partnership with Regional and Global Platforms: (a) Sharing knowledge for action; (b) Using a multidisciplinary approach to research; and (c) Building systems resilience through local, national, regional and international partnerships (Aitsi-Selmi et al. 2015).

Concrete recommendations to be enhanced by STAG and IRDR—as the representative of the Major Group on Science and Technology—to build a partnership to support the integration of science in Disaster Risk Reduction (DRR) include:

1. Establishment of national DRR science-policy councils/platforms or national focal points for science to support disaster risk reduction and management;
2. Understanding the root causes and underlying risk factors of disaster risk (see Oliver-Smith et al. 2016);
3. Carrying out a periodic review of knowledge needs, new science, and research gaps;
4. Using the expanding science and technology evidence base to support capacity building and interdisciplinary capacity development for disaster risk management;
5. Leveraging science for DRR by promoting the dialogue between decision makers and researchers to guarantee integrated disaster risk governance;
6. Supporting integrated and holistic approaches to the use of science and technology for DRR;
7. Increasing the role of social science from a multidisciplinary approach to understand behavior and decision making in DRR and the role of the wider societal context in disaster risk creation and reduction;
8. Promoting open access, multi-hazard data platforms and standardized perspectives and tools for mapping and the use of data and scenarios that make science meaningful to decision-makers and people;
9. Using participatory approaches for communities to work together in the co-production of risk knowledge;
10. Investigating and documenting the effects of disasters and DRR interventions, including the ethical dimension of scientific research;
11. Strengthening DRR science-policy and cross-sectoral dialogues to facilitate risk assessments, post-disaster reviews, data sharing, and decision-making;
12. Developing guidelines for evidence-based risk assessments and their implementation to support the practical application of risk assessment (Aitsi-Selmi et al. 2016) (Fig. 5).

Challenges for the Integration of Science into International Policy Development for Landslide Disaster Risk Reduction

In spite of the uncertainty for predicting landslide occurrence in time and space, landslide risk assessments offer valuable insights in terms of hazard evaluation, vulnerability analysis, and elements at risk, which can be useful for simple tasks such as identifying symptoms of slope instability in the field, in addition to landslide mapping, instrumentation, monitoring and for designing and implementing Early Warning Articulated Systems (EWAS's). The main attribute of EWAS's is the inclusion of the human dimension of disaster risk as the core of the system. It involves a process aiming at the comprehensive understanding of disaster risk by the exposed communities by exploring the underlying or root causes of disasters. Likewise, hazard assessment, the analysis of risk perception, and the different multi-dimensions of vulnerability and resilience to apprehend and communicate risk are integrated into the warning system. In consequence, EWAS's intend to provide a mechanism of awareness, knowledge-sharing, preparedness and capacity building for people to understand disaster risk as a social construct and its potential consequences, so that assessment of likely disaster scenarios, realistic structural and non-structural measures, risk management procedures, and response strategies and actions, can be properly enhanced (Alcántara-Ayala and Oliver-Smith 2017).

Investigations concerning landslide hazards and their associated vulnerability have been dominated by

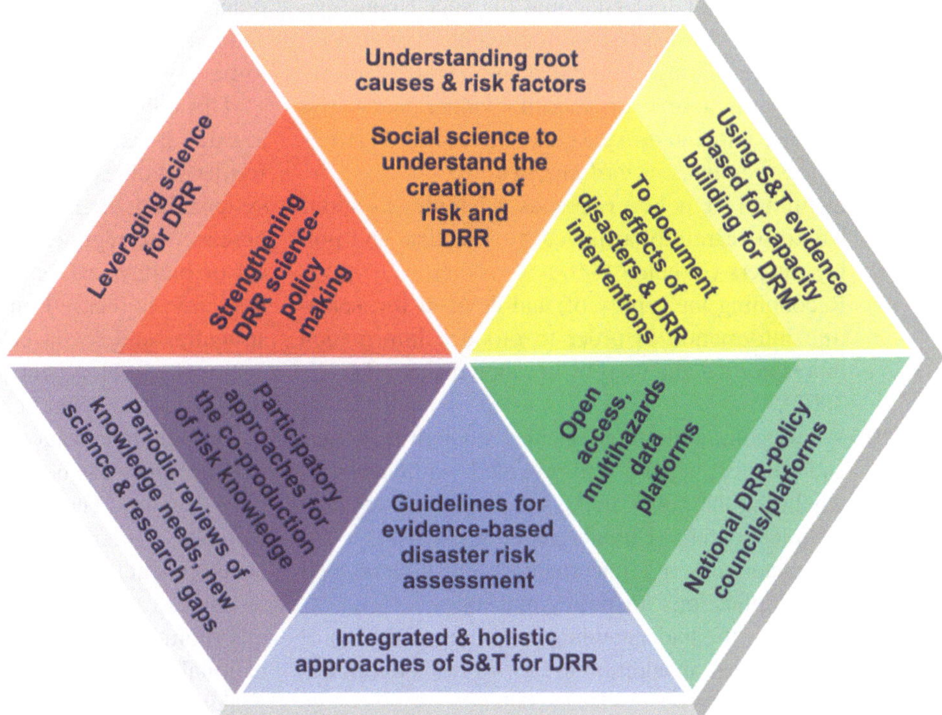

Fig. 5 Recommendations for building the interface between science and technology for policy informing guidance (adapted from STAG and IRDR)

mono-disciplinary perspectives, although quite recently multi-disciplinary and interdisciplinary research have also contributed to the understanding of landslide disasters. As in the case of other hazards, the work and new direction of work on landslide disaster risk, requires the need to move towards integrated research with a transdisciplinary angle, so that different stakeholders can be fruitfully engaged in the co-production of disaster risk knowledge. To do so in the best possible way, a series of challenges must be faced in terms of the approach towards disaster risk per se, and the impasse scenarios generated by the nature of policy making and practice, as follows:

(a) ***Landslide disaster risk approach***

(1) *Understanding disaster risk* as a social construct derived from the interaction between societies and the environment to build territories based on models of development that have been derived in the creation of vulnerability and exposure conditions to hazards (Blaikie et al. 1994; Wisner et al. 2004; Burton 2010, 2015; Oliver-Smith 2013; Oliver-Smith et al. 2016). Along the same line, the use of the term natural disasters should be avoided (Briceño 2015).

(2) Identifying and addressing the ***root or underlying causes*** of disaster risk and disasters (Blaikie et al. 1994; Wisner et al. 2004; Burton 2010, 2015; Oliver-Smith 2013; Oliver-Smith et al. 2016).

(3) Recognizing and dealing with ***dynamic pressures*** or ***drivers of disaster risk***, as they favor vulnerability and exposure and intensify unsafe conditions (Blaikie et al. 1994; Wisner et al. 2004; Burton 2010, 2015; Oliver-Smith 2013; Oliver-Smith et al. 2016). In the case of landslides, particular attention must be given to those aspects known as or associated with anthropogenic determining factors of hillslope instability (i.e., population growth and spread into areas with unstable slopes, the rural-urban interface, and urbanization processes, deforestation processes, mining, excavations, deficient drainage and sewage systems, land degradation, land-use changes, etc.).

(4) Anticipating low-frequency, high magnitude landslide events to try to provide the best possible reliable estimates, and preventing high-frequency, low magnitude landslides by addressing root causes and drivers of risk.

(5) Enhancing the availability of consistent landslide hazard, vulnerability and exposure data, and filling the gaps of landslide knowledge.

(6) Stimulating landslide community-based strategies and applying landslide risk reduction measures to those communities predicted to be at risk in the future (Anderson et al. 2014).

(7) Design and implementation of ***Early Warning Articulated Systems (EWAS's)*** as integrated

mechanisms for understanding the social construction of disaster risk through awareness, knowledge-sharing, preparedness and capacity building in terms of hazard, vulnerability and exposure (Alcántara-Ayala and Oliver-Smith 2017).

(b) ***Scientist, policy makers and practice***

(8) Definition of policy issues as scientific questions has to be carefully achieved by scientists and policy makers (Boehlert 2007).

(9) Reconciling the supply of, and demand for, scientific information in order to produce information that is needed and used by decision makers (McNie 2007).

(10) When communicating science, enough evidence and support should be provided without overwhelming the policy makers and practitioners with technical details (Uhlenbrock et al. 2014).

(11) Characterizing risk uncertainties and their associated implications for policy makers.

(12) Fostering the understanding and communication of uncertainties, in addition to face-to-face continuous interaction.

(13) Strengthening the relationships, trust and information flows necessary for full integration of scientific knowledge into the decision-making process (Jacobs 2002).

(14) Prioritizing livelihoods and well-being over economic interests and balancing public and private interests.

(15) Aiming at the co-production of knowledge as an iterative process, research agendas should be flexible to meet the needs of decision-makers (Lemos and Morehouse 2005).

(16) Expanding the role of chief scientific advisers and advisory boards at national, sub-national and desirably local levels (UNISDR 2015b; Cutter et al. 2015).

Considerations for a Way Forward

The impacts of landslides are just one example of the need for an all-hazard approach to disaster risk reduction. Indeed, progress in addressing disaster risk reduction relies not only on the adoption but also on the parallel implementation of targets, goals and actions derived from the four international agendas. These agendas are related to financing development, sustainability, limiting global temperature increase, enhancing adaptive capacity, strengthening resilience and reducing vulnerability to climate change and disaster risk, and avoiding the creation of future risks, that are considered to be key to reducing future harm.

This collective action can be portrayed as an 'open door' towards transformation and transformational strategies. The former involves 'fundamental changes in the attributes of a system, including value systems; regulatory, legislative, or bureaucratic regimes; financial institutions; and technological or biophysical systems' (O'Brien et al. 2012), whereas the latter, is focused on addressing risk derived from social structures and social behavior, including disaster risk management, development goals, policy, and practice (Nelson et al. 2007; O'Brien et al. 2012).

Despite the advance of science and technology in disaster risk, progress has been relatively limited in the many of the policy arenas. The notion of disaster risk reduction clearly includes a number of dimensions related to sustainable development, availability and mobilization of financing resources, and climate change adaptation that rely on the incorporation of multi-stakeholders at local, sub-national, national and global scales.

As noted in the IPPC "Managing the Risks of Extreme Events and Disasters to Advance Climate Change Adaptation" report (IPCC 2012), integration of disaster risk reduction and management and climate change adaptation with the international development agenda could lead to enormous benefit to policies and practices, at local, sub-national, national, regional and global scales.

The concept of science and technology and the link to develop an evidence base for policy is a complex one, and a promise to engage in a continued communication process towards specific commitments and targets is difficult when economic interests tend to override the sustainable use of the environment.

Regardless of the commitments made by authorities at member state level, the weighty work falls on the shoulders of local authorities, given that only DRR implementation at local scale can lead to desired regional and global deep-seated transformations.

With the confluence of the international initiatives, these historic milestones for the societies of tomorrow require a permanent dialogue among all stakeholders and a solid but transparent atmosphere of partnerships where commitment, responsibility and ethical values are top priority for implementing the 2015 UN Frameworks.

Acknowledgements For Irasema Alcántara-Ayala to lead on this work, thanks are due to CONACyT for the financial support kindly provided through the research project 156242.

References

AAAA (2015) Addis Ababa Action Agenda of the third international conference on financing for development (Addis Ababa Action Agenda), United Nations, New York. Available at http://www.un.org/esa/ffd/wp-content/uploads/2015/08/AAAA_Outcome.pdf

Aitsi-Selmi A, Blanchard K, Al-Khudhairy D, Ammann W, Basabe P, Johnston D, Ogallo L, Onishi T, Renn O, Revi A, Roth C, Peijun S, Schneider J, Wenger D, Murray V (2015) UNISDR STAG 2015 report: science is used for disaster risk reduction. Available at http://preventionweb.net/go/42848

Aitsi-Selmi A, Murray V, Wannous C, Dickinson C, Johnston D, Kawasaki A, Stevance AS, Yeung T, et al (2016) Reflections on a science and technology agenda for 21st century disaster risk reduction. Based on the scientific content of the 2016 UNISDR science and technology conference on the implementation of the Sendai framework for disaster risk reduction 2015–2030. Int J Dis Risk Sci 7(1):1–29. doi:10.1007/s13753-016-0081-x

Alcántara-Ayala I (2016) On the multi-dimensions of integrated research on landslide disaster risk, In: Aversa S, Cascini L, Picarelli L, and Scavia C (eds) Landslides and engineered slopes. Experience, theory and practice, proceedings of the 12th international symposium on landslides (Napoli, Italy, 12–19 June 2016). CRC Press

Alcántara-Ayala I, Oliver-Smith A (2017) The necessity of early warning articulated systems (EWASs): critical issues beyond response. In: Sudmeier-Rieux K, Fernandez M, Penna I, Jaboyedoff M, Gaillard JC (eds) Linking sustainable development, disaster risk reduction, climate change adaptation and migration. Springer (in press)

Alcántara-Ayala I, Altan O, Baker D, Briceño S, Cutter S, Gupta H, Holloway A, Ismail-Zadeh A, Jiménez Díaz V, Johnston D, McBean G, Ogawa Y, Paton D, Porio E, Silbereisen R, Takeuchi K, Valsecchi G, Vogel C, Wu G, Zhai P (2015) Disaster risks research and assessment to promote risk reduction and management. In: Ismail-Zadeh A, Cutter S (eds) Ad Hoc group on disaster risk assessment (ICSU-ISSC), Paris. Available from: http://www.icsu.org/science-for-policy/disasterrisk/documents/DRRsynthesisPaper_2015.pdf (online)

Anderson M, Holcombe E, Holm-Nielsen N, Della Monica R (2014) What Are the emerging challenges for community-based landslide risk reduction in developing countries? Nat Hazards Rev, pp 128–139. doi:10.1061/(ASCE)NH.1527-6996.0000125

Blaikie P, Cannon T, Davis I, Wisner B (1994) At risk: natural hazards, people's vulnerability and disasters. Routledge, New York

Boaz A, Hayden C (2002) Pro-active evaluators: enabling research to be useful, Usable and Used. Evaluation 8(4):44053

Boehlert SL (2007) The role of scientists in policymaking. AAAS-CSPA S&T policy review: highlights from the 2007 forum on S&T policy

Briceño S (2015) Looking back and beyond Sendai: 25 years of international policy experience on disaster risk reduction. Int J Disaster Risk Sci 6(1):1–7

Burton I (2010) Forensic disaster investigations in depth: a new case study model. Environ Sci Policy Sustain Dev 52(5):36–41

Burton I (2015) The forensic investigation of root causes and the post-2015 framework for disaster risk reduction. Int J Disaster Risk Reduct 12:1–2

Cutter SL, Ismail-Zadeh A, Alcántara-Ayala I, Altan O, Baker DN, Briceño S, Gupta H, Holloway A, Johnston D, McBean GA, Ogawa Y, Paton D, Porio E, Silbereisen RK, Takeuchi K, Valsecchi GB, Vogel C, Wu G (2015) Global risks: pool knowledge to stem losses from disasters. Nature 522:277–279

Dilley M, Chen RS, Deichmann W, Lerner-Lam AL, Arnold M (2005) Natural disaster hotspots: a global risk analysis. The World Bank, Washington

ICSU—LAC (2010) Science for a better life: developing regional scientific programs in priority areas for Latin America and the Caribbean. In: Cardona OD, Bertoni JC, Gibbs T, Hermelin M, Lavell A (eds) Understanding and managing risk associated with natural hazards: an integrated scientific approach in Latin America and the Caribbean, vol 2. ICSU—LAC/CONACYT, Rio de Janeiro and Mexico City, 88 pp

IPCC (2012) Summary for policymakers. In: Field CB, Barros V, Stocker TF, Qin D, Dokken DJ, Ebi KL, Mastrandrea MD, Mach KJ, Plattner GK, Allen SK, Tignor M, Midgley PM (eds) Managing the risks of extreme events and disasters to advance climate change adaptation. A special report of working groups I and II of the intergovernmental panel on climate change. Cambridge University Press, Cambridge, UK, and New York, NY, USA, pp 3–21

IRDR (2013) Integrated research on disaster risk strategic plan 2013–2017, Beijing, China. http://www.irdrinternational.org/wp-content/uploads/2013/04/IRDR-Strategic-Plan-2013-2017.pdf

Jacobs K (2002) Connecting science, policy, and decision making: a handbook for researchers and science agencies. National Oceanic and Atmospheric Administration, Office of Global Programs

Kahneman D (2011) Thinking, fast and slow (Ferrar, Straus, and Giroux, New York)

Kennedy I T R, Petley D N, Williams R, Murray V (2015) A systematic review of the health impacts of mass earth movements (landslides). In: PLOS currents disasters, 1 ed

Lavell A (1996) Degradación Ambiental, Riesgo y Desastre Urbano: Problemas y Conceptos. In: Fernández MA Ciudades en Riesgo. Lima, Perú: LA RED. USAID

Lavell A (2003a) Sobre la Gestión del Riesgo: Apuntes hacia una Definición. Available from: http://disaster-info.net/cepredenac/pdf/pnud/productos/Documentos/definicion.pdf (online)

Lavell A (2003b) La gestión local del riesgo, nociones y precisiones en torno al concepto y la práctica. CEPREDENAC–PNUD

Lemos MC, Morehouse B (2005) The co-production of science and policy in integrated climate assessments. Glob Environ Change 15:57–68

McNie EC (2007) Reconciling the supply of scientific information with user demands: an analysis of the problem and review of the literature. Environ Sci Policy 10:17–38

Nelson DR, Adger N, Brown K (2007) Adaptation to environmental change: contributions of a resilience framework. Ann Rev Environ Resour 32:395–419

O'Brien K, Pelling M, Patwardhan A, Hallegatte S, Maskrey A, Oki T, Oswald-Spring U, Wilbanks T, Yanda PZ (2012) Toward a sustainable and resilient future. In: Field CB, Barros V, Stocker TF, Qin D, Dokken DJ, Ebi KL, Mastrandrea MD, Mach KJ, Plattner GK, Allen SK, Tignor M, Midgley PM (eds) Managing the risks of extreme events and disasters to advance climate change adaptation. managing the risks of extreme events and disasters to advance climate change adaptation. A special report of working groups I and II of the intergovernmental panel on climate change. Cambridge University Press, Cambridge, UK, and New York, NY, USA, pp 437–486

Oliver-Smith A (2013) A matter of choice. Int J Disaster Risk Reduct 3 (1):1–3

Oliver-Smith A, Alcántara-Ayala I, Burton I, Lavell A (2016) Forensic investigations of disasters (FORIN): a conceptual framework and guide to research. IRDR FORIN Publication No. 2. Integrated Research on Disaster Risk, ICSU, Beijing, 56 pp

Petley DN (2012) Landslides and engineered slopes: protecting society through improved understanding. In: Eberhardt E, Froese C,

Turner K, Leroueil S (eds) Landslides and engineered slopes. CRC Press, Canada
Rio Declaration (1992) Report on the UN conference on environment and development, Rio de Janeiro, 13–14 June 1992, UN doc. A/CONF.151/26/Rev.1 (vols 1–III)
Uhlenbrock K, Landau E, Hankin E (2014) Science communication and the role of scientists in the policy discussion. In: Drake JL, Kontar YY, Rife GS (eds) New trends in earth-science outreach and engagement advances in natural and technological hazards research, vol 38. Springer, Enfield, pp 93–105
UNFAO (2015) (http://www.fao.org/3/a-i5128e.pdf)
UNFCCC (2015) Adoption of the Paris agreement, COP 21, Paris, France. Available at: https://unfccc.int/resource/docs/2015/cop21/eng/l09r01.pdf
UNISDR (2009) Terminology on disaster risk reduction. Available from: http://www.unisdr.org/files/7817_UNISDRTerminologyEnglish.pdf (online)
UNISDR (2015a) Sendai framework for disaster risk reduction 2015–2030, Geneva, Switzerland
UNISDR (2015b) Proposed updated terminology on disaster risk reduction: a technical review, background paper. The United Nations Office for Disaster Risk Reduction. Geneva, Switzerland. Available from: http://www.preventionweb.net/files/45462_backgoundpaperonterminologyaugust20.pdf (online)
UNISDR (2015c) Disaster Risk reduction and resilience in the 2030 agenda for sustainable development, Geneva, Switzerland
United Nations (2015) Transforming our world: the 2030 agenda for sustainable development, A/RES/70/1
von Winterfeldt D (2013) Bridging the gap between science and decision making. Proc Natl Acad Sci U S A 110(S3):14055–14061
Wisner B, Blaikie P, Cannon T, Davis I (2004) At risk: natural hazards, people's vulnerability and disasters (2nd ed). New York: Routledge

World Federation of Engineering Organizations (WFEO)—World Federation of Engineering Organizations Activities in Disaster Risk Reduction

Kenichi Tsukahara

Abstract

In the Sendai Framework, the importance of coordination mechanisms within and across sectors and with relevant stakeholders at all levels is stressed. Engineers are working for disaster risk management in the public sector, private sector, and community. Thus, the World Federation of Engineering Organizations (WFEO), which represents the engineering profession worldwide, should take an important role in achieving the goals of the Sendai Framework. This paper explains what WFEO and WFEO's committee on disaster risk management (CDRM) are, and activities of CDRM in achieving goals of the Sendai Framework.

Keywords

Sendai framework • Multisectoral/transdisciplinary approach • Implementation of DRR

Outline and Mission of the World Federation of Engineering Organizations

Outline

The World Federation of Engineering Organizations (WFEO) is an international, non-governmental organization representing the engineering profession worldwide. On March 4th, 1968, representatives of 50 scientific and technical associations from all over the world met under the auspices of the United Nations Educational, Scientific and Cultural Organization (UNESCO) in Paris to establish WFEO, whose charter as an international, non-governmental organization is to unite multidisciplinary engineering associations throughout the world. WFEO encourages all of its national and international members to contribute to global efforts to establish a sustainable, equitable and peaceful world by providing an international perspective and enabling mechanisms:

(1) To provide information and leadership to the engineering profession on issues of concern to the public or the profession.
(2) To serve society and to be recognized, by national and international organizations and the public, as a respected and valuable source of advice and guidance on the policies, interests and concerns that relate engineering and technology to the human and natural environment.
(3) To make information on engineering available to the countries of the world and to facilitate communication between its member nations.
(4) To foster peace, socioeconomic security and sustainable development among all countries of the world, through the proper application of technology.
(5) To facilitate relationships between governments, business and people by adding an engineering dimension to discussions on policies and investment (Fig. 1).

Mission of the WFEO

WFEO is the internationally recognized and chosen leader of the engineering profession and cooperates with national and

K. Tsukahara (✉)
Secretary of Committee on Disaster Risk Management of World Federation of Engineering Organizations, Paris, France
e-mail: tsukahara@doc.kyushu-u.ac.jp

K. Sassa et al. (eds.), *Advancing Culture of Living with Landslides*,
DOI 10.1007/978-3-319-59469-9_11

Fig. 1 WFEO general assembly at Kyoto in 2015

other international professional institutions in being the lead profession in developing and applying engineering to constructively resolve international and national issues for the benefit of humanity. Based on this mission, WFEO's mission is described as follows:

(1) To represent the engineering profession internationally, providing the collective wisdom and leadership of the profession to assist national agencies choose appropriate policy options that address the most critical issues affecting countries of the world.
(2) To enhance the practice of engineering—to make information on engineering available to the countries of the world and to facilitate communication between its member nations about the world's best practices in key engineering activities.
(3) To foster socio-economic security and sustainable development and poverty alleviation among all countries of the world, through the proper application of technology.
(4) To serve society and to be recognized by national and international organizations and the public, as a respected and valuable source of advice and guidance on the policies, interests and concerns that relate engineering and technology to the human and natural environment.
(5) To cooperate with funding agencies such as development banks.
(6) To encourage public–private partnerships by including the engineering dimension.
(7) To address the issue of what public policies need to be implemented.

Committee on Disaster Risk Management of WFEO

Background of CDRM

In general, disasters are caused by natural hazards and/or manmade incidents. The disasters referred to hereafter arise as consequence of natural hazards such as earthquakes, tsunamis, floods, droughts, cyclones, typhoons or hurricanes, landslides, forest fires, and volcanic eruptions; they are also caused by manmade incidents such as industrial accidents, power cuts, and air-pollution. These disasters affect human lives directly or indirectly, while having an impact on human societies and natural environments.

There are various phases of disaster risk management (DRM)—from identifying, evaluating, mitigating, and managing disaster risks to adapting to climate change. Risk management includes disaster preparedness programs, disaster response, disaster damage assessment, reconstruction, and rehabilitation. Note that there are various types of damage, including direct and indirect damage, physical and non-physical damage; monetary and non-monetary damages; short-term, mid-term and long-term damage; as well as economic damage and the loss of human life. Hence, approaches to mitigation and adaptation activities should be evaluated for each disaster in cost-benefit terms.

Based on this understanding, establishment of the Committee on Disaster Risk Management of WFEO (CDRM) as a standing committee was approved by the General Council of WFEO on November 2009. Dr. Yumio Ishii was appointed as the founding president of WFEO-CDRM.

Dr. Ishii served as the president of CDRM until 2013, and Dr. Toshimitsu Komatsu was appointed as the second president from 2013.

CDRM helps to develop and implement advanced expertise and practical training to achieve societies that are strong and resilient in the face of disaster, while mobilizing every available resource to reduce disaster risks and potential damage. The extent and level of damage and risk should be discussed and determined within the CDRM. The definition of "risk" is not necessarily restricted to likely dangers. "Risk" also has a broader meaning, related to the mitigation of and adaptation to situations that result from disasters.

It is also important to consider not only single hazards or risks, but also the interrelation between multiple hazards and risks. After the 2011 Tohoku earthquake and tsunami in Japan, "resilience" has also been an important key word in disaster risk management. It is particularly important to consider the role of resilience (in conjunction with disaster reduction) when considering global climate change, where the disaster cannot be prevented by means of current technology and infrastructure. Thus, the concept of DRM can include "resilience."

Vison and Mission of CDRM

The CDRM vision is to be a WFEO standing technical committee that professionally and internationally carries out activities to reduce natural disaster risks, from an engineering perspective. It must introduce and recommend useful practices and lessons, as well as engineering knowledge and innovative approaches to disaster damage reduction and the promotion of sustainable, sound development. The CDRM should contribute to all parts of the risk cycle, including identification, evaluation, the mitigation of, and the adaptation to disasters, which relate to all aspects of human life as well as to the economy, social activities, and the environment. The CDRM's priorities include developing comprehensive structural and non-structural measures to increase disaster resiliency and business continuity.

As a standing committee of WFEO, the CDRM aims to exchange, share, and transfer knowledge, technology, and expertise in order to reduce disaster risks. It also fosters research and investigations that relate to DRM, including examples of best practice, lessons, and their implementation. The CDRM will create advisory documents, policy papers, guidelines, reports, and booklets.

The CDRM will gather and disseminate DRM-related information that can help WFEO member countries, engineering societies, and leading engineers effectively mitigate risk and help societies adequately adapt to potential risks. The CDRM also helps to build the capacity of engineers working in these areas, by disseminating information through WFEO member country linkages to decision makers, governmental organizations, engineering societies, and leading engineers. The CDRM also coordinates international DRM efforts, and organizes conferences, presentations, and workshops related to DRM.

Activities of CDRM

The CDRM operates as a knowledge-hub that focuses on recognizing and promoting the worldwide contributions of engineers and practitioners, and on providing need-driven education and information to engineering communities around the world, especially in developing countries. The scientific analysis and advice on reducing risk (a combination of hazards and vulnerability) provided and distributed by the CDRM, provides an opportunity to improve societies and ways of living. In particular, understanding the scientific basis of a particular risk creates a crucial opportunity to make sustainable societal adjustments instead of unsustainable ones. Reducing disaster risk reduces damage that might otherwise impede continued economic development and environmental sustainability. Sharing best practice and lessons learned, as well as DRM networking and information-sharing, are two important operating principles of the CDRM. Natural disasters are categorized as either water-related disasters or earthquake-related disasters for further investigation and discussion in the CDRM.

Because there are a wide variety of disaster and risk management categories, diverse methods and technologies are needed to deal with them. During its initial two-year cycle of activities, the CDRM only considered water- and earthquake-related disasters. As a result of its good outcomes between 2009 and 2013, the committee was strongly advised to expand the scope of its activities to other disasters such as air pollution, volcanic explosions, forest fires, drought, desertification, and communicable diseases. Accordingly, the CDRM has upgraded its activities through the following three subcommittees: Water-related Disaster Risk Management (WDRM), Earthquake-related Disaster Risk Management (EQDRM), and Capacity Building for Natural Disaster Risk Management (CBNDRM), in order to deal with many different types of disasters.

WFEO/CDRM and Sendai Framework

In January 2015, the Tokyo Conference on International Study for Disaster Risk Reduction and Resilience was held in Tokyo. World leaders and top scientists participated in the conference prior to the Third UN World Conference on Disaster Risk Reduction to discuss and formulate how science and technology could help in disaster risk reduction, hence fostering sustainable development. Core members of CDRM participated in the conference and made substantive inputs from the viewpoint of engineering onto DRM. In particular, Dr. Komatsu, the chair of CDRM, was invited as a guest speaker and delivered a speech entitled "Implementation oriented technology for coping with natural disasters under the climate change."

In January 2016, the UNISDR Science and Technology Conference on the implementation of the Sendai Framework for Disaster Risk Reduction was held in Geneva and discussed how the science and technology community will best support the implementation of the Sendai Framework. Many engineers and engineering organizations actively participated in this conference and displayed how engineering supports the implementation of the Sendai Framework.

At the Asian Civil Engineering Coordinating Council (ACECC) 7th Civil Engineering Conference of Asian Region in August 2016, a Technical Committee 21 (TC21): Transdisciplinary Approach (TDA) for Building Societal Resilience to Disasters was launched to promote a trans-disciplinary approach for scientific knowledge-based decision-making for building societal resilience to disasters at national and local levels. As the Civil Engineering community is one of the most active areas to work with implementation of the Sendai Framework, CDRM has been working with TC21 to promote DRM activities of WFEO.

The World Engineering Conference and Convention 2015, Kyoto

WFEO holds a World Engineers' Convention (WEC) at intervals of four years. The 1st WEC took place in Germany in 2000, the 2nd in China in 2004, the 3rd in Brazil in 2008, and the 4th in Switzerland in 2011. The World Engineering Conference and Convention (WECC 2015), the 5th WEC, was held in Kyoto, Japan, from 29 November to 2 December 2015. This event showcased Japan as a nation built on scientific and technological innovation to achieve sustainable development in the 21st century. WECC2015 also provided a forum for sharing information focused on innovations required to make sustainable development a reality and advances in fundamental technologies to achieve this vision. Adopting both local and global perspectives, WECC2015 made a productive contribution to worldwide peace and socioeconomic progress.

The overall conference theme of WECC2015 was "Engineering: Innovation and Society", and 10 tracks were convened in WECC2015. Among the 10 tracks, "Resilient Infrastructure for Society" is placed as the first track of WECC 2015. This track dealt with several aspects of disaster risk management from the viewpoint of engineering. In this track, the following items were presented and discussed:

1.1 Reconstruction innovation
1.2 Land/city conservation and disaster mitigation
1.3 Robot technology used at disaster sites and its operating systems
1.4 Strengthening national interests and creating new industries using big data
1.5 Creating a resilient economy
1.6 Resilience in the manufacturing and energy sectors.

In addition to these tracks, several prominent speakers delivered important lectures at WECC2015. The Crown Prince of Japan delivered an address. In His Imperial Highness's address he said "I believe that it is of great importance to hold discussions at WECC2015 as to how we can realise innovations and solve those challenges that face communities, such as energy issues as well as natural disasters and the safety of people" HIH expressed his expectation of engineering innovation in disaster risk reduction. In addition, Dr. Han Seung-soo, the UN Secretary-General's Special Envoy on Disaster Risk Reduction and Water, and Former Prime Minister of the Republic of Korea, delivered a lecture "Science and Technology for Water in a Carbon-Constrained World". The lecture was delivered from the perspectives of climate change and disaster management, and centered on the importance of science and technology in resolving these issues. Dr. Han explained that the importance of evidence-based policy decisions through science and technology had been being emphasized in major conferences in 2015, such as the Third UN World Conference on Disaster Risk Reduction in Sendai, the 7th World Water Forum in Daegu, Republic of Korea, and the UN Sustainable Development Summit in New York.

Dr. Yumio Ishii, the founding chair of CDRM, served as the Executive Committee Chair and made an enormous effort over many years in WECC2015 preparation (Fig. 2).

At WECC2015, CDRM held an international symposium on "River Technologies for Innovations and Social Systems" (hereafter, "River Technology") attended by over 100 participants. This symposium comprised opening and closing remarks, as well as three sessions, covering river/watershed disaster mitigation, environmental conservation, and a panel discussion to enhance worldwide information exchange on and public awareness of river technologies. Speakers had varying backgrounds related to disaster risk reduction (Fig. 3).

Conclusion

This paper explained what WFEO and WFEO's committee on disaster risk management (CDRM) are, and the activities of CDRM in achieving the goals of the Sendai Framework. Both The International Consortium on Landslides (ICL) and WFEO are international

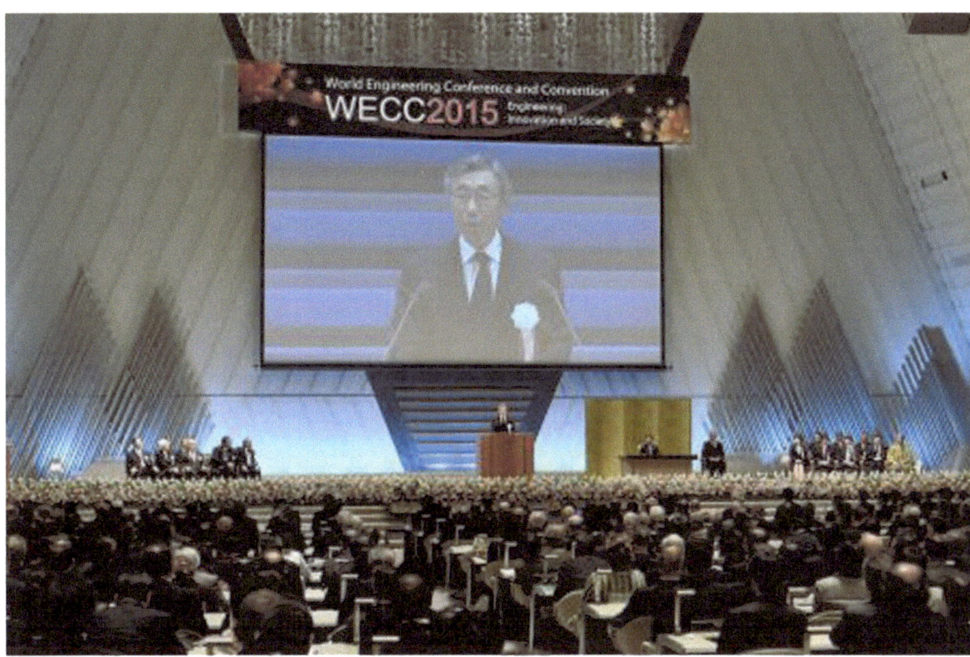

Fig. 2 Closing remarks—Dr. Ishii at WECC2015

Fig. 3 CDRM symposium at WECC2015

non-governmental and non-profit organizations that aim to utilize their member expertise to realize a better society, including reducing disaster risks. As stated in the paper, many important initiatives have been launched to support implementation of the Sendai Framework in the science and technology community. As ICL and CDRM have common objectives that utilize their scientific and engineering knowledge of DRR, close coordination of ICL and CDRM would make valuable inputs to implementing the Sendai Framework.

References

WFEO, Homepage "About". http://www.wfeo.org/about/

WFEO-CDRM, Homepage "CDRM Main Page". http://www.wfeo.org/stc_disaster_risk_management/

WFEO-CDRM, CDRM E-Newsletter—January 2016. http://www.wfeo.org/wp-content/uploads/stc-disaster_risk/CDRM_Newsletter_2016_Jan_No5_Final.pdf

The Organizing Committee of the World Engineering Conference and Convention, Report on the World Engineering Conference and Convention (WECC2015), March 2016. http://www.jfes.or.jp/wecc2015/common/pdf/wecc2015_report_e.pdf

International Union of Geological Sciences (IUGS)—Sendai—Foreseeable but Unpredictable Geologic Events—IUGS Reactions

Roland Oberhänsli, Yurijo Ogawa, and Marko Komac

Abstract

This paper gives an overview of the International Union of Geological Sciences (IUGS) activities that are related to the geologically related events—geohazards that pose risk to contemporary society. As geohazards are common events, IUGS has established an initiative with the aim to address the issue from a geological perspective and consequentially enable more holistic approach to the geohazds, including understanding processes, approaching them with the most effective solutions and educating public.

Keywords

Geohazard • Geological risk • IUGS

What Is IUGS?

The International Union of Geological Sciences (IUGS) is the international umbrella organisation for all geoscientists. It has represented all geoscientists at the highest international level since its formation in 1961 and supports both fundamental research and applied aspects of the Earth Sciences of an international and interdisciplinary nature. Through member countries and affiliate members it represents around a million geoscientists. Its main scientific forum is the International Geological Congress (IGC). IUGS promotes development of the Earth Sciences through:

- developing broad-based scientific studies relevant to the entire Earth system;
- applying the results of these and other studies to preserving Earth's natural environment, using all natural resources wisely and improving the prosperity of nations and the quality of human life;
- strengthening public awareness of geology and advancing geological education in the widest sense;
- developing/endorsing geoscientific standards such as stratigraphic charts, rock nomenclature, GIS data management and interoperability, isotope half-life times, as well as geochemical baselines;
- its new strategic initiative "Resourcing Future Generations".

IUGS Risk and Hazard Activities

The "Risk and Hazard" (R&H) theme, for the long term, was at the heart of the International Geophysical Union (IUGG), the home of seismology and volcanology. In IUGS main emphasis was placed on rock-fall, landslides and flooding. Because tectonic signs of R&H are obvious in the sedimentary record, in 2013 IUGS decided to place more focus on disaster risk reduction and hazards.

R. Oberhänsli (✉)
Institute for Earth & Environmental Sciences, Potsdam University, Karl-Liebknecht Strasse 24, 14467 Potsdam, Germany
e-mail: R.Oberhaensli.IUGS@geo.uni-potsdam.de

Y. Ogawa
University of Tsukuba, 1-1-1 Ten-Nodai, Tsukuba, 305-8572, Japan
e-mail: fyogawa45@yahoo.co.jp

M. Komac
OneGeology, IUGS—International Union of Geological Sciences, Ljubljana, Slovenia
e-mail: m.komac@telemach.net

K. Sassa et al. (eds.), *Advancing Culture of Living with Landslides*,
DOI 10.1007/978-3-319-59469-9_12

The basic phenomena of earthquakes and volcanic eruptions, for instance, are primarily geological, such as earthquakes being related to active fault movement, and volcanic activities being due to magma genesis and tectonic conditions. It is well known that earthquakes are triggers of liquefaction and land sliding (slope failure) and also, if occurring under the sea, of tsunami. Historic occurrences are recorded as geological events in landscapes (topography) or outcrops of sediments showing deformation, or both (as shown in Fig. 1).

While many affiliated organizations already had a focus on these themes, IUGS, with its former Secretary General Ian Lambert, decided to undertake an initiative on Hazard and Risk. The first step took place after the Tohoku Tsunami

Fig. 1 *Above* record of a major tectonic event (paleoearthquake) in sediment accreted in a volcanic arc, Bono peninsula, Japan (*above*); soft-sediment deformation (paleoearthquake?) in a fluvio-lacustrine sequence near Lake Van, Turkey (©R. Oberhänsli) (*below*)

2011. This led to an IUGS sponsored G-EVER meeting at Sendai, Japan. The outcome of the meeting was the Sendai agreement that was published as part of a special issue of EPISODES (Vol. 37(4) 2014). Following this first meeting, which was combined with an excursion to the areas damaged by the tsunami, the special issue was published. In a trench dug near Iwanuma, not only were the effects of the 2011 Tohoku tsunami visible, but also the 1611 Keicho event, as well as one before the 915 Towada volcanic eruption (Fig. 2).

Renewed attendance at the G-EVER meetings and a series of IUGS contributions at the "Capacity Building of Earth Sciences toward Decrease of Geohazards: Establishment of Global Networks for Decreasing Geohazards" Workshop at the 3rd United Nations World Conference on Disaster Reduction at Sendai, Japan finally led to the concept and establishment of a Task Group (TG) on Disaster Risk Reduction and Geo-hazards.

By convening this task group, IUGS expanded its responsibilities for dealing with societal impacts of geologic events. Effects of energy release in geologic systems and the socio-economic consequences of foreseeable but unpredictable events (Fig. 3) to society are taking an increasingly prominent position in modern geosciences.

This calls for new efforts, not only in specific geoscience education, but also for public literacy in earth science and increasing understanding at the administrative level of governance. This need led to the establishment of courses in geo-governance for political administrators at Potsdam University in Germany; an inspiring initiative for IUGS.

Of course, geologic phenomena related to the disaster risk and hazard themes had been worked on much earlier under the auspices of IUGS, but mostly out of scientific curiosity through basic investigations of geologic processes undertaken by:

- the Commission on Tectonics and Structural Geology (TekTask) focusing on geodynamic processes and their geologic expression such as thrusts and faults, paleo earthquakes, landslides, rock falls, etc.
- the Commission of Geosciences for Environmental Management (GEM) dealing with man-made strata and geo-pollution particularly in respect of liquefaction-fluidization.

Y. Ogawa' s observation with interpretation of the archaeological trench site (Stop 2) of Iwanuma City, Sendai Plain, Miyagi Prefecture (Oct. 21, 2013; Red and white band scale 5 cm long each; refer to HP of Iwanuma City report as http://www.city.iwanuma.miyagi.jp/kakuka/050300/050302/kikakuten.html)
Mud layer deposited during inundation stage of March 11, 2011.
Turbidite-like sandy deposits of normal grading of sands with upper parallel laminations during the first tsunami overwash, March 11, 2011.
Paddy field muddy layer probably cultivated during Edo period (17th C) to 2011 (Present upper part yields late 18th C potteries, the original upper part may be eroded by Tohoku Tsunami).
Yellow, fine-grained sandy deposits probably corresponding to 1611 Keicho tsunami. Upper part was dated 17th C, lower part 15th C.
Peat layer, suggesting swamp environments.
Ash layer from Towada (To-a) AD 915, convoluted into pinch-and-swell structure (by liquefaction?)
Jogan tsunami deposits of AD 869. Laminations or irregular layers may be roots of plant of the late peat stage??
Mud layer of lacustrine environments. Several tsunami deposits were recognized below in other trenches or cores, showing approx. 1000 yrs intervals.

Fig. 2 Trench in the Sendai Plain near Iwanuma city recording three major tsunamis: Tohoku 2011, Keicho 1611 and one before the Towada volcanic eruption 915. (Ogawa et al. 2014)

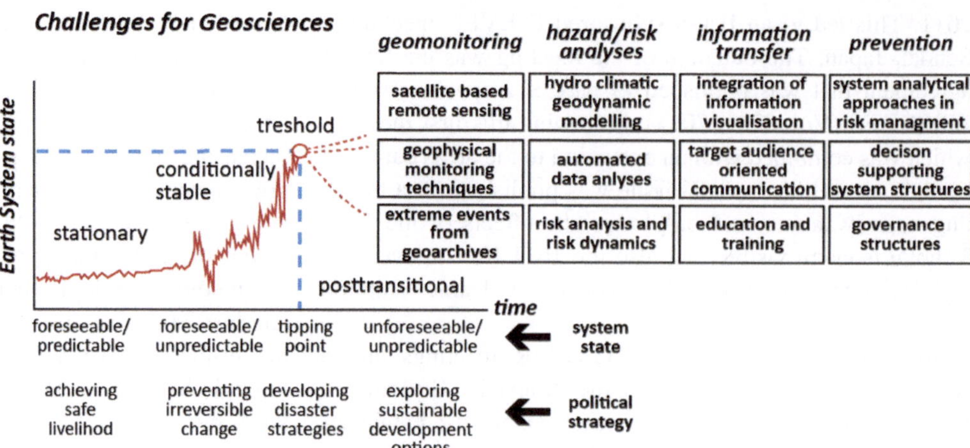

Fig. 3 Preparing for the unpredictable. Consequences for geoscience education and communication with a target audience (adapted from www.geogovernance.de and documents of Potsdam Earth Science Institutions; courtesy M. Strecker)

– the Commission of Management and Application of Geoscience Information (CGI). The efforts of this commission producing standards for IT-based maps are also relevant for topical maps on risk distribution.

The joint-research programs that IUGS runs in co-operation with UNESCO and the International Union of Geodesy and Geophysics (IUGG) have distinct foci and a long history in addressing problems of risk and hazard. With UNESCO, the IUGS has run for over 45 years the International Geoscience Program (IGCP) and with IUGG the International Lithosphere Program (ILP).

One of IGCP science themes is: "Geohazards: mitigating the risks". A selection of recent and active projects under this umbrella is listed below. These encompass historical as well as health aspects such as: earthquake archaeology; medical geology; the environmental and health impacts of major and abandoned mines in Sub-Saharan Africa; impacts of mining on the environment in Africa; preparing for coastal change; seismo-tectonics and seismic hazard in Africa; the significance of modern and ancient submarine slopes and landslides; and deformation and fissuring caused by exploitation of subsurface fluids. In addition to this plethora of geohazard-focused research projects in the past 45 years, several have also addressed the landslide issue. Some of such IGCP projects are Submarine Mass Movements and Their Consequences (IGCP-585 project): S4SLIDE—Assessing Geohazards, Environmental Implications and Economic Significance of Submarine Landslides across the World's Continental Margins (IGCP-640 project), E-MARSHAL—Submarine Mass Movements and Their Consequences (IGCP-511 project), M3EF3—Mechanisms, Monitoring and Modelling Earth Fissure generation and Fault activation due to subsurface Fluid exploitation (IGCP-641 project) and the Landslide Hazard Assessment and Cultural Heritage (IGCP-425 project) that resulted in a "spin-off" the International Consortium on Landslides (ICL) and it's very successful initiative, the International Programme on Landslides (IPL), which focuses on research and capacity development in the field of landslide themes.

All aspects of the IGCP program, which was the first scientific program to be started under the UNESCO logo jointly with IUGS and which served as a template for other UNESCO science programs, are compiled in "Tales Set in Stone" (Fig. 4). The program has been reshaped recently but keeps its important place and logo in the new structure 2015 of the Earth Science Division of UNESCO that now also includes a risk & hazard department.

A second joint program that IUGS runs with IUGG is devoted to scientific studies of the Lithosphere. Different task groups focused on seismology, magmatism, volcanism and geodynamics undertake curiosity-driven research relevant to disaster risk and hazard assessment. A flagship for modern research with high societal relevance is the International Continental Drilling Program ICDP (http://www.icdp-online.org) with the three main research areas: climate and ecosystems, sustainable geo-resources and natural hazards. The ICDP hazard theme focuses on faults, volcanic eruptions, impact structures and plate boundaries. From studies on faults, principles for understanding seismic processes and early warning systems are being developed (e.g. GONAF; http://www.gonaf.de/).

IGCP continues to be one of the IUGS tools that help different research fields with seed money in the first phases of geoscientific networking. ICL is a living proof that such an approach is successful. Therefore IUGS will continue to support interesting and relevant topics, including various geohazard research topics, into the future.

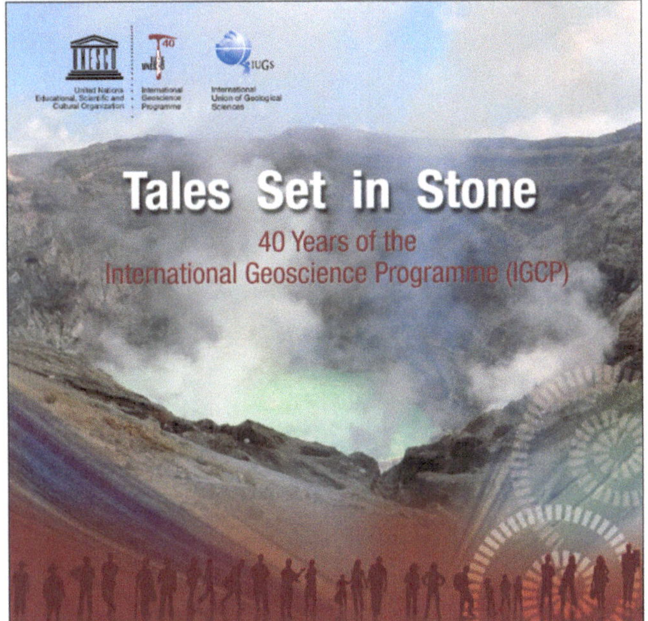

Fig. 4 UNESCO IGCP brochure, 2015 (*above*); Rock fall, 1978, in the northern Adula nappe, Switzerland (©R. Oberhänsli) (*below*)

Acknowledgements Authors would like to thank all geoscientists around the world that contribute to IUGS expertise and its bodies with their knowledge and expertise.

References

Derbyshire E (ed) (2012) Tales set in stone—40 Years of the International Geoscience Programme (IGCP).—Global earth observation section of the United Nations Educational Scientific an Cultural Organisation, 7, place de Fontenoy, 75352 Paris 07 SP, France

Lambert I, Oberhänsli R (2014) Towards more effective risk reduction: catastrophic tsunami. Episodes 37(4):227–228

Ogawa Y, Dile Y, Takarada S (2014) Episodes 37(4). http://www.episodes.org/journal/ArchiveArticle.do

Strecker et al. (2011): PROGRESS (Potsdam Research Cluster for Georisk Analyses; Environmental Change and Sustainability) internal report. University of Potsdam. 123 pp

The Sendai Agreement (2014) Episodes 37(4):329–331. http://g-ever.org/en/sendai/index.html

Tsukuda E, Takarada S, Kuwahara Y, Ishikawa Y, Koizumi N, Uchida T, Takada A, Shigematsu N, Furukawa R, Maruyama T, Ando R, Hara J, Bandibas J (2014) Report of the 2nd G-EVER international symposium and the 1st IUGS and SCJ international workshop on natural hazards and the "Sendai Agreement". Episodes 37(4):329–331

International Union of Geodesy and Geophysics (IUGG)—Integrating Natural Hazard Science with Disaster Risk Reduction Policy

Alik Ismail-Zadeh

Abstract

Science-driven approaches to disaster risk reduction and management can help communities and governments become more resilient and reduce the human and economic impacts of disasters. The International Union of Geodesy and Geophysics (IUGG) promotes international scientific research and cooperation in natural hazards and disaster risks, and contributes to development of sound scientific knowledge on hazards, based on monitoring of physical phenomena and integrated observations, analysis, and modeling. IUGG makes scientific information available to people, and bridges advanced science with policymaking via international and intergovernmental programs. This report describes the union's major activities in the area of hazard and risk research and considers potential contribution of IUGG to the Sendai Partnerships. The contribution could include assessments of landslide hazards and risks; development of a scientific background to high-precision early warning systems for landslides; geophysical and geodetic monitoring of landslides; analysis and modeling of landslides and other rapid land movements; and relevant science education and capacity building.

Keyword

Geohazards • Risk assessment • International cooperation • IUGG

Introduction

Disasters triggered by geophysical events (e.g. landslides, earthquakes, volcanoes, tsunamis) continue to grow in number and impact. In many regions, geohazards are becoming direct threats to national security because their impacts are amplified by rapid growth of population, and unsustainable development practices, both of which increase exposure and vulnerabilities of communities, capital, and environmental assets. Reducing disaster risk using scientific knowledge is a foundation for sustainable development (Cutter et al. 2015). For example, risk caused by landslides is evolving and growing despite considerable progress in understanding of mechanisms triggering landslides (e.g. Sassa and Canuti 2009; Reichenbach and Günther 2014). Our knowledge of geohazards and their interaction with human systems is lacking in some important areas and is being challenged by the unforeseen or unknown repercussions of a rapidly changing and increasingly interdependent world – one transformed by technological change, globalization of economic systems, and political and economic instability. In such a tightly coupled world, a disaster not only affects the immediate area where it occurs, but also has cascading impacts that can affect other nations near and far. Co-designed and co-productive integrated research on disaster risk and science-based disaster risk assessments coupled with political decisions could significantly reduce disasters (Ismail-Zadeh 2017).

A. Ismail-Zadeh (✉)
Institute of Earthquake Prediction Theory and Mathematical Geophysics, Russian Academy of Sciences, Profsoyuznaya Str. 84/32, Moscow 117997 Moscow, Russia
e-mail: Alik.Ismail-Zadeh@kit.edu

A. Ismail-Zadeh
Institute of Applied Geosciences, Karlsruhe Institute of Technology, Adenauerring 20b, Karlsruhe, 76131, Germany

K. Sassa et al. (eds.), *Advancing Culture of Living with Landslides*,
DOI 10.1007/978-3-319-59469-9_13

Since 2005, the International Consortium on Landslides (ICL) has promoted a holistic approach to research and learning on integrated Earth system risk analysis and sustainable disaster management via its International Programme on Landslides (IPL). At the Third World Conference on Disaster Risk Reduction (WCDRR) held in Sendai, Japan, in 2015, the ICL and its IPL contributed further to disaster reduction activities proposing the Sendai Partnerships 2015–2025 for Global Promotion of Understanding and Reducing Landslide Disaster Risk. Several international and intergovernmental organizations, which signed the Sendai Partnerships, agreed to mobilize their efforts "to pursue prevention, to provide practical solutions, education, communication, and public outreach to reduce landslide disaster risk" (Sassa 2015).

One of the signatory Partners of the Sendai Partnerships is the International Union of Geodesy and Geophysics (IUGG). The primary motivation for signing the document was that the partnerships may allow, via international scientific cooperation and inter-governmental decision-making, the best scientific knowledge about landslide hazards and risks to be integrated and implemented into the practice to reduce the risks and possible losses.

This paper outlines IUGG's activities related to promotion of studies in natural hazards, extreme events, and risks. It also discusses potential contributions of geophysical sciences and geodesy in support of the Sendai Partnerships.

Promoting Natural Hazard Research

IUGG (http://www.iugg.org) is a non-profit, non-governmental, scientific organization established in 1919. The IUGG mission is to initiate, promote and coordinate, through international cooperation, studies of the Earth and its environment in space for the benefit of humanity. Those physical, chemical, and mathematical studies related to natural hazards include geodynamics, tectonics, seismology, the generation of magmas, volcanism and lava flows, the hydrology, oceanic waves, hazards associated with atmosphere, cryosphere, ionosphere, and magnetosphere; and climatic and environmental changes influencing the frequency and severity of extreme events. IUGG is a vibrant modern scientific union of nations and individual scientists from all over the world promoting research, science education, and capacity building, and linking scientific knowledge to societal needs (Ismail-Zadeh and Beer 2009).

IUGG is composed of eight international, semi-autonomous, scientific associations promoting specific discipline(s) of geoscience: the International Association of Geodesy (IAG, maintaining observational services to monitor hazards and assist in disaster management), the International Association of Cryospheric Sciences (IACS, promoting research in snow avalanches and other types of cryospheric hazards and risks), the International Association of Geomagnetism and Aeronomy (IAGA, in extreme space weather), the International Association of Hydrological Sciences (IAHS, in flooding, severe precipitation, droughts and other hydrological hazards and risks), the International Association of Meteorology and Atmospheric Sciences (IAMAS, in hurricanes, storms, tornadoes, and other meteorological hazards and risks), the International Association for the Physical Sciences of the Oceans (IAPSO, in tsunamis) together with the International Association of Seismology and Physics of the Earth Interior (IASPEI, in surface ruptures, soil properties, earthquakes, and seismic hazard and risk assessment) and with the International Association of Volcanology and Chemistry of the Earth's Interior (IAVCEI, in volcano eruptions, lava, debris, lahar flows, landslides, and associated hazard and risk assessment). IUGG and its Associations operate through more than 110 scientific divisions, commissions, committees, working groups, and services. IUGG holds general and scientific assemblies, during which policies governing the union are discussed and adopted, and research programs requiring international participation are formulated, coordinated, and planned. At the assemblies, national delegates of IUGG Member countries pass resolutions on important scientific and science policy issues, particularly related to natural hazards and reduction of risks (Ismail-Zadeh 2016).

IUGG Associations work to set global standards for research, such as the International Terrestrial Reference Frame (ITRF 2014), the Manual of Seismological Observatory Practice (Bormann 2012), and the Guidelines for Professional Interaction during Volcanic Crises (IAVCEI Subcommittee for Crisis Protocols 1999). In particular, the International Terrestrial Reference Frame provides a realization (i.e., a set of coordinates of some points located on the Earth's surface) to the International Terrestrial Reference System, a world spatial reference system co-rotating with the Earth in its diurnal motion in space. World navigation systems are referenced either to a specific ITRF realization or to their own coordinate systems, which are then referenced to an ITRF realization. Precise navigation assists in monitoring natural hazards, particularly movements of landslides, lavas, lahar, and debris flows. IUGG Associations oversee eighteen geodetic and geophysical services, particularly those related to monitoring natural hazards. For example, IAG established an observing system – the Global Geodetic Observing System—which works with IAG's international services to provide the geodetic infrastructure necessary for monitoring the Earth system including hazards and disasters and for global change research (Plag and Pearlman 2009).

Strengthening International Cooperation

IUGG makes research visible to the international scientific community, to government agencies, to industry, and to the public in general through its education and outreach activities. Furthermore, science policy is an essential component of IUGG activities related to dissemination of scientific knowledge among the countries and the public (Ismail-Zadeh 2016).

IUGG has initiated and/or supported collaborative efforts that have led to highly productive worldwide interdisciplinary research and outreach programs related to natural hazards, disaster risks and sustainability. The International Hydrological Decade (1965–74), the Global Atmospheric Research Programme (1967–80), and the International Lithosphere Program (1980–present) have been promoting research in floods, hurricanes, earthquakes, volcanoes, and land movements and their hazard/risk assessments.

After the 2004 Indian Ocean earthquake and tsunami, IUGG scientists put forward an initiative about integrated studies on disaster risks, and convinced the International Council for Science (ICSU) to develop a major program on this topic to understand complicated natural and social processes and to contribute to disaster risk reduction. Since 2008, when the scientific program "Integrated Research on Disaster Risk" (IRDR) was established by ICSU and co-sponsored by the International Social Sciences Council (ISSC) and the U.N. Office for Disaster Risk Reduction, IUGG has been working closely with the IRDR Scientific Committee on the problems related to natural hazards and disaster risks (e.g. Ismail-Zadeh et al. 2014; Cutter et al. 2015).

IUGG has developed productive partnerships related to natural and human-induced hazards and disaster risk reduction with several intergovernmental organizations including the Group on Earth Observations, the International Civil Aviation Organization (ICAO), the Preparatory Commission for the Comprehensive Nuclear-Test-Ban Organization, U.N. Committee of Experts on Global Geospatial Information Management, U.N. Environment Programme, U.N. Education, Scientific and Cultural Organization (UNESCO), and the World Meteorological Organization (WMO). Particularly, IUGG has been cooperating for decades with the UNESCO International Hydrological Programme on topics related to hydrological hazards and risks, and with the Intergovernmental Oceanographic Commission on topics related to tsunami research and early warning systems. Since 1955, IUGG has been cooperating with WMO in meteorology, climate research, hydrology, cryosphere, space weather, volcanology, and related aspects of hazards and disaster risk analysis. For example, after the 2010 Eyjafjallajökull volcano eruption, and at a request of ICAO, WMO and IUGG established the joint Volcanic Ash Scientific Advisory Group, which provides scientific advice on volcanic ash to civil aviation.

Contributing to Disaster Risk Reduction

IUGG promotes scientific knowledge related to disaster risk reduction. The most recent activities of the Union have been related to understanding extreme events, their societal implications, and scientific assessment of disaster risks (e.g. Ismail-Zadeh and Takeuchi 2007; Ismail-Zadeh et al. 2014; Ismail-Zadeh and Cutter 2015).

In 2009, IUGG initiated a major project "Extreme Natural Hazards and Societal Implications—ENHANS" supported by ICSU, international scientific unions, and several international and intergovernmental bodies (http://www.icsu-geounions.org/enhans). The principal goals of the project were: to improve understanding of critical phenomena associated with extreme natural events; to analyze impacts of the natural hazards on sustainable development of society; to promote studies on the prediction of extreme events and on natural hazards mitigation; to disseminate scientific knowledge and data on natural hazards and disaster risks for the advancement of research and education in general and especially in developing countries; and to establish links and networks with the organizations involved in research on extreme natural hazards and their societal implications. The goals of the ENHANS project were achieved via scientific meetings and open forums that brought together research experts, decision makers, disaster managers, insurance agency practitioners, and mass-media representatives. The project placed special emphasis on the importance of research on extreme natural hazards and disaster risk mitigation in the most vulnerable regions of the world, particularly in Latin America and the Caribbean, Africa, Middle East, Asia and the Pacific. The project contributed to, and promoted, research activities in natural hazard and disaster risk reduction, in particular via implementation of the IRDR Programme.

The ENHANS Declaration called for the following actions: (i) to promote comprehensive holistic inter- and trans-disciplinary approaches to natural hazard and disaster risk research, which have to integrate knowledge from natural and social sciences, mathematics, engineering, disaster management, insurance sectors and other stakeholders dealing with disaster risk; (ii) to help in networking existing regional scientific and educational centers with the aim to establish a regional center of excellence in disaster risk research (e.g. in sub-Sahara Africa); and (iii) To negotiate on setting up a process of assessing and synthesizing the policy-relevant results of peer-reviewed published research on the understanding of the natural phenomena and the social vulnerability associated with disasters; on the capability of predictive systems to disseminate timely and accurate information needed for policy and decision making; on methodologies and approaches for reducing vulnerability and increasing resilience of societies; and on the overall ability of societies to reduce risk (prevent, mitigate and prepare for the increasing impact of natural events). The proposed assessments could contribute to enhancement of the knowledge on disaster risk at global, regional, and local levels, and to the awareness of the people living with risk. The scientific assessments would help policymakers to develop a policy for significant reduction of disasters (Ismail-Zadeh et al. 2014).

Following an IUGG intervention on the topic of disaster risk assessments at the 2011 ICSU General Assembly in Rome, Italy, an ad hoc group of scientific experts on disaster risk reduction was set up by ICSU, with an ISSC participation, to analyze the state of the art in disaster risk science and risk assessment and to prepare a synthesis report on the topic. The 2014 ICSU General Assembly in Auckland, New Zealand, urged its Members to work closely with international and intergovernmental bodies in order to integrate scientific knowledge on disaster risks and risk assessments into decision-making and actions, and invited ICSU National Members to actively encourage their governments to support the proposed intergovernmental disaster risk assessment process. The report called for a multidisciplinary unified risk assessment, and its summary for policymakers was distributed to representatives of national governments attending the WCDRR in Sendai in 2015 (Ismail-Zadeh and Cutter 2015).

Supporting Landslide Disaster Risk Reduction

A landslide is one of the major geological hazards, which may lead to a disaster in vulnerable areas. The areas of high landslide risk are inhabited by 66 million people occupying a land surface of about 820,000 km^2 (Dilley et al. 2005). The social impact of landslide disasters is significant, and it may continue to increase in the future due to the expansion of human settlements in mountainous areas, urbanization processes, and landscape modifications (Alcántara-Ayala 2014).

Understanding physical processes associated with landslides, predicting their extreme behavior, forecasting magnitude, time and place of landslide events as well as assessing vulnerability and risk are challenging problems. Landslides can be triggered by earthquakes, soil liquefaction, rainfalls, and volcanic eruption; submarine landslides in the ocean floor can trigger tsunami waves. These concatenated events may impact vulnerable communities and lead to disasters. Moreover, climate change may intensify the risk of landslides through an increase in the frequency and magnitude of rainfall (IPCC 2012).

IUGG promotes scientific studies of rainfalls, earthquakes, volcanic eruptions, and tsunamis which are linked to landslide hazards (e.g. Babayev et al. 2010; Gudmundsson 2012; Kremer et al. 2012; Yin et al. 2015; Emberson et al. 2016). Also the union encourages analysis, modeling, and risk assessments related to natural hazards including landslides, lahars, lava, and debris flows, avalanches and rock falls (e.g. Crutchley et al. 2013; Edwards et al. 2014; Korotkii et al. 2016).

For example, Crutchley et al. (2013) developed a marine seismic dataset collected over volcanic landslide deposits, acquired offshore of the Soufrière Hills volcano on the island of Montserrat in the Lesser Antilles. These data allow for detailed quantitative analysis of mass movements and of their manifestations, such as surface deformation and the distribution of transported blocks. Three-dimensional time-dependent numerical models of mass movements (Fig. 1) can assist in realistic simulations of the natural processes (e.g. Ismail-Zadeh et al. 2001; Tsepelev et al. 2016).

Global Positioning System (GPS), Global Navigation Satellite System (GNSS), and radar interferometry (InSAR) became useful tools in monitoring surface dynamics, and in detecting change in the motion through measurement of surface displacement (e.g. Scaioni 2015). The IAG's GGOS (Fig. 2) and its geodetic services provide the highest quality GPS/GNSS and InSAR data in support of scientific research on landslides, monitoring of land instability, early warnings, rescue, and damage assessment (Dow et al. 2009).

IUGG's associations such as IAG, IAHS, IASPEI, and IAVCEI as well as the IUGG Tsunami and GeoRisk Commissions can contribute to the Sendai Partnerships. IUGG Associations work with relevant engineering communities (e.g. the U.S. Army Corps of Engineers, the European Association of Earthquake Engineering) to reduce physical vulnerability through proper land use and urban planning, building codes, risk assessments, and early warning systems, and with insurance companies to mitigate disasters associated with natural hazards. Also the union promotes trans-disciplinary education in disaster risks (Declaration 2015), and hence could contribute to relevant capacity building activities.

Concluding Remarks

Sound scientific knowledge based on integrated observations, analysis and modeling, and scientific information available to policymakers along with interpretations of the knowledge in ways that accommodate diverse and unique local needs and expectations are important components in

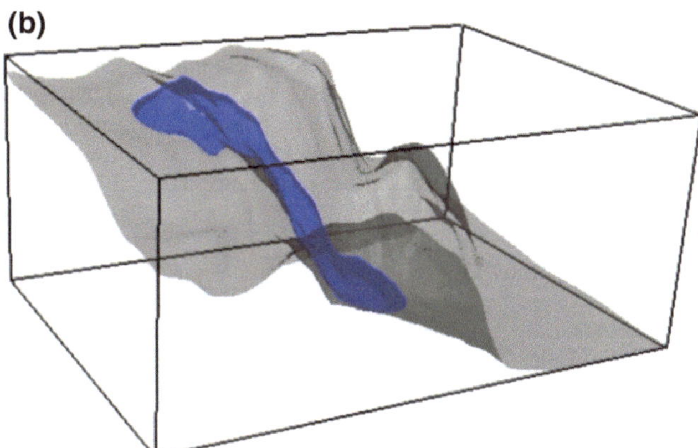

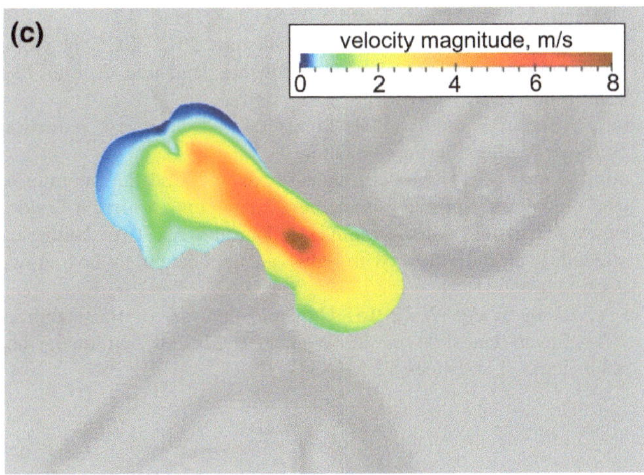

Fig. 1 Observed volcano landslide and modeled fluid flow. (**a**) An aerial view of the Soufriere Hills volcano on the island of Montserrat in the Lesser Antilles, which presents a volcanic dome collapse into the sea (*Credit* the National Oceanography Centre, Southampton, UK). (**b**) Snapshot of a modeled flow on a topographic surface and (**c**) the flow rates

managing future risks. IUGG bridges advanced science with policymaking via international and intergovernmental programs. Timely interventions and sustained efforts to support disaster risk management including scientific research,

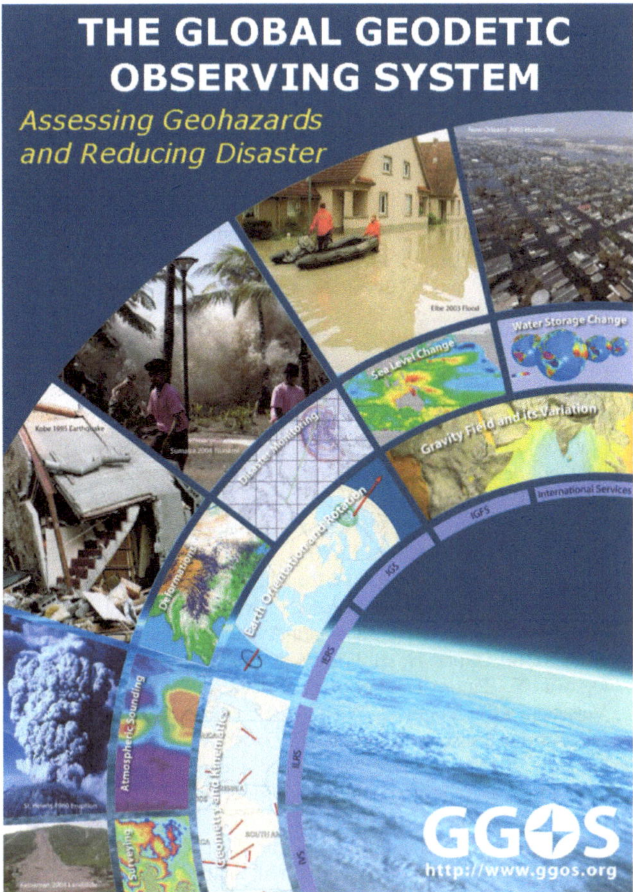

Fig. 2 Geodetic observations play a crucial role in assessment of natural hazards and disaster prevention/mitigation (image by IAG/GGOS)

management and resilience building can enhance sustainable development efforts.

A contribution of the IUGG to the Sendai Partnerships can include (but is not limited to) the following:

- scientific research related to understanding of landslide hazard and disaster risk;
- development of a scientific foundation for reliable prediction of landslides and for landslide early warning system of increased precision;
- landslide hazard and vulnerability assessments, and multi-hazard risk identification;
- improved technologies for monitoring, testing, and analysis of landslides and their analogue and computer simulations;
- teaching courses and tools on natural hazards.

Acknowledgements The author acknowledges the International Consortium on Landslides for its efforts in promoting research and capacity building and for its contribution to reduction of disasters caused by landslides. The research on lava-debris flow presented in Fig. 1 was supported by the Russian Science Foundation (grant 14-17-00520).

References

Alcántara-Ayala I (2014) The special-temporal dimensions of landslide disasters. In: Ismail-Zadeh A, Fucugauchi J, Kijko A et al (eds) Extreme natural events, disaster risks and societal implications. Cambridge Univ. Press, Cambridge, pp 113–125

Babayev G, Ismail-Zadeh A, Le Mouël J-L (2010) Scenario-based earthquake hazard and risk assessment for Baku (Azerbaijan). Nat Hazards Earth Sys Sci 10:2697–2712

Bormann P (ed) (2012) New manual of seismological observatory practice (NMSOP-2). IASPEI and GFZ German Research Centre for Geosciences, Potsdam, Germany

Crutchley GJ, Karstens J, Berndt C, Talling PJ, Watt SFL, Vardy ME, Hühnerbach V, Urlaub M, Sarkar S, Klaeschen D, Paulatto M, Le Friant A, Lebas E, Maeno F (2013) Insights into the emplacement dynamics of volcanic landslides from high-resolution 3D seismic data acquired offshore Montserrat, Lesser Antilles. Mar Geol 335:1–15

Cutter S, Ismail-Zadeh A, Alcántara-Ayala I, Altan O, Baker DN, Briceño S, Gupta H, Holloway A, Johnston D, McBean GA, Ogawa Y, Paton D, Porio E, Silbereisen RK, Takeuchi K, Valsecchi GB, Vogel C, Wu G (2015) Pool knowledge to stem losses from disasters. Nature 522:277–279

Declaration (2015) Future earth and space science education. Available at: http://www.icsu-geounions.org/files/FutureESSE_Declaration.pdf. Accessed on 25 Jul 2016

Dilley M, Chen RS, Deichmann W, Lerner-Lam AL, Arnold M (2005) Natural disaster hotspots: a global risk analysis. The World Bank, Washington D.C

Dow JM, Neilan RE, Rizos C (2009) The International GNSS Service in a changing landscape of Global Navigation Satellite Systems. J Geodesy 83:191–198

Edwards BR, Belousov A, Belousova M (2014) Propagation style controls lava-snow interactions. Nat Commun 5:5666. doi:10.1038/ncomms6666

Emberson R, Hovius N, Galy A, Marc O (2016) Chemical weathering in active mountain belts controlled by stochastic bedrock landsliding. Nat Geosci 9:42–45

Gudmundsson A (2012) Strengths and strain energies of volcanic edifices: implications for eruptions, collapse calderas, and landslides. Nat Hazards Earth Syst Sci 12:2241–2258

IAVCEI Subcommittee for Crisis Protocols (1999) Professional conduct of scientists during volcanic crises. Bull Volcan 60:323–334

IPCC (2012) Summary for policymakers. In: Managing the risks of extreme events and disasters to advance climate change adaptation In: Field CB et al., Barros V, Stocker TF, Qin D, Dokken DJ, Ebi KL, Mastrandrea MD, Mach KJ, Plattner G-K, Allen SK, Tignor M, Midgley PM (eds) A special report of working groups I and II of the intergovernmental panel on climate change. Cambridge University Press, Cambridge, UK, and New York, NY, USA, pp 1–19

Ismail-Zadeh A, Beer T (2009) International cooperation in geophysics to benefit society. EOS 90(51):493, 501–502

Ismail-Zadeh A, Cutter S (eds) (2015) Disaster risks research and assessment to promote risk reduction and management. ICSU-ISSC, Paris, France

Ismail-Zadeh A (2016) Geoscience international: the role of scientific unions. Hist Geo Space Sci 7:103–123

Ismail-Zadeh A, Takeuchi K (2007) Preventive disaster management of extreme natural events. Nat Haz 42:459–467

Ismail-Zadeh A, Cutter SL, Takeuchi K, Paton D (2017) Forging a paradigm shift in disaster science. Nat Hazards 86:969–988

Ismail-Zadeh A, Urrutia Fucugauchi J, Kijko A, Takeuchi K, Zaliapin I (eds) (2014) Extreme natural hazards. Cambridge University Press, Cambridge, UK, Disaster Risks and Societal Implications

Ismail-Zadeh AT, Korotkii AI, Naimark BM, Tsepelev IA (2001) Numerical simulation of three-dimensional viscous flows with gravitational and thermal effects. Comput Mathem Math Phys 41:1331–1345

ITRF (2014) The international terrestrial reference frame, 2014 release. Available at: http://itrf.ensg.ign.fr/ITRF_solutions/2014. Accessed on 25 Jul 2016)

Korotkii A, Kovtunov D, Ismail-Zadeh A, Tsepelev I, Melnik O (2016) Quantitative reconstruction of thermal and dynamic characteristics of lava from surface thermal measurements. Geophys J Int 205:1767–1779

Kremer K, Simpson G, Girardclos S (2012) Giant Lake Geneva tsunami in AD 563. Nat Geosci 5:756–757

Plag H-P, Pearlman M (eds) (2009) Global geodetic observing system. Springer-Verlag, Berlin-Heidelberg

Reichenbach P, Günther A (2014) Progress in landslide hazard and risk evaluation. Nat Hazards Earth Syst Sci 14:2711–2713

Sassa K (2015) ISDR-ICL Sendai Partnerships 2015–2025 for global promotion of understanding and reducing landslide disaster risk. Landslides 12:631–640

Sassa K, Canuti P (eds) (2009) Landslides—disaster risk reduction. Springer-Verlag, Berlin-Heidelberg

Scaioni M (ed) (2015) Modern technologies for landslide monitoring and prediction. Springer, Heidelberg-New York-Dordrecht-London

Tsepelev I, Ismail-Zadeh A, Melnik O, Korotkii A (2016) Numerical modeling of fluid flow with rafts: an application to lava flows. J Geodyn 97:31–41

Yin Y, Li B, Wang W (2015) Dynamic analysis of the stabilized Wangjiayan landslide in the Wenchuan Ms 8.0 earthquake and aftershocks. Landslides 12:537–547

Cabinet Office, Government of Japan (CAO)—Japan's International Cooperation on DRR: Mainstreaming DRR in International Societies

Setsuko Saya

Abstract

The Disaster Management Bureau of the Cabinet Office of Japan has a mandate to coordinate policies and systems for all phases of disaster risk reduction in Japan. The bureau took a key role to host the the Third UN World Conference on Disaster Risk Reduction (WCDRR), which was held in Sendai City, Miyagi Prefecture, 14–18 March, 2015. Japan has suffered from various disasters, including earthquakes, volcanic eruptions, floods, landslides, tsunamis and others and took active roles in international cooperation for disaster risk reduction at the World Conference on Natural Disaster Reduction in Yokohama, Japan in 1994, the World Conference on Disaster Reduction in Kobe, Japan in 2005 and the World Conference on Disaster Risk Reduction in Sendai, Japan in 2015. At the WCDRR in Sendai, Japan, the Government of Japan advocated the importance of "mainstreaming DRR (Disaster Risk Reduction)". The Cabinet office encouraged and supported the International Consortium on Landslides (ICL) to propose the ISDR-ICL Sendai Partnerships 2015–2025 as a voluntary commitment to the WCDRR.

Keywords

World conference on disaster risk reduction (WCDRR) • Sendai framework for disaster risk reduction (SFDRR) • Sendai cooperation initiative for disaster risk reduction (SCIDRR) • ISDR-ICL sendai partnerships • Cabinet office of japan

Mission of the Cabinet Office in Disaster Management

In Japan, the Disaster Management Bureau of the Cabinet Office has a mandate to coordinate policies and systems for all phases of disaster risk reduction, including disaster risk mitigation and preparedness, emergency response, and recovery and reconstruction. It also has responsibility for overall coordination among relevant ministries and agencies in responding to large-scale disasters. As part of the government reform of 2001, the Minister of State for Disaster Management was established in the Cabinet Office to integrate and coordinate disaster risk management policies and measures of ministries and agencies. In the Cabinet Office, which is responsible for securing cooperation and collaboration among related government organizations in wide-ranging issues, the Director-General for Disaster Management undertakes the planning of basic disaster management policies and response to large-scale disasters, as well as conducting overall coordination (Fig. 1).

The disaster management system has been continuously reviewed and revised following the lessons learned from large-scale disasters in the past. The Central Disaster Management Council, which consists of the Prime Minister as the Chair, all members of the Cabinet, heads of major public corporations and experts, develops the Basic Disaster Management Plan, establishes basic disaster management

S. Saya (✉)
International Cooperation Division, Cabinet Office Government of Japan, Tokyo, Japan
e-mail: setsuko.saya.e8f@cao.go.jp

S. Saya
Disaster Management Bureau, Dhaka, Bangladesh

K. Sassa et al. (eds.), *Advancing Culture of Living with Landslides*,
DOI 10.1007/978-3-319-59469-9_14

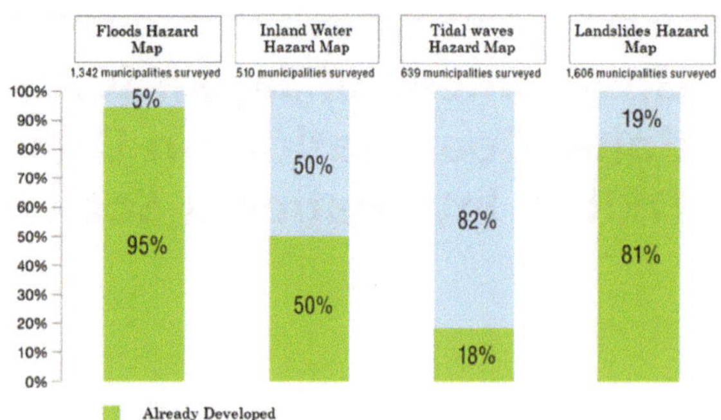

Landslides triggered by Typhoon No.26 (Izu-Oshima Island, 2013)

Fig. 1 Preparation of Hazard Maps (as of the end of March 2013) (Cabinet office 2015b). Hazard maps for landslides have been completed for 81% of the 1606 municipalities in Japan. *Right* photo shows recent landslides triggered by a typhoon (824 mm cumulative rainfall and 118.5 mm/hour peak hourly rainfall) in Izu-Oshima Island, in Tokyo. In all, 35 persons died, 4 persons were missing, and 86 houses were destroyed

policies, and plays a role of promoting comprehensive disaster countermeasures. The Council also deliberates important issues on disaster management upon requests from the Prime Minister or the Minister of State for Disaster Management.

In extreme disaster events that cannot be handled by the prefectural governments, the Major Disaster Management Headquarters led by the Minister of State for Disaster Management or the Extreme Disaster Management Headquarters led by the Prime Minister is established, depending on the scale and impact of a disaster. The Cabinet Office is responsible for collection and dissemination of accurate information, reporting to the Prime Minister, and establishment of the emergency activities system, including the Government's Disaster Management Headquarters, and overall cross-jurisdictional coordination among relevant ministries and agencies for disaster response (Fig. 2).

The Third WCDRR and the Cabinet Office

The Sendai Framework for Disaster Risk Reduction (SFDRR) 2015–2030, the successor framework to the Hyogo Framework for Action (HFA) 2005–2015, was adopted at the Third WCDRR, as the guideline for disaster risk reduction in countries around the world.

Over the coming 15-year period, promoting the SFDRR will be the responsibility of the international community, and of Japan as a country that has led the world in the field of disaster risk reduction (Fig. 3).

In accordance with the Sendai Cooperation Initiative for Disaster Risk Reduction (SCIDRR) announced by Prime Minister Shinzo Abe, Japan will contribute to the mainstreaming of disaster risk reduction in the international community by effectively combining the promotion of structural and non-structural supports, as well as global

Fig. 2 Programme of 3rd WCDRR (Cabinet office 2015a) The figure shows a list of the major meeting and sessions: Official Plenary Session, High-level partnership dialogues, Ministerial Roundtables, and Working Sessions, and their themes

Fig. I-2-3 **Major Meetings and Sessions During the WCDRR**

Session Title	Schedule	Theme
Official Plenary Session	Mar. 14-17 (Sat.-Tues.)	Official statements by Member State delegations, international organizations, and other major groups
High Level Partnership Dialogues	15:00-18:00 March 14 (Sat.)	Mobilizing Women's Leadership in Disaster Risk Reduction
	15:00-18:00 March 16 (Mon.)	Risk Sensitive Investment: Public-Private Partnerships
	10:00-13:00 March 17 (Tues.)	Inclusive Disaster Risk Management: Governments, Communities, and Organizations Acting Together
Ministerial Roundtables	10:00-13:00 March 15 (Sun.)	Reconstructing After Disasters: Building Back Better
	15:00-18:00 March 15 (Sun.)	International Cooperation in Support of a Post-2015 Framework for Disaster Risk Reduction
	10:00-13:00 March 16 (Mon.)	Governing Disaster Risk: Overcoming Challenges
	15:00-18:00 March 16 (Mon.)	Reducing Disaster Risk in Urban Settings
	15:00-18:00 March 17 (Tues.)	Public Investment Strategies for Disaster Risk Reduction
Working Sessions	Mar. 14-17 (Sat.-Tues.)	Sessions (34) were held on the following four themes: (1) Progress on the existing HFA Priorities for Action (2) Emerging Risks (3) Commitments to Post-HFA Implementation (4) Accelerating Post-HFA Implementation

Fig. I-2-4 List of Working Sessions

	Theme	Working Session Title	Schedule
1	Progress on Existing HFA Priorities for Action	Governance and Development Planning at National/Local Levels (Priority 1)	10:00-11:30 March 15 (Sun.)
2a		Risk Identification and Assessment (Priority 2)	15:00-16:30 March 14 (Sat.)
2b		Early Warning (Priority 2)	17:00-18:30 March 14 (Sat.)
3		Education and Knowledge in Building a Culture of Resilience (Priority 3)	10:00-11:40 March 16 (Mon.)
4		Underlying Risk Factors (Priority 4)	10:00-11:30 March 16 (Mon.)
5		Preparedness for Effective Response (Priority 5)	12:00-13:30 March 17 (Tues.)

Fig. 3 The working session "Underlying Risk Factors" (Cabinet office 2015a). The figure lists the initial 5 working session within 33 working sessions held in WCDRR. The working session "Underlying Risk Factors (Priority 4) was co-organized by the International Consortium on Landslides (ICL), UNESCO, the Japanese Ministry of Land, Infrastructure, Transport and Tourism (MLIT) and other pertinent organizations. "ISDR-ICL SENDAI PARTNERSHIPS 2015–2024 for global promotion of understanding and reducing landslide disaster risk, Tools for Implementing and Monitoring the Post-2015 Framework for Disaster Risk Reduction and the Sustainable Development Goals" was adopted in this session

cooperation and region-wide cooperation; contributing a total of four billion dollars to fields related to disaster risk reduction and providing human resource development for 40,000 people over the next four years.

Japan will also continue to promote international cooperation in DRR through measures that include multilateral cooperation through international agencies such as the UN, regional cooperation in Asia, and intergovernmental cooperation. The following sections discuss these activities.

Japan's Role in International Disaster Management

Japan has cultivated knowledge, systems and technologies in disaster countermeasures as a result of its numerous disaster experiences. Due to its geographical and climate conditions, Japan has long been vulnerable to all types of disasters, including earthquakes, tsunamis, floods, storm surges, high waves, slope failures, volcanic eruptions, debris flows, and heavy snows. Utilizing that expertise, Japan is cooperating in the efforts of disaster reduction in the world, making an important visible contribution to the international societies.

The number of disasters around the world is increasing, and disasters remain a major drawback to sustainable development. Reducing vulnerabilities to natural hazards and damage caused by them is an inevitable challenge in the international community. Almost every year, disasters hit worldwide and a great many people are killed and there is huge damage to the local and world economies. In the past 30 years, (1984–2013), more than 247 million people were killed and more than US$2.4 trillion was lost in damages. Approximately 80% of the casualties are concentrated in low- to middle income countries, making the vicious cycle of disasters and poverty another challenge.

The Government of Japan advocated the importance of "mainstreaming DRR" at the Third WCDRR. The primary meaning of "DRR mainstreaming" is making prior efforts in initiatives to mitigate damage from disasters in particular; in other words, to ensure that DRR efforts are reflected in all policies on a widespread basis. All sectors are affected once a disaster strikes, and DRR cannot be achieved unless advanced preparations are made in all types of policies. Japan has been building a disaster management system with relevant ministries and agencies, public agencies, local governments, and others under the Central Disaster Management Council, comprising all relevant Cabinet ministers (Fig. 4).

It is imperative that all stakeholders in countries around the world work to create systems to address DRR during ordinary times, before a disaster strikes. The SFDRR adopted at the Third WCDRR includes numerical DRR goals to advance this concept of mainstreaming DRR. It includes prior DRR investment and fundamental disaster prevention measures in the recovery phase after a disaster strikes, in adherence with the "Build Back Better" principle, and establishes a concept of DRR governance as the responsibility, not only of governments, but of all stakeholders, with diverse stakeholders fully performing their respective roles. Implementation of these measures in the international community will lead to the mainstreaming of DRR. To this end, Japan will continue to actively cooperate in the area of international DRR as a leader in the DRR field.

Fig. I-3-1 Outline of the Sendai Cooperation Initiative for Disaster Risk Reduction (SCIDRR)

Sendai Cooperation Initiative for Disaster Risk Reduction (SCIDRR)

1. Basic Concept

- Disasters are an obstacle to poverty eradication and sustainable development, and thus a threat to human security.
- **Mainstreaming of disaster risk reduction (DRR)** - introducing the DRR perspective in all development policy and planning - is important. Clearly positioning DRR in the post-2015 development agenda is important from the perspective of resource mobilization.
- High attention to the efforts for "adaptation" at the climate change negotiation where an agreement is required by the end of this year. Firm DRR efforts will contribute to the climate change negotiation.
- Japan will build with the international community a society that is resilient to disasters by sharing with the world its knowledge and technology as a country advanced in DRR.

2. Basic Policies

- Japan attaches particular importance to the three points in DRR policies outlined below, building on the experience of the past 10 years since the formulation of HFA.
  - **(1) Investment in DRR from the long-term perspective**
    Prior investment in DRR is more cost-effective than post-disaster emergency response and recovery and contributes to sustainable development.
  - **(2) Build Back Better**
    The post-disaster phase provides an opportunity to implement drastic measures to build countries and regions that are resilient to disasters.
  - **(3) Collaboration between the central governments and various actors**
    Addressing with networks including local governments, private companies, NGOs/CSOs, international organizations and regional organizations, with the central government taking the initiative.
- Japan will take the following perspectives into consideration in implementing cooperation.
  - (1) **The human security approach** and **promoting women's participation** (women, children, the elderly and persons with disabilities)
  - (2) Cooperation based on the perspective of **adaptation to the impacts of climate change**
  - (3) Utilizing **Japan's knowledge and technology**

  ⇒ Cooperation through effectively combining (i) non-material assistance, (ii) material assistance and (iii) global and region-wide cooperation.

3. Concrete Measures

DRR cooperation totaling to 4 billion US dollars and training of 40 thousand from 2015 to 2018

Non-material assistance

Assistance for establishing laws, institutions and systems, human resource development and other technical assistance

- Laws and regulations relating to DRR (basic acts on disaster countermeasures, laws and regulations on the use of land / building standards)
- Basic DRR plans, master plans for flood control, master plans for urban planning, land-use plans, urban planning
- Assistance to and strengthening setup of DRR branches in government
- Assistance to build and strengthen partnership systems among the public and private sectors and NGOs
- Disaster risk assessment (development hazard maps, research assistance for adaption to climate change, etc.)
- Technologies for disaster observation, prediction and warning (ICT, earth observation, geospatial information)
- Community-based DRR, disaster education
- Human resource development, training, technology transfer for DRR policy planning and emergency disaster relief
- Training to promote women's leadership in DRR

Material assistance

Economic and social infrastructure development with Japanese technology as prior investment in DRR ("quality growth")

- Countermeasures against flooding, debris flow, landslides and storm surges, forest improvement for disaster reduction
- Satellites necessary for disaster observation, perdition and warning, and information and communication infrastructure
- Improvement of buildings quality (earthquake resistance, wind resistance)
- Provision of equipment related to DRR
- Transportation, lifeline and public facilities resilient to disasters, DRR-related information and communication facilities
- Recovery and reconstruction assistance

Global and region-wide cooperation

Assistance for UNISDR and IRP, region-wide cooperation

- Assistance for the monitoring of the global targets and the improvement of its methods, as well as for the development of indicators
- Development of international disaster statistics
- Dissemination of information on good practices of "Build Back Better" including efforts from the Tohoku region
- Assistance for efforts to build region-wide institutions and systems (Sentinel Asia, Asian Disaster Reduction Center, AHA Centre, etc.)
- Assistance for countermeasures against climate change (including Green Climate Fund (GCF))
- Collaboration between regional cooperation of each region and Japan's bilateral cooperation

Fig. 4 Sendai cooperation Initiative for Disaster Risk Reduction (SCIDRR) (Cabinet office 2015a). Prime Minister Shinzo Abe, Japan, announced the Sendai Cooperation Initiative for Disaster Risk Reduction (SCIDRR) which consists of Basic Concept, Basic Policies, Concrete measures of non-material assistance, material assistance as well as global and region-wide cooperation by contributing a total of four billion dollars to fields related to disaster risk reduction and providing human resource development over the next four years

References

Cabinet Office, Government of Japan (2015a) White paper on disaster management 2015. Available at. http://www.bousai.go.jp/kaigirep/hakusho/pdf/WP2015_DM_Full_Version.pdf

Cabinet Office, Government of Japan (2015b) Disaster management in Japan. Available at. http://www.bousai.go.jp/1info/pdf/saigaipamphlet_je.pdf

Disaster Prevention Research Institute (DPRI), Kyoto University

Kaoru Takara

Abstract

This article describes an outline of the Disaster Prevention Research Institute (DPRI), which was established in Kyoto University in 1951, including its mission and objectives in terms of research, education and social contributions. Brief history of DPRI, as well as that of Research Centre on Landslides (RCL), is also given in relation with domestic and international activities such as the Natural Disaster Research Council (NDRC), designated COE programs, a Leading Graduate Schools Program (GSS), the International Decade for Natural Disaster Reduction (IDNDR), UNESCO-KU-ICL UNITWIN Program, UNESCO International Hydrological Program (IHP), Science and Technology Research Partnership for Sustainable Development (SATREPS) projects, Japan-ASEAN Science, Technology and Innovation Platform (JASTIP) and the Global Alliance of Disaster Research Institutes (GADRI).

Keywords

Center of excellence • Joint usage/research center • NDRC • GADRI • JASTIP

Introduction

Since its inception in 1951, the Disaster Prevention Research Institute (DPRI) of Kyoto University has been pursuing principles of natural disaster reduction, establishing integrated methodologies for disaster prevention on the basis of natural and social sciences, and educating students in related fields. The research staff members of the Institute are also affiliated with the Graduate Schools of Science, Engineering and Informatics of Kyoto University. Many graduate students come to the Institute to carry out their studies under supervision of its staff members.

Currently, DPRI consists of four research groups, which include five Research Divisions and six Research Centers. It is managing the Natural Disaster Research Council, which is a research network for natural disaster risk reduction, since 2001. In March 2015, it established the Global Alliance of Disaster Research Institutes (GADRI) as one of the actions for the Sendai Framework for Disaster Risk Reduction (SFDRR) 2015-2030. Kyoto University also agreed with the Implementation of the **ISDR-ICL Sendai Partnerships 2015–2025 for Global Promotion of Understanding and Reducing Landslide Disaster Risk**.

K. Takara (✉)
Disaster Prevention Research Institute, Kyoji University,
Uji, Kyoto, 611-0011, Japan
e-mail: takara.kaoru.7v@kyoto-u.ac.jp

K. Sassa et al. (eds.), *Advancing Culture of Living with Landslides*,
DOI 10.1007/978-3-319-59469-9_15

DPRI's Mission and Objectives

Today's society becomes ever more rapidly vulnerable to natural hazards and consequent disasters due to the concentration of populations in mega-cities. Additionally, changes in the global environment threaten us with the possibility of severe typhoons, floods, landslides, sea level rise, and droughts. Considering these rapid changes of ambient conditions, and to meet urgent research requirements in a more timely manner, in 1996 the Institute reorganized itself into five research divisions and five research centers; namely, Integrated Management of Disaster Risk; Earthquake Disaster Prevention; Geo-Disasters; Fluvial and

Marine Disasters; Atmospheric Disasters; Research Center for Disaster Environment; Research Center for Earthquake Prediction; Sakurajima Volcano Research Center; Water Resources Research Center; Research Center for Disaster Reduction Systems, and Research Centre on Landslides.

The Division of Integrated Management of Disaster Risk and the Research Center for Disaster Reduction Systems have been set up by rearranging and increasing the number of staff members who have been involved in disaster study from human, social, and planning aspects. The Research Center for Disaster Environment unifies experimentation stations and observatories located in distant places; namely, the Ujigawa Hydraulics Laboratory, the Shionomisaki Wind Effect Laboratory, the Hodaka Sedimentation Observatory, the Shirahama Oceanographic Observatory, the Ogata Wave Observatory, and the Tokushima Landslide Observatory. The Center carries out synthetic observational and experimental research projects, collaborating not only with other staff members but also with researchers outside the Institute.

Although the Institute belongs to Kyoto University, it has been open since 1996 to all researchers from other universities around the country who are concerned with investigations of disasters. Collaboration is maintained through joint research projects and research meetings. Researchers from both inside and outside of the Institute can submit proposals, which are assessed and approved through the peer-review by the Collaboration Committee, a group consisting of members from both outside and inside the Institute. The Advisory Board will advise the DPRI Director on the policy of the institute's operation. The Advisory Board is composed of two members from the Institute and several prominent professors from outside such as Deans and Institute Directors.

Its objectives are summarized as follows:

Research

(1) DPRI enhances international cooperative research on natural disasters, as well as cooperative research in Japan, through the Joint Usage/Research Center for natural disaster reduction research, as designated by MEXT (Japanese Government). This work will be done through the use of research facilities of DPRI, development of databases for natural disaster information, such as the Natural Disaster Resource Data (SAIGAI), and organizing reconnaissance teams for natural disaster events.

(2) DPRI pursues the principles of natural hazard reduction through basic research focusing on the changes of modern natural disasters associated with the evolution of the natural environment and human society.

(3) DPRI promotes practical research for disaster reduction to meet the urgent needs of society, by integrating various research fields related to disaster reduction.

Education

(1) DPRI welcomes young researchers and students from overseas, as well as Japanese students, in order to meet the high demand of research and educational needs for disaster reduction around the world.

(2) DPRI fosters researchers of the next generation, by conducting seminars and preparing educational materials. This will be done through the Joint Usage/Research Center for natural disaster reduction research and with the cooperation of other related organizations in Japan and around the world.

(3) DPRI promotes education for regional disaster mitigation planners at the working-level, as well as young researchers, through collaborations of the cooperative studies in the Joint Usage/Research Center.

Social Contributions

(1) DPRI transfers research results and specialized knowledge regarding disasters and disaster reduction to society, and help people understand disaster reduction methods. DPRI will also provide disaster reduction strategies to national and local governments.

(2) DPRI promotes international cooperative research and education by establishment of a world leading Joint Usage/Research Center.

Administration

DPRI strengthens the function of the Joint Usage/Research Center by providing well-organized support and evaluation for schedules, research and education.

History of DPRI

Kyoto University established the Disaster Prevention Research Institute (DPRI) with three chairs (three full-professors) in April 1951. DPRI's mission was to promote science on disasters and its application. It has been adding many research sections and centers, having 34 full-professors at its restructuring in May 11, 1996 immediately after the 1995 Kobe Earthquake. DPRI's mission has been changed as to promote science on disasters and studies on comprehensive disaster prevention and mitigation. In 1997, the Japanese Ministry of Education, Science and Culture (Monbusho) had designated DPRI as a center of excellence (COE) by. In 2002, DPRI was again designated as a 21st Century COE by the Ministry of Education, Culture, Sports, Science and Technology of Japan (MEXT: Monbukagakusho). In April 2003, the DPRI established Research Centre on Landslides. In April 2005, the DPRI restructured its organization with four research groups including five research divisions and six research centers.

The DPRI promotes joint research programs with other Japanese universities and research organization through the Natural Disaster Research Council (NDRC), which was established in DPRI in 2001. The NDRC's mission includes:

(1) To plan scientific strategy to promote natural disaster science,
(2) To organize the Integrated Symposium on Natural Disaster Science every year, and
(3) To arrange research projects and teams for emergent investigation on disaster events taking place in Japan and abroad

The MEXT started a system to encourage such joint usage/research organizations since 2009. The DPRI was designated as a Joint Usage/Research Center of Excellence (2010–2015) and again renewed further six years (2016–2021).

The DPRI is also committing interdisciplinary graduate school education. It obtained a Global COE Program: Sustainability/Survivability Science for a Resilient Society Adaptable to Extreme Weather Conditions (GCOE-ARS; Takara, Asian Journal of Environment and Disaster Management, 2011). The GCOE-ARS conducted various research topics such as extreme weather, resultant meteorological and hydrological disasters (floods and droughts), landslides, and oceanographic/coastal disasters, and produced 35 doctors who took this GCOE-ARS course from Graduate Schools of Science, Engineering, Agriculture, and Global Environmental Studies in Kyoto University during its program period (2009–2013) and its follow-up period (2014–2017).

Another educational program that the DPRI is implementing is the Inter-Graduate School Program for Sustainable Development and Survivable Societies (GSS = Global Survivability Studies) for seven years (2011–2018) under the MEXT's Leading Graduate School Program. The GSS program is implemented in cooperation with nine graduate schools (25 departments) and three research institutes of Kyoto University, having more than 80 students.

The DPRI has been contributing international activities such as the International Decade of Natural Disaster Reduction (IDNDR), UNESCO's International Hydrological Programme (IHP), UNESCO-KU-ICL UNITWIN Network Programme, Integrated Research on Disaster Risk (IRDR) and Future Earth. It also contributed to the establishment of the Integrated Disaster Risk Management Society (IDRiM),

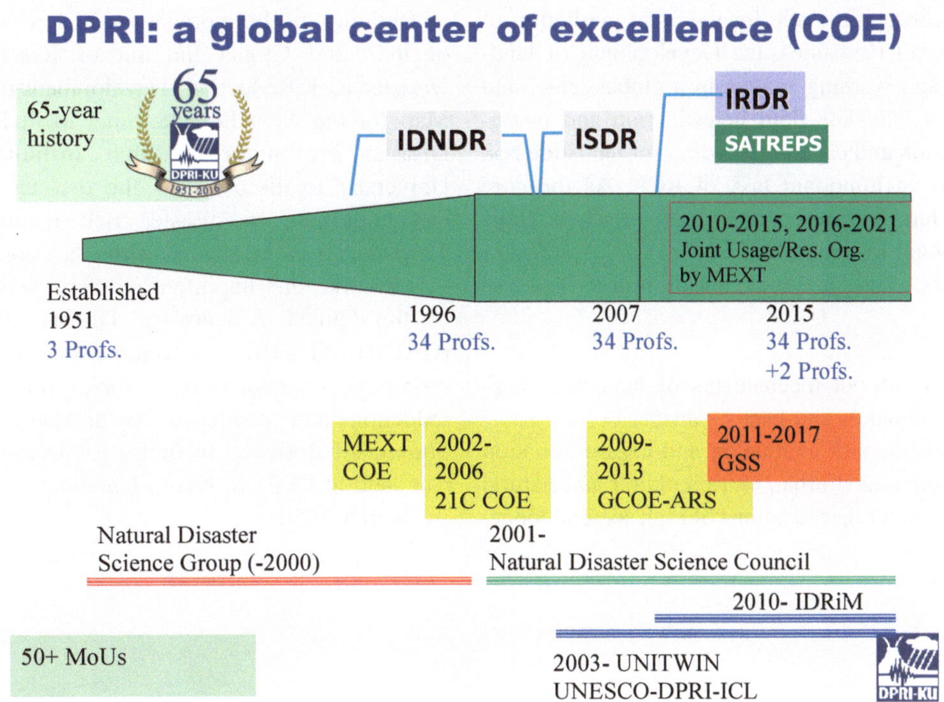

an international academic society, in October 2009. Recently, it is implementing SATREPS projects for Croatia (2008–2012), Vietnam (2011–2016), Indonesia (2013–2018), Bangladesh (2013–2018) and Mexico (2015–2020). Currently about 60 MoU's are concluded with renowned overseas universities and research organizations.

Research Centre on Landslides

Two thirds of the Japanese Archipelago is mountainous areas. A hundred million people are forced to live and work in mountain slopes or around slopes. Landslides cause disasters on such slopes. In 1959, Landslide Research Section was founded in the DPRI, Kyoto University and was reorganized into the Research Centre on Landslides (RCL) in April 2003. This Research Centre on Landslides is very unique because it is the only one centre specialized for landslide research in national universities supported by the Ministry of Education, Culture, Sports, Science and Technology of Japan (MEXT). The Tokushima Landslide Observatory (TLO) was founded in Tokushima Prefecture, Shikoku Island, Japan in 1969 as a field of landslide monitoring and investigation in DPRI. Since then, both RCL and TLO have done landslide field and experimental studies, cooperating with the Slope Conservation Section and Mountain Hazards Section in the Research Division of Geohazards, DPRI, Kyoto University.

The RCL aims to pursue research for protecting human lives, properties, and cultural and natural heritages from landslides. RCL conducts research on the mechanisms of initiation and motion of landslides triggered by earthquakes and rainstorms. Efforts are made for the areal prediction of rapid and long-travel landslides, the Development of landslide monitoring and warning system in a global scale, and new techniques of landslide field investigation and instrumentation. Education and capacity building for landslide risk mitigation is also an important task of RCL. As the core centre of global landslide research network, RCL is coordinating international programmes.

The main topics of research and education in RCL are as follows:

(1) Initiation and run-out mechanisms of landslides triggered by earthquakes and heavy rains;
(2) Reliable landslide risk evaluation and hazard zonation for densely populated urban areas, cultural and natural heritage sites, and other locations of high societal value;
(3) Development of high-precision and reliable monitoring system of landslides from a local scale to a global scale;
(4) Field investigation and development of instrumentations for landslide research; and
(5) Education and capacity building to reduce landslide disasters in developing countries.

The RCL had been one of the main actors of the International Consortium on Landslides (ICL) and the UNESCO-DPRI-ICL UNITWIN Programme.

Global Alliance of Disaster Research Institutes (GADRI)

Following the Hyogo Framework for Action (HFA), advances have been made in the application of damage reduction principles, but many challenges remain as seen by the continued increase in disaster losses. Also, the new Sendai Framework for Disaster Risk Reduction 2015–2030 (SFDRR 2015–2030) with support from the Scientific and Technical Advisory Group of the United Nations Office for Disaster Risk Reduction (UNISDR), has pointed out the need to better link sustainable development and climate change efforts towards a common goal of harmonious living with nature and our planet. The Framework explicitly calls for coordinated efforts by the scientific community to deepen the understanding of disaster risks, promote evidenced-based implementation of disaster risk reduction strategies, and transfer and disseminate scientific knowledge and technologies in support of decision making processes.

Building on the momentum of these efforts, participants of the "2nd Global Summit of Research Institutes for Disaster Risk Reduction: Development of a Research Road Map for the Next Decade," met to in March 2015 at the Disaster Prevention Research Institute (DPRI), Kyoto University to discuss how the disaster research institutes can contribute to disaster risk reduction in the next 15 years, in accordance with the goals of the Sendai Framework. One important outcome was the establishment of the Global Alliance of Disaster Research Institutes (GADRI). GADRI is a forum for sharing knowledge and promoting collaboration on topics related to disaster risk reduction and resilience to disasters. The 3rd Global Summit of Research Institutes for Disaster Risk Reduction was held at DPRI in Kyoto University Uji Campus on 19–21 March 2016.

GADRI also welcomes any landslide-related research institutes and organizations from all over the world.

JASTIP Disaster Prevention

Jastip

In 2015, Kyoto University has launched "Japan-ASEAN Science, Technology and Innovation Platform (JASTIP): Promotion of Sustainable Development Research" within the framework of the Collaboration Hubs for International Research Program (CHIRP) funded by the Strategic International Collaborative Research Program (SICORP) of the Japan Science and Technology Agency (JST).

Under the JASTIP project, joint research between Japan and ASEAN countries is ongoing for achieving remarkable science and technology outcomes. In addition, the joint research promotes a number of researchers in leading positions and helps train the researchers who have become subsequent leaders in the field. The research project goal is to develop a broader understanding in Japan and in the ASEAN society with the academic achievements for which the community should be proud, and to accelerate research that will be a driving force in our shared goal to attain sustainable development under a closer collaboration between Japan and the ASEAN countries. The JASTIP project promotes research that can be used to resolve shared local issues to ultimately build a sustainable society in Japan and in ASEAN countries. In addition, our aim through these activities is for Japan to "show face" and "have face be seen" in the ASEAN region, and to build an inclusive Japanese-ASEAN cooperative platform related to science, technology and innovation.

Disaster Prevention in JASTIP

The JASTIP includes three research fields: Energy & Environment, Bioresources & Biodiversity, and Disaster prevention. The DPRI is leading this Disaster Prevention field by setting up satellite sites in ASEAN countries. The Malaysia-Japan International Institute of Technology (MJIIT) located at University Technology Malaysia (UTM) Kuala Lumpur Campus is one of the satellite sites that promote JASTIP international cooperation research projects. Another satellite site has been established at Thuy Loi University in Ho Chi Minh City, Vietnam in 2016.

The main activities performed at this site are as follows:

(1) Joint research in Malaysia and Vietnam related to heavy rain, flooding, and landslide disasters
(2) Joint research in Thailand related to ground foundation disasters due to heavy rain, flooding, landslide disasters, and earthquakes
(3) Joint research in Indonesia related to earthquakes, volcanoes, and landslides
(4) Joint research in the Philippines related to earthquakes, volcanoes, landslides, and typhoons
(5) Advanced technology development of early warning systems to reduce the risk of large-scale natural disasters, a common issue for all the involved countries
(6) Practical implementation of the developed early warning systems
(7) Development of curriculum for human resource cultivation programs at MJIIT (including a disaster prevention curriculum at the graduate level)
(8) Building a research network with leading universities in ASEAN countries for disaster prevention research, including Chulalongkorn University, Asian Institute of Technology, Gadjah Mada University, University of the Philippines, and Thuy Loi (Vietnam Water Resources) University
(9) Launching the "ASEAN Disaster Prevention Joint Research and Human Resource Cultivation Program"
(10) Establishing the ASEAN Disaster Prevention Research Cooperation Framework Concept

Many landslide issues are serious in ASEAN region, as indicated above. The JASTIP also promotes joint research into comprehensive disaster prevention useful for sustainable development and collaboration on disaster prevention projects of SATREPS (Science and Technology Research Partnerships for Sustainable Development) such as Indonesian volcanoes and Myanmar flooding.

Master Program in MJIIT

The MJIIT established the Disaster Preparedness and Prevention Centre (DPPC) and started a graduate school course "Masters Disaster Risk Management" since 2016. This course is a one-year master program for practitioners working at governmental and public offices for disaster management. The DPRI dispatches several lecturers to this course to encourage higher education in disaster risk management for Malaysia and surrounding countries. Further details of DPPC and Masters Disaster Risk Management course can be seen at:

http://mjiit.utm.my/dppc/organization-chart/
http://mjiit.utm.my/dppc/masters-in-disaster-risk-management/

Understanding and Reducing Landslide Disaster Risk: Challenges and Opportunities for Italian Civil Protection

Pagliara Paola, Onori Roberta, and Ambra Sorrenti

Abstract

This work provides an overview of the hydraulic and hydrogeological warning system, starting with a description of the tasks carried out by the Italian Civil Protection Department and in an Italian context. The Italian early warning system was put in place with the intent of following a specific approach, also confirmed by the Sendai Framework, that has shifted its focus towards Disaster Risk Management (DRM) as opposed to Disaster Management. The aim of the paper is to stress the challenge and the relevance of the approach to reducing landslide risk, which requires the involvement of many actors, including scientists and decision makers, as well as international, national, local, governmental, and non-governmental institutions, to find, develop and share new and best practices in the technical-scientific and regulation fields in order to make the necessary tools and instruments available to carry out the challenging tasks defined in the Sendai Framework toward real Disaster Risk Reduction.

Keywords

Early warning system • Disaster risk reduction • Landslide

Italian Civil Protection Department

Outline and Mission

The Italian national territory is exposed to a broad range of natural hazards, including landslides, that every year cause a significant number of casualties and considerable economic damage. In some cases, the vulnerability of the population and the environment is increased by human activities.

In this perspective, within the National Civil Protection Service (Law n. 225/92 and all the amendments and modifications of Law n. 100/12), the Italian Civil Protection Department (DPC) is responsible for a wide array of sectors pertaining to civil protection, ranging from prevention, forecast and assessment, early warning and alert systems to emergency response and recovery from emergency.

The DPC is a structure of the Presidency of the Council of Ministers. It is responsible for coordinating the National Civil Protection Service, which includes local authorities, research institutions, private companies, volunteer associations and all Italian operational forces. During major emergencies, it ensures horizontal (line Ministries) and vertical (central-local) coordination. It plays a leading role, in cooperation with regional and local governments, to ensure risk prevention, forecasting and monitoring activities, as well as emergency preparedness and intervention procedures in case of on-going or upcoming disaster events.

P. Paola (✉)
Manager of the National Centre for Forecasting and Surveillance for Hydrologic and Hydraulic Risk, Italian Civil Protection Department (DPC), Rome, Italy
e-mail: paola.pagliara@protezionecivile.it

O. Roberta
Italian Civil Protection Department (DPC), National Centre for Forecasting and Surveillance for Hydrologic and Hydraulic Risk, Rome, Italy
e-mail: roberta.onori@protezionecivile.it

A. Sorrenti
International Relation Unit, Italian Civil Protection Department (DPC), Rome, Italy
e-mail: ambra.sorrenti@protezionecivile.it

K. Sassa et al. (eds.), *Advancing Culture of Living with Landslides*,
DOI 10.1007/978-3-319-59469-9_16

For this purpose, it relies on a well-developed network of risk monitoring and forecasting centres (one in each region, besides the central one located in the DPC Headquarter premises, which has a coordinating role). They work closely with the local branches of the National Civil Protection Service and the scientific community, as well as from a national inter-institutional operational room, where all the operational forces are present 24/7.

DPC promotes drills, national and international training projects and activities that contribute to spreading the culture of civil protection based on prevention and preparedness rather than just response.

In the framework of its guiding role in the phase of emergency preparedness, the DPC:

- develops and implements contingency plans for major events, as well as awareness-raising campaigns, jointly with local governments and volunteer associations throughout the country;
- provides technical support to local government for the elaboration/updating of contingency plans, including testing activities (such as simulations and field exercises);
- issues guidelines, standard operating procedures aimed at regions, provinces and municipalities, to prepare and implement prediction and prevention programs based on risk scenarios;
- promotes field exercises for testing multi-level co-ordination (international, national and local), including host nation support aspects.

The DPC operates at an international level aimed at advancing work on disaster risk management and disaster risk reduction by promoting and exchanging knowledge and experience in these domains.

In particular, the DPC is involved in a consistent number of initiatives of international cooperation and capacity-building related to different types of risk in the framework of programs and projects, mainly financed by the European Union, involving the EU countries, the Balkans, North Africa and Middle Eastern countries.

Due to its role in coordinating the National Civil Protection Service, the DPC is the focal point of the Union Civil Protection Mechanism since its establishment in 2001. The Mechanism, revised in 2013, aims to strengthen cooperation between the Union and the Member States and to facilitate coordination in the field of civil protection in order to improve the effectiveness of systems in prevention, preparedness and response to natural and man-made disasters. In this regard the DPC has been strongly engaged in contributing to advance the work on these three pillars, both from policy and operational points of view.

Furthermore, in cooperation with components of the Civil Protection System, the Department is committed to the design, establishment and testing of national and multi-national resources and capacities to be offered and deployed upon request from a country hit by a disaster through the Mechanism.

Among the valuable achievements in the scientific field, it is worthwhile to mention that the DPC has actively contributed, through its ten-year participation in the GMES projects that led to the current operational service, to the implementation of the newly established Copernicus Emergency Mapping Service (EMS).

Relations between the DPC and the international organisations are very fruitful as well. Through Inter-governmental processes and initiatives, the DPC contributes in advance to DRR and DRM topics at international level as well as at regional and national ones.

In this regard the collaboration with the United Nation Office for Disaster Risk Reduction UNISDR has been constant over the years, offering opportunities to foster disaster risk reduction policy, strategies and plans.

After the adoption of the Hyogo Framework for Action, Italy established the National Platform for Disaster Risk Reduction under the coordination of the DPC. The Italian commitment in DRR has been further renewed with the Sendai Framework for Disaster Risk Reduction 2015–2030.

The Department participates actively in several initiatives promoted by UNISDR in different domains, such as 'Making cities resilient', 'Indicators and Terminology related to DRR', 'School Safety and Science' and technology.

The Italian Hydraulic and Hydrogeological Early Warning System

Role and Responsibilities

In Italy, the separation of responsibilities is clearly stated in the text of Law 225/1992, and further specified in the Decree-Law of 15 May 2012, n. 59, converted with amendments by Law 12 July 2012, n.100.

In particular, regarding the "prevention" of hydraulic and hydrogeological risk, there are two major categories (http://www.protezionecivile.gov.it/):

- *Structural prevention* is based on the implementation of structural interventions aimed at mitigating the risk conditions through the reduction of danger and the probability of occurrence of a hydrological or hydraulic phenomena in a given time interval. In this context, measures include works of soil conservation measures such as the consolidation of the slopes, embankments, dams, spillways, etc. Planning, programming and implementation of these works is not, however, due to the expertise of the National System of Civil Protection, but to other institutions such as the Ministry of Environment, Regions and Basin Authority and the local authorities, on the basis of provisions mostly established at regional level.

- *Non-structural prevention* refers to actions and measures in the first place to mitigate risk by the reduction of exposure. The early warning system for hydraulic and hydrogeological risk, the regional offices, the emergency plans, the provision of information to the population, etc. are all examples of non-structural prevention The main purpose of these interventions, which refer to the competence of the different components of the national civil protection system, essentially consists in ensuring the safety and protection of human lives. In addition, territorial and urban planning, which are important for a complete and efficient soil protection policy, fall exclusively within the institutional powers of local governments. Since then, the crosscutting nature of soil defence and the fragmentation of responsibilities have clearly stressed the need for closer integration between the different institutions, for various reasons, demanding a wider level of competence, including forecasting, prevention and mitigation of geological risk.

National Centre for Forecasting and Surveillance for Hydraulic and Hydrogeological Risk

To address its mandate, the Civil Protection Service has organized a comprehensive system that includes a great number of both local and centralized resources (Boni et al. 2015). In particular, for hydraulic and hydro-geological risk, a national alert system is run by the DPC and regional authorities built around a network of Functional Centres (Boni et al. 2015) providing service in two phases—forecast of expected flooding and landslides and then monitoring and observation of current weather, flooding and geological conditions. This system, defined by the Directive of the President of the Council of Ministers 27 February 2004, was the result of a long sequence of disasters.

For these purposes and for the decision-making and the consequent assumption of responsibility, the Functional Centre is organized as a network consisting of 21 regional or decentralized Centres, based in the Regions or Autonomous Provinces, and by a Central Functional Centre, based at the Department of the National Civil Protection (Fig. 1).

Each Functional Centre is in charge of carrying out forecasting, real-time monitoring and surveillance of meteorological phenomena, with the consequent evaluation of the expected effects on people and things in a certain territory (alert zone).

The forecast phase consists in the evaluation, supported by appropriate numerical modelling, of weather, snow, hydrogeological and geomorphological conditions and the corresponding potential impact that these conditions may have on the integrity of life, property, settlements and environment. The phase of monitoring and surveillance is articulated in the qualitative and quantitative observation, direct and instrumental, of the meteorological and hydrogeological event in progress and in a short prediction of its effects by the measures collected in real time.

Rainfall thresholds are used to measure the expected risk scenario of geological and flood events in the alert zone. The thresholds identify the precipitation critical values and the effects expected if the events overcome the thresholds described in hydrogeological or hydraulic type event scenarios, in relation to the different levels of criticality defined (ordinary criticality, moderate, high).

Like the establishment of the alert areas, rainfall thresholds also were defined by the Regions and Autonomous Provinces of Understanding with the Department of National Civil Protection.

The use of thresholds is however not the only factor to be taken into account. The previous precipitation determines the degree of humidity of the soil and can affect the assessment of the expected risk scenario. The water levels of the main rivers have to be also considered, as well as sudden increases of temperature that can cause melting of the snowpack, factors such as landslides that are already active, and the geomorphological characteristics of the specific area.

On the basis of the criticality level, evaluated by the Functional Centre, the regional civil protection in charge may define corresponding alert levels, using three colour codes: yellow, orange and red, identifying from the lower to the maximum alert level. Each Municipality, taking into account the alert level, has to activate the corresponding operational phase (attention, pre alarm and alarm), as defined in each emergency plan. The alert system reaches its goal when the citizen is aware of self-protection rules and what to do in relation to the event that is taking place (Fig. 2).

In order to make this system effective, specific information campaigns for the general public take place both yearly ("I don't take risks" campaign, carried out in more than one hundred municipalities with the support of local volunteers, ad hoc trained) (http://iononrischio.protezionecivile.it) and when an event occurs, after which the population needs to be informed on what happened, what they can do and what they can expect to happen in the near future.

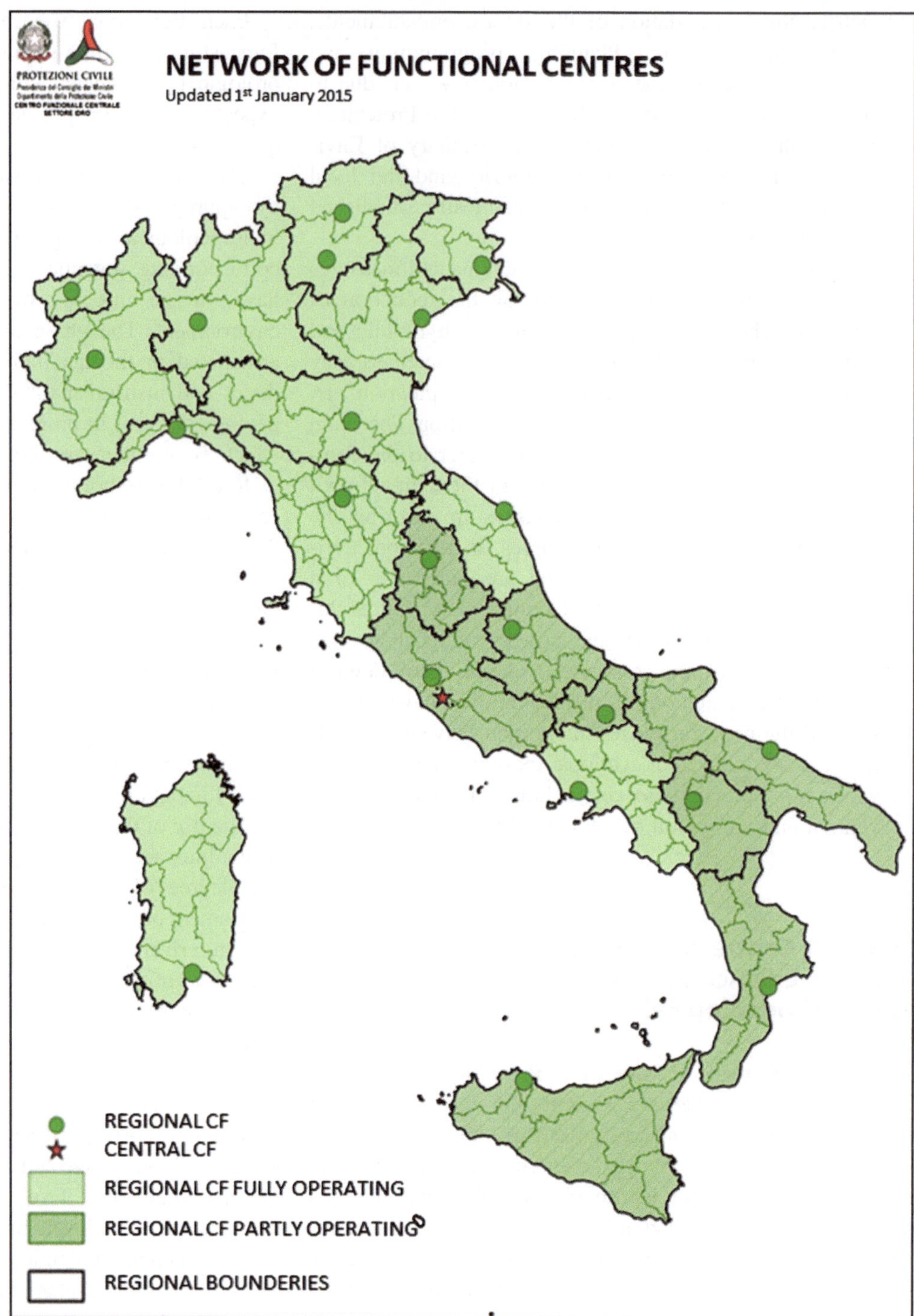

Fig. 1 Network of Italian functional centre for forecasting and surveillance for hydraulic and hydrogeological risk

Understanding and Reducing Landslide Disaster Risk

The widespread distribution of landslides, the occurrence of a number of destructive events, and the cost in terms of human lives and economic resources have led the scientific community and the institutions responsible for land management to a change in their analysis and management of risks related to hydrogeological hazards. After becoming conscious that it was no longer sufficient to tackle the problem during emergencies, providing assistance and restoring the damage, it was realised that instead there is a need to put in place a forecasting policy (identifying the areas at greater risk and estimates of the expected effects) and prevention (risk mitigation), so as to reduce as much as possible the damage expected as a result of landslides. This

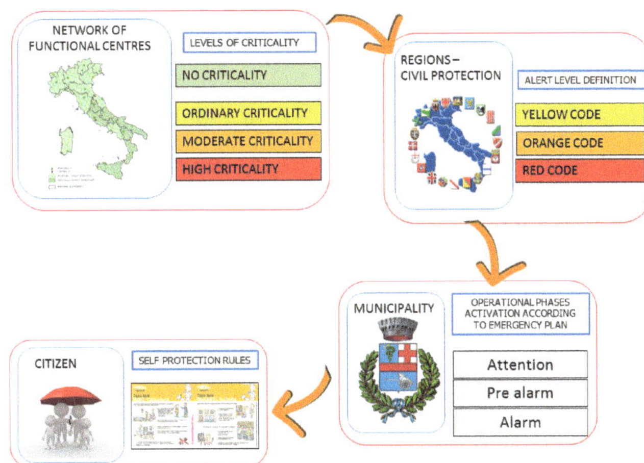

Fig. 2 General diagram of the early warning system for hydraulic and hydrogeological risk

vision is totally coherent with the Sendai Framework position, which shifts the emphasis towards Disaster Risk Management (DRM) as opposed to Disaster Management.

The cost of the damage caused by landslides must take into account the indirect damage associated with loss of productivity, the reduction of real estate asset value, the reduction in tax revenue and other economic impacts, for a more complete and realistic estimate. An understanding of the spatial and temporal evolution of landslides is thus of primary importance for hazard assessment (Guzzetti et al. 2005; Van Westen et al. 2006), risk management (Reichenbach et al. 2005), and the definition, design, and implementation of effective prevention and mitigation strategies, which can be effectively reached if long-term monitoring data are made available (Calòa et al. 2014).

The Role of Science and Research

Within the National Civil Protection Service, the Department of Civil Protection supports research efforts on the assessment of vulnerability and exposure of the population, buildings and critical infrastructure to landslides, and also is involved in national and international research projects (Bianchini et al. 2012; Cigna et al. 2013, 2011; Guzzetti 2000). With regard to deferred-time activities in the pre-event phase, the DPC provides strong support for knowledge application activities concerning landslide risk through a network of Italian Competence Centres (Centres for technological and scientific services) (Bianchini et al. 2012; Cigna et al. 2013, 2011; Guzzetti 2000).

Competence Centres play an important role in supporting the Department in the implementation and development of an Early Warning System (EWS), as well as in daily efforts for its updating and improvement. The role of science in supporting the activities of the Department, in particular concerning the EWS, is to provide support in terms of competences and tools to make the activities that have to be carried out daily more effective. In addition, the opportunity for close cooperation with scientists can address studies to understand the needs of EWS through "ad hoc" applied research (http://drmkc.jrc.ec.europa.eu—March 2016 Newsletter #2).

The DPC works at a national scale, also providing unified approaches and guidance to be followed at local level for risk mitigation. Therefore, we often support regional and local authorities in integrating science into their policy-making processes. The scientific community, of course, also directly supports the local authorities. In this respect, the role of the universities is very important, as some of them are very active at the local level. Considering the wide variety of scientific findings provided by different researchers on the same subject, it is also very important to favour and support the networking of those universities that conduct research activities and projects in the field of civil protection. In this sense, it is crucial to develop applied research programs and scientific applications that involve the entire scientific community. Moreover, particular attention is dedicated to achieving the objective that the entire national system of civil protection be coordinated, working on the basis of shared and reliable scientific information (http://drmkc.jrc.ec.europa.eu—March 2016 Newsletter #2).

But, for other risks, in addition to landslide risk reduction, effort has to be made to share the knowledge, and not only at a National level. In this framework the direct involvement of the DPC in several national but also international research projects is a way to improve and increase understanding of the best practices to be developed, both from a technical point of view and based on procedural approaches, to reduce the impact of landslides on the territory and on the population.

A common EU European approach also allows everyone to benefit from the achievements and practices of the entire scientific community in DRM. The European Commission started with a new initiative addressed to arrange the Disaster Risk Management Knowledge Centre (DRMKC) at EU level. The DRMKC are scientific institutions that provide tools, information, data, processing, and best practices in risk assessment and management, as well as technical-scientific advice on topics relevant to civil protection. This could be useful to facilitate the sharing of best practices, analyses and methodologies among the Member States and to support them for specific requests. It could also play a significant coordinating role in identifying and promoting the application of the most advanced scientific methods to support the countries to meet the goals foreseen in the European Union and Sendai Framework. (http://drmkc.jrc.ec.europa.eu/ March 2016—Newsletter #2).

From a regulation point of view, to support the transition from traditional flood defence strategies to a flood risk management approach at the basin scale in Europe, the EU has adopted the Flood Directive (2007/60/EC) at the end of 2007. One of the major tasks which Member States must carry out in order to comply with this Directive is to map flood hazards and risks in their territory, which will form the basis of future flood risk management plans (de Moel et al. 2009). This Directive represents a strength and a good starting point to share efforts to improve the management of Flood Risk and its reduction. The topic of landslides is included in the Thematic Strategy for Soil Protection (EC 2006a), adopted by the European Commission on 22 September 2006, and also in the legislative package is foreseen the development of a Soil Framework Directive which could include the identification of the risk areas and the development of risk management strategies (Van Den Eeckhaut and Hervás 2012). Like the Flood Directive, a Directive for the landslides could be useful in developing a common approach to risk reduction and improving the knowledge and expertise among different countries.

The Sendai Partnerships 2015–2025 for Global Promotion of Understanding and Reducing Landslide Disaster Risk

The Sendai Partnerships 2015–2025 for Global Promotion of Understanding and Reducing Landslide Disaster Risk represents the way to improve and share the efforts "to pursue prevention, to provide practical solutions, education, communication, and public awareness raising to reduce landslide disaster risk" (Sassa 2015).

The objective is to reach an ideal Disaster Risk Management decision-making process in which the scientific community is a key actor who contributes by providing a quantitative evaluation of the risk and a cost-benefit evaluation of the possible risk-mitigating actions, supporting the definition of strategies and policies and the implementation of the consequent measures in a timely, effective and efficient way. Their contribution is also relevant in the framework of risk communication and dissemination of a civil protection culture.

The partnership is relevant to facilitating the sharing of best practices, analyses and methodologies to meet the goals foreseen in the Sendai Framework, as well as making better use of existing knowledge at all stages of the DRM cycle, from prevention, reduction and preparedness to response and recovery, at all levels—local, national, European and global—and is therefore a priority both in policy-making and operations.

The direct involvement of Civil Protection Authorities in the Sendai Partnerships 2015–2025 for Global Promotion of Understanding and Reducing Landslide Disaster Risk is a strong opportunity for strengthening the use of science and technology in policymaking and is also relevant to address efforts to proceed in the direction of compliance with the perspective of DRR and to develop services, procedures and innovations that can be easily integrated in the operational workflow.

For the Civil Protection authorities, the protection of communities and cities is not only a duty, it is also an opportunity for reducing the social and economic losses related to disasters, and to progress towards building more resilient and equal societies.

This perspective is strictly connected with the pillars of the Sendai Framework, as well as with the Seven Global Targets. The following objectives have to be reached as a starting point to improve and increase resilience:

- close collaboration with technical and scientific communities to promote the development of effective monitoring tools and facilitate the use and transfer of new technologies to improve the knowledge of territories and risks;
- the promotion of awareness of risks among the population, developing an effective information system;
- the improvement of training of experts and decision makers;
- identify and promote, in the framework of the Platform, for Disaster Risk Reduction, global and national, a strategic fruitful cooperation

The achievements of the goals require an effort at national and international levels to improve the collaboration and, in this direction, the Sendai Partnerships 2015–2025 for Global Promotion of Understanding and Reducing Landslide Disaster Risk is a key initiative.

Acknowledgements The authors acknowledge the International Consortium on Landslides for its efforts in supporting the authorities in disaster risk reduction; the scientific and research communities that work and support the activities of civil protection with a special effort addressed to cope the user needs and operative requirements, and all the institutions, organizations and the actors involved who spend resources and effort to promote and improve the actions towards a strategy addressed to disaster risk reduction.

References

Boni G, Pulvirenti L, Silvestro F, Squicciarino G, Pagliara P, Onori R, Proietti C, Candela L, Pisani A R, Zoffoli S (2015) Flood mapping by Italian civil protection. Satellite earth observations in support of disaster risk reduction—special 2015 WCDRR edition

Bianchini S, Cigna F, Righini G, Proietti C, Casagli N (2012) Landslide hotspot mapping by means of persistent scatterer Interferometry. Environ Earth Sciences

Calò F, Ardizzone F, Castaldo R, Lollino P, Tizzani P, Guzzetti F, Lanari R, Angeli M-G, Pontoni F, Manunta M (2014) Enhanced landslide investigations through advanced DInSAR techniques: the Ivancich case study, Assisi, Italy. Remote Sens Environ 142:69–82

Cigna F, Del Ventisette C, Liguori V, Casagli N (2011) Advanced radar-interpretation of InSAR time series for mapping and characterization of geological processes. Nat Hazards Earth Syst Sci 11:865–881

Cigna F, Bianchini S, Casagli N (2013) How to assess landslide activity and intensity with Persistent Scatterer Interferometry (PSI): the PSI-based matrix approach. Landslides 10:267–283

de Moel H, van Alphen J, Aerts JCJH (2009) Flood maps in Europe—methods, availability and use. Nat Hazards Earth Syst Sci 9:289–301

Guzzetti F, Reichenbach P, Cardinali M, Galli M, Ardizzone F (2005) Probabilistic landslide hazard assessment at the basin scale. Geomorphology 72(1–4):272–299

Guzzetti F (2000) Landslide fatalities and evaluation of landslide risk in Italy. Eng Geol 58:89–107

http://www.protezionecivile.gov.it/

http://iononrischio.protezionecivile.it

http://drmkc.jrc.ec.europa.eu/ March 2016—Newsletter #2

Sassa K (2015) ISDR-ICL Sendai Partnerships 2015–2025 for global promotion of understanding and reducing landslide disaster risk. Landslides 12:631–640

Van Den Eeckhaut M, Hervás J (2012) Landslide inventories in Europe and policy recommendations for their interoperability and harmonization A JRC contribution to the EU-FP7 SafeLand project

Van Westen CJ, Van Asch TWJ, Soeters R (2006) Landslide hazard and risk zonation—why is it still so difficult? Bull Eng Geol Environ 65:167–184

Landslide Dynamics: ISDR-ICL Landslide Interactive Teaching Tools (LITT)

Kyoji Sassa, Fausto Guzzetti, Hiromitsu Yamagishi, Željko Arbanas, Nicola Casagli, Binod Tiwari, Ko-Fei Liu, Alexander Strom, Mauri McSaveney, Eileen McSaveney, Khang Dang, and Hendy Setiawan

Abstract

The International Consortium on Landslides (ICL) and ICL supporting organizations jointly established the ISDR-ICL Sendai Partnerships 2015–2025 which is the voluntary commitment to the Sendai Framework for Disaster Risk Reduction 2015–2030. As the core activity of the Sendai Partnerships, ICL has created "Landslide Dynamics: ISDR-ICL Landslide Interactive Teaching Tools", which are always updated and continuously improved, based on responses from users and lessons during their application. This paper describes the aim, outline, the contents of Text tools, PPT tools for lectures and PDF tools including already published reference papers/reports, guidelines, etc. Core parts of two fundamentals of the Teaching Tools, namely 1. Landslide types: description, illustration and photos, and 2. Landslide Dynamics for Risk Assessment are introduced.

Keywords

International consortium on landslides (ICL) • International strategy for disaster risk reduction (ISDR) • Landslide dynamics • Landslide types • Capacity development

K. Sassa (✉)
International Consortium on Landslides (ICL), Kyoto, Japan
e-mail: sassa@iclhq.org

F. Guzzetti
Istituto di Ricerca per la Protezione Idrogeologica, Consiglio Nazionale delle Ricerche, via Madonna Alta 126, 06128 Perugia, Italy
e-mail: Fausto.Guzzetti@irpi.cnr.it

H. Yamagishi
Asian Institute of Spatial Information, Shiroishi-Ku, Hongodori 2chome Kita 3-10, Sapporo, 003-0025, Japan
e-mail: hiromitsuyamagishi88@gmail.com

Ž. Arbanas
Faculty of Civil Engineering, University of Rijeka, Radmile Matejčić 3 Rijeka 51000, Rijeka, Croatia
e-mail: zeljko.arbanas@gradri.uniri.hr

N. Casagli
Department of Earth Sciences, University of Firenze, Via La Pira 4, 50212 Florence, Italy
e-mail: nicola.casagli@unifi.it

B. Tiwari
Civil & Environmental Engineering Department, California State University, Fullerton, 800 N State College Blvd., E-419, Fullerton, CA 92831, USA
e-mail: btiwari@fullerton.edu

K. Liu
National Taiwan University, No.1, Roosevelt Road, Chinese Taipei, 10617, Taiwan, Republic of China
e-mail: kfliu@ntu.edu.tw

M. McSaveney · E. McSaveney
GNS Science, PO Box 30368, Lower Hutt, 5040, New Zealand
e-mail: m.mcsaveney@gns.cri.nz

E. McSaveney
e-mail: e.mcsaveney@gns.cri.nz

H. Setiawan
Disaster Prevention Research Institute, Kyoto University, Uji, 611-0011, Japan
e-mail: hendy@flood.dpri.kyoto-u.ac.jp

A. Strom
Geodynamics Research Centre—branch of JSC "Hydroproject Institute", Volokolamsk Highway 2, Moscow, 125993, Russia
e-mail: strom.alexandr@yandex.ru

K. Dang
International Consortium on Landslides, Kyoto, Japan
e-mail: khangdq@gmail.com

K. Dang
VNU University of Science, Hanoi, Vietnam

K. Sassa et al. (eds.), *Advancing Culture of Living with Landslides*,
DOI 10.1007/978-3-319-59469-9_17

Aim of Landslide Dynamics: ISDR-ICL Landslide Interactive Teaching Tools

The International Consortium on Landslides (ICL) proposed the ISDR-ICL Sendai Partnerships 2015–2025 for global promotion of understanding and reducing landslide disaster risk at a session of "Underlying risk factors" of the 3rd WCDRR on the morning of 16 March 2015. The partnership was proposed as a voluntary commitment to the World Conference on Disaster Risk Reduction, Sendai, Japan, 2015, and also as tools for implementing and monitoring the Post-2015 Framework for Disaster Risk Reduction and the Sustainable Development Goals. It was approved and signed by 16 global stakeholders in the afternoon of the same day in Sendai, Japan, and the Secretary-General Mr. Petteri Taalas of the World Meteorological Organization (WMO) signed it on 16 April 2016. The number of current ICL members (as of 30 November 2016) that are a part of the Sendai Partnerships is 64. The number will be updated every year. The signatory organizations may increase at the high-level panel discussion and the round-table discussion during the Fourth World Landslide Forum in Ljubljana, Slovenia in 2017. The Sendai Partnerships is being updated during the period.

The Sendai partnerships acknowledge that

- At a higher level, social and financial investment is vital for understanding and reducing landslide disaster risk, in particular social and institutional vulnerability, through coordination of policies, planning, research, capacity development, and the production of publications and tools that are accessible, available free of charge and are easy to use for everyone in both developing and developed countries.

Landslide science and technologies have continuously been developed to be more reliable, precise or cost-effective for landslide disaster risk reduction over the world. However, this scientific and technological progress has not been shared equally over the world. The gap between the available level of science and technologies and the practical use of those in many countries, regions and communities is very wide. To fill this gap, ICL has created Landslide Interactive Teaching Tools, which are always updated and continuously improved, based on responses from users and lessons during their application. All text books gradually become outdated. To avoid this problem, ICL plans to upload the latest teaching tools in the WEB of Teaching Tools and print text tools periodically.

Landslide Dynamics

A landslide is a downslope movement of rock, soil or both (Cruden 1991, 1996). Landslide disasters are caused by exposure to **hazardous motions of soil and rock** that threaten vulnerable human settlement in mountains, cities, coasts, and islands, as stated in the Sendai Partnerships. Understanding "Landslide dynamics" is the very basis of landslide disaster risk reduction.

Organizations Contributing Teaching Tools

Each teaching tool will be submitted by the teaching tool contributing organization as shown in the list of contribution organizations. Each organization has its own Teaching Tool Identifying Number consisting of telephone number of the country and the registered number within the country (Table 1). The involvement of organization as well as individual researcher is better to keep quality and updating of each tool.

Outline of the ISDR-ICL Landslide Interactive Teaching Tools

The teaching tools are classified in five major parts. The part number is included in each teaching tool identifier.

0. Fundamentals
   (1) Landslide Types: Description, illustrations and photos
   (2) Landslide Dynamics for risk assessment
1. Mapping and Site Prediction
   (1) Basic Mapping
   (2) Site Prediction Using GIS
   (3) Field Guidelines
2. Monitoring and Early Warning
   (1) Remote Sensing Techniques for Landslide Monitoring
   (2) Monitoring System Instrumentation
   (3) Rainfall Threshold for Landslide Prediction
   (4) Landslide Time Prediction from Pre-failure Movement Monitoring
   (5) Guidelines for Landslide Monitoring and Early Warning Systems
3. Risk Assessment
   (1) Numerical Modeling and Simulation
   (2) Physical and Mathematical Modeling
   (3) Laboratory Soil Testing for Landslide Analysis
   (4) Analysis and Assessment of Landslides

Table 1 List of contributing organizations with identifier number and email of leader

Identifier No.	Organizations and email of leaders
001-1	Department of Civil and Environmental Engineering, California State University, Fullerton, 800 N. State College Blvd., E-419, Fullerton, CA 92831 Binod Tiwari: <btiwari@fullerton.edu>
001-2	U.S. Geological Survey (Denver Federal Ctr., Denver, CO 80225 USA) Lynn Highland: <highland@usgs.gov> Geological Survey of Canada (601 Booth St., Ottawa, Ontario, Canada KIA 0E8) Peter Bobrowsky: <peter.bobrowsky@canada.ca>
007-1	Sergeev Institute of Environmental Geoscience RAS, Ulansky per., 13, PB 145, Moscow 101000, Russia Valentina Svalova <inter@geoenv.ru>
020-1	Mining Department, Faculty of Engineering, Cairo University, Giza – 12613, Egypt Yasser ELSHAYEB <yasser.elshayeb@eng.cu.edu.e.g.>
034-1	Department of Civil and Environmental Engineering, Universitat Politècnica de Catalunya BarcelonaTech. Jordi Girona l-3. 08034 Barcelona, Spain. Jordi Corominas: <jordi.corominas@upc.edu>
034-2	Catalan Institute for Water Research (ICRA), Emili Grahit 101, 17003 Girona, Spain. Formerly at Sediment Transport Research Group (GITS), Department of Hydraulic, Marine, and Environmental Engineering, Universitat Politècnica de Catalunya—BarcelonaTech (UPC) Francesco Bregoli: <fbregoli@icra.cat>
039-1	Research Institute for Geo-Hydrological Protection, CNR, via Madonna Alta 126, 06128 Perugia, Italy Fausto Guzzetti <F.Guzzetti@irpi.cnr.it>
039-2	ISPRA – Italian National Institute for Environmental Protection and Research, Geological Survey of Italy Claudio Margottini: <claudio.margottini@isprambiente.it>
039-3	Department of Earth Sciences, University of Firenze, Via La Pira 4, Firenze, Italy Nicola Casagli: <nicola.casagli@unifi.it>
039-4	Department of Informatics, Modelling, Electronics and System Engineering, University of Calabria – Ponte Pietro Bucci, 41B Building, 5th Floor, 87036 Arcavacata di Rende (CS)—Italy Pasquale Versace: <pasquale.versace@unical.it>
043-1	Institute of Mountain Risk Engineering, University of Natural Resources and Life Sciences, Peter-Jordan-Straße 82, Vienna, A-1190, Austria Johannes Hübl: <johannes.huebl@boku.ac.at>
044-1	Department of Civil & Structural Engineering, University of Sheffield, Mappin St, Sheffield, United Kingdom. Elisabeth T. Bowman: <e.bowman@sheffield.ac.uk>
052-1	Institute of Geography, National Autonomous University of Mexico (UNAM), Circuito Exterior, Ciudad Universitaria, 04510, Coyoacán, Mexico City, Mexico Irasema Alcántara-Ayala: <irasema@igg.unam.mx>
052-2	Instituto de Geografía, Universidad Nacional Autónoma de México, México Gabriel Legorreta Paulíni: <legorretag@hotmail.com>
060-1	Slope Engineering Branch, Jabatan Kerja Raya MALAYSIA
062-1	Faculty of Engineering, Universitas Gadjah Mada, Indonesia Dwikorita Karnawati: <dwiko@ugm.ac.id> Teuku Faisal Fathani< tfathani@ugm.ac.id>
064-1	GNS Science, New Zealand Mauri McSaveney <m.mcsaveney@gns.cri.nz> Chris Massey <c.massey@gns.cri.nz>
066-1	Asian Disaster Preparedness Center (ADPC),Thailand NMSI Arambepola <arambepola@adpc.net>
081-1	ICL Headquarters, Japan Kyoji Sassa <sassa@iclhq.org>
081-2	Tohoku Gakuin University, Japan Toyohiko Miyagi <miyagi@izcc.tohoku-gakuin.ac.jp>

(continued)

Table 1 (continued)

Identifier No.	Organizations and email of leaders
081-3	Erosion and Sediment Control Department, Ministry of Land, Infrastructure, Transport and Tourism (MLIT), Japan
081-4	Asian Institute of Spatial Information, Shiroishi-ku, Hongodori 2chome kita 3-10 003-0025, Japan Hiromitsu Yamagishi <hiromitsuyamagishi88@gmail.com>
081-5	Center for Earth Information Science and Technology, Japan Agency for Marine-Earth Science and Technology, 3173-25 Showamachi Kanazawa-ku, Yokohama Kanagawa 236-0001, Japan Keiko Takahashi <takahasi@jamstec.go.jp>
0.81-6	National Research and Development Agency, Public Works Research Institute (PWRI), Tsukuba, Japan Jie Dou <j-dou@pwri.go.jp/douj888@gmail.com>
081-7	Geosphere Engineering & Disaster Management Office, International Consulting Operations, NIPPON KOEI CO., LTD., 1-14-6 Kudan-kita Chiyoda-ku, Tokyo 102-8539, Japan Kiyoharu Hirota <sbhirota@gmail.com>
081-8	Department of Disaster prevention, Meteorology and Hydrology, Forestry and Forest Products Research Institute, 1 Matsunosato, Tsukuba, Ibaraki 305-8687, Japan Shiho Asano <shiho03@ffpri.affrc.go.jp>
081-9	Okuyama Boring Co., Ltd., Akita Japan Shinro Abe <abeshinro@gmail.com>
084-1	VNU University of Science, Vietnam DUC Do Minh<ducgeo@gmail.com>
084-2	Vietnam Institute of Transport Science and Technology, Hanoi, Vietnam Dinh Van Tien <dvtien.gbn@gmail.com>
084-3	Vietnam Institute of Geosciences and Mineral Resources, No. 67 Chien Thang street, Van Quan ward, Ha Dong district, Hanoi City, Vietnam Le Quoc Hung <le.quoc.hung@vigmr.vn; hunglan@gmail.com>
086-1	Northeast Forestry University, Harbin, Heilongjiang, China Wei Shan <shanwei456@163.com>
380-1	ITGIS NASU (13, Chokolivsky Blvd., Kyiv, 03186, Ukraine) Oleksander Trofymchuk <itelua@kv.ukrtel.net>
385-1	Croatian Landslide Group from Faculty of Civil Engineering, Rijeka University and Faculty of Mining, Geology and Petroleum Engineering, University of Zagreb Željko Arbanas <zeljko.arbanas@gradri.uniri.hr> Snježana Mihalić Arbanas <snjezana.mihalic@rgn.hr>
386-1	Faculty of Civil and Geodetic Engineering, University of Ljubljana, Jamova c. 2, Ljubljana, SI-1000, Slovenia Matjaž Mikoš <matjaz.mikos@fgg.uni-lj.si> Nejc Bezak <nejc.bezak@fgg.uni-lj.si>
386-2	Independent researcher and Associated Professor at the University of Nova Gorica, Slovenia Marko Komac < m.komac@telemach.net>
504-1	Honduras Earth Science Institute. National Autonomous University of Honduras. Ciudad Universitaria, Boulevard Suyapa E-1. Tegucigalpa, Honduras. Lidia Torres Bernhard <torres.lidia@unah.edu.hn>
504-2	Faculty of Engineering, Central American Technological University, Zona Jacaleapa, Tegucigalpa, Honduras. Rigoberto Moncada López <rigoberto.moncada@unitec.edu>, <rigoberto.moncada.lopez@gmail.com>
886-1	National Taiwan University, Department of Civil Engineering, Chinese Taipei Ko-Fei Liu <kfliu@ntu.edu.tw>
886-2	Department of Civil Engineering, National Chiao Tong University (Hsin Chu, 30010, Taiwan) An-Bin Huang<huanganbin283@gmail.com>
886-3	Soil and Water Conservation Bureau, Council of Agriculture (Nantou, 54044, Taiwan) Hsiao-Yuan Yin <sammya@mail.swcb.gov.tw>
886-4	Department of Hydraulic and Ocean Engineering, National Cheng Kung University, No. 1 University Road, Tainan 70101, Taiwan Chyan-Deng Jan <cdjan@mail.ncku.edu.tw>

(continued)

Table 1 (continued)

Identifier No.	Organizations and email of leaders
886-5	Socio-Economic System Division, National Science & Technology Center for Disaster Reduction, Taiwan, R.O.C. Hsin-Chi Li <hsinchi@ncdr.nat.gov.tw>
886-6	Department of Soil and Water Conservation, National Chung Hsing University, (Taichung, 402, Taiwan, R.O.C.) Su-Chin Chen <scchen@dragon.nchu.edu.tw>
886-7	Water Environment Research Center, National Taipei University of Technology, 1, Sec. 3, Chung-Hsiao E. Rd., Taipei 106, Taiwan Chia-Chun Ho <ccho@ntut.edu.tw>

4. Risk Management and Country Practices
   (1) Landslide Risk Management
   (2) Community Risk Management
   (3) Country Practices.

The teaching tools consist of three types of tools.

1. The first type are text-tools consisting of original texts with figures. The first edition includes two volumes of books.
2. The second type are PPT-tools consisting of PowerPoint files and video tools made for visual lectures.
3. The third type are PDF-tools consisting of already published reference papers/reports, guidelines, laws and others.
   The second and the third type of tools are supplementary tools of the text tools (text books).
   Each teaching tool has its own identifier. The identifier of each tool consists of three parts:

1. the number of the part of the tool box in which it appears (Parts 0–4);
2. the country telephone code and an assigned unique number for each contributing organization (for example 081-1 signifies Japan-ICL headquarters, and 081-3 signifies Japan – Erosion and Sediment Control Department, Ministry of Land, Infrastructure, Transport and Tourism);
3. the last part of the identifier is a consecutive number assigned to the teaching tool by its contributing organization.

The following tables (Tables 1, 2, 3 and 4) present the list of contributing organizations and the list of contents of the teaching tools.

Fundamentals of the ISDR-ICL Landslide Interactive Teaching Tools (LITT)

All tools include visual explainations with many full color illustrations and photos. The tools start from two fundamental tools (TXT Tool 0.001-2.1 and TXT Tool 0.081-1.1). To present examples and illustration and photos used in LITT and also present an overview of the content of two fundamental aspects, some selected illustrations and photos are presented below.

Landslide Types: Descriptions, Illustrations and Photos

All figures and captions are copied from TXT Tool 0.001-2.1. Fig. 1.

The definitions of landslides were not uniform around the world before the United Nations International Decade for Natural Disaster Reduction (IDNDR) 1990–2000. Landslide disasters are one of the major disasters to be tackled in IDNDR. The united definition of landslides forms the basis for investigations and statistics of landslide disasters, through IDNDR as its base. The International Geotechnical Societies and UNESCO Working Group for World Landslide Inventory (Chair: David Cruden) was established. The landslide was then defined to be "the movement of a mass of rock, debris or earth down a slope". The types were explained in detail in "Landslides-investigation and mitigation", edited by A Keith Turner and Robert L, Schuster, Special Report 247 of the National Research Council (U.S.) Transportation Research Board in 1996. In order to disseminate this definition of landslides, including debris flows, rock falls and others, Lynn Highland and Peter Bobrowsky implemented the IPL 106 Best Practice handbook for landslide hazard mitigation (2002–2007) to create a handbook on landslides and published its result as "The Landslide Handbook—A guide to Understanding Landslides" (USGS Circular 1325) (Highland and Bobrowsky 2008). This book was well evaluated and translated into several languages. This project was awarded an "IPL Award for Success" at the Second World Landslide Forum at FAO Headquarters, Rome, October 2011. This definition, which includes debris flows, rock falls and other different types of landslides, is the basis of landslide science and it is the base of the International Journal "Landslides" founded in 2004. TXT Tool 0.001-2.1 (Highland and Bobrowsky 2017) presents many photos and illustrations to explain this definition by IDNDR,

Table 2 Contents of text tools

Landslide dynamics: ISDR-ICL landslide interactive teaching tools Vol. 1 fundamental, mapping and monitoring Kyoji Sassa, Fausto Guzzetti, Hiromitsu Yamagishi, Zeljko Arbanas, Nicola Casagli, Mauri McSaveney, Khang Dang Editors			
No.	Identifier	Title	Author
Fundamentals			
1.	TXT-Tool 0.001-2.1	Landslide types: descriptions, illustrations and photos	Lynn Highland, Peter Bobrowsky
2.	TXT-Tool 0.081-1.1	Landslide dynamics for risk assessment	Kyoji Sassa, Khang Dang
Part 1. Mapping and site prediction			
I.1	*Basic mapping*		
1.	TXT-Tool 1.081-2.1	Landslide mapping through the interpretation of aerial photographs	Toyohiko Miyagi
2.	TXT-Tool 1.081-2.2	Landslide mapping through the interpretation of aerial photographs and topographic maps	Eisaku Hamasaki et al.
3.	TXT-Tool 1.081-3.1	Landslide recognition and mapping using aerial photographs and Google Earth	Hiromitsu Yamagishi, Rigoberto Moncada Lopez
4.	TXT-Tool 1.039-1.1	Very-high resolution stereoscopic satellite images for landslide mapping	Francesca Ardizzone et al.
5.	TXT-Tool 1.504-1.1	Landslide Inventory educational methodology derived from experiences in Latin America	Rigoberto Moncada Lopez, Hiromitsu Yamagishi
I.2	*Site prediction using GIS*		
6.	TXT-Tool 1.052-1.1	GIS using landslides Inventory mapping for Volcanoes	Gabriel Legorreta Paulin
7.	TXT-Tool 1.052-1.2	GIS using landslides susceptibility mapping model for Volcanoes	Gabriel Legorreta Paulín, Michael Polenz, Trevor Contreras
8.	TXT-Tool 1.504-1.1	Landslide susceptibility assessment method	Marko Komac, Jernej Jež
9.	TXT-Tool 1.386-2.1	A comparative study of the binary logistic regression (BLR) and artificial neural network (ANN) models for GIS-based spatial predicting landslides at a regional scale	Jie Dou et al.
10.	TXT-Tool 1.386-2.2	Practical approach to assessing the factors influencing landslide susceptibility modelling—a case of Slovenia	Marko Komac
11.	TXT-Tool 1.084-3.1	Landslide susceptibility mapping at regional scale in Vietnam	Le Quoc Hung et al.
12.	TXT-Tool 1.039-1.2	Bedding attitude information through the interpretation of stereoscopic aerial photographs and GIS modeling	Ivan Marchesini et al.
13.	TXT-Tool 1.086-1.1	Distribution of Isolated Patches Permafrost in the Lesser Khingan Mountains of Northeast China using Landsat7 ETM + Imagery	Chunjiao Wang, Wei Shan
I.3	*Field guidelines*		
14.	TXT-Tool 1.081-7.1	Investigating landslides in the field Using Google Earth and PowerPoint: a case study of Altos de Loarque in Honduras	Kiyoharu Hirota
15.	TXT-Tool 1.064-1.1	Field guide for the identification and assessment of landslide and erosion features and related hazards affecting pipelines	Chris Massey et al.
Part 2. Monitoring and early warning			
II.1	*Remote sensing techniques for landslide monitoring*		
1.	TXT-Tool 2.039-3.1	Satellite remote sensing techniques for landslides detection and mapping	Nicola Casagli et al.
2.	TXT-Tool 2.039-3.2	Ground-based remote sensing techniques for landslides detection, monitoring and early warning	Nicola Casagli, Stefano Morelli et al.
3.	TXT-Tool 2.386-2.1	SAR interferometry as a tool for detection of landslides in early phases	M. Komac, T. Peternel, M. Jemec
4.	TXT-Tool 2.039-3.3	Ground-based radar interferometry for landslide monitoring	Nicola Casagli et al.
II.2	*Monitoring system instrumentation*		
5.	TXT-Tool 2.062-1.1	A Landslide monitoring and early warning system	Teuku Faisal Fathani, Dwikorita Karnawati
6.	TXT-Tool 2.007-1.1	Monitoring alarm system of landslide and seismic safety for potentially hazardous objects	Alexander Ginzburg et al.

(continued)

Table 2 (continued)

Landslide dynamics: ISDR-ICL landslide interactive teaching tools Vol. 1 fundamental, mapping and monitoring Kyoji Sassa, Fausto Guzzetti, Hiromitsu Yamagishi, Zeljko Arbanas, Nicola Casagli, Mauri McSaveney, Khang Dang Editors			
No.	Identifier	Title	Author
7.	TXT-Tool 2.007-1.2	Landslide and seismic monitoring system on the base of unified automatic equipment	Alexander Ginzburg et al.
II.3	*Rainfall threshold for landslides prediction*		
8.	TXT-Tool 2.039-1.1	Italian national landslide warning system	Mauro Rossi et al.
9.	TXT-Tool 2.084-3.1	Rainfall thresholds for triggering Geohazards in Bac Kan Province (Vietnam)	Nguyen Duc Ha et al.
10.	TXT-Tool 2.039-1.2	Rainfall thresholds for the possible initiation of shallow landslides in the Italian Alps	Maria Teresa Brunetti et al.
11.	TXT-Tool 2.039-1.3	Topographic and pedological rainfall thresholds for the prediction of shallow landslides in central Italy	Silvia Peruccacci, Maria Teresa Brunetti et al.
12.	TXT-Tool 2.039-4.1	FLaIR model (Forecasting of Landslides Induced by Rainfalls)	Pasquale Versace et al.
13.	TXT-Tool 2.386-2.1	A system to forecast rainfall induced landslides in Slovenia	Mateja Jemec Auflič et al.
14.	TXT-Tool 2.886-1.1	Early warning criteria for debris flows and their application in Taiwan	Chyan-Deng Jan et al.
15.	TXT-Tool 2.081-5.1	High-resolution rainfall prediction for early warning of landslides	Ryo Onishi et al.
16.	TXT-Tool 2.386-1.1	Intensity-Duration Curves for Rainfall-Induced Shallow Landslides and Debris Flows Using Copula Functions	Nejc Bezak et al.
17.	TXT-Tool 2.039-1.5	An algorithm for the objective reconstruction of rainfall events responsible for landslides	Massimo Melillo et al.
II.4	*Landslide time prediction from pre-failure movement monitoring*		
18.	TXT-Tool 2.385-1.1	A comprehensive landslide monitoring system: the Kostanjek landslide, Croatia	Snježana Mihalić Arbanas et al.
19.	TXT-Tool 2.385-1.2	Landslide comprehensive monitoring system: the grohovo landslide case study, Croatia	Željko Arbanas et al.
20.	TXT-Tool 2.062-1.2	A monitoring and early warning system for debris flows in rivers on Volcanoes	Teuku Faisal Fathani, Djoko Legono
21.	TXT-Tool 2.380-1.1	Monitoring and early warning system of the building constructions of the Livadia Palace, Ukraine	O. Trofymchuk, I. Kaliukh, O. Klimenkov
22.	TXT-Tool 2.039-4.2	LEWIS project: An Integrated System For Landslides Early Warning	Pasquale Versace et al.
23.	TXT-Tool 2.039-3.4	Methods to improve the reliability of time of slope failure predictions and to setup alarm levels based on the inverse velocity method	Tommaso Carlà et al.
II.5	*Guidelines for landslides monitoring and early warning systems*		
24.	TXT-Tool 2.886-1.2	Guidelines for landslide monitoring systems	An-Bin Huang, Wen-Jong Chang
25.	TXT-Tool 2.386-1.2	Practice guideline on monitoring and warning technology for debris flows	Johannes Hübl, Matjaž Mikoš
26.	TXT-Tool 2.886-1.3	Debris flow monitoring guidelines	Hsiao-Yuan Yin, Yi-Min Huang
Back material			
1.		List of PDF-tools and PPT-tools	
2.		ICL structure	

Landslide dynamics: ISDR-ICL landslide interactive teaching tools Vol. 2 testing, risk management and country practice Kyoji Sassa, Binod Tiwari, Kofei Liu, Mauri McSaveney, Eileen McSaveney Alexander Strom, Hendy Setiawan editors			
No.	Identifier	Title	Author
Part 3. Risk assessment			
III.1	Numerical modeling and simulation		
1.	TXT-Tool 3.081-1.1	An integrated model simulating the initiation and motion of earthquake and rain induced rapid landslides and its application to the 2006 Leyte landslide	Pham Van Tien, Kyoji Sassa, Khang Dang
2.	TXT-Tool 3.385-1.1	Application of Integrated Landslide Simulation Model LS-Rapid to the Kostanjek Landslide, Zagreb, Croatia	Karolina Gradiški et al.

(continued)

Table 2 (continued)

Landslide dynamics: ISDR-ICL landslide interactive teaching tools Vol. 2 testing, risk management and country practice Kyoji Sassa, Binod Tiwari, Kofei Liu, Mauri McSaveney, Eileen McSaveney Alexander Strom, Hendy Setiawan editors			
No.	Identifier	Title	Author
3.	TXT-Tool 3.081-1.2	Simulation of landslide induced tsunami (LS-Tsunami) based on the landslide motion predicted by LS-RAPID	Khang Dang, Kyoji Sassa
4.	TXT-Tool 3.081-1.3	A hypothesis of the Senoumi submarine megaslide in Suruga Bay in Japan based on the undrained dynamic-loading ring shear tests and computer simulation	Hendy Setiawan, Kyoji Sassa, Bin He
5.	TXT-Tool 3.081-1.4	Initiation Mechanism of Rapid and Long Runout Landslide and Simulation of Hiroshima Landslide Disasters using the Integrated Simulation Model (LS-RAPID)	Loi Doan Huy et al.
6.	TXT-Tool 3.385-1.2	Deterministic landslide susceptibility analyses using LS-Rapid software	Sanja Dugonjić Jovančević et al.
7.	TXT-Tool 3.886-1.1	Debris2D tutorial	Ko-Fei LIU, Ying-Hsin WU
8.	TXT-Tool 3.081-1.5	Manual for the LS-RAPID software	Kyoji Sassa, Hendy Setiawan et al.
9.	TXT-Tool 3.034-2.1	A debris flow regional fast hazard assessment tool	F. Bregoli et al.
10.	TXT-Tool 3.386-1.1	Two-dimensional debris-flow modelling and topographic data	Jošt Sodnik, Matjaž Mikoš
11.	TXT-Tool 3.886-1.2	Simulation for the debris flow and sediment transport in a large-scale watershed	Ying-Hsin Wu et al.
12.	TXT-Tool 3.385-1.3	Landslide occurrence prediction in the Rječina River Valley as a base for an early warning system	Martina Vivoda et al.
III.2	*Physical and mathematical modeling*		
13.	TXT-Tool 3.001-1.1	Physical modelling of rain-induced landslides	Binod Tiwari, Beena Ajmera
14.	TXT-Tool 3.001-1.2	Physical modeling of earthquake-induced landslides	Beena Ajmera and Binod Tiwari
15.	TXT-Tool 3.044-1.1	The runout of chalk cliff collapses in England and France—case studies and physical model experiments	E. T. Bowman, W. A. Take
16.	TXT-Tool 3.007-1.1	Mechanical-mathematical modeling and monitoring for landslide processes	Svalova Valentina
III.3	*Laboratory soil testing for landslide analysis*		
17.	TXT-Tool 3.081-1.6	Manual for undrained dynamic-loading ring shear apparatus	Hendy Setiawan et al.
18.	TXT-Tool 3.081-1.7	Undrained dynamic-loading ring shear apparatus and its application to landslide dynamics	Hendy Setiawan et al.
19.	TXT-Tool 3.081-1.8	A new high-stress undrained ring-shear apparatus and its application to the 1792 Unzen–Mayuyama megaslide in Japan	Khang Dang et al.
20.	TXT-Tool 3.001-1.3	Laboratory measurement of fully softened shear strength and its application for landslide analysis	Binod Tiwari, Beena Ajmera
III.4	*Analysis and assessment of landslides*		
21.	TXT-Tool 3.081-1.9	Dynamic properties of earthquake induced large-scale rapid landslides within past landslide masses	Pham Van Tien, Kyoji Sassa
22.	TXT-Tool 3.001-1.4	Using excel tools for slope stability analysis	Beena Ajmera, Binod Tiwari
23.	TXT-Tool 3.034-1.1	A textural classification of argillaceous Rocks and their durability	J. Corominas, J. Martínez-Bofill, A. Soler
24.	TXT-Tool 3.039-1.1	Landslide-related WPS Services	Ivan Marchesini, Massimiliano Alvioli, Mauro Rossi
Part (4) Risk management and country practices			
IV.1	*Landslide risk management*		
1.	TXT-Tool 4.084-1.1	Soil slope stability analysis	Do Minh DUC
2.	TXT-Tool 4.081-2.1	Risk Evaluation using the Analytic Hierarchy Process (AHP)—Introduction to the process concept	Eisaku HAMASAKI, Toyohiko MIYAGI
3.	TXT-Tool 4.039-3.1	Terrestrial laser scanner and geomechanical surveys for the rapid evaluation of rock fall susceptibility scenarios	Gigli, G. et al.

(continued)

Table 2 (continued)

Landslide dynamics: ISDR-ICL landslide interactive teaching tools Vol. 2 testing, risk management and country practice Kyoji Sassa, Binod Tiwari, Kofei Liu, Mauri McSaveney, Eileen McSaveney Alexander Strom, Hendy Setiawan editors			
No.	Identifier	Title	Author
4.	TXT-Tool 4.039-3.2	How to assess landslide activity and intensity with Persistent Scatterer Interferometry (PSI): the PSI-based matrix approach	Cigna, F., Bianchini, S., Casagli, N.
5.	TXT-Tool 4.034-1.1	Quantitative rockfall risk assessment for roadways and railways	Olga Mavrouli, Jordi Corominas
6.	TXT-Tool 4.084-1.2	Landslide Vulnerability assessment: a case study of Backan Town, Northeast Vietnam	Do Minh Duc et al.
7.	TXT-Tool 4.086-1.1	Landslide investigations in the northwest section of the Lesser Khingan Range in China using combined HDR and GPR methods	Zhaoguang Hu, Wei Shan, Hua Jiang
8.	TXT-Tool 4.052-1.1	Landslide risk perception	Irasema Alcántara Ayala
9.	TXT-Tool 4.886-1.1	Taiwan Typhoon Loss Assessment System (TLAS Taiwan) Web Tool	Hsin-Chi Li, Yi-Chen Chen, Mei-Chun Kuo
10.	TXT-Tool 4.385-1.1	Method for prediction of landslide movements based on random forests	Martin Krkač et al.
IV.2	*Community risk management*		
11.	TXT-Tool 4.062-1.1	Community hazard maps for landslide risk reduction	Dwikorita Karnawati et al.
12.	TXT-Tool 4.039-1.1	Definition and use of empirical rainfall thresholds for possible landslide occurrence	Silvia Peruccacci, Maria Teresa Brunetti
13.	TXT-Tool 4.062-1.2	A socio-technical approach for landslide mitigation and risk reduction	Dwikorita Karnawati et al.
14.	TXT-Tool 4.066-1.1	Community-based Landslide risk management approaches	Asian Disaster Preparedness Centre (ADPC)
15.	TXT-Tool 4.039-2.1	On the protection of cultural heritages from landslides	Claudio Margottini
IV.3	*Country practices*		
16.	TXT-Tool 4.886-1.2	Procedures for constructing disaster evacuation maps: guidelines and standards	Su-Chin Chen, Lien-Kuang Chen
17.	TXT-Tool 4.886-1.3	Emergency post-landslide disaster documentation	Lien-Kuang Chen
18.	TXT-Tool 4.386-1.1	State-of-the-art overview on landslide disaster risk reduction in Slovenia	Matjaž Mikoš et al.
19.	TXT-Tool 4.086-1.2	Shallow landslides and plant protection in seasonal frozen regions	Ying Guo et al.
20.	TXT-Tool 4.886-1.4	Ecological countermeasure guidelines and case histories in Taiwan	Chia-Chun HO et al.
21.	TXT-Tool 4.086-1.3	The impact of climate change on landslides in Southeastern high-latitude permafrost regions of China	Wei Shan, Zhaoguang Hu, Ying Guo
22.	TXT-Tool 4.052-1.2	Landslide risk communication	Irasema Alcántara Ayala
23.	TXT-Tool 4.504-1.1	How to make a database of landslides in Tegucigalpa, Honduras	Lidia Torres Bernhard et al.
24.	TXT-Tool 4.039-3.3	Debris flows modeling for hazard mapping	Massimiliano Nocentini et al.
25.	TXT-Tool 4.039-4.1	Landslide investigations and risk mitigation. The Sarno, Italy, case study	Giovanna Capparelli, Luciano Picarelli, Pasquale Versace
26.	TXT-Tool 4.081-8.1	Landslide monitoring for early warning in the Hai van station landslide in Vietnam	Shiho Asano, Hirotaka Ochiai, Huynh Dang Vinh
27.	TXT-Tool 4.081-1.1	Mechanism of large-scale deep-seated landslides induced by rainfall in gravitationally deformed slopes: A case study of the Kuridaira landslide in Kii Peninsula	Pham Van Tien et al.
28.	TXT-Tool 4.081-9.1	Rotary sampling drilling technology to extract a core of high quality by using sleeve-incorporating core barrel and polymer mud-From the cases of drilling in the landslide area of Japan and Vietnam	Shinro Abe et al.
29.	TXT-Tool 4.081-1.2	Mechanism of the Aratozawa large-scale landslide induced by the 2008 Iwate-Miyagi Earthquake	Hendy Setiawan et al.
Back material			
1.		List of PDF-tools and PPT-tools	
2.		ICL structure	

Table 3 Contents of PPT tools and video tools

List of PPT tools & video tools			
No.	Identifier	Title	Author
1.	PPT-tool 1.039-1.1	Remote Sensing data and methodology for event landslide recognition and mapping	Alessandro Mondini
2.	PPT-tool 1.052-1.1	Logisnet Manual and Quick-start Tutorial	Gabriel Legorreta Paulín, Marcus I. Bursik
3.	PPT-tool 2.039-1.1	Italian National Landslide Warning System (29 pages)	Mauro ROSSI et al.
4.	PPT-tool 2.062-1.1	Landslide Monitoring and Early Warning System	Teuku Faisal FATHANI, Dwikorita KARNAWATI
5.	PPT-tool 2.062-1.2	Monitoring and Early Warning System for Debris Flows in Rivers on Volcanoes (37 pages)	Teuku Faisal FATHANI, Djoko LEGONO
6.	PPT-tool 2.385-1.2	Landslide Comprehensive Monitoring System: The Grohovo Landslide Case Study, Croatia	Željko ARBANAS et al.
7.	PPT-Tool 3.385-1.3	Landslide Occurrence Prediction in the Rječina River Valley as a Base for an Early Warning System	Martina VIVODA PRODAN et al.
8.	PPT-tool 2.886-1.2	Landslide Monitoring and Warning	An-Bin HUANG, Wen-Jong Chang
9.	PPT-tool 3.039-1.1	Landslide Hazards and Risk Assessment	Fausto GUZZETTI
10.	PPT-tool 3.039-1.2	Landslide-related WPS services	Ivan MARCHESINI
11.	PPT-tool 3.039-1.3	Probabilistic approach to physically based landslide modeling	Massimiliano ALVIOLI et al.
12.	PPT-tool 3.039-1.4	Advanced 2D Slope stability Analysis by LEM by SSAP software: a full freeware tool for teaching and scientific community	Lorenzo BORSELLI
13.	PPT-tool 3.886-1.1	Debris-2D Tutorial	Ko-Fei LIU, Ying-Hsin WU
14.	PPT-tool 4.039-1.1	Definition and Use of Empirical Rainfall Thresholds for Possible Landslide Occurrence	Maria Teresa BRUNETTI, Silvia PERUCCACCI
15.	PPT-tool 4.039-1.2	Landslide Risk to the Population of Italy	Paola SALVATI, Cinzia BIANCHI
16.	PPT-tool 4.062-1.1	Socio-Technical Approach for Landslide Mitigation and Risk Reduction	Dwikorita KARNAWATI et al.
17.	PPT-tool 4.062-1.2	Community Hazard Maps for Landslide Risk Reduction	Dwikorita KARNAWATI et al.
18.	PPT-tool 4.066-1.1	Course on Landslide Disaster Risk Reduction for Local Government Level Stakeholders	Asian Disaster Preparedness Centre
19.	PPT-tool 4.886-1.1	Typhoon Loss Assessment System (TLAS) Taiwan Web Tool	Hsin-Chi Li et al.
20.	PPT-tool 4.886-1.2	Assessment Social Impact of debris flow disaster by Social Vulnerability Index	Ko-Fei Liu et al.
21.	PPT-tool 4.886-1.3	Tutorial: Procedures for Constructing Disaster Evacuation Maps (56 pages)	Su-Chin Chen, Lien-Kuang Chen
22.	Video-tool 3.084-2.1	Manual for undrained dynamic-loading ring shear apparatus	Lam Huu Quang

as the major contribution from the global landslide community.

The following four types of landslides: Falls (falls and topples), Slides (rotational and translational), Spreads and Flows (debris flow and debris avalanches) are explained using illustrations and photo examples in Figs. 2, 3, 4 and 5.

Landslide Dynamics for Risk Assessment (TXT-Tool 0.081-1.1)

All figures and captions are copied from TXT-Tool 0.081-1.1.

Table 4 Contents of PDF tools

List of PDF Tools			
No.	Identifier	Title	Author
1.	PDF-tool 1.064-1.1	Field guide for the identification and assessment of Landslide and Erosion features and hazards affecting pipelines	Chris Massey, Graham Hancox, Mike Page
2.	PDF-tool 3.081-1.1	An integrated model simulating the initiation and motion of earthquake and rain induced rapid landslides and its application to the 2006 Leyte landslide	Kyoji Sassa
3.	PDF-tool 3.081-1.3	A hypothesis of the Senoumi submarine megaslide in Suruga Bay in Japan based on the undrained dynamic-loading ring shear tests and computer simulation	Kyoji Sassa, Bin He
4.	PDF-tool 3.081-1.5	Manual for the LS-RAPID software	Kyoji Sassa, Hendy Setiawan et al.
5.	PDF-tool 3.081-1.6	Manual for undrained dynamic-loading ring shear apparatus	Hendy Setiawan et al.
6.	PDF-tool 3.081-1.7	Undrained dynamic-loading ring shear apparatus and its application to landslide dynamics	Kyoji Sassa et al.
7.	PDF-tool 3.081-1.9	Dynamic properties of earthquake-induced large-scale rapid landslides within past landslide masses	Kyoji Sassa
8.	PDF-tool 4.091-1.1	Guidelines for landslides management in India	Surya Parkash
9.	PDF-tool 4.091-1.2	Training module on comprehensive landslide risk management	Surya Parkash
10.	PDF-tool 0.001-2.1	The landslide handbook: a guide to understanding landslides	Lynn Highland, Peter Bobrowsky
11.	PDF-tool 4.064-1.1	Guidelines for assessing planning policy and consent requirements for landslide-prone land	Wendy Saunders and Phillip J. Glassey
12.	PDF-tool 4.064-1.2	Shut happens—building hazard resilience for businesses in NZ	Resilient Organisations
13.	PDF-tool 4.064-1.3	Working from the same page—consistent messages for CDEM: PART B: Hazard-specific information—landslides	Ministry of Civil Defence and Emergency Management
14.	PDF-tool 4.081-3.1	Japanese laws, codes, guideline and standard procedure regarding disaster prevention and risk reduction in Japan	Erosion and Sediment Control Department, Ministry of Land, Infrastructure, Transport and Tourism, Japan and International Sabo Network
15.	PDF-tool 4.007-2-1	Summer school guidebook	Alexander Strom

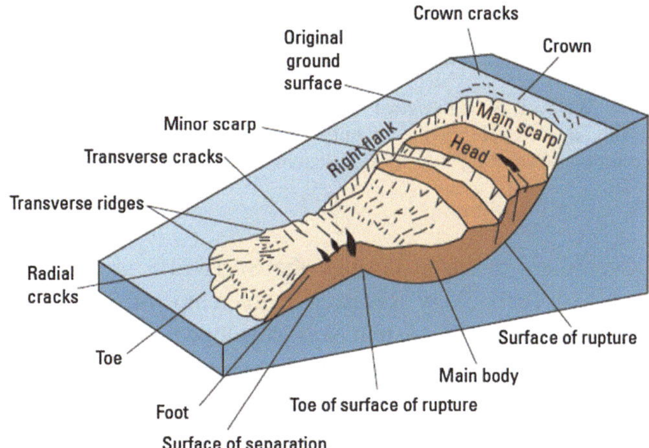

Fig. 1 This graphic illustrates commonly-used labels for the parts of a landslide. The image shows a rotational landslide that has evolved into an earthflow (modified from Varnes, 1978)

Figure 6 presents three major types of tests to measure the shear strength of soils: (1) Direct shear tests (shear box tests) in which a sample is sheared until failure in the drained condition in a shear speed control test, (2) Triaxial compression tests in which a sample is compressed until failure, either in either a drained condition or undrained condition, in either a stress control or speed control test, 3. Ring shear test in which a sample is sheared until a residual strength is obtained after failure in the drained condition during a speed control test.

Figure 7 presents the new undrained dynamic loading ring shear test which can simulate the initiation and the motion of landslides, by loading normal and shear stress in the field, including seismic stress loading and pore-pressure increase during rainfall. The most important feature of this new apparatus is the ability of to maintain an undrained condition and the measurement of pore-water pressure changes near the sliding surface (zone). The most important

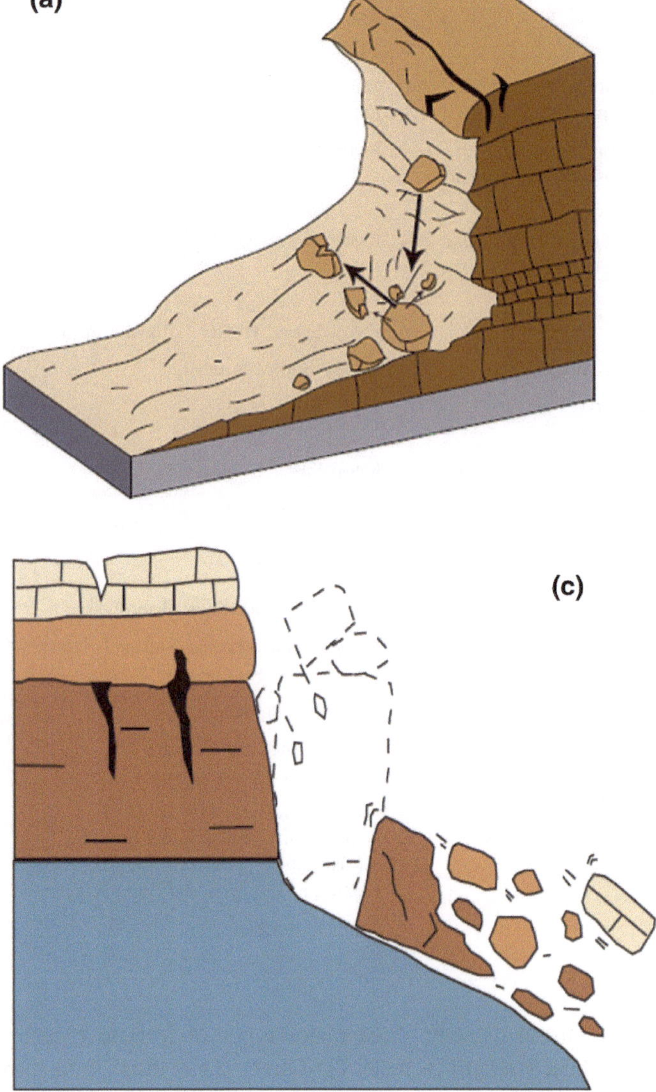

Fig. 2 Illustrations and photos of an example of Falls. **a** Schematic illustration of rockfall. Note that rocks may roll and bounce at potentially great distances depending on a number of factors. **b** A large rockfall due to the May, 2008 Wenchuan, China Earthquake. Photo by Dave Wald, U.S. Geological Survey. **c** Schematic illustration of a topple. **d** A topple in the vicinity of Jasper National Park, British Columbia, Canada. Photo by G. Bianchi Fasani

factor for the landslide disaster risk assessment is the estimation of velocity and the travel distance, and the moving area of landslides. The velocity and hazard area of a landslide is controlled by shear resistance mobilized in the sliding surface of the landslide. The resistance is regulated by the pore water pressure generated during initiation and motion, as well as seismic shaking. The water leakage from the gap between the upper shear box and the lower shear box is prevented by a rubber edge. The contact stress of the rubber edge to the upper ring shear box is controlled to be higher than the generated pore water pressure in the servo-control system. Two sets of new apparatuses (ICL-1 and ICL-2) were donated to Croatia and Vietnam from the Government of Japan through the SATREPS project (Science and Technology Research Partnership for Sustainable Development). The system was developed to be practical toly maintain,ed even in developing countries.

Figure 8 illustrates the great effect of pore water pressure. During rainfalls, the ground water level is increased in a soil layer on the bed rock (stable) layer of the slope. The mobilized shear resistance on the sliding surface is affected by the effective weight of the soil layer.

B illustrates a block inside a pool. A necessary horizontal force to move this block is decreased when a the water table will increases and the effective weight of the block is decreased by the buoyant force due to water. If the density of

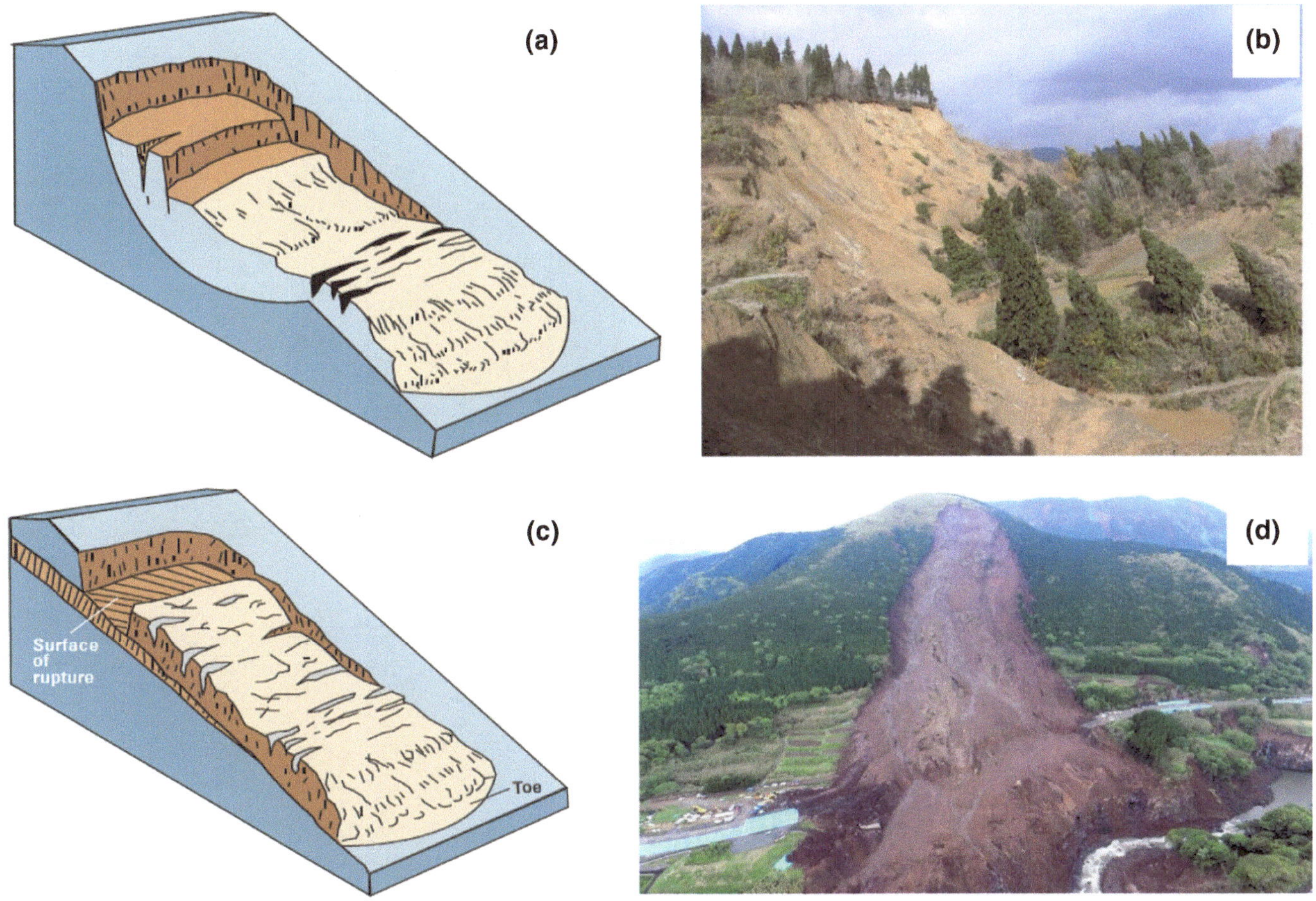

Fig. 3 Illustrations and photos of examples of slides. **a** Schematic of a rotational landslide. **b** A photo of a rotational landslide, showing the Dainichi-san landslide triggered by the October 23, 2004 Mid-Niigata Prefecture earthquake (Sassa 2005). **c** Schematic of a translational (parallel to slope) landslide. **d** The Minami-Aso Landslide shows a translational landslide in a steep slope. The landslide was triggered by the Kumamoto Earthquake of 2016 in Japan (Dang et al. 2016). Photo taken from UAV by Khang Dang and Kyoji Sassa

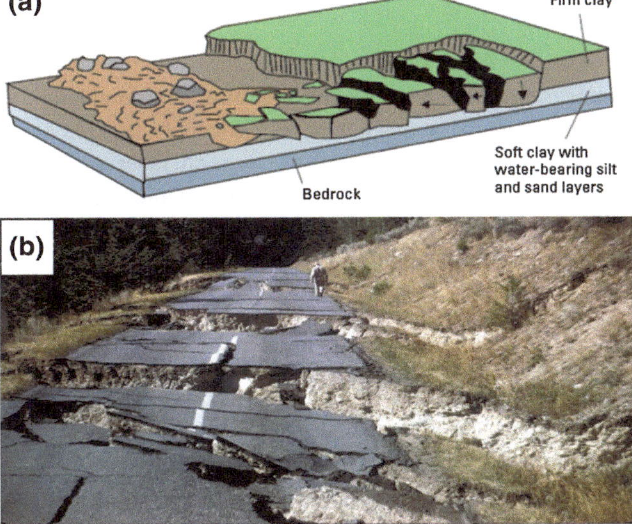

Fig. 4 Illustration and photo of an example of a spread. **a** Schematic of a lateral spread. **b** Lateral spreads at Hebgen Lake near West Yellowstone, Montana (USA), due to the effects of the Magnitude 7.3 Hebgen Lake earthquake, on August 18, 1959. Photo by R.B. Colton, U.S. Geological Survey

the block is similar to water, the block is almost floated floating in the water, and the necessary horizontal force to move this block is zero. Namely, the shear resistance between the block and the bottom of this pool is zero. The most difficult part is to know the function of pore water pressure.

Figure 9 illustrates the landslide initiation mechanism. Two figures show the normal stress and the shear stress relationships working on the sliding surface of a potential landslide. The left top shows a soil column within a slope. The mass of the soil column imparts a load (mass and gravity—mg) to the sliding surface. The shear stress component is mg·sinθ and the normal stress component is mg·cosθ. From this relationship, the initial stress is expressed as point I in this normal stress-shear stress chart, assuming no pore water pressure. When the ground water table is increased during rain, pore water pressure (u) is increased. In this case, the effective normal stress (normal stress minus pore water pressure) is decreased. Namely, the stress point in this chart will move to the left direction by u from the point I. When the stress point reaches the failure

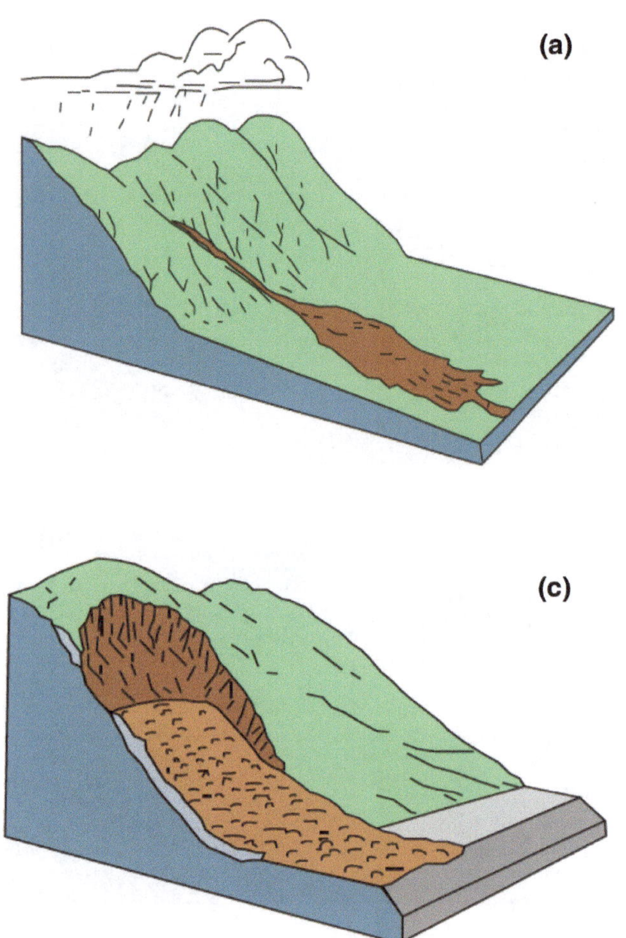

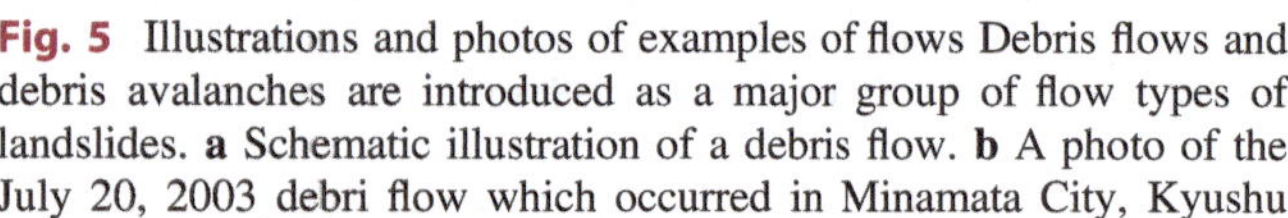

Fig. 5 Illustrations and photos of examples of flows Debris flows and debris avalanches are introduced as a major group of flow types of landslides. **a** Schematic illustration of a debris flow. **b** A photo of the July 20, 2003 debri flow which occurred in Minamata City, Kyushu Island, Japan, resulting in 14 deaths and 15 houses destroyed (Sassa et al. 2004). **c** Schematic of a debris avalanche. **d** A debris avalanche that buried a village in Guinsaugon, Southern Leyte, Philippines, in February, 2006 (Photo by University of Tokyo Geotechnical Team)

line of this soil, the soil will fail. This is the initiation of landslide by the mechanism of rainfall.

The right figure illustrates the initiation of landslide by pore water pressure plus earthquake loading. In the case of slope layer that includes a certain height of ground water table, as shown in Fig. 8, the initial stress before the earthquake is located at A in Fig. 9. When an earthquake strikes this area, seismic stress is loaded. The direction and the stress level will differ depending on the earthquake acceleration and its direction, but the stress point moves from A to somewhere. If the stress point reaches the failure line, the soil layer will fail and a landslide is initiated. This is the mechanism of an earthquake-induced landslide. Whether the landslide will be initiated or not can be simulated using the undrained ring shear apparatus by loading the seismic stress (an example test result is shown in Fig. 17).

Figure 10 illustrates two cases of the undrained dynamic loading ring shear tests for the initiation of a landslide and the movement of a the landslide. A sample will be taken from the potential shear zone or the soil layer or a layer which is estimated to have the same mechanical properties. The ring shear test will be conducted to determine whether a landslide will be initiated or not as shown in Fig. 9. The initiated landslide mass will move to the lower slope or onto the alluvial deposit, as shown in the right figure of Fig. 10. The shear surface will be formed within the deposit. A sample will be taken from the deposit on which the landslide mass now loadsrests on. A dynamic stress simulating the undrained loading by the moving landslide mass is given applied to the sample in the ring shear testing. The stress necessary stress to shearing the deposit and the generated pore water pressure and mobilized shear resistance

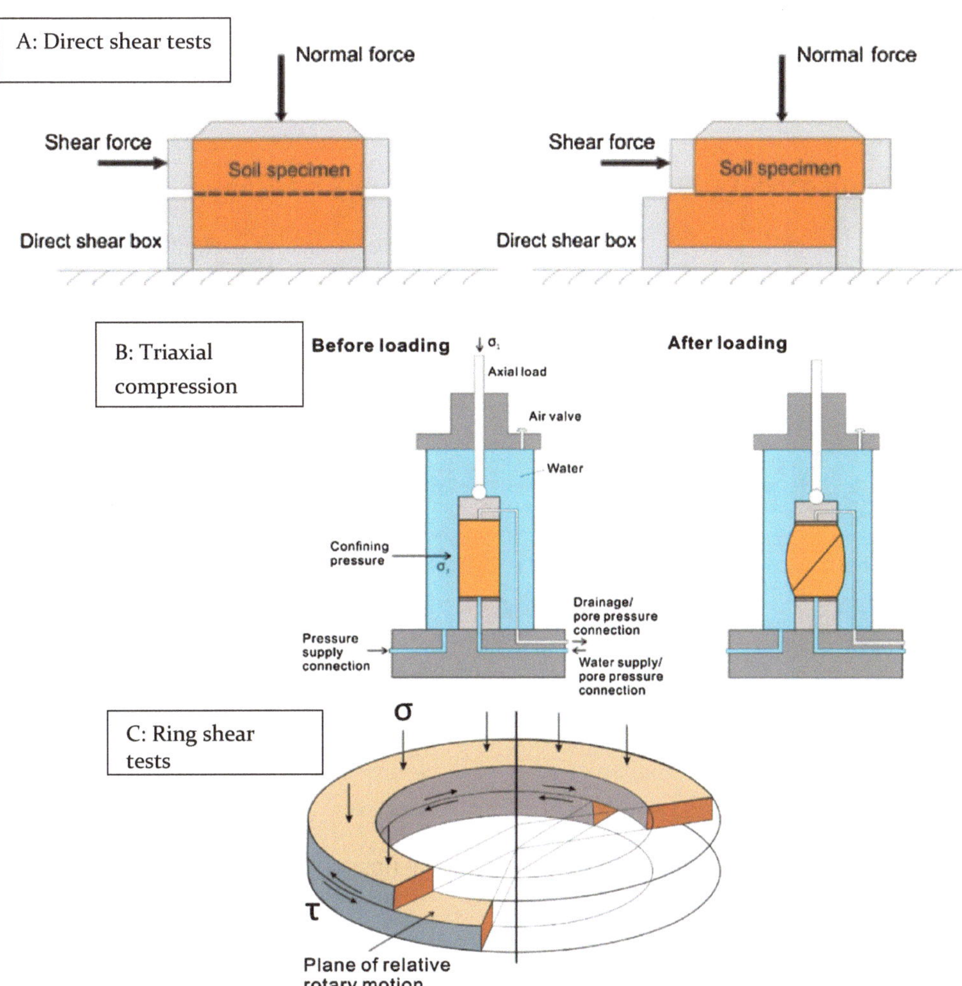

Fig. 6 Major three types of shear test to measure the shear strength of soils. *A* and *B* (Direct shear and Triaxial compression tests) are to measure the shear strength at failure. *C* (Ring shear test) is to measure the residual shear strength after failure

during loading and motion will be measured. Normally two tests are necessary to assess the initiation of landslide and the motion of landslides.

Figure 11 shows the setup (Nos. 1–6) of the undrained ring shear apparatus of ICL-2. Within this photo:

No. 1 is the computer and its two monitors (one for the test control system and one for the recording system).

No. 2 is the control unit, including the amplifiers for various monitoring sensors and the four servo-control amplifiers (Normal stress, shear stress, gap control and pore water pressures).

No. 3 is the main body of the undrained ring shear apparatus, including loading shear stress and speed control motor and gap control motor, normal stress loading system, the vertical and shear displacement measuring sensors, and pore pressure sensors (shown in Fig. 7).

No. 4 is the electricity supply and control system box.

No. 5 is pore pressure supply and control system.

No. 6 is the de-aired water supply system with a vacuum pump, vibration control system, and a vacuum tank to produce the sample fully saturated by de-aired water.

Figure 12 illustrates the concepts of the integrated landslide simulation model (LS-RAPID). A soil column is taken from the landslide mass, and all forces (self-weigh of soil column, the seismic forces, the lateral pressure, shear resistance on the bottom and the normal stress on the bottom) acting on this column are summed. The sum of force should accelerate the soil mass of the column. The change of pore water pressure and the resulting shear resistance on the bottom during the seismic loading, dynamic loading and shearing are obtained from testing using the undrained dynamic loading ring shear apparatus. The development of the undrained dynamic loading ring shear test has enabled the development of an integrated landslide simulation model applicable from the initiation of the motion until the termination of motion. Examples of the simulation results are shown in Figs. 13, 14 and 15.

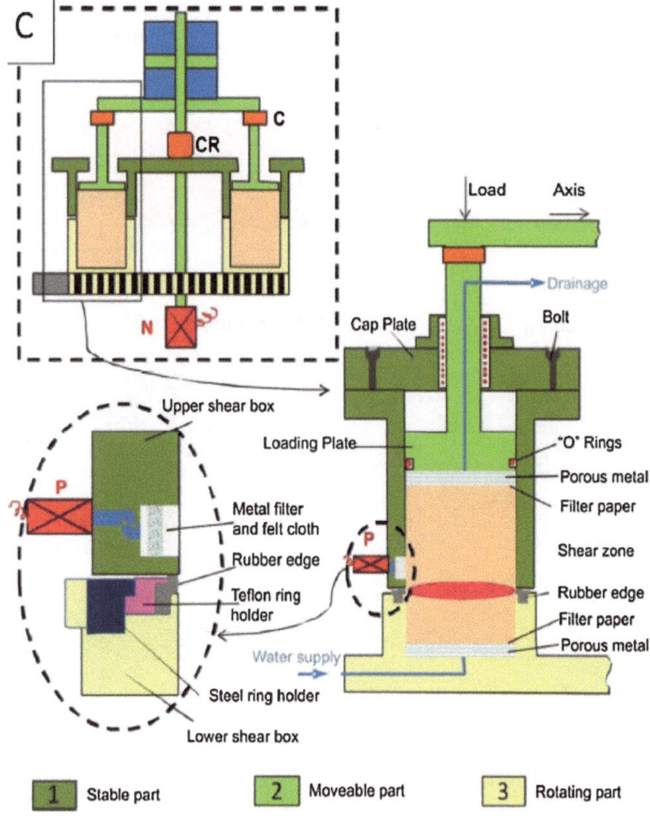

Schematic figure of the undrained dynamic-loading ring shear apparatus

Photo of shear box and normal stress loading piston and shear stress sensor

Fig. 7 An undrained dynamic-loading ring shear apparatus. The *left* figure presents the schematic figure of the latest version of undrained dynamic-loading ring shear apparatus (Sassa et al. 2016). The *right* photo shows the shear box of the apparatus. *A* Shear box; *B* Normal stress loading piston; *C* A pair of two shear stress sensors; *D* Loading cap; *E* Hanging frame to lift the loading cap; *F* Pore-pressure sensor; *G* Connection to the pore-pressure control system; *H* One-touch plug for the water drainage from the shear box; *I* One-touch plug for the de-aired water supply from the bottom of the shear box

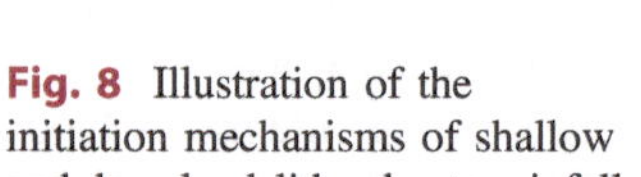

Fig. 8 Illustration of the initiation mechanisms of shallow and deep landslides due to rainfall

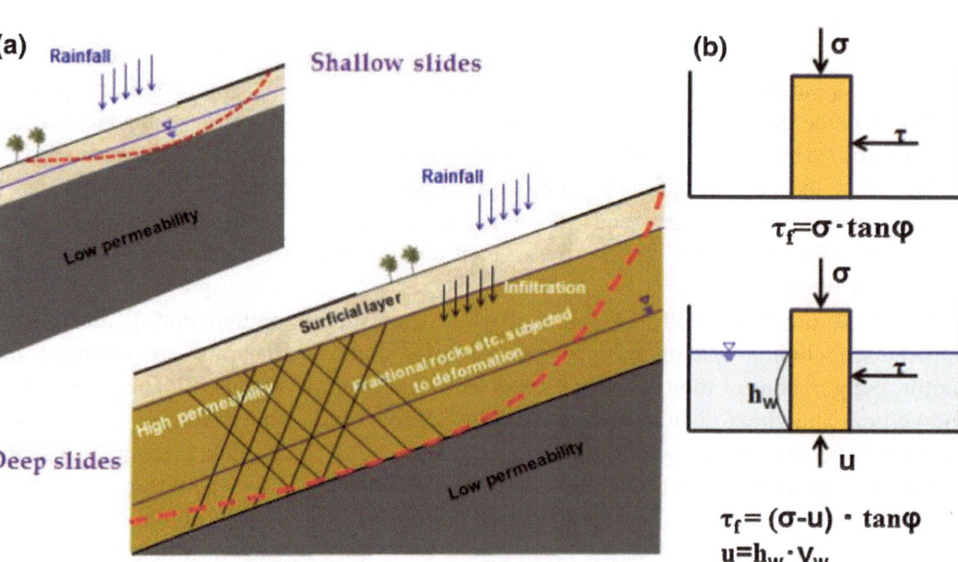

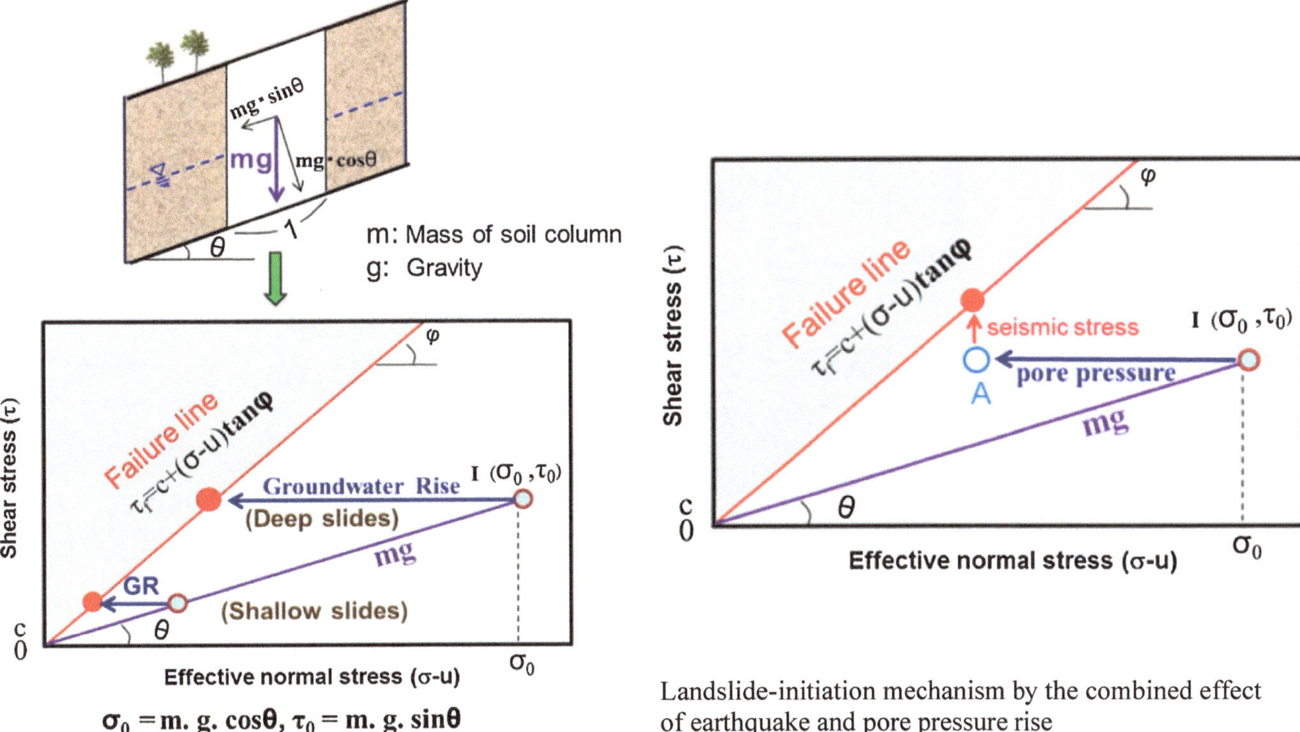

Fig. 9 Illustration of the mechanism to trigger failure within a slope by the triggering factors of rain (*left*) and the pore pressure plus earthquakes (*right*)

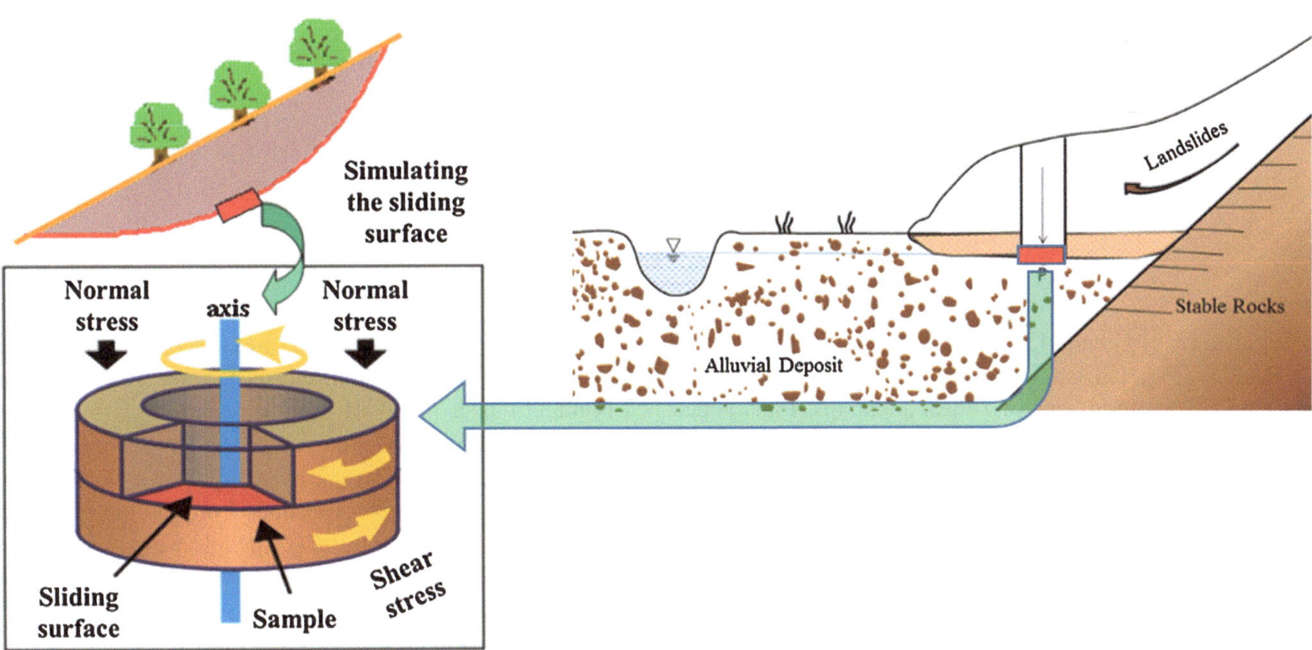

Fig. 10 Schematic figure of concept of an undrained dynamic-loading ring-shear apparatus

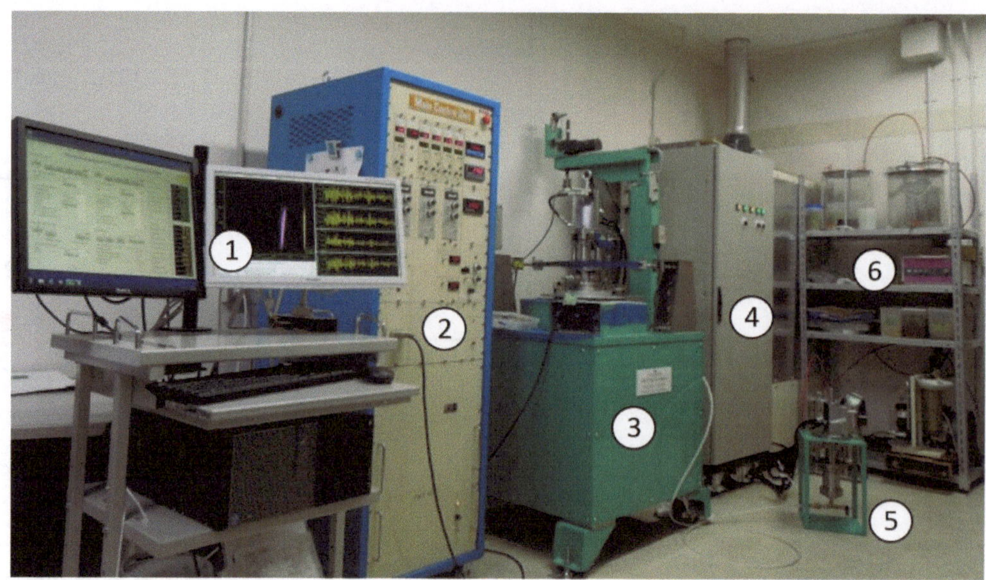

Fig. 11 Photo showing the setup of the ICL-2 apparatus

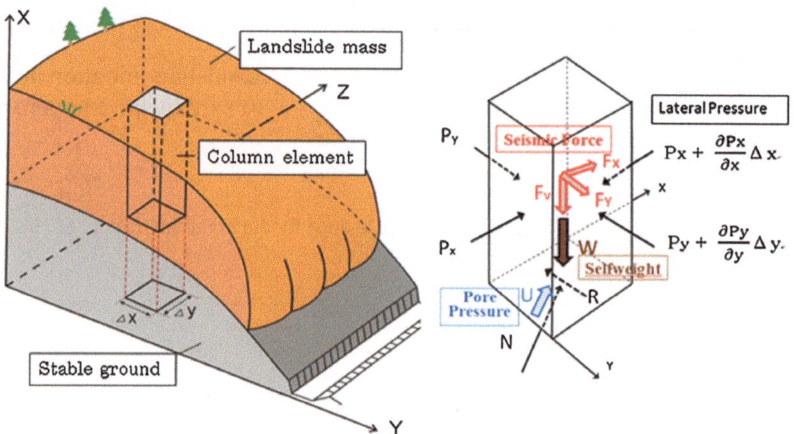

Fig. 12 Concept of the integrated landslide simulation model (LS-RAPID)

The landslide mass (m) will be accelerated by an acceleration (a) from the sum of all forces acting on the soil column.

Driving force = (self-weight + seismic forces) + lateral pressure + shear resistance

$$am = (W + Fv + Fx + Fy) + (\frac{\partial P_x}{\partial x}\Delta x + \frac{\partial P_y}{\partial y}\Delta y) + R$$

Figures 13, 14 and 15 are show the application of LS-RAPID (Fig. 12) using the test results of the from an undrained dynamic-loading ring shear apparatus (Figs. 8 and 11). Figure 13 is shows the application of the model to the Leyte landslide in Guinsaugon in the southern Leyte, Philippines in February 2006 (introduced in the bottom of Fig. 5). The landslide was triggered by a small nearby earthquake after a long period of rainfall (Sassa et al. 2010). Figure 13 and its caption explains the simulation result in the time series figure. For the initial two figures of A and B, the loaded initial pore water ratio due to rain and the loaded seismic stress calculated from the seismic record are shown.

Figure 14 is a photo of the Yagi and Midorii area of Hiroshima city, Japan (Doan et al. 2016). Both areas and the surrounding areas were struck by debris flows which came from the many of shallow landslides from the top of slopes during a local heavy rain. In all 74 people were killed in the urbanized residential areas. Figure 15 shows the result of the LS-RAPID computer simulation using the landslide dynamics parameters obtained from the testing using the

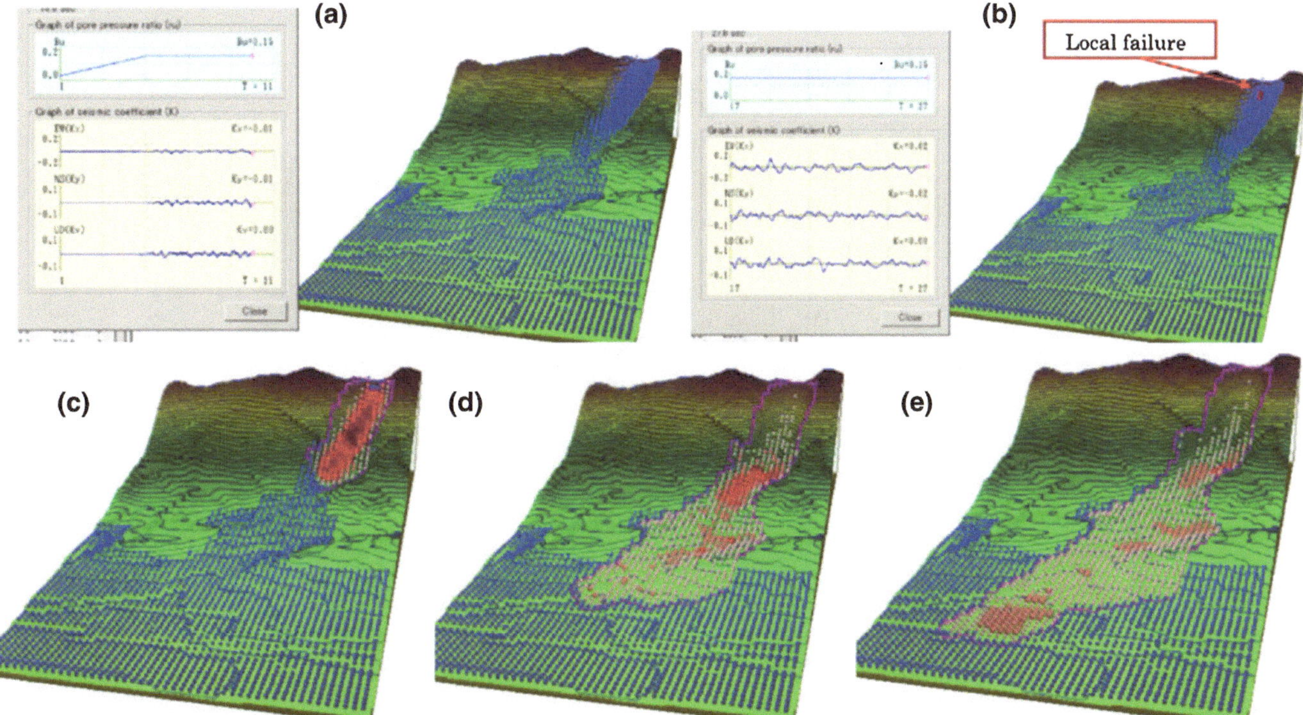

Fig. 13 Simulation results of the Leyte landslide. *A* r_u rises to 0.15 and the earthquake will start, but there is no motion. *B* Continued earthquake loading triggers a local failure, as shown by the *red color* mesh, *C* An entire landslide block is formed and moving, *D* The *top* of landslide mass moves onto alluvial deposits. *E* Deposition. Mesh size is 40 m; the area is 1960 × 3760 m; contour interval is 20 m; there is 3 m of unstable deposits on the alluvial deposit area (*blue balls*)

Fig. 14 Photo of the Yagi-area and Midorii in the 2014 Hiroshima landslide-debris flow disaster

undrained dynamic-loading ring shear apparatus. For the triggering factors, the 10-minute rainfall record was used. The initiation and the motion of landslides and the landslide hazard areas are reasonably reproduced by this simulation.

Figures 16, 17, and 18 present a study of the 1792 Unzen Mayuyama landslide using undrained ring shear testing and the integrated computer simulation. Figure 16 shows the Google view of the landslide. The urban area is Shimabara city and the sea is the Ariake Sea on Kyushu island of Japan. The mountain is a part of Unzen volcano. This 1792 Unzen Mayuyama landslide-and-tsunami-induced disaster is both the largest landslide disaster and also the largest volcanic

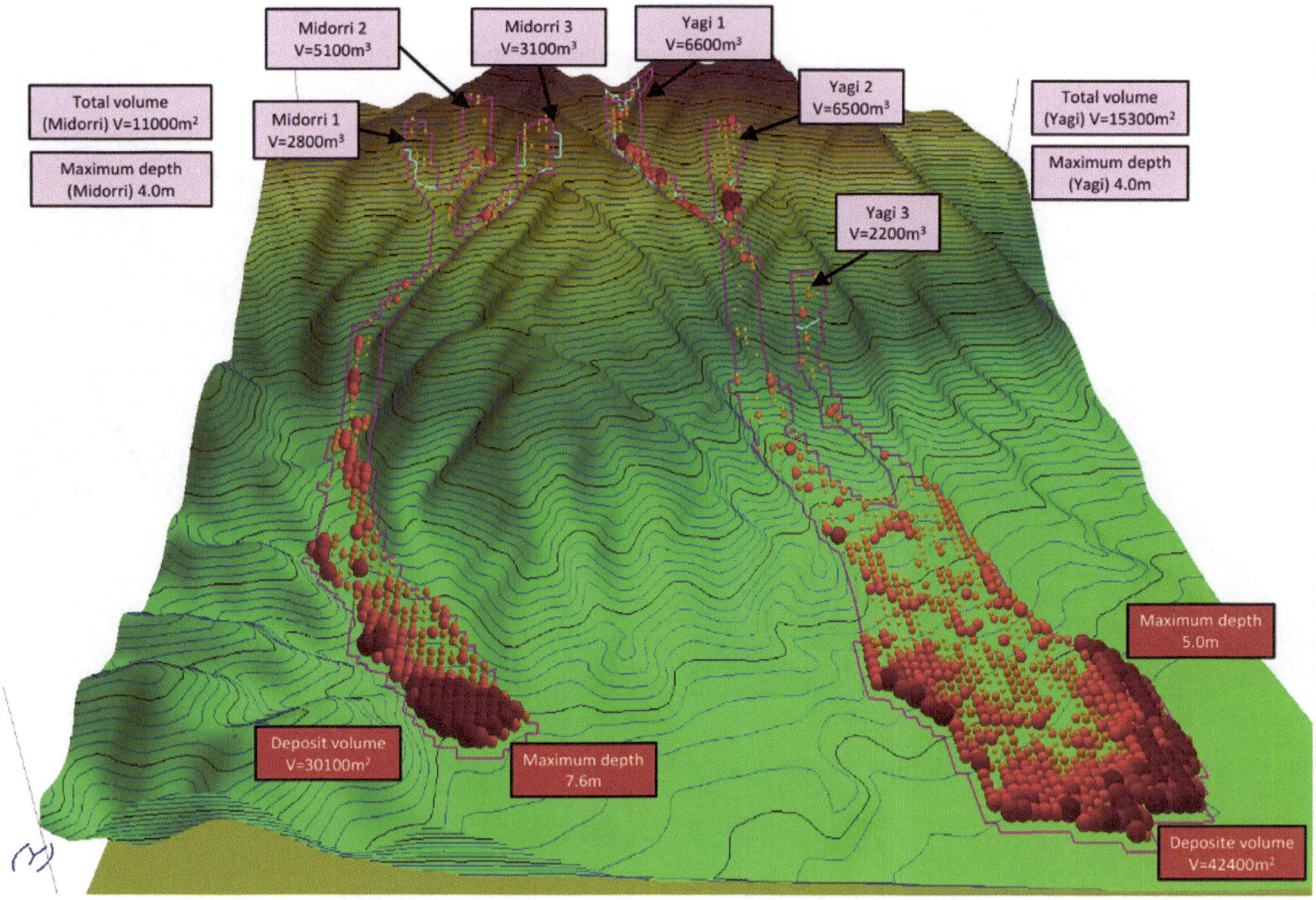

Fig. 15 Result of LS-RAPID simulation (Sassa et al. 2014)

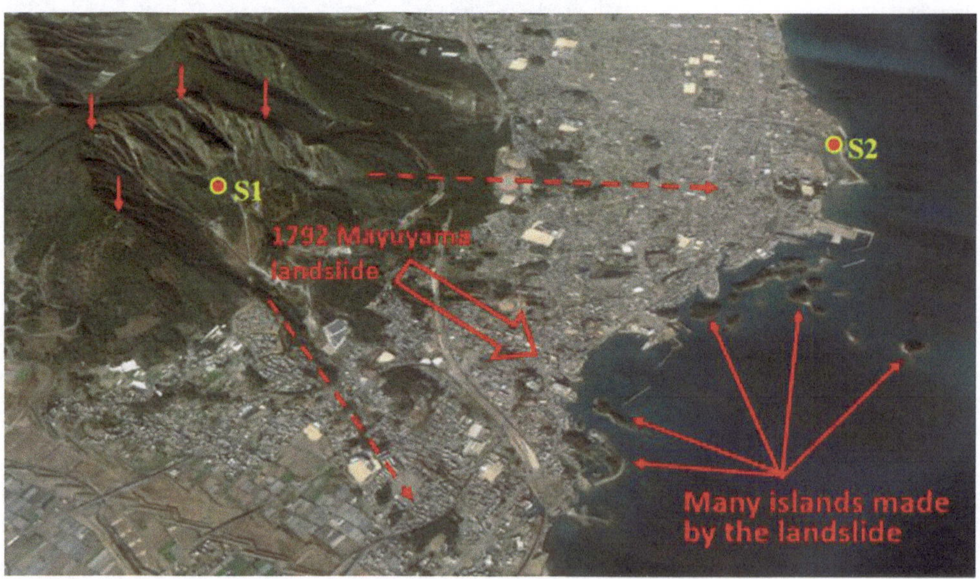

Fig. 16 Overview of the 1792 Unzen-Mayuyama landslide

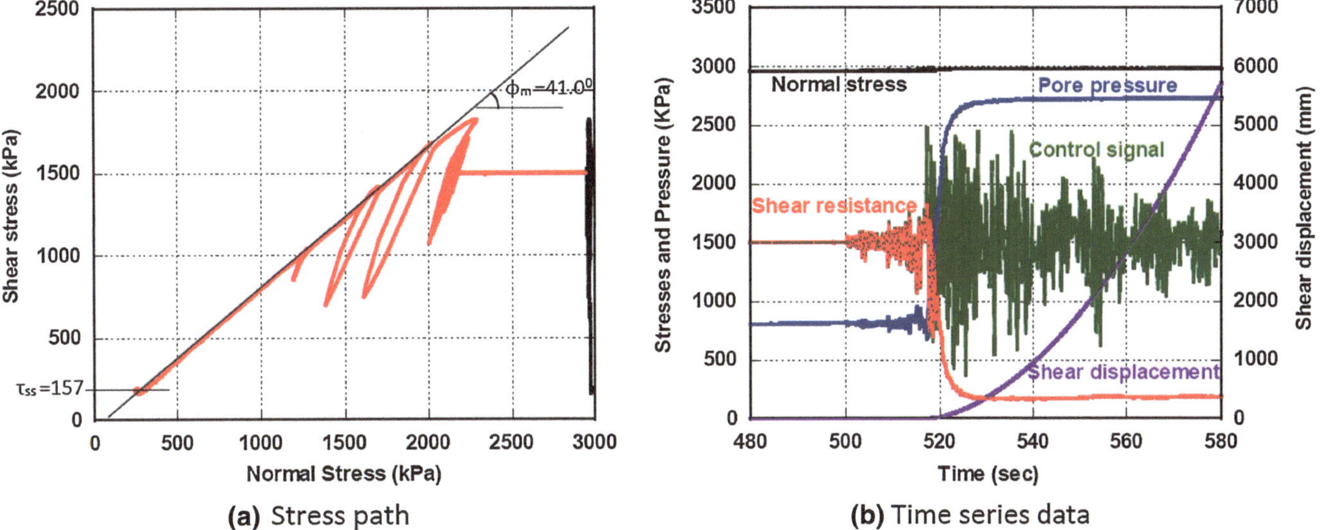

Fig. 17 Undrained seismic loading test on Sample 1 (S1). B_D = 0.94, Seismic wave: 2008 Iwata-Miyagi Earthquake record, 5 times slower speed

disaster in Japan, and also one of greatest tsunami disasters in Japan. The landslide mass entered into the Ariake Sea. Currently there are still some islands which are parts of the landslide mass deposited in the Ariake Sea. S1 is the sampling point to study landslide initiation behavior and S2 is the sampling point to study the motion of the landslide. All the area of the moving landslide mass area is now covered by heavily developed urban building. A sample was taken from the outside of the landslide moving area. Figure 17 is an example of the undrained ring shear testing, which involved (1) loading the initial shear stress and the normal stress, (2) loading pore water pressure before the earthquake, (3) loading the seismic stress using the seismic acceleration record of the 2008 Iwate-Miyagi Earthquake, which triggered a large-scale landslide (67 million cubic meters) (Miyagi et al. 2010). The earthquake was not recorded in 1792, but the acceleration was estimated based on detailed investigation of the damage to the houses and the tomb stones in Shimabara city. The test results indicate a steady-state shear resistance of 157 kPa and a friction angle during motion of 41°. Pore water pressure is built up during seismic loading and the pressure was very much increased in the progress of shearing.

Figure 18 is the simulation result of this landslide from its initiation to the motion into the sea. The initial landslide started from the middle of the source area (17 s) and the progressive failure expanded to the top of the landslide source area (26 s) and the total mass moved into the sea (64 s) and stopped after 226 s. The length of the deposit area from the simulation is rather close to the area determined by field investigation by the Unzen Restoration Office.

Figures 19, 20, and 21 present the most advanced study of the landslide-induced tsunami, which was published online (April 2016) and in print (Sassa et al., Landslides Vol. 13, No. 6, 2016). Figure 19 presents the historical record of the landslide-induced tsunami disaster in the 1792 Unzen Mayuyama landslide and tsunami. In this disaster, 15,153 people were killed.

The central disaster is the concept of the landslide-induced-tsunami simulation model (LS-Tsunami). The basic concept of this model is that the landslide-induced tsunami will be simulated using the well-established and widely used model (Intergovernmental Oceanographic Commission (IOC) (1997). The basic equation is shown below in Fig. 20. As the triggering factor of the tsunami, the landslide simulation results of LS-RAPID are used. Two steps are completely separated. Interactive shear forces between the landslide mass and the water is neglected. It is assumed that the vertical uplift of the sea floor by the moving landslide mass lifts the water mass above the landslide mass vertically. Figure 21 is the result of the landslide simulation, landslide-induced tsunami simulation, and the

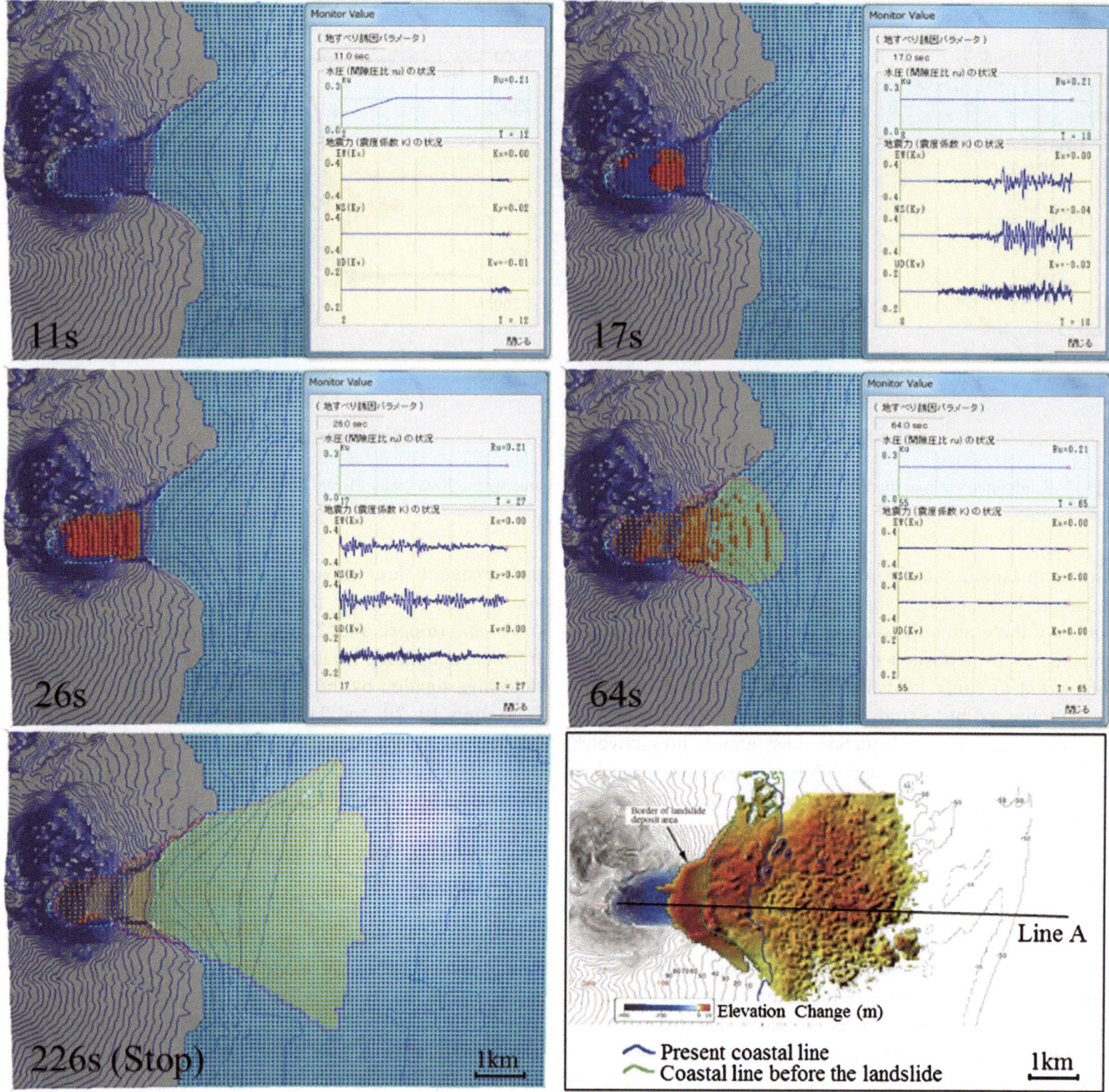

Fig. 18 LS-RAPID simulation result of the 1792 Unzen-Mayuyama landslide

tsunami motion over the sea, including reflection from the opposite shore. The first figure at 0 m 20 s shows the initiation of landslide. The second figure at 0 m 35 s shows the landslide mass reaching the coast, the third figure at 1 m 25 s shows the tsunami wave induced by the landslide mass, the fourth figure and the fifth figures at 5 m 55 s and 10 m 45 s show the expansion of the tsunami wave. The final two figures present the reflected wave from the opposite bank (Kumamoto Prefecture) attached the Shimabara Peninsula again. The red color tsunami wave is more than 5 m above sea level and the blue color tsunami wave is more than 5 m below sea level. The detailed tsunami height records at 5

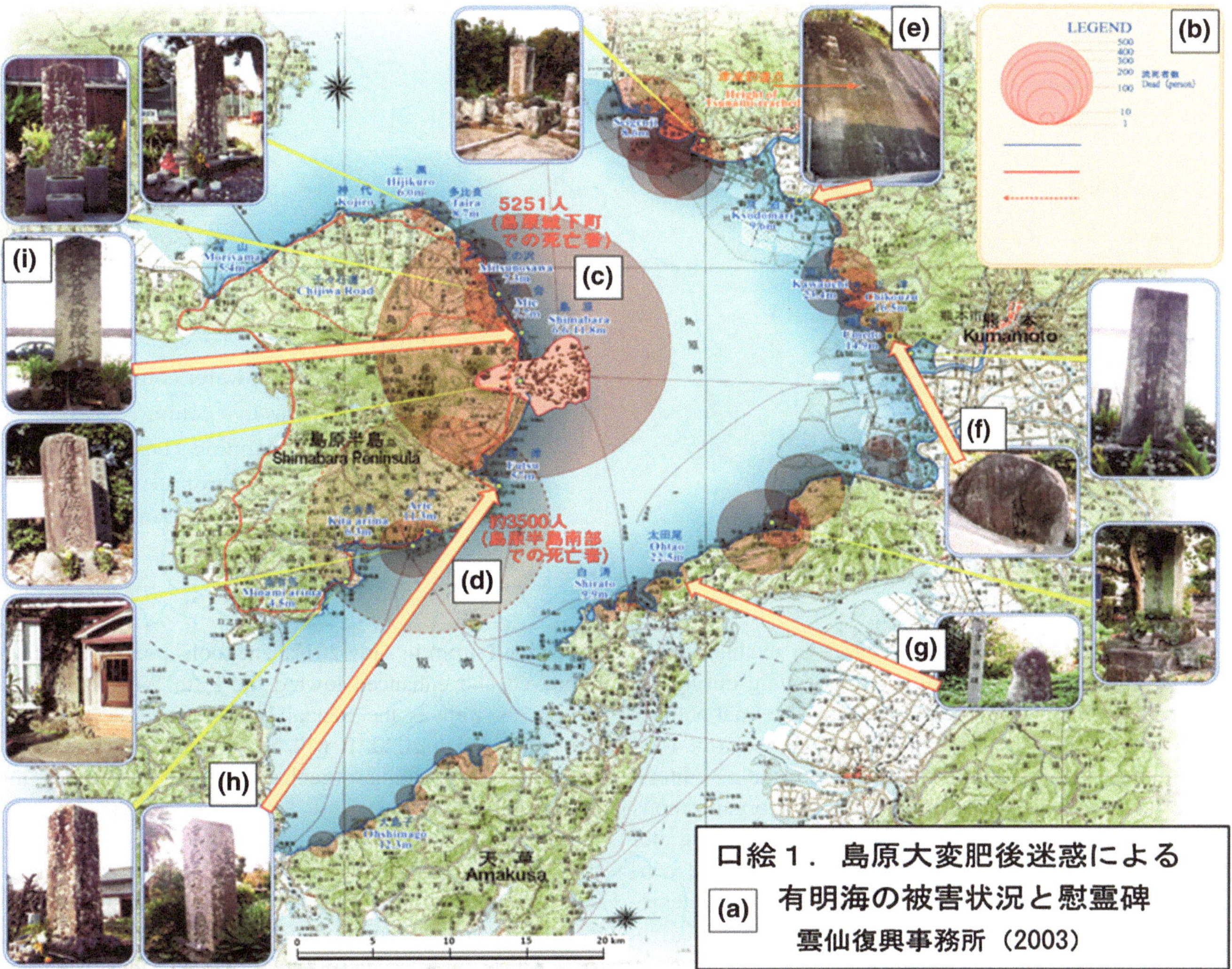

Fig. 19 Records of the Unzen landslide-and-tsunami disaster (by Unzen Restauration Office 2003). The total number of deaths is 15,153 persons. The size of *circles* is proportional to the number of human fatalities in the area. The legend for the number of deaths is show in the *right-top corner*. *A* Disasters around Ariake Sea and monuments by the Shimabara-Taihen, Higo-Meiwaku. The "Catastrophe" in Shimabara Area and "Annoyance" or "adversely affected" in Higo (Kumamoto) Region. *B* The numbers of deaths are shown in the circles (the largest is 500 persons). *C* The greatest number of deaths are in Shimabara town around the castle (5251 persons). *D* The second largest number of deaths are in the southern part of Shimabara Peninsula (around 3500 persons). *E*, *F* and *G* Tsunami-Dome-Ishi (*A* stone showing the tsunami reaching that point) were set to record the tsunami by the communities in Kyodomari (*E*), Umedo (*F*) and Otao (*G*) of the Higo (Kumamoto) Han area. The Tsunami-Dome-Ishi in Kyodomari was moved for the construction of a road, but its former location is marked on the road retaining wall (by the regional education committee). The Tsunami-Dome-Ishi is limited in Higo (Kumamoto) Han area. These tsunami records are reliable. *H*, *I* Stone pillars for memorial services for deaths by tsunami in Futsu (*H*) and Mie (*I*) in Shimabara Han area

locations on land and those estimated by this computer simulation were compared. The values are rather similar in 4 locations.

ICL called for cooperation for the ISDR-ICL Landslide Interactive Teaching Tools soon after the World Conference on Disaster Risk Reduction (WCDRR) in Sendai Japan and the establishment of the ISDR-ICL Sendai Partnerships for global promotion of understanding and reducing landslide disaster risk 2015–2025 on 16 March 2015. Many ICL members have offered their cooperation and contributed

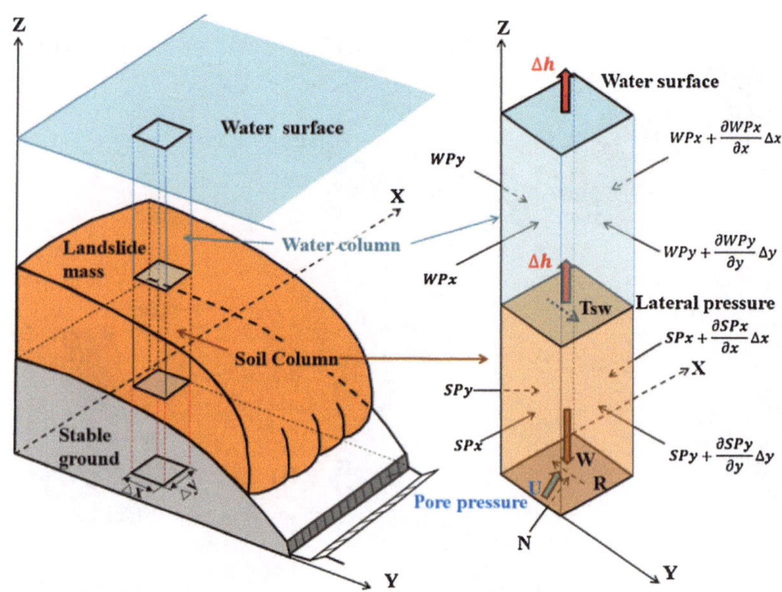

$$am_w = \frac{\partial P_x}{\partial x} \Delta x + \frac{\partial P_y}{\partial y} \Delta y + R(Manning)$$

Where

a: Acceleration of water column

m_w: Water mass in a column

R (Manning): Manning's basal resistance between water and ground

Fig. 20 Basic principles of the landslide-induced tsunami simulation model

many teaching tools. ICL asked editors to evaluate those submitted tools. Tools and editors are changing during the process of producing these teaching tools. The final number of accepted teaching tools are 97 in two volumes (the total page number is 1700) and there are 11 cooperating editors. Firstly ICL appreciates all authors and their organizations which contributed teaching tools. The planned teaching tool set is not fixed, but continually evolving—it will be continuously updated, improved and enhanced by the interaction between authors and users.

The initial version of ISDR-ICL Landslide Interactive Teaching Tools will be published before the Fourth World Landslide Forum in May 2017. The tools are expected to be improved and enhanced toward the Fifth World Landslide Forum, as well as the Sendai Partnerships mid-term conference in Japan 2020. It will be very much appreciated if voluntary contributing organizations and individuals join this initiative.

All ICL member organizations and all World Centres of Excellence for Landslide Disaster Reduction (WCoEs) and non-ICL cooperating organizations are requested to contribute to capacity building using the ISDR-ICL teaching tools and to improve these living tools as better, wider and more practical resources for landslide disaster risk reduction.

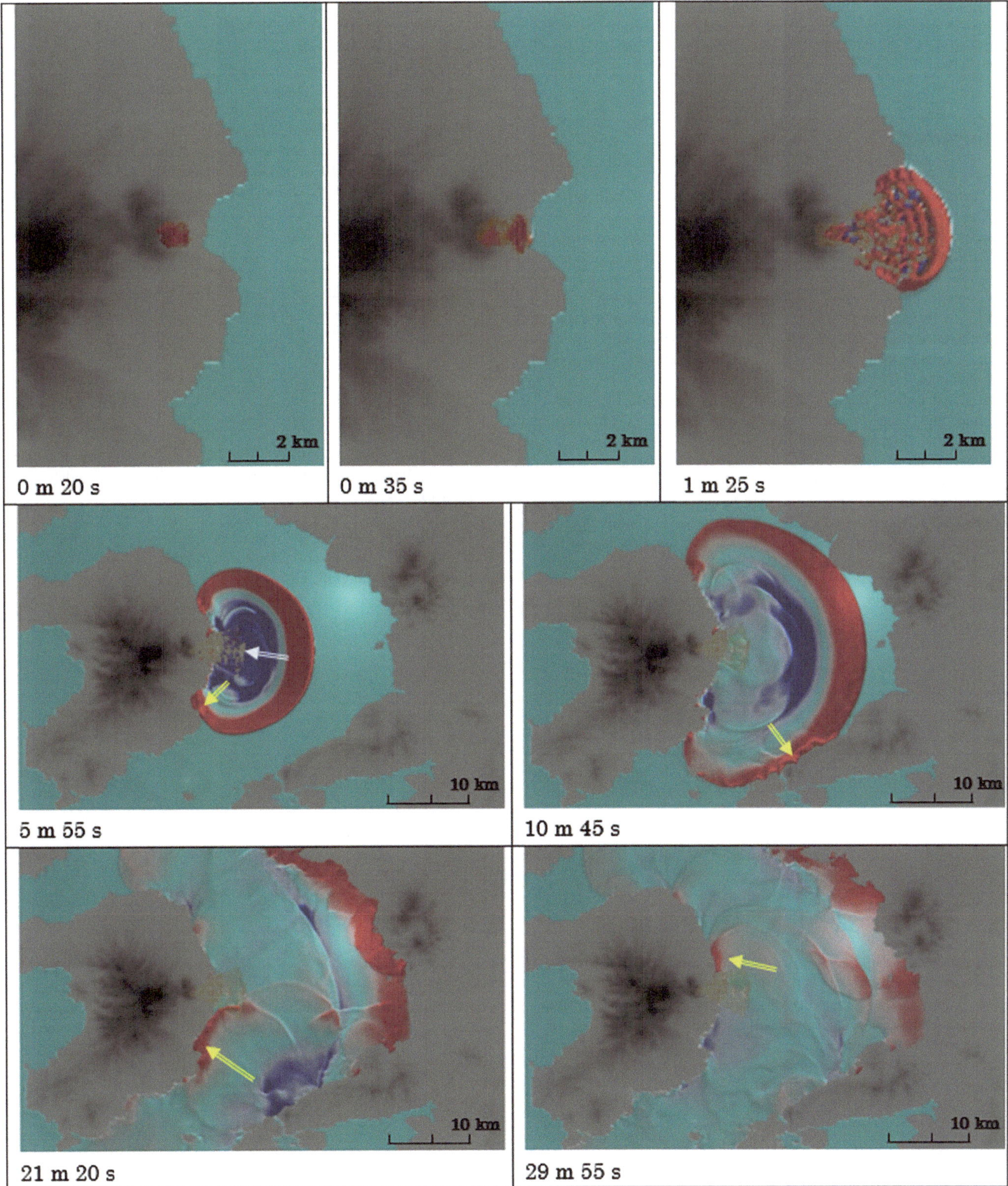

Fig. 21 LS-Tsunami simulation result for the 1792 Unzen-Mayuyama landslide-induced tsunami disaster

Acknowledgements Acknowledgement and call for cooperation for the development of the ISDR-ICL Sendai Partnerships 2015–2025 contributing to the Sendai Framework for Disaster Risk Reduction 2015–2030.

References

Cruden DM (1991) A simple definition of a landslide. Bull Int Assoc Eng Geol 43:27–29

Cruden DM, Varnes DJ (1996) Landslide types and processes. In: Turner AK, Schuster RL (eds) Landslides investigation and mitigation. Transportation Research Board, US National Research Council. Special Report 247, Washington, DC, Chapter 3: 36–75

Dang K, Sassa K, Fukuoka H et al (2016) Mechanism of two rapid and long-runout landslides in the 16 April 2016 Kumamoto earthquake using a ring-shear apparatus and computer simulation (LS-RAPID). Landslides 13(6):1525–1534

Doan HL, Sassa K, Fukuoka H et al. (2016) Initiation mechanism of rapid and long runout landslides and simulation of hiroshima landslide disasters using the Integrated Simulation Model (LS-RAPID). TXT-Tool 3.081-1.4. Landslide dynamics: ISDR-ICL landslide interactive teaching tools, vol 1 fundamental, mapping and monitoring. Accepted

Highland LM, Bobrowsky P (2008) The landslide handbook: a guide to understanding landslides. U.S. Geological Survey Circular 1325, 129 p. http://pubs.usgs.gov/circ/1325/pdf/C1325_508.pdf

Highland LM, Bobrowsky P (2017) Landslide types: descriptions, illustrations and Photos. TXT-Tool 0.001-2.1. Landslide Dynamics: ISDR-ICL landslide interactive teaching tools, vol 1 fundamental, mapping and monitoring. Accepted

Miyagi T, Yamashina S, Esaka F, Abe S (2010) Massive landslide triggered by 2008 Iwate-Miyagi inland earthquake in the Aratozawa Dam area, Tohoku, Japan. Landslides 8:99–108

Sassa K (2005) Landslide disasters triggered by the 2004 Mid-Niigata Prefecture earthquake in Japan. Landslides 2:135–142

Sassa K, Dang K (2016) Landslide dynamics for risk assessment. TXT-Tool 0.081-1.1. Landslide dynamics: ISDR-ICL landslide interactive teaching tools, vol1 fundamental, mapping and monitoring. Accepted

Sassa K, Nagai O, Solidum R et al (2010) An integrated model simulating the initiation and motion of earthquake and rain induced rapid landslides and its application to the 2006 Leyte landslide. Landslides 7(3):219–236

Sassa K, Dang K, He B, Takara K, Inoue K, Nagai O (2014) Development of a new high-stress undrained ring shear apparatus and its application to the 1792 Unzen-Mayuyama megaslide in Japan. Landslides 11(5):827–842

Sassa K, Dang K, Yanagisawa H et al (2016) A new landslide-induced tsunami simulation model and its application to the 1792 Unzen-Mayuyama landslide-and-tsunami disaster. Landslides 13 (6):1405–1419

Varnes DJ (1978) Slope movement types and processes. In: Schuster RL, Krizek RJ (eds) Landslides, analysis and control, special report 176: Transportation Research Board. National Academy of Sciences, Washington, DC., pp 11–33

Progress of the World Report on Landslides

Biljana Abolmasov, Teuku Faisal Fathani, KoFei Liu, and Kyoji Sassa

Abstract

The IPL World Reports on Landslides (WRL) database is created as a cooperation platform for sharing landslide case studies and the best practice in the global landslide community. ICL and IPL wishes to promote and publish global landslide information using the ICL/IPL network for the ISDR-ICL Sendai Partnership 2015–2025 and the Sendai Framework for Disaster Risk Reduction 2015–2030 through WRL activities were assigned as one of priority action. World Report on Landslides data base contains 40 submitted reports on landslide cases over the world. The best rating reports are accessible for world-wide landslide community as open access data, as well as all basic reports. In this paper results of ICL/IPL World Report on Landslides Commeetee members and related activities from 2010 to 2016 are presented.

Keywords

Landslides • ICL • IPL • Sendai partnership • UNISDR

Introduction

The International Consortium on Landslides (ICL) and International Program on Landslides (IPL) created web data base and web cooperation platform for sharing information about landslide case studies in the global landslide community. The idea of World Report on Landslides (WRL) was first examined in the ICL-IPL Secretariat meeting held on January 2010 in Kyoto, Japan. The WRL web portal and the instruction for authors were launched before the ICL-IPL meeting in November 2010. The portal and Instruction for authors were further examined in the ICL-IPL meeting during 2nd World Landslide Forum in Rome, Italy, 2011. The idea of Sendai Partnership 2015–2030 at 3rd World Conference on Disaster Risk Reduction (WCDRR) as well as Beijing Declaration during 3rd World Landslide Forum 2014 were built and ICL coordinator for WRL was elected. After ICL Steering Committee meeting KoFei Liu started to manage WRL with Faisal Fathani as web moderator in October 2014. At the 3rd WCDRR which was convened by the United Nations and hosted by Japan in Sendai from 14 to 18 March 2015, the ICL and its IPL, besides *The Sendai Partnership 2015–2030 for Global Promotion of Understanding and Reducing Landslide Disaster Risk*, contributed further to the UN International Strategy for Disaster Reduction with additional three main types of activities:

- Publication of ICL *Landslides Journal* to communicate frontier of Landslide Science and Technology

B. Abolmasov (✉)
Faculty of Mining and Geology, University of Belgrade, Djusina 7, 11000 Belgrade, Serbia
e-mail: biljana.abolmasov@rgf.bg.ac.rs

T.F. Fathani
University of Gadjah Mada, Yogyakarta, Indonesia
e-mail: tfathani@gmail.com

K. Liu
Department of Civil Engineering, National Taiwan University, Chinese Taipei, Taiwan
e-mail: kfliu@ntu.edu.tw

K. Sassa
International Consortium on Landslides, 138-1, Tanaka Asukaicho, Sakyo-Ku, Kyoto, 606-8226, Japan
e-mail: sassa@iclhq.org

K. Sassa et al. (eds.), *Advancing Culture of Living with Landslides*,
DOI 10.1007/978-3-319-59469-9_18

- Publication of *Landslide Dynamics: ISDR-ICL Landslide Interactive Teaching Tools* for Education and Capacity development and
- IPL web portal for *World Report on Landslides* to inform global landslide community about world-wide landslide case studies, thereby promoting research cooperation within ICL members though IPL web platform.

ICL and IPL wishes to promote and publish global landslide information using the ICL and IPL network for the ISDR-ICL Sendai Partnership 2015–2025 and the Sendai Framework for Disaster Risk Reduction 2015–2030 through WRL activities were assigned.

Progress of WRL activities was reported on ICL-IPL Kyoto Conference in March 2016 and followed by activities with Kyoji Sassa during March–October 2016. The Chair of WRL Committee was replaced by Biljana Abolmasov (University of Belgrade, Serbia) and Deputy Chairs are KoFei Liu (National Taiwan University, Chinese Taipei) and Teuku Faisal Fathani (University of Gadjah Mada, Indonesia). Core members of WRL are Khang Dang (ICL research promotion officer, Kyoto University, Japan) and Miloš Marjanović (University of Belgrade, Serbia). The WRL activities were presented and examined during ICL-IPL UNESCO Conference 15–18 November 2016 within ICL members.

In this paper results of ICL/IPL World Report on Landslides Commeetee members and related activities from 2010 to 2016 are presented.

World Report on Landslides

The IPL World Reports on Landslides (WRL) database is created as a cooperation platform for sharing landslide case studies and the best practice in the global landslide community. The reports may be used for research and capacity development in landslide risk reduction activities in the frame of the United Nations International Strategy for Disaster Reduction (UNISDR). Generally, everyone is welcome to contribute landslide case studies in this WRL database as the reporter. Landslide experts of ICL members and colleagues are invited to contribute landslide cases over the World Reports on Landslides at IPL web page http://iplhq.org/ls-world-report-on-landslide/ (Fig 1).

Instruction for Authors

This instruction provide explanation of items in Data Sheet for World Report on Landslides. Every report should be prepared according to the unified data information:

1. Landslide Case Identifier (LCI)—Each registered landslide case has its own Identifier number. Number consists of three letter country code (ISO 3166-1 alpha-3) and year/date/time of submission (Example: JPN1512101230).
2. Location of landslides—Unit of latitude and longitude should be in degree/minute/second. It is possible to find the latitude and longitude of reporting landslide by using the Google earth.
3. Authors—Office and Address are necessary to identify the authors.
4. Landslide Types—Types of movements and material involved are illustrated in The Landslide Handbook—A Guide to Understanding Landslides (Highland and Bobrowsky 2008), IPL Project 106.
5. Velocity—velocity range which is estimated or guessed as the maximum speed range of reporting landslide (extremely rapid, very rapid, rapid, moderate, slow, very slow and extremely slow, unknown) has to be selected.
6. Slope—range of slope which is measured or guessed for the case (extremely steep, very steep, steep, moderate, gentle, very gentle and almost flat, unknown) has to be selected.
7. Depth—depth range which is measured, estimated or guessed for the reporting landslide (extremely deep, very deep, deep, moderate, shallow, very shallow, extremely shallow, unknown) has to be selected.
8. Volume—volume range which is estimated or guessed for landslide reporting (extremely large, very large, large, large-moderate, moderate-small, small, very small, unknown) has to be selected.
9. Activity—data about landslide occurrence has to be filled in if known: Date of occurrence: Day/Month/Year and state of activity (currently active, active in the past) or unknown. If the landslide is moving slowly or repeatedly moves, it can be classified as currently active.
10. Triggering factors—one or more triggers (rainfall, earthquake, snow melting, erosion, human activities, others, unknown) have to be selected.
11. Damage—Damage is presented as number of death(s) and/0r missing persons, houses and other structural damage and economical loss if estimated.
12. Land use—Land use should be explained for source area and run-out/deposition area (forest, farming, pasture, wildland, urban area, human settlement, industrial use, road, railways, sea/lake, river, cultural heritage site).
13. Description—It has to be provided summary for reporting landslide and reference paper.

In addition, the reporter is asked to provide further information on landslide case as attachment files such as photo, map, plan and cross section, figures or simulation video.

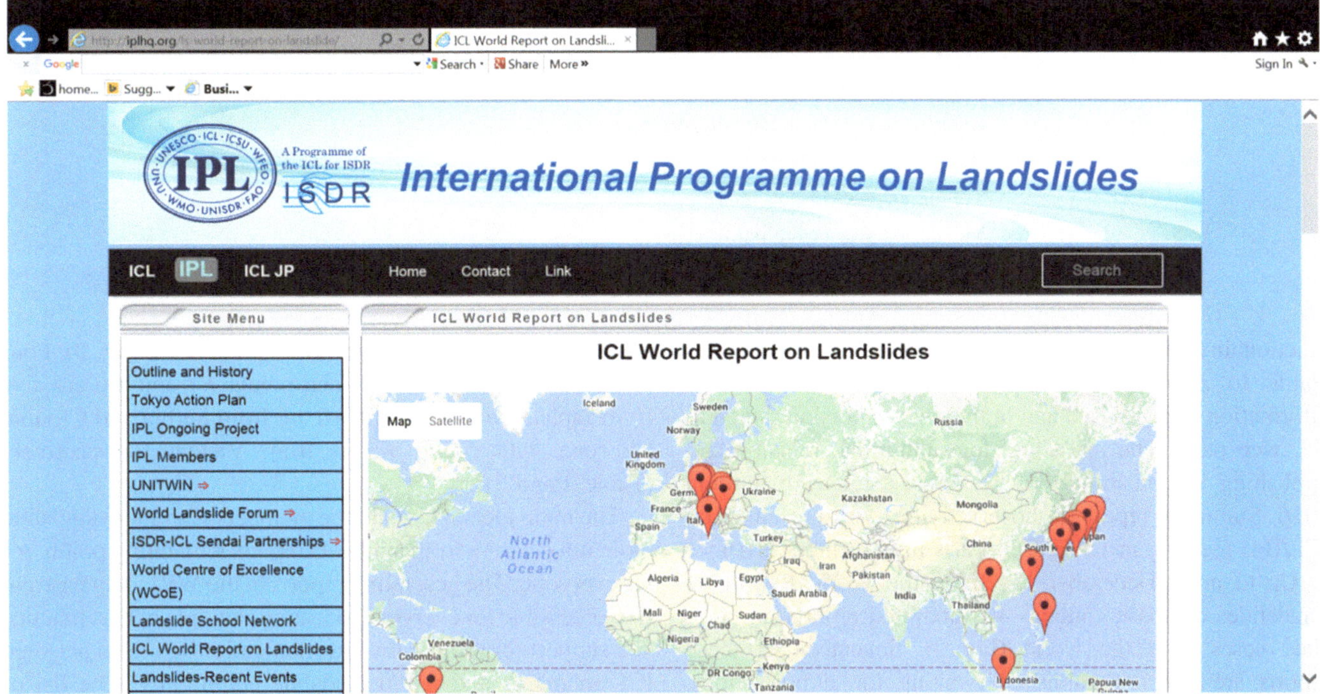

Fig. 1 Screenshot of world report on landslides web page http://iplhq.org/ls-world-report-on-landslide/

Instruction for authors is available at http://iplhq.org/icl/wp-content/uploads/2016/02/1.-Instruction-for-WRL-Final.doc.

Guide to Submit Report

Before reporting landslide cases, authors need to register with the registration form. The system will send ID and password via email. After registration, authors need to login to input data for reporting landslide. After login, reporter will be redirected to landslide report submission form. To submit landslide report, reporter have to fill in the Report Form personal information (Email address, First Name, Last Name, Country and Institution) and landslide data (Landslide name, Landslide Case Identifier and other relevant data). After filling in the necessary fields, reporter have to submit report to save into database. On successful submission reporter will see success notification within green box on top of the page.

Guidelines on report submission are available at http://iplhq.org/icl/wp-content/uploads/2016/02/2.-Guide-to-submit-reports.docx.

Rating and Accessibility

Each report is rated by 7 components (basic information, plan/section, reference, graphics and other resources) attached to the report (Table 1). WRL report is rated by the total rating points from 1 to 7, so that each component brings 1 point. Each report is categorized from ★1 to ★7, depending on the sum of rating points. ICL earning points are then calculated as a total sum of all accepted reports given by one reporter. Reporter earns accessibility to other reports by earning sufficient amount of these ICL points. However, ★1 report is open for everyone, while access to other reports from ★2 to ★7 requires ICL earning points obtained by accepted reports, except open access reports (high category reports assigned by the WRL Committee). Report can be submitted by one or maximally two reporters, and their names are shown in the ICL-WRL web portal. In the case of two authors, both will obtain the same amount of ICL points (from one to seven) for their report. It is important to notice that is necessary to obtain signed authorisation sheet from both reporters to avoid any plagiarism of WRL report.

Rating and accessibility of WRL is available at http://iplhq.org/icl/wp-content/uploads/2016/10/3.-Rating-and-Accessibility-of-WRL-16.10.20.doc.

Progress of the World Report on Landslides

Web platform and data base of IPL World Report on Landslides are functionally operated since 2010, but during 2016 additional improvements of visibility and more

Table 1 Rating points of each component on landslide report

Components	Rating point
Basic information (location/reporting items/description/photo/Google earth kmz. file	1
Plan of landslide pdf, image	1
Cross section of landslide pdf, image	1
Reference (paper/report) link	1
Testing graphics in pdf, image	1
Monitoring graphics in pdf, image	1
Video of moving landslides including 3D simulation link	1

functionality are established. The Instruction for Authors, Guide to Submit Report and Rating and Accessibility information were revised and in the revised version linked to IPL web page. The procedure for submitting, revision and publishing responsibilities were discussed during autumn 2016. Summary report of those activities was presented on ICL/IPL meeting in UNESCO, Paris in November 2016.

Up to now (December 15 2016), the World Report on Landslides database contains 40 submitted reports on landslide cases over the world (Fig. 2). Seventeen of these reports are still not published (waiting for authorization), while 23 of them are available on the web platform (Table 2). The most of reports are rated as more than ★5, which means that almost all components are included in the report. One published report is rated as ★1 which means that only basic landslide information is provided. The simple analysis of reported landslides and their distribution according to the obtained rating points are given in Fig. 3.

The analysis of reports by country origin shows that the highest number of WRL (uploaded or published) originated from Japan (45%), (http://www.iplhq.org/) and (Fig. 3). Four of reports with highest obtained points ★7 and ★6 are also from Japan, two with point ★6 are from Serbia and Croatia, and one with rating ★5 is from Vietnam, respectively (Table 2 and 3).

The main idea of Committee members is to share landslide basic information from landslide reports with rating points ★1 for everyone. The best rating reports in the WRL data base are free accessible for everyone too, as an example of best practice and support for promotion and cooperation activities between ICL members and world landslide community. The list of open access WRL reports are listed in Table 3.

An example of free access of World Report on Landslides is 1792 Unzen-Mayuyama landslide which killed around 15.000 people by landslide and landslide induced tsunami. The following Figures shows geographical position of landslide (Fig. 4); cross section of landslide (Fig. 5); testing data from samples taken from landslide (Fig. 6) and Screenshot of video simulation results (Fig. 7) uploaded on YouTube https://www.youtube.com/watch?v=GwAWjdXXNbk.

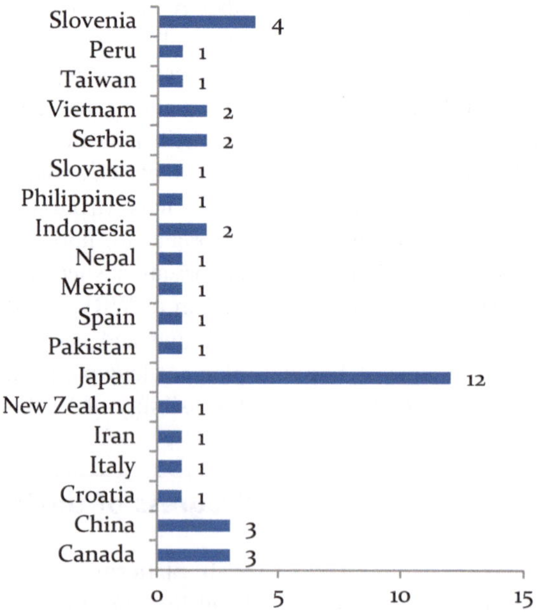

Fig. 2 Total of uploaded reports distribution by country

Table 2 The list of published world report on landslides on IPL web platform

	ID	Title	Reporters	Country	Points
1.	2305	Kalitelaga landslide	Bintri Simbolon, Faisal Fathani	Indonesia	★3
2.	2306	Ledoksari landslide	Tasdiq Hasan, Dwikorita Karnawati	Indonesia	★3
3.	3299	Leyte landslide	Pham Tien, Kyoji Sassa	Philippines	★4
4.	3321	Unzen-Mayuyama megaslide	Khang Dang, Kyoji Sassa	Japan	★7
5.	3322	Minamata debris flow	Khang Dang, Kyoji Sassa	Japan	★4
6.	3323	Spis castle landslide	Khang Dang, Jan Vlcko	Slovakia	★4
7.	3324	Tsukidate landslide	Khang Dang, Hiroshi Fukuoka	Japan	★2
8.	3325	Qianjiangping landslide	Khang Dang, Fa-Wu Wang	China	★4
9.	3326	Nikawa landslide	Khang Dang, Kyoji Sassa	Japan	★6
10.	3403	Leva Reka landslide	Milos Marjanovic, Biljana Abolmasov	Serbia	★1
11..	3405	Kostanjek landslide	Martin Krkac, Snjezana Mihalic Arbanas	Croatia	★6
12.	3406	Takanodai landslide	Khang Dang, Kyoji Sassa	Japan	★7
13.	3407	Aso-ohashi landslide	Khang Dang, Kyoji Sassa	Japan	★7
14.	3408	Umka landslide	Biljana Abolmasov, Svetozar Milenkovic	Serbia	★6
15.	3423	Ha Long landslide	Doan Loi, Quang Lam	Vietnam	★5
16.	3448	1792 Unzen landslide and Tsunami disaster	Khang Dang, Kyoji Sassa	Japan	★7
17.	3497	Haivan station landslide	Khang Dang, Quang Lam	Vietnam	★6
18.	3500	Juan Grijalva	Victor Manuel Hernandez-Madrigal, Juan Carlos Mora-Chaparro	Mexico	★4
19.	3476	Abbotsford landslide	Ha Nguyen Duc, Graham Hancox	New Zealand	★3
20.	3315	Aratozawa	Ha Nguyen Duc, Hendy Setiawan	Japan	★3
21.	3297	Kuridaira landslide dam	Pham Tien	Japan	★5
22.	3404	Akatani landslide dam	Pham Tien	Japan	★5
23.	3499	Sunkoshi landslide	Pham Tien	Nepal	★3

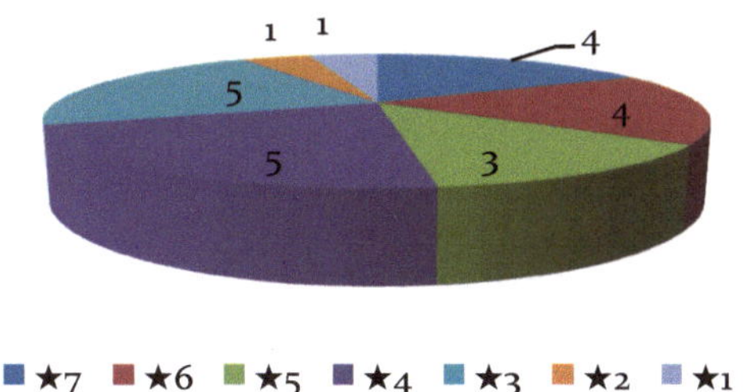

Fig. 3 Number of published reports and their rating

Table 3 The list of open access world report on landslides on IPL web platform

Landslide name	Location	Reporter(s)	Date of occurence	Latitude	Longitude
1792 Unzen landslides and tsunami disaster LCI: JPN1611131539 Rating: ★7	Shimabara, Nagasaki, Japan	Khang Dang, Kyoji Sassa	May 21, 1792	32:46:13.24 N	130:30:0.82 E
Unzen-Mayuyama megaslide LCI: JPN1607121655 Rating: ★7	Shimabara, Nagasaki, Japan	Khang Dang, Kyoji Sassa	May 21, 1792	32:45:51.51 N	130:20:54.16 E
Kostanjek landslide LCI: HRV20160913 Rating: ★ 6	Zagreb, City of Zagreb, Croatia	Martin Krkac, Snjezana Mihalic Arbanas	Dec 02, 1963	45:49:20 N	15:51:24 E
Takanodai landslide LCI: JPN1609161455 Rating: ★6	Kawayo, Kumamoto, Japan	Khang Dang, Kyoji Sassa	Apr 16, 2016	32:53:4.66 N	131:0:15.14 E
Umka landslide LCI: SRB1616092150 Rating: ★6	Belgrade, Belgrade area, Serbia	Biljana Abolmasov, Svetozar Milenković	unknown	44:40:18.46 N	20:17:53.35 E
Ha Long landslide LCI: VNM161020844 Rating: ★5	HaLong, QuangNinh, Vietnam	Doan Loi, Quang Lam	Jul 28, 2015	20:57:29.91 N	107:6:10.26 E

Fig. 4 Google earth kmz. file of Unzen landslide area, Japan

Fig. 5 Cross section of 1792 Unzen-Mayuyama landslide

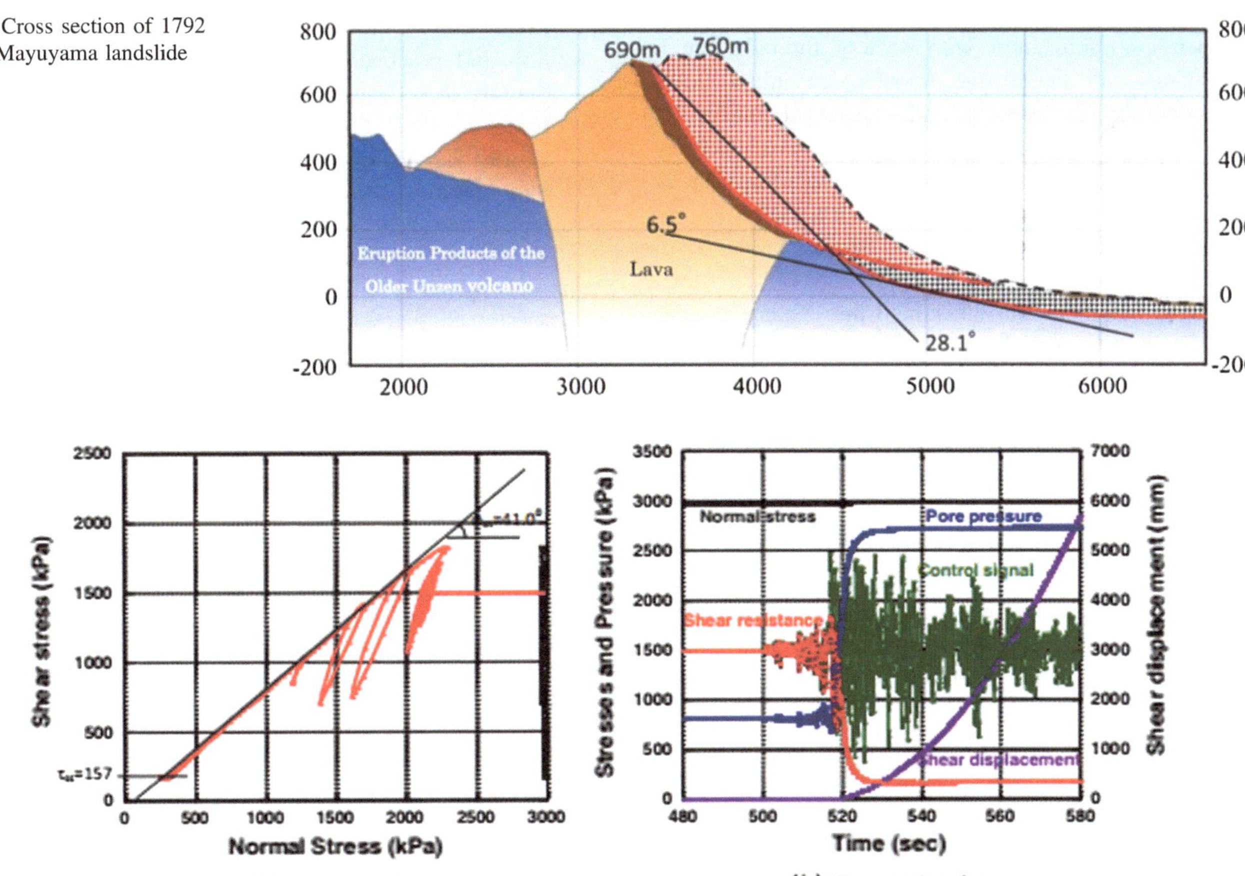

Fig. 6 Testing results from samples taken from 1792 Unzen-Mayuyama landslide

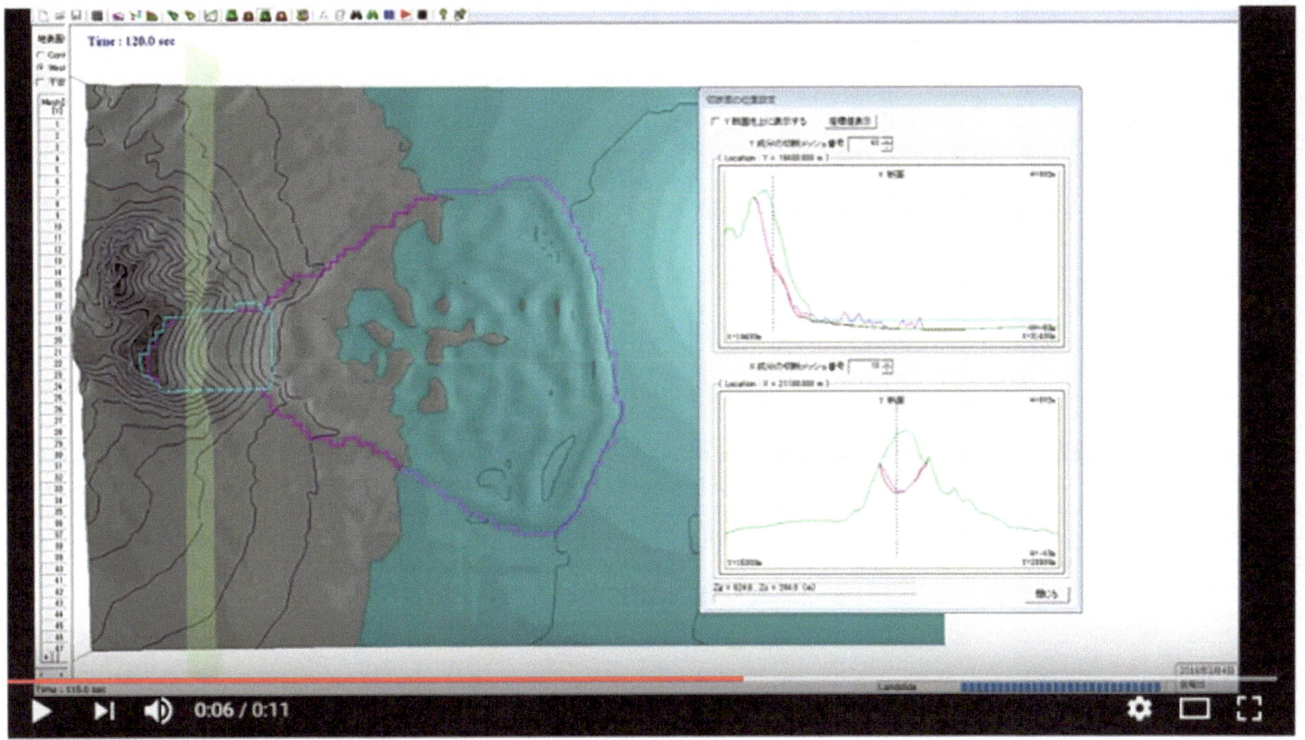

Fig. 7 Screenshot of simulation results video from 1792 Unzen-Mayuyama landslide uploaded on https://www.youtube.com/watch?v=GwAWjdXXNbk

Conclusion

Open communication with society through integrated research and knowledge transfer are initial fields of cooperation in research and capacity building in the UNISDR-ICL Sendai partnership 2015-2025 Resolution. The ICL/IPL WRL data base and web platform are one of three major types of global ICL/IPL activities within voluntary commitment to the Sendai Framework for Disaster Risk Reduction.

ICL members and other landslide experts colleagues are invited to contribute landslide cases over the World Reports on Landslides at IPL web page.

Committee members support landslide reporting with a goal to:

- Increasing number of reports from over the world, generally (to share global information)
- Improving quality of reports in scientific and technical content (to share knowledge)
- Improving number of reports with high rating points (to share best practice data and examples) as free-open access data.

ISDR-ICL Partnership is focused on delivering information and practical results within landslides community that are directly related to the implementation of Sendai Framework for Disaster Risk Reduction. That ICL/IPL World Report on Landslide activities are promoting integrated research, knowledge transfer, education activities and capacity building in the coming years.

Acknowledgements The authors are thankful to ICL members for comments and suggestion and reporters for their contribution in IPL WRL activities. The authors would also like to thank Khang Dang, Ha Nguyen Duc and Pham Tien from Kyoto University, DPRI, Japan, and Miloš Marjanović from University of Belgrade, Serbia for their extensive help.

References

Highland LM, Bobrowsky P (2008) The landslide handbook-a guide to understanding landslides: Reston, Virginia, U.S. Geological Survey Circular 1325, 129 p

http://www.iplhq.org/

https://www.youtube.com/watch?v=GwAWjdXXNbk

Part II

International Programme on Landslides (IPL)

International Programme on Landslides (IPL): Objectives, History and List of World Centres of Excellence and IPL Projects

Qunli Han, Kyoji Sassa, Feng Min Kan, and Claudio Margottini

Abstract

The initial stage of IPL project which was managed by ICL started in 2002 at the same time of ICL foundation. The first IPL project was publication of International Journal of Landslides at this stage. The current second stage of IPL was defined by 2006 Tokyo Action Programme on Landslides as an international programme managed by IPL Global Promotion Committee consisting of ICL and ICL supporting organizations (UNESCO, UNISDR and others). IPL includes IPL Projects conducted by ICL member organizations, the triannual World Landslide Forum and the World Centres of Excellence on Landslide Risk Reduction (WCoE). This paper describes those activities and the list of WCoE since 2008 and the list of IPL projects both in the initial stage of IPL projects (2002–2008) and the second stage of IPL projects (2008–present).

Keywords

International programme on landslides (IPL) • World centres of excellence on landslide risk reduction (WCoE) • International consortium on landslides (ICL) • International strategy for disaster risk reduction (ISDR)

Objectives and Management

The objectives of the International Programme on Landslides (IPL) are to provide international authorization and advice to the projects proposed and implemented by ICL members. ICL was initially planned to create an International Programme on Landslides, namely a version focusing on the study on landslides of the UNESCO-IUGS joint project, International Geological Correlation Programme (IGCP).

The implementation of IGCP-425 "Landslide Hazard Assessment and Mitigation for Cultural Heritage Sites and Other Locations of High Societal Value" provided a higher evaluation and greater fund raising for most of the thirty-one subproject members. They are very active in research, but suffer from lower evaluations and funding because landslide research is a minor or a very marginal study within the organizations they belong to, and also within established scientific fields in funding programmes. The IGCP project will terminate within 5 years. Continuation is not easy, and a

Q. Han (✉)
Ecological and Earth Sciences, UNESCO, Paris, France
e-mail: q.han@unesco.org

Q. Han · F.M. Kan · C. Margottini
IPL Global Promotion Committee, Kyoto, Japan
e-mail: kanf@un.org

C. Margottini
e-mail: claudio.margottini@gmail.com

K. Sassa
International Consortium on Landslides (ICL), Kyoto, Japan
e-mail: sassa@iclhq.org

F.M. Kan
UNISDR Asia-Pacific Office in Bangkok, Bangkok, Thailand

C. Margottini
ISPRA-Italian Institute Form Environmental Protection and Research, Varese, Italy

K. Sassa et al. (eds.), *Advancing Culture of Living with Landslides*,
DOI 10.1007/978-3-319-59469-9_19

number of landslide projects will not be adopted in IGCP (a very wide Geoscience Programme). The funding of IGCP425 was around 4000 USD per year, which was shared among 31 subproject leaders as partial travel support fees. The aim of the International Programme on Landslides (IPL) is not to support research funding itself, but provide partial travel fees for members to attend ICL-IPL meetings, propose projects, evaluate projects, report project results, and terminate or continue the projects. The membership fee funds the sustainable and basic financial resources. All management of projects, including proposal, evaluation, and decision on ongoing status is implemented by the IPL Global Promotion Committee of ICL (IPL-GPC). The committee consists of all ICL member organizations and IPL partner organizations (UNESCO, UNISDR, WMO, FAO, UNU, ICSU, WFEO and IUGS). During the first phase from 2008 to 2013, the Chair of IPL-GPC was Salvano Briceno (Former Director of UNISDR for 2001–2011), and the Deputy Chairs were Badaoui Rouhban (Former Director of Section for Disaster Reduction, UNESCO) and Kyoji Sassa (Executive Director of ICL). For the current phase from 2014 to the present, the Chair of the Committee is Qunli Han, Director of Ecological and Earth Sciences, UNESCO, and Deputy chairs are Feng Min Kan, Chief of UNISDR Asia-Pacific Office in Bangkok and Claudio Margottini (Vice President of ICL).

Development of International Programme on Landslides (IPL)

The initial stage of IPL was from the First Board of Representatives (BOR) meeting in 2002 to the First World Landslide Forum in 2008. When the initial thirty-three ICL members gathered in this first BOR meeting, IPL projects were classified in two categories. One category is coordinating projects which are planned by ICL multiple members, e.g., the whole ICL project to found a new landslide journal: IPL C100: Landslides: Journal of the International Consortium on Landslides, and the project succeeding the IGCP-412: C101: Landslide risk evaluation and mitigation in cultural and natural heritage sites. One subproject of C101 was C101-1: Landslide investigation and capacity building in Machu Pichu—Aguas Calientes area. Another category is member projects proposed by a single member organization together with its partners.

Initially the ICL board decided to allocate 2000 USD for IPL projects proposed by developing countries, funded by membership fees from developed countries. However, the research result was not cost-effective for the initial three years' experiences. In the Steering Committee meeting in Kyoto just before the UN World Conference on Disaster Reduction in Kobe, 2005, how to use the limited funds effectively was initially examined. The concept was to provide research funds, but to give authorization to each project, and ICL-IPL will evaluate each project at each World Landslide Forum and give an IPL Award for Success in one to three projects. The Award (3000 USD) for one project will be given to its leader. The IPL Award for Success is not an award for the best IPL project, as funding, research facilities and research infrastructure differ in various countries. The award will be given to those projects with success within the given social, technical and financial conditions.

This award was approved in Tokyo, 2008 and applied in the Second World Landslide Forum in Rome, 2011.

The IPL Awards for Success were given at the Second World Landslide Forum at FAO Headquarters, Rome, October 2011, to:

(1) Lynn Highland: U.S. Geological Survey, and Peter Bobrowsky: Geological Survey of Canada (Publication)
IPL 106 Best Practice handbook for landslide hazard mitigation (2002–2007)
(2) Farrokh Nadim and Bjørn Kalsnes: International Centre for Geohazards, Norway (Research)
IPL 102 Assessment of global high-risk landslide disaster hotspots (2002–2004)
IPL 144 Changing pattern of landslide risk and strategies for its management (2009–present)
(3) Dwikorita Karnawati and Faisal Fathani: Universitas Gadjah Mada, Indonesia (Capacity Building)
IPL 158 Development of Community-based Landslide Early Warning System (2009–present)
IPL 159 Development of Education Program for Sustainable Development in Landslide Vulnerable Area through Student Community Service (2009–present).

The IPL Awards for Success were given at the Third World Landslide Forum at China National Convention Center, Beijing, China, 2014, to:

(1) Wei Shan: Research Center of Cold Regions Landslide, China
IPL-132 Research on vegetation protection system for highway soil slope in seasonal frozen regions (2008-present)
IPL-167 The effect of freezing-thawing on the stability of ancient landslide of North-Black highway (2009–present)
(2) Ogbonnaya Igwe: Department of Geology, University of Nigeria, Nsukka, Nigeria
IPL-150 Capacity building and the impact of

climate-driven changes on regional landslide distribution, frequency and scale of catastrophe (2010–present)
IPL-183 Landslides in West Africa: impacts, mechanism and management (2012–present).

The Second Stage of the International Programme on Landslides (IPL)

The Round Table Discussion, Strengthening Research and Learning on Earth System Risk Analysis and Sustainable Disaster Management within UN-ISDR as Regards "Landslides", was co-organized by ICL, UNESCO, WMO, FAO, UNISDR, UNEP, UNU and Kyoto University at United Nations University, Tokyo, Japan on 18–20 January 2006. This discussion aimed to implement the 2005 Letter of Intent agreed at thematic session 3.8 New International Initiatives for Research and Risk Mitigation of Foods (IFI) International programme on Landslides (IPL) at the 2nd United Nations World Conference on Disaster Reduction in 2005. During this Round Table Discussion, the 2006 Tokyo Action Plan on Landslides was adopted. This plan defined the International Programme on Landslides (IPL) as a programme of ICL for the International Strategy for Disaster Reduction (ISDR), which is managed by the IPL Global Promotion Committee consisting of ICL and ICL partners. To promote the 2006 Tokyo Action Plan and to implement the International Programme on Landslides, ICL exchanged Memorandums of Understanding with UNESCO, WMO, FAO, UNISDR, UNU, ICSU and WFEO in 2006. The IPL logo then included those organizations.

New initiatives under the Tokyo Action Plan were to identify the World Centre of Excellence on Landslide Risk Mitigation and to organize a triannual World Landslide Forum. In addition, two categories of IPL (Coordinating Project and Member project) were combined and it was termed as IPL-101, 150, and 200.

World Landslide Forum

In order to report and disseminate the activities and achievements of ICL, a World Landslide Forum (WLF) shall be convened once every three years by inviting ICL Members and ICL Supporting organizations, individual members within those organizations, and all levels of co-operating organizations and individual researchers, engineers and administrators. The World Landslide Forum will receive reports on ICL's activities and provide a forum for open discussion and new initiatives from all participants. This role of the meeting was defined as a General Assembly. The first General Assembly was organized at Keck Center of the National Academy of Sciences in Washington DC on 12–14 October 2005. The second General Assembly was organized as the First World Landslide Forum.

World Centres of Excellence

The World Centres of Excellence on Landslide Risk Reduction were established by the 2006 Tokyo Action Plan; the Global Promotion Committee (GPC) of the International Programme on Landslides (IPL) will identify at the World Landslide Forum, organized every 3 years, eligible organizations such as universities, institutes, NGOs, government ministries and local governments, contributing to "Risk Reduction for Landslides and Related Earth System Disasters". An independent Panel of Experts, set up by the Global Promotion Committee of International Programme on Landslides (IPL-GPC), endorses the WCoEs. Twelve World Centres of Excellence (WCoEs) 2008–2011 were identified at the First World Landslide Forum in November 2008 at UNU in Tokyo, Japan. Fifteen WCoEs for 2011–2014 were identified at the Second World Landslide Forum in October 2011 at FAO, Rome, Italy. Fifteen WCoEs for 2014–2017 were identified at the Third World Landslide Forum in June 2014 at Beijing, China. Certificates of WCoEs were awarded in Beijing to each leader of the WCoEs by Ms. Irina Bokova, UNESCO Director General.

Objectives of WCoE

To strengthen the International Programme on Landslides (IPL) and IPL Global Promotion Committee;

To create “A Global Network of entities contributing to landslide risk reduction”;
To implement the ISDR-ICL Sendai Partnerships 2015–2025 for global promotion of understanding and reducing landslide disaster risk; and
To improve the global recognition of “Landslide Risk Reduction” and its social-economic relevance, and entities contributing to this field.

Criteria for WCoE Candidates

Governmental and non-governmental entities such as universities, agencies, and other institutions, and their subsidiary entities (faculties, departments, centres, divisions or others) which meet the following two conditions:

- Contributing to “Landslide Disaster Risk Reduction”; and
- Willing to support IPL intellectually, practically and financially by either joining ICL or/and contributing to IPL-GPC and promote “landslide research and disaster risk reduction” on a regional and/or global scale in a mutually beneficial manner.

Guidelines for WCoE

(1) Candidates of WCoEs must submit the application form to the Secretariat of the IPL Global Promotion Committee.
(2) Candidates will be evaluated from their achievements and current activities (scientific, technical and educational capacity, training courses, publications, dissemination of knowledge and information) and planned activities contributing to IPL.
(3) WCoEs will be identified at every World Landslide Forum (held every 3 years). The status as a WCoE will be given for 3 years until the next Forum.
(4) Each WCoE must submit an annual activity report each year which is uploaded in the IPL web and also report its activities in Landslides: Journal of the International Consortium on Landslides or/and a World Landslide Forum.
(5) The status as a WCoE may be extended for another 3 years in the same topics or in a revised topic by the IPL Global Promotion Committee, based on the activities carried out as a WCoE.

Procedure for Identification of WCoEs

(1) The application form from an eligible entity will be submitted to the Secretariat of the IPL-GPC (IPL World Centre).
(2) Preliminary screening of received applications will be conducted according to the Criteria by the secretariat of IPL World Centre. Feasible applications will be passed on for in-depth evaluation.
(3) Candidates of WCoEs passing to in-depth evaluation will be invited to orally present their proposals in the IPL Global Promotion Committee. Revision of application may be advised by the committee to avoid duplication of activities with other proposals or to improve the proposals when necessary.
(4) Final application forms will be evaluated by the technical evaluation committee of the IPL World Centre.
(5) Summarizing all evaluation values, appropriate candidates will be selected, and a list of recommendations will be submitted to the Independent Panel of Experts which consists of experts outside of ICL.
(6) The Independent Panel of Experts will review the recommendation, and endorse appropriate WCoEs.
(7) The Endorsement of WCoEs by the Independent Panel of Experts will be reported to the IPL Global Promotion Committee. WCoEs endorsed by the Panel must be approved by the vote of more than half of the participating members of the IPL Global Promotion Committee.
(8) The WCoE for the next 3 years will be announced at the World Landslide Forum.

Acknowledgements Organizations and individuals supporting the development of ICL and IPL: A programme of ICL were acknowledged in this book at the end of Section 1.3-1 International Consortium on Landslides, although it could not cover all cooperating organizations and individuals. At the end of this article, the record of activities which could be implemented thanks to their strong support is attached with thanks as an Appendix.

Appendix

Figure 1 and Tables 1, 2.

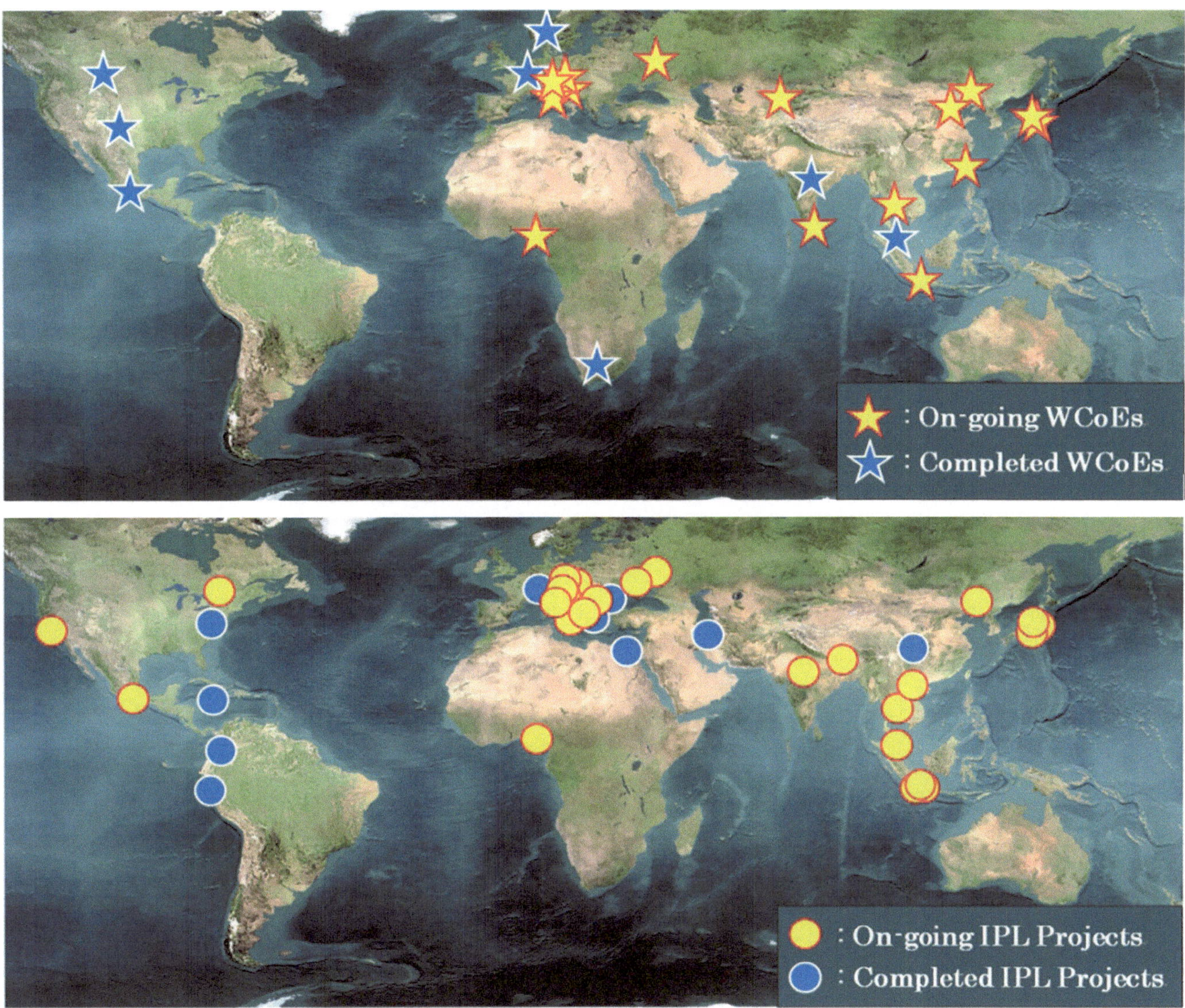

Fig. 1 The geographical distribution of World Centres of Excellence and IPL projects

Table 1 The chronological order of all World Centres of Excellence, including both past and ongoing WCOEs

List of World Centres of Excellence					
No.	Title	Leader	Country	Organization	Status
World Centre of Excellence 2008–2011					
1.	Scientific research for mitigation, preparedness and risk assessment of landslides	Yuepin Yin	China	China Geological Survey	Completed
2.	Landslide field research and capacity building through international collaboration	Vit Vlimek	Czech Republic	Faculty of Science, Charles University in Prague	Completed
3.	Earth observation advanced technologies for landslide monitoring, management and mitigation	Nicola Casagli	Italy	Department of Earth Science, University of Florence	Completed
4.	Research and development of advanced technology for landslide hazard analysis	Alberto Presitininzi/Gabriele Scarascia-Mugnozza	Italy	Research Centre on Prediction Prevention and Control of Georisks of Rome University "La Sapienza"	Completed
5.	Development of methodology for risk assessment of the earthquake-induced landslides	Hideaki Marui	Japan	The Japan Landslide Society	Completed
6.	Implementation of National Slope Master Plan	Ashaari Mohamad/Che Hassandi bin Abdullah	Malaysia	Slope Engineering Branch, Public Works Department of Malaysia	Completed
7.	Research on mitigation of landslide risk and training of specialists.	Farrokh Nadim	Norway	International Centre for Geohazards (ICG) at NGI,	Completed
8.	International Summer School on Rockslides and Related Phenomena in the Kokomeren River basin, Kyrgyzstan	Alexandar Strom	Russia and Kyrgyz	Institute of Geospheres Dynamics of Russian Academy of Science (IDG RAS) & Kyrgyz Institute of Seismology (KIS)	Completed
9.	Mechanisms of landslides in over-consolidated clays and flysch	Bojan Majes/Matjaž Mikoš	Slovenia	University of Ljubljana, Faculty of Civil and Geodetic Engineering (UL FGG)	Completed
10.	Landslide inventorization and susceptibility mapping in South Africa	S. Diop/SG Chiliza	South Africa	Engineering Geoscience Unit, Council for Geoscience	Completed
11.	Promoting Knowledge sharing, Innovations and Institutions with South-South focus network on Landslide Risk Reduction in Asia	N.S.M.I. Arambepola	Thailand	Asian Disaster Preparedness Center	Completed
12.	Conduct landslide hazard assessments and develop early warning systems	Peter Lyttle	USA	U.S. Geological Survey	Completed
World Centre of Excellence 2011–2014					
13.	Canadian Landslide Loss Risk Reduction Strategy and Implementation	Peter Bobrowsky	Canada	Geological Survey of Canada	Completed
14.	Risk Assessment and Disaster Mitigation Code for Long Run-out Landslides	Yueping Yin	China	China Geological Survey	Completed
15.	Scientific research for landslide risk analysis and international education for mitigation and preparedness	Vit Vilimek	Czech Republic	Charles University, Faculty of Science	Completed
16.	Research on landslide risk management harmonisation in	Javier Hervás	European Commission	Joint Research Centre, European Commission	Completed

(continued)

Table 1 (continued)

List of World Centres of Excellence					
No.	Title	Leader	Country	Organization	Status
	support to European Union policy making				
17.	Training, Research and Documentation on Landslides Risk Management	Surya Parkash	India	National Institute of Disaster Management	Completed
18.	Development of Community-based and Most Adaptive Technology for Landslide Risk Reduction	Dwikorita Karnawati	Indonesia	Universitas Gadjah Mada	Completed
19.	Advanced Technologies for Landslides	N. Casagli, F. Catani	Italy	Department of Earth Science, University of Florence, Italy	Completed
20.	Development of a methodology for risk reduction of earthquake-induced landslides	Daisuke Higaki	Japan	The Japan Landslide Society, Japan	Completed
21.	Risk identification and land-use planning for disaster mitigation of landslides	Hideaki Marui	Japan	Niigata University, Institute for Natural Hazards and Disaster Recovery	Completed
22.	Landslide monitoring and community based early warning systems	Irasema Alcántara-Ayala	Mexico	National Autonomous University of Mexico	Completed
23.	Research on mitigation of landslide risk and training of specialists	Farrokh Nadim	Norway	International Centre for Geohazards	Completed
24.	Annual Summer School on Rockslides and Related Phenomena in Kyrgyzstan	Alexander Strom	Russia and Kyrgyz	Inst. of Geospheres Dynamics of Russian Academy of Sciences & Kyrgyz Institute of Seismology	Completed
25.	Mechanisms of landslides in over-consolidated clays and flysch	Bojan Majes	Slovenia	University of Ljubljana, Faculty of Civil and Geodetic Engineering	Completed
26.	Promoting Knowledge, Innovations and Institutions with South-South focus through a Regional network of Landslide Risk Reduction	N.S.M.I. Arambepola	Thailand	Asian Disaster Preparedness Center	Completed
27.	Scientific Research for Landslide Hazard Analysis, U.S. Geological survey	Peter Lyttle	USA	U.S. Geological Survey Landslide Programme	Completed
World Centre of Excellence 2014–2017					
28.	Formation Mechanism Research, Disaster Warning and Universal Education of Cold Regions Landslide	Wei Shan	China	Research Center of Cold Regions Landslide, China	On-going
29.	Scientific Research for Mitigation, Preparedness and Risk Assessment of Landslides	Wang Min	China	China Geological Survey, China	On-going
30.	Scientific Research for Landslide Risk Analysis, Modeling, Mitigation and Education	Liang-Jenq Leu	Taiwan	Department of Civil Engineering, National Taiwan University, China, Taiwan	On-going
31.	Landslide Risk Reduction in the Adriatic-Balkan Region through the Regional Cooperation	Željko Arbanas/Snježana Mihalić Arbanas	Croatia	Croatian Landslide Group, Croatia	On-going
32.	Landslide Risk Assessment and Development Guidelines for Effective Risk Reduction	Josef Stemberk	Czech Republic	Institute of Rock Structure and Mechanics Czech Academy of Sciences & Charles University, Faculty of Science	On-going

(continued)

Table 1 (continued)

List of World Centres of Excellence					
No.	Title	Leader	Country	Organization	Status
33.	Development of Community-based and Most Adaptive Technology for Landslide Risk Reduction	Dwikorita Karnawati	Indonesia	Universitas Gadjah Mada, Yogyakarta, Indonesia	On-going
34.	Advanced Technologies for LandSlides (ATLaS)	Nicola Casagli	Italy	Department of Earth Science, University of Florence, Italy	On-going
35.	Emergency Response Support System for Large-scale Landslide Disasters	Satoshi Tsuchiya	Japan	The Japan Landslide Society (JLS), Japan	On-going
36.	Risk Identification and Land-use Planning for Disaster Mitigation of Landslides	Hiroshi Fukuoka	Japan	Niigata University, Institute for Natural Hazards and Disaster Recovery, Japan	On-going
37.	Implementation of National Slope Master Plan	Che Hassandi Abdullah	Malaysia	Slopes Engineering Branch, Public Works Department of Malaysia, Malaysia	On-going
38.	Building Human Capacities and Expertise in Landslide Disaster Risk Managements	Ogbonnaya Igwe	Nigeria	Department of Geology, University of Nigeria, Nsukka, Nigeria	On-going
39.	International Summer School on Rockslides and Related Phenomena in the Kokomeren River Valley, Tien Shan, Kyrgyzstan	Alexander Strom	Russia and Kyrgyz	Geodynamics Research Center—branch of JSC "Hydroproject Institute" & Institute of Seismology of National Academy of Sciences of Kyrgyz Republic	On-going
40.	Mechanisms of Landslides and Creep in Over-consolidated Clays and Flysch	Ana Petkovšek	Slovenia	University of Ljubljana, Faculty of Civil and Geodetic Engineering (UL FGG), Ljubljana, Slovenia	On-going
41.	Developing Model Policy Frameworks, Standards and Guidelines	Nihal Rupasinghe A A Virajh Dias	Sri Lanka	Central Engineering Consultancy Bureau, Colombo, Sri Lanka	On-going
42.	Promoting Knowledge, Innovations and Institutions with South-South focus through a Regional network of Landslide Risk Reduction in Changing Climate Scenario in Asia	N.M.S.I. Arambepola	Thailand	Asian Disaster Preparedness Center (ADPC), Thailand	On-going

Table 2 The chronological order of all IPL Projects, including past and ongoing projects

IPL Projects (2002–2016)						
No.	Title	Organization	Country	Leader	Year	Status
The initial stage of the International Programme on Landslides (IPL: 2002–2008) under support from UNESCO Coordinating projects (C by multiple members) and Member projects (M by a single member)						
C100	Landslides: Journal of the International Consortium on Landslides	Kyoto University, DPRI, RCL/ICL	Japan	Kyoji Sassa	2002/2008–2010	Completed
C101	Landslide risk evaluation and mitigation in cultural and natural heritage sites	Kyoto University, DPRI, RCL/ICL	Japan	Kyoji Sassa and Paolo Canuti	2002/2008–2010	Completed
C101-1	Landslide investigation and capacity buildng in Machu Pichu-Aguas Calientes area	Kyoto University, DPRI, RCL/ICL	Japan	Kyoji Sassa	2002/2008–2011	Completed
C101-1-1	Low environmental impact technologies for slope monitoring by radar interferometry: application to Machu Picchu site	CIVITA Consortium/ENEA	Italy	Claudio Margottini	2002–2006	Completed
C101-1-2	Expressions of risky geomorphologic processes as well as paleogeographical evolution of the area of Machu Picchu	Charles University, Research Center of Earth Dynamic	Czech Republic	Vit Vilimek, Jiri Zvelebil	2002–2007	Completed
C101-1-3	Shallow geophysics and terrain stability mapping techniques applied to the Urubamba Valley, Peru: Landslide hazard evaluation	Instituto Geologico Minero y Metalurgico	Peru	Romulo Mucho, Peter Bobrowsky	2004–2006	Completed
C101-1-4	A proposal for an integrated geophysical study of the Cuzco region	Instituto Nazionale Di Oceanografia E Di Geofisica Sperimentale-OGS	Italy	Daniel Nieto Yabar	2004–2006	Completed
C101-1-5	UNESCO-Italian-ESA Satellite monitoring of Machu Picchu	University of Firenze, Earth Sciences Department	Italy	Paolo Canuti, Claudio Margottini,	2004–2006	Completed
C101-3	The geomorphological instability of the Buddha niches and surrounding cliff in Bamiyan valley (Central Afghanistan)	CIVITA Consortium/ENEA/ISPRA	Italy	Claudio Margottini	2002/2008–2015	Completed
C101-4	Stability assessment and prevention measurement of Lishan Landslide, Xian, China	Lishan Landslide Observatory, Xian Municipal Government	China	Qing Jin Yang	2002–2007	Completed
C101-5	Environment protection and disaster mitigation of rock avalanches landslides and debris flow in Tianchi Lake region and natural preservation area of Changbai Mountains, Northeast China	Jilin University, Environmental Geological Disaster Research Institute	China	Binglan Cao	2002–2007	Completed
C101-6	Conservation of Masouleh Town	Building and Housing Research Center	Iran	S. H. Tabatabaei	2002–2007	Completed
C101-7	Cultural and natural heritage threatened by landslides in the region of Iassy, Romania	Proexrom S.R.L., Technical University, Civil Engineering Faculty	Romania	Nicolae Botu	2005–2007	Completed

(continued)

Table 2 (continued)

IPL Projects (2002–2016)						
No.	Title	Organization	Country	Leader	Year	Status
The initial stage of the International Programme on Landslides (IPL: 2002–2008) under support from UNESCO Coordinating projects (C by multiple members) and Member projects (M by a single member)						
C102	Assessment of global high-risk landslide disaster hotspots	International Centre for Geohazards, (ICG) in Oslo	Norway	Farrokh Nadim	2002–2004	Completed
C103	Global landslide observation strategy	Kyoto University, DPRI, Flood Section	Japan/Italy	Kaoru Takara, Nicola Casagli	2004/2008–2012	Completed
C104	World Landslide Database	Kyoto University, DPRI, RCL	Japan/Italy	Hiroshi Fukuoka, Nicola Casagli	2006/2008–2012	Completed
C105	Early Warning of Landslides	Kyoto University, DPRI, RCL/ICL	Japan	Kyoji Sassa	2007-2013	Completed
C106	Capacity building and outreach	ISPRA-Italian Institute for Environmental Protection and Research	Italy/Russia	Claudio Margottini, Alexander Strom	2008–2012	Completed
C106-1	Landslide museum in Civita di Bagnoregio	ENEA/ISPRA	Italy	Claudio Margottini	2006/2008–2012	Completed
M101	Areal prediction of earthquake and rain induced rapid and long-travelling flow phenomena (APERITIF)	Kyoto University, DPRI, RCL	Japan	Hiroshi Fukuoka	2002/2008–2012	Completed
M102	Disaster evaluation and mitigation of the giant Jinnosuke-dani Landslide in the Tedori water reservoir area, Japan	Kanazawa University	Japan	Tatsunori Matsumoto	2002–2004	Completed
M103	Capacity building on management of risks caused by landslides in Central American countries	International Centre for Geohazards, (ICG) in Oslo	Norway	Farrokh Nadim	2002–2007	Completed
M104	A global literature study on the use of critical rainfall intensity for warning against landslide disasters	International Centre for Geohazards, (ICG) in Oslo	Norway	Haakon Heyerdal	2002–2004	Completed
M105	Hurricane-flood-landslide continuum: a forecast system	U. S. Geological Survey	USA	Randall Updike	2002–2006	Completed
M106	A best practices handbook for landslide hazard mitigation	Geological Survey of Canada	Canada	Lynn Highland, Peter Bobrowsky	2002–2007	Completed
M107	Landslide risk assessment in landslide prone regions of Slovakia—modelling of climatic changes impact	Comenius University, Faculty of Natural Sciences	Slovakia	Rudolf Hozer	2002–2006	Completed
M108	Disaster evaluation and mitigation of landslides in the Three-Gorge water reservoir area, China	Chongqing Seismological Bureau	China	Renjie Ding	2002–2005	Completed
M109	Recognition, mitigation and control of landslides of flow type in Greater Kingston and adjoining parishes in Eastern Jamaica, including public education on landslide hazard	University of the West Indies	Jamaica	Rafi Ahmad	2002–2006	Completed
M110	Capacity Building in Landslide Hazard Management and Control for Mountainous Developing Countries in Asia	Niigata University, Research Institute for Hazards in Snowy Areas	Japan	Hideaki Marui	2002–2007	Completed

(continued)

Table 2 (continued)

IPL Projects (2002–2016)						
No.	Title	Organization	Country	Leader	Year	Status
The initial stage of the International Programme on Landslides (IPL: 2002–2008) under support from UNESCO Coordinating projects (C by multiple members) and Member projects (M by a single member)						
M111	Detail study of the internal structure of large rockslide dams in the Tien Shan and international field mission	"Institute hydroproject"/ Institute of the Geospheres Dynamics, Russian Academy of Sciences	Russia	Alexander Strom	2002–2006	Completed
M113	Zone risk map: Towards harmonized, intercomparable landslide risk assessment and risk maps	Cairo University, Faculty of Engineering	Egypt	Yasser Elshayeb	2002–2005	Completed
M114	Landslide hazard assessment along Tehran-Caspian seaside corridors	Soil Conservation and Watershed Mangement Research Institute/Agri. Research and Education Organization	Iran	Zieaoddin Shoaei	2002–2007	Completed
M115	Establishment of a regional network for disaster mitigation, disaster education, and disaster database system in Asia	Ehime University, Faculty of Engineering	Japan	Ryuichi Yatabe	2003–2007	Completed
M116	Standardization of terminology, integration of information and the development of decision support software in the area of landslide hazards	Geological Survey of Canada	Canada	Catherine Hickson	2003–2006	Completed
M117	Geomorphic Hazards from landslide dams	Swiss Federal Institute for Snow and Avalanche Research SLF	Switzerland	Oliver Korup	2003–2006	Completed
M118	Development of an expert DSS for assessing landscape impact mitigation works for cultural heritage at risk	ENEA(Italian Agency for New technologies Energy and Environment)	Italy	Giuseppe Delmonaco	2003–2007	Completed
M119	Slope instability phenomena in Korinthos county	Institute of Geology and Mineral Exploration (IGME)	Greece	Nikos Nikolaou	2002–2005	Completed
M120	Landslide hazard zonation in Garwal using GIS and geological attributes	Indian Institute of Technology	India	Ashok Kumar Pachauri	2003–2004	Completed
M121	Integrated system of a new generation for monitoring of dynamics of unstable rock slopes and rock fall early warning	Charles University, Research Center of Earth Dynamic	Czech Republic	Jiri Zvelebil, Vit Vilimek	2003–2007	Completed
M122	Inka cultural heritage and landslides: detailed studies in Cusco and Sacred Valleys, Peru	Grudec Ayar	Peru	Raul Carreno	2004–2007	Completed
M123	Cusco regional landslide hazard mapping and preliminary assessment	Grudec Ayar	Peru	Raul Carreno	2004–2007	Completed
M124	The influence of clay mineralogy and ground water chemistry on the mechanism of landslides	Russian Academy of Sciences, Institute of Environmental Geoscience	Russa	Viktor Osipov	2004–2007	Completed
M125	Landslide mechanisms on volcanic soils	Universidad Nacional de Colombia	Colombia	Carlos Edusrdo Rodriguez	2004–2007	Completed

(continued)

Table 2 (continued)

IPL Projects (2002–2016)						
No.	Title	Organization	Country	Leader	Year	Status
The initial stage of the International Programme on Landslides (IPL: 2002–2008) under support from UNESCO Coordinating projects (C by multiple members) and Member projects (M by a single member)						
M126	Compliation of landslide/rockslide inventory of the Tien Shan mountain system	Institute of the Geospheres Dynamics, Russian Academy of Sciences	Russia	Alexander Strom	2004–2007	Completed
M127	Development of low-cost detector of slope instability for individual use	University of Tokyo, Geotechnical Engineering Group, Civil engineering	Japan	Ikuo Towhata	2004–2007	Completed
M128	Development of sounding methodology for a root-reinforced landslide mass	University of Tokyo, Inst. Industrial Science	Japan	Kazuo Konagai	2004–2007	Completed
M129	Evaluation of natural hazards associated with rapid glacial retreat in Cordillera Blanca (Peru)	Charles University, Research Center of Earth Dynamic	Czech Republic	Vit Vilimek	2005–2007	Completed
M131	Technology development for landslide monitoring in China	China Geological Survey/Geological Survey of Canada	China/Canada	Yueping Yin and Peter Bobrowsky	2006–2007	Completed
M132	Research on vegetation protection system for highway soil slope in seasonal frozen regions	Northeast Forestry University	China	Wei Shan, Fawu Wang	2006/2008–2015	Completed
M133	Establishment of rainfall-soil chart for erosion induced landslide prediction	Mara University of Technology	Malaysia	Roslan Abidin	2006–2007	Completed
M134	Large-scale rockslides in coarse-bedded carbonate rocks in the Apennines (Italy), Caucasus (Russia) and Zagros (Iran): evaluation of possible triggers and hazard assessment	Institute of the Geospheres Dynamics, Russian Academy of Sciences	Russia	Alexander Strom	2007–2011	Completed
M135	Landslide hazard assessment in Changunarayan hill of Kathmandu, Nepal—Geotechnical investigation and preventive plan-	Ehime University, Faculty of Engineering	Japan	Ryuichi Yatabe	2008–2011	Completed
M136	Shear behaviour and mechanics of Megaslides and their nearby faults in Hittian Balla, Pakistan and Shaolin, Taiwan	University of Tokyo, Inst. Industrial Science	Japan	Kazuo Konagai, Kyoji Sassa	2008–2012	Completed
M137	Italian Landslide Inventory (IFFI Project)	Italian Agency for New technologies Energy and Environment (ENEA)	Italy	Alessandro Triglia	2008–2011	Completed
M138	Long run out and Catastrophic Landslides study: Yigong Landslide, Tibet China.	China Geological Survey	China	Yin Yueping	2008–2011	Completed
M140	Landslide and multi geohazards mapping for community empowerment in Indonesia	Gadjah Mada University	Indonesia	Dwikorita Karnawati	2008–2011	Completed
M141	Geo-Risks Management for Third World Countries–Mapping and Assessment of Risky Geo-factors for Land Use (e.g. in Ethiopia)	Charles University, Research Center of Earth Dynamic	Czech Republic	Jiří Zvelebil	2008–2012	Completed

(continued)

Table 2 (continued)

IPL Projects (2002–2016)						
No.	Title	Organization	Country	Leader	Year	Status
The initial stage of the International Programme on Landslides (IPL: 2002–2008) under support from UNESCO Coordinating projects (C by multiple members) and Member projects (M by a single member)						
International Programme on Landslides (IPL)-A programme of ICL for ISDR (2008–Present)—MoUs with UNESCO, WMO, FAO, UNISDR, UNU, ICSU and WFEO						
IPL-101-2	Landslides monitoring and slope stability at selected historic sites in Slovakia	Comenius University, Faculty of Natural Sciences	Slovakia	Jan Vlcko	2008	On-going
IPL-106-2	International Summer School on Rockslides and Related Phenomena in the Kokomeren River Valley, Tien Shan, Kyrgyzstan	Institute of the Geospheres Dynamics, Russian Academy of Sciences/JSC "Hydroproject Institute"	Russia	Alexander Strom	2008	On-going
IPL-112	Landslide mapping and risk mitigation planning in Thailand	Ministry of Agriculture and Cooperatives, Land Development Department	Thailand	Saowanee Prachansri	2008	On-going
IPL-139	Development of low-cost early warning system of slope instability for civilian use	University of Tokyo, Geotechnical Engineering Group	Japan	Ikuo Towhata/Taro Uchimura	2008	On-going
IPL-142	Seismic landslide hazards mapping in Sichuan	China Geological Survey	China	Yuepin Yin	2009–2011	Completed
IPL-143	Evaluation of sensitivity of the combined hydrological model (dynamic) for landslide susceptibility risk mapping in Sri Lanka	Central Engineering Consultancy Bureau (CECB)	Sri Lanka	A.A. Virajh Dias	2008–2012	Completed
IPL-144	SafeLand—Living with landslide risk in Europe: Assessment, effects of global change, and risk management strategies	International Centre for Geohazards, (ICG) in Oslo	Norway	Bjørn Kalsnes	2009–2012	Completed
IPL-145	Preparation of landslide risk map in Taleghan Area–Iran	Building and Housing Research Center	Iran	S.H. Tabatabaei	2009–2011	Completed
IPL-146	Spatial monitoring of joint influence of an atmospheric precipitation and seismic motions on formation of landslides in Uzbekistan (Central Asia)	Institute Hydroingeo, State Committee of Geology of Uzbekistan	Uzbekistan	Rustam Niyazov	2010–2012	Completed
IPL-147	Study on Debris Flow Controlling Factors and Triggering Mechanism in Peninsular Malaysia	Slope Engineering Branch, Public Works Department of Malaysia	Malaysia	Che Hassandi Abdullah	2010–2011	Completed
IPL-148	Geo-evaluation of the stability of slopes around crater lakes in Cameroon: The cases of lakes Nyos, Barombi, Mbo and Awing	University of Buea	Cameroon	Ntasin Edwin Bongsiysi	2010–2011	Completed
IPL-149	Canadian Landslide Best Practice Manual	Geological Survey of Canada	Canada	Peter Bobrowsky	2009–2015	Completed
IPL-150	Capacity building and the impact of climate-driven changes on regional landslide distribution, frequency and scale of catastrophe	Department of Geology, University of Nigeria, Nsukka	Nigeria	Ogbonnaya Igwe	2009	On-going

(continued)

Table 2 (continued)

IPL Projects (2002–2016)						
No.	Title	Organization	Country	Leader	Year	Status
The initial stage of the International Programme on Landslides (IPL: 2002–2008) under support from UNESCO Coordinating projects (C by multiple members) and Member projects (M by a single member)						
IPL-151	Soil matrix suction in active landslides in flysch—the Slano Blato landslide case	University of Ljubljana, Faculty of Civil and Geodetic Engineering	Slovenia	Bojan Majes	2010–2012	Completed
IPL-152	Assessment of coastal landslides risk by innovative remote sensing techniques.	University of Roma “La Sapienza”	Italy	Gabriele Scarascia Mugnozza	2010–2011	Completed
IPL-153	Landslide hazard zonation in Kharkov region of Ukraine using GIS	Insutitute of Telecommunication and Global Information Space	Ukraine	Oleksandr M. Trofymchuk	2010–2011 2012–2014	Completed
IPL-154	Development of a methodology for risk assessment of the earthquake-induced landslides	Japan Landslide Society	Japan	Satoshi Tsuchiya	2009	On-going
IPL-155	Determination of soil parameters of subsurface to be used in slope stability analysis in two different precipitation zones of Sri Lanka	Central Engineering Consultancy Bureau	Sri Lanka	A.A. Virajh Dias	2009	On-going
IPL-156	Best practices for early warning of landslides in a changing climate scenarios	Asian Disaster Preparedness Center (ADPC)	Thailand	N.M.S.I. Arambepola	2009–2012	Completed
IPL-157	Dynamics of subaerial and submarine megaslides	ICL	Japan	Kyoji Sassa	2009	On-going
IPL-158	Development of Community-based Landslide Early Warning System	Gadjah Mada University	Indonesia	Teuku Faisal Fathani	2009	On-going
IPL-159	Development of Education Program for Sustainable Development in Landslide Vulnerable Area through Student Community Service.	Gadjah Mada University	Indonesia	Dwikorita Karnawati	2009	On-going
IPL-160	Landslides and floods under extreme weather condition and resilient society	Kyoto University, DPRI, RCL	Japan	Hiroshi Fukuoka	2009–2012	Completed
IPL-161	Risk identification and land-use planning for disaster mitigation of landslides and floods in Croatia	Kyoto University, DPRI/Niigata University, R. Institute for Natural Hazards and Disaster Recovery	Japan	Hiroshi Fukuoka/Hideaki Marui	2009	On-going
IPL-163	Mechanical-mathematical modeling and monitoring for landslide processes	Russian Academy of Sciences, Institute of Environmental Geoscience	Russia	Svalova Valentina	2009	On-going
IPL-165	Development of community-based landslide hazard mapping for landslide risk reduction at the village scale in Java, Indonesia	Gadjah Mada University	Indonesia	Dwikorita Karnawati	2010	On-going
IPL-167	Landslides Mechanism and the Subgrade Stability Controlling Measures in Island Permafrost Area	Northeast Forestry University	China	Wei Shan	2010	On-going
IPL-168	Engaging U.S. citizens in Landslide Science through the website, “Did You See It? Report a Landslide”	U. S. Geological Survey	USA	Rex Baum	2010–2013	Completed

(continued)

Table 2 (continued)

IPL Projects (2002–2016)						
No.	Title	Organization	Country	Leader	Year	Status
The initial stage of the International Programme on Landslides (IPL: 2002–2008) under support from UNESCO Coordinating projects (C by multiple members) and Member projects (M by a single member)						
IPL-169	Landslide hazard and risk assessment in Geyser Valley (Kamchatka)	Laboratory of Engineering Geodynamics, Geological Faculty, Moscow State University	Russia	Oleg V. Zerkal	2010–2012	Completed
IPL-170	Landslide susceptibility and landslide hazard zonation in volcanic terrains using Geographic Information System (GIS): A case study in the Río Chiquito-barranca Del Muerto watershed	Institute of Geography, UNAM	Mexico	Gabriel Legorreta Paulín	2010–2013	Completed
IPL-171	Study of the geotechnical characteristics of an unstable urban area of Barranquilla (Colombia) severely affected for slope instabilities and soil volume changes	Universidad Nacional de Colombia	Colombia	Guillermo Ávila	2010–2015	Completed
IPL-172	Documentation, Training and Capacity Building for Landslides Risk Management	National Institute of Disaster Management, New Delhi	India	Surya Parkash	2011–2014	Completed
IPL-173	Croatian virtual landslide data center	Faculty of Mining, Geology and Petroleum University of Zagreb	Croatia	Snjezana Mihalic	2011	On-going
IPL-175	Development of landslide risk assessment technology and education in Vietnam and other areas in the Greater Mekong Sub-region	ICL/Institute of Transport Science and Technology	Japan, Vietnam	Kyoji Sassa/Nguen Xuan Khang	2011	On-going
IPL-176	Slope Data Acquisition along Highways in Sabah State for hazard assessment and mapping	Slope Engineering Branch, Public Works Department of Malaysia	Malaysia	Che Hassandi Abdullah	2012–2013	Completed
IPL-177	Study on geological disasters focusing on landslides in and around Tegucigalpa City, Honduras	Universidad Politécnica de Ingeniería, UPI	Honduras	Aníbal Godoy	2012–2013	Completed
IPL-179	Database of Glacial Lake Outburst Floods (GLOFs)	Charles University, Research Center of Earth Dynamic	Czech Republic	Adam EMMER/Vit Vilimek	2012	On-going
IPL-180	Introducing Community-based Early Warning System for Landslide Hazard Management in Cox's Bazaar Municipality, Bangladesh	Asian Disaster Preparedness Center (ADPC)	Thailand	N.M.S.I. Arambepola	2011–2013	Completed
IPL-181	Study of slow moving landslide Umka near Belgrade, Serbia	University of Belgrade, Faculty of Mining and Geology	Serbia	Biljana Abolmasov	2012	On-going
IPL-182	Characterization of landslides mechanisms and impacts as a tool to fast risk analysis of landslides related disasters in Brazil	CENACID—UFPR (Center for Scientific Support in Disasters—Federal University of Parana)	Brazil	Renato Eugenio de Lima	2012–2014	Completed

(continued)

Table 2 (continued)

IPL Projects (2002–2016)						
No.	Title	Organization	Country	Leader	Year	Status
The initial stage of the International Programme on Landslides (IPL: 2002–2008) under support from UNESCO Coordinating projects (C by multiple members) and Member projects (M by a single member)						
IPL-183	Landslides in West Africa: impacts, mechanism and management	Department of Geology, University of Nigeria, Nsukka	Nigeria	Igwe Ogbonnaya	2012	On-going
IPL-184	Study of landslides in flysch deposits of North Istria, Croatia: sliding mechanisms, geotechnical properties, landslide modeling and landslide susceptibility	Faculty of Civil Engineering University of Rijeka	Croatia	Željko Arbanas	2012	On-going
IPL-185	Design and Validation of an Early Warning System for Landslides—DeVEL	Technische Universitat Darmstadt, Institute and Laboratory of Geotechnics	Germany	Rolf Katzenbach	2013	Completed
IPL-186	Rock-fall hazard assessment and monitoring in the archaelogical site of Petra, Jordan	ISPRA-Italian Institute for Environmental Protection and Research	Italy	Claudio Margottini	2013	On-going
IPL-187	Landslide hazards assessment and modeling and sediment yield	Institute of Geography, UNAM	Mexico	Gabriel L.Paulin	2013	On-going
IPL-188	Study of slow moving landslide Potoška Planina (Karavanke Mountain, NW Slovenia)	Geological Survey of Slovenia	Slovenia	Marko Komac	2013	On-going
IPL-190	Landslide risk identification and resilience study in tectonically active mountains and sea floors	Niigata University, Research Institute for Natural Hazards and Disaster Recovery	Japan	Hiroshi Fukuoka	2015	On-going
IPL-191	Landslide hazard zonation in Carpathian region of Ukraine using GIS	Institute of Telecommunication and Global Information Space	Ukraine	Yakovliev Yevhenii/Oleksandr Trofymchuk	2015	On-going
IPL-192	Development of post-earthquake rainfall induced landslide (PERIL) hazard mitigation framework	California State University, Fullerton	USA and Nepal	Binod Tiwari	2015	On-going
IPL-193	Integrated systems for landslides monitoring, early warning and risk mitigation along motorways	University of Calabria, DIMES, CAMILAB	Italy	Pasquale Versace	2015	On-going
IPL-194	Public awareness and education programme for landslides management in Malaysia	Slope Engineering Branch, Public Works Department of Malaysia	Malaysia	Che Hassandi Abdullah	2015	On-going
IPL-195	Study for mitigation and recovery of mud eruption disaster in East Java and modeling for risk reduction mudflow hazards	Parahyangan Catholic University	Indonesia	Paulus P. Rahardjo	2015	On-going
IPL-196	Development and applications of a multi-sensors drone for geohazards monitoring and mapping	University of Firenze, Earth Sciences Department	Italy	Veronica Tofani	2015	On-going
IPL-197	Low frequency, high damaging potential landslide events in "low risk" regions—challenges for hazard and risk management	Institute of Rock Structure and Mechanics Academy of Sciences	Czech Republic	Jan Klimeš	2015	On-going

(continued)

Table 2 (continued)

IPL Projects (2002–2016)						
No.	Title	Organization	Country	Leader	Year	Status
The initial stage of the International Programme on Landslides (IPL: 2002–2008) under support from UNESCO Coordinating projects (C by multiple members) and Member projects (M by a single member)						
IPL-198	Multi-scale rainfall triggering models for Early Warning of Landslides (MUSE)	University of Firenze, Earth Sciences Department	Italy	Filippo Catani	2015	On-going
IPL-199	The effect of root systems in natural slope erosion protection in the hill country of Sri Lanka	Central Engineering Consultancy Bureau	Sri Lanka	Pvip Perera	2015	On-going
IPL-200	An assessment of the rock fall susceptibility based on cut slopes adjacent to highways and railways	Central Engineering Consultancy Bureau	Sri Lanka	H.M.J.M.K. Herath	2015	On-going
IPL-201	Landslide inventory and Susceptibility map in Durres and Kavaja region	Albanian Geological Survey	Albania	Hasan Kulici	2016	On-going
IPL-202	Ripley landslide monitoring project (Ashcroft, BC, Canada)	Geological Survey of Canada	Canada	Peter Bobrowsky/Claudio Margottini	2016	On-going
IPL-203	Analysis and identify of landslides based on species distribution and surface temperature difference	Institute of Cold Regions Science and Engineering, Northeast Forestry University	China	Ying Guo	2016	On-going
IPL-204	A study on socio-economic and environmental impacts of landslides	National Institute of Disaster Management, New Delhi	India	Surya Prakash	2016	On-going
IPL-205	Integrated systems for landslides monitoring, early warning and risk mitigation along motorways	University of Calabria, DIMES, CAMILAB	Italy	Pasquale Versace/Giovanna Capparelli	2016	On-going
IPL-206	Towards improved landslide mapping and forecasting	Istituto di Ricerca per la Protezione Idrogeologica of the Italian National Research Council	Italy	Fausto Guzzetti/Mario Parise	2016	On-going
IPL-207	Evaluation on social research approach In determining "acceptable risk" and "tolerable risk" in landslide risk areas in Malaysia	Slope Engineering Branch, Public Works Department of Malaysia	Malaysia	Che Hassandi Bin Abdullah	2016	On-going
IPL-208	Landslide disaster risk communication in mountain areas	Institute of Geography, UNAM	Mexico	Irasema Alcántara Ayala	2016	On-going
IPL-209	Landslides and related sediment disaster project covering the entire South-East Nigeria, West Africa	Department of Geology, University of Nigeria, Nsukka	Nigeria	Igwe Ogbonnaya	2016	On-going
IPL-210	Massive landsliding in Serbia following Cyclone Tamara in May 2014	University of Belgrade, Faculty of Mining and Geology	Serbia	Biljana Abolmasov	2016	On-going
IPL-211	Development of wireless sensor network for monitoring and earlier warning of shallow and deep landslides (WISE-LAND)	Research Center for Geotechnology, Indonesian Institute of Sciences	Indonesia	Adrin Tohari	2016	On-going

UNESCO-KU-ICL UNITWIN Cooperation Programme for Landslides and Water-Related Disaster Risk Management

Kaoru Takara and Kyoji Sassa

Abstract

UNITWIN is the abbreviation for the university twinning and networking scheme. This UNESCO programme was established in 1992. During ICL foundation meeting in January 2002, participants from UNESCO advised to link the planned International Programme on Landslides (IPL) to one of UNESCO Programme for the promotion and the authorization. Then, ICL applied for UNITWIN programme to UNESCO soon after the foundation of ICL in 2002. UNITWIN-UNESCO/KU/ICL Landslides Mitigation for Society and Environment Cooperation Programme was established in 2003 at Kyoto University, Kyoto, Japan. In 2010, the UNESCO-KU-ICL UNITWIN Cooperation Programme was extended to "Landslide and Water-Related Disaster Risk Management" to include more participants dealing with rainfall-induced landslides on slopes, as well as flood, sediment and debris flows in river systems. This paper describes its progress and the activities of capacity development including the list of students and post-doctoral researchers within this programme.

Keywords

UNESCO • UNITWIN programme • International programme on landslides (IPL) • International consortium on landslides (ICL) • Kyoto university

Introduction

During the International Decade for Natural Disaster Reduction (IDNDR) (1990–1999), which was a UN-designated decade for promotion of disaster prevention and mitigation activities, the Disaster Prevention Research Institute (DPRI) of Kyoto University conducted a number of international cooperative research projects with Asian countries such as China, Indonesia and the Philippines. One of them was a research project on the Lishan Landslide in Xi'an, China, to assess the landslide hazard and protect a cultural heritage site, the Lishan Resort Palace from the Tan Dynasty (618-907), located at the foot of Lishan Slope. No landslide has occurred in this slope since the Tan Dynasty, but landslide experts in Japan and China indicated the necessity of detailed monitoring and testing, by mobilizing the most advanced technologies and establishing new landslide hazard assessment technology to protect this cultural

K. Takara (✉)
UNITWIN Cooperation Programme,
Kyoto University, Kyoto, Japan
e-mail: takara.kaoru.7v@kyoto-u.ac.jp

K. Takara
Disaster Prevention Research Institute,
Kyoto University, Kyoto, Japan

K. Sassa
UNITWIN Cooperation Programme from ICL,
Kyoto University, Kyoto, Japan
e-mail: sassa@iclhq.org

K. Sassa
International Consortium on Landslides, 138-1 Tanaka-Asukai
Cho, Sakyo-Ku, Kyoto, 606-8226, Japan

K. Sassa et al. (eds.), *Advancing Culture of Living with Landslides*,
DOI 10.1007/978-3-319-59469-9_20

heritage. Based on the results of this Japan-China joint research, in parallel, UNESCO-IUGS Joint Programme IGCP (International Geological Correlation Programme) sought the development of IGCP from "Geological Correlation" to "Geoscience for the Society". Our project was evaluated to be a good example of "Geoscience for the Society". This project was submitted to IGCP, and IGCP-425: "Landslide Hazard Assessment and Cultural Heritage" (1998–2004) was approved. Note that it is renamed as the International Geoscience Programme but still keeps its original abbreviation IGCP; currently it is a part of the International Geoscience and Geoparks Programme (IGGP), which was approved officially on 17 November 2015 at the General Conference of UNESCO.

As a part of the activities of IGCP-425, the DPRI, Kyoto University (KU), organized a UNESCO-Kyoto University Joint Symposium in Kyoto, Japan in 2002, which included participants from UNESCO, UNISDR, WMO, and Japanese ministries such as MOFA (Ministry of Foreign Affairs) and MEXT (Ministry of Education, Culture, Sports, Science and Technology). At this occasion, the International Consortium on Landslides was officially established and UNESCO proposed the establishment of a UNESCO-KU-ICL UNITWIN Network to support the International Programme on Landslides (IPL), which is considered a scientific programme contributing to UNISDR. This manuscript outlines the development of the UNESCO-KU-ICL UNITWIN Network since its official establishment in March 2003.

UNESCO UNITWIN Programme

The UNITWIN programme is explained in the UNITWIN-UNESCO-Chair Programme WEB at: http://en.unesco.org/unitwin-unesco-chairs-programme as follows:

"UNITWIN" is the abbreviation for the university twinning and networking scheme. This UNESCO programme was established in 1992, in accordance with a resolution adopted by the General Conference of UNESCO at its 26th session (1991). The UNITWIN/UNESCO Chairs Programme consists of the establishment of UNESCO Chairs and UNITWIN Networks in higher education institutions. This UNESCO programme serves as a prime means of building the capacities of higher education and research institutions through the exchange of knowledge and sharing, in a spirit of international solidarity. Thus it promotes North-South, South-South and triangular cooperation as a strategy to develop institutions. These institutions work in partnership with NGOs, foundations and public and private sector organizations and play an important role in the field of higher education. The UNITWIN/UNESCO Chairs Programme opens avenues for the higher education and research community to join forces with UNESCO to contribute to the implementation of its programme and the achievement of the Millennium Development Goals (MDGs). The UNITWIN/UNESCO Chairs Programme covers training, research and exchange of academics and offers a platform for information sharing in all fields within the competence of UNESCO. The majority of the projects are interdisciplinary and intersectoral and involve all the programme sectors of UNESCO, with the active cooperation of its field Offices, Institutes and Centres. National Commissions play an important role by helping to promote the programme nationally, facilitating its execution and evaluating its impact. Because it is totally multidisciplinary in nature, the UNITWIN/UNESCO Chairs Programme is one of the Organization's most intersectoral programmes".

Currently 50 UNITWIN networks from 24 countries are ongoing as of 1st December 2016. UNITWIN Networks in the fields of Disaster Risk Reduction (only two) and sustainable development (six) are introduced below from the UNITWIN-UNESCO-Chair Programme WEB (many French titled networks are included, but they are not introduced here).

1. UNITWIN-UNESCO/KU/ICL International Consortium on Landslides Mitigation for Society and Environment Cooperation Programme (605), established in 2003 at Kyoto University, Kyoto, Japan
2. The Network on Emergency Preparedness and Responses (704), established in 2005 at Waseda University, Tokyo, Japan
3. UNESCO Interdisciplinary Chair/Network for Sustainable Development (6), established in 1995 at Universidad Católica de Cuyo, La Plata
4. UNESCO Chair/Network on Global Economics and Sustainable Development (42), established in 1996, Colegio do Brasil, Rio de Janeiro, RJ, Brasil.
5. UNESCO Chair/International Network of Water-Environment Centres for the Balkans on "Sustainable Management on Water and Conflict Resolution" (618), established in 2003 at the Aristotle University of Thessaloniki, Thessaloniki, Greece
6. Network on International Cooperation and Development (1114), established in 2015 at the University of Pavia and the University of Bethlehem University, Italy
7. International UNESCO Chair/Network on Transfer of Technologies for Sustainable Development (TTSD) (203), established in 1993 at the International Centre for Educational Systems, Moscow, Russia
8. UNITWIN Cooperation Programme in Marine Biology and Sustainable Development for East Africa (854), established in 2009 at the University of Dar-Es- Salaam and the Bangor University (U.K.), United Republic of Tanzania.

Each UNITWIN cooperation programme has its own logo. Figure 1 is the logo of the UNESCO-KU-ICL UNITWIN Cooperation Programme.

The First Phase of the UNESCO-KU-ICL UNITWIN Cooperation Programme

In March 2003, the UNESCO-KU-ICL UNITWIN Network was agreed upon and signed at the Office of the President of Kyoto University among UNESCO, Kyoto University (KU) and ICL (Fig. 2); the name chosen for the network was the "Landslide Risk Mitigation for Society and the Environment Cooperation Programme." The joint network for landslide risk mitigation of UNESCO/KU/ICL is to carry out education, research and capacity building in the field of landslide risk mitigation. The activities are jointly conducted by Kyoto University (especially, the Research Centre on Landslides, which was established in 2003 in the Disaster Prevention Research Institute, and other groups in Kyoto University), ICL and ICL Supporting Organizations. The Integrated Disaster Risk Management group (NEXUS IDRiM), established in 2005 with its secretariat in DPRI, Kyoto University, has joined this network.

The UNESCO-KU-ICL UNITWIN Cooperation Programme was established in 2003 to be hosted in Kyoto University. The application to construct the UNITWIN headquarters building as its secretariat was approved by Kyoto University. ICL organized its international symposium on landslide risk mitigation and protection of cultural and natural heritage on 21–24 January in 2004. During this symposium, a memorial meeting of the establishment of the headquarters of UNESCO-KU-ICL UNITWIN Cooperation Programme was organized, inviting UNESCO, UNISDR, MEXT, Japan Meteorological Agency representing the WMO, and a representative from the Italian Embassy, as well as ICL members, in the main building of Kyoto University, together with President Kazuo Oike (bottom left photo of Fig. 3). The left panel of Fig. 3 is a symbol of the UNITWIN headquarters building. Prof. Oike composed a haiku—"Eyebrow moon in the sky of Andes, Realm of Landslides". The main IPL project at the beginning of ICL was "Landslides threatening World Heritage Machu Picchu, Peru". The background of this figure is the Machu Picchu citadel. The Andes Mountains are similar to the Japanese archipelago; both are affected by active faults and earthquakes and are full of landslides. All haiku include a key word symbolizing one of four seasons as a rule. Eyebrow moon is a key word of Autumn, and also symbolizes an initial stage of development to the full moon, with a wish for the development of ICL and IPL and UNITWIN. The panel is displayed in the UNITWIN headquarters building (top-right photo of Fig. 3). The building has three rooms, for joint research, the journal and a meeting room. The right-bottom photo presents the opening ceremony of UNITWIN Headquarters Building held in the central meeting room on 3 September 2004.

The President of Kyoto University (Dr. Kazuo Oike), three delegates from UNESCO (Dr. Wolfgang Eder, Director of the Division of Earth Sciences, Ms. Winsome Gordon, Section Head of Higher Education, Dr. Badaoui Rouhban, Section Head of Disaster Reduction), the Rector of UNU (Dr. Hans van Ginkel), officers from the Japanese Cabinet Office (Dr. Satoru Nishikawa) and the MEXT (Mr. Nakamura and Mr. Kasuo Akiyama), and the Peruvian Ambassador (Luis J. Macchiavello,), as well as DPRI colleagues, including Director at that time Kazuya Inoue, attended its opening ceremony (Fig. 3).

The ICL organized the 2006 Tokyo Round Table Discussion "Strengthening Research and Learning on Earth System Risk Analysis and Sustainable Disaster Management within UN-ISDR as Regards Landslides" towards a dynamic global network of the International Programme on Landslides (IPL) at the United Nations University (UNU), Tokyo, Japan on 18th to 20th January 2006. This international conference adopted the 2006 Tokyo Action Plan on the International Programme on Landslides (IPL). It proposed the global cooperation network of the IPL, and established the IPL Global Promotion Committee and the IPL World Centre as its secretariat to coordinate and support

Fig. 1 Logo of the UNESCO-KU-ICL UNITWIN cooperate programme

Fig. 2 Signing at the office of Kyoto University President

Fig. 3 UNITWIN headquarters building constructed in Uji Campus, Kyoto University in 2004

ICL has exchanged MoUs with UNESCO, WMO, FAO, UN/ISDR, UNU, ICSU, WEFO to promote the 2006Tokyo Action Plan

Mr. Koïchiro Matsuura (UNESCO Director General) and Mr. Kyoji Sassa (ICL President) exchanging MoU, 22 August 2006 in Tokyo. Vice President of KU (S. Nishimura) attended the signing ceremony.

Fig. 4 Signing ceremony between UNESCO and ICL in Tokyo on 22 August 2006

implementation of the IPL to formulate a framework for cooperation and to identify focus areas for reducing landslide disaster risk worldwide. It was further agreed that the 2006 Tokyo Action Plan would be implemented within the scope of the Hyogo Framework for Action (HFA) 2005–2015, "Building the Resilience of Nations and Communities to Disasters", adopted at the United Nations World Conference on Disaster Reduction held in Kobe, Hyogo, Japan in January 2005.

The ICL exchanged Memorandums of Understanding (MoUs) with a number of international organizations, such as UNESCO, WMO, FAO, UNISDR, UNU, ICSU and WFEO, to promote the 2006 Tokyo Action Plan. Figure 4 shows a snapshot at the signing ceremony between Mr. Koïchiro Matsuura, the Director-General of UNESCO, and Mr. Kyoji Sassa, the President of ICL. The UNESCO-KU-ICL UNITWIN Cooperation Programme has strengthened its cooperation. Based on such activities, the ICL organized the First World Landslide Forum in Tokyo on 18-21 November 2008, of which its objectives were:

(1) Promotion of research and exchange of experience through open Forums, Symposia and Workshops;
(2) Advances and achievements of IPL; and
(3) Designation of World Centres of Excellence on Landslide Risk Reduction.

This World Landslide Forum (WLF) was succeeded by the Second WLF in Rome in 2011 and the Third WLF in Beijing in 2014. The Fourth WLF will be held in the Cultural and Congress Centre in Ljubljana, Slovenia in 2017. The Fifth WLF is currently planned to assemble at the International Conference Center, Kyoto on 2–6 November 2020.

The Second Phase of the UNESCO-KU-ICL UNITWIN Cooperation Programme

In 2010, the UNESCO-KU-ICL UNITWIN Cooperation Programme was extended from its original research area "Landslide Risk Mitigation" to "Landslide and Water-Related Disaster Risk Management" to include more participants dealing with rainfall-induced landslides on slopes, as well as flood, sediment and debris flows in river systems. UNESCO, Kyoto University and ICL renewed the MoU for the UNITWIN programme as an Amendment to the Agreement, which was made on 18 March 2003, at the UNESCO Headquarters in Paris on 16 November 2010. Figure 5 shows the signing by Mr. Paolo Canuti (ICL President), Ms. Sonia Bahri (UNESCO Higher Education Division) and Mr. Norio Okada (Kyoto University DPRI Director) with witnesses: Mr. Kyoji Sassa (ICL Former President), Mr. Badaoui Roubhan (UNESCO Disaster Management Section) and Mr. Salvano Briceno (UNISDR Director-General).

In this second phase, Prof. Kaoru Takara has been playing an important role in capacity building. As show in Table 1, he introduced many international scientists and graduate students to his laboratory at the Disaster Prevention Research Institute (DPRI), Uji Campus Kyoto University. During 2010–2016, he produced eighteen Ph.D.s from nine countries (Brazil 2, China 4, Croatia 2, India, Indonesia 2, Japan 3, Laos PDR, Malaysia 2, Vietnam) at the Graduate School of Engineering, Kyoto University, in which Prof. Takara also has a position as professor in the Civil and Earth Resources Engineering Department of the graduate school.

International Journal "Landslides"

One of the most important activities of this UNITWIN Cooperation Programme is the publication of an international journal. In cooperation with Springer, a world-renowned publisher, in 2004 the ICL started a journal "Landslides" in full color.

The Springer home page (http://link.springer.com/journal/10346) introduces this journal as follows.

> As catastrophic events, landslides can cause human injury, loss of life and economic devastation, and destroy construction works and cultural and natural heritage. The journal Landslides is the common platform for publication of integrated research on all aspects of landslides. The journal publishes research papers, news of recent landslide events and information on the activities of the International Consortium on Landslides.
>
> Coverage includes landslide dynamics, mechanisms and processes; volcanic, urban, marine and reservoir landslides; related tsunamis and seiches; hazard assessment and mapping; modeling, monitoring, GIS techniques; remedial or preventive

UNITWIN cooperation programme was extended from landslide risk mitigation to landslide and water-related disaster risk management at UNESCO in 2010. Mr. Norio Okada, DPRI Director, Mr. Paolo Canuti and Ms Sonia Bahri, Education Sector of UNESCO signed the Amendment to the Agreement on 18 March 2003.

Fig. 5 Signing ceremony among UNESCO, Kyoto University and ICL in Paris on 16 November 2010

measures; early warning and evacuation and a global landslide database.

Landslides has been accepted at Thompson ISI for coverage in Science Citation Index Expanded, Current Contents/Physical Chemical and Earth Sciences and Current Contents/Engineering Computing and Technology.

The journal already has published 13 volumes (52 issues) since 2004, including 839 articles as of December 2016. Its Impact Factor (IF) is 3.049, which indicates the quality of this journal. In addition to this regular journal, the ICL publishes books at the occasions of the World Landslide Forums as irregular publications, also from Springer. The editorial work on these publications is implemented at the UNITWIN Headquarters Building at the Disaster Prevention Research Institute (DPRI) in Uji Campus, Kyoto University. Further information is reported in the article "Landslides: Journal of International Consortium on Landslides" by Sassa and Arbanas in this volume (Fig. 6).

UNITWIN Laboratory in the Main Campus of Kyoto University, Kyoto Japan

The UNESCO-KU-ICL UNITWIN Network group proposed two SATREPS (Science and Technology Research Partnership for Sustainable Development) projects, jointly funded by the Japan Science and Technology Agency (JST) and the Japan International Cooperation Agency (JICA). New undrained dynamic-loading ring shear apparatuses (ICL-1 and ICL-2) were developed by these two projects.

1. Risk identification and land-use planning for disaster mitigation of landslides and floods in Croatia (2009–2014) with development of a transportable compact undrained dynamic-loading ring shear apparatus.
2. Development of Landslide Risk Assessment Technology along Transport Arteries in Viet Nam (2011–2017) with development of high-stress undrained dynamic-loading ring shear apparatus to simulate the landslide initiation and the motion of landslides deeper than 100 m.

UNESCO-KU-ICL UNITWIN Project has applied for and accepted the use of one laboratory, UNITWIN Laboratory, since 2011. The two undrained ring shear apparatuses that were developed are installed in this laboratory and invited researchers and students have used these two apparatuses to study landslides.

Figure 7 shows the developed ICL-1 transportable undrained dynamic-loading ring shear apparatus. The counterparts of the Croatia project are three universities in Rijeka, Zagreb and Split, and also the Geological Survey of Croatia. The undrained ring shear apparatus developed in the Disaster Prevention Research Institute, Kyoto University, was difficult to transport because of the difficulty of setting it up for undrained capability, as well as its size and weight. To maintain an undrained condition, the central axis must be completely vertical and shear boxes must be completely horizontal in a single plane. Once the apparatus was moved, reproducing the undrained condition was not easy. It took some months to succeed in maintaining the undrained state during rotation. ICL-1 was designed to be transportable by a SUV car, with no necessity of resetting for the undrained condition.

UNITWIN students and researchers who used this apparatus are:

Maja Ostric (Croatia Water)
School: Graduate School of Engineering, Kyoto University, Japan
Title of Doctoral Dissertation: Development of portable undrained ring shear apparatus and its application
Date of certification: 23 September, 2013

Vivoda Prodan M—Doctor of Philosophy (Ph.D.)
School: Faculty of Civil Engineering, University of Rijeka, Rijeka, Croatia.
Title of Doctoral Dissertation: The influence of weathering process on residual shear strength of fine grained lithological flysch components.
Date of certification: 16 September 2016

Krkač M—Doctor of Philosophy (Ph.D..)
School: Faculty of Mining, Geology and Petroleum Engineering, University of Zagreb, Zagreb, Croatia.
Title of Doctoral Dissertation: A phenomenological model of the Kostanjek landslide movement based on the landslide monitoring parameters.
Date of certification: 17 July 2015

Table 1 Students and post-doctoral researchers participating in UNITWIN in the period of 2010–2016

Name	Years	Status at UNITWIN	Current position	Country
Kenichiro Kobayashi	2009–2013 Ph.D.	P.D.	Assoc. Prof., Kobe U	Japan
He Bin	2010–2013	P.D.	Prof., CAS	China
Apip	2008–2011 Ph.D. 2011–2013	Ph.D. student P.D.	Researcher, LIPI	Indonesia
Roberto Valmir Da Silva	2008–2011 Ph.D.	Ph.D. student	Asst. Prof., UFSS	Brazil
Tutao Oizumi	2008–2012 Ph.D.	Ph.D. student	Researcher, JAMSTEC	Japan
Mohd Remy Rozainy bin Mohd Arif Zainol	2009–2012 Ph.D.	Ph.D. student	Asst. Prof., USM	Malaysia
Luo Pigping	2009–2012 Ph.D. 2012–2014	Ph.D. student JSPS research fellow	Prof., Changan University	China
Pedro Luiz Borges Chaffe	2009–2012 Ph.D.	Ph.D. student	Asst. Prof., UFSC	Brazil
Netrananda Sahu	2009–2012 Ph.D.	Ph.D. student	Asst. Prof., U. Delhi	India
Tomoko Teramoto	2009–2015 Ph.D.	Ph.D. student	Broad Band Tower	Japan
Maja Ostric	2010–2013 Ph.D.	Ph.D. student	Researcher, Croatia Water	Croatia
Shin Young-A	2011–2013	M.Sc. student	Researcher, K-Water	R. of Korea
Kang Eunbi	2011–2013	M.Sc. student		R. of Korea
Nor Eliza binti Alias	2011–2014 Ph.D.	Ph.D. student	Asst. Prof., UTM	Malaysia
Duan Weili	2011–2014 Ph.D.	Ph.D. student	Asst. Prof., CAS	China
Hendy Setiawan	2012–2014 2014–2017 Ph.D.	M.Sc. student Ph.D. student	Student in Kyoto University	Indonesia
Xue Han	2012–2014 2014–2017 Ph.D.	M.Sc. student Ph.D. student	Student in Kyoto University	China
Vilaysane Bounhieng	2012–2015 Ph.D.	Ph.D. student	Lecturer, National U. Laos	Lao PDR
Dang Quang Khang	2012–2015 Ph.D.	Ph.D. student	Research promotion officer, ICL	Vietnam
Josko Troselj	2012–2016 Ph.D.	Ph.D. student	P.D., Kyoto University	Croatia
Hu Maochuan	2013–2016 Ph.D.	Ph.D. student	P.D., Kyoto University	China
Pham Hong Nga	2012–	JSPS RONPAKU Fellow	Lecturer, Vietnam Water Resources University	Vietnam
Doan Huy Loi	2013–2015	M.Sc. student	Institute of Transport Science and Technology, Vietnam	Vietnam
Pham Van Tien	2013–2015 2015–	M.Sc. student Ph.D. student	Student in Kyoto University	Vietnam
Eva Mia Siska	2014–	Ph.D. student	Student in Kyoto University	Indonesia
Khai Lin Chong	2014–	Ph.D. student	Student in Kyoto University	Malaysia
Karlina	2015–	Ph.D. student	Student in Kyoto University	Indonesia
Shi Yongxue	2014–2016 2016–	M.Sc. student Ph.D. student	Student in Kyoto University	China
Adnan Artyunov	2014–2016 2016–	M.Sc. student Ph.D. student	Student in Kyoto University	U.S.A.
Toma Stoyanov	2016–	Ph.D. student	Student in Kyoto University	Bulgaria
Nguyen Duc Ha	2016–	Ph.D. student	Student in Kyoto University	Vietnam
Jamila Rajabi	2016–	M.Sc. student	Student in Kyoto University	Afganistan
Saima Riaz	2016–	DPRI Research Fellow	Asst. Prof., University of Engineering & Technology Lahore	Pakistan

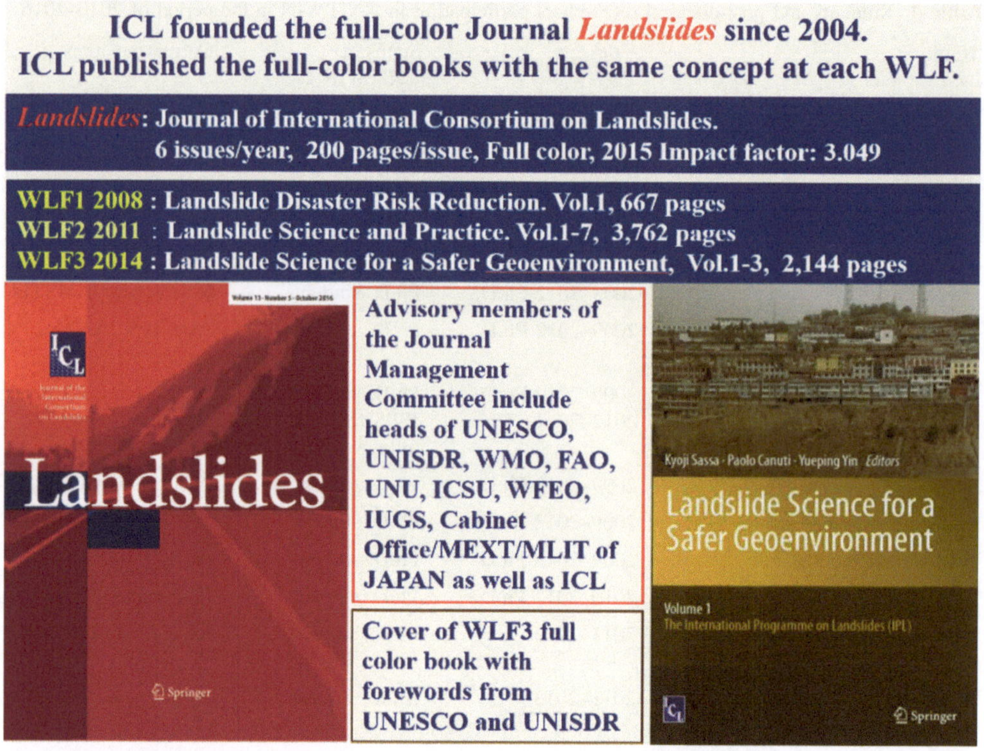

Fig. 6 Cover pages of the journal "Landslides" and an ICL book for the WLF

ICL founded the full-color Journal *Landslides* since 2004.
ICL published the full-color books with the same concept at each WLF.

Landslides: Journal of International Consortium on Landslides.
6 issues/year, 200 pages/issue, Full color, 2015 Impact factor: 3.049

WLF1 2008 : Landslide Disaster Risk Reduction. Vol.1, 667 pages
WLF2 2011 : Landslide Science and Practice. Vol.1-7, 3,762 pages
WLF3 2014 : Landslide Science for a Safer Geoenvironment, Vol.1-3, 2,144 pages

Landslides
Springer

Advisory members of the Journal Management Committee include heads of UNESCO, UNISDR, WMO, FAO, UNU, ICSU, WFEO, IUGS, Cabinet Office/MEXT/MLIT of JAPAN as well as ICL

Cover of WLF3 full color book with forewords from UNESCO and UNISDR

Kyoji Sassa · Paolo Canuti · Yueping Yin *Editors*
Landslide Science for a Safer Geoenvironment
Volume 1
The International Programme on Landslides (IPL)
Springer

Fig. 7 Photo of developed transportable undrained dynamic-loading *ring shear* apparatus in the factory of the producing company. (At the *center front* is late Takeshi Nakasono, chief designer of the apparatus, at the *center in the back* is Maja Ostric, invited from Croatia, who obtained her Ph.D. based on the development of this apparatus)

Figure 8 shows the developed ICL-2 "High-stress undrained dynamic-loading ring shear apparatus" in use in the UNITWIN Laboratory. A new student who passed the Japanese Government Scholarships examination in the Japanese Embassy in Hanoi started his doctoral course study from 1st October 2016. ICL-2 was very much modified and developed to avoid trouble and damage to the apparatus by miss-handling, because severe damage occurred to break the

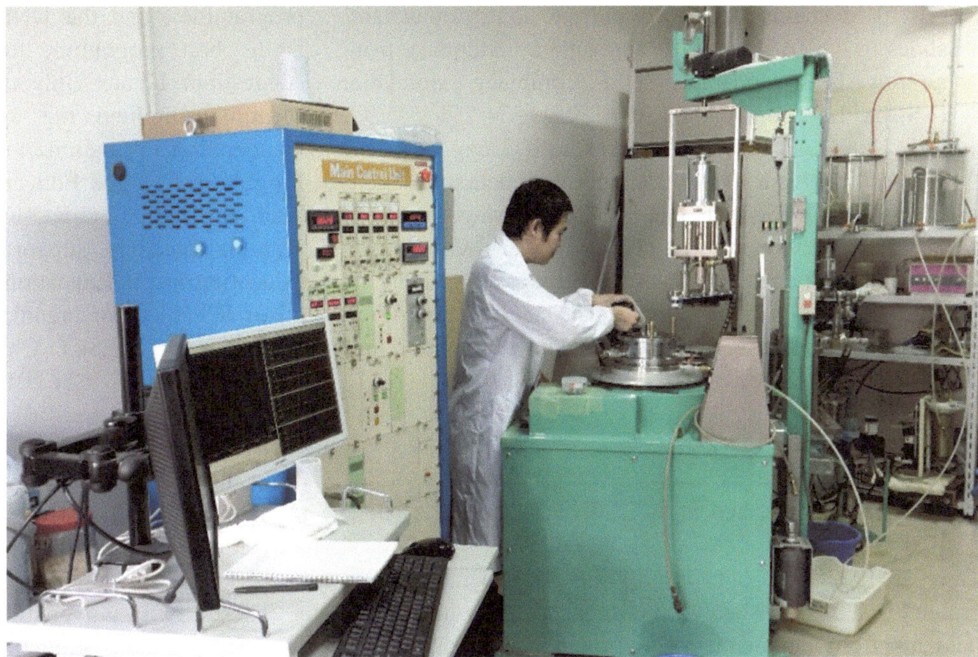

Fig. 8 ICL-2 High-stress undrained dynamic-loading ring shear apparatus (3 MPa) in the UNITWIN laboratory, Kyoto University

central axis of 3 tons under high-stress testing by the servo-control systems for normal stress, shear stress and gap control during the training.

The apparatus was designed for use in developing countries that need equipment with low maintenance costs and technology. The apparatus was donated to the Institute of Transport Science and Technology of the Ministry of Transport in Vietnam (ITST). ITST has made a very good video manual for ICL-2. Many engineers, even without training in Japan, can handle this complicated apparatus. In September 2016, two researchers were invited to write doctoral papers for a Ph.D. During the discussion, they noted that further testing was needed. They asked engineers in ITST in Hanoi to test samples in the fully saturated state. Then, they conducted the undrained dynamic-loading ring shear test by remote control through the internet. It means that the apparatus of ICL-2 can be used from anywhere in the world, if users know how to test though the UNESCO-KU-ICL UNITWIN network.

UNITWIN students and researchers who have used and currently using this apparatus are:

Dang Quang Khang—Doctor of Philosophy (Ph.D.)
School: Graduate School of Engineering, Kyoto University, Japan
Title of Doctoral Dissertation: Development of a new high-stress dynamic-loading ring-shear apparatus and its application to large-scale landslides
Date of certification: 23 September, 2015

Hendy Setiawan
School: Graduate School of Engineering, Kyoto University, Japan
Title of Doctoral Dissertation: Landslide Hazard Assessment on the Upstream of Dam Reservoir
Date of certification: The doctoral thesis has been submitted in December 2016, and is under examination.
Pham Van Tien (ITST, Vietnam) and Nguyen Duc Ha (VIGMR, Vietnam) are currently studying using the ICL-2 and ICL-1 to obtain Ph.D. degrees in the Graduate School of Engineering, Kyoto University, Japan
Lam Huu Quang (ITST, Vietnam) is testing soils taken from Vietnam in the ICL-2 that was donated from Japan to Vietnam. He studied and supervised in the UNTWIN programme in Japan and in Vietnam.

Recent papers published or submitted to the journal *Landslides* using ICL-2 apparatus include the following.

1. Sassa K, Dang K, He B, Takara K, Inoue K, Nagai O (2014) A new high-stress undrained ring-shear apparatus and its application to the 1792 Unzen–Mayuyama megaslide in Japan. Landslides 11(5):827–842.
2. Sassa K, Dang K, Yanagisawa H, He B (2016) A new landslide-induced tsunami simulation model and its application to the 1792 Unzen-Mayuyama landslide-and-tsunami disaster. Landslides 13(6):1405–1419
3. Dang K, Sassa K, Fukuoka H, Sakai N, Sato Y, Takara K, Lam HQ, Doan HL, Pham VT, Nguyen DH

(2016) Mechanism of two rapid and long runout landslides in the 16 April 2016 Kumamoto earthquake using a ring-shear apparatus and computer simulation (LS-RAPID). Landslides 13(6):1525–1534
4. Lam HQ, Doan HL, Sassa K, Takara K, Dang K, Abe S, Asano S (2016) Risk Assessment of a Precursor Stage of Landslide Threatening the Haivan Railway Station in Vietnam (Submitted to Landslides, under revision).
5. Doan HL, Lam HQ, Sassa K, Takara K, Dang D, Nguyen KT, Pham VT (2016) The 28 July 2015 rapid landslide at Ha Long city, Quang Ninh, Vietnam (Submitted to Landslides).

Acknowledgements and Call for Cooperation

The UNESCO-KU-ICL UNITWIN Cooperation programme was established at the suggestion, from six participants from the Science Sector and Cultural Sector of UNESCO at the ICL foundation meeting in January 2002, that the International Programme on Landslides (IPL), to be created by the International Consortium on Landslides (ICL), should be authorized by one of UNESCO programmes, and the UNITWIN Cooperation Programme is the best programme to support IPL. Because of this suggestion, Kyoto University and the International Consortium submitted the application form of UNITWIN Cooperation Programme to the UNITWIN/UNESCO Chair Programmes of the Education Sector of UNESCO. Thereafter, the UNESCO-KU-ICL UNTWIN Cooperation Programme obtained support from the Section of Higher Education in the Education Sector of UNESCO, as well as all of ICL member organizations.

Two new UNESCO Chairs were established within this UNITWIN Network in 2016:

- Prevention and Mitigation of Geo-hydrological Hazards at the University of Florence, Italy.
- UNESCO Chair for Water-Related Disaster Risk Reduction at the University of Ljubljana, Slovenia.

These two universities, which have hosted these UNESCO Chair programmes, are active ICL member organizations. We extend our sincere appreciation to all the involved organizations and individuals and wish the successful further development of this network with strong cooperation from all those partners.

Landslides: Journal of the International Consortium on Landslides

Kyoji Sassa and Željko Arbanas

Abstract

The international journal ***Landslides: Journal of the International Consortium on Landslides*** was established in April 2004 as the core project of the International Programme on Landslides and a joint initiative of the International Consortium on Landslides and the United Nations and other global organizations. The aims of ***Landslides*** are to promote landslide science, technology, and capacity building, and to strengthen global cooperation for landslide risk reduction within the United Nations International Strategy for Disaster Risk Reduction (ISDR). The importance of landslide occurrences, as a one of the main global hazards increasing under global climate change in recent years, focused the scientists, engineers and stakeholders all over the world, especially in regions threatened by landslides, on landslide risk reduction research, with the aim of reducing their consequences. The landslide scientists recognized ***Landslides*** as the most important scientific journal in the fields of natural hazards, engineering geology, geotechnics and civil engineering related to any type of landslide research. Results of the most significant landslide research conducted last year were submitted and published in ***Landslides***. The increasing number and quality of published manuscripts in the last years has resulted in a continuous rise of the ***Landslides*** journal impact, as expressed by the Thompson Reuters Impact Factor. The Thompson Reuters Impact Factor 2015 is 3.049; ranking No.1 in the category of Engineering, Geological journals. The aims of the ***Landslides*** Editorial Board are further improvements of manuscript quality, speed-up of the peer review process and faster publication of landslide science achievements.

Keywords

Landslides • Journal • Science • Impact factor • Publication

Introduction

The International Consortium on Landslides (ICL) was established in January 2002 to promote landslide research and capacity building, primarily in developing countries, for the benefit of society and the environment, through the

K. Sassa
International Consortium on Landslides, 138-1 Tanaka Asukai-cho, Kyoto, 606-8226, Japan
e-mail: sassa@iclhq.org

Ž. Arbanas (✉)
Faculty of Civil Engineering, University of Rijeka, Radmile Matejčić 3, 51000 Rijeka, Croatia
e-mail: zeljko.arbanas@gradri.uniri.hr

K. Sassa et al. (eds.), *Advancing Culture of Living with Landslides*,
DOI 10.1007/978-3-319-59469-9_21

establishment of the International Programme on Landslides (IPL) (Sassa 2004). The concept of IPL was developed through discussions at the United Nations World Conference on Disaster Reduction held in Kobe 2005, followed by a roundtable discussion resulting to the 2006 Tokyo Action Plan (Sassa 2006). Landslide research and science have been developed and conducted within the various disciplines and fields of natural, engineering and social sciences outlined in the 2006 Tokyo Action Plan, but until 2004 there was no international journal strictly focused on landslides. The Japan Landslide Society had published an international newsletter, ***Landslide News***, annually from 1987 to 2003 with Kyoji Sassa as chairperson of the publishing committee and editor-in-chief and with the support of Robert L. Schuster as international chief editor (Sassa et al. 2009).

At the first session of the Board of Representatives of ICL, held at the UNESCO Headquarters in Paris, 19–21 November 2002, Kyoji Sassa proposed ***Landslides: Journal of the International Consortium on Landslides***, as the core project of the IPL which was approved as the first IPL coordinating project (IPL-C100) (Sassa et al. 2009). The proposal included development of the international newsletter ***Landslide News*** into an international journal on landslides published in full color, with no color-print charges for authors. After negotiation with several international publishers for publication of a full-color scientific journal, agreement was reached with Springer in 2003. Until ***Landslides***, there was no full-color scientific journal in the world (Sassa et al. 2009). Figure 1 is a cover of ***Landslides*** 2016. The design has remained the same from the foundation issue of the Journal in April 2004.

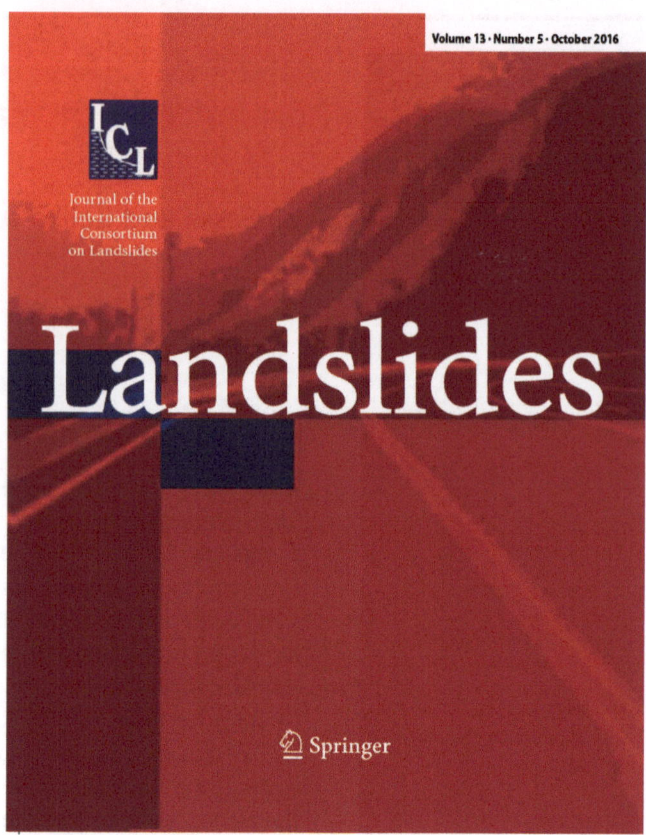

Fig. 1 The cover of ***Landslides: Journal of the International Consortium on Landslides***

The first issue of ***Landslides: Journal of the International Consortium on Landslides*** was published in April 2004 and appeared in journal citations of Thomson Reuters (Institute for Scientific Information, ISI) in 2005. The first Thomson Reuters ISI Web of Knowledge Impact Factor, released in 2008, was 0.986. After 2008 ***Landslides*** has been recognized by landslide scientists, researchers, professional engineers, international organizations, national and regional governments and other stakeholders as a high-quality specialized journal related to the scientific, technical, social and all other aspects of landside investigation and remediation in the frame of landslide risk reduction. The number of submitted and published manuscripts has continuously increased, providing ***Landslides'*** readers better and better insight into all aspects of landslide research around the world. The rate of manuscripts accepted for publishing is almost the same from the first volume until today, ensuring the established quality level of the journal, with the consequent continuous rise of the quality indicators and impact on the scientific community expressed in its Thomson Reuters ISI Web of Knowledge impact factor, which reached 3.049 for 2015 and the ranking No. 1 in geological engineering journals.

The achievements of ***Landslides*** from 2004 to 2009 were presented by Sassa et al. (2009) and from 2009 to 2013 by Sassa et al. (2013) in ***Landslides***. In this paper we will present some achievements of ***Landslides*** from the first issue in April 2004 until the writing of this manuscript at the end of September 2016, based on data from Thomson Reuters (Institute for Scientific Information) and publisher Springer Nature.

Impact of the *Landslides* Journal

Impact Factor

One of the most widely accepted impact indexes of a publication to the society is Impact Factor (IF), which is annually reported by Thomson Reuters (Institute for Scientific Information, ISI). The Impact Factor is calculated by dividing the number of current year citations in ISI databases to the source items published in that journal during the previous two years. The IF value is reflected by authors/scientists of published papers, not always by the practitioners/users of published papers.

The Impact Factor of ***Landslides*** after its first release in 2008 has shown an almost continuous rise, especially in the last three years, from 0.986 to 3.049 in 2015. The Impact Factor and related data necessary for Impact Factor calculation from 2008 to 2015 are presented in Table 1. There is also a clearly visible rise of citable articles in 2013, when started bimonthly publication of the ***Landslides*** journal with 100 pages per issue (Vol. 10, Issues No. 1-4) and 150 pages per issue (Vol. 10. Issues No. 4-5), and again in 2014 when increased the size to 200 pages per issue (from Vol. 11 Issue No.3). The consequence of the rise of published citable articles is also an increasing citation of published articles of 40% in 2013, 23% in 2014 and 40% in 2015.

Citations

Impact Factors are calculated based on the number of citations of each article published in a journal during the previous two years, or five years for the five-year Impact Factor. The most cited articles published in ***Landslides*** journal, especially in years following the publication of an article, are the most important for a value of the Impact Factor. The five-year Impact Factor indicates the general quality of published articles and it is more appropriate for journals in certain fields such as landslides, because the body of citations may not be large enough to make reasonable comparisons, publication schedules may be consistently late, or it may take longer than two years to disseminate and respond to published works. Table 2. presents the 10 most cited manuscripts published in ***Landslides*** to the end of 2015 according to Institute for Scientific Information, ISI.

The most cited article is "The rainfall intensity-duration control of shallow landslides and debris flows: an update" prepared by Fausto Guzzetti, Silvia Peruccacci, Mauro Rossi and Colin P. Stark; Landslides Vol. 5. Issue No. 1 (2008) with 256 citations.

The Journal Impact

The influence and impact of a journal in the world of science is quantifiable according to positon of the journal in the Thompson Reuters Journal Citation Reports® related to the achieved Impact Factor. Thompson Reuters Journal Citation Reports® offers a systematic, objective means to critically evaluate the world's leading journals, with quantifiable, statistical information based on citation data. By compiling articles' cited references, JCR helps to measure research influence and impact at the journal and category levels, and shows the relationship between citing and cited journals (http://ipscience.thomsonreuters.com/product/journal-citation-reports).

Landslides journal, according to the Thompson Reuters Journal Citation Reports®, is ranked at the top of 35 journals in the category Engineering, Geological (1/35) and 32nd of 184 journals in category of Geosciences, Multidisciplinary (32/184) (see Table 3) (http://ipscience.thomsonreuters.com/product/journal-citation-reports).

The category Engineering, Geological, according to the Thompson Reuters Journal Citation Reports®, includes multidisciplinary resources that encompass the knowledge and experience drawn from both geosciences and various engineering disciplines (primarily civil engineering). Resources in this category cover geotechnical engineering, geotechnics, geotechnology, soil dynamics, earthquake engineering, geotextiles and geomembranes, engineering geology and rock mechanics. The much wider category Geosciences, Multidisciplinary covers resources having a general or interdisciplinary approach to the study of the Earth and other planets. Relevant topics include geology, geochemistry/geophysics, hydrology, paleontology, oceanography, meteorology, mineralogy, geography, and energy and fuels. Resources having a primary focus on geology, or geochemistry and geophysics are placed in their own categories (http://ip-science.thomsonreuters.com/mjl/scope/scope_scie).

Landslides journal has been on the top of the list of journals in the category Engineering, Geological continuously for the

Table 1 Impact factor, 5-year impact factor, number of total citable articles, total cites and cites per volume

Year	Impact factor	5-year Impact factor	Total citable articles	Total cites	Cites per volume
2015	3.049	3.616	90	1839	35
2014	2.870	3.205	85	1310	139
2013	2.814	3.045	59	1067	228
2012	2.093	2.358	41	760	243
2011	2.216	1.841	45	535	190
2010	1.625	1.938	41	461	223
2009	1.703	2.374	33	460	213
2008	0.754	N/A	39	231	202
2007	0.986	N/A	35	155	182

Table 2 Ten most cited manuscript published in ***Landslides*** to the end of 2015 according to Institute for Scientific Information, ISI

No	Article title	Authors	Vol	No	Year	Citations
1	The rainfall intensity-duration control of shallow landslides and debris flows: an update	Fausto Guzzetti et al.	5	1	2008	256
2	Landslide hazard mapping at Selangor, Malaysia using frequency ratio and logistic regression models	Saro Lee and Biswajeet Pradhan	4	1	2007	212
3	Landslide hazards triggered by the 2008 Wenchuan earthquake, Sichuan, China	Yueping Yin et al.	6	2	2009	147
4	Landslide susceptibility mapping using GIS-based weighted linear combination, the case in Tsugawa area of Agano River, Niigata Prefecture, Japan	Lulseged Yimam Ayalew et al.	1	1	2004	146
5	Regional landslide susceptibility analysis using back-propagation neural network model at Cameron Highland, Malaysia	Biswajeet Pradhan and Saro Lee	7	1	2010	115
6	An approach for GIS-based statistical landslide susceptibility zonation - with a case study in the Himalayas	Ashis K. Saha et al.	2	1	2005	113
7	Global landslide and avalanche hotspots	Farrokh Nadim et al.	3	2	2006	108
8	The 12 May Wenchuan earthquake-induced landslide lakes: distribution and preliminary risk evaluation	Peng Cui et al.	6	3	2009	95
9	Survey and monitoring of landslide displacements by means of L-band satellite SAR interferometry	Tazio Strozzi et al.	2	3	2005	93
10	The Varnes classification of landslide types, an update	Oldrich Hungr et al.	9	1	2014	88

Table 3 Thompson Reuters Journal Citation Reports® ranking of ***Landslides*** in the period 2007–2015

Year	Engineering, geological		Geosciences, multidisciplinary	
	Rank	Quartile	Rank	Quartile
2015	1/35	Q1	32/184	Q1
2014	1/32	Q1	30/175	Q1
2013	1/33	Q1	31/174	Q1
2012	2/32	Q1	56/172	Q2
2011	1/30	Q1	38/170	Q1
2010	5/30	Q1	65/167	Q2
2009	3/27	Q1	51/155	Q2
2008	12/25	Q2	109/144	Q4
2007	6/26	Q1	76/137	Q3

last three years and the difference in Impact Factor between the ***Landslides*** and the second ranked ***Acta Geotechnica*** is more than 0.6. The top 20 journals, their Impact Factors and total cites in 2015 in the category Engineering, Geological are listed in Table 4. In the much wider category Geosciences, Multidisciplinary, ***Landslides*** journal is ranked as 32nd after ***Journal of Glaciology*** (IF2015 = 3.109) and before ***Journal of Hydrology*** (IF2015 = 3.043) and ***Geomorphology*** (IF2015 = 2.813) (https://jcr.incites.thomsonreuters.com/JCRJournalHomeAction). It is important that ***Landslides*** journal has ranked in quartile Q1 for the last three years in the category Geosciences, Multidisciplinary because the position in quartile Q1 is another indicator of the quality and influence of a journal, especially in Europe. In the category Engineering, Geological the ***Landslides*** journal has been continuously in quartile Q1 from 2007, except 2008.

Table 4 Top 20 journals ranked by Impact Factor in the category Engineering, Geological in 2015 (http://ipscience.thomsonreuters.com/product/journal-citation-reports)

Rank	Journal title	Publisher	IF 2015	Total cites
1	Landslides	Springer Heidelberg	3.049	1839
2	Acta Geotechnica	Springer Heidelberg	2.426	649
3	Rock Mechanics and Rock Engineering	Springer Vienna	2.386	2487
4	Geotextiles and Geomembranes	Elsevier Science BV	2.366	1851
5	Earthquake Spectra	Earthquake Engineering Research Inst	2.298	3068
6	Engineering Geology	Elsevier Science BV	2.196	7398
7	Earthquake Engineering and Structural Dynamics	Wiley-Blackwell	2.127	6379
8	Geosynthetics International	ICE Publishing	2.066	725
9	Bulletin of Earthquake Engineering	Springer	2.036	1468
10	International Journal of Rock Mechanics and Mining Sciences	Pergamon-Elsevier Science Ltd	2.010	9728
11	Geotechnique	ICE Publishing	2.000	7384
12	Canadian Geotechnical Journal	Canadian Science Publishing, NRC Research Press	1.877	6107
13	International Journal for Numerical and Analytical Methods in Geotechnics	Wiley-Blackwell	1.758	3261
14	Computers and Geotechnics	Elsevier Science BV	1.705	3067
15	Journal of Geotechnical and Geoenvironmental Engineering	ASCE-Amer Soc Civil Engineers	1.696	6304
16	Soil Dynamics and Earthquake Engineering	Elsevier Science BV	1.481	3663
17	International Journal of Geomechanics	ASCE-Amer Soc Civil Engineers	1.387	994
18	Bulletin of Engineering Geology and Environment	Springer Heidelberg	1.252	1237
19	Soils and Foundation	Elsevier Science BV	1.238	2492
20	Geomechanics and Engineering	Techno Press	1.085	188

Publications

The number of articles submitted and accepted for publication in the ***Landslides*** journal continuously increased from the first year of publishing until now. The ***Landslides*** journal started in 2004 as a three-monthly journal (four issues per volume) with approximately 100 pages per issue. An increased number of accepted articles accompanied a change from three to bi-monthly publication of the ***Landslides*** journal in 2013 with 100 pages per issue (Vol. 10, Issues No. 1-4). The further rise of accepted articles resulted from increasing the number of the pages per issue to 150 pages from Vol. 10. Issues No. 4-5 and to 200 pages per issue from Vol. 11 Issue No. 3 in 2014. A rise in the number of submitted articles followed the increasing Impact Factor and the number of accepted manuscripts in the second half of 2016 indicates a necessary new increase in printed pages per issue.

Categories of Articles

Landslides published four major categories of articles:

- Original Papers (6–12 pages): original research and investigation results;
- Technical Note (less than 6 pages): research notes, review notes, case studies, progress of technology, and best practices;
- Recent Landslides (generally less than 6 pages): reports of recent landslides, including location (latitude/longitude), plan, section, geology, volume, movement, mechanism, and disasters within the available extent; in monitoring, testing, investigation, and mitigation measures; and
- International Consortium on Landslides (ICL)/International Programme on Landslides (IPL) Activities (length depending on the content): progress of IPL projects and ICL Committee activities.

The categories of Original paper and Technical note are the same as in other scientific journals. The category of Recent Landslides is unique to ***Landslides*** and carries on the tradition begun by the Landslide News (1987–2003), an international newsletter published by the Japan Landslide Society (Sassa et al. 2009, 2015). The International Consortium on Landslides aims to contribute to the United Nations International Strategy for Disaster Reduction through developing landslide sciences, technology, and capacity building, and strengthening global cooperation for landslide risk reduction within developed and developing countries. The ICL established the International Programme on Landslides (IPL), together with ICL supporting organizations (UNESCO, UNISDR, WMO, FAO, UNU, ICSU, WFEO, and IUGS). These activities are reported in ICL/IPL Activities (Sassa et al. 2015).

Classification of Articles

The submission of a manuscript to the ***Landslides*** journal should be only electronic through the Landslides Editorial Manager (EM) managed by Springer Nature. When an article needs to be submitted through the ***Landslides*** Editorial Manager, the author is requested to classify the manuscript based on a list of article classifications (see Table 5). Starting from Vol. 10 (2013), the current article classification has four major classes: Background Science,

Table 5 Classification of articles

Classification of articles
10: Background Science
010: Geology
020: Geomorphology
030: Geotechnology
040: Geophysics
050: Hydrology & Meteorology
20: Methodology
010: Field investigation and ground exploration
020: Monitoring
030: Material testing
040: Physical modeling
050: Numerical simulation
060: GIS
070: Remote sensing
080: Planning and design
30: Application
010 Hazard and risk mapping
020: Early warning
030: Risk assessment
040: Remedial measures & prevention works
050: Risk reduction strategy
060: Database
070: Capacity development
40 Types of landslides
010: Debris flows
020: Rock falls
030: Earthquake-induced landslides
040: Rain-induced landslides
050: Landslides in cultural/natural heritage sites
060: Anthropogenic landslides
070: Landslides in urban areas

Table 6 Editorial status summary 2013-2015 (Schwarz and Mannsperger 2016)

Submissions	2013	2014	2015
Total Submitted	189	242	301
Total Decisions	188	208	279
Accept	76	87	117
Reject	112	121	162
Acceptance Rate (%)	40	42	42
Rejection Rate (%)	60	58	58
Average Days to First Decision	65	79	49
Average Days to Final Disposition Accept	228	214	205
Average Days to Final Disposition Reject	59	76	68

Methodology, Application and Types of Landslides. An author can chose one or more classes for the submitted manuscript.

Editorial Process

Each article submitted to the ***Landslides*** journal should be uploaded in the Editorial Manger (EM) as a *New assignment* article. New assignment articles are passed to Executive Editors, who are requested to upload their opinions of whether the article should be passed to in-depth review or rejected without in-depth review, and whether contributed category and page length are appropriate or not.

Executive Editors organize editorial meetings every week by *Skype* and decide the assignment of a handling Editor for each article passed for in-depth review. The handling Editors are assigned from a database of around 100 registered Editors in Editorial Manager (EM), identifying classifications attached to their research area. The handling Editor will assign one or two Reviewers, depending on the article's category, from the database of around 950 registered Reviewers in Editorial Manager, with their personal classifications. Reviewers are searched by classification matching. The handling Editor's recommendation, based on results of conducted reviews for each article, will be uploaded to EM. The Editor-in-Chief will then make a final decision, mostly following the handling Editor's recommendations. During the peer review process, submitted manuscripts go through one or more revision stages leading up to final acceptance or rejection. The editorial status summary is presented in Table 6 (Schwarz and Mannsperger 2016). The table summarizes the activity for the journal office between January 1st and December 31st of each year, but Only "Original Submissions" have been taken into account. From Table 6 it is clearly visible that there has been a continuous rise in submitted manuscripts and also that the acceptance and rejection rates have been almost identical during the last three years (and before). The rejection rate for a year is calculated as the number of rejected manuscripts this year compared to the total number of decisions in the year, which is defined here as the number of rejected manuscripts plus the number of accepted manuscripts. The term Reject is used for the calculation of the acceptance and rejection rates, which includes all terms that may exist for rejection decisions: Reject before review; Reject after review; Reject, but resubmit; Reject, out of scope; and so forth. Only the papers for which the Final Disposition Date has been set are taken into account. Final disposition date means that a manuscript is fully completed (Schwarz and Mannsperger 2016).

One of important tasks for the Editorial Board is to reduce the peer review process to enable publishing of an article as soon as possible. The longest period from submission to the final decision was a maximum of around one year. The

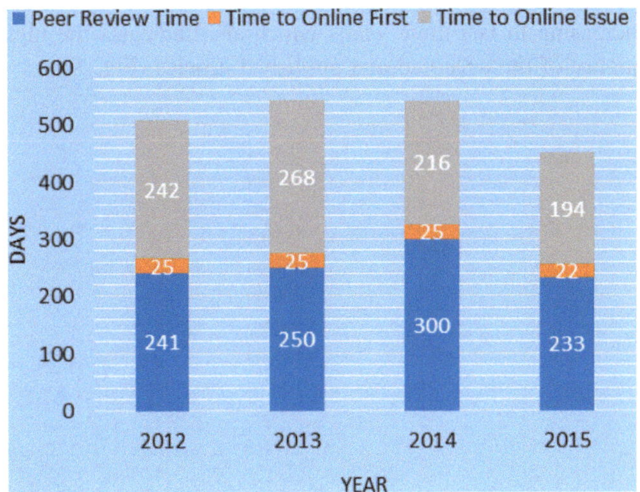

Fig. 2 Average time between submission of a manuscript and publication in an Online Issue 2013–2015 (Schwarz and Mannsperger 2016)

Editorial Board is making efforts to reduce this period to 6 months. The average time between submission of the article and publication is presented in Fig. 2 (Schwarz and Mannsperger 2016).

It is clear in Fig. 1 that peer review time has decreased continuously during the last three years, but the time for publishing an article after Springer's Online First® is very long, although this period is becoming shorter.

The publication of accepted articles via Springer's Online First® is very important and enables internet readers to view the article soon after final acceptance. Articles published via Springer's Online First® service are final articles published online after an author has reviewed proofs and all corrections have been carried out. Metadata is sent to all relevant bibliographic services for inclusion in abstracting and indexing databases immediately after online publication. Articles are published on the SpringerLink platform in PDF format and only final pagination and the citation line are later added in the printed version. Articles are fully citable by their DOI (Digital Object Identifier) and the official publication date is the online publication date. Publication of papers through Springer's Online First® helps shorten the time between publication and citation (Schwarz and Mannsperger 2016).

Article Downloading

While accepted articles are available to the readers after publishing via Springer's Online First® service, article downloading is an important way for disseminating article results. The downloading of ***Landslides*** articles is rapidly increasing in last three years, which also indicates the high quality of the accepted and published articles (Fig. 3).

The top 10 most downloaded ***Landslides'*** articles in 2015, according to Institute for Scientific Information, ISI, are listed in Table 7.

Landslides' Best Paper Award

The Best Paper Award for the best paper published in ***Landslides: Journal of the International Consortium on Landslide*** has been given annually, beginning with the year 2004 for the first volume (Vol. 1) of the journal. The selection of the Best Paper Award is carried out by the Best Paper Award Subcommittee. The judging and ranking of papers were based on a numerical grading system that involved three elements in the final score of the paper: (i) Scientific and technical quality of the paper (up to 50%), (ii) Impact on the profession and society (up to 30%), and (iii) Quality of figures and tables (up to 20%). The proposal of Best Paper Award should be approved by the Board of Representatives of the International Consortium on Landslides. The ***Landslides*** Best Paper Awards from the Vol. 1 to Vol. 12 are listed in Table 8.

Acknowledgements ***Landslides***' Editorial Board deeply appreciate the work of all Editors and Reviewers, for their voluntary contributions to editing and reviewing articles submitted to the ***Landslides*** journal. In the last 5 years, the number of articles submitted and published in ***Landslides*** has increased more than three times from 100 pages per issue and 4 issues per year to 200 pages per issue and 6 issues per year and is still increasing, and may result in more pages per issue and/or possible monthly publication of ***Landslides***. These circumstances would increase the editorial work for Landslides' Editors and Reviewers. The main goals of the Landslides Editorial Board, Editors, Reviewers and all others included in the editorial process are further improvements of manuscript quality, speeding-up of the peer review process and faster publication of landslide science achievements to enable all types of ***Landslides***' readers quick access to new knowledge in landslide science.

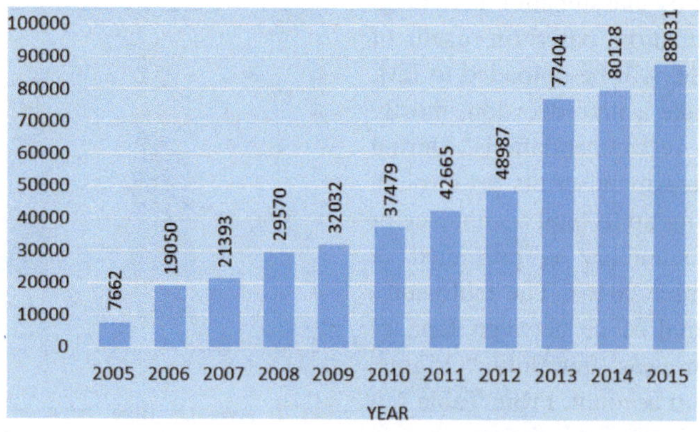

Fig. 3 Downloads of ***Landslides*** articles 2012–2015 (Schwarz and Mannsperger 2016)

Table 7 Top 10 most downloaded ***Landslides*** articles in 2015 according to Institute for Scientific Information, ISI

Rank	Article Title	Authors	Vol	No	Year	Download
1	The Varnes classification of landslide types, an update	Oldrich Hungr et al.	11	2	2014	2053
2	Landslide susceptibility mapping using multi-criteria evaluation techniques in Chittagong Metropolitan Area, Bangladesh	Bayes Ahmed	12	6	2015	1131
3	The rainfall intensity–duration control of shallow landslides and debris flows: an update	Fausto Guzzetti et al.	5	1	2008	1030
4	Three (nearly) complete inventories of landslides triggered by the May 12, 2008 Wenchuan Mw 7.9 earthquake of China and their spatial distribution statistical analysis	Chong Xu et al.	11	3	2014	694
5	Landslide hazards triggered by the 2008 Wenchuan Earthquake, Sichuan, China	Yueping Yin et al.	6	2	2009	634
6	Landslide susceptibility mapping using GIS-based multi-criteria decision analysis, support vector machines, and logistic regression	Taskin Kavzoglu et al.	11	3	2014	630
7	Generation of a Landslide Risk Index Map for Cuba using spatial multi-criteria evaluation	Enrique Armando Castellanos Abella et al.	4	4	2007	590
8	New insights into the temporal prediction of landslides by a terrestrial SAR interferometry monitoring case study	Paolo Mazzanti et al.	12	1	2015	536
9	Spatial prediction models for shallow landslide hazards: a comparative assessment of the efficacy of support vector machines, artificial neural networks, kernel logistic regression, and logistic model tree	Dieu Tien Bui	13	2	2016	528
10	Integration of rainfall thresholds and susceptibility maps in the Emilia Romagna (Italy) regional-scale landslide warning system	Samuele Segoni et al.	12	4	2015	491

Table 8 The *Landslides* best paper awards from the Vol. 1 to Vol. 12 (2004–2015)

Authors	Article Title	Vol.	No	Year
Margottini C.	Instability and geotechnical problems of the Buddha niches and surrounding cliff in Bamiyan Valley, central Afghanistan	1	5	2004
Baum R.L., Coe J.A., Godt J.W., Harp E.L., Reid M.E., Savage W.Z., Schulz W.H., Brien D.L., Chleborad A.F., McKenna J.P. and Michael J.A.	Regional landslide-hazard assessment for Seattle, Washington, USA	2	4	2005
Nadim F., Kjekstad O., Peduzzi P., Herold C., and Jaedicke C.	Global landslide and avalanche hotspots	3	2	2006
Leynaud D., Sultan N., and Mienert J.	The role of sedimentation rate and permeability in the slope stability of the formerly glaciated Norwegian continental margin: the Storegga slide model	4	4	2007
Prochaska A.B., Santi P.M., Higgins J.D., and Cannon S.H.	A study of methods to estimate debris flow velocity	5	4	2008
Lundström K., Larsson R., and Dahlin T.	Mapping of quick clay formations using geotechnical and geophysical methods	6	1	2009
Massey C.I., Manville V., Hancox G.H., Keys H.J., Lawrence C., and McSaveney M.	Out-burst flood (lahar) triggered by retrogressive landsliding, 18 March 2007 at Mt Ruapehu,New Zealand—a successful early warning	7	3	2010
Brideau M.A., Pedrazzini A., Stead D., Froese C., Jaboyedoff M. and van Zeyl D.	Three-dimensional slope stability analysis of South Peak, Crowsnest Pass, Alberta, Canada	8	2	2011
Pinyol N.M., Alonso E.E., Corominas J. and Moya J.	Canelles landslide: modelling rapid drawdown and fast potential sliding	9	1	2012
Sosio R., Crosta G.B. and Hungr, O.	Numerical modeling of debris avalanche propagation from collapse of volcanic edifices	9	3	2012
Staley D.M., Kean J.W., Cannon S. H., Schmidt K.M. and Laber J.L.	Objective definition of rainfall intensity – duration thresholds for the initiation of post-fire debris flows in southern California	10	5	2013
Hungr O., Leroueil S. and Picarelli L.	The Varnes classification of landslide types, an update	11	2	2014
Huang D., Cen D., Ma G., and Huang R.	Step-path failure of rock slopes with intermittent joints	12	5	2015

References

Abella EAC, van Westen CV (2007) Generation of a landslide risk index map for Cuba using spatial multi-criteria evaluation. Landslides 4(4):311–325

Ahmed B (2015) Landslide susceptibility mapping using multi-criteria evaluation techniques in Chittagong metropolitan area, Bangladesh. Landslides 12(6):1077–1095

Ayalew LY, Yamagishi H, Ugawa N (2004) Landslide susceptibility mapping using GIS-based weighted linear combination, the case in Tsugawa area of Agano River, Niigata Prefecture, Japan. Landslides 1(1):73–81

Baum RL, Coe JA, Godt JW, Harp EL, Reid ME, Savage WZ, Schulz WH, Brien DL, Chleborad AF, McKenna JP, Michael JA (2005) Regional landslide-hazard assessment for Seattle, Washington, USA. Landslides 2(4):266–279

Brideau MA, Pedrazzini A, Stead D, Froese C, Jaboyedoff M, Zeyl D (2011) Three-dimensional slope stability analysis of South Peak, Crowsnest Pass, Alberta, Canada. Landslides 8(2):139–158

Cui P, Zhu Y, Han Y, Chen X, Zhuang J (2009) The 12 May Wenchuan earthquake-induced landslide lakes: distribution and preliminary risk evaluation. Landslides 6(3):209–233

Guzzetti F, Peruccacci S, Rossi M, Stark CP (2008) The rainfall intensity-duration control of shallow landslides and debris flows: an update. Landslides 5(1):3–17

Huang D, Cen D, Ma G, Huang R (2015) Step-path failure of rock slopes with intermittent joints. Landslides 12(5):911–926

Hungr O, Leroueil S, Picarelli L (2014) The Varnes classification of landslide types, an update. Landslides 11(2):167–194

Kavzoglu T, Sahin EK, Colkesen I (2014) Landslide susceptibility mapping using GIS-based multi-criteria decision analysis, support vector machines, and logistic regression. Landslides 11(3):425–439

Lee S, Pradhan B (2007) Landslide hazard mapping at Selangor, Malaysia using frequency ratio and logistic regression models. Landslides 4(1):33–41

Leynaud D, Sultan N, Mienert J (2007) The role of sedimentation rate and permeability in the slope stability of the formerly glaciated Norwegian continental margin: the Storegga slide model. Landslides 4(4):297–309

Lundström K, Larsson R, Dahlin T (2009) Mapping of quick clay formations using geotechnical and geophysical methods. Landslides 6(1):1–15

Margottini C (2004) Instability and geotechnical problems of the Buddha niches and surrounding cliff in Bamiyan Valley, central Afghanistan. Landslides 1(1):41–51

Massey CI, Manville V, Hancox GH, Keys H, Lawrence C, McSaveney M (2010) Out-burst flood (lahar) triggered by retrogressive landsliding, 18 March 2007 at Mt Ruapehu, New Zealand—a successful early warning. Landslides 7(3):303–315

Mazzanti P, Bozzano F, Cipriani I, Prestininzi A (2015) New insights into the temporal prediction of landslides by a terrestrial SAR interferometry monitoring case study. Landslides 12(1):55–68

Nadim F, Kjekstad O, Peduzzi P, Herold C, Jaedicke C (2006) Global landslide and avalanche hotspots. Landslides 3(2):159–173

Pinyol NM, Alonso EE, Corominas J, Moya J (2012) Canelles landslide: modelling rapid drawdown and fast potential sliding. Landslides 9(1):33–51

Pradhan B, Lee S (2010) Regional landslide susceptibility analysis using backpropagation neural network model at Cameron Highland, Malaysia. Landslides 7(1):13–30

Prochaska AB, Santi PM, Higgins JD, Cannon SH (2008) A study of methods to estimate debris flow velocity. Landslides 5(4):413–444

Saha AK, Gupta RP, Sarkar I, Arora MK, Csaplovics E (2005) An approach for GIS-based statistical landslide susceptibility zonation-with a case study in the Himalayas. Landslides 2(1):61–69

Sassa K (2004) The international programme on landslides (IPL). Landslides 1:95–99

Sassa K (2006) "2006 Tokyo Action Plan"—strengthening research and learning on landslides and related earth system disasters for global risk preparedness. Landslides 3:361–369

Sassa K, Tsuchiya S, Ugai K, Wakai A, Uchimura T (2009) Landslides: a review of achievements in the first 5 years (2004–2009). Landslides 6(4):275–286

Sassa K, Tsuchiya S, Fukuoka H, Mikoš M, Doan L (2015) Landslides: review of achievements in the second 5-year period (2009–2013). Landslides 12(2):213–223

Schwarz J, Mannsperger C (2016) Landslides—2015 Publisher's Report. Springer Nature, Heidelberg. 48p

Segoni S, Lagomarsino D, Fanti R, Moretti S, Casagli N (2015) Integration of rainfall thresholds and susceptibility maps in the Emilia Romagna (Italy) regional-scale landslide warning system. Landslides 12(4):773–785

Sosio R, Crosta GB, Hungr O (2012) Numerical modeling of debris avalanche propagation from collapse of volcanic edifices. Landslides 9(3):315–334

Staley DM, Kean JW, Cannon SH, Schmidt KM, Laber JL (2013) Objective definition of rainfall intensity—duration thresholds for the initiation of post-fire debris flows in southern California. Landslides 10(5):547–562

Strozzi T, Farina P, Corsini A, Ambrosi C, Thüring M, Zilger J, Wiesmann A, Wegmüller U, Werner C (2005) Survey and monitoring of landslide displacements by means of L-band satellite SAR interferometry. Landslides 2(3):193–201

Tien Bui D, Tuan T, Klempe H, Pradhan B, Revhaug I (2016) Spatial prediction models for shallow landslide hazards: a comparative assessment of the efficacy of support vector machines, artificial neural networks, kernel logistic regression, and logistic model tree. Landslides 13(2):361–378

Xu C, Xu X, Yao X, Dai F (2014) Three (nearly) complete inventories of landslides triggered by the May 12, 2008 Wenchuan Mw 7.9 earthquake of China and their spatial distribution statistical analysis. Landslides 11(3):441–461

Yin Y, Wang F, Sun P (2009) Landslide hazards triggered by the 2008 Wenchuan Earthquake, Sichuan, China. Landslides 6(2):139–152

Advanced Technologies for Landslides (WCoE 2014–2017, IPL-196, IPL-198)

Nicola Casagli, Veronica Tofani, Filippo Catani, Sandro Moretti, Riccardo Fanti, and Giovanni Gigli

Abstract

The Earth Sciences Department of the University of Firenze (DST-UNIFI) since 2002 has been a member of the International Consortium on Landslides (ICL) and three times it has been awarded status as a World Centre of Excellence (WCoE) for Landslide Risk Reduction (2008–2010, 2011–2013, 2014–2016). Since 2016, DST-UNIFI has established a UNESCO Chair on Prevention and sustainable management of geo-hydrological hazards. In this paper we describe the activities carried out by DST-UNIFI as a member of ICL and as WCoE in the framework of landslide risk reduction, landslide prevention and management.

Keywords

Italy • Landslides • Monitoring • Remote sensing • Landslide prediction • Multi-sensors drone

Introduction

The Earth Sciences Department of the University of Firenze (DST-UNIFI) has, since 2002, been a member of the International Consortium on Landslides (ICL) and it has been awarded status as a World Centre of Excellence (WCoE) for Landslide Risk Reduction three times (2008–2010, 2011–2013, 2014–2016).

N. Casagli (✉) · V. Tofani · F. Catani · S. Moretti · R. Fanti · G. Gigli
Department of Earth Sciences, University of Firenze, Via La Pira 4, 50212 Florence, Italy
e-mail: nicola.casagli@unifi.it

V. Tofani
e-mail: veronica.tofani@unifi.it

F. Catani
e-mail: filippo.catani@unifi.it

S. Moretti
e-mail: sandro.moretti@unifi.it

R. Fanti
e-mail: riccardo.fanti@unifi.it

G. Gigli
e-mail: giovanni.gigli@unifi.it

K. Sassa et al. (eds.), *Advancing Culture of Living with Landslides*,
DOI 10.1007/978-3-319-59469-9_22

DST-UNIFI is one of the largest centres for scientific and technological services on geohazards in Italy, and currently comprises 46 full-time employees. The group participates in research and technological development projects in several areas of the world, often in active collaboration with international, national and regional organizations and agencies. The main objective of the group is to focus on landslide studies at all scales, with an emphasis that in recent years has moved towards the application and development of new technologies for landslide disaster prevention, monitoring and early warning, with a special emphasis on remote sensing and regional landslide forecasting models.

In June 2016, the DST-UNIFI successfully established an UNESCO Chair on prevention and sustainable management of geo-hydrological hazards. Since 2002 DST-UNIFI has coordinated or has been involved in several ICL/IPL projects.

In this paper we describe the research activities carried out by DST-UNIFI as WCoE (2014–2017) and the research outcomes of two ongoing ICL/IPL projects (IPL196: Development and applications of a multi-sensors drone for geohazards monitoring and mapping, IPL198: Multi-scale rainfall triggering models for Early Warning of Landslides (MUSE)) led by DST-UNIFI.

WCoE (2014–2017) Activities

Research Activities

The objective of the WCoE has been to develop new methodologies and advanced technologies for landslide risk reduction. DST-UNIFI has carried out research and development for the prevention and management of landslides, in order to support policies and actions of risk reduction. In particular, research activity has focused on the development of advanced methodologies useful for applications to landslides as such and has been structured in the following three work packages (WPs):

- WP_1: Ground-based SAR interferometry for landslide monitoring and development of reliable procedures and technologies for early warning.
- WP_2: EO (Earth Observation) data and technology to detect, map, monitor and forecast ground deformations.
- WP_3: Coupling of short-term weather forecasting with geotechnical modeling for shallow landslide prediction.

Concerning WP_1 the DST-UNIFI performed monitoring activities of unstable slopes in order to estimate the deformational evolution of the landslide events (in space and time) and to implement the most suitable operational early warning systems (EWS) based on different critical situations (Carlà et al. 2016; Agostini et al. 2014; Bardi et al. 2014; Intrieri et al. 2013). Several sites have been monitored in Italy (Fig. 1 reports the example of the Calatabiano landslide, which occurred in November 2016), all of them were monitored using advanced sensors and portable instruments such as ground-based synthetic aperture radar interferometer (GB-InSAR), terrestrial laser scanning (LIDAR), satellite interferometry (PS-InSAR), GPS antennas and infrared thermography and traditional instrumentation (e.g., strain gauges, inclinometers, piezometers) (Fig. 1) (Nocentini et al. 2015; Margottini et al. 2015; Di Traglia 2014a, b; Gigli et al. 2014a, b).

Each time the best-integrated uses and synergetic approaches were evaluated. Using all these monitoring activities, the DST-UNIFI performed and validated a rapid procedure (called Landslide Triage) for the unambiguous assessment of landslide risk through a specific evaluation sheet, which considers all the hazardous conditions associated with landslides.

The goal of this procedure is to assign, as quickly and objectively as possible, a level of criticality, which allows speeding up the selection of actions to take.

In WP_2 the research was focused on the development of a satellite surveillance system, exploiting satellite data for the identification, mapping, monitoring and analysis of risk scenarios associated with landslides, from local to regional scale (Tofani et al. 2017; Bianchini et al. 2014; Lu et al. 2014). Its implementation was carried out by:

- The application of satellite radar interferometry in monitoring and emergency support to areas affected by hydro-geologic instabilities threatening or involving the population and its activities;
- The creation of geodatabases for each critical area, arranged to contain all the available data (both optical and radar) and new specific calculations in a homogenized framework;
- The integration of susceptibility models and interferometric data acquired by the satellite platform;
- Stitching (merged analysis) between RADARSAT-1 and RADARSAT-2 images in order to ensure continuity of data monitoring.

Through these issues we tried to use satellite data to increase the number, quality and accuracy of products that can be readily utilized to reduce risk in areas affected by slope instability. Some of these products are: (i) ground deformation velocity maps, (ii) damage assessments, (iii) susceptibility and (iv) risk zonation maps (Ciampalini et al. 2016a).

In particular, the development of a geodatabase dedicated to landslides allowed us to include different types of data such as pre-existing inventories (ancillary records), auxiliary data (e.g., thematic maps, DEMs, VHR optical photos) and new remote sensing SAR (Synthetic Aperture Radar) data (e.g. from COSMO-SkyMed constellation) and subsequently to properly link them. This makes possible the generation of high quality map products at various levels of specificity (Bianchini et al. 2016; Ciampalini et al. 2016b; Raspini et al. 2015; Tofani et al. 2014; Del Ventisette et al. 2014).

On the other hand, using remotely sensed data acquired through different SAR satellites (e.g., ERS1/2, RADARSAT1/2, ENVISAT, COSMO-SkyMed), the slope deformations of the study areas were monitored in order to promptly undertake alert procedures and to address suitable countermeasures in case of high criticalities. The long time series of displacement provided by the Persistent Scatterer SAR Interferometry (PSInSAR) technique allowed us to define the most critical unstable sectors, to update the extension of landsliding phenomena and to redefine their state of activity (Fig. 2). This was possible by comparing the ground deformation velocities measured at different times over the same target area. Usability of remote sensing data, comprising satellite, optical and ground-based data (e.g.,

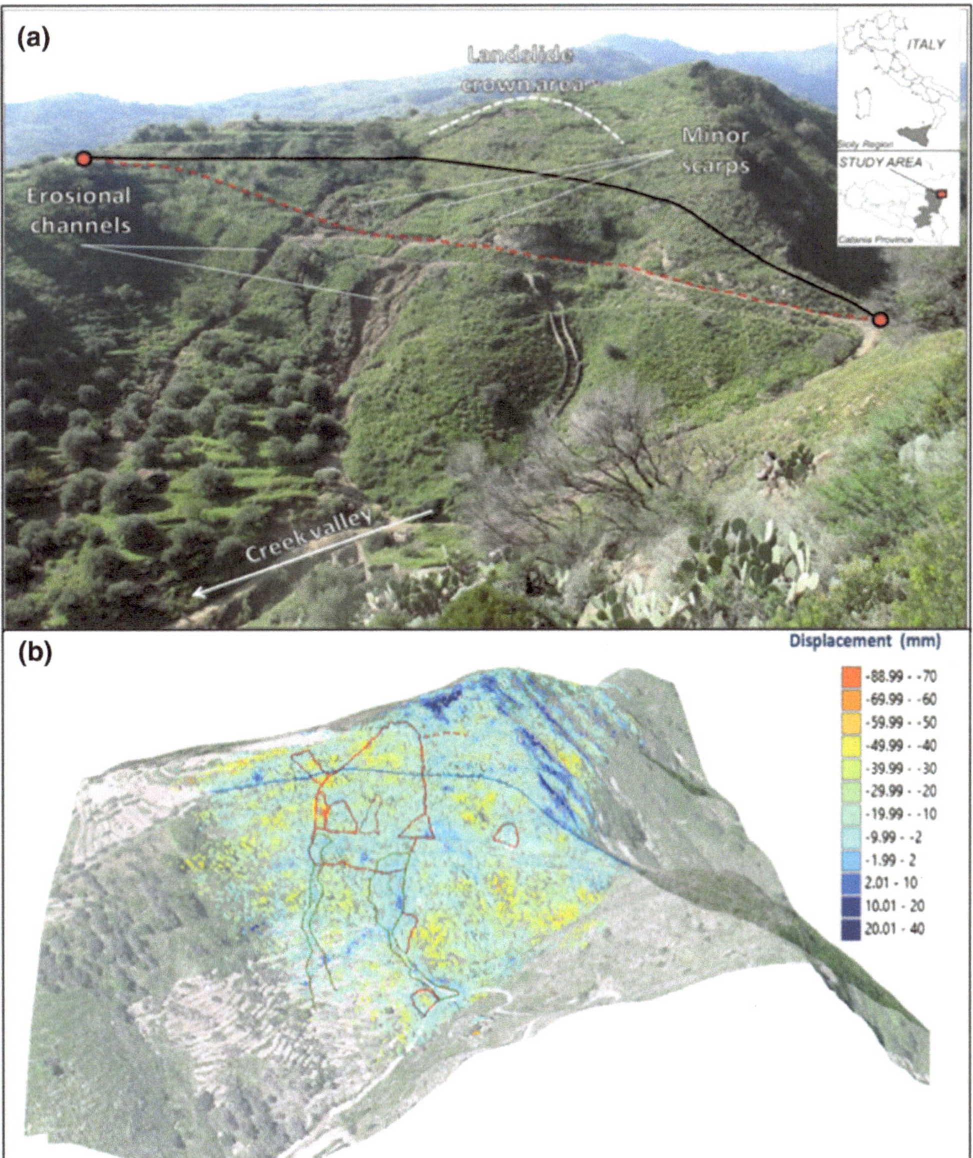

Fig. 1 Calatabiano landslide occurred in October 2016 in Soukhern Italy and was caused by a damaged water pipe. An integrated monitoring system has been installed just after the event. **a** study area and landslide. **b** Displacements map derived from GB-Insar measurements from 11/11/2015 to 31/12/2015. (modified from Lombardi et al. 2017)

GB-InSAR) was also exploited during all the phases of the emergency management cycle: prevention, crisis, and post-crisis (Raspini et al. 2014).

In WP_3 the research activities include the development of a physically-based model, called HIRESSS (Rossi et al. 2013; Mercogliano et al. 2013), for the real time prediction of shallow landslides induced by rain on large areas, predictive statistical models (SIGMA and MACUMBA models) (Martelloni et al. 2012; Lagomarsino et al. 2013; Segoni et al. 2014a, b; Rosi et al. 2016) based on the identification of rainfall thresholds for triggering landslides, and a nowcasting system which includes a forecasting and alert system valid at the national level for landslides with fast kinematics (not monitorable by satellite).

During the last year, the system of collection, storage and querying of meteorological data was implemented with the creation of an efficient database. The two statistical models have been extended to the whole national territory in order to find a peculiar rainfall threshold for the landslides initiation in each alert zone (137) outlined by the Italian Civil Protection. In particular, the thresholds calculated by the software MACUMBA were essentially improved by applying the new rainfall collecting platform created, then they were subjected to a rigid manual quality control (Segoni et al. 2014c).

The integrated nowcasting system is mainly formed by two levels of forecasting: one is managed by the MACUMBA model in its real-time and one is derived from the output of the HIRESSS physically based model (Fig. 3).

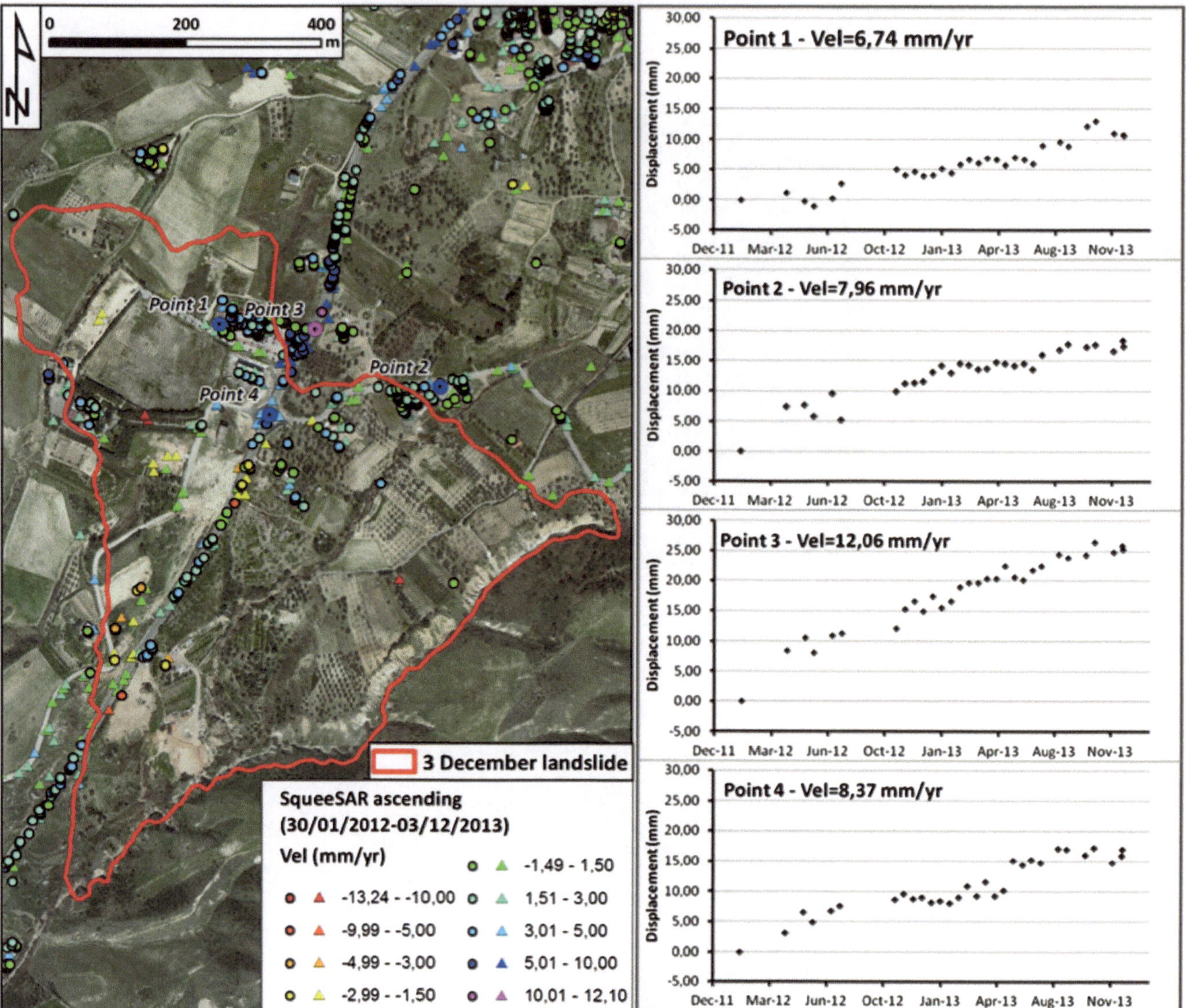

Fig. 2 The Montescaglioso landslides (Basilicata Region, southern Italy) occurred on 3 December 2013, caused by prolonged and intense rainfalls and rapid slope failure, and producing a ground displacement of several meters. In the figure is an analysis of the pre-event SqueeSAR data (Raspini et al. 2015)

Capacity Development and International Cooperation

In March 2015, the DST-UNIFI successfully applied for an UNESCO Chair on prevention and sustainable management of geo-hydrological hazards (www.unesco-geohazards.unifi.it).

The mission of the Chair is to promote research and development for the prevention and management of geo-hydrological hazards, in order to support policies and actions of risk reduction.

In particular, the Chair aims at the implementation of the Sendai Partnership 2015B2025, launched at the World Conference on Disaster Risk Reduction (WCDRR) in Sendai by the International Strategy for Disaster Reduction (ISDR) and by the International Consortium on Landslides (ICL), for global promotion of understanding and reducing landslide disaster risk, which was also signed by UNESCO and the Italian Government, among other partners and UN organizations. Applied research for a Safer Society will be the main keyword of the Chair.

The specific objectives of the Chair are:

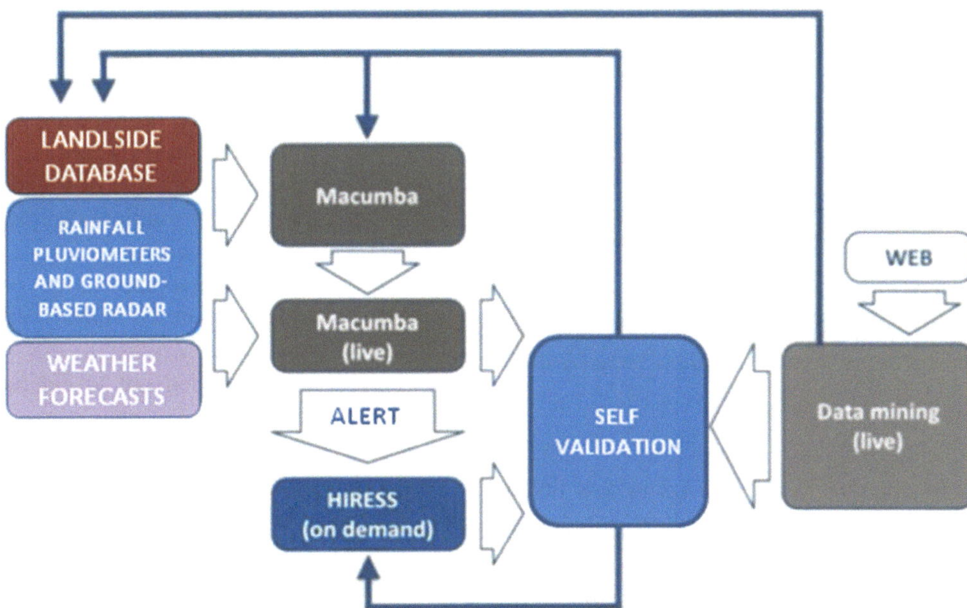

Fig. 3 Block diagram illustrating the mode operation of the integrated nowcast system

(1) to promote the development of innovative technologies for the prevention and mitigation of geo-hydrological hazards;
(2) to develop tools and procedures for supporting risk reduction policies and emergency management for the safety of human life;
(3) to promote the protection of cultural heritage threatened by geo-hydrological hazards;
(4) to promote research at an international level by offering scientific facilities to postgraduate students and visiting researchers.

Several types of activities will be carried out in the framework of the Chair, such as a postgraduate teaching programme, short-term training, research, a visiting professorship, scholarships, and institutional development. The target beneficiaries will be students, professionals, and governmental organizations.

ICL/IPL Projects Research Activities

DST-UNIFI is currently involved into two IPL projects, namely:

- IPL 196: Development and applications of a multi-sensors drone for geohazards monitoring and mapping
- IPL198: Multi-scale rainfall triggering models for Early Warning of Landslides (MUSE).

Development and Applications of a Multi-sensors Drone for Geohazards Monitoring and Mapping (IPL 196)

The objective of the proposed project is to test the applicability of a multi-sensors drone to the mapping and monitoring of geohazards. The project has two specific objectives: (i) development of the drone, sensors, safety and automation and (ii) application of the drone as a platform for integrated sensors (multispectral sensor, visible light camera, infrared camera and LIDAR) for the mapping and monitoring of geohazards.

The project activities are related to the two work packages:

WP_1: Development of the multicopter drone, sensors, safety and automation.
WP_2: Application of the multi-sensors drone for rapid mapping, 3D surface reconstruction, monitoring.

In WP_1 the activities have concerned the research and development of the chassis and the flight model to improve flight range and transportable instruments weight limits. Recently a new chassis drone (SATURN) has been designed, after over 70 h of flight of the first prototype. This new chassis is stiffer and more resistant, while maintaining the same weight as the previous one, ensuring more stability and precision during flight, even with heavier instruments (Fig. 4).

Fig. 4 SATURN drone

In order to avoid damage or loss of costly instruments after a possible electrical failure, a redundant safety system has been developed. This system includes redundant flight electronics, and in particular we focused on the reliability of the power supply and the GPS signal.

The multicopter drone has been equipped with a visible light and a multispectral NIR camera, while a very light hyperspectral sensor (visible and near infrared) has been specifically designed to be installed on the drone and is currently under development.

WP_2 has mainly focused on the application of multi-sensor drones to selected case studies and on the validation of the multi-sensor drone acquisitions with data from ground-based instrumentation for selected test sites. In particular the activity has dealt with the application of the multicopter drone equipped with visible light camera for mapping and 3D surface reconstruction of landslides. 3D surface reconstruction with optical sensors has been carried out during multiple campaigns in order to define the volume variation in time.

The first application of the drone is in the village of Ricasoli, in Tuscany (Italy), which is strongly affected by landslide activity. In this site three survey campaigns have been carried out to determine the capability of this rising technology to characterize and to monitor landslides. High temporal frequency DEMs have been derived by the DST drone in Ricasoli village. In order to obtain a detailed reconstruction of the topography of the site, three aerial photogrammetric survey were conducted on July 2015, March 2016 and April 2016, the second and the third ones being conducted after two landslide events.

The DEMs derived from the three photogrammetric surveys were analyzed and compared with each other in order to detect any possible area affected by slope deforming processes. The volume mobilized among the three acquisitions was also computed (Fig. 5). Further information on the analysis and comparison of DEMs derived from the drone surveys of Ricasoli landslides can be found in Tanteri et al. (2017).

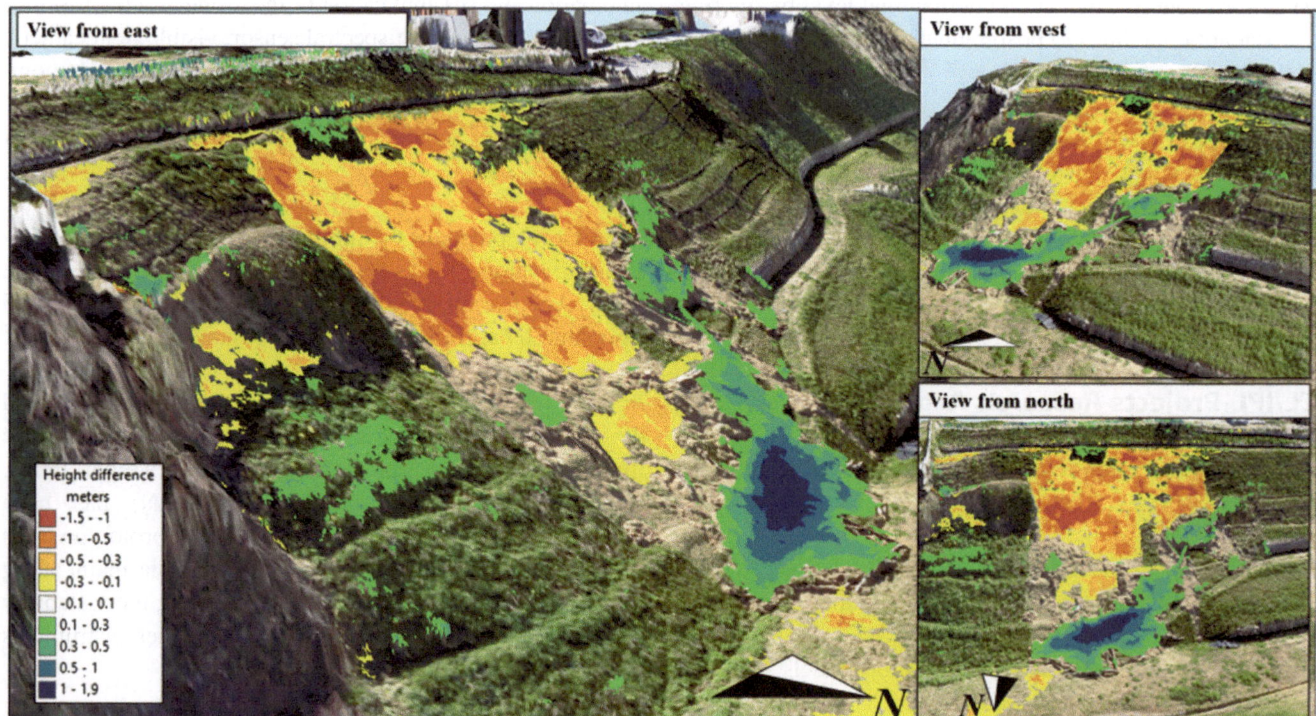

Fig. 5 Elevation difference between the second (March 2016) and first (July 2015) drone photogrammetric surveys in Ricasoli village

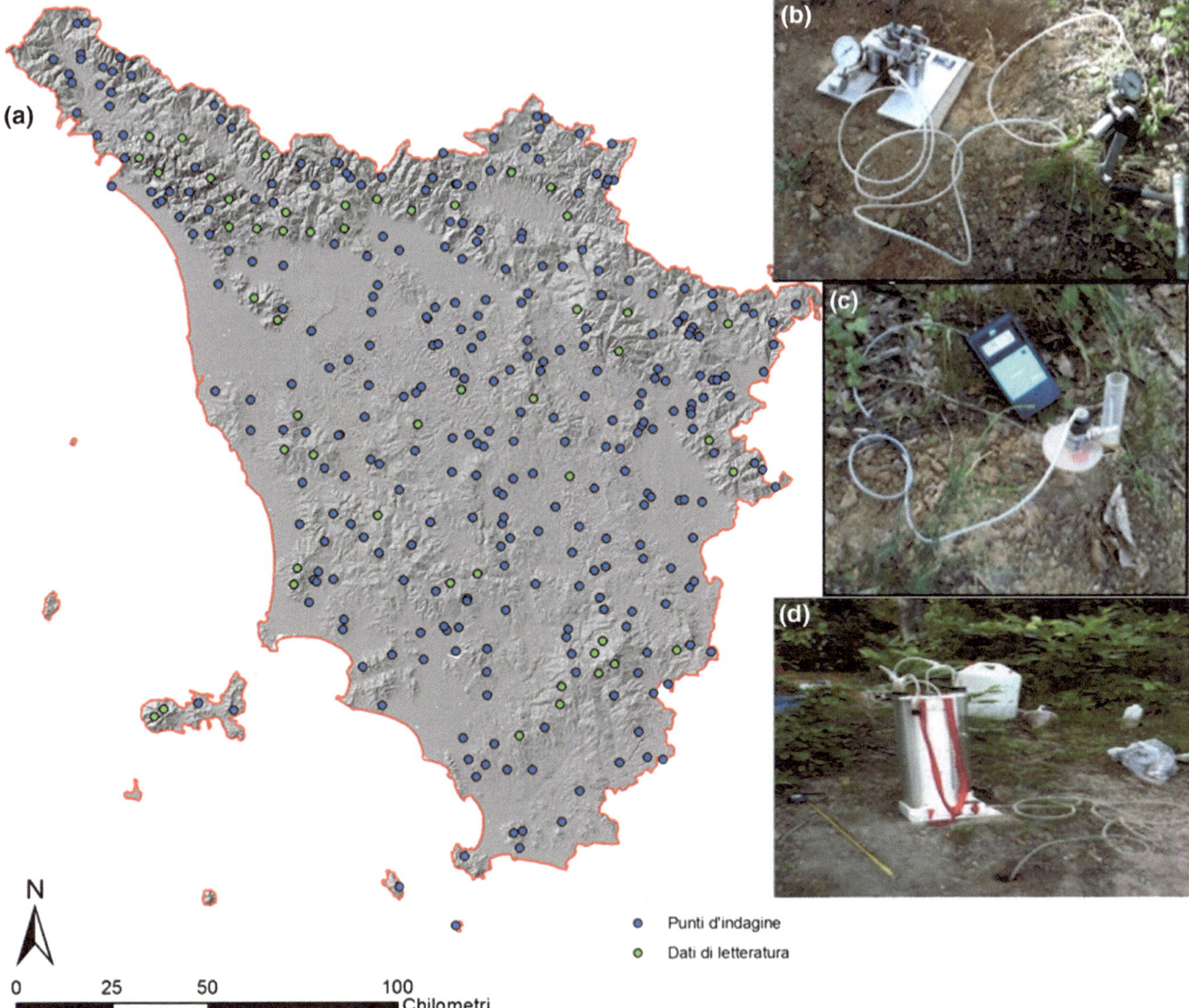

Fig. 6 **a** Distribution of survey points (*blue* and *green*) in Tuscany (Italy), **b** Borehole Shear Test elevation difference, **c** Tensiometer, **d** constant-head well permeameter Amoozemeter

Multi-scale Rainfall Triggering Models for Early Warning of Landslides (MUSE) (IPL 198)

The main objective of this IPL project is the enhancement of knowledge and methods to integrate landslide prediction models at different scales to build an effective operational multi-scale system for real-time early warning of rainfall-triggered mass movements. The project is structured into two work packages:

WP_1: Soil properties variability study.
WP_2: Analysis and integration of rainfall data and rainfall forecasts.

In WP_1 the activity has focused on the characterization of geotechnical properties of the soil cover of the hillslopes in Tuscany (Central Italy) in order to improve the reliability of deterministic model output such as the HIRESSS (HIgh REsolution Slope Stability Simulator; Rossi et al. 2013). This software makes use of a Monte Carlo simulation technique to manage the uncertainty typical of geotechnical and hydrological input parameters, which is a common weakness of deterministic models. Uncertainty in the geotechnical input parameter values can be solved by using a frequency distribution model for the parameters. This helps to restrict the number of in situ and laboratory measurements and is crucial for reducing computation time within

HIRESSS and other similar distributed numerical models at the basin scale (Fig. 6).

In the Tuscany region (22,994 km^2) 130 survey points were selected. The following in situ measurements were performed at each survey point (Fig. 6): (i) shear strength under natural conditions by means of a Borehole Shear Test (Lutenegger and Halberg 1981) jointly with the (ii) matric suction obtained with a tensiometer, (iii) saturated hydraulic conductivity (ksat) by means of a constant-head well permeameter Amoozemeter (Amoozegar 1989) and (iv) sampling of two aliquots ($\sim$2 kg each) of the material for laboratory tests. The Borehole Shear Test allows the measure of shear strength parameters under natural in situ conditions.

In addition to the in situ measures, the grain size distribution and the relationships of phases (porosity, dry unit weight γ_d) are determined in the laboratory following the ASTM (American Society for Testing and Materials) recommendations (ASTM D422-63 2007, ASTM D2217-85 1998 and ASTM D-4318 2010).

The geotechnical parameters measured have been statistically analyzed and linked to the main lithologies of the bedrock in order to define the input hydrological and geotechnical parameters for the HIRESSS model. The HIRESSS code has been applied in a selected study area located in Northern Tuscany in order to test the performance of the models with the measured geotechnical and hydrological parameters. Further applications will be carried out to test the sensitivity of the model to the variation of the model results (Tofani et al. in review).

In WP_2 the system of collection, storage and querying of meteorological data was implemented with the creation of an efficient database that allows the storage of both historical records and real-time data coming from the National Radar Network and also from the rainfall network belonging to the National Civil Protection. In this way, new data can be continuously added to the already stored files in the same format through the same indexing procedures. Therefore, this reference platform was necessary to transfer the physically-based model (HIRESSS) from first an experimental scientific software to an operative system for real-time applications at a regional scale.

References

Agostini A, Tofani V, Nolesini T, Gigli G, Tanteri L, Rosi A, Cardellini S, Casagli N (2014) A new appraisal of the Ancona landslide based on geotechnical investigations and stability modelling. Q J Eng GeolHydrogeol 47(1):29–43. doi:10.1144/qjegh2013-028

Amoozegar A (1989) Compact constant head permeameter for measuring saturated hydraulic conductivity of the vadose zone. Soil Sci Soc Am J 53:1356–1361

Bardi F, Frodella W, Ciampalini A, Bianchini S, Del Ventisette C, Gigli G, Fanti R, Moretti S, Basile G, Casagli N (2014) Integration between ground based and satellite SAR data in landslide mapping: The San Fratello case study. Geomorphology 223:45–60. doi:10.1016/j.geomorph.2014.06.025

Bianchini S, Ciampalini A, Raspini F, Bardi F, Di Traglia F, Moretti S, Casagli N (2014) Multi-temporal evaluation of landslide movements and impacts on buildings in San Fratello (Italy) by means of C-band and X-band PSI data. Pure appl Geophys 172(11):3043–3065. doi:10.1007/s00024-014-0839-2

Bianchini S, Raspini F, Ciampalini A, Lagomarsino D, Bianchi M, Bellotti F, Casagli N (2016) Mapping landslide phenomena in landlocked developing countries by means of satellite remote sensing data: the case of Dilijan (Armenia) area. Geomatics, Natural Hazards and Risk, pp 1–17

Carlà T, Intrieri E, Di Traglia F, Nolesini T, Gigli G, Casagli N (2016) Guidelines on the use of inverse velocity method as a tool for setting alarm thresholds and forecasting landslides and structure collapses. Landslides. doi:10.1007/s10346-016-0731-5

Ciampalini A, Raspini F, Lagomarsino D, Catani F, Casagli N (2016a) Landslide susceptibility map refinement using PSInSAR data. Remote Sens Environ 184:302–315

Ciampalini A, Raspini F, Frodella W, Bardi F, Bianchini S, Moretti S (2016b) The effectiveness of high-resolution LiDAR data combined with PSInSAR data in landslide study. Landslides 13(2):399–410. doi:10.1007/s10346-015-0663-5

Del Ventisette C, Righini G, Moretti S, Casagli N (2014) Multitemporal landslides inventory map updating using spaceborne SAR analysis. Int J Appl Earth Obs Geoinf 30(1):238–246. doi:10.1016/j.jag.2014.02.008

Di Traglia F, Cauchie L, Casagli N, Saccorotti G (2014a) Decrypting geophysical signals at Stromboli Volcano (Italy): Integration of seismic and Ground-Based InSAR displacement data. Geophys Res Lett 41(8):2753–2761. doi:10.1002/2014GL059824

Di Traglia F, Intrieri E, Nolesini T, Bardi F, Del Ventisette C, Ferrigno F, Frangioni S, Frodella W, Gigli G, Lotti A, Tacconi Stefanelli C, Tanteri L, Leva D, Casagli N (2014b) The ground-based InSAR monitoring system at Stromboli volcano: linking changes in displacement rate and intensity of persistent volcanic activity. Bulletin of Volcanology, vol 76(2). doi:10.1007/s00445-013-0786-2

Intrieri E, Gigli G, Casagli N, Nadim F (2013) Brief communication landslide early warning system: toolbox and general concepts. Nat Hazards Earth Syst Sci 13(1):85–90. doi:10.5194/nhess-13-85-2013

Gigli G, Frodella W, Garfagnoli F, Morelli S, Mugnai F, Menna F, Casagli N (2014a) 3-D geomechanical rock mass characterization for the evaluation of rockslide susceptibility scenarios. Landslides 11(1):131–140. doi:10.1007/s10346-013-0424-2

Gigli G, Morelli S, Fornera S, Casagli N (2014b) Terrestrial laser scanner and geomechanical surveys for the rapid evaluation of rock fall susceptibility scenarios. Landslides, vol 11(1). doi:10.1007/s10346-012-0374-0

Lagomarsino D, Segoni S, Fanti R, Catani F (2013) Updating and tuning a regional-scale landslide early warning system. Landslides 10(1):91–97. doi:10.1007/s10346-012-0376-y

Lombardi L, Nocentini M, Frodella W, Nolesini T, Bardi F, Intrieri E, Carlà T, Solari L, Dotta G, Federica Ferrigno F, Casagli N (2017) The Calatabiano landslide (Southern Italy): preliminary GB-InSAR monitoring data and remote 3D mapping. Landslides J 14(2):685–696

Lu P, Catani F, Tofani V, Casagli N (2014) Quantitative hazard and risk assessment for slow-moving landslides from Persistent Scatterer Interferometry. Landslides 11(4):685–696. doi:10.1007/s10346-013-0432-2

Lutenegger AJ, Halberg GR (1981) Borehole shear test in geotechnical investigation. Special technical Publ. Am Soc Test Mater 740:566–578

Margottini C, Antidze N, Corominas J, Crosta GB, Frattini P, Gigli G, Giordan D, Iwasaky I, Lollino G, Manconi A, Marinos P, Scavia C, Sonnessa A, Spizzichino D, Vacheishvili N (2015) Landslide hazard, monitoring and conservation strategy for the safeguard of

Vardzia Byzantine monastery complex. Georgia Landslides 12 (1):193–204
Martelloni G, Segoni S, Fanti R, Catani F (2012) Rainfall thresholds for the forecasting of landslide occurrence at regional scale. Landslides 9(4):485–495. doi:10.1007/s10346-011-0308-28-2
Mercogliano P, Segoni S, Rossi G, Sikorsky B, Tofani V, Schiano P, Catani F, Casagli N (2013) Brief communication "A prototype forecasting chain for rainfall induced shallow landslides". Nat Hazards Earth Syst Scie 13(3):771–777. doi:10.5194/nhess-13-771-2013
Nocentini M, Tofani V, Gigli G, Fidolini F, Casagli N (2015) Modeling debris flows in volcanic terrains for hazard mapping: the case study of Ischia Island (Italy). Landslides. doi:10.1007/s10346-014-0524-7
Raspini F, Moretti S, Fumagalli A, Rucci A, Novali F, Ferretti A, Prati C, Casagli N (2014) The COSMO-SkyMed constellation monitors the Costa Concordia wreck. Remote Sens 6(5):3988–4002. doi:10.3390/rs6053988
Raspini F, Ciampalini A, Del Conte S, Lombardi L, Nocentini M, Gigli G, Casagli N (2015) Exploitation of amplitude and phase of satellite SAR images for landslide mapping: the case of Montescaglioso (South Italy). Remote Sens 7(11):14576–14596
Rosi A, Peternel T, Jemec-Auflič M, Komak M, Segoni S, Casagli N (2016) Rainfall thresholds for rainfall-induced landslides in Slovenia. Landslides. doi:10.1007/s10346-016-0733-3
Rossi G, Catani F, Leoni L, Segoni S, Tofani V (2013) HIRESSS: a physically based slope stability simulator for HPC applications. Nat Hazards Earth Syst Sci 13(1):151–166. doi:10.5194/nhess-13-151-2013
Segoni S, Rossi G, Rosi A, Catani F (2014a) Landslides triggered by rainfall: a semi-automated procedure to define consistent intensity–duration thresholds. Comput Geosci 63:123–131. doi:10.1016/j.cageo.2013.10.009
Segoni S, Battistini A, Rossi G, Rosi A, Lagomarsino D, Catani F, Moretti S, Casagli N (2014b) Technical Note: An operational landslide early warning system at regional scale based on space-time variable rainfall thresholds. Nat Hazards Earth Syst Sci Discuss 2:6599–6622. doi:10.5194/nhessd-2-6599-2014
Segoni S, Rosi A, Rossi G, Catani F, Casagli N (2014c) Analysing the relationship between rainfalls and landslides to define a mosaic of triggering thresholds for regional scale warning systems. Nat Hazards Earth Syst Sci Discuss 2:2185–2213. doi:10.5194/nhessd-2-2185-2014
Tanteri L, Rossi G, Tofani V, Vannocci P, Moretti S, Casagli N (2017a) Multitemporal UAV survey for mass movement detection and monitoring. In: Proceedings of Fourth World Landslide Forum, Ljubljana, Slovenia, May 29–June 2, 2017
Tofani V, Del Ventisette C, Moretti S, Casagli N (2014) Integration of remote sensing techniques for intensity zonation within a landslide area: a case study in the Northern Apennines, Italy. Remote Sens 6 (2):907–924. doi:10.3390/rs6020907
Tofani V, Bicocchi G, Rossi G, Segoni S, D'Ambrosio M, Catani F, Casagli N (2017) Soil characterization for shallow landslides modeling: a case study in the Northern Apennines (Central Italy). Landslides J 14(2):755–770

Mechanisms of Landslides and Creep in Over-Consolidated Clays and Flysch (WCoE 2014–2017)

Matjaž Mikoš, Janko Logar, Matej Maček, Jošt Sodnik, and Ana Petkovšek

Abstract

The Faculty of Civil and Geodetic Engineering of the University of Ljubljana (UL FGG), Slovenia, Europe, was voted in 2014 at the 3rd World Landslide Forum in Beijing, China to be one of the 15 new World Centres of Excellence (WCoE) in Landslide Disaster Reduction for the period 2014–2017. This successful nomination followed the period 2011–2014, in which UL FGG successfully fulfilled the role as one of the WCoEs for the second time. The title of the activities of the WCoE in this third term was slightly modified to be "Mechanisms of Landslides and Creep in Over-Consolidated Clays and Flysch". We can divide the activities of the WCoE at UL FGG into international and national research activities. The international ones consisted of the ICL related activities with the main task of being the main organizer of this 4th World Landslide Forum 2017, international cooperation, European research activities, and bilateral cooperation. The national ones consisted of the national projects and the national research program "Water Science and Technology, and Geotechnics". In the paper, these activities of the WCoE at UL FGG are elaborated in more detail, with a comprehensive list of publications to show the dissemination and capacity building efforts.

Keywords

Creep • Debris flows • Flysch • Laboratory tests • Landslides • Slovenia • Suction • WCoE

Introduction

The Faculty of Civil and Geodetic Engineering of the University of Ljubljana (UL FGG), Slovenia, Europe, was elected at the 3rd World Landslide Forum in Beijing, China to be one of the 15 new World Centres of Excellence (WCoE) in Landslide Disaster Reduction for the period 2014 to 2017. The title of the activities of the WCoE at UL FGG was slightly changed from the WCoE 2011–2014 to a new title "Mechanisms of landslides and creep in over-consolidated clays and flysch".

In this paper we present the WCoE research activities in the domain of international and national activities, respectively.

M. Mikoš (✉) · J. Logar · M. Maček · J. Sodnik · A. Petkovšek
Faculty of Civil and Geodetic Engineering, University of Ljubljana, Jamova C. 2, 1000 Ljubljana, Slovenia
e-mail: matjaz.mikos@fgg.uni-lj.si

J. Logar
e-mail: janko.logar@fgg.uni-lj.si

M. Maček
e-mail: matej.macek@fgg.uni-lj.si

J. Sodnik
e-mail: jost.sodnik@fgg.uni-lj.si

A. Petkovšek
e-mail: ana.petkovsek@fgg.uni-lj.si

K. Sassa et al. (eds.), *Advancing Culture of Living with Landslides*,
DOI 10.1007/978-3-319-59469-9_23

International Research Activities

ICL-Related Activities

Among the ICL/IPL related activities of the WCoE at UL FGG we can name:

- Cooperation in the ICL regional "Adriatic-Balkan Network", where we actively supported the 2nd Regional Symposium on Landslides in the Adriatic-Balkan Region (2nd ReSyLAB; http://resylab2015.rgf.rs/) in May 2015 in Belgrade, Serbia, and
- Cooperation in the ICL thematic network "Landslide Monitoring and Warning Thematic Network—LaMaWaTheN", in which we were one of the initiators of this network (Maček et al. 2014).
- We intend to co-host, together with the Geological Survey of Slovenia, the 3rd Regional Symposium on Landslides in the Adriatic-Balkan Region (3rd ReSyLAB) in September/October 2017 in Ljubljana, Slovenia. The initial idea to hold this symposium as a WLF4 side event was abandoned in favor of a separate event with post-conference proceedings.
- We support the ICL activities also by serving in ICL/IPL working bodies, i.e. as the elected ICL Vice President (Mikoš 2015–2017), and chairing the IPL Evaluation Committee: (http://icl.iplhq.org/category/icl/structure-and-officers/).
- Our different experiences in the field of landslide disaster risk reduction have been submitted as text teaching tools (TXT-Tools 2017a, b, c) to a new Springer reference work entitled "Landslide Dynamics: ISDR-ICL Landslide Interactive Teaching Tools".
- The main ICL-related activities were dedicated to the preparation of the 4th World Landslide Forum in 2017 in Ljubljana (www.wlf4.org), for which the University of Ljubljana (jointly Faculty of Civil and Geodetic Engineering and Faculty of Natural Sciences and Technology) is the major organizer in Slovenia (Forum Chair), together with the Geological Survey of Slovenia. At UL FGG we have prepared a 3-day WLF4 Post-Forum Technical (Study) Tour to Slovenia entitled "Living with slope mass movements in Slovenia and its surroundings". We have been personally involved in the preparation of the WLF4 Landslide Photo Contest entitled "Landslides and Mankind", collecting photographs in three categories, and in the WLF4 Student Award Contest. Together with the Minisitry of the Environment and Spatial Planning, we are working on organizing an one-day national workshop as a side event to WLF4 on Wednesday, May 31, 2017.
- We have represented ICL at the General Assembly of IUGG in Prague in June 2015.

International Research Cooperation

We have been actively involved into the 3rd WCDRR in Sendai in March 2015, and have disseminated its results to stakeholders (Mikoš 2015a, c, 2016).

In 2016, we succeeded in establishing a UNESCO Chair on Water-Related Disaster Risk Reduction at the University of Ljubljana (www.unesco-floods.eu). Among the supporting partners were the International Consortium on Landslides (ICL) and one of its members, Niigata University, Japan. The next step is already envisaged, i.e., the establishment of a UNESCO Category 2 Center at the University of Ljubljana.

Rainfall-induced landslides and debris flows in mountainous and hilly areas are seen as parts of sediment–related disasters, i.e., closely associated if not covered by the water-related disasters.

European Research Activities

In 2014, we collaborated in the framework of the European Alpine Space project START_it_up "State-of-the-Art in Risk Management Technology: Implementation and Trial for Usability in Engineering Practice and Policy" (ended in December 2014; project leader S. Rusjan).

As a part of the project deliverables, we have prepared an overview of legislation in Slovenia in the field of landslide hazard and risk assessment (Mikoš et al. 2014).

Using as a theoretical base of how monitoring is connected to perception of natural hazards and to hazard warnings (Fig. 1), and practical experiences gained in installing debris-flow monitoring sites in the Alps (see Fig. 2 for an example), an overview of available debris-flow monitoring techniques has been prepared (Hübl and Mikoš 2014).

Bilateral Research Cooperation

In years 2014 and 2015, we collaborated with the University of Rijeka, Croatia (member of the Croatian Landslide Group, also an ICL member), on a bilateral research project SoLiFlyD "Study of landslides in flysch deposits: sliding mechanisms and geotechnical properties for landslide modeling and landslide mitigation" (Project leader M. Mikoš). Using staff exchange (early stage researchers) and via short visits (see Fig. 3), we compared geotechnical (Maček et al. 2015) and geological aspects (Peternel et al. 2015) of landslides in flysch in Slovenia and the Croatian Istrian Peninsula.

The project goals were to collect, unify and exchange data and knowledge in the field of landslide investigation in

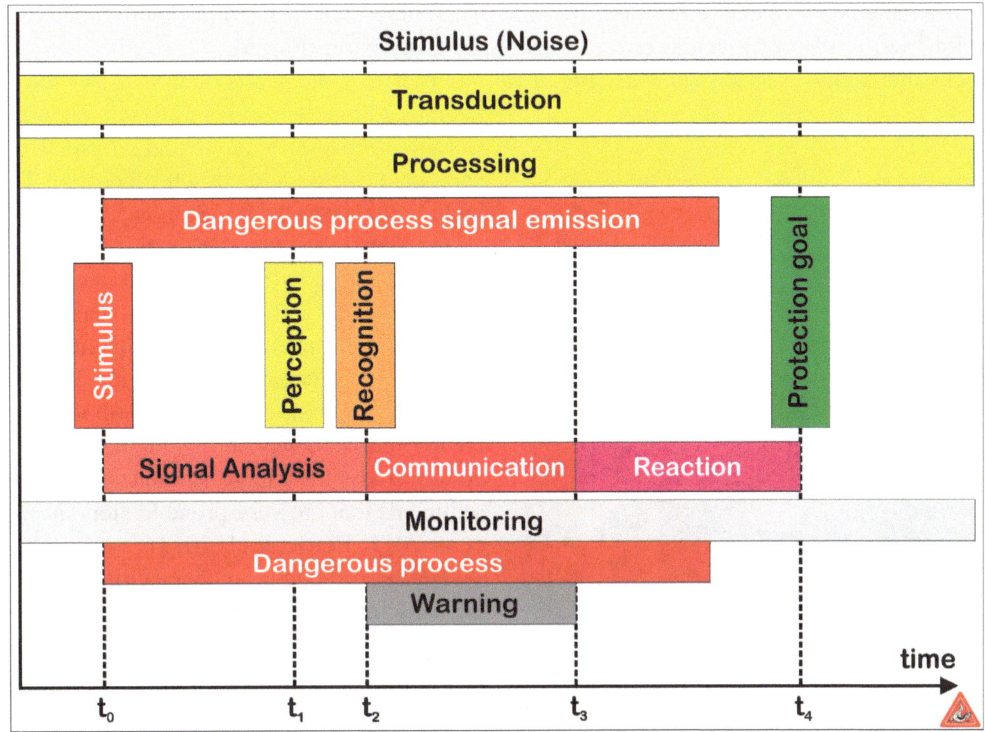

Fig. 1 From perception via monitoring to warning against natural hazards (from Hübl and Mikoš 2014)

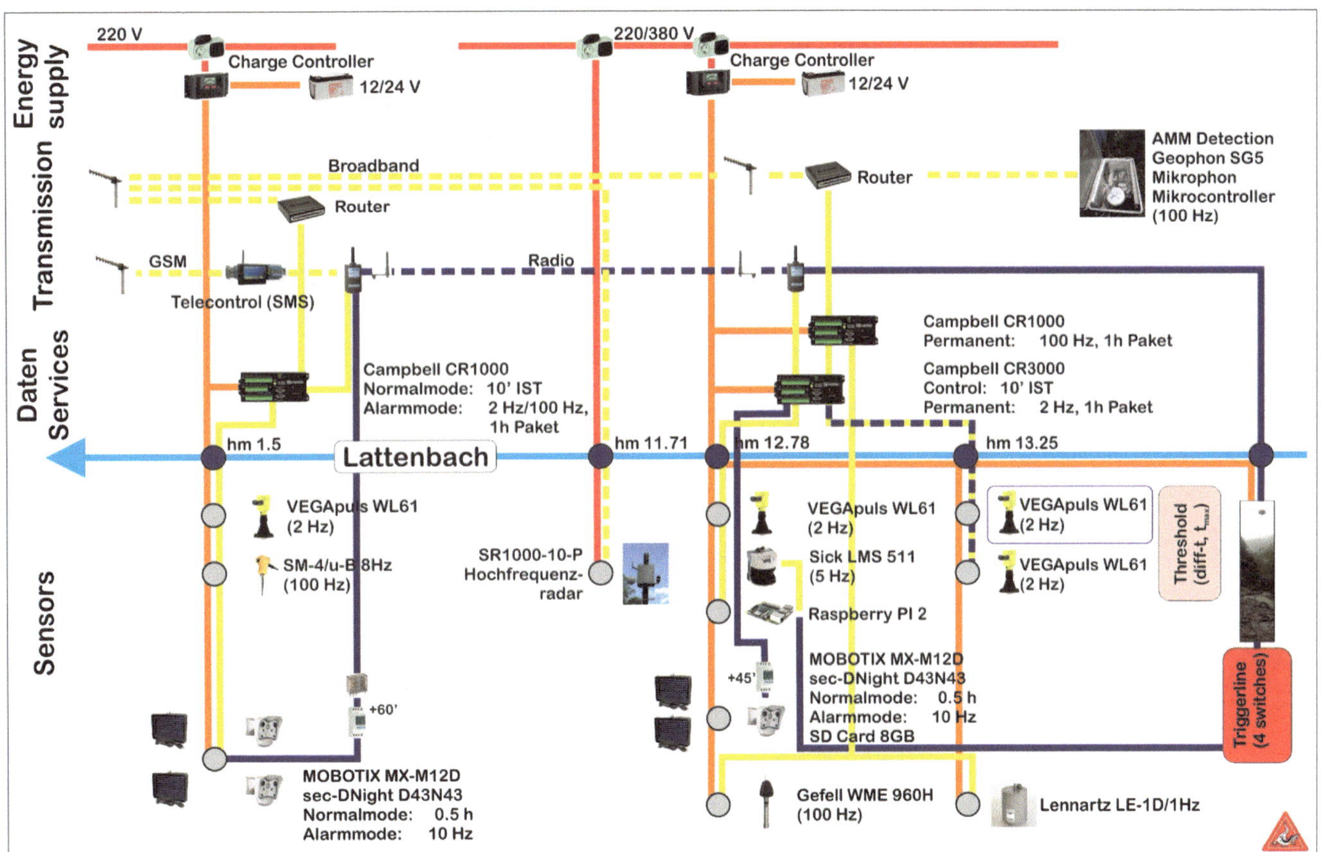

Fig. 2 A conceptual scheme of the debris-flow monitoring site at Lattenbach, Austria, showing sensors, communication, and energy supply (from Hübl and Mikoš 2014)

flysch deposits in Slovenia and Croatia. The study area in Slovenia (~1050 km^2 i.e. ~5% of Slovenia) covers a territory of clastic rocks that includes flysch deposits (Fig. 4).

Fig. 3 A group of Slovenian and Croatian researchers visiting landslides in flysch in the Istrian Peninsula in Croatia (Photo: Tina Peternel, September 25, 2014)

The study area in Croatia (approximately 550 km^2) lies in the Grey Istria (Fig. 5).

From a geological point of view, the common characteristic of large landslides in flysch deposits in Slovenia and Croatia is that they are mainly formed on the contact of carbonate rocks and flysch deposits. Typical landslides include the Slano Blato, Stogovce and Grohovo landslides.

In each case the Mesozoic carbonate rocks (mainly limestone) are over-thrust on the Eocene flysch deposits. The consequence of active tectonics is that the flysch deposits are folded and ruptured and consequently very prone to fast weathering to depth (Logar et al. 2005; Petkovšek et al. 2011; Benac et al. 2014). Furthermore, the carbonate rocks are also prone to weathering, and as a result the flysch slopes are covered with a large amount of talus material and slope sediments that are very prone to slope instability (Petkovšek et al. 2011; Benac et al. 2014).

The second common characteristic of landslides in flysch deposits in Slovenia and Croatia is the complexity of the sliding phenomena, which is connected with the softening of clay-bearing rock layers and that is mostly activated or reactivated by extreme weather events (Petkovšek et al. 2011). In the last decades a large number of landslides have

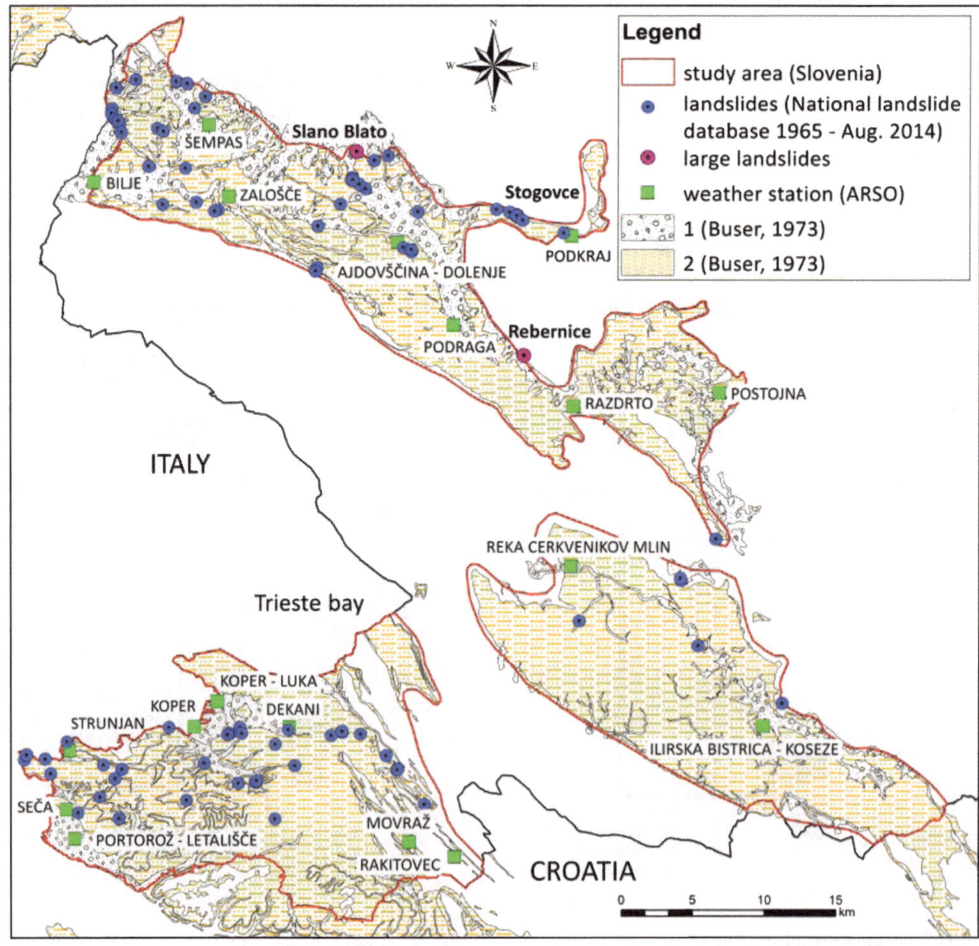

Fig. 4 The study area in Slovenia with the locations of landslides (according to Slovenian National landslide database—Status on August 2014) shown on the Geological map of Slovenia: 1 Alluvium; 2 Eocene Flysch deposits (adopted from basic geological map of SFRJ sheets: Gorica, Postojna, Trst, Ilirska Bistrica)

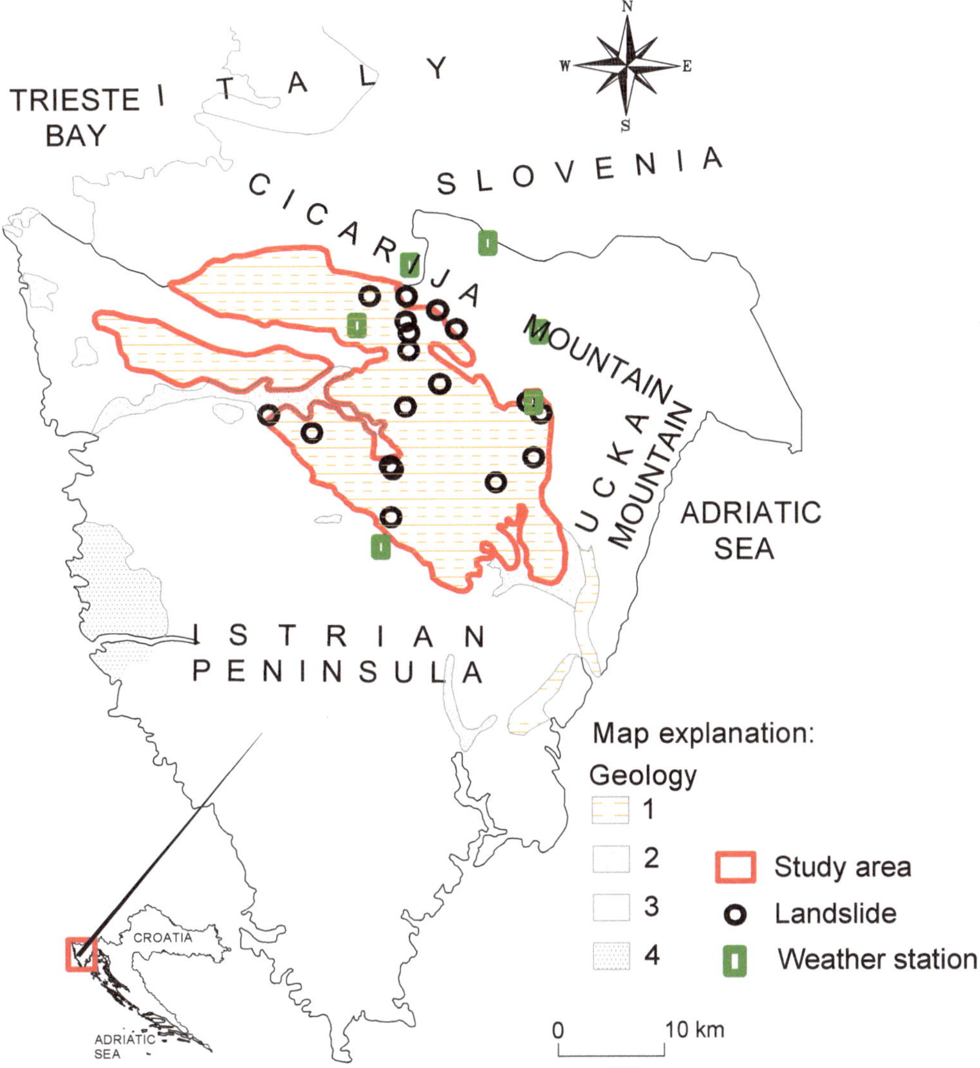

Fig. 5 The study area in the Grey Istria with the locations of landslides shown on the simplified Geological map of Istria: *1* Paleogene flysch, *2* Alluvium, *3* Cretaceous Limestone, *4* Jurassic Limestone (after Velić et al. 1995)

occurred in Slovenia and Croatia in flysch deposits, mostly triggered by prolonged rainfall or short intensive rainfall events.

In Slovenia several landslides in flysch deposits are also related to earthquakes, which are quite a common phenomenon (Logar et al. 2005; Petkovšek et al. 2011).

For rainfall-induced landslides, there is a difference between shallow landslides, which are generally triggered by short duration and intense rainfall, and deep-seated (and often also large scale) landslides, mainly triggered during or after prolonged rainfall. This was confirmed by a study of rainfall-triggered landslides in Slovenia in the last 25 years (Bezak et al. 2015a, b), which included both landslides in flysch and in other geological strata.

The results shown in Fig. 6 indicate that when using empirical rainfall-threshold curves as part of an Early Warning System in Slovenia, different empirical curves (Caine 1980; Clarizia et al. 1996; Aleotti 2004; Guzzetti et al. 2008) should be applied using a rainfall measuring network with an appropriate high density. This step was recently partially carried out by dividing Slovenia into four regional units, and by using rainfall data from 41 pluviometers (Rosi et al. 2016). Both studies confirmed that No Rain Gap (Rosi et al. 2016), respectively Inter-Event Time (Bezak et al. 2015a), as a parameter indicating the number of consecutive hours without rain, has a large impact on rainfall intensity and rainfall duration values.

This bilateral research project was evaluated as successful, since a new bilateral research project between Slovenia and Croatia was confirmed for the period 2016–2017, entitled "Laboratory investigations and numerical modelling of landslides in flysch deposits in Croatia and Slovenia" (Project leader M. Maček). With the Croatian side we will exchange knowledge and will cooperate in the field of

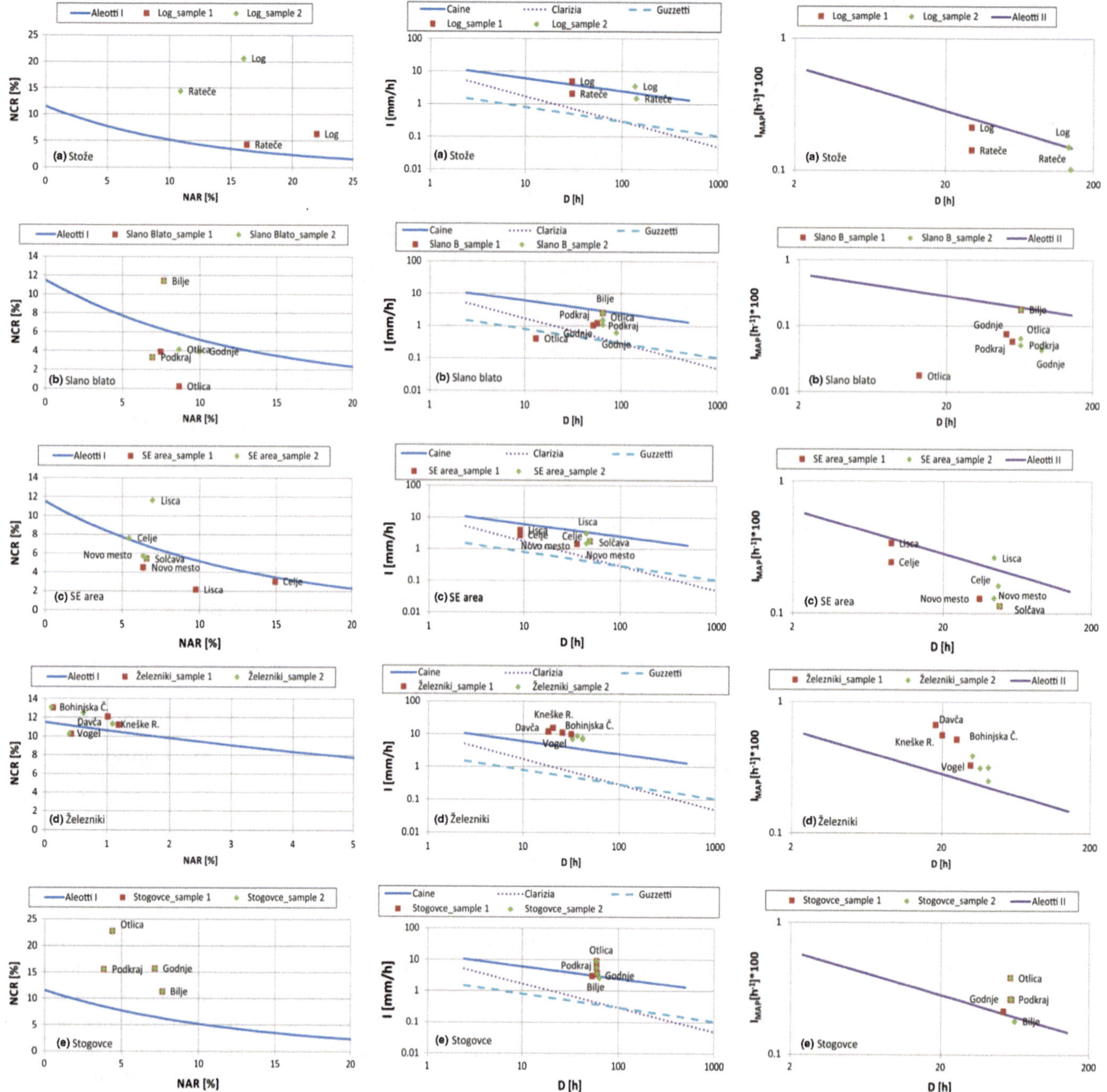

Fig. 6 Evaluation of a few empirical rainfall-threshold curves for selected landslides (shallow and deep-seated landslides and debris flows), which occurred in Slovenia in the last 25 years (from Bezak et al. 2015a)

laboratory research on rheology parameters of potential debris flows as a part of a debris flow hazard assessment on torrential fans.

National Research Projects

In 2014, we finished a national research project on the protection efficiency of beech-dominated forests against debris-flow hazards in a narrow Alpine valley in NW Slovenia of the Sava Bohinjka River gorge called Soteska, with the main state road and railway (Project leader M. Mikoš). The study area was a part of the Soteska gorge, where we assessed the starting points of potential debris flows based on a small-scale field geological survey that produced a geological map of the study area in the scale 1:5000 (produced by the Geological Survey of Slovenia—GSS). General lithological and structural geological data

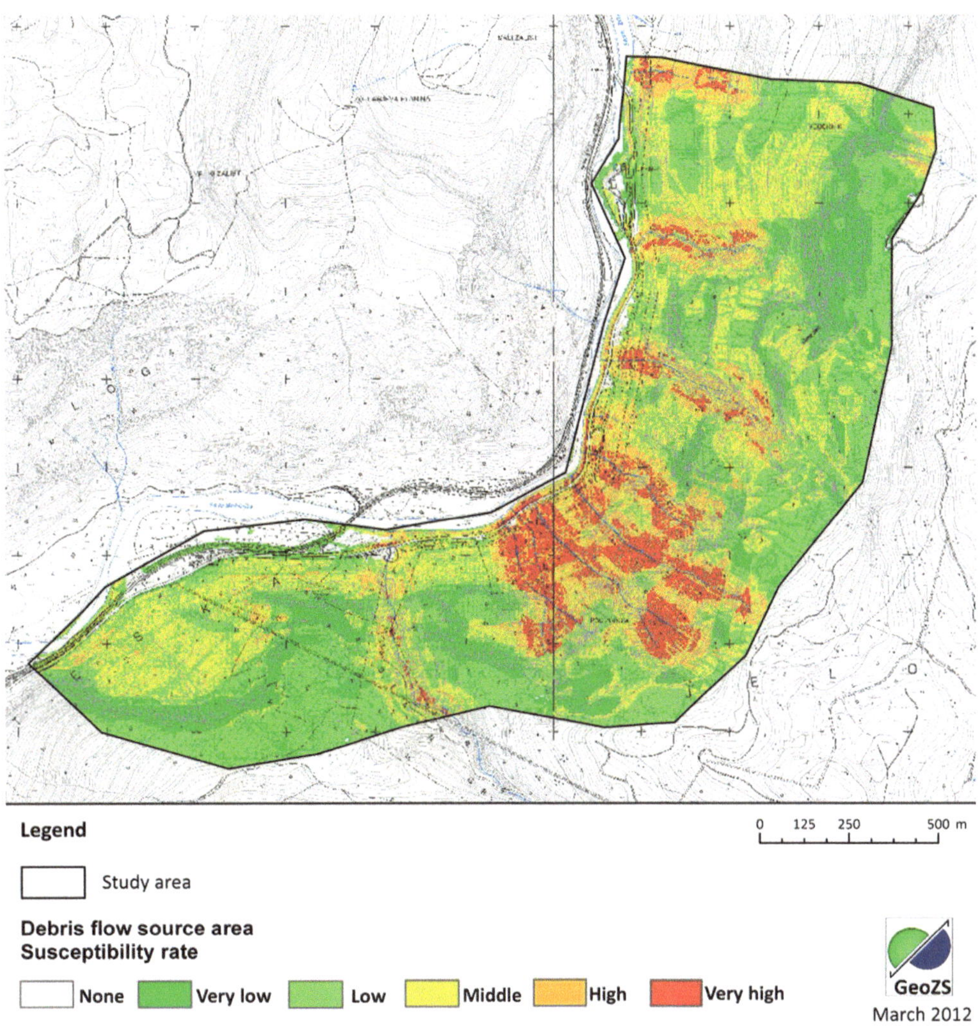

Fig. 7 The debris-flow susceptibility map of the study area in the Sava Bohinjka River gorge called Soteska

were taken into account in the creation of the geological map, with special emphasis on the identification of unconsolidated sediments such as scree deposits that can be involved in mass movement processes.

The debris flow susceptibility map (Fig. 7) was created using methods that the GSS developed for different spatial resolutions and different types of mass movements (e.g., landslides, mass-flows, rock falls). The methodology consists of four consecutive phases: a synthesis of archived data, geostatistical modeling with the GSS algorithm (Komac 2005), elaboration of a geohazard map, and field verification of the most susceptible areas. In addition to data on lithology, crushed tectonic zones, and distance from structural elements, the impact analysis and creation of the susceptibility model included elevation data, slope and curvature, distance to surface waters, energy potential of streams, and 48-hour rainfall intensity.

Furthermore, we modelled debris flows with the Top Run Debris Flow (TopRunDF) model, version 1.1. (Scheidl 2009). The model is a tool for the two-dimensional run-out simulation of the debris flow deposit phase on debris cones. TopRun DF produces two estimates: (1) an inundated simulation area, combined with overflow probability of each related cell (this was used as a debris flow warning map—Fig. 8); and (2) a deposited area and the deposition height of each cell. The goal was to identify debris-flow hazard areas on the debris cones in the valley hitting the railroad. The details of the study can be found in Fidej et al. (2015).

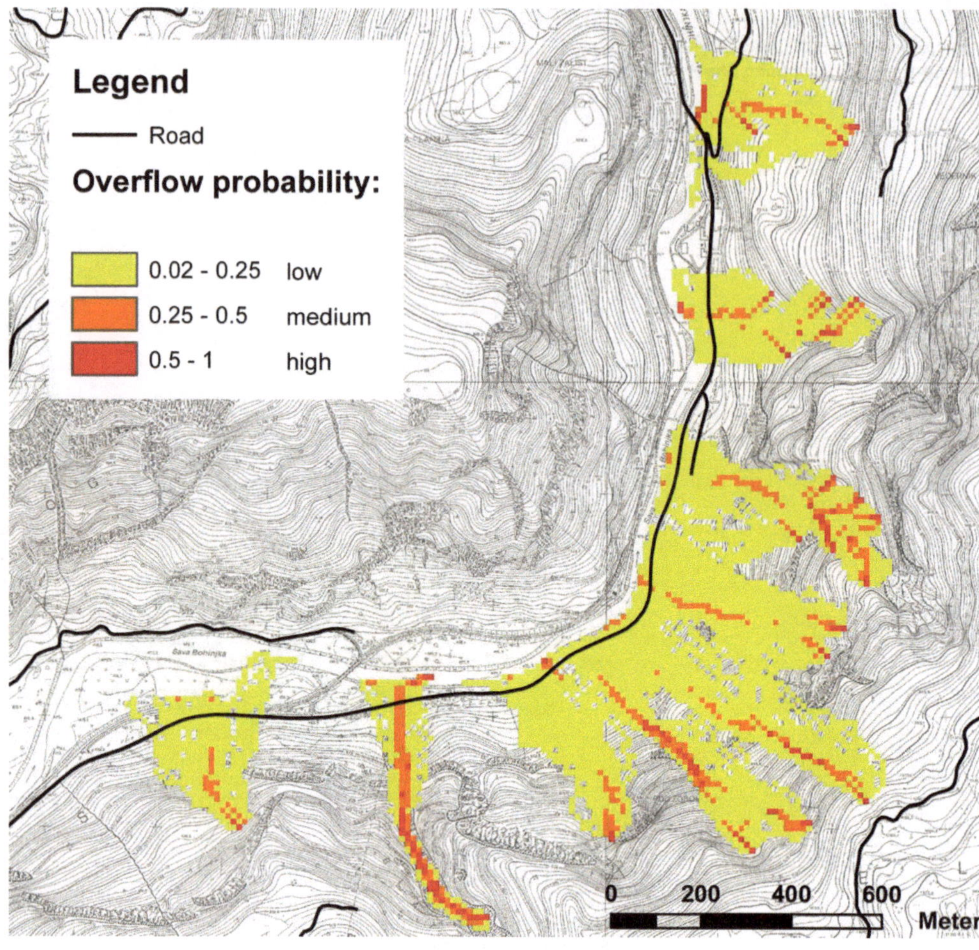

Fig. 8 The debris-flow warning map at a scale of 1:15,000, prepared with the TopRunDF model. The color chart shows the debris-flow overflow probability for the stretch of the regional railroad Jesenice-Nova Gorica

In the field of landslide terminology, we published a paper in the national journal dedicated to natural and other hazards (Mikoš 2014), and at the XIIth IAEG Congress in Turin we reported on the investigation on torrent check dams as debris-flow sources (Sodnik et al. 2015b). We published a research paper in the journal *Landslides* on the executed mitigation measures on two large deep-seated landslides in Slovenia (Pulko et al. 2014).

On the Slano blato landslide, 11 dowels or shafts (Fig. 9) were constructed for both drainage of the upper part of landslide and as retaining works. The arch of 11 dowels is seen in Fig. 10. The 3D finite element method has been applied for the calculation of landslide stability and for the internal forces in the shafts.

At the Slano blato landslide a drop of suction was one possible reason for landslide instabilities (Petkovšek et al. 2009). In 2007 suction monitoring started at three different locations (MS on Fig. 10). One of the results of monitoring was dry and wet envelopes of pore pressures during monitoring periods and changes in the factor of safety due to water lever variation (Fig. 11). A scientific report on these six years of suction monitoring was published in an Italian geotechnical journal (Maček et al. 2016).

National Research Program

In 2015, we studied empirical rainfall thresholds for rainfall-induced landslides in Slovenia (Bezak et al. 2015a; 2016), tailings dam failure related risk management (Petkovšek and Pulko 2015), and the effects of over-consolidation ratios and shear rate on the shear strength of soils in a direct shear apparatus (Šelekar 2015). The in situ investigations on the Slano blato landslide were compared to laboratory tests. The results show that at the Slano blato landslide soil suction measurements cannot be used as an

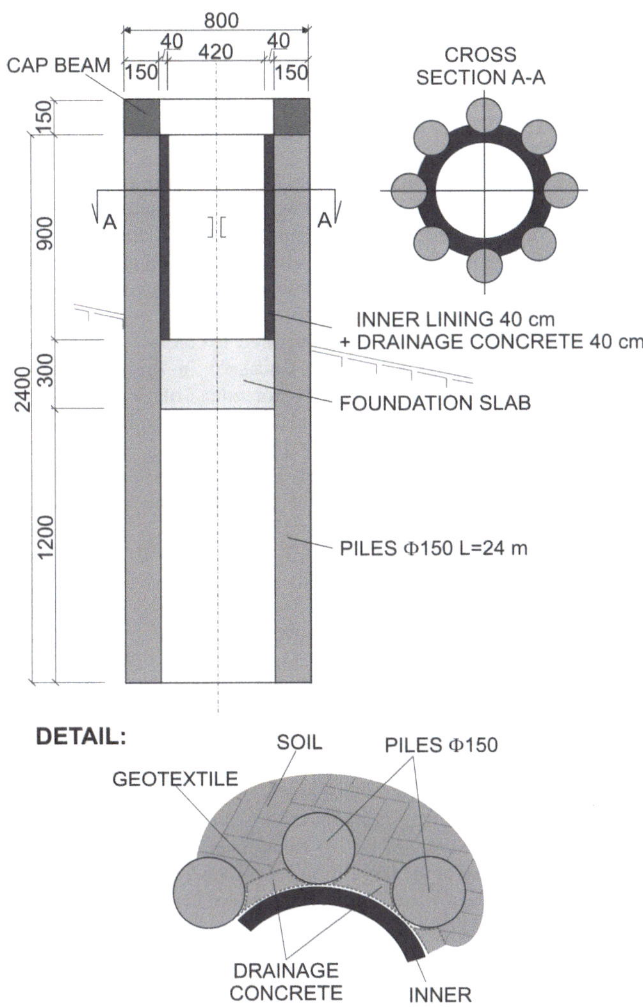

Fig. 9 Vertical and horizontal cross section of the modified dowel design at Slano blato landslide (from Pulko et al. 2014)

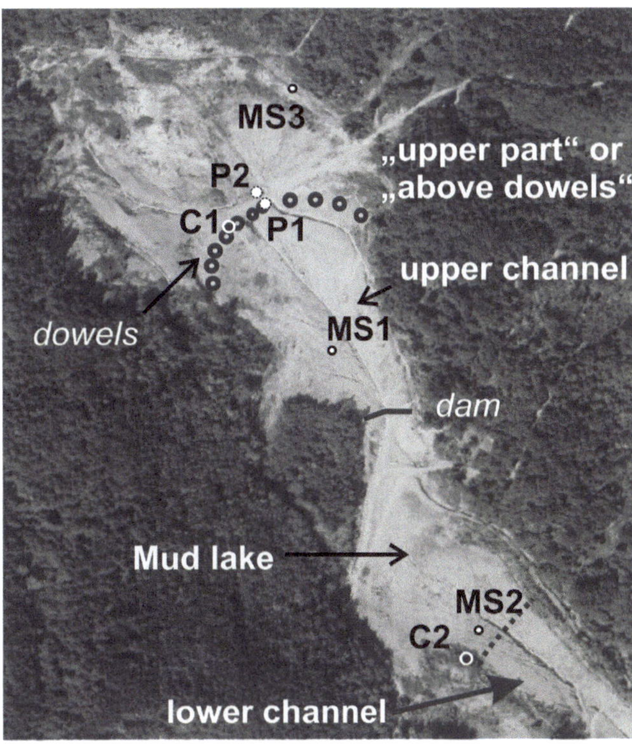

Fig. 10 Monitoring of the Slano blato landslide: *C1* TV camera, *P1* Piezometer, and *MS1* suction measurement station (Maček et al. 2016)

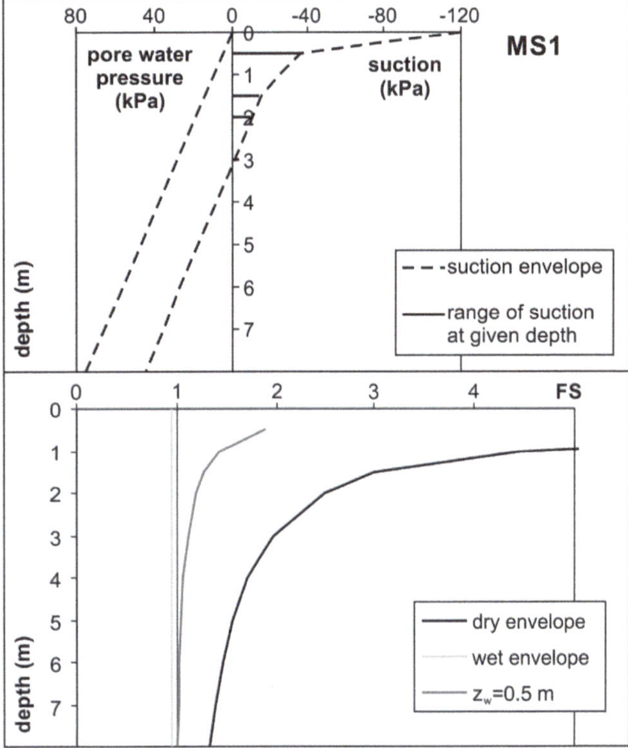

Fig. 11 Pore water pressure envelopes for measurement profile MS1 and factor of safety against depth for the dry and wet envelopes, and for the case of ground water at a depth of 0.5 m

indicator of new earth flow occurrences, however they may serve as an indicator of the periods with the lowest safety, when the new instabilities may appear (Maček et al. 2016).

We have compared the investments into water infrastructure (including landslide risk mitigation) in Slovenia and Austria (Sodnik et al. 2015a). We helped with the bibliometric analysis of the achievements of the ICL journal *Landslides* (Sassa et al. 2015). We currently support the activities of the Slovenian National Platform on Disaster Risk Reduction, established in 2014.

At UL FGG, four chairs established a new research institute called Research Institute for Geo and Hydro Threats (RIGHT; www.right.si) that will help to coordinate research work in the field of natural disaster risk reduction at the faculty level between hydrology, hydraulic engineering, geological engineering, remote sensing, and geodetic engineering.

Acknowledgements Our research activities in the field of landslide research have been mainly financed by the Slovenian Research Agency (ARRS) through the national Research Programme P2-0180 "Water Science and Technology, and Geotechnics". This financial support is greatly acknowledged.

References

Aleotti P (2004) A warning system for rainfall-induced shallow failures. Eng Geol 73(3–4): 247–265

Benac Č, Oštrić M, Dugonjić Jovančević S (2014) Geotechnical properties in relation to grain-size and mineral composition: The Grohovo landslide case study (Croatia). Geologia Croatica J Croatian Geol Surv Croatian Geol Soc 67(2):127–136

Bezak N, Šraj M, Mikoš M (2015a) Analysis of rainfall-triggered extreme landslide events in Slovenia in the last 25 years. In: Abstract Proceedings of the 2nd ReSyLAB—Regional Symposium on Landslides in the Adriatic-Balkan Region, 14–16 May 2015, Belgrade, Serbia, pp 123–126

Bezak N, Šraj M, Brilly M, Mikoš M (2015b) Empirical rainfall thresholds and copula-based IDF curves for shallow landslides and flash floods. Geophysical research abstracts 17. http://meetingorganizer.copernicus.org/EGU2015/EGU2015-5171.pdf

Bezak N, Mikoš M, Šraj M (2016) Copula-based IDF curves and empirical rainfall thresholds for flash floods and rainfall-induced landslides. J Hydrology. doi:10.1016/j.jhydrol.2016.02.058

Caine N (1980) The rainfall intensity-duration control of shallow landslides and debris flows. Geogr Ann A 62(1–2): 23–27

Clarizia M, Gullà G, Sorbino G (1996) Sui meccanismi di innesco dei soil slip. In: International conference Prevention of hydrogeological hazards: the role of scientific research, vol 1, pp 585–597 (in Italian)

Fidej G, Mikoš M, Rugani T, Jež J, Kumelj Š, Diaci J (2015) Assessment of the protective function of forests against debris flows in a gorge of the Slovenian Alps. IForest. 8(1):73–81

Guzzetti F, Peruccacci S, Rossi M, Stark CP (2008) The rainfall intensity-duration control of shallow landslides and debris flows: an update. Landslides 5(1):3–17

Hübl J, Mikoš M (2014) Debris flow monitoring. Wildbach- und Lawinenverbau. 78(173):50–66

Komac M (2005) Verjetnostni model napovedi nevarnih območij glede na premike pobočnih mas—primer občine Bovec (Probabilistic model of slope mass movement susceptibility—a case study of Bovec municipality, Slovenia). Geologija. 48(2):311–340

Logar J, Fifer Bizjak K, Kočevar M, Mikoš M, Ribičič M, Majes B (2005) History and present state of the Slano Blato landslide. Nat Hazards and Earth System Sci 5:447–457

Maček M, Petkovšek A, Majes M, Mikoš M (2014) Landslide monitoring techniques database. Landslide science for a safer geoenvironment—vol 1. In: Sassa K, Canuti P, Yin Y (eds) The international programme on landslides. Springer, Heidelberg, 493 p

Maček M, Petkovšek A, Arbanas Ž, Mikoš M (2015) Geotechnical aspects of landslides in flysch in Slovenian and Croatian. In: Abstract Proceedings of the 2nd ReSyLAB—Regional Symposium on Landslides in the Adriatic-Balkan Region, 14–16 May 2015, Belgrade, Serbia, pp 136–138

Maček M, Majes B, Petkovšek A (2016) Lessons learned from 6 years of suction monitoring of the Slano blato landslide. Rivista Italiana di Geotecnica 16(1):21–33

Mikoš M (2014) On Terms such as accident, natural disaster, natural catastrophe and natural cataclysm. Ujma 28:306–310

Mikoš M (2015a) ISDR-ICL partnerships 2015–2025 for global promotion of understanding and reducing landslide disaster risk. Ujma 29:424-429. http://www.sos112.si/slo/tdocs/ujma/2015/424_429.pdf

Mikoš M (2015b) The 4th world landslide forum—call for abstracts. Landslides 12(6):1235–1240. doi:10.1007/s10346-015-0656-4

Mikoš M (2015c) From natural disasters protection to culture of living with them. In: Proceedings of the 26th Mišičev vodarski dan 2015, 8 December 2015. Maribor, Slovenia, pp 7–13. http://www.mvd20.com/LETO2015/R6.pdf

Mikoš M (2016) Slovenia and the UN 3rd World Conference on Disaster Risk Reduction, Sendai, Japan, 2015. Ujma. 30:309–316

Mikoš M, Čarman M, Papež J, Janža M (2014) Legislation and procedures for the assessment of landslide, rockfall and debris flow hazards and risks in Slovenia. Wildbach- und Lawinenverbau. 78 (174):212–221

Peternel T, Mikoš M, Đomlija P, Dugonjić Jovančević S, Arbanas Ž (2015) Geological conditions of landslides in flysch deposits in Slovenian and Croatian. In: Abstract Proceedings of the 2nd ReSyLAB—Regional Symposium on Landslides in the Adriatic-Balkan Region, 14–16 May 2015, Belgrade, Serbia. pp. 140-142

Petkovšek A, Maček M, Kočevar M, Benko I, Majes B (2009) Soil matric suction as an indicator of the mud flow occurence. In: Proceedings of the 17th international conference on soil mechanics and geotechnical engineering, vol 3, pp 1855–1860

Petkovšek A, Fazarinc R, Kočevar M, Maček M, Majes M, Mikoš M (2011) The Stogovce landslide in SW Slovenia triggered during the September 2010 extreme rainfall event. Landslides. 8:499–506

Petkovšek A, Maček M, Mikoš M, Majes B (2013) Mechanisms of active landslides in Flysch. In: Sassa K, Rouhban B, Briceño S, McSaveney M, He B (eds) Landslides: global risk preparedness. Springer, Heidelberg, pp 149–164. ISBN 978-3-642-22086-9

Petkovšek A, Pulko B (2015) Tailings dams—operational reliability and failure related risk management. Ujma 29:293–304. http://www.sos112.si/slo/tdocs/ujma/2015/293_304.pdf

Pulko B, Majes B, Mikoš M (2014) Reinforced concrete shafts for the structural mitigation of large deep-seated landslides: an experience from the Macesnik and the Slano blato landslides (Slovenia). Landslides 11(1):81–91

Rosi A, Peternel T, Jemec-Auflič M, Komac M, Segoni S, Casagli N (2016) Rainfall thresholds for rainfall-induced landslides in Slovenia. Landslides. doi:10.1007/s10346-016-0733-3

Sassa K, Si Tsuchiya, Fukuoka H, Doan L, Mikoš M (2015) Landslides: review of achievements in the second 5-year period (2009–2013). Landslides 12(2):213–223. doi:10.1007/s10346-015-0567-4

Scheidl C (2009) English manual for using TopRunDF (v. 1.0), p 9. http://www.debris-flow.at/index.php/en/download-en/category/6-toprundf-doc

Sodnik J, Kogovšek B, Mikoš M (2015a) Investments into water infrastructure in Slovenia and in Austria. Gradbeni vestnik 64(1):3–12

Sodnik J, Martinčič M, Kryžanowski A, Mikoš M (2015b) Are Torrent Check-Dams Potential Debris-Flow Sources? Engineering geology for society and territory—vol 2. In: Lollino G, Giordan D, Crosta G, Corominas J, Azzam R, Wasowski J, Sciarra N (eds) Landslide processes. Springer, Heidelberg, 2177p. ISBN 978-3-319-09056-6

START_it_up Partnership (2014) START_it_up—Common Strategic Paper and Final Booklet. 76 p

Šelekar N (2015) Effects of overconsolidation ratios and the shear rate on the shear strength of soils in the direct shear apparatus. Graduation Thesis, University of Ljubljana FGG, Ljubljana, Slovenia. http://drugg.fgg.uni-lj.si/5318/

TXT-tool 1.386-1.1 (2017a) Intensity-duration-frequency curves for rainfall-induced shallow landslides and debris flows using copula functions. Landslide dynamics: ISDR-ICL landslide interactive teaching tools (in print)

TXT-tool 2.386-1.1 (2017b) Practice guidelines on monitoring and warning technology for debris flows. Landslide dynamics: ISDR-ICL landslide interactive teaching tools (in print)

TXT-tool 5.386-1.1 (2017c) State-of-the-Art Overview on Landslide Disaster Risk Reduction in Slovenia. Landslide Dynamics: ISDR-ICL Landslide Interactive Teaching Tools (in print)

Velić I, Tišljar J, Matičec D, Vlahović I (1995) General review of the geology of Istria. Vlahović I, Velić I (eds) Excursion guide-book, Institute of Geology, Zagreb, pp 5–20

Research on Heavy-Rainfall-Induced and Hydraulic-Driven Geological Hazards in China (WCoE 2014–2017)

Yueping Yin, Yongqiang Xu, and Wenpei Wang

Abstract

China Geological Survey (CGS) is one of the 15 new World Centres of Excellence (WCoE) in Landslide Disaster Reduction for the period 2014–2017. The title of the activities of the WCoE has been "Scientific research for mitigation, preparedness and risk assessment of landslides" since 2008, due to the complex conditions for the occurrence of geohazards in China. The Center of Geohazards Emergency of Ministry of Land Resources, which is directly under CGS, is responsible for the emergency response to major geohazards nationwide, including survey and investigation, monitoring and warning, risk assessment, prevention, training, and information systems. In this paper, the important activities of CGS are elaborated. The research project entitled "Research Project on the Early Recognition and Warning on Heavy Rainfall-Induced and Hydraulic-Driven Geological Hazards in China" was conducted from 2011 to 2015. From 2015 to 2017, a new project on mechanisms and hazards patterns of hydraulic-driven landslides has been conducted by the Center of Geohazards Emergency, China Geological Survey.

Keywords

Landslide • Heavy rainfall • Hydraulic-driven

Introduction

Affected by global extreme climate change and ill-advised human activity, serious geological hazards induced by heavy rainfall still occur frequently in China. The catastrophic landslides that kill dozens of people and cause huge loss of property are still too difficult to recognize and mitigate. The research project entitled "Research Project on the Early Recognition and Warning on Heavy Rainfall-Induced Geological Hazards in China" was conducted from 2011 to 2015, co-funded by the Ministry of Sciences and Technology, the PRC and the China Geological Survey, and hosted by Prof. Yin Yueping of the Center of Geohazards Emergency, China Geological Survey. From 2015 to 2017, a new project on the mechanisms and patterns of hydraulic-driven landslides is being conducting by Center of Geohazards Emergency, China Geological Survey. The research projects focus on three aspects: the forming mechanisms of rain-storm triggered landslides, early recognition and warning methodology and technology, and risk assessment, through the implementation of five sub-projects. The projects selected the Southeast China typhoon-rainstorm coast, the Wenchuan earthquake region, the Three Gorges reservoir, and other typical areas where catastrophic landslides frequently happened because of heavy rain and hydraulic factors to conduct the research. This includes the mechanisms of occurrence of geological hazards in heavy rain areas and the key methodology and technology of early recognition, emergency response, monitoring and early warning of geological hazards, and risk assessments based

Y. Yin (✉) · Y. Xu · W. Wang
Center of Geohazards Emergency, China Geological Survey, Haidian, Beijing, 100081, China
e-mail: yyueping@mail.cigem.gov.cn

Y. Xu
e-mail: Xuyq@mail.cigem.gov.cn

W. Wang
e-mail: wangwp@mail.cigem.gov.cn

K. Sassa et al. (eds.), *Advancing Culture of Living with Landslides*,
DOI 10.1007/978-3-319-59469-9_24

on the occurrence mechanisms of the geological disasters. Finally, codes and guidelines were formulated to improve the capacity and technological level of mitigation of rainfall-triggered geologic hazards.

Some research advances on typical landslides due to rainfall and hydraulic factors from 2011 to 2015 are briefly introduced below.

Heavy-Rainfall-Induced Rockslide-Debris Flows After the Wenchuan Ms 8.0 Earthquake

During July 7–12 of 2013, heavy rainfall occurred in the Wenchuan Ms 8.0 Earthquake area. The 5 days cumulative precipitation in the Dujiangyan area of Sichuan Province was up to 1129 mm, which was rare in previous years and exceeded the mean annual precipitation of 1110 mm over the twenty-five years from 1987 to 2012. The heavy rainfall triggered a rapid and long run-out rockslide in Sanxicun village. Its debris had a volume of 1.9 million m^3, and it slid a distance of up to 1200 m, killing 166 people in the village. The Sanxicun landslide was located in Zhongxing Town, Dujiangyan City of Sichuan Province. The site is located at (E103°33'48", N30°54'55"), 68 km from Chengdu City to the east, and 16 km from the Dujiangyan urban district (Fig. 1) (Yin et al. 2016a).

This area was famous for its agreeable climate and beautiful natural mountain scenery, lakes, and abundant vegetation. The typical natural conditions at resorts and vacation spots often concealed the potential for rockslides, rockfalls, and debris flows. This resulted in the misconception of a lower risk of geological hazards. This area enjoyed a high reputation stemming partly from the Qincheng Mountain-Dujianyan Scenic Resort and attracted many tourists every summer.

Therefore, research on rainfall-induced post-earthquake geologic hazards in these areas is significant for the reconstruction of towns and villages and tourism safety after the earthquake. The rockslide dynamic response characteristics triggered by the Wenchuan Ms 8.0 Earthquake and post-earthquake debris flow disaster have received much attention from researchers (Yin et al. 2009; Tang et al. 2011). Tang et al. reported a heavy-rainfall-induced debris flow disaster on September 24 of 2008 at the earthquake-ruined Beichuan town in the central part of the area affected by the Wenchuan Ms 8.0 Earthquake and presented the possibility that due to the ground shaking, the critical amount of accumulated precipitation and the hourly rainfall intensity necessary to initiate debris flows were reduced compared with values before the earthquake. Recently, Zhou and Tang (2013) summarized eleven rainfall events that induced debris flows between 2008 and 2012

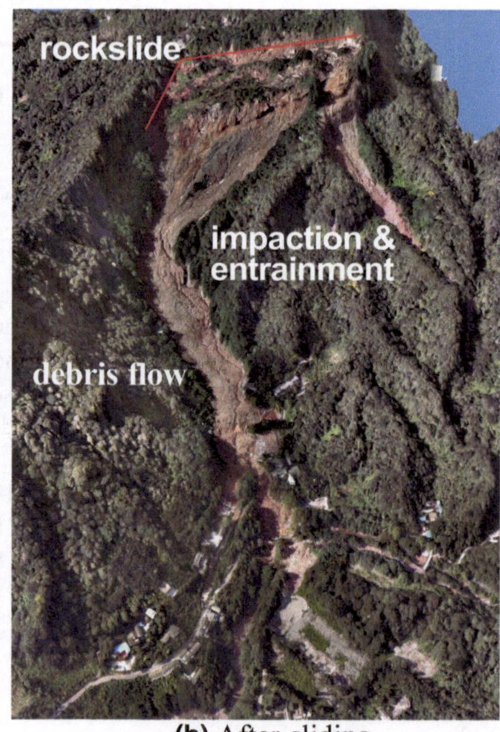

Fig. 1 3D images showing catastrophic rockslide-debris flows at Sanxicun, in the Wenchuan earthquake region (*red dashed line* indicates the small rockfall due to Ms 8.0 earthquake)

after the Wenchuan earthquake, and discussed the rainfall thresholds for debris flow occurrence. However, research on rainfall-induced rockslides after the earthquake and their transformation into long run-out debris flows has made little significant progress.

The parent rock of the Sanxicun rockslide is a slightly inclined, nearly horizontal strata sandstone formation with intact rock constituents (Fig. 2). This rockslide was located southwest of the Wenchuan Ms 8.0 Earthquake fault and was only 12 km away from the epicentral region. The seismic intensity in the Sanxicun area was up to degree IX.

Based on the emergency investigation after the Wenchuan earthquake, three small-scale rockfalls were induced. Two rockfalls were at the margin of the slope of the rockslide source area and moved in the direction of NNE10°, and the other was at the southeastern corner and moved in the direction of NWW288°. The rockfalls disturbed or even damaged the slope. Research on this rockslide is significant for post-earthquake geological hazard prevention. Therefore, this project uses the Sanxicun landslide as an example to study the stability of fractured slopes after the Wenchuan Ms 8.0 Earthquake and the formation mechanism of rapid and long run-out rockslides triggered by heavy rainfall. The adverse conditions of the geological features prior to the rockslide are investigated, which provides a scientific basis for classifying hazard areas that should be subject to monitoring and prevention. The slope site was affected by the Wenchuan Ms 8.0 Earthquake in 2008. The sliding involved the thick fractured and layered rockmass with a gentle dip plane at Sanxicun. The formation processes included: (1) toppling due to shear failure at a high-level position, (2) bulldozing of the accumulative layer below, (3) formation of a debris flow of the highly weathered bottom rockmass, and (4) flooding downward along the valley (Fig. 3).

After the shear failure of rock occurred at a high-level position, the rockslide moved for approximately 47 s downward along the valley with a maximum velocity of 35 m/s. This is typical for a rapid and long run-out rockslide. Finally, the research concludes that the identification of the potential geological hazards at the Wenchuan mountain area is crucial to prevent catastrophic rockslides triggered by heavy rainfall. The identified geological hazards should be properly considered in the town planning of the reconstruction works.

Catastrophic Long-Runout Landslide in Zhenxiong, Yunnan

On 11th January 2013 a catastrophic long-runout landslide (located at latitude 27°33'5" and longitude 104°59'15") triggered by a prolonged low-intensity rainfall occurred in Zhenxiong, Zhaotong, Yunnan, southwestern China. This landslide destroyed the village of Zhaojiagou, killing 46 people, and burying over 60 houses in the distal part of its path (Fig. 4).

In the project, we selected the Zhenxiong landslide to examine the possible mechanisms and behaviour of this type of landslide from initiation to final deposition We conducted a series of ring shear tests on samples collected from the sliding surface and runout path to examine their shear behaviour under saturated conditions (Fig. 5). To determine

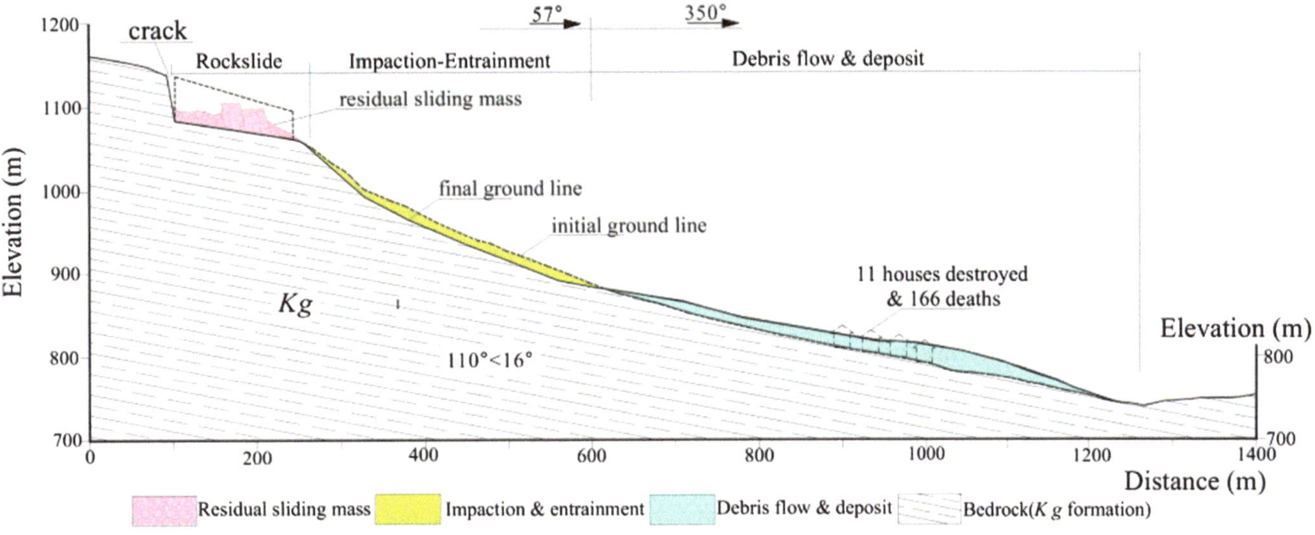

Fig. 2 Longitudinal profile of a rockslide-debris flow at Sanxicun, Wenchuan Ms 8.0 earthquake region

Fig. 3 The debris flow, showing the height difference of about 15 m between the maximum level and the accumulation

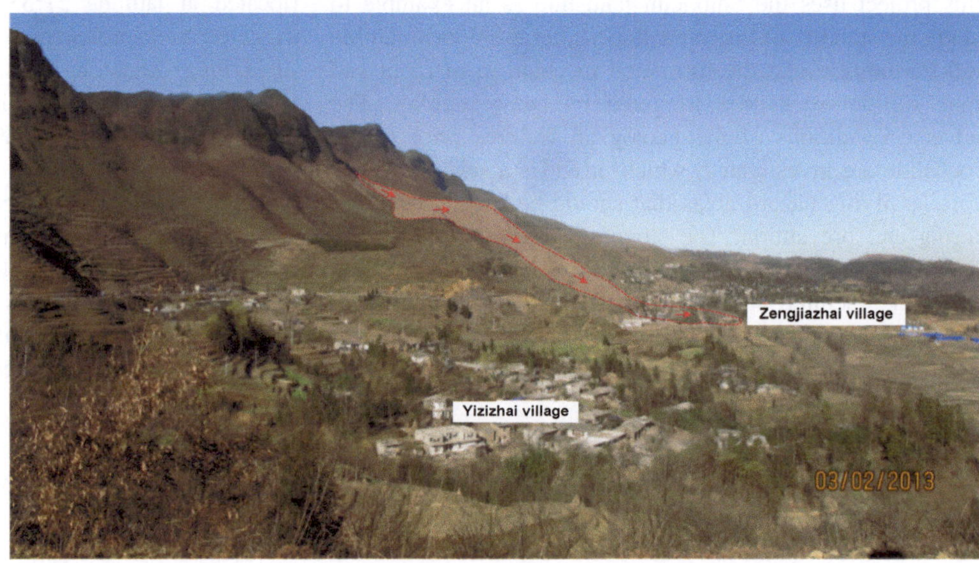

Fig. 4 Aerial view of the Zhenxiong landslide taken on 12 January 2013

the appropriate rheological relationships and parameters to numerically model the landslide from initiation to final deposition, it is important to investigate the shear behaviour of source and runout path materials. Recently, a series of ring shear apparatus was developed and improved by the Disaster Prevention Research Institute (DPRI), Kyoto University (Sassa et al. 2010). These apparatuses enable shearing with different types of loadings under either drained or undrained conditions. We sampled the source (S1) and runout path material (S2) (Figs. 4 and 5e) and examined the shear behaviour of the samples using the ring shear apparatus (DPRI-5). Also, for hazard assessment of this type of landslide in the same area, it is essential to estimate its runout behavior, including travel distance and velocity. Based on the results of ring shear tests, we analyzed the post-failure behaviour of the landslide with the DAN-W model developed by Hungr (1995). It is expected that these models and parameters can improve the accuracy of hazard assessment for those areas with similar geo-environmental settings to the Zhenxiong landslide area.

The following conclusions can be drawn from this study (Yin et al. 2016b):

- The Zhenxiong landslide is a rapid and long-runout catastrophic landslide triggered by prolonged low-intensity rainfall. The undrained ring shear tests on the samples taken from the sliding surface and runout path reveal that the shear resistance of the samples can be decreased to a small value, and this enables the initiation and rapid movement of the landslide.

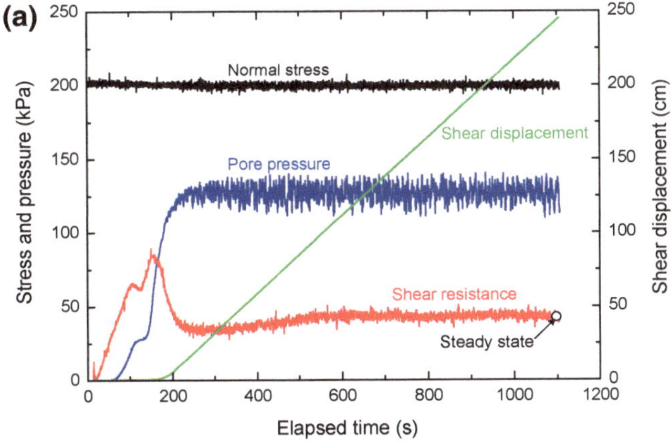

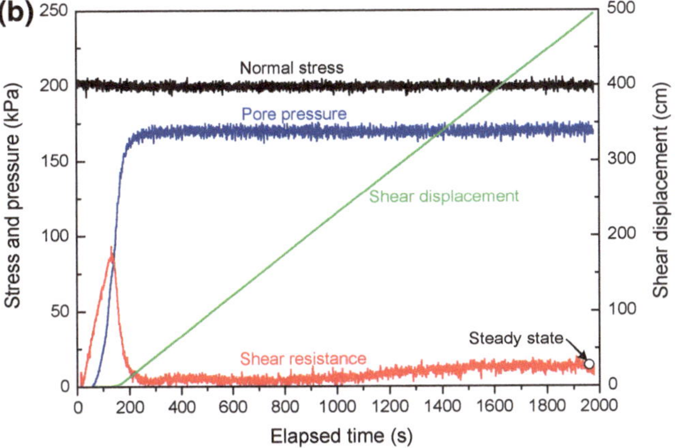

Fig. 5 Results of undrained ring-shear tests on the saturated samples taken from the source (**a**) and runout path (**b**)

- The partially drained test suggests that the shear strength of samples from the sliding surface is less affected by shear rate, while the shear rate has a negative effect on the shear strength of runout path material.
- The DAN-W model was used to simulate the post-failure behaviour based on ring shear test results. The results show that the mobilized friction angle obtained from ring shear tests provided the best performance in simulating this landslide. The total duration of the landslide is 48 s for an average velocity of 16.7 m/s. There is good agreement between the observed and simulated results, suggesting that this model, using parameters obtained through ring shear tests and back analyses, could be a useful tool for predicting post-failure behaviour in the same area, and the information used to mitigate this type of landslide hazard.

Reservoir-Induced Landslides at Three Gorges Reservoir

The Three Gorges region in China was basically a geohazard-prone area even prior to construction of the Three Gorges Reservoir. In September 2008, a trial impoundment of 175 m ASL commenced in the reservoir. The water level increased by about 100 m from the original water level of about 75 m. In addition, a 30-m annual variation in water level was specified by reservoir regulations. As a result of the combined effect of the large water level variation, relocation, and other factors, the geological conditions in the Three Gorges area have been significantly changed, and challenges in terms of geological hazards prevention have been encountered in the Yangtze River (Yin et al. 2016c).

This project first presents the spatiotemporal distribution of landslides in six periods of 175 m ASL trial impoundments from 2008 to 2014. The results show that the number of landslides sharply decreased from 273 at the initial stage to less than ten at the second stage of impoundment.

Figure 6 shows the locations of landslides induced by six periods of 175 m asl trial impoundments. From a geographical perspective, reservoir-induced landslides are mainly distributed along the main streams and some branches of the Yangtze River. Because of the marked differences in geological environment conditions of slopes along the Yangtze River, major differences in landslide occurrences in various sections were observed.

Based on this, the reservoir-induced landslides in the Three Gorges region can be roughly classified into five failure patterns, i.e., accumulation landslides, dip-slope landslides, reversed bedding landslides, rockfalls, and karst breccia landslides. The accumulation landslides and dip-slope landslides account for more than 90%. The Shuping accumulation landslides (a sliding mass volume of 20.7 × 106 m^3) in Zigui County and the Outang dip-slope landslide (a sliding mass volume of about 90 × 106 m^3) in Fengjie County are two typical cases; the mechanisms of reactivation of the two landslides were analyzed (Figs. 7 and 8). The monitoring data and factor of safety (FOS) calculations show that the accumulation landslide is dominated by water level variation in the reservoir, as most of the mass body is under 175 m ASL, and that the dip-slope landslide is controlled by the coupling effect of reservoir water level variation and precipitation, with an extensive recharge area of rainfall from the rear, and a front mass below 175 m ASL.

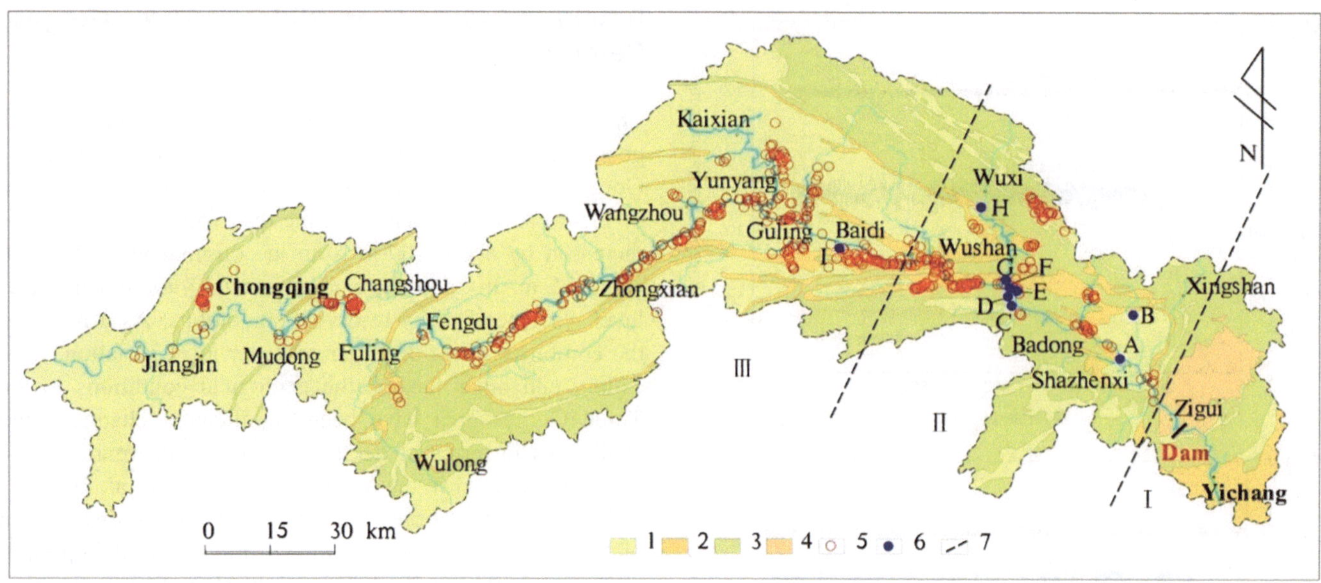

Fig. 6 Distribution map of landslides induced by six periods of 175 m asl trial impounding (Data from Sep. 1, 2008 to Aug. 31, 2014). Geology and locations: **1** sandstone, shale and coal strata, **2** mudstone mixed with sandstone and shale, **3** limestone, dolomite interbedded with shale, **4** granite, **5** six reservoir-induced landslides during 175 m asl trial impounding, **6** nine landslides described in the paper, **7** the section line of the reservoir region. Major landslides: **A** Shuping landslide, **B** Ni'erwang landslide, **C** Qingshi landslide, **D** Ganjingzi landslide, **E** Wangxia collapse, **F** Gongjiafang landslide, **G** Hongyanzi landslide, **H** Chuanzhu landslide, **I** Liangshuijing landslide

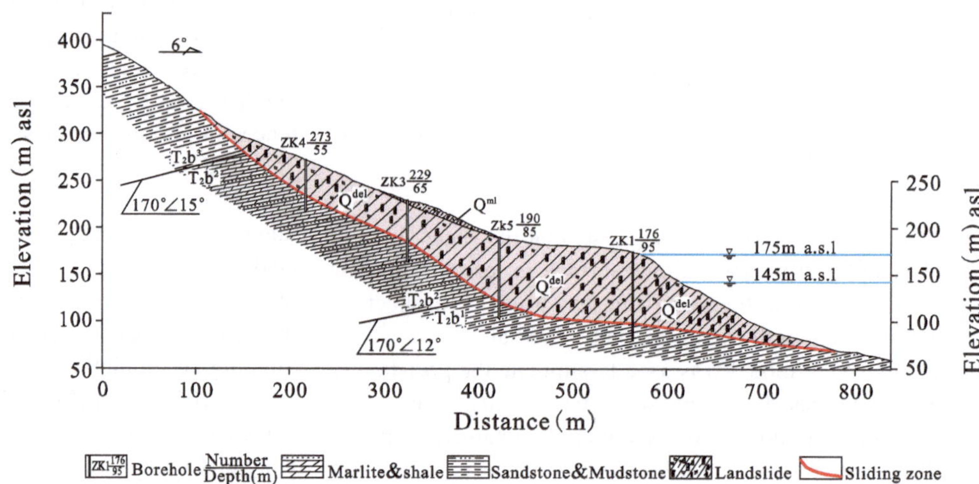

Fig. 7 Engineering geologic section for Shuping landslide in Zigui County, Hubei Province

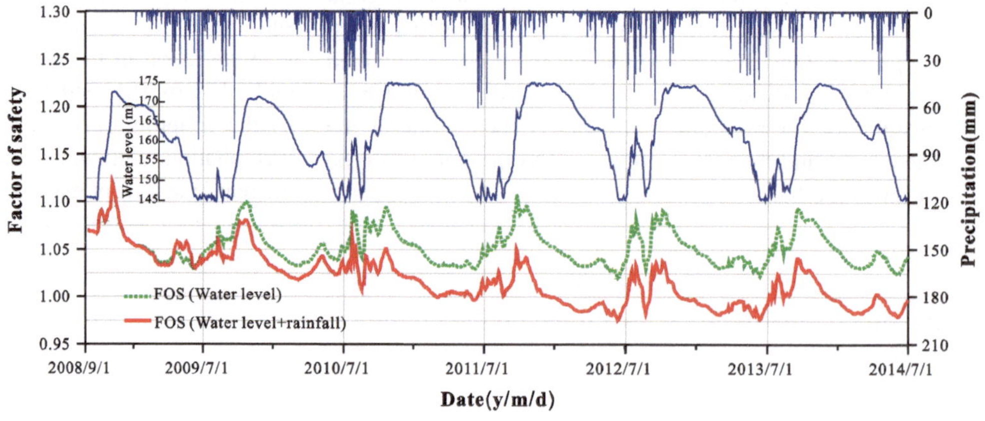

Fig. 8 Variations in the FOS of Shuping Landslide in Zigui County Hubei Province under the effects of water level and precipitation (data from Sep. 1, 2008 to Jul. 1, 2014)

The characteristics of landslide-induced impulsive wave hazards after and before reservoir impoundment were studied, and the probability of occurrence of landslide-induced-impulsive waves has increased in the reservoir region.

Simulation results of the Ganjingzi landslide in Wushan County indicate the strong relationship between landslide-induced surges and water variation, with high potential risk to shipping and residential areas. With regard to reservoir regulation in the Three Gorges, when using a single index, i.e., 1-d water level variation, water resources are not rationally utilized, and there has also been potential risks of disasters since 2008. In addition, various indices such as 1-d, 5-d, and 10-d water level variations are proposed for reservoir regulation. Finally, taking reservoir-induced landslides in June 2015 as an example, the feasibility of the optimizing indices of water level variations is verified.

Catastrophic Landslide at the Shenzhen Landfill

On December 20, 2015, a large landslide occurred at the Hong'ao Village construction solid waste (CSW) landfill in the Guangming New District of Shenzhen, Guangdong, China (103°33'48"E, 30°54'55"N) (Fig. 9). The landslide involved 2.73 million m^3 of construction waste, and had a length of about 1100 m, making it the largest landfill slope failure in the world. Because the landslide killed 77 people and destroyed 33 houses within an industrial zone of Shenzhen, the State Council of China immediately organized an investigation team to look into the causes of the landslide. Prof. Yin Yueping (first author of this paper) was the head of the expert group within this investigation team. The CGS collectively produced the cause-analysis report for the State Council and conducted research on the failure mechanism of this landslide based on field investigations, unmanned aerial vehicle (UAV) drone 3D photogrammetry, the dynamic analysis of multiphase remote-sensing images, in situ and laboratory physical-mechanics tests, computer simulation, relevant archives, and witness interviews. With increasing volumes of waste and constant changes in a landfill's slope structures, the physical and mechanical properties and the hydrologic performance of a given landfill changes over time. Therefore, in contrast to analysis of natural landslides, the study of landfill slope failures must employ a dynamic analysis that probes into boundary conditions and stability. For this landslide investigation, we adopted a multistage modeling technique to study the influence of soil mass structure and hydrologic performance changes on the stability of the landfill landslide during different phases of placement. The dynamics of rapid long run-out sliding triggered by liquefaction after slope failure was simulated with LS-RAPID software. Finally, taking typical landfill landslides worldwide into consideration, this paper discusses controls on the geotechnical risks of urbanization (Yin et al. 2016d).

The analysis indicated that the Shenzhen landfill could be divided into a frontal unit (the landfill slope with low moisture content) and a rear unit (the placing unit with ponding and a high moisture content). This dual structure had two effects: first, surface-water infiltration, in that externally generated pressures at the back pond of fresh landfilled waste were higher than would be the case under normal hydrostatic conditions in an existing landfill slope; and second, consolidation seepage, in that the externally generated load from upper placement of waste in steps leads to excess pore pressure in the lower saturated soil body. Therefore, groundwater seeped from the rear unit into the front unit, causing a decrease of stability of the front slope, and inducing the landslide.

We applied a multistage modeling technique to study the various characteristics of CSW landfill's slope structure during the five stages of CSW placement, and used non-steady fluid flow theory to analyze the groundwater seepage affecting the landfill. The results show that the landfill could be divided into two units: the frontal unit (the landfill slope), with low water content, and the rear unit (fresh waste), with ponded water and high moisture levels. This structure created two effects to trigger the landslide at the landfill—surface water infiltration and consolidation seepage. We also used soil dynamic parameters of the landslide from cone penetration, triaxial, and ring-shear tests to simulate the characteristics of a flowing slide with a long run-out due to the liquefaction effect. The result suggested that the landslide had a maximum speed of about 30 m/s with a maximum width of about 23 m, and traveled about 610 m in 130 s, at an apparent friction angle of 6° (Fig. 10).

Figure 11 shows the simulation result from LS-RAPID of the landslide at the Shenzhen landfill, showing the movement and accumulation at different times. The red grids represent the moving landslide and the green grids represent the stable mountain and plain. Figure 11a at 0 s represents the initial state of the landslide, showing the morphological characteristics before the landslide failure. Figure 11b at 5 s represents the launch of the landslide. The whole slope failure starts to move downward, with a speed of 14 m/s and a movement distance of 480 m. Figure 11c at 30 s represents the moving downward stage. The slide mass exits and the rear and its periphery collapse to the middle, with a maximum speed of 18 m/s and a movement distance of 560 m. Figure 1d at 65 s represents the stage of continuous

Fig. 9 Comparison of images showing the Hong'ao Village location, landfill, and landslide. **a** Remote-sensing image taken in December, 2013 (three months prior to CSW placement); **b** remote-sensing-based image taken on December 18, 2015 (two days before the landslide); and **c** aerial view taken on December 21, 2015 (one day after the landslide) from a UAV

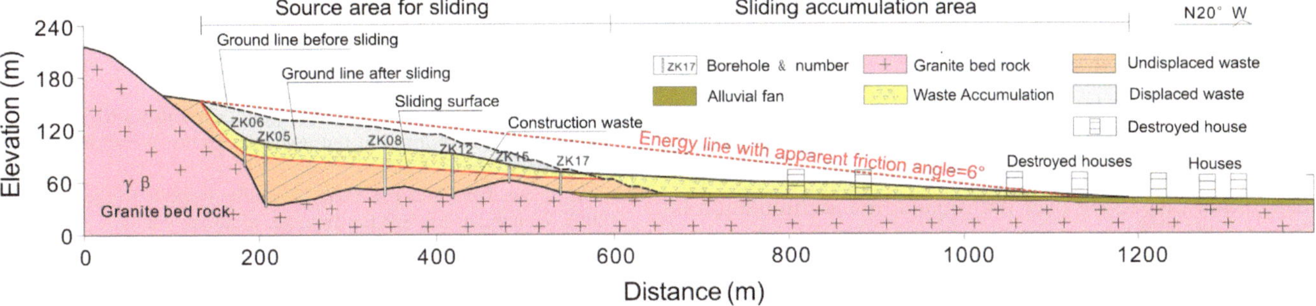

Fig. 10 Longitudinal profile of the landslide at the Shenzhen landfill. The energy line is indicated from scar of landslide to the front of accumulation with an angle of 6° and the drilling holes after sliding reveal that the slip zone is translational with a angle of 4°. The landslide exits at the terrace T1

Fig. 11 Simulation result of the long run-out of the landslide at the Shenzhen landfill with LS-RAPID

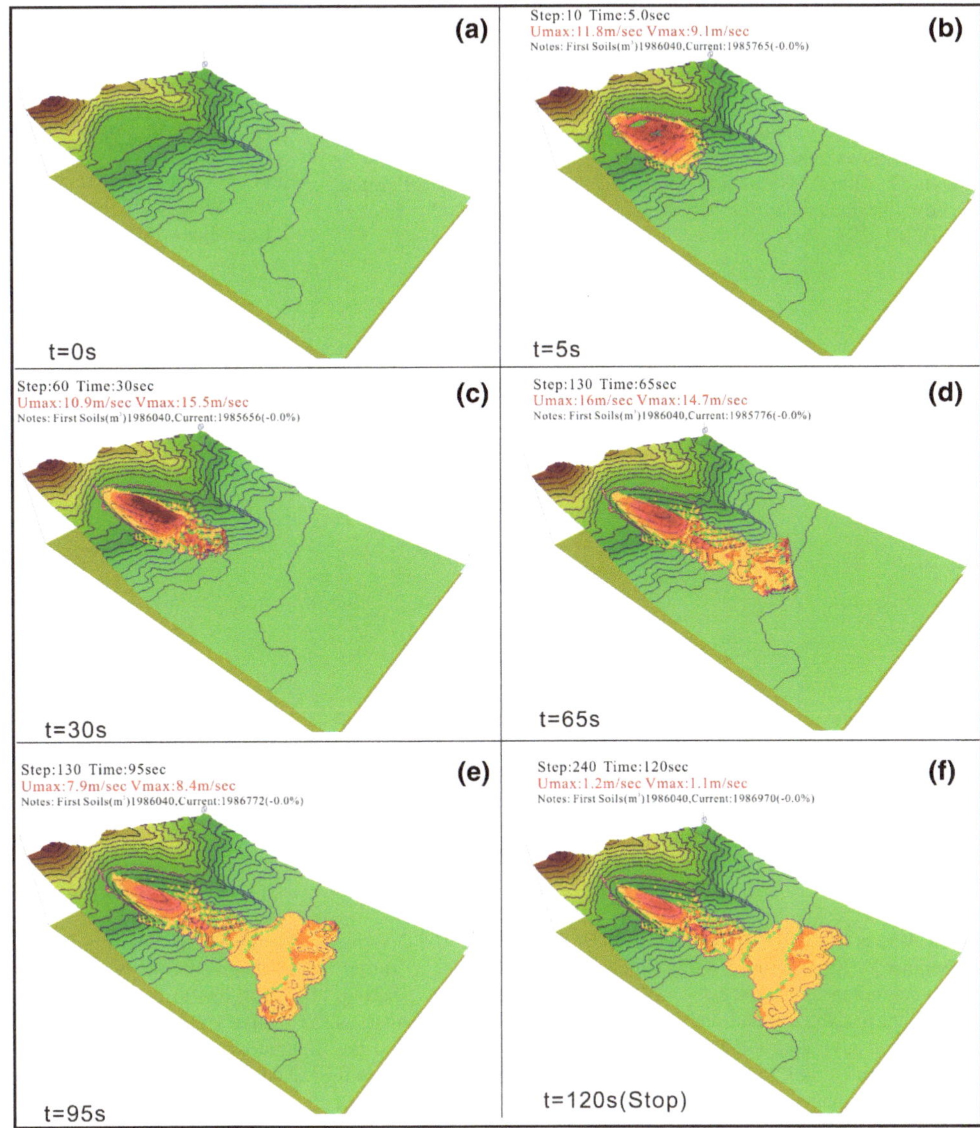

speed forward. The slide mass gradually disperses, with a maximum speed of 21.7 m/s and a movement distance of 720 m. Figure 21e at 95 s represents the decelerated motion stage. The accumulation is changing constantly, with a maximum speed of 11.5 m/s and a movement distance of 920 m. Figure 11f at 120 s represents the front accumulation stage. After dispersing over flat ground, the accumulation is still changing, with a maximum speed of 1.4 m/s and a movement distance of 1060 m.

Acknowledgements Our research activities in the field of landslide research have been mainly financed by the Ministry of Science and Technology, the Ministry of Land Resources and China Geological Survey. This financial support is greatly acknowledged.

References

Hungr O (1995) A model for the runout analysis of rapid flow slides, debris flows, and avalanches. Can Geotech J 32(4):610–623

Hungr O (2009) Numerical modeling of the motion of rapid, flow-like landslides for hazard assessment. KSCE J Civ Eng 13 (4):281–287

Sassa K, Nagai O, Solidum R, Yamazaki Y, Ohta H (2010) An integrated model simulating the initiation and motion of earthquake and rain induced rapid landslides and its application to the 2006 Leyte landslide. Landslides 7(3):219–236

Tang C, Zhu J, Ding J (2011) Catastrophic debris flows triggered by a 14 August 2010 rainfall at the epicenter of the Wenchuan Ms 8.0 Earthquake. Landslides 8:485–497

Yin YP, Wang FW, Sun P (2009) Landslide hazards triggered by the 2008 Wenchuan Ms 8.0 Earthquake, Sichuan, China. Landslides 6 (2):139–152

Yin YP, Cheng YL, Liang JT, Wang WP (2016a) Heavy-rainfall-induced catastrophic rockslide-debris flow at Sanxicun, Dujiangyan, after the Wenchuan Ms 8.0 earthquake. Landslides 13(1):9–23

Yin YP, Xing AG, Wang GH et al. (2016b) Experimental and numerical investigations of a catastrophic long-runout landslide in Zhenxiong, Yunnan, Southwestern China. Landslides. doi:10.1007/s10346-016-0729-z

Yin YP, Huang BL, Wang WP et al (2016c) Reservoir-Induced landslides and risk control in the Three Gorges Project, the Yangtze River, China. J Rock Mech Geotech Eng 8(5):577–589

Yin YP, Li B, Wang WP et al (2016d) Mechanism of the December 2015 catastrophic landslide at the Shenzhen landfill and controlling geotechnical risks of urbanization. Engineering 2 (2):230–249

Zhou W, Tang C (2013) Rainfall thresholds for debris flow initiation in the Wenchuan earthquake-stricken area, southwestern China. Landslides 11(5):877–887

Landslide Risk Reduction in Croatia: Scientific Research in the Framework of the WCoE 2014–2017, IPL-173, IPL-184, ICL ABN

Snježana Mihalić Arbanas, Željko Arbanas, Martin Krkač, Sanja Bernat Gazibara, Martina Vivoda Prodan, Petra Đomlija, Vedran Jagodnik, Sanja Dugonjić Jovančević, Marin Sečanj, and Josip Peranić

Abstract

In this paper scientific activities of the Croatian Landslide Group (CLG), World Centre of Excellence on Landslide Risk Reduction (WCoE) of the International Consortium on Landslide (ICL) for the period 2014–2017, are shortly described. The results of scientific research are presented through the fields of landslide science: landslide identification and mapping, landslide investigation and testing, landslide monitoring, landslide modelling and landslide stabilization and remediation. It is concluded that the resulting landslide inventory maps, regional empirical rainfall intensity-duration thresholds, kinematic landslide models and soil strength parameters, landslide movement prediction models, numerical models and simulations and behavior of geotechnical construction for landslide stabilization provide necessary information for landslide risk management in Croatia. Besides applied scientific research, the general objectives of ICL WCoE are achieved in the framework of two Croatian IPL Projects and regional ICL Adriatic-Balkan Network.

Keywords

Croatian landslide group • ICL WCoE • Landslide science • Landslide risk reduction

S. Mihalić Arbanas (✉) · M. Krkač · S. Bernat Gazibara · M. Sečanj
Faculty of Mining, Geology and Petroleum Engineering, University of Zagreb, Pierottijeva 6, Zagreb, 10000, Croatia
e-mail: smihalic@rgn.hr

M. Krkač
e-mail: mkrkac@rgn.hr

S. Bernat Gazibara
e-mail: sbernat@rgn.hr

M. Sečanj
e-mail: msecanj@rgn.hr

Ž. Arbanas · M. Vivoda Prodan · P. Đomlija · V. Jagodnik · S. Dugonjić Jovančević · J. Peranić
Faculty of Civil Engineering, University of Rijeka, Radmile Matejčić 3, Rijeka, 51000, Croatia
e-mail: zeljko.arbanas@gradri.uniri.hr

M. Vivoda Prodan
e-mail: martina.vivoda@gradri.uniri.hr

P. Đomlija
e-mail: petra.domlija@gradri.uniri.hr

V. Jagodnik
e-mail: vedran.jagodnik@gradri.uniri.hr

S. Dugonjić Jovančević
e-mail: sanja.dugonjic@gradri.uniri.hr

J. Peranić
e-mail: josip.peranic@gradri.uniri.hr

K. Sassa et al. (eds.), *Advancing Culture of Living with Landslides*,
DOI 10.1007/978-3-319-59469-9_25

Introduction

The Croatian Landslide Group (CLG) from University of Rijeka and the University of Zagreb encompass scientists from the fields of engineering geology and geotechnics dealing with scientific research on landslides, i.e., landslide science. Scientists from the CLG started joint research in 2009 in the framework of the bilateral Japanese-Croatian scientific SATREPS FY2008 Project (2009–2014) aimed at enhancement of landslide research in Croatia through collaboration with scientists from Kyoto and Niigata universities. Based on outcomes of the SATREPS FY2008 Project, CLG become a World Centre of Excellence on Landslide Risk Reduction (WCoE) of the International Consortium on Landslide (ICL) for the period 2014–2017. The general objectives of ICL WCoE are: (1) to strengthen the International Programme on Landslides (IPL) and IPL Global Promotion Committee; (2) to create "A Global Network of entities contributing to landslide risk reduction"; and (3) to improve the global recognition of "Landslide Risk Reduction" and its social-economic relevance, and entities contributing to this field.

This article presents scientific research of the CLG in the period 2014–2017, related to the following topics: landslide identification and mapping, landslide investigation and testing, landslide monitoring, landslide modelling and landslide stabilization and remediation. Here are also incorporated IPL and ICL activities of the CLG, IPL Projects 173 and 184, and the regional ICL network, the Adriatic-Balkan Network (ICL ABN). The scientific research of CLG is also considered in terms of its application to landslide risk reduction on a national and international level.

Landslide Identification and Mapping

Extreme weather conditions in winter and spring of 2013 (re) activated more than 900 shallow landslides in North-Western Croatia, as it is recorded by the Croatian National Protection and Rescue Directorate (DUZS). Bernat Gazibara et al. (2016) compiled a catalog of 85 precipitation events that have caused landslides in the continental part of Croatia, within an area of approx. 10,500 km^2 (Fig. 1), and covering the period between June 2006 and October 2014. Three counties show a high average frequency of 2.4–4.8 precipitation events per year. The paper presents a preliminary analysis of precipitation conditions and a comparison with global and national ID (Intensity-Duration) thresholds. For example, annual precipitation in 2010, 2013 and 2014 was 20–40% higher than the Mean Annual Precipitation (883.6 mm) in the last 152 years, according to data from the Zagreb-Grič meteorological station, located in the central part of NW Croatia. Analysis of data showed that during January, February and March 2013, cumulative monthly precipitation was 130–190% higher than the average monthly values for the same period from 1862 to 2014. The cumulative precipitation for a 3-month period in 2013 has the highest value in the last 150 years (Bernat et al. 2014c). A comparison between precipitation conditions that caused landslides in NW Croatia and published global ID thresholds shows that precipitation conditions are lower than the threshold curves proposed by Caine (1980), Crosta and Frattini (2001), and Guzzetti et al. (2008). The best fit to the analyzed precipitation data shows the global ID threshold for soil slips according to Clarizia et al. (1996). A catalog of precipitation events will serve as the basis for further more comprehensive spatial and temporal probabilistic analysis for defining regional empirical threshold for shallow soil slides formed in Upper Miocene and Quaternary sediments of NW Croatia. Precipitation events analyses performed in this study were based on limited landslide inventory data (Mihalić Arbanas et al. 2016a). Therefore, more reliable analyses require a more detailed landslide inventory of small (<10^5 m^3) superficial to moderately shallow (<20 m) landslides, characteristic of the study area.

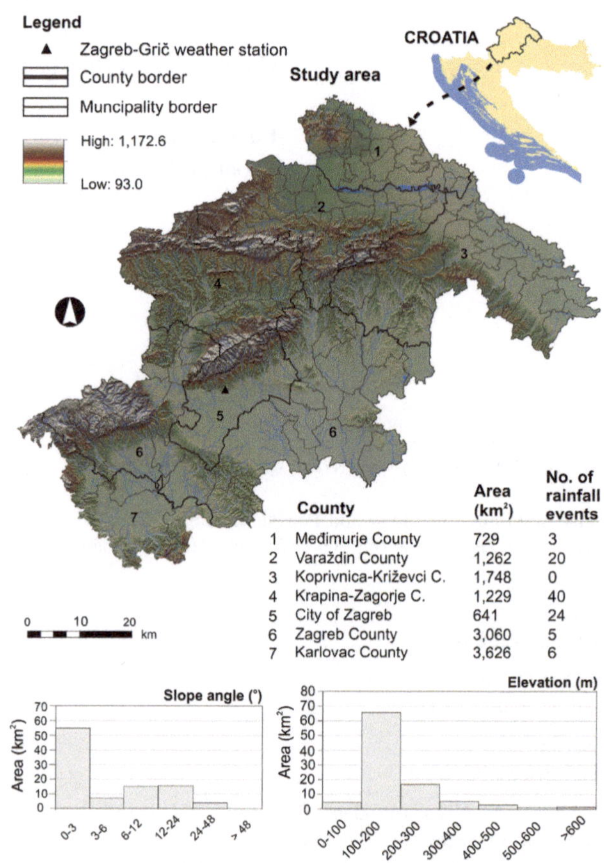

	County	Area (km²)	No. of rainfall events
1	Međimurje County	729	3
2	Varaždin County	1,262	20
3	Koprivnica-Križevci C.	1,748	0
4	Krapina-Zagorje C.	1,229	40
5	City of Zagreb	641	24
6	Zagreb County	3,060	5
7	Karlovac County	3,626	6

Fig. 1 Relief map of NW Croatia with the number of precipitation events in catalog (Bernat Gazibara et al. 2016)

A seasonal landslide inventory for the hilly area of the Medvednica Mt. (City of Zagreb in central part of NW Croatia), compiled based on records of reported landslide events from January to April 2013, contains 55 landslides (Bernat et al. 2014b). All recorded landslides are soil slides; 51% are superficial landslides (depth <1 m, area <200 m^2) formed in colluvial deposits overlaying engineering soil and soft rocks (marls); and 49% are shallow landslides (estimated depth 3–12 m, area <14,000 m^2) developed in stratified Upper Miocene (silty and sandy soils) and Quaternary sediments (heterogeneous mixtures of unfoliated, mostly impermeable fine grained soils). Figure 2 shows part of the geomorphological inventory compiled for the test area of the Medvednica Mt. hills (21 km^2) based on visual interpretation of airborne LiDAR (Light Detection and Ranging) DTM data from December 2013, with a spatial resolution of 15 × 15 cm (Bernat Gazibara et al. 2017). Landslide identification resulted in a landslide inventory map indicating the contours of 783 landslides, which implies an average landslide density of 37.28 landslides per square kilometer. Seventy-five percent of the landslides had an area between 159 and 2018 m^2. The area of the smallest identified landslide in the test area is 43 m^2. Only a minor number of landslides were reactivated in 2013 and two of them, showed in Fig. 2, are fully or partially located inside a zone of agricultural areas.

Podolszki (2014) also performed conventional visual interpretations of stereoscopic aerial photographs from 1964 at a scale of 1:8000 for the hilly area of the Medvednica Mt. He derived a landslide inventory map with an area of 54.14 km^2 for 963 landslides, which gives an average landslide density of 17.8 landslides per square kilometer. Stereoscopic analysis of historical aerial photographs from 1964 over a large scale enabled the identification of landslides over a range from 78 m^2 to 281,886 m^2. The landslide areas of most of the landslides (90.6%) range from 200–3600 m^2, which is in accordance to abovementioned results of application of new innovative LiDAR technologies. The reliability of aerial photo identification is estimated as very low, because only 50% of all mapped landslides were evaluated as reliable, based on certainty of identification, which implies only 9 reliably identified landslides per square km. Moreover, the quality of mapped landslide contours is very low because of forested terrain that makes it difficult or even impossible to map small and shallow landslides.

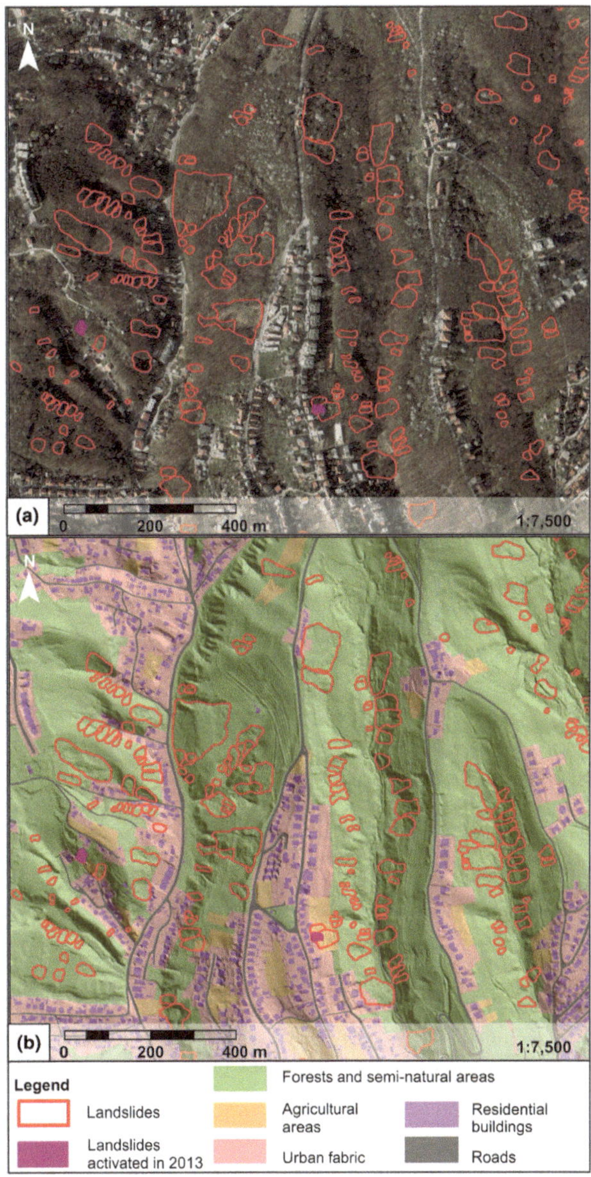

Fig. 2 Geomorphological inventory compiled for the test area of the Medvednica Mt. hills, based on visual interpretation of airborne LiDAR DTM data from December 2013 with a spatial resolution of 15 × 15 cm (Bernat Gazibara et al. 2017): **a** Landslide contours depicted on the orthophoto map at the scale 1:5000; **b** Landslide contours depicted on the official land use map of the City of Zagreb

The geomorphological environment composed of carbonate and flysch rocks, located in a coastal part of Croatia, are prone to a variety of landslide types. Research on landslide identification in the Vinodol Valley (Primorsko-Goranska County) within the wider zone of the northern Adriatic coast, has been undertaken in the framework of doctoral research by Đomlija (2017). Đomlija et al. (2016) identified geomorphological processes causing slope instabilities in the Vinodol Valley, fluvial erosion and landslides. There are three types of fluvial erosion processes: planar, rill and gully erosion. Figure 3a illustrates linear erosion processes on steep slopes composed of hard carbonate rocks. Figure 3b shows denudational slope in the central part of the Valley composed of sylicyclastic soft

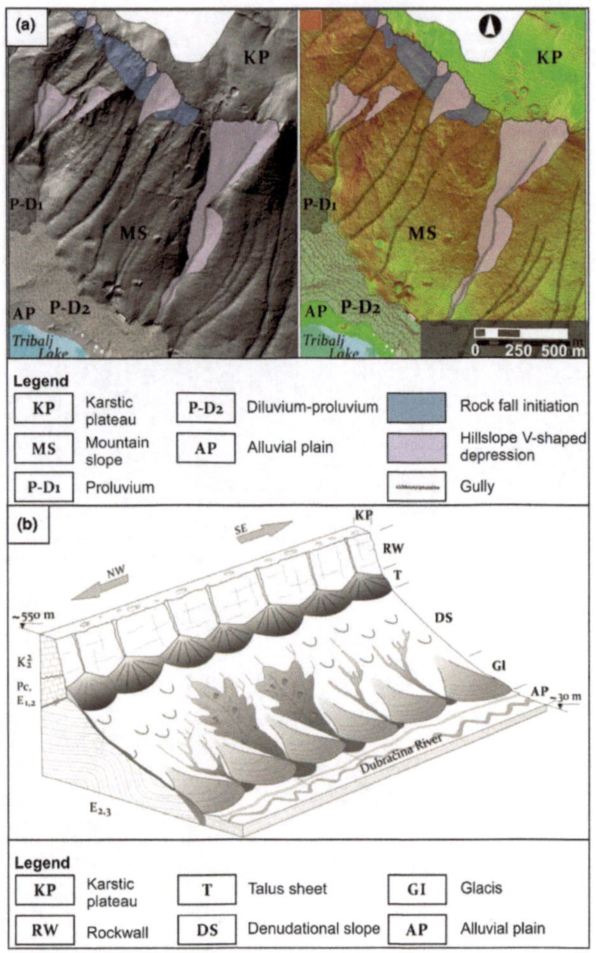

Fig. 3 Identified geomorphological phenomena and geomorphological units in the Vinodol Valley according to Đomlija et al. (2016): **a** Rock falls and gullies on the steep slopes composed of carbonate rocks; **b** Simplified model depicting geomorphological units in the central part of the Vinodol Valley

rocks in which excessive erosion is dominant, especially in some sub-basins, resulting in badland relief type. Bernat et al. (2014a) gave review of literature dealing with causes of excessive erosion in the Salt Creek catchment in the Vinodol Valley, related to mineralogical composition of flysch bedrock. Table 1 lists processes and phenomena in the central part of the mineralogical composition of flysch bedrock. Table 1 Vinodol Valley. Landslides classified according to classification by Hungr et al. (2014) encompass rock falls, breccia falls, boulder/debris/silt falls, rock block topples, gravel/sand/silt topples, rock slides (wedge, planar, rotational), debris slides, clay/silt slides (planar, rotational), breccia spreads, rock avalanches, mud flows, rotational earthflows, soil creep and slope deformation. Đomlija et al. (2014) present an applied methodology of identification and mapping, which is a visual interpretation of morphological features of instability phenomena on LiDAR-derived imagery from March 2012 with a 1 m-resolution. The methodology proved to be useful for identifying and mapping of variety of landforms and phenomena resulting from different types of hazardous geomorphological processes developed in carbonate rock masses, flysch and its derivative materials, classed as superficial deposits. The research shows the necessity of identification of geomorphological units, as a base for compilation of a landslide inventory and susceptibility and hazard analyses. Moreover, conditions in geological and geomorphological environments composed of carbonate and flysch rocks are complex, requiring analysis of the interrelationships of landslides and erosion hazards. Đomlija et al. (2016) showed that spatial density of landslides in zones of excessive erosion is extremely high, resulting in necessity of identification of landslide and erosion phenomena.

Landslide Investigation and Testing

This section presents research which results in quantitative characterization of the landslide site, including identification of the geometry of relatively homogenous zones of soil/rock, as well as the constructive properties of the material within the zones. Constructive properties are those parameters that allow the prediction of a material's strength, deformation, or permeability, in response to changes over time due to stress or other environmental conditions (Dowding 1979).

Surface observations and engineering geologic mapping were undertaken at the area of the Valići Lake landslide (Rječina River Basin in Primorsko-Goranska County), as a part of urgent measures, immediately after landslide reactivation in February 2014 (Mihalić Arbanas et al. 2016b). The purpose of the detailed mapping of this active landslide formed in weathered flysch (siltstones and sandstones), as well as the adjacent region that could contribute to causes of movement, was to provide the foundation for numerical modelling of the landslide (Arbanas et al. 2016), planning subsurface investigations and locating instrumentation for landslide monitoring. Collection of surface data included access to existing information, i.e., use of topographic maps, and field site inspection. Airborne LiDAR-derived imagery from March 2012 with a spatial resolution of 1 × 1 m was used for a comprehensive overall view of the landslide site. Figure 4a shows contours of the large deep historical landslide (landslide area of 1.3×10^6 m^3, landslide depth of 28.5 m) interpreted using a high-resolution bare-earth DEM. The estimated dimensions of the landslide body of the

Table 1 Geomorphological units in the central part of the Vinodol Valley with attributed hazardous processes and phenomena (Đomlija et al. 2016)

Geomorphological unit		Movement processes	Movement phenomena	Erosion processes	Erosion phenomena
KP	Karstic plateau	Slope deformation	Rock slope deformations	Sheet washing Gullying	Gullies
MS	Mountain slope	Falling	Rock falls		
		Toppling	Rock block topples		
		Sliding	Rock planar slides Rock wedge slides Rock rotational slides		
		Slope deformation	Rock slope deformations		
RW	Rockwall	Falling	Rock falls		
		Toppling	Rock block topples		
		Sliding	Rock planar slides Rock wedge slides		
		Flowing	Dry debris flow		
T	Talus sheet	Flowing	Dry debris flows	No processes	No phenomena
Br	Talus breccias	Falling	Breccia fall	Sheet washing	No phenomena
		Spreading	Breccia spreads		
DS	Denudational slope	Falling	Boulder/debris/silt falls	Sheet washing Rilling Gullying	Rills Gullies
		Toppling	Gravel/sand/silt topples		
		Sliding	Debris slides Clay/silt rotational slides Clay/silt planar slides		
		Flowing	Mud flows		
		Complex	Rotational slide-earthflows		
		Creeping	Soil creep		
Gl	Glacis	Sliding	Clay/silt rotational slides Clay/silt planar slides		
		Complex	Rotational slide-earthflows		
AP	Alluvial plain	No processes	No phenomena	No processes	No phenomena

historical landslide are length of 350 m and width of 135 m. The toe of the historical landslide reached the bank of the Valići reservoir 250 m upstream of the Valići dam. During construction of the Valići dam and lake in 1960, the historical landslide was partially remediated by a surface and subsurface drainage system. Figure 4b presents a predictive model for reactivation of the historical landslide in 2014 (landslide volume of 3.68×10^5 m^3, landslide depth of 20 m) and its relative position to Valići Lake, which is endangered by potential movement of the landslide. The movement of the reactivated landslide of approx. 12–15 m down the slope caused substantial damage to the local road. The trigger for landslide reactivation in 2014 was a 15-day rainfall event with 302.6 mm of cumulative rainfall from 30 Jan to 13 Feb 2014.

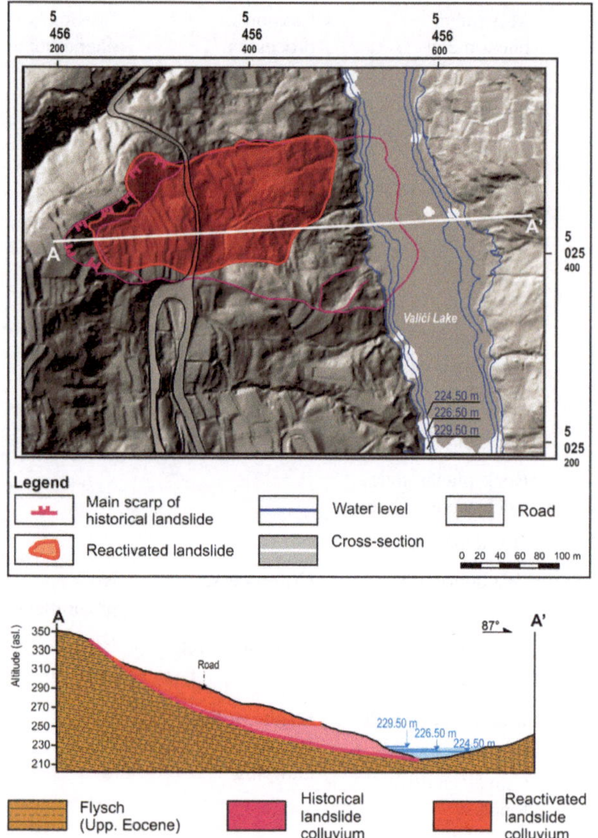

Fig. 4 Valići Lake landslide reactivated in February 2014 (Mihalić Arbanas et al. 2016b): **a** Landslide map with contours of an historical landslide visually identified using airborne LiDAR DTM data from May 2012, 1 m-resolution, and landslide contours of the reactivated landslide; **b** Engineering geological cross-section showing the historical and reactivated landslides

Identification of digital surface models derived from the point cloud data obtained with terrestrial and airborne laser scanning (TLS and ASL) was applied to rock slope characterization for the purpose of stability analyses and design of remedial measures. Sečanj et al. (2017) presented visual and semi-automatic identification of rockfall-prone areas on the steep slopes above the historical town of Omiš (Splitsko-Dalmatinska County), situated on the southern part of the Adriatic Coast. The ancient town of Omiš is a tourist destination, cultural and historical heritage site. The residential part of Omiš, along with its historical landmarks, are located at the toe of the steep slopes of Omiška Dinara Mt., composed of a massive, blocky to very blocky carbonate rock mass. People and material goods are highly endangered by rock falls from steep slopes with a total area of approx. 15 ha and a maximum slope height of 277 m. To identify rockfall-prone areas on steep slopes, discontinuities and unstable rock blocks were mapped in detail using remote sensing. Potentially unstable rock blocks were identified based on kinematic analysis. The volume of hazardous rock blocks varies in size from 0.1 to 1235 m^3. The same approach in investigation and interpretation of remote sensing data was used for rock slope characterization of road slope cuts along county roads in National Park Krka on the north side of the Brljan Lake (Šibensko-Kninska County), cut in a very blocky and disintegrated carbonate rock mass. Total length of the slope cuts along county road is 900 m, maximal height of the slopes is 65 m and slope angles vary from 30° to 90°. The road is placed just in the bottom of the slope cuts, without a rock fall catchment zone, which makes the road very risky, especially for people. Identification of columnar rock blocks in the base of ancient Momjan Castle from the 13th century (Istrian County in the north Adriatic Coast) built on a cliff was conducted by interpretation of point cloud data obtained using terrestrial laser scanning (TLS). Vertical columnar rock blocks have been continuously forming along multiple sets of vertical discontinuities intersecting a mega-bed (thickness of about 10 m) of calcareous rock in flysch, underlain by marl. Weathering along vertical discontinuities in the hard rock, as well as weathering of soft rock in the base of columnar blocks, is the main preparatory causal factor of rock toppling. These instabilities in the foot of the Castle foundations seriously endanger this cultural heritage site with the threat of complete ruination of the Castle.

Research on the changes of engineering properties of flysch rock mass due to weathering in the wider zone of the northern Adriatic coast (Rječina River and Vinodol Valley in Primorsko-Goranska County and Istrian Peninsula) have been undertaken in the framework of doctoral research by Vivoda Prodan (2016). Vivoda Prodan et al. (2016) analyzed the effects of weathering on the shear strength of siltstones from the flysch rock mass. Six weathering grades of siltstone member were identified, based on material color, state of discontinuities, presence or absence of the original rock structure, and uniaxial strength. X-ray analyses revealed significant increase of clay mineral (phyllosilicates) content, especially chlorite and illite, with increasing weathering grade and a consequent increase in cation exchange capacity (CEC) and water adsorption. Identification of physical and mechanical properties and the residual strength of different weathering grades of siltstones were performed on the soil-like material produced in laboratory conditions by crushing samples to sand-sized particles. The testing procedure, using ring shear device ICL-1 (described in Oštrić et al. 2014) and direct shear devices, enabled simulation of slip surface development in remolded material from flysch slopes

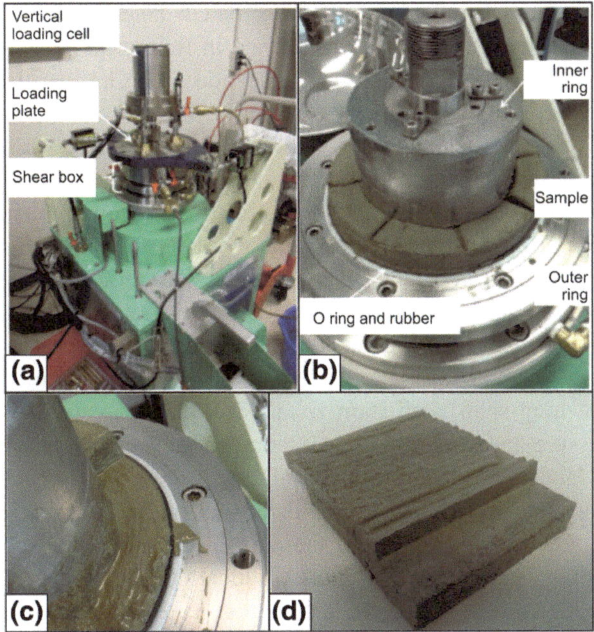

Fig. 5 Siltstone sample in a ring shear device: **a** during and **b** after testing. The sliding surface is shown after shearing in **c** the ring shear device and **d** the direct shear device (Vivoda Prodan et al. 2016)

(Fig. 5). The test results indicate increase of liquid limits, plasticity indexes and decrease of peak and residual shear strength with increase in weathering grade of siltstones from flysch rock masses. Vivoda Prodan and Arbanas (2016) also performed a series of drying–wetting cycles to simulate natural conditions of the weathering process involved in the disintegration of the rock mass material into sand-sized and smaller particles to establish the relationship of geotechnical properties and durability with weathering of siltstones from a flysch rock mass in the Istria Peninsula. According to the results of the study, weathering has a significant influence on the plasticity of soil-like remolded material of siltstones in terms of increase of liquid limit and plasticity index with increasing weathering grade. The uniaxial compressive strength determined with a Schmidt hammer and the Point Load Test (PLT) decreases with increasing siltstone weathering grade. The study showed that determination of durability of siltstones requires calculation of a degradation ratio and the modified degradation ratio as additional parameters that indicate the manner of disintegration of siltstones of different weathering grades. According to the classification based on the slake durability index, the siltstone samples of different weathering grades from Istria are classified in a higher durability class than in classifications based on the fragmentation during the slake durability test. The obtained laboratory results indicate that weathering has a significant influence on the plasticity, uniaxial compressive strength, durability characteristics and shear strength of the siltstones. Application of the described research results are demonstrated by use of different physical-mechanical parameters of different weathering states of siltstones in numerical simulations of behavior of the Valići Lake landslide (Vivoda Prodan and Arbanas 2017).

Landslide Monitoring

Integrated automated monitoring system of the large and deep Kostanjek landslide (City of Zagreb) in soft rocks of Upper Miocene age in the continental part of Croatia was established in the period of 2011–2013 (Krkač et al. 2014c). External triggers at the Kostanjek landslide have been measuring with a rain gauge and accelerometers, displacements at the surface by GNSS sensors and extensometers, subsurface displacement have been measured by vertical extensometers and an inclinometer; while hydrological monitoring consists of groundwater level (GWL) measurements and discharge measurements. Krkač et al. (2014b) reviews the monitoring parameters which have been continuously and discontinuously observed at the Kostanjek landslide. The Kostanjek landslide presents a hazard and risk for approx. 290 buildings (mostly residential houses) in an area of 1 km^2 in the urban part of the city of Zagreb. Automated data transmission from all sensors makes the monitoring system suitable for the establishment of an early warning system (EWS), which is planned in collaboration with City administration responsible for emergency management (Krkač et al. 2014a).

In a doctoral thesis, Krkač (2015) presents analysis of a temporal data series of movement and GWL captured by the Kostanjek landslide monitoring system in the period from January 2013 to January 2015. Analysis of movement measurements showed that 90% of total displacement (426.09 mm) happened during five periods of accelerated displacement. Statistical analysis revealed that the periods of accelerated displacements are a consequence of nine periods of GWL increase. Analysis of precipitation data proved that changes in GWL happened during wet periods, i.e., the same amount of precipitation will have a different influence on GWL in dry and wet periods. For this reason, modeling of the relation between precipitation and GWL, and GWL and landslide movement, are performed separately. Statistical methods used for precipitation and GWL and GWL and landslide movement modeling were multiple linear regression and random forests. The purpose of the modeling was prediction of GWL based on precipitation data, as well as prediction of movement based on GWL data. Based on

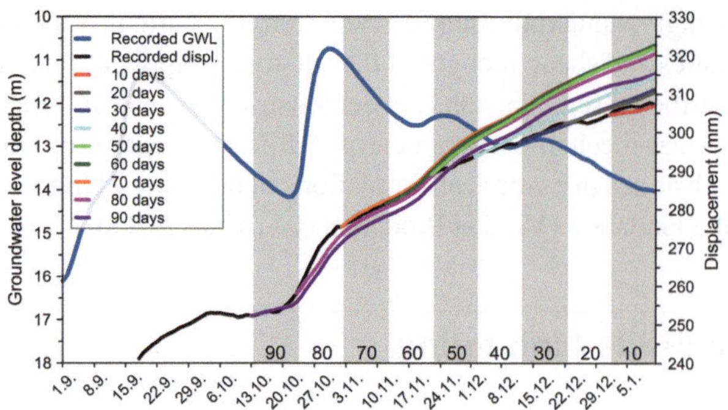

Fig. 6 Prediction of landslide displacement by random forests from groundwater level data (Krkač et al. 2016)

comparison of multiple linear regression and random forests models, performed using mean square error and correlation coefficient between modeled and measured values, as well as k-fold cross validation and validation, it is concluded that better results have been achieved using random forests. Krkač et al. (2016, 2017) presents a methodology for prediction of landslide movements using random forests, a machine learning algorithm based on regression trees. The process of landslide movement prediction is divided into two separate models: (1) a model for prediction of GWL from precipitation data; and (2) a model for prediction of landslide movements from GWL data. In a GWL prediction model, 75 parameters were used as predictors, calculated from precipitation and evapotranspiration data. In the landslide movement prediction model, 10 parameters calculated from GWL data were used as predictors. Model validation was performed through the prediction of GWL and prediction of landslide movements for periods from 10 to 90 days (Fig. 6). The validation results show the capability of the model to predict the evolution of daily displacements from predicted variations of GWL for the period of up to 30 days. A practical contribution of the developed method is the possibility of automated predictions of landslide movement, updated and improved on a daily basis, which would be an important source of information for decisions related to crisis management in the case of risky landslide movements.

Integrated automated monitoring system of the large and moderate shallow landslide Grohovo (Rječina Valley in the City of Rijeka) in flysch rocks was established in the period of 2010–2011 (Arbanas et al. 2014b, c). The Grohovo landslide is the largest active landslide (landslide area of 10 ha; landslide volume of 3×10^6 m^3) along the Croatian part of the Adriatic coast, formed at the contact of fresh and weathered siliciclastic rock (i.e., flysch; Benac et al. 2014) covered by debris with limestone boulders. The landslide represents the reactivated part of a much larger historical landslide of unknown age (estimated landslide area of 40 ha), with the main scarp formed in limestone rock mass visible as a cliff at the top of landslide. The Grohovo landslide is complex landslide which exhibits two types of movement, rock falls and debris sliding. Landslide reactivation in 1996 shifted attention to the high level of landslide hazard and necessity of its monitoring. The comprehensive monitoring system of the Grohovo landslide consists of geodetic and geotechnical monitoring. Geodetic monitoring includes geodetic surveys with a robotic total station and displacement measurements of GPS sensors. Equipment for the geotechnical monitoring includes vertical inclinometers, and long-span and short-span wire extensometers. Pore pressure gauges and weather station are aimed at monitoring of landslide causal factors. All monitoring equipment is connected in one system, with continuous monitoring and automated transmission of the data. The main disadvantage of the system is power supply. Ljutić et al. (2014) describes problems related to the installation and working of the system, as well as solutions which should improve the system for successful continuous work.

Landslide Modeling

Numerical modeling of Croatian landslides has been performing using LS-RAPID Software since 2013. LS-RAPID Version 2.03 Beta10, developed in 2013, is described in He et al. (2014). This landslide simulation software is developed to reproduce the initiation process and the runout process from stable state until deposition within the same model. It can simulate the initiation and motion of landslides triggered by earthquakes, rainfall or the combined effects of rainfall and earthquakes. The modeling is based on the following key parameters: the shear resistance at the steady state, the peak friction angle, cohesion, the shear displacements at peak and the onset of steady state and the lateral pressure ratio with physical meaning.

Landslide modeling using LS-RAPID software was performed for the Grohovo landslide (Vivoda et al. 2014) and the Valići Lake landslide (Arbanas et al. 2016, 2017) in the Rječina River Valley, and the Kostanjek landslide in the City of Zagreb (Gradiški et al. 2013, 2014). Vivoda et al. (2014) implemented the deterministic 3D landslide stability analyses of the wider area (approx. 40 ha) of the Grohovo landslide to found out possible landslide reactivation under unfavorable groundwater conditions inside the area of the historical landslide. The topography of the Grohovo landslide was determined using 10 m-resolution DEM data. The limestone rock mass is situated at the top of the slopes, while the siliciclastic rocks and flysch are situated on the lower slopes and the bottom of the valley. Depth of the sliding mass varies from 3 to 10 m and the slip surface is formed in flysch bedrock. The final result of the modeling was a failure probability map, converted from the simulation model. The long-term rainfalls and consequent ground water level rise were detected as the main triggering factor for the Grohovo landslide reactivation. Arbanas et al. (2016) used LS-RAPID software to determine possible scenarios for the Valići Lake landslide movement for four scenarios with different reservoir water levels at the foot of the landslide, which correspond to full reservoir, dam overflow level, and two lower safety levels. The foot of the landslide is submerged in the reservoir and the magnitude of motion, run off sliding path, and deposition of the sliding mass will significantly depend on the reservoir water level. Analyses showed that in a case of a high reservoir water level, the sliding mass would fill the reservoir, causing the water level to rise and waves (tsunamis) which would overflow the Valići dam, resulting in significant damage downstream along the Rječina River channel. Numerical simulation results enabled an estimation of landslide risk for the cases of different scenarios of reservoir water level (Arbanas et al. 2017). Gradiški et al. (2013, 2014) performed simulation of the Kostanjek landslide motion using LS-RAPID software in a case of rising ground water level and a possible earthquake. It was found that the eastern and central parts of the landslide are more unstable than the western part. An earthquake with an acceleration of 0.344 cm/s^2 will cause movements of the whole existing landslide body. The LS-RAPID simulation software was used for deterministic landslide susceptibility analyses in the flysch area of North Istria after identification of typical landslide phenomena on flysch slopes (Arbanas et al. 2014a; Peternel et al. 2016). This region is identified as an area of frequent, small and shallow instability phenomena, mostly triggered by rainfall. The deterministic landslide susceptibility analyses were also performed for the part of the City of Buzet area (Dugonjić Jovančević 2013; Dugonjić Jovančević et al. 2014a, b, 2016) to identify the zones susceptible to sliding caused by rising ground water levels. Ground water level is taken into account through the pore pressure ratio, which gradually increases due to superficial water infiltration. Žic et al. (2015) analyzed propagation of a possible mudflow from the Grohovo landslide downstream through the Rječina watercourse to the City of Rijeka using a smoothed-particle hydrodynamics (SPH) model. Based on conducted computational simulations, it can be concluded that the potential mudflow propagation is unlikely to threaten the urban part of the city of Rijeka and that it is unlikely to cause substantial effects on the environment or lead to loss of human lives.

Table 2 Landslides and designed remedial measures in the period 2013–2017 by the CLG

Landslide name	Landslide type	Landslide area (m^2)	Remedial measures
Gabrovica	Debris slide	83	Gabion wall
Cerina 1	Debris slide	100	Gabion wall
Muellerov Brijeg	Debris slide	198	Pile wall
Črnica	Rock fall	200	Rock anchoring
Cerina 2	Debris slide	210	Gabion wall
Sovinjak	Soil slide	351	Concrete wall
Sv. Martin	Soil slide	500	Gabion wall
Marovići	Soil slide	525	Gabion wall
Braslovje	Soil slide	532	Gabion wall
Gradišće	Soil slide	709	Gabion wall
Cerina 3	Debris slide	760	Gabion wall
Grohovski put	Soil slide	850	Pile wall
Remete 1	Soil slide	877	Gabion wall
Galgovo-Zagrebačka	Debris slide	1.200	Gabion wall
Grdanjci	Rock slide	1.269	Gabion wall
Grohovo-HEP	Debris slide	1.547	Pile wall
Momjan Castle	Rock topple	1.600	Rock anchoring
Havišće	Rock falls	2.000	Rock anchoring
Brljan	Rock falls	20.000	Rock anchoring, rock fall barriers
Omiš	Rock falls	150.000	Rock anchoring, rock fall barriers

Landslide Stabilization and Remedial Measures

To improve understanding of remedial constructions on slope stability, two research studies based on in situ-testing, long-term monitoring and numerical analyses were conducted. Jagodnik, in the framework his Ph.D. thesis (Jagodnik 2014) tested behavior of horizontally loaded piles on a testing site constructed in natural sandy gravels. Individual piles, group of piles and pile walls are very often applied as effective remediation measures of landslides (Popescu 2001). The research resulted with *p-y* curves of piles exposed to controlled horizontal loads in the head of a pile (Jagodnik and Arbanas 2015a, b) that would enable better control of forces in piles used as a remedial measure for landslides. Grošić in his Ph.D. thesis (Grošić 2014) analyzed time-dependent deformation of reinforced cuts in flysch rock masses based on long-term monitoring results. The analysis results have indicated significant time-dependent deformation in slopes built of weak rock masses and consecutive redistribution of the forces in rock mass reinforcement elements over time that can lead to the progressive failure of a slope (Grošić and Arbanas 2014). This finding pointed to the necessity of long-term analyses of remedial constructions at slopes in weak rock masses.

Field investigation, soil and rock laboratory testing and landslide remediation designs of several recent landslides were also conducted, as a part of CLG's activities in landslide risk reduction. Most of the investigated landslides occurred as sliding processes which affected roads in the North-Western part of Croatia, Istria and outback of the City of Rijeka, while rock fall phenomena endangered cultural heritage and urban areas in coastal parts of Croatia (Istria and Dalmatia). The investigated landslides and developed remediation measure design methods in the period 2013–2017 are listed in Table 2.

Discussion and Conclusions

The scientists from the Croatian Landslide Group (CLG) deal with basic and applied landslide science in the field of engineering geology and geotechnics (civil engineering), working at the two Croatian universities. As a stakeholder of the "Sendai Framework for Disaster Risk Reduction 2015–2030", which involves academic, scientific and research entities, the CLG focuses on disaster risk factors and scenarios, including emerging disaster risks in the medium and long term; increases research for regional, national and local applications; supports action by local communities and authorities; and supports the interface between policy and science for scientifically based options for decision making.

Described research of disaster risk factors undertaken by the CLG encompasses:

(i) Compilation of landslide hazard inventories in the form of geomorphological inventories of shallow landslides and of different types of landslides associated with erosion phenomena (multi-hazard approach);
(ii) Landslide susceptibility assessment to map initiation and runout zones of rock falls;
(iii) Hazard assessment by compilation of seasonal landslide inventories and catalogs of precipitation events to determine the size and probability of potential landslide occurrences for a given return period;
(iv) Determination of environmental risk factors by analyzing a landslide material's strength, deformation, or permeability in response to changes over time due to other environmental conditions, i.e., weathering.

Research of disaster risk scenarios was performed by modelling of the initiation and motion of the most dangerous landslides: reactivation of the Grohovo landslide under unfavorable groundwater conditions inside the area of a historical landslide and propagation of a possible mudflow from the landslide; initiation and runout of the Valići Lake landslide for different reservoir water levels and possible filling of the lake; and reactivation of the urban Kostanjek landslide due to rising ground water level and possible earthquake. Emerging landslide risk has been determined in the form of most frequent landslide initiations and reactivations as a consequence of higher frequency of precipitation events identified as triggering factors. Results of landslide hazard identification and assessment are important for medium risk analysis as well as for long-term risk assessment necessary for adaptation to climate changes and extreme weather conditions.

Scientific activities focused on landslide risk factors listed from (i)–(iii) are implementing in the framework of the IPL Project 173. Research on landslide risk factors described under (iv) and landslide risk scenarios are part of IPL Project 184. All described research in the fields of landslide identification and mapping, landslide investigation and testing, landslide monitoring, landslide modelling and landslide stabilization and remediation are aimed at local, national and regional applications for landslide risk reduction. Additionally, the CLG is leading member of the regional scientific network of landslide scientists, the ICL ABN. Network activities include joint activities related to landslide risk reduction with the scientific and academic institutions from Croatia, Slovenia and Serbia, scientific institutions from Albania and Slovenia, professional association from Bosnia and Herzegovina and local government from Croatia.

Research results in the form of landslide hazard maps, prognostic models and landslide remediation designs support action by local communities and authorities managing landslide risk. Scientific CLG activities support the interface between policy and science for scientifically based options for decision making in the systems of land-use planning, construction, civil protection and environmental protection.

References

Arbanas Ž, Dugonjić Jovančević S, Vivoda Prodan M, Mihalić Arbanas S (2014a) Study of landslides in flysch deposits of North Istria, Croatia: landslide data collection and recent landslide occurrences. In: Proc of the 3rd World Landslide Forum, vol 1: Methods of Landslide Studies, 2–6 June 2014. Springer, Cham, pp 837–842

Arbanas Ž, Jagodnik V, Ljutić K, Vivoda M, Dugonjić Jovančević S, Peranić J (2014b) Remote monitoring of a landslide using an integration of GPS, TPS and conventional geotechnical monitoring methods. In: Proc of the 1st Reg Symp on Landslides in the Adriatic-Balkan Region, 6–9 March 2013. Croatian Landslide Group, Zagreb. pp 39–44

Arbanas Ž, Sassa K, Nagai O, Jagodnik V, Vivoda M, Dugonjić Jovančević S, Peranić J, Ljutić K (2014c) A landslide monitoring and early warning system using integration of GPS, TPS and conventional geotechnical monitoring methods. In: Proc of the 3rd World Landslide Forum, Vol 2: Methods of Landslide Studies, 2–6 June 2014. Springer, Cham, pp 631–636

Arbanas Ž, Vivoda M, Peranić J, Sečanj M, Bernat Gazibara S, Krkač M, Mihalić Arbanas S (2016a) Analysis of a reservoir water level impact on landslide reactivation. In: Proc of the 2nd Reg Symp on Landslides in the Adriatic-Balkan Region, 14–15 May 2015. Beograd (in press)

Arbanas Ž, Mihalić Arbanas S, Vivoda M, Peranić J, Sečanj M, Bernat Gazibara S, Krkač M (2017) Preliminary investigations and numerical simulations of a landslide reactivation. In: Proc of the 4th World Landslide Forum, 29 May-2 June 2017. Springer, Cham (in prep)

Benac Č, Oštrić M, Dugonjić Jovančević S (2014) Geotechnical properties in relation to grain-size and mineral composition: the Grohovo landslide case study (Croatia). Geologia Croatica 67 (2):127–136

Bernat S, Đomlija P, Mihalić Arbanas S (2014a) Slope movements and erosion phenomena in the Dubračina River Basin: a geomorphological approach. In: Proc of the 1st Reg Symp on Landslides in the Adriatic-Balkan Region, 6–9 March 2013. Croatian Landslide Group, Zageb, pp 79–84

Bernat S, Mihalić Arbanas S, Krkač M (2014b) Landslides triggered in the continental part of Croatia by extreme precipitation in 2013. In: Proc of the XII IAEG Congress, vol 2, Landslide Processes, 15–19 September 2014. Springer, Heidelberg, pp 1599–1603

Bernat S, Mihalić Arbanas S, Krkač M (2014c) Inventory of precipitation triggered landslides in the winter of 2013 in Zagreb (Croatia, Europe). In: Proc of the 3rd World Landslide Forum, vol 2: Methods of Landslide Studies, 2–6 June 2014. Springer, Cham, pp 829–836

Bernat Gazibara S, Mihalić Arbanas S, Krkač M (2016) Catalog of precipitation events that triggered landslides in northwestern Croatia. In: Proc of the 2nd Reg Symp on Landslides in the Adriatic-Balkan Region, 14–15 May 2015. Beograd (in press)

Bernat Gazibara S, Mihalić Arbanas S, Krkač M (2017) Identification and mapping of shallow landslides in the City of Zagreb (Croatia) using LiDAR-based terrain model. In: Proc of the 4th World Landslide Forum, 29 May-2 June 2017. Springer, Cham (in prep)

Caine N (1980) The rainfall intensity-duration control of shallow landslides and debris flows. Geografiska Annaler, Series A, Physical Geography (1/2):23–27

Clarizia M, Gulla G, Sorbino G (1996) Sui meccanismi di innesco dei soil slip. In: Int. Conf. prevention of hydrogeological hazards: the role of scientific research, vol 1. L'Artistica Savigliano pub, Alba, pp 585–597 (in Italian)

Crosta GB, Frattini P (2001) Rainfall thresholds for triggering soil slips and debris flow. In: Proc of the 2nd EGS Plinius Conf. on Mediterranean Storms. Siena, Italy, pp 463–487

Dowding CH (1979) Perspective and challenges of site investigation. In: Dowding CH (ed) Site characterization and exploration. ASCE, New York, pp 10–35

Dugonjić Jovančević S (2013) Landslide hazard assessment on flysch slopes. In: Ph.D. thesis, Faculty of Civil Engineering, University of Rijeka, Zagreb, Croatia (in Croatian)

Dugonjić Jovančević S, Nagai O, Kyoji S, Arbanas Ž (2014a) Deterministic landslide susceptibility analyses using LS Rapid software. In: Proc of the 1st Reg Symp on Landslides in the Adriatic-Balkan Region, 6–9 March 2013. Croatian Landslide Group, Zagreb, pp 73–77

Dugonjić Jovančević S, Vivoda M, Arbanas Ž (2014b) Landslide susceptibility assessment on slopes in flysch deposits: a deterministic approach. In: Proc of the XII IAEG Congress, vol 2, Landslide Processes, 15–19 September 2014. Springer, Heidelberg, pp 1615–1618

Dugonjić Jovančević S, Arbanas Ž, Vivoda Prodan M, Peranić J, Đomlija P (2016) Landslide hazard and risk assessment in Istria, Croatia. In: Proc of the 2nd Reg Symp on Landslides in the Adriatic-Balkan Region, 14–15 May 2015. Beograd (in press)

Đomlija P (2017) Identification of landslides and erosion phenomena through the visual interpretation of airborne laser scanning digital elevation model of the Vinodol Valley. In: Ph.D. thesis, Faculty of Mining, Geology and Petroleum Engineering, University of Zagreb, Zagreb, Croatia, In Croatian (in prep)

Đomlija P, Bernat S, Mihalić Arbanas S, Benac Č (2014) Landslide inventory in the area of Dubračina River Basin (Croatia). In: Proc of the 3rd World Landslide Forum, vol 2. Methods of Landslide Studies, 2–6 June 2014. Springer, Cham, pp 837–842

Đomlija P, Bočić N, Mihalić Arbanas S (2016) Identification of geomorphological units and hazardous processes in the Vinodol Valley. In: Proc of the 2nd Reg Symp on Landslides in the Adriatic-Balkan Region, 14–15 May 2015. Beograd (in press)

Gradiški K, Krkač M, Mihalić Arbanas S, Bernat S (2013) Slope stability analyses of the Kostanjek Landslide for extreme rainfalls in the winter of 2013. In: Abstract Proc of the 4th Workshop of the Japanese-Croatian SATREPS FY2008 Project, 12–14 December 2013. University of Split, Split, pp 15–16

Gradiški K, Sassa K, He B, Arbanas Ž, Mihalić Arbanas S, Krkač M, Kvasnička P, Oštrić M (2014) Application of integrated landslide simulation model LS-Rapid to the Kostanjek Landslide, Zagreb, Croatia. In: Proc of the 1st Reg Symp on Landslides in the Adriatic-Balkan Region, 6–9 March 2013. Croatian Landslide Group, Zagreb, pp 15–16

Grošić M (2014) Time-dependent deformation of flysch rock mass. In: Ph.D. thesis, Faculty of Civil Engineering, University of Rijeka, Rijeka, Croatia (in Croatian)

Grošić M, Arbanas Ž (2014) Time-dependent behaviour of reinforced cuts in weathered flysch rock masses. Acta Geotechnica Slovenica 11(1):4–17

Guzzetti F, Peruccacci S, Rossi M, Stark CP (2008) The rainfall intensity–duration control of shallow landslides and debris flows: an update. Landslides 5:3–17

He B, Sassa K, Nagai O, Takara K (2014) Manual of LS-RAPID Numerical Simulation Model for Landslide Teaching and Research. In: Proc of the 1st Reg Symp on Landslides in the Adriatic-Balkan Region, 6–9 March 2013. Croatian Landslide Group, Zagreb, pp 5–10

Hungr O, Leroueil S, Picarelli L (2014) The Varnes classification of landslide types, an update. Landslides 11(2):167–194

Jagodnik V (2014) Behavior of laterally loaded piles in natural sandy gravels. In: Ph.D. thesis, Faculty of Civil Engineering, University of Rijeka, Rijeka, Croatia (in Croatian)

Jagodnik V, Arbanas Ž (2015a) Testing of laterally loaded piles in natural sandy gravels. Int J Phys Model Geotechn 15(4):191–208

Jagodnik V, Arbanas Ž (2015b) Comparison of deflections of laterally loaded pile obtained from test results and analyses using mixed finite element approach. In: Proc of the XVI Europ Conf on Soil Mechanics and Geotechnical Engineering, 14–15 May 2015. ICE Publishing, Edinburgh, pp 3923–3928

Krkač M (2015) A phenomenological model of the Kostanjek landslide movement based on the landslide monitoring parameters. In: Ph.D. thesis, Faculty of Mining, Geology and Petroleum Engineering. University of Zagreb, Zagreb, Croatia (in Croatian)

Krkač M, Mihalić Arbanas S, Arbanas Ž, Bernat S, Špehar K (2014a) The Kostanjek landslide in the City of Zagreb: forecasting and protective monitoring. In: Proc of the XII IAEG Congress, vol 5, Urban Geology, Sustainable Planning and Landscape Exploitation, 15–19 September 2014. Springer, Heidelberg, pp 715–719

Krkač M, Mihalić Arbanas S, Arbanas Ž, Bernat S, Špehar K, Watanabe N, Nagai O, Sassa K, Marui H, Furuya G, Wang C, Rubinić J, Matsunami K (2014b) Review of monitoring parameters of the Kostanjek Landslide (Zagreb, Croatia). In: Proc of the 3rd World Landslide Forum, vol 2: Methods of Landslide Studies, 2–6 June 2014. Springer, Cham, pp 637–645

Krkač M, Mihalić Arbanas S, Nagai O, Arbanas Ž, Špehar K (2014c) The Kostanjek landslide—monitoring system development and sensor network. In: Proc of the 1st Reg Symp on Landslides in the Adriatic-Balkan Region, 6–9 March 2013. Croatian Landslide Group, Zagreb, pp 27–32

Krkač M, Špoljarić D, Bernat S, Mihalić Arbanas S, (2016) Method for prediction of landslide movements based on random forests. Landslides (in review)

Krkač M, Mihalić Arbanas S, Arbanas Ž, Bernat Gazibara S, Sečanj M (2017) Prediction of Kostanjek landslide movements based on monitoring results using Random Forests technique. In: Proc of the 4th World Landslide Forum, 29 May-2 June 2017. Springer, Cham (in prep)

Ljutić K, Jagodnik V, Vivoda M, Dugonjić Jovančević S, Arbanas Ž (2014) The Grohovo landslide monitoring system—experiences from 18 months period of monitoring system operating. Proc of the 1st Reg Symp on Landslides in the Adriatic-Balkan Region, 6–9 March 2013. Croatian Landslide Group, Zagreb, pp 45–50

Mihalić Arbanas S, Krkač M, Bernat S (2016a) Application of advanced technologies in landslide research in the area of the City of Zagreb (Croatia, Europe). Geologia Croatica 69/2:231–243

Mihalić Arbanas SM, Sečanj M, Bernat Gazibara S, Krkač M, Arbanas Ž (2016b) Identification and mapping of the Valići Lake landslide (Primorsko-Goranska County, Croatia). In: Proc of the 2nd Reg Symp on Landslides in the Adriatic-Balkan Region, 14–15 May 2015. Beograd (in press)

Oštrić M, Sassa K, Ljutić K, Vivoda M, He B, Takara K (2014) Manual of transportable ring shear apparatus, ICL-1. In: Proc of the 1st Reg Symp on Landslides in the Adriatic-Balkan Region, 6–9 March 2013. Croatian Landslide Group, Zagreb, pp 1–4

Peternel T, Mikoš M, Đomlija P, Dugonjić Jovančević S, Arbanas Ž (2016) Geological conditions of landslides in flysch deposits in Slovenia and Croatia. In: Proc of the 2nd Reg Symp on Landslides in the Adriatic-Balkan Region, 14–15 May 2015. Beograd (in press)

Podolszki L (2014) Stereoscopic analysis of landslides and landslide susceptibility on the southern slopes of Medvednica Mt. In: Ph.D. thesis, Faculty of Mining, Geology and Petroleum Engineering, University of Zagreb, Zagreb, Croatia (in Croatian)

Popescu ME (2001) A suggested method for reporting landslide remedial measures. Bull Eng Geol Environ 60(1):69–74

Sečanj M, Mihalić Arbanas S, Krkač M, Bernat Gazibara S, (2017) Identification of rock fall prone areas on the steep slopes above the town of Omiš, Croatia. In: Proc of the 4th World Landslide Forum, 29 May-2 June 2017. Springer, Cham (in prep)

Vivoda M, Dugonjić Jovančević S, Arbanas Ž (2014) Landslide occurrence prediction in the Rječina River Valley as a base for an early warning system. In: Proc of the 1st Reg Symp on Landslides in the Adriatic-Balkan Region, 6–9 March 2013. Croatian Landslide Group, Zagreb, pp 85–90

Vivoda Prodan M (2016) The influence of weathering process on residual shear strength of fine grained lithological flysch components. In: Ph.D. thesis, Faculty of Civil Engineering, University of Rijeka, Rijeka, Croatia (in Croatian)

Vivoda Prodan M, Arbanas Ž (2016) Weathering influence on properties of siltstones from Istria, Croatia. Advances in materials science and engineering. 2016, ID 3073202, 1–15. doi:10.1155/2016/3073202

Vivoda Prodan M, Mileusnić M, Mihalić Arbanas S, Arbanas Ž (2016) Influence of weathering processes on the shear strength of siltstones from a flysch rock mass along the northern Adriatic coast of Croatia. Bull Eng Geol Environ (in press) doi:10.1007/s10064-016-0881-7

Vivoda Prodan M, Arbanas Ž, (2017) Parametric analysis of weathering effect on possible reactivation of the Valići landslide, Croatia. In: Proc of the 4th World Landslide Forum, 29 May-2 June 2017. Springer, Cham (in prep)

Žic E, Arbanas Ž, Bićanić N, Ožanić N (2015) A model of mudflow propagation downstream from the Grohovo landslide near the city of Rijeka (Croatia). Nat Hazards Earth Sys Sci 15(1):293–313

Shapes and Mechanisms of Large-Scale Landslides in Japan: Forecasting Analysis from an Inventory (WCoE 2014–2017)

S. Ogita, W. Sagara, Daisuke Higaki, and Research Committee on Elucidating Mechanisms of Large-Scale Landslides

Abstract

Large-scale landslides with widths and lengths of 1 km or more have been reported in many parts of the world. Occurrences of large-scale landslides have recently tended to increase due to climate change and frequent seismic activity. To conduct research on proper measures for large-scale landslides, elucidation of the occurrence mechanism, for which there are as yet many unclear points, will be required in future. The Japan Landslide Society established a research committee that worked from 2011 to 2014 to elucidate the occurrence mechanisms of large-scale landslides. Analysis of examples of large-scale landslides collected from members of the research committee showed that a volume of moving body larger than 1×10^6 m^3 and a maximum landslide thickness of more than 30 m are appropriate as the definition of a large-scale landslide. The shape of a large-scale landslide depends on the geology and age of the landslide site, and landslide activity and history affect the symmetry of the shape of a landslide. This paper presents some results of the WCOE (2014–2017) project titled "Emergency response support system for large-scale landslide disasters" by the Japan Landslide Society.

Keywords

Large-scale landslide • Kinematic structure • Landslide topography

Introduction

Many large-scale landslides, which have brought disaster to lives and properties, have occurred in Japan.

The Jizukiyama landslide (Fig. 1), which occurred in Nagano City in 1985, for example, was 350 m wide and 280 m long, with a maximum thickness of 60 m and a volume of 2.5×10^6 m^3. The landslide buried facilities for the disabled and housing-development apartments, and 26 lives were lost (National Research Institute for Earth Science and Disaster Prevention (NIED) 1986; Shinshu University Research Committee 1986). The Dozan river landslide, which occurred in Yamagata-prefecture in 1996, was 1100 m wide and 1200 m long. Its maximum thickness was 180 m, the volume was 1.0×10^8 m^3, and the event caused damage to national roads and the Dozan River.

S. Ogita (✉)
Okuyama Boring Co. Ltd, 10-39 Shimei-Cho, 013-0046 Yokote-Shi, Akita-Ken, Japan
e-mail: ogita@okuyama.co.jp

W. Sagara
Sabo and Landslide Technical Center, 4-8-21, Kudan-Minami, 102-0074 Chiyoda-Ku, Tokyo, Japan
e-mail: sagara@stc.or.jp

D. Higaki
Hirosaki University, 3 Bunkyo-Cho, 036-8561 Hirosaki-Shi, Aomori-Ken, Japan
e-mail: dhigaki@hirosaki-u.ac.jp

Research Committee on Elucidating Mechanisms of Large-Scale Landslides
Japan Landslide Society, 5-26-8 Shinbashi, 105-0004 Minato-Ku, Tokyo, Japan

K. Sassa et al. (eds.), *Advancing Culture of Living with Landslides*,
DOI 10.1007/978-3-319-59469-9_26

The Sumikawa landslide, which occurred in Akita Prefecture in 1997, was 350 m wide and 650 m long, with a maximum thickness of 70 m. Some of the moving body, with a volume of 6.0×10^6 m^3 (Furuya and Enokida 2004), partially moved down the river, damaging hot-spring facilities and national roads (Oyagi and Ikeda 1988).

Landslides have occurred at Akadani and Nagadono in the Kii Peninsula in 2011. The volume of the moving body at Akadani is estimated as 9.4×10^6 m^3 (Sakurai 2013); that of Nagadono is 5.0–6.0×10^6 m^3 (Sakurai et al. 2013).

In consideration of the occurrence of large-scale landslides such as those mentioned above, the Japan Landslide Society established the "Research Committee on Elucidating Mechanisms of Large-Scale Landslides" (hereinafter referred to as the Committee) in 2011 to contribute to the prediction of landslide occurrence, the planning of measures, and the development of analysis techniques, with a focus on large-scale landslides.

In the Committee, the main areas of research were the definition of large-scale landslides, characteristics of topography and geology, time-series kinetic structure, effect of groundwater at great depth, shear strength, and stability analysis techniques. A questionnaire was provided to each member to collect examples of large-scale landslides. The instances include ones in which several landslide blocks are aligned (Furuya 2003). Here, 30 instances for which the landslide mechanism has been elucidated are collected to consider the characteristics of landslide kinetic structure and the shapes of landslide sites. For the kinetic structure, the "distribution pattern of displacement, velocity, and acceleration vectors of landslide moving body" defined by Oyagi (2004) is used.

Questionnaire Results

A total of 30 instances of landslides were collected using the questionnaire: 16 in rocks of Cenozoic age; the other 14 in rocks of Mesozoic and Paleozoic age. The landslides were assigned to groups to summarize the scale and the gradient of landslide sites in the occurrence areas (Fig. 2 and Table 1). Figure 2 shows landslide positions. In Table 1, the maximum landslide width (h), length (L), relative height (rh), gradient of landslide slope (θ), maximum thickness (t), volume (v), and gradient of sliding surface (α) are listed.

Figure 3 shows the dimension of landslides obtained from the questionaire.

Fig. 1 Full view of the Jizukiyama landslide site (Shinshu University Research Committee 1986)

The maximum landslide width ranges from 190 to 2000 m. For landslides with a maximum width of 1000 m or longer, four instances in Cenozoic strata and one in Mesozoic and Paleozoic rocks are reported (Fig. 3a).

Landslide length ranges from 200 to 2100 m. There are nine instances in Cenozoic rocks and four instances in Mesozoic and Paleozoic strata of landslides that are 1000 m or more in length (Fig. 3b).

The relative height from the top of a main cliff to the bottom is between 45 and 500 m. For landslides with relative height of less than 100 m, four instances in Cenozoic rocks and one example in Mesozoic and Paleozoic rocks are reported. In contrast, for landslides with a relative height of 300 m or more, there is one instance in Cenozoic rocks and five in Mesozoic and Paleozoic rocks (Fig. 3c).

The gradient of landslide slope is 6–40°. There are seven examples in Cenozoic rocks and one example in Mesozoic and Paleozoic rocks of landslides with a slope smaller than 10°; in contrast, for slopes 20° or more, there are three reported instances in Cenozoic rocks and nine in Mesozoic and Paleozoic strata. The examples in Mesozoic and Paleozoic rocks display larger gradients (Fig. 3d).

The maximum landslide thickness is 30–180 m. For landslides with a maximum thickness of 100 m or more, there are six instances in Cenozoic strata and three in Mesozoic and Paleozoic strata; thus, there are slightly more occurrences in Cenozoic rocks than in older strata (Fig. 3e).

The volume of the moving body is a minimum of 7.0×10^5 m^3 and a maximum of 1.0×10^8 m^3. All 16 instances in Cenozoic rocks and 13 of 14 examples in Mesozoic and Paleozoic rocks have a volume of 1.0×10^6 m^3 or more (Fig. 3f).

The gradient of the sliding surface is the slope immediately under the center of gravity of the landslide section; this parameter varies from 2° to 31°. There are seven examples in Cenozoic rocks and one example in Mesozoic and Paleozoic

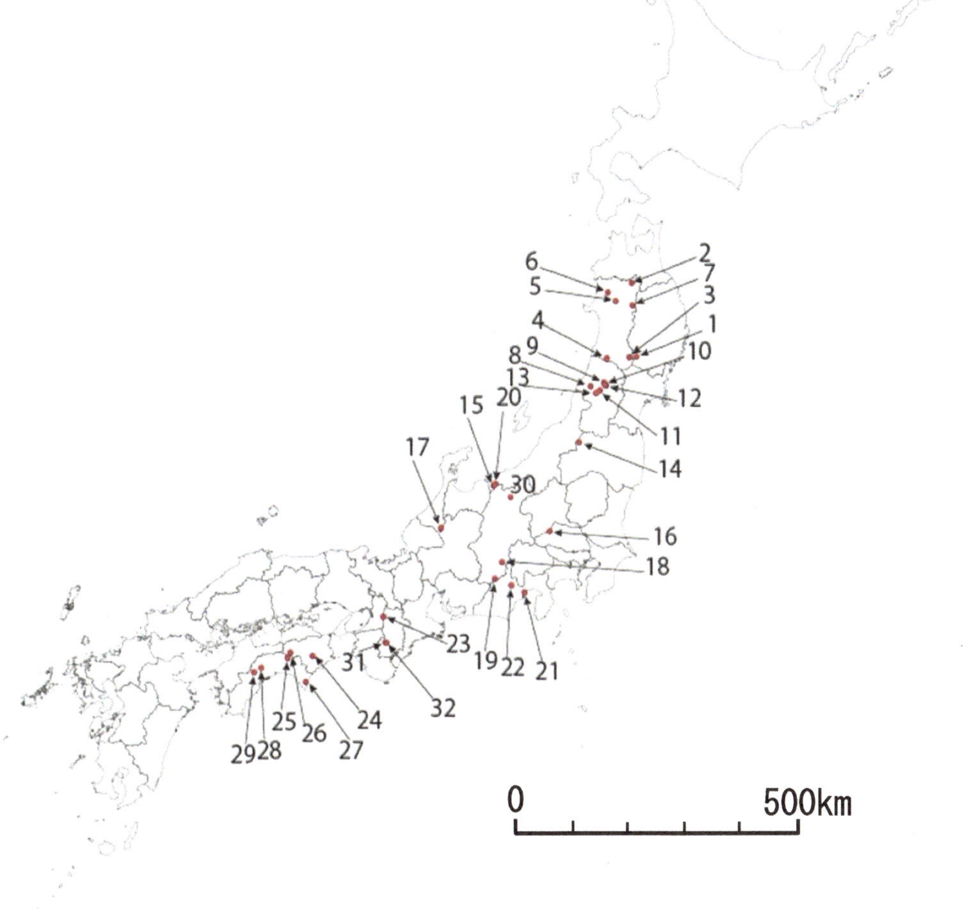

Fig. 2 Instances of landslides obtained by using the questionnaire

Table 1 Landslide scale based on the questionnaire results

Geologicla age	No.	Generating area						
		h (m)	L (m)	rh (m)	θ (°)	t (m)	v (m^3)	α (°)
Cenozoic era	1.	250	450	130	14.0	80	9.0E + 06	7.5
	2.	240	200	90	28.0	45	1.0E + 06	31.0
	3.	950	1200	200	6.5	55	1.0E + 07	11.0
	4.	1200	1000	250	15.0	165	6.0E + 07	15.0
	5.	800	1100	230	12.0	90	1.5E + 07	
	6.	280	250	45	25.0	45	1.5E + 06	
	7.	350	650	120	1.5	70	6.0E + 06	11.5
	8.	840	1140	150	8.0	122	4.9E + 07	4.5
	9.	470	750	100	13.0	100	1.0E + 07	
	10.	500	1000	50	6.0	45	1.3E + 07	4.0
	11.	500	750	70	6.5	40	1.0E + 07	5.5
	12.	2000	1500	350	10.5	150	8.0E + 07	14.0
	13.	1100	1200	190	8.0	180	1.0E + 08	2.0
	14.	1300	2100	150	8.5	140	4.8E + 07	3.5
	15.	250	1500	250	9.0	60	1.0E + 07	4.0
	30.	350	280	100	22.0	60	2.5E + 06	10.0
Mesozoic–Paleozoic era	16.	600	450	190	23.0	50	8.0E + 06	17.5
	17.	400	1200	500	21.0	130	2.1E + 07	14.0
	18.	270	300	90	12.5	30	1.0E + 06	11.0
	19.	350	450	170	20.0	48	5.4E + 06	11.0
	20.	200	500	130	12.5	60	6.0E + 06	12.0
	21.	350	400	200	28.0	65	3.6E + 06	24.5
	22.	400	900	260	23.0	60	2.0E + 07	15.0
	23.	400	650	100	4.0	70	9.0E + 06	12.5
	24.	190	390	180	40.0	30	7.0E + 05	18.0
	25.	700	900	440	26.5	65	1.7E + 07	25.0
	26.	400	350	100	16.5	35	2.0E + 06	12.0
	27.	640	1290	460	20.0	100	7.0E + 06	
	28.	1000	1050	450	20.0	160	1.0E + 08	13.0
	29.	400	1150	360	13.0	58	1.2E + 07	8.0

Blank fields for slide surface slope indicate places for which longitudinal section views were unavailable

rocks of landslides with a slope smaller than 10°; in contrast, for slopes 20° or more, there is one reported instance in Cenozoic rocks and two in Mesozoic and Paleozoic strata (Fig. 3g).

Discussion

Topography, Geology, and Scale

From the results of the questionnaire, in 29 of the 30 examples of landslides, the volume of the moving body is 1.0×10^6 m^3 or more. Similarly, for maximum landslide thickness, all instances had a thickness of 30 m or more (Table 1). A volume of 1.0×10^6 m^3 or larger was also treated as a large-scale landslide by Fujita (2004). These values, therefore, were regarded as a criterion for the definition of large-scale landslides by the Committee.

Shape of Landslide

From plane views of longitudinal and cross sections collected in the questionnaire, the shapes of landslide bodies in

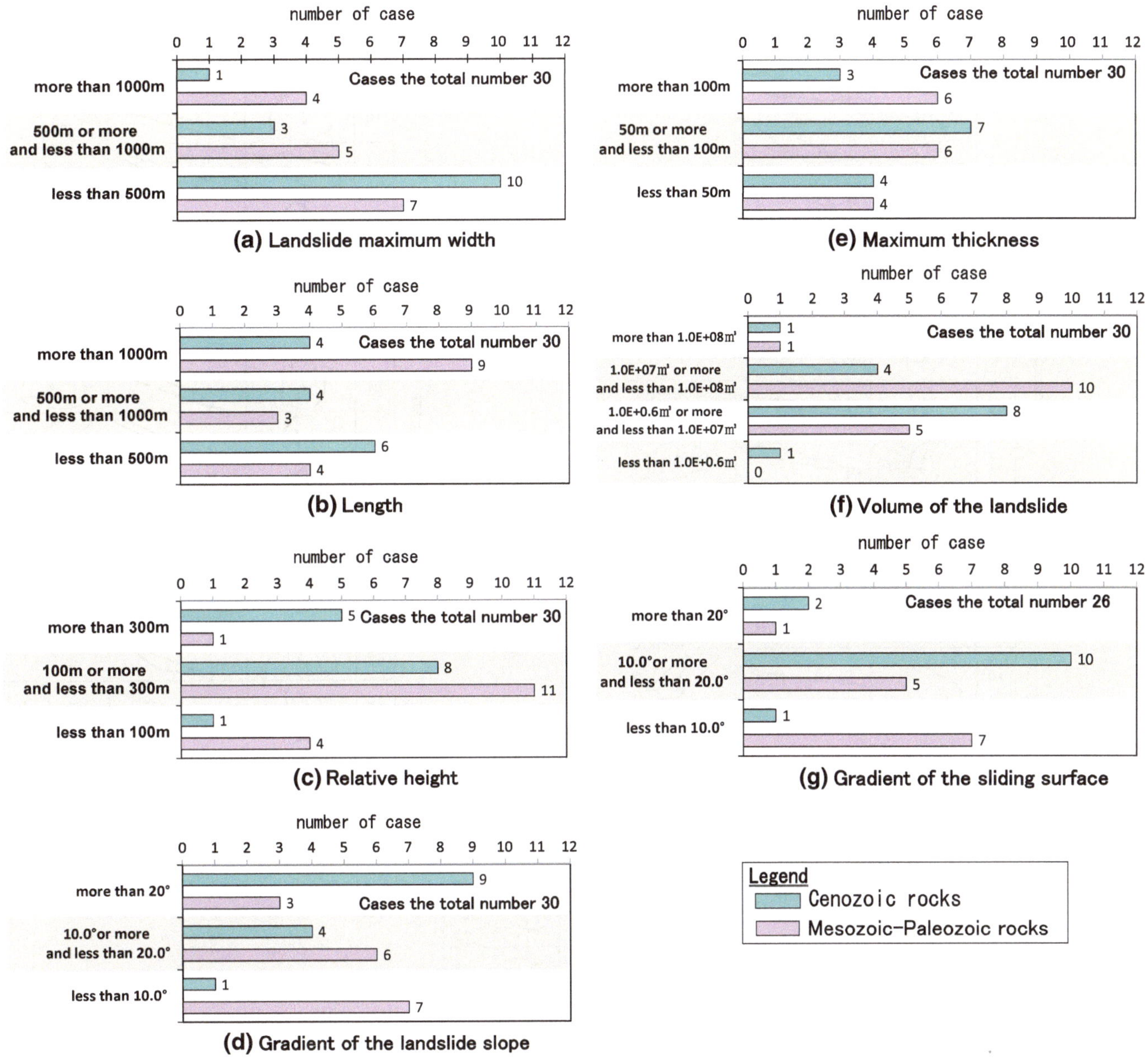

Fig. 3 Landslide dimensions from the questionnaire results

terms of geology were divided into symmetric and asymmetric (Fig. 4).

For plane shapes, a straight landslide direction was considered as symmetric and a bent landslide direction as asymmetric. For longitudinal-section shapes, cases in which the gradient of landslide slope and that of the sliding surface are equal on average were considered to be symmetric; those in which the thickness of the moving body at the top of landslide is different from that at the bottom and in which the dip of sliding surface at the toe part is opposite to that of the landslide slope were regarded as asymmetric. For cross-sectional shapes, a ship-bottom shape and an isosceles-triangle shape were regarded as symmetric, and cases in which the landslide thickness slants to one side were deemed asymmetric. Landslides for which the shape was unclear were indicated by a question mark. When information became available, the section shape was added near the question mark.

For landslides in Cenozoic rocks, 11 of the 12 instances for which information is available are identified as asymmetric. Of these, 7 examples are asymmetric in planar view,

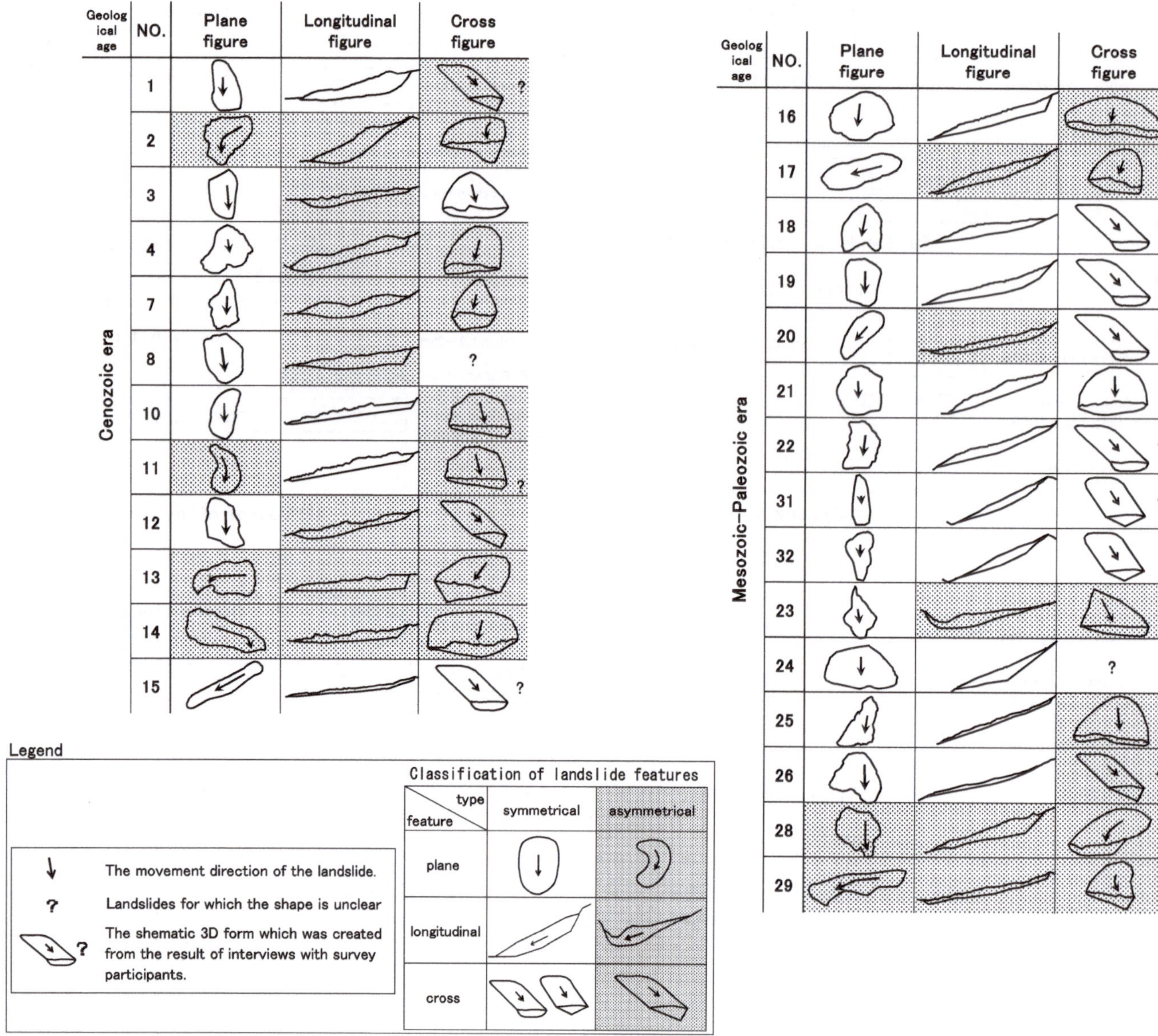

Fig. 4 Classification of landslide shape

in longitudinal section, and/or in cross-section. On the other hand, for landslides in Mesozoic and Paleozoic strata, 8 of the 15 examples are classified as asymmetric. Of these, 4 examples are asymmetric in shape, fewer than the number in Cenozoic strata.

Instances No. 7 in Cenozoic rocks and Nos. 31 and 32 in Mesozoic and Paleozoic rocks in Fig. 3 are landslides in which part or all of the landslide moving body slid downward. No. 7 is asymmetric in the longitudinal section and cross-section; Nos. 31 and 32 are symmetric in all views. Although the gradient of sliding surface of No. 2 is 31°, a steep slope, and the displacement over the last two years has been 1 m, this example does not slide downward. It is suggested that a landslide with many asymmetric elements tends not to slide down easily and shows higher resistance, compared to landslides in which the planar, longitudinal section, and cross-sectional shapes are symmetric.

For example, Fig. 5 is a bird's-eye view from the northeast of the Tozawa landslide in Akita Prefecture (No.4 landslide in Table 1). It occurs in Cenozoic strata; is 1.2 km wide and 1.0 km long; has a maximum thickness of 165 m; and the gradient of its sliding surface is 15° (Table 1). This landslide is symmetrical in plan view and asymmetrical in longitudinal section and cross-section (Figs. 6 and 7).

Fig. 5 Full view of the Tozawa landslide site

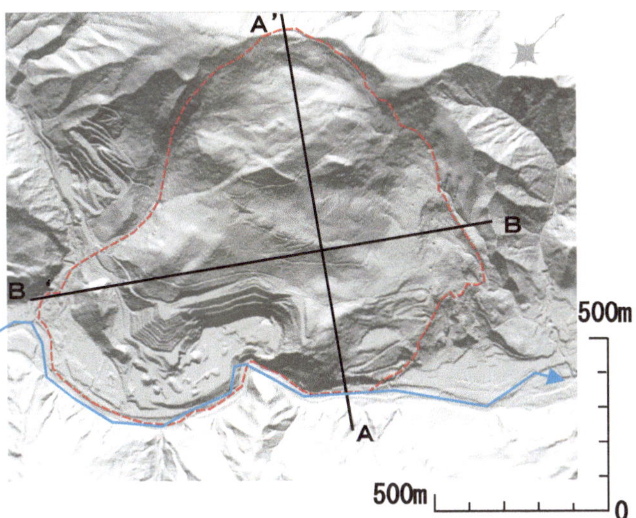

Fig. 6 Plan view of the Tozawa landslide site

For the movement state of the Tozawa landslide, the height of the sliding cliff is approximately 20 m, and the observed landslide movement was 0.4 m/year for 2002–2003. The river at the bottom of the landslide is now raised, but no downward sliding movement has occurred.

As an example of a landslide in Mesozoic and Paleozoic rock, Fig. 8 shows a view of landslide No. 25 in Table 1. The landslide is 700 m wide and 900 m long, and has a maximum thickness of 65 m and a slide surface of 25° in gradient. This landslide is symmetric in plan view and longitudinal section, but asymmetric in cross-section (Fig. 9). For the landslide, the difference in height of the sliding cliff is approximately 5 m, the observed landslide movement rate is 3 cm/year for 2007–2009, and the landslide is currently not sliding.

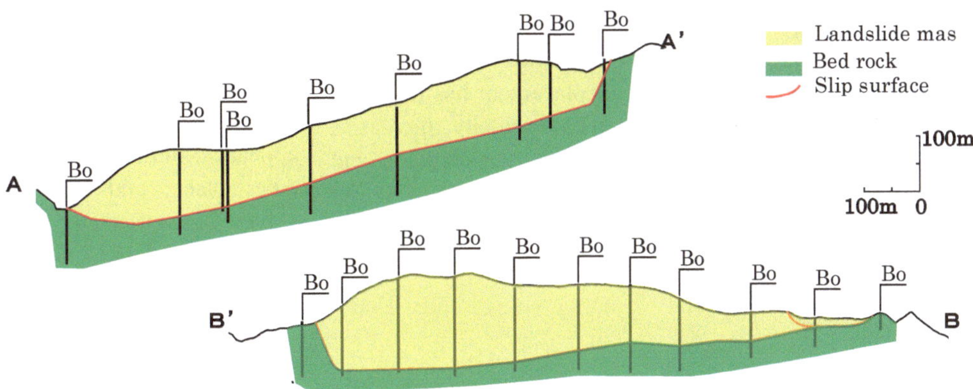

Fig. 7 Cross section of the Tozawa landslide site

Fig. 8 Overall view of landslide No. 25

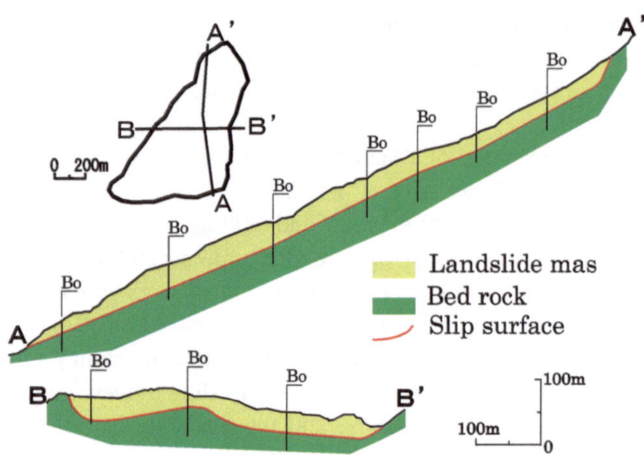

Fig. 9 Cross-section view of landslide No. 25

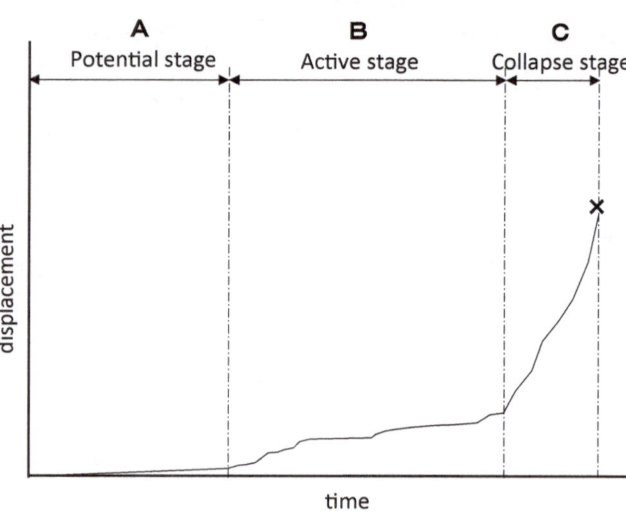

Fig. 10 Three stages of landslide displacement

Landslide Kinematic Structure

Time-Series Displacement of Landslide

For some landslides, after small displacement has occurred for a certain time, the displacement velocity increases, leading to downward sliding; for others, the displacement velocity tends to increase and decrease, not leading to further sliding downward. Taking this information into consideration, landslide movement can be divided in a time-series manner into: period A, in which small landslide displacement that can be observed with creep by monitoring equipment occurs; period B, in which displacement that appears differently on the ground and within the structure takes place; and period C, in which the displacement velocity increases and part or all of the moving body slides down and collapses. Herein, period A is called the potential

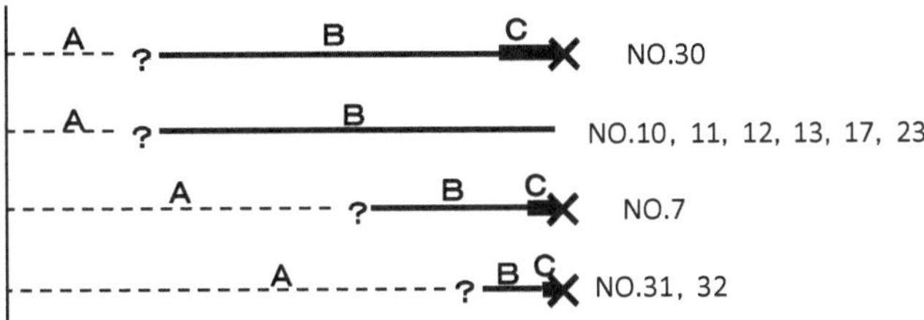

Fig. 11 Time-historical changes in the displacement stage of each landslide (**A**, **B**, **C** is shown in Fig. 10 No.7–32 is shown in Fig. 1)

stage of a large-scale landslide, period B is called the active stage, and period C is called the collapse stage (Fig. 10).

Examples of Landslide Movement History

As shown in Fig. 10, when the kinematic structure of a landslide is expressed as periods A, B, and C, in the case of landslide No. 30 in Table 1, after a period B lasting several years period C commenced, and part of the moving body slid down and flowed. For landslides Nos. 10, 11, 12, 13, 17, and 23 in Table 1, displacement increases during snow-melt and heavy rain; however, as these landslides have not collapsed at present, they are in period B. For the example of landslide No. 7 in Table 1, which occurred in a volcanic region, a period B of several years was confirmed (Oyagi and Ikeda 1988). After a period C lasting nine days, part of the moving body collapsed and moved for 1.5 km as a debris avalanche (Ogawauchi et al. 1988). Landslides Nos. 31 and 32 in Fig. 1 exhibited a period A, in which double ridges and a sliding cliff were formed, and a short period B; subsequently, sliding down and flowing out occurred (Sakurai 2013; Sakurai et al. 2013) (Fig. 11).

Summary

The results of the Committee's research can be summarized as follows.

- In the questionnaire to members of the Committee, which consists of technical and research experts, details of large-scale landslides in which the volume of the moving body was 1.0×10^6 m^3 or larger were obtained for 29 instances, excluding one. A large-scale landslide, therefore, is defined herein as one in which the volume of the moving body is 1.0×10^6 m^3 or larger and the maximum landslide thickness is 30 m or larger.
- There is a trend for the relative height, gradient of landslide slope, and that of sliding surface of large-scale landslides to be smaller in Cenozoic strata than in Mesozoic and Paleozoic rocks.
- One of the factors controlling why some large-scale landslides do not slide down but displacement continues is asymmetry of shape (in any or all of plan view, longitudinal section, and cross-section) of the landslide moving body. This trend is particularly strong in landslides in Cenozoic rock. On the other hand, one of the factors causing Mesozoic and Paleozoic rock to sliding down relatively easily is that these landslides have a relatively symmetric shape.

Further analysis incorporating additional examples is required; there is a possibility that the characteristics of large-scale landslides deduced herein are also common in areas outside of Japan that have similar geology, topography, and evolution of geological structure. For large-scale landslides in which the scale of measures is apt to be large, monitoring of landslides and emergency measures suited to the characteristics of the ground of occurrence sites must be taken. It is necessary, therefore, for landslide experts to share detailed information on large-scale landslides.

References

Fujita T (2004) Scale, landslide—topography and geology knowledge and terms. Japan landslide society, land form and geology terms committee on landslide, pp 16–28 (in Japanese)

Furuya T (2003) Hazard map with landslide disaster. Geography 48 (9):25–28(in Japanese)

Furuya T, Enokida M (2004) Slope carte and description example, landslide—topography and geology knowledge and terms. Japan landslide society, land form and geology terms committee on landslide, pp 285–295 (in Japanese)

Ogawauchi Y, Yamazaki T, Yamasaki K, Kikuchi A, Obara Y, Hosaka K (1988) Occurrence mechanism of the Sumikawa landslide—result of survey on the Sumikawa landslide and outline of measures. Landslide 35(2):38–45 (in Japanese with English abstract)

Oyagi N (2004) Landslide structure—topography and geology knowledge and terms. Japan landslide society, land form and geology terms committee on landslide, pp 29–45 (in Japanese)

Oyagi N, Ikeda H (1988) Sumikawa landslide from the viewpoint of landslide structure and broad field. Landslide 35(2):1–10 (in Japanese with English abstract)

Sakurai W (2013) Measures for large-scale landslide disaster caused by typhoon No. 12 in September 2011. Kansai Branch of Japan Landslide Society, collected papers on deep layer collapse and channel blockage, pp 7–32 (in Japanese)

Sakurai W, Tokunaga H, Amino K, Kato A, Suzuki S (2013) Survey on deep layer collapse and channel blockage caused by typhoon No. 12 in 2011. Kansai branch of Japan landslide society, collected papers on deep layer collapse and channel blockage, pp 33–70 (in Japanese)

Science and Technology Agency: National Research Institute for Earth Science and Disaster Prevention (1986) Report on disaster caused by the Jizukiyama landslide in Nagano-city on July 26, 1985, No. 26, p 45 (in Japanese)

Shinshu University Research Committee on natural disaster (1986) The Jizukiyama landslide. Japan Geotechnical Consultant Association, geology-related information Web: https://www.zenchiren.or.jp/tikei/saigai.html. Accessed 8 June 2016 (in Japanese)

Retrospective and Prospects for Cold Regions Landslide Research (2012–2016) (WCoE 2014–2017, IPL-132, IPL-167, IPL-203, CRLN)

Wei Shan and Ying Guo

Abstract

For nearly 100 years, the average temperature of the global surface has showed a consistent warming trend. Climate change and extreme weather events causing landslides are rising, especially landslides in cold regions, and the topic has become a hot issue in landslide research. With the support of ICL and the Chinese government, based on highway construction projects in Heilongjiang Province (China), Prof. Shan and his group (Institute of Cold Regions Science and Engineering, North East Forestry University, China) conducted thematic studies focusing on environmental and engineering geology problems in cold regions in the context of climate change, such as IPL132, IPL167, and IPL203. These studies attracted the interest of international colleagues, then Chinese colleagues, together with researchers from Russia, Canada, Japan, Italy and Czech Republic, together organized the ICL-cold regions landslide network (ICL-CRLN). In ICL-CRLN researchers could exchange research information and results, and so promote the development of landslide research in cold regions. In 2014, IPL-GPC approved the establishment of IPL-WCoE: Research Center of Cold Regions Landslide, so landslide research in cold regions came into a new stage of development. This article is a summary and outlook of these activities.

Keywords

Cold region landslides • IPL projects • ICL-CRLN • WCoE

Introduction

For nearly 100 years, the average temperature of the surface of the globe has been showing a consistent warming trend. The rate of warming in the past 50 years is almost double that in the last 100 years (IPCC 2007). Since the early 1980s, the temperature of most permafrost regions has increased. In some places in northern Alaska, the observed warming rate reached 3 °C, in the northern European region of Russia it reached 2 °C during 1975–2005, and it has been observed that the thickness and scope of permafrost has been significantly reduced (IPCC 2013). Today, climate change and its effects is a topic of widespread concern (Fig. 1).

The numbers of landslides caused by climate change and extreme weather events are rising (Blunden et al. 2011), and gradually gaining more attention of governments and international academic organizations (EU-FP7 2008; ICL 2014). Landslides in cold regions in particular have become a hot issue in landslide research (ICL 2012; Guo et al. 2013).

The mechanisms and evolution of landslides in mountainous areas are closely related to the geological conditions and environmental factors, and are controlled not only by geological forces, lithologic structure, and other crustal internal factors; but also by topography, land cover, precipitation, changes in human activities, and environmental conditions. The spatial and temporal distribution of

W. Shan (✉) · Y. Guo
Institute of Cold Regions Science and Engineering, North East Forestry University, No.26 Hexing Road, Harbin, 150020, China
e-mail: shanwei456@163.com

Y. Guo
e-mail: samesongs@163.com

K. Sassa et al. (eds.), *Advancing Culture of Living with Landslides*,
DOI 10.1007/978-3-319-59469-9_27

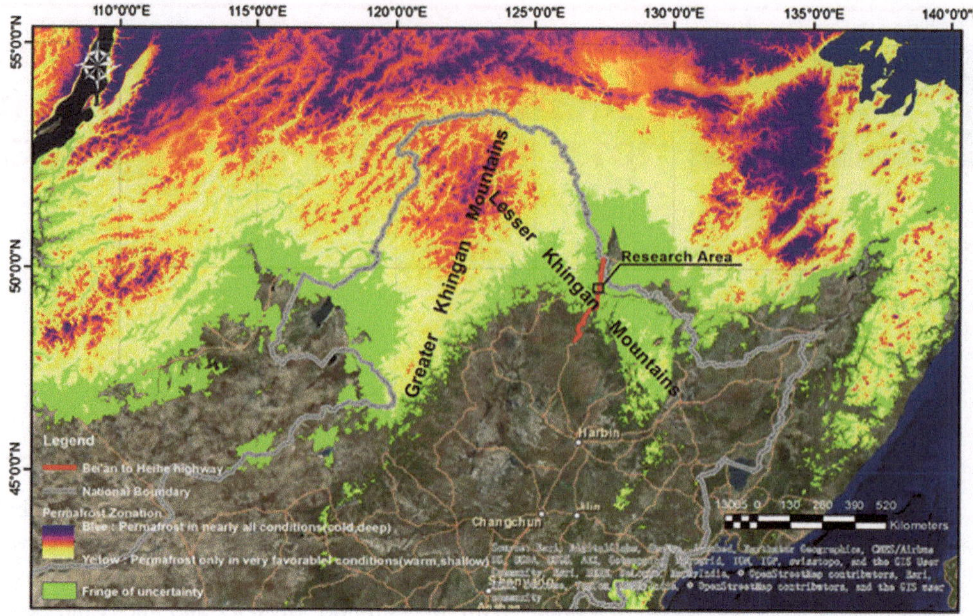

Fig. 1 High-latitude permafrost distribution in Northeast China (data from Zurich University, Switzerland), and location of study area

landslides is uncertain, sporadic in nature, continuous, and irreversible. Landslides are the result of geological and environmental changes and also drastically change the geological environment (Shan et al. 2014b). Scientists have used varying methods to analyze the relationships between climate change and landslide mechanisms and their evolution in cold areas (Shan et al. 2014a). The impact of landslides induced by glacier and permafrost degradation in cold regions on the topography, geological environment, water resources, and on biodiversity; and the role of climate, a main factor influencing landslide movement, in the evolution of landslides in cold areas have been studied (Fischer et al. 2013; Haeberli 2013; Kliem et al. 2013; Starnberger et al. 2013; Ballantyne et al. 2014; Graband Linde 2014; Nussbaumer et al. 2014). However, due to a lack of monitoring data, the majority of these studies were broad scale (Stoffel et al. 2014). Currently, there have been no reports on the mechanisms, movement characteristics, and patterns of landslides induced by the combined effect of permafrost thawing and extreme weather events due to climate change and geological conditions (Fig. 2).

Relevant Organizations of ICL and Its Working Results

With the support of ICL and the Chinese government, utilising highway construction projects in Heilongjiang Province(China), Prof. Shan and his group (Institute of Cold Regions Science and Engineering, North East Forestry University, China) conducted a thematic study that focused on environmental and engineering geology problems in cold regions in the context of climate change, such as "Research on vegetation protection system for highway soil slope in seasonal frozen regions", "Landslides mechanism and the subgrade stability controlling measures in island permafrost area", "Analysis and identify of landslides based on species distribution and surface temperature difference"; those also are IPL projects. These studies attracted the interest of international and Chinese colleagues, with researchers from Russia, Canada, Japan, Italy and the Czech Republic together organizing an ICL-cold regions landslide network (ICL-CRLN), so researchers could exchange research information and results, and promote the development of landslide research in cold regions. In 2014, IPL-GPC approved the establishment of IPL-WCoE: Research Center of Cold Regions Landslide, which fostered a new stage of development for landslide research in cold regions (Fig. 3).

Results of Regional Projects (IPL132, IPL167, IPL203)

Project IPL132 focuses on shallow landslides caused by seasonal freeze-thaw in northeastern China, Project IPL167 focuses on landslides caused by permafrost degradation in permafrost regions of northeast China, and Project IPL203 focuses on environmental and ecological problems caused

Fig. 2 Different forms of landslides around the world in cold regions

by permafrost degradation in permafrost regions of northeast China.

IPL-132: Research on Vegetation Protection System for Highway Soil Slope in Seasonal Frozen Regions (Duration: 2008–2015)

Project leader:
Prof. Wei Shan, Cold Regions Science and Engineering, Northeast Forestry University, Harbin, China.
Core members of the Project:
Dr. Fawu Wang, Shimane University, Japan
Dr. Ying Guo, Northeast Forestry University, China
Dr. Chengcheng Zhang, Northeast Forestry University, China

Objectives:

The study area is located along a major highway in China's Heilongjiang Province, the Kiamusze to Harbin sector of the Tong-San Expressway. This study aims to establish a comprehensive vegetation protection system for highway soil slopes in seasonally frozen regions. The system will aid in greening the regional environment along the highway, and in highway slope stabilization.

Training Student: One doctor, Five Masters

(1) Doctor Ying Guo. Mechanisms of slope freeze-thaw instability of cuttings and their protection by vegetation in high-latitude frozen regions (2013)
(2) Master Chengcheng Zhang. Research on stability and protection of shallow side slopes of soil road cuttings in seasonal frozen regions (2012)
(3) Master Chunming Liu. Research on the slope instability mechanism on freezing and thawing of the K560 +300 to K561+000 section of the Tongsan-Sanya highway (2013)
(4) Master Fuliang Wang. The evaluation to plant comprehensive protection systems along a high-grade highway in a seasonal frost area (2008)

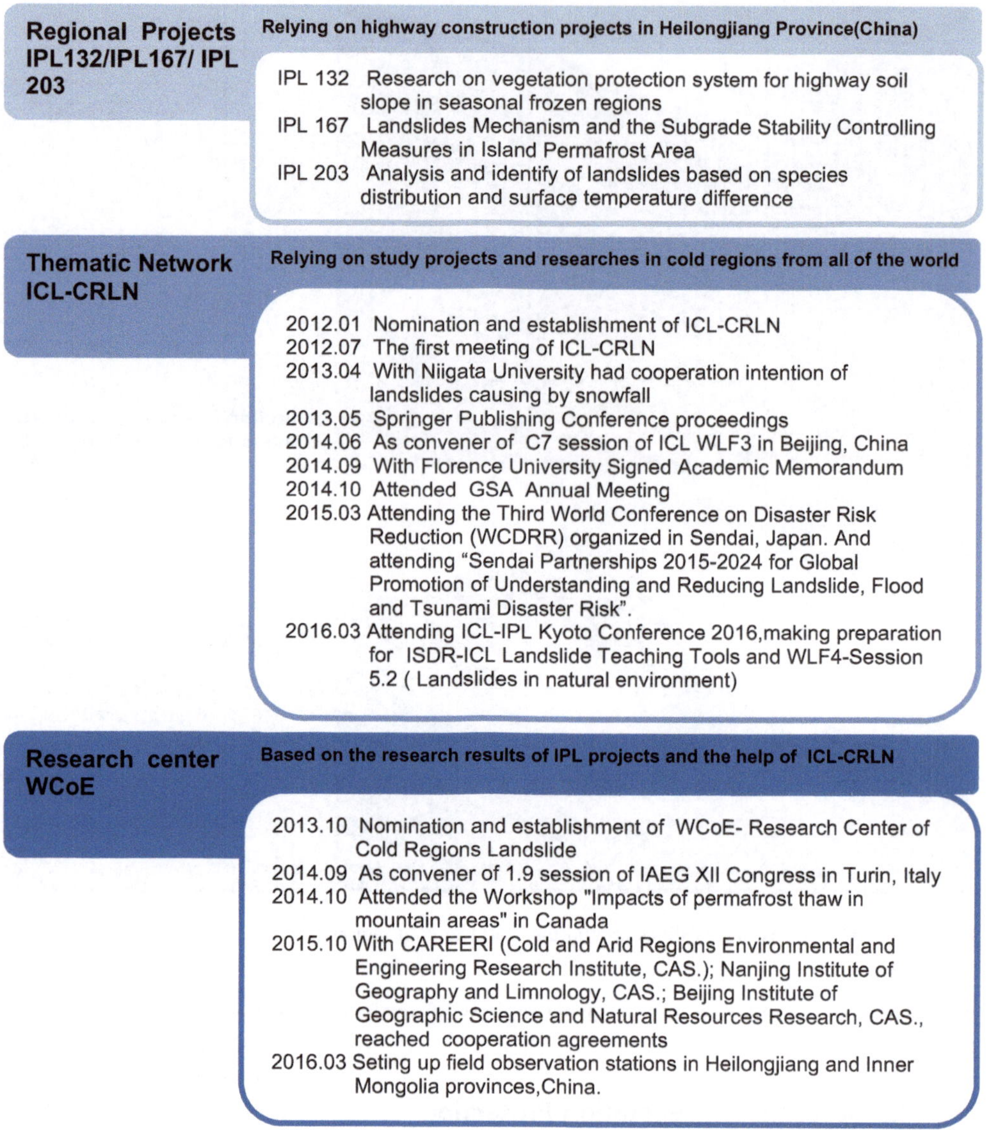

Fig. 3 Main research activities in cold regions

Regional Projects IPL132/IPL167/ IPL 203
Relying on highway construction projects in Heilongjiang Province(China)
IPL 132 Research on vegetation protection system for highway soil slope in seasonal frozen regions
IPL 167 Landslides Mechanism and the Subgrade Stability Controlling Measures in Island Permafrost Area
IPL 203 Analysis and identify of landslides based on species distribution and surface temperature difference
Thematic Network ICL-CRLN
Relying on study projects and researches in cold regions from all of the world
2012.01 Nomination and establishment of ICL-CRLN
2012.07 The first meeting of ICL-CRLN
2013.04 With Niigata University had cooperation intention of landslides causing by snowfall
2013.05 Springer Publishing Conference proceedings
2014.06 As convener of C7 session of ICL WLF3 in Beijing, China
2014.09 With Florence University Signed Academic Memorandum
2014.10 Attended GSA Annual Meeting
2015.03 Attending the Third World Conference on Disaster Risk Reduction (WCDRR) organized in Sendai, Japan. And attending "Sendai Partnerships 2015-2024 for Global Promotion of Understanding and Reducing Landslide, Flood and Tsunami Disaster Risk".
2016.03 Attending ICL-IPL Kyoto Conference 2016,making preparation for ISDR-ICL Landslide Teaching Tools and WLF4-Session 5.2 (Landslides in natural environment)
Research center WCoE
Based on the research results of IPL projects and the help of ICL-CRLN
2013.10 Nomination and establishment of WCoE- Research Center of Cold Regions Landslide
2014.09 As convener of 1.9 session of IAEG XII Congress in Turin, Italy
2014.10 Attended the Workshop "Impacts of permafrost thaw in mountain areas" in Canada
2015.10 With CAREERI (Cold and Arid Regions Environmental and Engineering Research Institute, CAS.); Nanjing Institute of Geography and Limnology, CAS.; Beijing Institute of Geographic Science and Natural Resources Research, CAS., reached cooperation agreements
2016.03 Seting up field observation stations in Heilongjiang and Inner Mongolia provinces,China.

(5) Master Yuying Sun. Study on mechanisms of plant protection of soil slopes on Heilongjiang Road (2008)
(6) Master Yao Liu. Establishment and experimental study of a frozen soil resistivity model (2015).

Publications:

1. Wei Shan, Wang Fawu, Liu Hongjun (2008) Cause Analysis on the Shallow Landslide of Highway Soil Cutting Slopes in Seasonally Frozen-Ground. First World Landslide Forum, 2008.
2. Wei Shan, Hongjun Liu, Lin Yang, Ying Guo (2008) Changes of soil moisture of Shallow Slope in seasonal frozen cutting slope. Rock and Soil Mechanics.
3. Wei Shan, Hongjun Liu, Ying Guo (2009) Effect of freeze-thaw on strength and microstructure of silty clay. Journal of Harbin Institute of Technology.
4. Wei Shan (2008) "Highway greening project" Northeast Forestry University Press.
5. Hongjun Liu, Ying Guo, Wei Shan et al. (2011) Instability of soil cutting slopes caused by freeze-thaw and reinforcement mechanism by vegetation Chinese Journal of Geotechnical Engineering. 08: 1197–1203.
6. Ying Guo, Wei Shan (2011) "Monitoring and Experiment on the Effect of Freeze-Thaw on Soil Cutting Slope Stability" Procedia Environmental Sciences 10: 1115–1121.
7. Ying Guo, Wei Shan et al. (2011) Landslides and Moisture—temperature for Cutting Slope soil in Freeze

—thaw Cycles. Proceedings of the Second World Landslide Forum 3–7 October 2011, Rome.

8. Ying Guo, Wei Shan (2013) The Effect of Freeze–Thaw and Moisture on Soil Strength Index of Cutting Slope. Progress of Geo-Disaster Mitigation Technology in Asia, Environmental Science and Engineering, DOI:10.1007/978-3-642-29107-4_20, Springer-Verlag.
9. Wei Shan, Fawu Wang et al. (2013) Shallow Slope Failure and Protection Method Along a Highway in a Seasonally Frozen Area in China. Progress of Geo-Disaster Mitigation Technology in Asia, Environmental Science and Engineering, DOI:10.1007/978-3-642-29107-4_25, Springer-Verlag.
10. Shan Wei, Guo Ying et al. (2012) Effects of Soil Water Content and Density on Slope Reinforcement by Plant Roots Journal of Northeast Forestry University. 40(12): 111–113.
11. Wei Shan, Chengcheng Zhang, Ying Guo (2012) Mechanism of shallow slide on soil road cutting slope during spring in seasonal frozen region. Applied Mechanics and Materials Vols. 178–181: 1258–1263.
12. Ying Guo, Wei Shan et al. (2014) Landslide Mechanism and Shallow Soil Moisture of Soil Cut Slopes in Seasonally Frozen Regions. Engineering Geology for Society and Territory-Volume 1 Climate Change and Engineering Geology, Springer.
13. Chengcheng Zhang, Wei Shan et al. (2014) The Impact of the Shrub Roots on the Stability of Soil Cut Slope in Seasonal Frozen Regions–Landslide Science for a Safer Geoenvironment-Volume 3 Targeted Landslides, Springer (2014).

IPL-167: Landslides Mechanism and the Subgrade Stability Controlling Measures in Island Permafrost Area (Duration: 2008-Now)

Project leader:
Prof. Wei Shan, Cold Regions Science and Engineering, Northeast Forestry University, Harbin, China.
Core members of the Project:
Dr. Ying Guo, Northeast Forestry University, China
Dr. Hua Jiang, Northeast Forestry University, China
Dr. Chunjiao Wang, Northeast Forestry University, China
Dr. Zhaoguang Hu, Northeast Forestry University, China

Objectives:

The study area is located along the Bei-Hei Expressway Extension Project K160～K182 Section. Under the permafrost, landslides and other complex geological conditions investigations, design, construction and monitoring technical aspects of expressway expansion project.

Training Student: Two doctors, Three Masters

(1) Doctor Chunjiao Wang. Land surface deformation research of permafrost degradation area in northeast china based on D-InSAR (2015)
(2) Doctor Hua Jiang. Formation laws of the landslides and their effect on the subgrade stability in permafrost degradation region (2015)
(3) Master Qingbin Sun. BeiHei Expressway K178 +530 Landslide stability analysis (2012)
(4) Master Zhaoguang Hu. Movement mechanism of permafrost landslide based on gps and resistivity surveying (2012)
(5) Master Shuliang Zhang. The BeiHei black highway island permafrost degradation effect on subgrad stability research (2012)
(6) Master Wei Zhang. Bei'an-Heihe Expressway K177 +550 landslide section of experimental study of soil physical and mechanical indexes (2013)

Publications

1. Wei Shan, Zhaoguang Hu (2011) Application of Geological Drilling combined with High-density resistance in Island Structure Permafrost Survey. [2011 International Conference on Electronics Communications and Control. 08 1898–1904.
2. Wei Shan, Hua Jiang (2012) Formation Mechanism and Stability Analysis of Bei'an-Heihe Expressway Expansion Project K178 Landslide. Advanced Materials Research. 368–373: 953–958.
3. Wei Shan, Hua Jiang, Gaohang Cui (2012) "Formation Mechanism and Characteristics of the Bei'an to Heihe Expressway K177 Landslide" [C]. Advanced Materials Research. 422: 663–668.
4. Wei Shan, Hua Jiang et al. (2011) "Formation Mechanism and Characteristics of the Bei'an to Heihe Expressway K177 Landslide" Proceedings of the Second World Landslide Forum 3–7 October 2011, Rome.
5. Shan Wei, Jiang Hua et al. (2012) Island Permafrost Degrading Process and Deformation Characteristics of Expressway Widen Subgrade Foundation. Disaster Advances. Vol. 5 (4): 827–832.
6. Wei Shan, ChunJiao Wang et al. (2012) Expressway and Road Area Deformation Monitoring Research Based on

InSAR Technology in Isolated Permafrost Area. Remote Sensing, Environment and Transportation Engineering (RSETE), 2nd International Conference. DOI:10.1109/RSETE.2012.6260574.
7. Wei Shan, Zhaoguang et al. (2013) Mechanism of Permafrost Landslide Based on GPS and Resistivity Surveying. Progress of Geo-Disaster Mitigation Technology in Asia, Environmental Science and Engineering, DOI:10.1007/978-3-642-29107-4_18, Springer-Verlag.
8. Hua Jiang, Zhaoguang Hu, et al. (2013) Cut Layer Rocky Landslide Development Mechanism in Lesser Khingan Mountain. Progress of Geo-Disaster Mitigation Technology in Asia, Environmental Science and Engineering, DOI:10.1007/978-3-642-29107-4_19, Springer-Verlag.
9. Zhaoguang Hu; Wei Shan et al. (2014) The Deformation Monitoring of Superficial Layer Landslide in the Northern Part of Lesser Khingan Mountains of China. Engineering Geology for Society and Territory-Volume 1 Climate Change and Engineering Geology, Springer.
10. Hua Jiang; Wei Shan et al. (2014) Freeway Extension Project Island Permafrost Section Foundation Deformation Characteristics. Engineering Geology for Society and Territory-Volume 1 Climate Change and Engineering Geology, Springer.
11. Chunjiao Wang; Wei Shan et al. (2014) Permafrost Distribution Research Based on Remote Sensing Technology in Northwest Section of lesser Khingan Range in China. Engineering Geology for Society and Territory-Volume 1 Climate Change and Engineering Geology, Springer.
12. Wei Shan et al. (2014) Environment and Engineering Geology Problems in Permafrost Section of China Bei'an to Heihe Expressway under the Background of Climate Change. Engineering Geology for Society and Territory-Volume 1 Climate Change and Engineering Geology, Springer.
13. Zhaoguang Hu; Wei Shan et al. (2014) Landslide deformation monitoring and analysis of influence factors at K178+530 of the Bei'an to Heihe Expressway. Landslide Science for a Safer Geoenvironment-Volume 3 Targeted Landslides, Springer.
14. Hua Jiang; Wei Shan et al. (2014) Formation Mechanism and Deformation Characteristics of Cut Layer Rock Landslide in Island Permafrost Region. Landslide Science for a Safer Geoenvironment-Volume 3 Targeted Landslides, Springer.
15. Chunjiao Wang; Wei Shan et al. (2014) Permafrost Distribution Study Based on Landsat ETM+ Imagery of the Northwest Section of the Lesser Khingan Range, China. Landslide Science for a Safer Geoenvironment-Volume 3 Targeted Landslides, Springer.
16. Wei Shan et al. (2014) The Impact of Climate Change on the Stability of Embankment and Slope of Bei'an Highway in Permafrost Regions. Landslide Science for a Safer Geoenvironment-Volume 1 The International Programme on Landslides (IPL), Springer.
17. Shan W, Hu Z, Guo Y (2015) The monitoring of soil pore water pressure and soil temperature in cutting slope before and after aufeis. Proceeding of the 68th Canadian Geotechnical Conference and 7th Canadian Permafrost Conference.
18. Shan W, Hu Z, Guo Y, Wang. C (2015). Environmental and Engineering Geology of the Bei'an to Heihe Expressway in China with a Focus on Climate Change, Engineering Geology for Society and Territory—Volume 1, DOI:10.1007/978-3-319-09300-0_51, Springer.
19. Shan W, Hu Z, Guo Y, Zhang C, Wang C, Jiang H, Liu Y, Xiao J (2015) The impact of climate change on landslides in southeastern of high-latitude permafrost regions of China. Front. Earth Sci. 3:7. doi:10.3389/feart.2015.00007.
20. Hu Z, Shan W (2015) Landslide investigations in the northwest section of the lesser Khingan range in China using combined HDR and GPR methods. Bulletin of Engineering Geology and the Environment. Open access at Springerlink.com doi:10.1007/s10064-015-0805-y.

IPL-203 New: Analysis and Identify of Landslides Based on Species Distribution and Surface Temperature Difference (Duration: 2016-Now)

Project leader:
Dr. Ying Guo, Cold Regions Science and Engineering, Northeast Forestry University, Harbin, China.
Core members of the Project:
Dr. Ying Guo, Northeast Forestry University(NEFU), China
Dr. Zhaoguang Hu, NEFU, China
Dr. Chunjiao Wang, NEFU, China
Dr. Cengcheng Zhang, NEFU, China
Dr. Hua Jiang, NEFU, China

Objectives:

In the Northeast high latitude permafrost zone of China, forest is extensively distributed, and landslide distribution is associated with tree species distribution. Using radar data and small UAV images, combined with manual investigation, to finally obtain landslide distributions within the whole forest area.

Table 1 The members of the ICL landslides in cold regions network until August 2012

ICL-CRLN Member Organization	Country	ICL-CRLN Board and Deputy Member	ICL Member
Northeast Forestry University, China	China	Wei Shan, Ying Guo	Yes
JSC "Hydroproject Institute", Russia	Russia	Alexander Strom	Yes
[a]Research Institute for Natural Hazards and Disaster Recovery, Niigata University, Japan	Japan	Hideaki Marui	Yes
[a]Geological Survey of Canada	Canada	Baolin Wang	Yes
[a]Department of Earth Sciences, University of Firenze, Italy	Italy	Filippo Catani	Yes
Department of Geosciences, Shimane University, Japan	Japan	Fawu Wang	No
College of Construction Engineering, Jilin University, China	China	Lei Nie	No
College of Geology Engineering and Geomatics, Chang'an University, China	China	Tonglu Li	No
Department of Geography, University of Zurich, Switzerland	Switzerland	Stephan Gruber	No
Ministry of Forests, Lands and Natural Resource Operations, Canada	Canada	Marten Geertsema	No
Earth Cryosphere Institute SB RAS, Russia	Russia	Marina Leibman	No
Georadar Division, IDS Ingegneria Dei Sistemi S. p.A. Italy	Italy	Paolo Farina	No

[a](WCoE) in the period 2011-2014

ICL-CRLN and Its Outcome

The International Consortium on Landslides (ICL) is an international non-governmental and non-profit scientific organization, which was established in January 2002 by worldwide universities, institutes, government organizations, academic societies and other entities. Now the ICL has 63 member institutions from 33 countries. The consortium promotes landslide research for the benefit of society and the environment, capacity building and education. At the 10th Session of ICL Board of Representatives, held in Rome, Italy on October 5, 2011, the ICL-Cold Region Landslides Network (ICL-CRLN) was established (Table 1).

The main goal of ICL-CRLN is to promote cooperation among scientists studying landslides in permafrost regions and regions with extreme weather conditions. It will support joint comprehensive investigations carried out by geographers, geologists, geocryologists and meteorologists from different countries and regions, study of landslide mechanisms, and distinguishing landforms, provision of landslide hazard assessment and elaboration of early warning systems. Such cooperation will enhance our understanding of hazardous phenomena in cold regions and, the safety of people living there and their properties and infrastructure. The following is detailed information on events.

(a) Approved by Executive Director of ICL Prof. Kyoji Sassa, the first meeting was held of the ICL Landslides in Cold Regions Network (ICL-CRLN) and First Symposium on Landslides in Cold Regions in Harbin, China on July 23–27, 2012. In total, 51 scientists, technical experts and government officials from Canada, China, Italy, Japan and Russia attended the meeting and leading scientists presented academic reports. The meeting received 12 papers from five countries, which were edited and printed in a full color Proceedings "Landslides in Cold Regions in the Context of Climate Change" (ISSN 1431-6250), which was published by Springer after the meeting. The Meeting determined the purpose of the ICL-CRLN and set up the first Special Committee of ICL-CRLN as the executive body of ICL-CRLN. The meeting elected Prof. Wei Shan from Northeast Forestry University, China as the ICL-CRLN Special Committee Chairman, and elected Dr. Alexander Strom from Geodynamic Research Center, Russia, and Dr. Hideaki Marui from Niigata University, Japan as vice-chairmen of the Special Committee. The Meeting determined the official secretariat of the ICL-CRLN, located at the Northeast Forestry University, China, that has the responsibility of necessary operations and management (Fig. 4).

Fig. 4 The first meeting of ICL-CRLN (Harbin, China, July 2012)

Fig. 5 Intention of cooperating with Niigata University on landslides caused by snowfall

(b) There is an intention to cooperate with the Research Institute for Natural Hazards and Disaster Recovery, Niigata University, Japan, with a focus on the landslides causing by snowfall (Fig. 5).

(c) In early June, 2014, Dr. Ying Guo, as the convener, hosted C7 session “Landslide in Cold Regions” at the Third World Landslide Forum in Beijing. The session received 16 papers, and made 7 oral reports. The second ICL-CRLN meeting was held during the Third World Landslide Forum. The papers of the C7 session (16 papers) were published in 3nd World Landslide Forum Proceeding, Landslide Science for a Safer Geoenvironment - Volume 3 Targeted Landslides (Springer, ISBN 978-3-319-04995-3). Dr. Ying Guo is the editor of Part 7 of this book (Fig. 6).

Fig. 6 The 2nd meeting of ICL-CRLN and the c7 session of WLF3, Beijing (June, 2014)

(d) On 13–14 September, 2014, Prof. Wei Shan visited Florence University, and signed a "Memorandum of Understanding for Academic Cooperation and Exchange Between the Department of Earth Sciences of the University of Florence, Italy and the Institute of Cold Regions Science and Engineering of Northeast Forestry University, China." The two colleges will cooperate in joint training of graduate students and on permafrost distribution, the impact of permafrost degeneration on landscape change and so on. This cooperation will combine field survey data, monitoring data with spacebourne radar data, which maybe open a new milestone for the study of landforms and environmental change in high-latitude permafrost regions (Fig. 7).

(e) Prof. Wei Shan and Dr. Ying Guo attended Geological Society of America (GSA) Annual Meeting held 19–22 October 2014 in Vancouver, Canada, and made two poster reports.

(f) On 14–18 March 2015, Prof. Wei Shan attended the Third World Conference on Disaster Risk Reduction (WCDRR) organized in Sendai, Japan, and attended "Sendai Partnerships 2015–2024 for Global Promotion

Fig. 7 Memorandum of understanding for academic cooperation and exchange with Florence University

of Understanding and Reducing Landslide, Flood and Tsunami Disaster Risk".

(g) With Network member Dr. Bailin Wang, Prof. Wei Shan attended the 68th Canadian Geotechnical Conference and 7th Canadian Permafrost Conference (Quebec, 20–25 Sept 2015) and made an oral report "The monitoring of soil pore water pressure and soil temperature in cutting slope before and after aufeis".

(h) Attended ICL-IPL Kyoto Conference 2016 in March 2016, making preparations for ISDR-ICL Landslide Teaching Tools and WLF4-Session 5.2 (Landslides in natural environment).

Publications

1. Ying Guo, Paolo Canuti, et al. (2012) The First Meeting of ICL Landslides in Cold Regions Network, Harbin, Landslides, DOI 10.1007/s10346-012-0369-x.
2. Giorgio Lollino, Andrea Manconi, John Clague, Wei Shan, Marta Chiarle. Engineering Geology for Society and Territory-Volume 1 Climate Change and Engineering Geology, Part V (Environmental and engineering geological problems in permafrost regions in the context of a warming climate. 21 papers) Springer, ISBN 978-3-319-09299-7 ISBN 978-3-319-09300-0 (eBook) DOI 10.1007/978-3-319-09300-0.
3. Kyoji Sassa, Paolo Canuti, Yueping Yin. Landslide Science for a Safer Geoenvironment-Volume 3 Targeted Landslides, Part VII (Landslide in Cold Regions, 16 papers) Springer, ISBN 978-3-319-04995-3 ISBN 978-3-319-04996-0 (eBook) DOI 10.1007/978-3-319-04996-0.
4. Kyoji Sassa, Paolo Canuti, Yueping Yin. Landslide Science for a Safer Geoenvironment-Volume 1, Part II (1 paper). The International Programme on Landslides (IPL), Springer, ISBN 978-3-319-04998-4 ISBN 978-3-319-04999-1 (eBook) DOI 10.1007/978-3-319-04999-1.
5. Shan W, Hu Z, Guo Y (2015)The monitoring of soil pore water pressure and soil temperature in cutting slope before and after aufeis. Proceedings of the 68th Canadian Geotechnical Conference and 7th Canadian Permafrost Conference, Sept 2015.

6) Wei Shan, Ying Guo, Fawu Wang, Hideaki Marui, Alexander Strom Landslides in Cold Regions in the Context of Climate Change. Springer, ISSN 1431-6250. ISBN 978-3-319-00866-0 ISBN 978-3-319-00867-7 (eBook) DOI 10.1007/978-3-319-00867-7.

WCoE and Its Outcome

Based on the research results of IPL projects and the help of ICL-CRLN, the Institute of Cold Regions Science and Engineering, North East Forestry University, China. proposed to IPL-GPC to build a Research Center of Cold Regions Landslide and it was approved. The following work has been carried out since the center was established.

(a) At the opening Ceremony of WLF3, "Research Center of Cold Regions Landslide, China." has been approved and designated as a "World Centre of Excellence on landslide disaster reduction, 2014–2017". Prof. Wei Shan, the leader of our network was approved, as the leader of IPL Award for success for IPL project.

Fig. 8 WCoE approved as "World Centre of Excellence on landslide disaster reduction, 2014–2017" and Prof. Wei Shan presented with an IPL Award for success for IPL project

Fig. 9 Session of IAEG XII Congress in Turin, Italy

Fig. 10 Workshop: Impacts of permafrost thaw in mountainous areas of Canada and beyond

Fig. 11 Academic cooperation and exchange with Chinese researchers

UNESCO Director-General Irina Bokova presented the award personally (Fig. 8).

(b) On September 16, 2014, while participating in IAEG XII Congress, held in Turin, Italy, Prof. Wei Shan convened and presided over session 1.9 on "Environmental and engineering geological problems in permafrost regions in the context of a warming climate" and received 21 papers. Professor Wei Shan, as a convener and compiler of a conference papers volume, was rewarded by Professor Giorgio Lollino, the President of the Assembly. The papers of session 1.9 (21 papers) was published in IAEG XII Congress Proceeding, Engineering Geology for Society and Territory-Volume 1 Climate Change and Engineering Geology (Springer, ISBN 978-3-319-09299-7). Prof. Wei Shan is one of the editors of this book (Fig. 9).

(c) On 22–25 October, 2014, Prof. Wei Shan and Dr. Ying Guo attended the Workshop "Impacts of permafrost thaw in mountain areas" organized by the Canadian Engineering Research Council (NSERC), the Canadian Forest Service and Ottawa University, to gain more knowledge of permafrost research in Canada. 25 people attended the meeting, which included academics from Canada (Geological Survey of Canada, University of British Columbia, Ottawa University, Carleton University), engineers from Canada, and foreign experts (Washington State University, Zurich University). Prof. Shan and Dr. Guo Ying, attended as foreign experts and representatives of ICL-CRLN. Each expert described relevant research reports from their own academic point of view, these report covered detection technology, theory methods, management, public awareness, databases and other aspects (Fig. 10).

(d) ICL-CRLN reached a cooperation agreement on joint research with CAREERI (Cold and Arid Regions Environmental and Engineering Research Institute, CAS.); Nanjing Institute of Geography and Limnology, CAS.; Beijing Institute of Geographic Science and Natural Resources Research, CAS. They will focus on landslides in high latitude permafrost regions in the context of climate change (Fig. 11).

Prospects

At the 3rd United Nations World Conference on Disaster Risk Reduction in Sendai, Japan in March 14–18, 2015, the ICL proposed the ISDR-ICL Sendai Partnerships 2015–2025 for Global Promotion of Understanding and Reducing Landslide Disaster Risk. It was accepted and signed by 16 United Nations and international stakeholders, and national organizations in 2015. For the effective implementation of "The Sendai Partnerships 2015–2025" cooperation work, at ICL-IPL Kyoto Conference 2016, ICL called for ICL Networks and IPL-WCoE to strengthen concrete collaboration within regional ICL members, as well as collaboration within non-ICL organization. The forms could be bilateral and multilateral project, thematic workshops and conferences, courses and summer schools, life-long learning, training, papers, brochures, proceedings, books and so on.

Based on the above activity plan, "ICL-CRLN" and "IPL-WCoE" will further strengthen exchanges between members of the network, as well as promote the development of cold regions landslide research, also make preparations for 4th World landslide Forum to be held in Slovenia in 2017.

References

Ballantyne CK, Sandeman GF, Stone JO, Wilson P (2014) Rock-slope failure following Late Pleistocene deglaciation on tectonically stable mountainous terrain. Quatern Sci Rev 86:144–157

Blunden J, Arndt DS (2011) State of the climate in 2011. In: American Meteorological Society (http://www.ncdc.noaa.gov/bams-state-of-the-climate/)

Eu-Fp7 (2008) ACQWA: Assessing climate impacts on the quantity and quality of water—a large integrating project under EU Framework Programme 7 (FP7), A summary for policymakers. (www.acqwa.ch)

Fischer L, Hugge C, Kääb Haeberli W (2013) Slope failures and erosion rates on a glacierized high-mountain face under climatic changes Earth Surf. Process Land 38:836–846. doi:10.1002/esp.3355

Grab SW, Linde JH (2014) Mapping exposure to snow in a developing African context: implications for human and livestock vulnerability in Lesotho. Nat Hazards 71:1537–1560. doi:10.1007/s11069-013-0964-8

Guo Y, Canuti P, Strom A, Hideaki M, Shan W (2013) The first meeting of ICL landslides in cold regions network, Harbin, 2012. Landslides 10:99–102. doi:10.1007/s10346-012-0369-x

Haeberli W (2013) Mountain permafrost—research frontiers and a special long-term challenge. Cold Reg Sci Technol 96:71–76

ICL (International Consortium on Landslides) (2012) International consortium on landslides strategic plan 2012–2021—to create a safer geo-environment. (http://iplhq.org/category/home/)

ICL (2014) The 2014 Beijing declaration landslide risk mitigation: toward a safer geo-environment (http://iplhq.org/category/home/)

IPCC (Intergovernmental Panel on Climate Change) (2007) Summary for policymakers. Climate change 2007: the physical science basis. Contribution of working group I to the fourth assessment report of the intergovernmental panel on climate change. Cambridge University Press, Cambridge

IPCC (2013) Summary for policymakers. Working group I contribution to the IPCC Fifth assessment report climate change 2013: the physical science basis. Cambridge University Press, Cambridge

Kliem P, Buylaert JP, Hahn A, Mayrd C, Murray AS, Ohlendorf C, Veres D, Wastegård S, Zolitschka B, Team TPS (2013) Magnitude, geomorphologic response and climate links of lake level oscillations at Laguna Potrok Aike, Patagonian steppe. Quatern Sci Rev 71:131–146

Nussbaumer S, Schaub Y, Huggel C, Nat AW (2014) Risk estimation for future glacier lake outburst floods based on local land-use changes. Hazards Earth Syst Sci 14:1611–1624. doi:10.5194/nhess-14-1611-2014

Shan W, Guo Y, Wang F, Marui H, Strom A (2014a) Landslides in cold regions in the context of climate change. Environmental Science and Engineering. Springer, Berlin. ISBN 978-3-319-00866-0. ISBN 978-3-319-00867-7 (eBook). doi:10.1007/978-3-319-00867-7

Shan W, Guo Y, Zhan C, Hu Z, Jiang H, Wang C (2014b) Climate-change impacts on embankments and slope stability in permafrost regions of Bei'an-Heihe Highway. Landslide Sci Safer Geoenviron 1:155–160. doi:10.1007/978-3-319-04999-1_18

Starnberger R, Drescher SR, Reitner JM, Rodnight H, Reimer PJ, Spötl C (2013) Late Pleistocene climate change and landscape dynamics in the Eastern Alps: the inner-alpine Unterangerberg record (Austria). Quatern Sci Rev 68:17–42

Stoffel M, Tiranti D, Huggel C (2014) Climate change impacts on mass movements—case studies from the European Alps. Sci Total Environ 2014:1255–1266

Large-Scale Rockslide Inventories: From the Kokomeren River Basin to the Entire Central Asia Region (WCoE 2014–2017, IPL-106-2)

Alexander Strom and Kanatbek Abdrakhmatov

Abstract

Large-scale bedrock landslides are among the most hazardous natural phenomena posing a threat to communities living in mountainous regions and in the river valleys therein. Their study requires regular mapping of past features and compilation of uniform and representative inventories. This paper presents the main activities of the World Center of Excellence on Landslide Disaster Reduction of the Geodynamics Research Center—branch of JSC "Hydroproject Institute" (Moscow, Russia) and of the Kyrgyz Institute of Seismology (Bishkek, Kyrgyzstan). Their activities include compilation of a landslide inventory for the Kokomeren River Basin in Central Tien Shan, where the annual Kokomeren Summer School on Rockslides has been carried out since 2006, and of the uniform inventory of large-scale bedrock landslides (rockslides) for the entire Central Asia region, including the Djungaria, Tien Shan and Pamir mountain systems. Basic principles of rockslides identification and the structure of the database are described in brief.

Keywords

Rockslide • Inventory • Kokomeren • Djungaria • Tien shan • Pamir • Central Asia

Introduction

Compilation of the landslide inventories is the first and critically important step in understanding the spatial and temporal distribution of these hazardous natural phenomena, in revealing factors governing their distributions and in correct assessing landslide susceptibility and, further, landslide hazard and risk.

The Central Asia region (Fig. 1), embracing the Djungaria, Tien Shan and Pamir mountain systems, is one of the global landslide hotspots (Nadim et al. 2006).

Besides numerous landslides in soft sediments, widely developed in the foothills and often covered by loess, this region is extremely rich in large-scale bedrock landslides, most of which are prehistoric. The majority of the historical case studies, including the world's largest non-volcanic slope failure—the 2.2 km^3 Usoi rockslide, were triggered by large earthquakes, although several non-seismic slope failures have been reported from this region as well.

Despite an arid climate favorable for the long-term preservation of landslides and facilitating their identification on remote sensing data, no complete rockslide inventory has been compiled up to now for the entire Central Asia region that includes six states—Afghanistan, China, Kazakhstan, Kyrgyzstan, Tajikistan and Uzbekistan. Detailed inventories of the landslides in soft sediments were compiled for large parts of the study region (Niyazov 2009; Havenith et al. 2015a, b) while rockslides were mapped and studied in detail more locally (see, e.g., Fedorenko 1988).

Although bedrock landslides are, luckily, rather rare phenomena compared to slope failures in soft sediments,

A. Strom (✉)
Geodynamics Research Centre—Branch of JSC "Hydroproject Institute", Volokolamskoe Shosse, 2, 125993 Moscow, Russia
e-mail: strom.alexandr@yandex.ru

K. Abdrakhmatov
Institute of Seismology, National Academy of Science, Asanbay 52/1, Bishkek, 720060, Kyrgyzstan
e-mail: kanab53@yandex.ru

K. Sassa et al. (eds.), *Advancing Culture of Living with Landslides*,
DOI 10.1007/978-3-319-59469-9_28

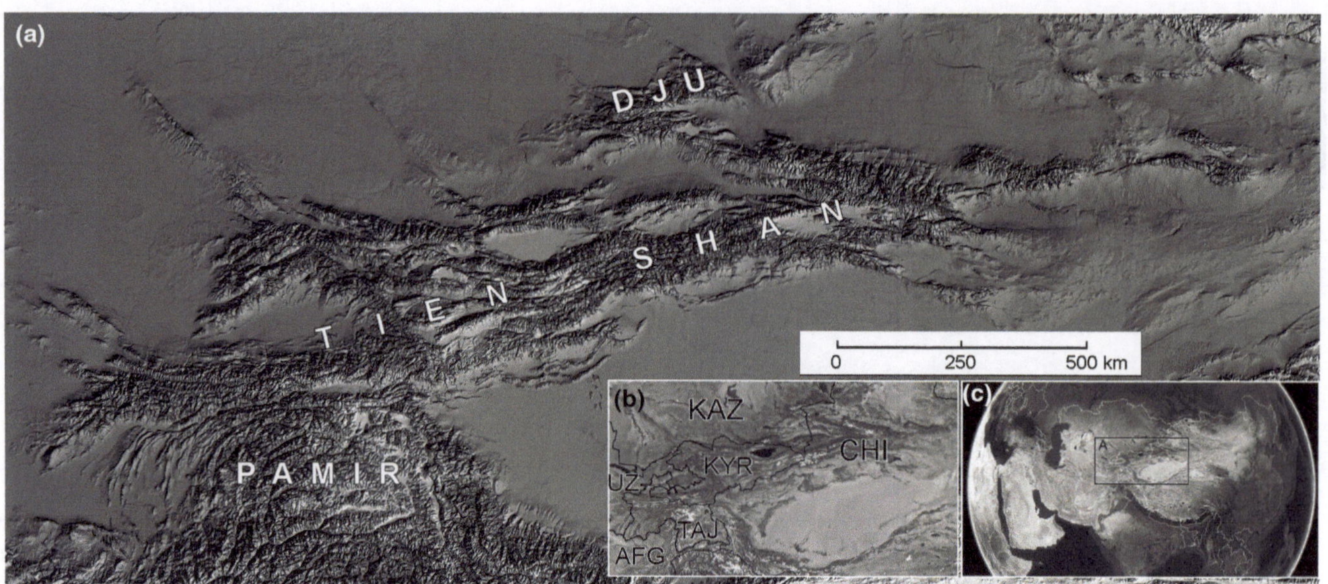

Fig. 1 **a** the 30" SRTM DEM of the study region; **b** political boundaries in the Central Asia region (from Google Earth); **c** location of the study region (*outlined*) in Asia. *DJU* Djungarian Range, *AFG* Afghanistan, *CHI* China, *KAZ* Kazakhstan, *KYR* Kyrgyzstan, *TAJ* Tajikistan, *UZ* Uzbekistan

their effects could be extremely disastrous, due to their abnormal mobility and ability to form large natural dams.

Study of such features in the Kokomeren River basin in Central Tien Shan (Kyrgyzstan) has been supported by several IPL Projects since 2005 (the last one is IPL106-2). The main goal of these projects was the organization of the annual International Summer School on Rockslides and Related Phenomena (the Kokomeren Summer School) that has been carried out since 2006, supported by the International Consortium on Landslides (ICL) and International Program on Landslides (IPL). Acknowledging the success of these activities, the Geodynamics Research Center—branch of the JSC "Hydroproject Institute" (ICL Member since 2003) and Kyrgyz Institute of Seismology were awarded status as a World Center of Excellence on Landslide Disaster Reduction (WCoE) for the periods of 2008–2011, 2011–2014 and 2014–2017.

Based on the experience gained in this region, we started compilation of a rockslide inventory of the entire Central Asia region.

Rockslides in the Kokomeren River Basin

The Kokomeren River basin (41.6°–42.4°N, 73.9°–74.6°E) (Fig. 2) is extremely rich in impressive bedrock landslides and rock avalanches of different types (see Fig. 2) ranging in volume from a few million up to ca. 1.5 billion cubic meters. They originated on slopes composed of different types of bedrock—Paleozoic granites, gneiss, sedimentary terrigenous rocks metamorphosed to a varying extent and Neogene conglomerates, sandstones and siltstones. About 25 rockslides were identified within a limited area of about 80 × 50 km, some of which formed natural dams while other transformed into highly mobile rock avalanches up to ~6 km long.

Since most of these features are located close to roads and are attainable during one-day trips from the base camp, this region was chosen for the two-week long annual field training course—the Kokomeren Summer School, which started in 2006 (Strom and Abdrakhmatov 2009). This area can be considered as a field museum of large-scale rockslides, not masked by thick vegetation and not disturbed by excessive human activities. It provides an excellent opportunity for students, young researchers and engineers to learn more about such phenomena, their distribution, origin, emplacement mechanisms and geological and geomorphic conditions favorable for their formation. Besides young professionals and students several experienced researchers participated in the Kokomeren Summer School too (Fig. 3).

The full-color Guidebook was printed and provided to Kokomeren Summer School participants and also has been uploaded on the IPL website for general use http://iplhq.org Download GUIDEBOOK.

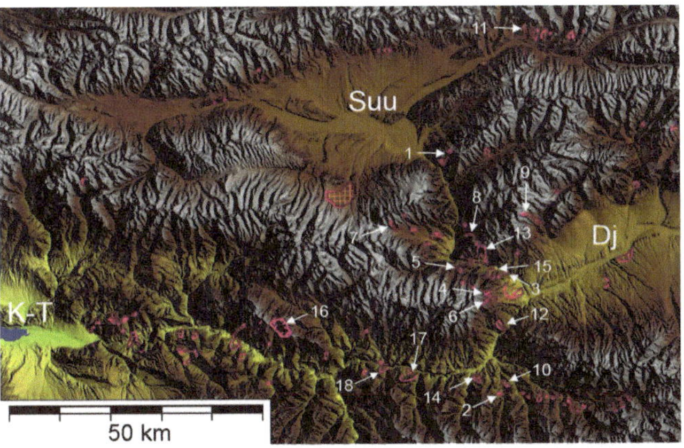

Fig. 2 Large landslides, rock avalanches, and caldera-like collapses in the Kokomeren River basin and adjacent part of the Naryn River basin. Suu, Dj, and K-T—the Suusamyr, the Djumgal, and the Ketmen-Tiube intermountain depressions. Selected features, most of which are studied during the training course: *1* Seit, *2* Ak-Kiol, *3* Mini-Köfels, *4* Kashkasu, *5* Northern Karakungey, *6* Southern Karakungey, *7* Chongsu, *8* Sarysu, *9* Ming-Teke, *10* Lower Ak-Kiol, *11* Snake-Head, *12* Lower-Aral, *13* Kokomeren, *14* Ornok, *15* Displaced Peneplain, *16* Kyzylkiol, *17* Karachauli, and *18* Lower Kokomeren

Fig. 4 Participants of the 2009 Kokomeren Summer School on top of the Seit rock avalanche; its headscarp can be seen in the background

Since 2006 about 100 students and landslide researchers from Argentina, Austria, Belgium, China (including Hong Kong), Czech Republic, France, Germany, Great Britain, Italy, Japan, Kyrgyzstan, New Zealand, Russia, Switzerland, Spain, Taiwan, Tajikistan and USA have attended it (Figs. 3, 4 and 5). Such joint work, apart from everything else, facilitates future cooperation of young landslide researchers.

Most of the case studies identified in this area were described in Abdrakhmnatov and Strom (2006), Hartvich et al. (2008), Strom (1994, 1996, 2006, 2010, 2013a, b, c, d, 2015), Strom and Groshev (2009), Strom and Korup (2006), Strom and Stepanchikova (2008), and Strom and Zhirkevich (2013).

However, comparison of some suspicious sites in this relatively well-studied river basin with some didactic case studies from other parts of the Central Asia region, allowed the identification of a new, previously unknown ~6 km long rock avalanche in the upper reaches of the Kashkasu river valley (compare Figs. 6 and 7). It highlights the importance of considering case studies from a much broader region even if the study area is limited to a particular river basin.

Compilation Methods for the Central Asia Rockslides Database

The arid climate typical of the Central Asian region facilitates long-term preservation of landforms created by bedrock slope failures and their expression on remote sensing data available at resources such as Google Earth, SAS Planet and Arc GIS Earth.

Fig. 3 Participants of the 2012 Kokomeren summer school at the outcrop of an active fault north of Chaek town. *First* from the *left* is Dr. Roman Lahodynsky (BUKU University), *second* from the *left* is Dr. James McCalpin—author of the famous “Paleoseismology”

Fig. 5 Participants of the 2016 Kokomeren Summer School at the headscarp crown of the Ak-Kiol rockslide, which had dammed the lake visible in the background

Fig. 6 The 5.5 km long Urmochdara rock avalanche in the Gunt River basin (South-western Pamir) with the striated microrelief of the deposits indicating flow-like debris motion. RG—rock glacier with a totally different morphology

These resources were used for the regular surveying of the study area, with the aim of identifying landforms typical of both the source zones and the rockslide deposits. More than 800 features exceeding ca. 1 million cubic meters in volume have been identified up to now.

Their boundaries, digitized on the above mentioned resources, were imported in the Global Mapper GIS and overlaid on the 3" SRTM DEM that allowed measurement of their basic parameters, such as headscarp height and width,

Fig. 7 A rock avalanche about 6-km long, with evidence of flow-like motion of debris, in the upper reaches of the Kashkasu River, Central Tien Shan. It was identified based on its similarity to the Urmochdara rock avalanche. The ~500 m high headscarp is shown in the *inset*

deposit area, width, and runout, vertical distance between headscarp crown and the tip of the deposits, etc.

Besides the well-developed bedrock landslides that had already failed, evidence of deep-seated gravitational slope deformations (DSGSD) that can, under certain conditions, convert into catastrophic rock slope failure (Fig. 8) are identified, digitized and included in the database.

The volume of the deposits was calculated for each case individually. Considering the unknown morphology of both the source zone and depositional area prior to slope failure, the accuracy of the volume assessment is estimated as ±30 – 50%. However, such accuracy seems to be sufficient for the statistical analysis of basic relationships between the magnitude of the phenomena (characterized by its volume) and its main geometrical parameters (runout, area, H/L ratio) and such parameter as the DBI—the Dimensionless Blockage Index (Ermini and Casagli 2003). It is higher than the scatter of the landslide volume versus area relationships, which usually are about 1 order larger (Honious et al. 1997; Guzetti et al. 2008).

Considering the significant scatter of volume estimates, some smaller features were included in the database even though, after being identified and outlined on space images, they were found to be smaller. In addition, small rockslides that form clusters accompanying larger slope failures were included in the database and described, considering their importance for revealing possible triggering factors of large-scale rockslides formation (Strom 2013b, c).

Besides rockslides and the DSGSDs, landslide-dammed lakes, both existing, filled, and breached, were mapped, along with evidence of outburst floods that were identifiable at several sites.

Structure of the Database

The rockslide database is compiled as an Excel spreadsheet. Parameters characterizing each landslide are grouped in several blocks: General information, Headscarp geology and parameters, Overall parameters of the rockslide deposits, Rock avalanche parameters, Landslide dams and dammed lake parameters. They are listed in the following tables.

Each landslide has its ID, which includes the Country name (Afghanistan, China, Kazarhstan, Kyrgyzstan, Tajikistan, Uzbekistan), its coordinates (for reference and easy search), individual feature name and date of the event, if available. Some comments describing the most peculiar characteristics of the features in question are added (Table 1).

The geology and morphology of the source zone is described in Table 2. Quantitative parameters presented here

Fig. 8 Deep-seated gravitational slope deformation in the Betdjin Creek valley (Chinese Tien Shan) at 41.146°N, 78.354°3D Google Earth view

Table 1 General information

Rockslide ID	Date (if known) and name (if given)	Comments	Latitude	Longitude
1	2	3	4	5

Table 2 Headscarp geology and parameters

Rock age	Rock type:	Rock sequence from top to bottom	Headscarp height (km)	Mean slope angle (°)	Headscarp width at its base (km)
6	7	8	9	10	11

Table 3 Overall parameters of the rockslide deposits

Area of the deposits (km^2)	Total area affected (km^2)	Volume (10^6 m^3)	Total length (km)	Maximal deposits' width (km)	Maximal height drop (km)	Height drop to the tip (km)	Confinement	Runup (km)
12	13	14	15	16	17	18	19	20

Table 4 Rock avalanche parameters

Rock avalanche type	Volume		Rock avalanche Length (km)	Rock avalanche height drop (km)	Secondary scar/jump height (km)
	(10^6 m^3)	% of total rockslide volume			
21	22	23	24	25	26

Table 5 Dammed lakes parameters

Dam parameters			Lake parameters				
Effective height (m)	Dams' width (km)	The lowermost part of the dams' crest position	Type (existing, breached, filled)	Maximal depth (m)	Length (m)	Mean width (m)	Volume (10^6 m^3)
27	28	29	30	31	32	33	34

and hereafter were measured based on the Google Earth data and 3" SRTM DEM.

The third block (Table 3) includes parameters characterizing each rockslide as a whole. As mentioned above, rockslide volume—one of the most critical parameters—was estimated on an individual basis for each feature, approximated by a combination of simple 3D geometrical figures—prisms, pyramids, cones.

Next block (Table 4) includes parameters of the mobile avalanche-like parts of rockslides classified as Jumping and Secondary rock avalanches (Strom 1996, 2006).

Considering the significant hazard presented by rockslides damming rivers and by their outburst floods, quantitative and qualitative data characterizing rockslide-dammed lakes are presented in the special block of the database (Table 5).

Conclusions

Compilation of the more or less uniform database of the Central Asian large-scale bedrock landslides and of associated features such as landslide-dammed lakes and DSGSDs will form grounds for a more rigorous regional rockslide susceptibility analysis and hazard assessment. Such analysis should be performed with due regard to the variability of the geological and geomorphic conditions of these mountain systems that predetermine formation of large-scale bedrock landslides, as well as to the variability of the tectonic (earthquakes) and climate-related (rainstorms, snow melt) triggering factors.

Compilation of the rockslide database will provide a valuable set of data for better understanding regularities in the occurrence and evolution of such phenomena, their spatial and temporal distribution over the entire region and of mechanism of rockslide and rock avalanche motion.

Such a database could be used by landslide researchers interested in new impressive examples of large-scale catastrophic bedrock landslides and in new data supplementing those derived from case studies from other regions. It can be used by emergency experts working on landslide and outburst flood hazard assessment and by researchers and practitioners in the fields of paleoseismology and seismic hazard assessment, since large-scale bedrock landslides are often considered as being triggered by large past earthquakes. The database will aid hydraulic engineers working on dam projects in the Tien Shan and Pamir mountainous rivers, considering the high potential of the Central Asian rivers for hydraulic energy production and the significant threat posed by large landslides to the safety of hydraulic schemes.

The results of this research project that, in addition to the Kokomeren Summer School, became the second main goal of our WCoE, will be presented in a book entitled: "Large landslides in Central Asia: distribution, impacts, and hazard assessment" that is planned to be published by Elsevier Publishers in 2017.

Acknowledgements Authors want to express sincere gratitude to ICL, IPL and to the Executive Director of ICL Prof. Kyoji Sassa for permanent support provided to organize the Kokomeren Summer School.

References

Ermini L, Casagli N (2003) Prediction of the behavior of landslide dams using a geomorphological dimensionless index. Earth Surf Proc Land 28:31–47

Fedorenko VS (1988) Rockslides and rock falls and their prediction. Moscow State University Publishing House, Moscow, 214 pp. (in Russian)

Guzzetti F, Ardizzone F, Cardinali M, Galli M, Reichenbach P, Rossi M (2008) Distribution of landslides in the Upper Tiber River basin, central Italy. Geomorphology 96(1–2):105–122. doi:10.1016/j.geomorph.2007.07.015

Hartvich F, Mugnai F, Proietti C, Smolkova V, Strom, A (2008) A reconstruction of a former rockslide-dammed lake: the case of the Kokomeren River valley (Tien Shan, Kyrgyzstan). In: Poster presentation at the EGU conference, Vienna

Havenith H-B, Strom A, Torgoev I, Torgoev A, Lamair L, Ischuk A, Abdrakhmatov K (2015a) Tien Shan geohazards database: Earthquakes and landslides. Geomorphology 249:16–31

Havenith H-B, Torgoev A, Schlögel R, Braun A, Torgoev I, Ischuk A (2015b) Tien Shan geohazards database: landslide susceptibility analysis. Geomorphology 249:32–43

Hovius N, Stark CP, Allen PA (1997) Sediment flux from a mountain belt derived by landslide mapping. Geology 25:801–804

Nadim F, Kjekstad O, Peduzzi P, Herold C, Jaedicke C (2006) Global landslide and avalanche hotspots. Landslides, 3:159-173

Niyazov RA (2009) Landslides in Uzbekistan (tendencies of evolution at the turn of the XXI Century). Tashkent, Gidroingeo, 206 pp. (in Russian)

Strom A (1994) Mechanism of stratification and abnormal crushing of rockslide deposits. In: Proceeding 7th International IAEG Congress 3, Rotterdam, Balkema, pp 1287–1295

Strom A (1996) Some morphological types of long-runout rockslides: effect of the relief on their mechanism and on the rockslide deposits distribution. In: Senneset K (ed) Landslides. Proceeding of the Seventh International Symposium on Landslides, 1996, Trondheim, Norway, Rotterdam, Balkema, pp 1977–1982

Strom A (2006) Morphology and internal structure of rockslides and rock avalanches: grounds and constraints for their modelling. In: Evans SG, Scarascia Mugnozza G, Strom A, Hermanns RL (eds) Landslides from massive rock slope failure. NATO Science Series: IV: Earth and Environmental Sciences 49:305–328

Strom A (2010) Evidence of momentum transfer during large-scale rockslides' motion. In: Williams AL, Pinches GM, Chin CY, McMorran TG, Massei CI (eds) Geologically active. Proceeding of the 11th IAEG Congress, Auckland, New Zealand, 5–10 September 2010, Taylor & Francis Group, London, pp 73–86

Strom A (2013a) Geological prerequisites for landslide dams' disaster assessment and mitigation in Central Asia. In: Wang F, Miyajima M, Li T, Fathani TF (eds) Progress of geo-disaster mitigation technology in Asia. Springer, Heidelberg, pp 17–53

Strom A (2013b) Use of indirect evidence for the prehistoric earthquake-induced landslide identification. In: Ugai K, Yagi H, Wakai A (eds) Earthquake-induced landslides. Springer, Berlin, pp 21–30

Strom AL (2013c) Constraints and promises of earthquake-triggered landslides discrimination. In: Grützner C, Rudersdorf A, Pérez-López R, Reicherter K (eds). Proceeding 4th International INQUA Meeting on Paleoseismology, Active Tectonics and Archeoseismology (PATA), 9–14 October 2013, Aachen, Germany, pp 267–270

Strom A (2013d) Rockslides and rock avalanches in the Kokomeren River valley (Kyrgyz Tien Shan). In: Arbanas SM, Arbanas Ž. (eds) Landslide and flood hazard assessment. Proceedings of the 1st Regional Symposium on Landslides in the Adriatic-Balkan Region, Zagreb, pp 245–250

Strom A (2015) Possible causes of accelerating landslide motion in confined environment. In: Lollino G, Giordan D, Crosta GB, Corominas J, Azzam R, Wasowski J, Sciarra N (eds) Engineering geology for society and territory—vol. 2, doi:10.1007/978-3-319-09057-3_151, © Springer International Publishing Switzerland, pp 883–885

Strom A, Abdrakhmatov K (2009) International summer school on rockslides and related phenomena in the Kokomeren River Valley, Tien Shan, Kyrgyzstan. In: Sassa K, Canuti P (eds) Landslides. Disaster Risk Reduction. Springer, Heidelberg, pp 223–227

Strom A, Groshev M (2009) Mysteries of rock massifs destruction. In: Abbie M, Bedford JS (eds) Rock mechanics: new research. Nova Science Publishers, New York, pp 211–231

Strom A, Korup O (2006) Extremely large rockslides and rock avalanches in the Tien Shan, Kyrgyzstan. Landslides 3:125–136

Strom A, Stepanchikova P (2008) Seismic triggering of large prehistoric rockslides: Pro and Con case studies. In: Proceedings of the International Conference on Management of Landslide Hazard in the Asia-Pacific Region (Satellite symposium of the First World Landslide Forum, Tokyo), Sendai, 11th–12th November 2008, pp 202–211

Strom A, Zhirkevich A (2013) "Remote" landslide-related hazards and their consideration for the hydraulic schemes design. In: Genevois R, Prestinnizi A (eds) Proceeding. International Conference on Vajont—1963–2013. Italian Journal of Engineering Geology and Environment, Book series No 6. pp 295–303

Interventions for Promoting Knowledge, Innovations and Landslide Risk Management Practices Within South and Southeast Asia (WCoE 2014–2017)

Peeranan Towashiraporn and N.M.S.I. Arambepola

Abstract

Asia is a dynamic nexus of economic and social change, with population growth, industrialisation and urbanisation playing a large part in shaping the exposure of communities to hydro-meteorological as well as geologic hazards. Among them, landslides have become most widespread and commonly observed events throughout South and Southeast Asia. Recent incidents triggered by South Asian monsoons affected by El Niño resulted in devastating landslides in many countries in Asia. Realizing the essential need for landslide risk reduction, the Asian Disaster Preparedness Center (ADPC) implements an umbrella program, namely the Asian Program for Regional Capacity Enhancement for Landslide Impact Mitigation (RECLAIM) to undertake various measures for landslide risk mitigation such as pilot demonstration projects, capacity building programs for the stakeholder institutions through regional and national level training courses, networking events for experience sharing. This paper presents some of the needs and gaps in the region and initiatives undertaken by the Asian Disaster Preparedness Center (ADPC) through its programs to address them. In acknowledgement of the initiatives being undertaken for landslide disaster risk reduction in Asia, the International Consortium of Landslides (ICL) has recognized ADPC to be one of the World Centers of Excellence (WCoE).

Keywords

Mandated institutions • Regional experience sharing

Introduction

Landslides are becoming one of the most widespread and frequent natural disaster events in many countries in South and Southeast Asia. Landslides are not limited to mountain settlements in a few countries, but are seen across sub-regions, and are being exacerbated by climate change. Landslide incidents that have occurred recently, triggered either by monsoon precipitation affected by El Niño or a relatively high level of shaking due to earthquakes, or due to the combined effects of both, have resulted in devastating impacts to respective national economies. Landslides not only kill people or destroy the houses, infrastructure and other lifeline facilities within urban and rural settlements, but they are also becoming a pressing environmental problem. An escalation of landslide events is becoming clear due to changing landscapes, and a growing population and increasing vulnerability due to their undesirable actions. The resultant slope destabilization and downslope movement of soil masses are responsible for serious degradation of mountain slopes, and higher sediment load transport, which demands speedy and sustainable solutions by authorities (Fig. 1).

Under its regional program for Regional Capacity Enhancement for Landslide Impact Mitigation (RECLAIM), the Asian Disaster Preparedness Center (ADPC) and its partners undertook various measures to help countries in

P. Towashiraporn (✉) · N.M.S.I.Arambepola
SM Tower, 24th Floor, 979/69 Paholyothin Road, Samsen Nai Phayathai, Bangkok 10400, Thailand
e-mail: peeranan@adpc.net

N.M.S.I.Arambepola
e-mail: arambepola@adpc.net

K. Sassa et al. (eds.), *Advancing Culture of Living with Landslides*,
DOI 10.1007/978-3-319-59469-9_29

Fig. 1 Meeriyabedda landslide, Sri Lanka (*image source* National Building Research Organization, Sri Lanka)

Asia manage landslide risk. Those measures had an impact at various levels through local pilot demonstration studies, national level capacity building programs, and regional knowledge-sharing events. The subsequent sections describe in detail those interventions under RECLAIM, as well as several other related projects.

Interventions for Building the Capacity of Mandated Institutions

One of the felt needs, to effectively deal with the landslide problem in a region, is development of the capacity of professionals and scientists involved in landslide services and research studies. It is beneficial to them if they can often share common challenges, problems and potential solutions to make the communities living in landslide prone areas more resilient. Realizing such needs, the Asian Disaster Preparedness Center (ADPC), with technical assistance from the Norwegian Geotechnical Institute (NGI), initiated an umbrella program, namely the Asian Program for Regional Capacity Enhancement for Landslide Impact Mitigation (RECLAIM) to build the capacity of stakeholder institutions through regional and national level training courses, risk information generation, experience sharing, etc. In addition, the program successfully developed a network of institutions mandated to deal with landslide risk management, conduct scientific studies, and engage in related research and development interventions.

The program, which received funding support from the Royal Norwegian Government, frequently organized annual networking events for member institutions to share experiences on good and sound practices of landslide risk management, as well as to discuss challenges and share lessons.

Such networking events have served as a platform to further enhance the scientific understanding of landslide processes and occurrence, share experience in early warning, discuss challenges in instrumentation and monitoring the active landslides, find sound solutions to improve preparedness at the local level, promote options for reducing the risk, and for capacity building and public awareness. Each participating country made presentations at the annual meetings organized by ADPC, providing an overview of the experience of the respective countries in landslide risk reduction. Participants attending a national level networking event in Myanmar are shown in Fig. 2.

Promoting Sound Practices for Landslide Early Warning

Use of early warning systems for natural hazards such as floods and typhoons have advanced fairly rapidly in the recent past and the successful use of such systems has helped to save lives of the people residing in the affected areas, for instance in Bangladesh, China, Thailand and the Philippines. However, use of effective early warning for alerting people in landslide-prone areas is lagging behind. There is a need for a joint effort in the Asian Region to share experiences in this matter and to establish good practices for efficient landslide early warning systems. The IPL 156- “Best practices for early warning of landslides in a changing climate scenarios”, has highlighted relevant ADPC-initiated activities that have been taken to promote sound practices for early warning of landslides in a changing climate. Improving landslide early warning mechanisms in high risk areas has contributed in providing valuable inputs to the multi-hazard

Fig. 2 Networking events are frequently organized among mandated institutions to discuss common challenges in landslide risk management. Field visits are a common feature during regional networking events

early warnings platforms that are currently operational in countries within Asia too.

Studies for Developing Precipitation Thresholds for Triggering Landslides

Landslide hazard prediction, for warning purposes, is still a difficult problem and an increasingly hot topic in the research on landslide hazards. Precipitation is a main triggering factor that controls the initiation of slope failures, and landslides are often triggered by weather events that are in excess of some thresholds.

Despite the benefits of advanced technologies for landslide early warning developed and used by developed countries, it is still not the main focus in developing countries. Various countries in Asia still rely on simple thresholds, such as the amount of rainfall, as triggering parameters of landslides. It is indeed necessary to investigate the possibility of interpreting landslide events in terms of the amount of rainfall and its patterns immediately preceding the slide event. The rainfall threshold values vary from region to region or district to district due to differences in not only the climatological patterns, but also existing soil characteristics. As a result, such approaches can be further enhanced by taking the geological formation into account in developing precipitation thresholds too.

Therefore, ADPC has undertaken a comprehensive study of rainfall patterns and landslide records in landslide-prone areas of different countries in Asia. Interpreting such data has helped derive reasonable threshold values of rainfall and use them as a tool for landslide forecasting. In addition, collection of critical rainfall values with sufficient geological data in the sub-region, and developing a possible correlation between such attributes was also found to be vital for predicting landslides accurately in highly vulnerable locations.

As a result of the study, several empirical relationships have been deduced for different landslide-prone districts of partner countries, based on the assumption that there exists a direct relationship between the occurrence of landslides, geological formations and the rainfall events characterized in terms of intensity, duration of rainstorm events (short-term, e.g., hourly rainfall or 24-hour cumulative rainfall, for examples) and antecedent rainfall (3 day cumulative, 7 day cumulative, etc.).

Nevertheless, some technical gaps that could deter the use of such threshold approach remain. Some of the networks of rain gauges maintained by the meteorological departments in countries were outdated and even non-functional in some cases. This has to be addressed through capacity building of the relevant departments mandated in maintaining the networks. At the same time, there are also operational issues such as automation of precipitation data collection and recording, as well as gaps in institutional communication. Such operational problems can be resolved through a dialogue among multi-stakeholder institutions. In this regard, ADPC has facilitated several discussions between mandated institutions and meteorological agencies (see Fig. 3) in partner countries to discuss common challenges and possible ways of addressing such challenges.

Initiatives being carried out by ADPC in this connection complements the ongoing international initiatives on setting up threshold values of precipitation for initiation of landslides.

Initiatives for Establishing Community-Based Systems for Landslide Warning

Devastating landslides have been frequently reported in settlements located within mountain districts. Most of them occur in communities in remote locations. Adverse climatic trends variations are increasingly making such landslide-prone, built-up areas more insecure for living. This was the case in hilly district cities of Bangladesh such as Chittagong, Cox Bazaar, and Teknaf.

In such rural areas, timely dissemination of data on landslide early warning to the communities at risk is a serious problem, as the data flow is not direct and often slow. Therefore it is difficult to provide location-specific landslide early warning through existing local networks. At the moment, ADPC is working with mandated institutions in Sri Lanka, Bangladesh, Myanmar, and Nepal in reviewing the rainfall monitoring networks and to expand the network to areas where rainfall data for landslide monitoring and early warning operations do not exist or are limited.

The Project interventions, which have been reported under IPL 180—"Introducing Community-Based Early Warning System for Landslide Hazard Management in Cox's Bazaar Municipality, Bangladesh" aims at developing a local-level landslide hazard risk management strategy in Bangladesh. Such interventions can be replicated elsewhere, so that other vulnerable settlements in Asia will be able to utilize similar practices for reducing future losses due to landslide events. In Fig. 4, a rain gauge set up at a community level is shown with a community member who was trained to observe precipitation thresholds to identify common symptoms of landslide initiation. These trained community members will promptly alert the communities and take appropriate actions to move away when landslide initiation is expected.

Framework for Country-Wide Landslide Inventory

ADPC submitted a technical paper on "Developing a framework for assessment of landslide hazard risk associated with critical national and provincial roads" in the ICL symposium held in Kyoto in November 2013. As described in the paper, the purpose of the proposed framework for a country-wide risk–based landslide inventory is to facilitate the taking of suitable proactive cost-effective measures by the road sector in order to reduce disruptions to transportation due to occurrences of landslides during monsoon periods. It is suggested that a country-wide landslide inventory be developed as a key feature in the long-term strategy and records maintained of all landslides recorded on critical national and provincial roads in a GIS environment. In the pilot study in Lao PDR, necessary training was provided to road authorities for field data collection, inventory preparation, maintenance and upkeep. The framework also will serve as a guideline for designing new slope stabilization measures where necessary during implementation of road

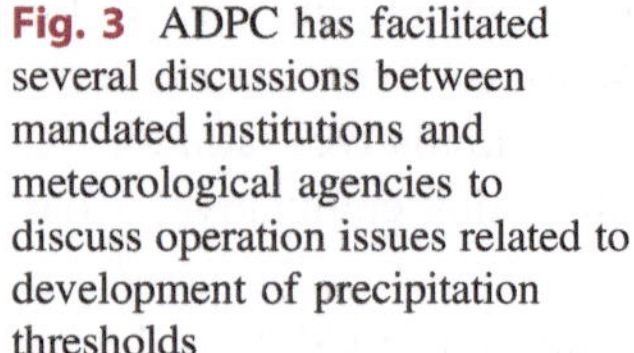

Fig. 3 ADPC has facilitated several discussions between mandated institutions and meteorological agencies to discuss operation issues related to development of precipitation thresholds

Fig. 4 Rain gauges set up in a community for observing the established thresholds

expansion and/or improvement projects. At present, road engineers in Lao PDR do not take into consideration landslide hazard and risk potential during implementation of road improvement projects and road expansion-associated excavations, which at times led to subsequent slope destabilization. Recommendations have been included in the framework for identifying likely high-risk landslide locations or landslide-prone areas with high hazard, which need special attention during excavations, as well in the implementation of priority mitigation measures. During the project period, actions have been taken to provide resilient landslide protection for some of the vulnerable locations identified during the project for providing safer access to public services (see Fig. 5: Resilient landslide protection measures in Bountai District, Lao PDR).

Landslide Hazard Zonation Mapping and Assessment of Risk

ADPC has carried out multi-hazard risk assessment projects in countries such as Bangladesh, Cambodia, Nepal, Lao PDR, Myanmar, and Timor Leste. In each project, national, provincial, and/or city-level landslide hazard maps were prepared (as shown in Fig. 6) and provided to the National Disaster Management Office (NDMO) or provincial

Fig. 5 Resilient landslide protection measures in Bountai District (Lao PDR) provided communities with safer access to public services

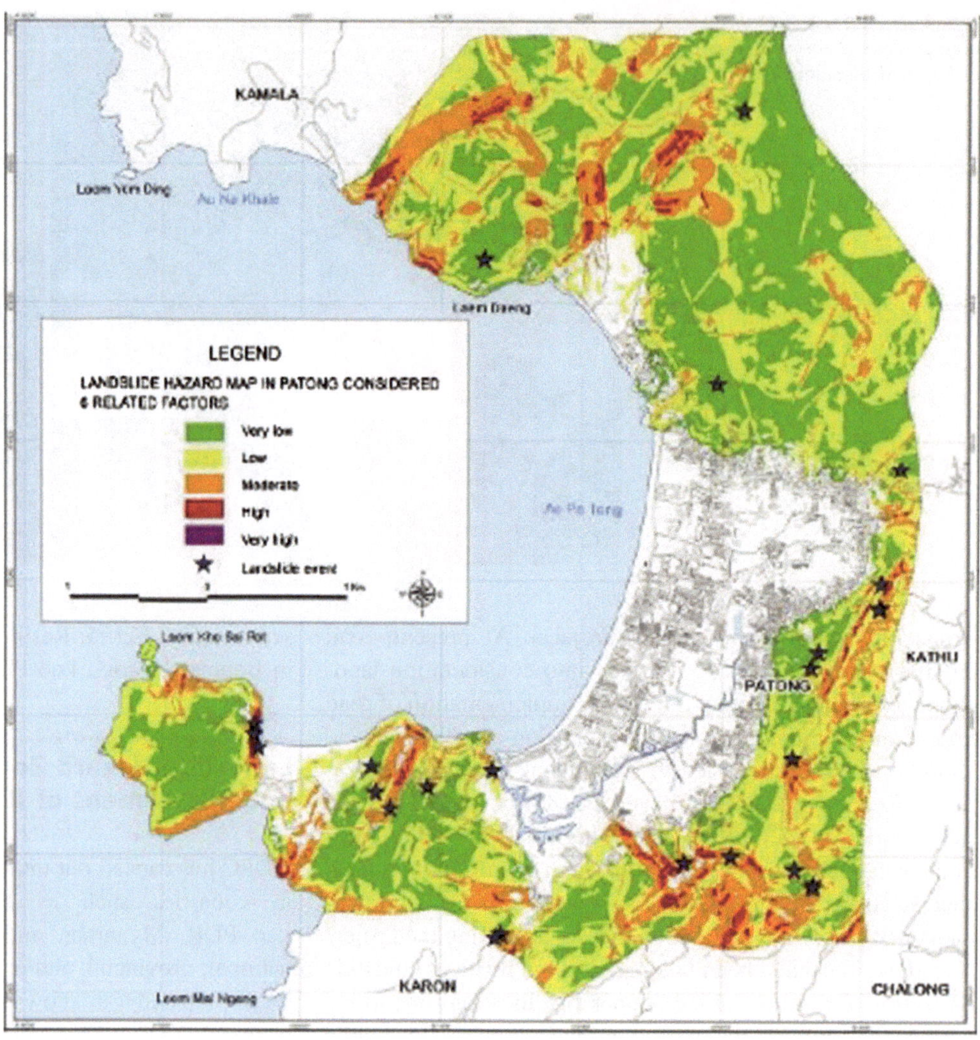

Fig. 6 Landslide hazard map of Patong city, Thailand

and city authorities, as well as to sector Ministries and line agencies. Using these maps, authorities can easily identify the areas that are prone to landslides, derive related vulnerabilities, and assess associated risk, and they can use such information in physical planning and undertaking risk mitigation interventions.

In the ICL-IPL Sendai Partnership Conference held on 13th March 2015, ADPC presented a paper on "Challenges in landslide risk management in Asia" capturing its experience in Asia on implementing structural and non-structural interventions for landslide risk management. One of the main challenges observed is the lack of interdisciplinary cooperation, including the communities, which is essentially needed to ensure that appropriate preparedness, response and mitigation actions are planned and owned by the communities. ADPC undertook regional and national level capacity building programs and awareness programs frequently to improve the knowledge and understanding by professionals of the landslide phenomena and interventions for prevention and mitigation of future disasters. Landslide hazard maps often were a useful tool in those capacity-building and awareness-raising activities.

Guidelines for Safer and Resilient Construction in Landslide Prone Areas

ADPC initiated pilot demonstration projects for landslide risk mitigation and to promote safer construction practices in landslide-prone areas in different places in Asia. The projects undertaken under this initiative covered resilient design and hands-on training in construction practices for dwelling houses, community infrastructure, and slope stabilization measures along major highways, as well as construction of critical facilities such as schools and hospitals (Fig. 7).

For example, under the program on "Mainstreaming Disaster and Climate Risk Management into Investment Decisions in Lao PDR", funded by the World Bank and the Japan Policy and Human Resources Development Fund, ADPC provided technical services to the Ministry of

Fig. 7 Landslide mitigation, Lao PDR

Planning and Investment, the Ministry of Public Works and Transport, and the Ministry of Agriculture and Forestry to develop resilient infrastructure for landslide and flood-affected areas. Previously, some critical sections of the roads in hilly terrain were usually washed out by flash flooding or affected by landslides. It often left the villagers in the lurch, with no ready access to basic services such as transportation, health, and education. The slope stabilization measures, culvert improvements for specific road patches, and landslide mitigation measures for affected sections of the major highways were some of the outcomes of the technical skills imparted to the government officials, so that such infrastructure facilities are no longer affected by floods or landslides in the future.

Teaching Tools

Certain urban and infrastructure developments can amplify the effect of landslide trigger mechanisms. There is a direct correlation between geo-environmental and physical factors associated with development, so control of land use is needed in order to bring about a reduction in the devastating impacts of landslides. In Asian countries, responsibility for local-level development control is vested with the local governments. It is therefore important to provide knowledge and appropriate skills in construction to the local governments in landslide-prone areas. ADPC has developed and delivered a training tool on "Landslide Disaster Risk Reduction for Local Government Level Stakeholders". This has been recorded as one of the tools under the ICL landslide teaching tools project. This course tries to educate the stakeholders within urban local government areas on the issues that need to be taken into consideration in safer development practice in hilly areas. The course also provides an opportunity for local government officials, technocrats, professionals, practitioners and other stakeholders in urban local government areas to discuss the measures that can be taken to mainstream landslide disaster risk reduction measures in urban planning and development and to regulate the land development in landslide-prone areas in order to reduce the impacts due to future devastating events of landsides.

Conclusions

A crucial aspect for the success and sustainability of landslide disaster risk reduction interventions in any country is institutionalization and mainstreaming of such approaches to become routine practices from local to national levels. In line with this, it is also important to make efforts for capacity building of mandated national level institutions for landslide risk management. Other efforts such as inter-agency collaboration, development of close linkage between agencies involved in disaster management and local government units, and the active participation of affected communities also play a vital role, among others. Experience sharing through regular consultative meetings among the mandated institutions of partner countries could provide significant inputs in improving the knowledge base on sound and cost effective practice.

Acknowledgements The authors gratefully acknowledge the contributions made in numerous ways by the representatives of the mandated institutions for landslide risk management in partner countries.

Appreciations are due for them in particular in helping ADPC to undertake pilot projects in various parts of South and Southeast Asia to demonstrate the effectiveness of local level actions for effective landslide risk management. The authors wish to note the contributions made by other stakeholder institutions, landslide professionals, academia, and staff from local governments in providing assistance in successful implementation of respective programs.The contributions made by the Norwegian Government in supporting the implementation of the Asian Program for Regional Capacity Enhancement for Landslide Impact Mitigation (RECLAIM) by ADPC in the region since 2004 and technical assistance by the Norwegian Geotechnical Institute for implementing the program are also acknowledged and greatly appreciated. The authors wish to record their gratitude to the management of the Asian Disaster Preparedness Center (ADPC), Thailand for the encouragement provided and to ADPC colleagues involved in implementation of landslide projects for the support extended in developing this paper.

Bibliography

ADPC (2012) Report on establishment of early warning mechanisms within the identified high risk areas in Bangladesh. Asian Disaster Management News (2008) ADPC, May–August 2008

ADPC, Project completion report on Community Based Landslide Early Warning System development in Cox's Bazar and Teknaf Municipalities in Bangladesh

Arambepola NMSI, Proposal for South-East Asian Network for Landslide Risk Management. http://icl.iplhq.org/category/icl/icl-networks/

Arambepola NMSI, Krishna Devkota Developing a framework for assessment of landslide hazard risk associated with critical national and provincial roads in Lao PDR. In: Proceedings of the International Geotechnical Conference (ICGE-Colombo 2017) organized by Sri Lanka Geotechnical Society

Arambepola NMSI and Krishna Chandra Dakota Developing a framework for assessment of landslide hazard risk associated with critical national and provincial roads. In: Proceedings of the ICL conférence in Kyoto, November 2013

Arambepola NMSI (2014a) Efforts in landslide risk reduction in Asia. In: Proceedings of the WLF 3, Beijing, China

Arambepola NMSI (2014b) Tools for Capacity Building at Local Government Level for Landslide Disaster Risk Reduction. In: Proceedings of the WLF 3, Beijing, China

Bandara RMS (2008) Landslide early warning in Sri Lanka, Regional seminar on experience of geotechnical investigations and mitigation for landslides, Bangkok, Thailand, 13–14 October 2008

Bandara RMS, Bhasin RK, Kjekstad O, Arambepola NMSI (2011) Cost effective practices for landslide monitoring for early warning in selected countries in Asia. In: Proceedings of the WLF 2, Rome, Italy

Basnayke S (2014) Study of the status of landslide early warning in selected countries in Asia. In: Proceedings of the WLF 3, Beijing, China

Billedo EB, Bhasin RK, Kjekstad O, Arambepola NMSI (2011) An appraisal on ongoing practices for landslide early warning systems in selected South and East Asian Countries. In: Proceedings of the WLF 2, Rome, Italy

Devkota KC (2014) Landslide problems in road sector of Lao PDR. In: Proceedings of the WLF 3, Beijing, China

ICL Web site, ICL webpage: http://icl.iplhq.org/category/icl/icl-networks/ South-East Asian Network for Landslide Risk Management

Maskrey A (2011) Revisiting community-based disaster risk management. Environ Hazards 10:42–52

National Building Research Organization (NRBO) Internal Publication (for restricted circulation) (1995) Landslide Instrumentation—general consideration and measurement of landslide activity. Landslide Hazard Mapp Proj, SRL 89(001):1–93

Soralump S, Chotikasathien W (2007) Integration of geotechnical engineering and rainfall data into landslide hazard map in Thailand. In: Proceeding of GEOTHAI'07 international conference on geology of Thailand: towards sustainable development and sufficiency economy, pp 125–131

Tatong T (2009) The best practices for landslide early warning system in Thailand. In: Proceedings of RECLAIM partner meeting

UNISDR (2006) the international early warning programme—the four elements of effective early warning systems –brochure, platform for the promotion of early warning (PPEW), 4 pp, Available at: http://www.unisdr.org/2006/ppew/iewp/IEWP-brochure.pdf

Promoting a Global Standard for Community-Based Landslide Early Warning Systems (WCoE 2014–2017, IPL-158, IPL-165)

Teuku Faisal Fathani, Dwikorita Karnawati, and Wahyu Wilopo

Abstract

The implementation of early warning systems is in line with the Sendai Framework for Disaster Risk Reduction (SFDRR) 2015–2030. One of the four priorities of the Sendai Framework for Action emphasizes the improvement of preparedness in response to a disaster by carrying out a simple, low-cost early warning system and improving its dissemination. A new proposal of a standard for community-based landslide early warning systems has been promoted to the International Organization for Standardization (ISO) by Universitas Gadjah Mada, in corporation with the Indonesian Standardization Agency and the Disaster Management Authority. The standard will serve to empower individuals and communities who are vulnerable to landslides to act in sufficient time in appropriate ways to reduce the possibility of injuries, loss of life and damage to property and the environment. It is designed to encourage communities to play a much more active role in their own protection. The guidelines adopted the concept of people-centered early warning system by UN-ISDR (Developing an early warning system: a checklist. Bonn, Germany, 2006) and will be used by communities vulnerable to landslides, and by government agencies and non-governmental organizations at central, provincial, municipality/district, sub-district, and village levels. The recommendations include: (1) Risk assessment; (2) Dissemination and communication; (3) Formation of disaster preparedness and response teams; (4) Development of evacuation maps; (5) Development of standard operating procedures; (6) Monitoring, early warning, and evacuation drills; (7) Commitment of the local authority and community to maintain the system. The standard will be developed by ISO/TC 292 *Security and resilience*, with the participation of 43 countries in the committee's work and another 14 as observers. The basic concept of this global standard has been initiated since 2007 through the Asian Joint Research on Early Warning of Landslides proposed by International Consortium on Landslides (ICL) and Disaster Prevention Research Institute (DPRI) Kyoto University and funded by JST and implemented in Indonesia, China, Korea and Japan. This paper describes the achievements and the current activities of the World Centre of Excellence (WCoE) on Landslide Risk Reduction (2014–2017), IPL Project (IPL-158) "Development of community-based landslide early warning system", and IPL Project (IPL-165) "Development of community-based landslide hazard mapping for landslide risk reduction at the village scale in Java, Indonesia".

T.F. Fathani (✉)
Civil and Environmental Engineering Department, Faculty of Engineering, Universitas Gadjah Mada, Jalan Grafika No. 2, Yogyakarta, 55281, Indonesia
e-mail: tfathani@ugm.ac.id

D. Karnawati · W. Wilopo
Geological Engineering Department, Faculty of Engineering, Universitas Gadjah Mada, Yogyakarta, 55281, Indonesia
e-mail: dwiko@ugm.ac.id

W. Wilopo
e-mail: wilopo_w@ugm.ac.id

K. Sassa et al. (eds.), *Advancing Culture of Living with Landslides*,
DOI 10.1007/978-3-319-59469-9_30

Keywords
Landslide monitoring • Early warning • Community development • Evacuation • ISO

Introduction

Landslides occur in different topographic and geologic settings and cause great social and economic losses. Landslide occurrence may increase due to human expansion into unstable hill-slope areas. Landslide mitigation efforts may be carried out both structurally and non-structurally. Structural mitigation includes the arrangement of slope geometry, reinforcement and protection of slopes, and improvement of drainage systems, all of which would incur a high cost. Meanwhile, relocation usually is not possible for residents living in areas prone to landslides. Under these conditions, the most effective disaster risk reduction effort is non-structural mitigation through the improvement of the community's preparedness by implementing early warning systems.

Currently many countries already have landslide and debris flow early warning systems (Baum and Godt 2009), including Indonesia (Fathani et al. 2014). The implementation of early warning systems is in line with the Sendai Framework for Disaster Risk Reduction (SFDRR) 2015–2030, with four priorities in disaster risk reduction. The fourth priority emphasizes the improvement of preparedness to respond effectively to a disaster by implementing a simple, low-cost early warning system and improving the dissemination of information about early warning signs of natural disasters at local and national levels.

According to UN-ISDR (2006), a complete and effective early warning system is composed of four interrelated key elements, i.e., (1) risk knowledge, (2) monitoring and warning service, (3) dissemination and communication, and (4) response capability. The implementation of a community-based early warning system must take into account the correlation between a strong bond and effective communication channels among all of these elements. The goal of the development of people-centered early warning system is to empower individuals and communities vulnerable to hazards to act in sufficient time and in appropriate ways to reduce the possibility of injuries, loss of life, and damage to properties and the environment.

Demographic, social, economic, and cultural aspects in general are often left out in the implementation of early warning systems, compared to other technical aspects (Karnawati et al. 2011). The experts, researchers, and the decision-makers at regional, national, provincial, municipality/district or village levels should follow up the training in warnings and actions to be taken addressed for the community. Consequently, early warning system guidelines encourage the roles of the community and social aspects.

Referring to the four key elements of people-centered early warning systems (UN-ISDR 2006), it is crucial to develop a global standard for landslide early warning systems to have uniformity in the effort to carry out early warning systems and improve the preparedness of the community and related stakeholders in landslides prone areas.

Standard for Landslide Early Warnings

This proposed standard specifies the requirements for a landslide early warning system, which include the definition, understanding, stage of procedures, implementation, as well as types of activities. This standard should be used by government agencies and non-governmental organizations, at central, provincial, municipality/district, sub-district, and village levels, and by communities vulnerable to landslides.

The community-based landslide early warning system was derived from four key elements of people-centered early warning system (UN-ISDR 2006) and constitutes of seven main sub-systems as elaborated below (Fathani et al. 2016):

1. Risk assessment;
2. Dissemination and communication;
3. Establishment of disaster preparedness and response teams;
4. Development of evacuation routes and maps;
5. Development of standard operating procedures;
6. Monitoring, early warning, and evacuation drills;
7. Commitment of the local government and community to the operation and maintenance of the whole system.

The components of the seven sub-systems of the community-based landslide early warning system are elaborated in this following section.

Risk Assessment

Risk assessment is carried out with technical (geomorphology, geology, and geotechnics), institutional, and social-economy-cultural surveys of vulnerable communities. Technical surveys are conducted to understand the geological conditions in certain areas, especially to determine landslide susceptibility and stable zones (Lacasse and Nadim 2009). The survey is also carried out to collect data regarding mass movement, such as cracks and subsidence, appearance of springs, fractures in structures, and tilting of poles and trees, to figure the placement of the landslide early warning instruments. During the technical survey, information on lithology types and distribution, slope soil composition, geological structure types and their orientation distribution, slope cracks, and slope inclination should also be included.

An institutional survey is carried out to find out if there is an institution to monitor and mitigate landslide hazards in a disaster-prone area. Meanwhile, a cultural-economic survey is performed to collect data on the demographics of the community, such as population (household), age, education, financial situation (vehicles and livestock), and culture. Information on potentially vulnerable inhabitants and infrastructure is important to determine the risk level in a certain area.

A social survey is performed to figure the community's perception of landslide risks and landslide disaster risk reduction efforts and applied technology. As a prerequisite for formulating strategies for risk reduction programs suitable for the local social conditions, the community must be eager and motivated to actively participate in the program. Therefore, it is critical to increase people's awareness; and training and education are two of the programs to increase people's awareness. This activity imparts knowledge and increases people's capability to decide the actions to prevent landslides and protect themselves from landslides.

Dissemination and Communication

Dissemination aims to impart knowledge and understanding to the community, particularly for landslide disasters, and to understand the community's aspirations. Methods and materials to aid in dissemination are tailored to the performed preliminary data of the risk assessments. This understanding includes landslide definition, their types, mechanisms of occurrence, factors controlling and triggering them, signs, and both structural and non-structural mitigation of landslides, including early warning systems, warning levels, and signs. The dissemination and communication increases the community understanding about landslides, ranging from the mechanisms and the signs, to how to minimize risks, especially with the early warning system to be installed. In addition, it is necessary to determine the key people with a strong commitment as initiators in the formation of a disaster preparedness team.

Establishment of Disaster Preparedness and Response Team

A disaster preparedness and response team is formed based on the community meetings facilitated by the government or related agencies. The appointment of the team is based on the abilities of the member in preparedness, prevention, mitigation, and post-disaster management of landslide disaster. The team (at least) consists of a chairperson, data and information division, evacuee mobilization division, first aid division, logistics division, and security division (Fig. 1). The number of divisions in the disaster preparedness and response team is determined based on the needs of the community and should be in agreement with the purpose of early warning systems. Each division comprises 3 people at the minimum or in proportion to its population. Each division should consist of a chief and members who are aware of and responsible for their duties. Moreover, the division should be comprised of the people living at the location where the early warning system is placed. An example of a disaster preparedness and response team is shown in Fig. 1.

The disaster preparedness and response team is tasked with conducting all activities related to preparedness, such as

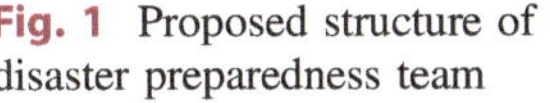

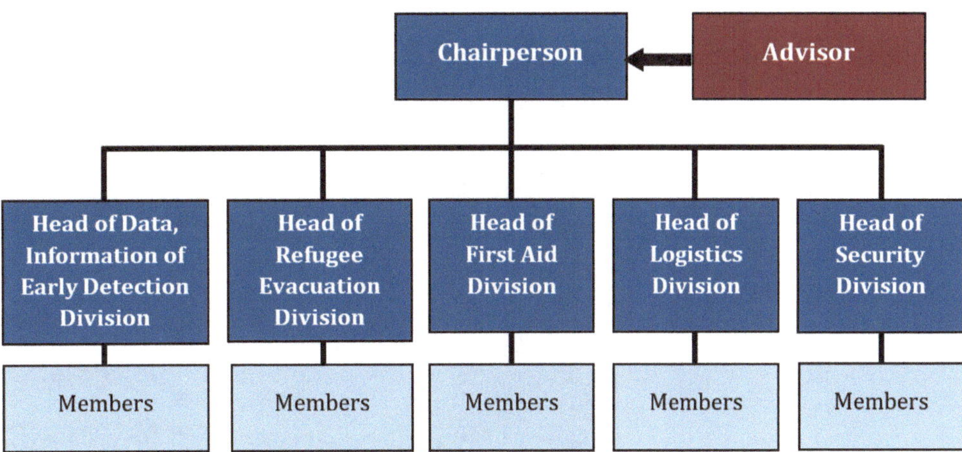

Fig. 1 Proposed structure of disaster preparedness team

organizing the community to support the technical system effectively. The team is responsible for deciding landslide risk zones and evacuation routes that are verified by the local authority or institution officials or experts, and arrange the people to evacuate before the landslide hits. The disaster preparedness and response team is in charge of operating and maintaining monitoring tools and the landslide early warning system that has been installed.

Development of an Evacuation Map

An evacuation map includes landslide risk zones and evacuation routes, which provide information on the zones that are safe or unsafe during a landslide emergency, safe evacuation routes for the residents to evacuate and secure locations for assembly points (Fig. 2). The disaster preparedness team, community, and stakeholders uses the landslide risk zones and evacuation routes as the operational guidelines to gather at the assembly point and subsequently evacuate by following the evacuation route that has been determined. The zones are created by the disaster preparedness team who have gained a basic knowledge of communication, and have been verified by experts. The map/ground plan provides the following information (as suggested by Karnawati et al. 2013b; Fathani et al. 2016):

(a) High-risk and low-risk (safe) zones;
(b) Homes, which include the homeowners;
(c) Important facilities: school, places of worship, community health center, offices, and landmarks;
(d) Streets and alleys;
(e) Installation point of early warning system;
(f) Alert post;
(g) Evacuation routes;
(h) Assembly Points;
(i) Refugee camps.

Development of Standard Operating Procedures (SOP)

Standard operating procedures (SOP) serve as a guide for the disaster preparedness team and the community when facing all hazard levels, i.e., Caution (Level 1), Warning (Level 2), and Evacuate (Level 3). The SOP contains the procedures for responses by the disaster preparedness team and the community to the alert issued by the control center in the field or by local authorities (Table 1). The SOP is developed in regards to the discussions of each division and based on the direction of relevant stakeholders to follow the flow of warning information delivery mechanisms and evacuation commands.

Monitoring, Early Warning and Evacuation Drill

Early detection devices are placed in areas that have the highest risk and the greatest number of affected population. Determination of the locations is based on the identification of landslide risk zones. Installation of the equipment is

Fig. 2 Example of an evacuation route map installed at a village prone to landslides

Table 1 The proposed Standard Operating Procedures (SOP) for evacuation

Status/alert level	Criteria/sign	Action/response by the community	Action by the local authority
Caution (Level 1)	**Criteria**: determined by rainfall measurement **Sign**: "blue" lamps and/or siren that sounds "caution, high rainfall" or another sound showing the lowest threat level or subject to the local conditions	• The data and information division checks the condition of the monitoring equipment, collects data of the community, informs the alert level, and encourages the residents to prepare essential items to bring. • The Disaster Preparedness and Response Team submits recurring reports to the team leader	• Receives report from the team leader of disaster preparedness team • Examines the field condition the and keeps the coordination with the disaster preparedness team
Warning (Level 2)	**Criteria**: determined by the increase in rainfall or slope hydrology, and landslide indications **Sign**: "orange" lamps and sirens that sounds "warning, evacuation" or another sound showing the increase of threat level to warning or subject to the local conditions	• The data and information division reassess the condition of soil movement and the monitoring equipment and collects data of the community • The team leader orders the vulnerable group to evacuate to the determined assembly point, assisted by the evacuee mobilization division and the security division • The data section ensures that the vulnerability group has been evacuated by collecting their data • The security division is responsible for the security of the residents' homes and environment	• Receives report from the team leader of disaster preparedness team • Examines the field condition the and keeps the coordination with the disaster preparedness team • Provides support to the evacuated vulnerable group
Evacuate (Level 3)	**Criteria:** determined by the increase in rainfall or slope hydrology and rate of landslide movement **Sign:** "red" lamps and sirens that sounds "evacuate" or another sound showing the highest level of threat or subject to the local conditions	• The team leader orders all residents to evacuate to the determined assembly point, assisted by of the evacuee mobilization division and the security division • The data and information section examines the early detection equipment and collects data of the residents in the shelter	• Receives report from the team leader of disaster preparedness team • Examines the field condition the and keeps the coordination with the disaster preparedness team • Provides emergency support to the evacuees

done with the community, aiming to increase the sense of ownership and responsibility for the equipment's condition to guarantee people's safety. The type of early detection and danger levels should be appropriate to the geological conditions and the size of the territory. The monitoring devices that must be installed to support early warning systems are:

(a) **rain gauge**: to measure the intensity of rainfall within a certain period.
(b) **surface deformation instruments**: to monitor the deformation on the land surface in a certain period. The common devices used are *extensometers* (to monitor the relative distance between two points on the crack) as shown in Fig. 3 and *tiltmeters* (to measure the change of slope inclination)

To improve measurement accuracy, additional devices can be used, as follows:

(a) **underground deformation meter**: to measure the deformation through the movement of the sliding plane within a certain period of time (*inclinometer*, *pipe strain gauge*, *multi-layer movement meter,* etc.).
(b) **groundwater level meter**: to measure changes in groundwater level in the landslide zones, mounted inside the borehole.
(c) **pore water pressure sensor**: to measure changes in pore water pressure in the land mass, installed in the borehole.
(d) **soil moisture sensor**: to measure the changes of water level in the land mass.
(e) **survey stake**: to monitor mass movement by using wooden sticks or bamboo.

The monitoring and early detection devices should utilize effective technology to deliver a community-based landslide early warning system. The technology is in the form of a radar monitoring module or conventional monitoring module i.e., in situ geotechnical instrumentation (Barla and Antolini 2015). The minimum landslide early warning devices to be installed are the mass movement gauge and rain gauge (Fig. 3). The landslide monitoring and early warning scheme are shown in Fig. 4.

Fig. 3 Example of monitoring and warning devices: **a** extensometer; **b** rain gauge and local server

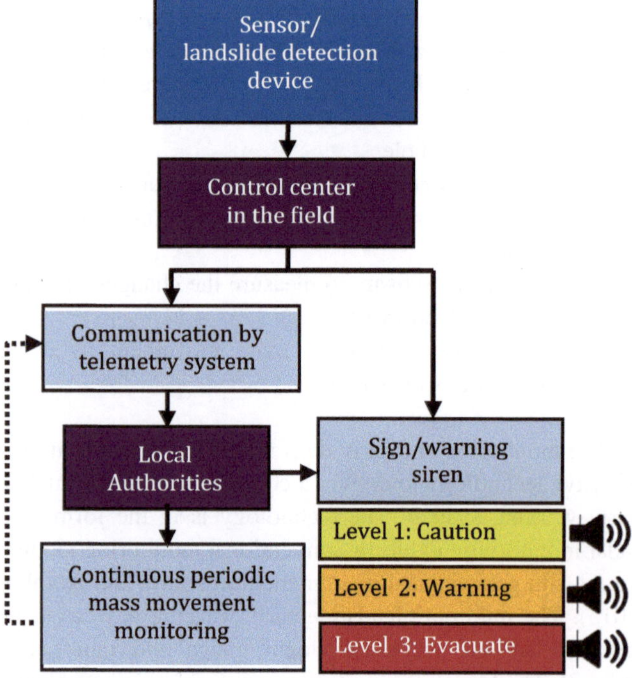

Fig. 4 Scheme of landslide monitoring and early warning system (Fathani et al. 2016)

After the devices are placed, the disaster preparedness team is established, the evacuation routes are determined, and the evacuation SOP is developed, an evacuation drill to ensure the functionality of the devices and the community's responses to anticipate disaster is carried out. The drill is conducted using a SOP-based scenario. It serves to instill vigilance, preparedness, and responsibility in the disaster preparedness team when the devices show the indication of a potential landslide. Furthermore, the evacuation drill is also aimed at introducing and familiarizing the local community with the sound of sirens of every early detection device, and familiarizing them with evacuation in emergency situations. The evacuation drill must be conducted at least once a year at the end of the dry season.

Commitment of the Local Government and Community to the Operation and Maintenance of the System

To ensure that all stages in the SOP is run well, the commitment of the local government and the community is vital for the operation and maintenance of the system. The government, the community, and the private sector must acknowledge the duty and responsibility including the ownership, installation, operation, maintenance, and security of the early warning system that are adjusted to the conditions of each location.

Discussions

The global standard includes one of the priorities of Sendai Framework for Disaster Risk Reduction (SFDRR) 2015–2030, which is described in four elements of a people-centered early warning system. The elements are then developed into seven sub-systems of a landslide early warning system. To assist the implementation of landslide early warning systems in Indonesia where the trial of this standard has been conducted, a hybrid socio-technical approach is carried out. Both approaches (technical and social), supported with education and research, are expected to be able to involve all of the related stakeholders, reduce the cost of system implementation, and maintain its sustainability (Karnawati et al. 2011, 2013a). The monitoring and warning service equipment installed in 20 provinces since 2012 is still in excellent condition due to the implementation of the above proposed methods. The description above shows that the proposed methodology or standard is successfully implemented in Indonesia.

The primary issue that the implementation of this system tries to address is that the implementation of only the technical approach is not effective for sustainable disaster prevention. This is often the case when the local authority/third party installs early warning system devices without involving local community. As a consequence, the community does not have a proper response capacity when the devices are triggered. The establishment and implementation of the seven sub-systems as a global standard for landslide-prone countries will support the disaster risk reduction effort. Moreover, with community involvement, the operation, maintenance, and sustainability of an entire system are secured early. At the end of 2015, this proposed standard is being nominated as an international standard for landslide early warning systems by ISO/TC 292 *Security and resilience.*

Acknowledgements The authors gratefully acknowledge the International Consortium on Landslides (ICL) and Disaster Prevention Research Institute (DPRI) Kyoto University for initiating the Asian Joint Research on Early Warning of Landslides, which is funded by JST and has been implemented in Indonesia, China, Korea and Japan. We acknowledge the Indonesian National Disaster Management Authority (BNPB) and the National Standardization Agency of Indonesia (BSN) for supporting and promoting this global standard to the International Organization for Standardization: ISO/TC 292, *Security and resilience* and for the participating countries and observers during the establishment of this new standard.

References

Barla M, Antolini F (2015) An integrated methodology for landslides' early warning systems. Landslides 13: Springer, Berlin, Heidelberg

Baum RL, Godt JW (2009) Early warning of rainfall-induced shallow landslides and debris flow in the USA. Landslides 7: Springer, Berlin, Heidelberg

Fathani TF, Karnawati D, Wilopo W (2016) An integrated methodology to develop a standard for landslide early warning system. J Nat Hazards Earth Sys Sci. doi:10.5194/nhess-2016-209

Fathani TF, Karnawati D, Wilopo W (2014) An adaptive and sustained landslide monitoring and early warning system. Landslide Science for a Safer Geoenvironment. Springer International Publishing, pp 563–567

Karnawati D, Fathani TF, Wilopo W, Setianto A, Andayani B (2011) Promoting the hybrid socio-technical approach for effective disaster risk reduction in developing countries. In: Brebbia CA, Kassab AJ, Divo EA (eds) Disaster management and human health risk II. WIT Press, Southampton, UK, pp 175–182

Karnawati D, Ma'arif S, Fathani TF, Wilopo W (2013a) Development of socio-technical approach for landslide mitigation and risk reduction program in Indonesia. ASEAN Eng J Part C 21(7):22–47

Karnawati D, Fathani TF, Wilopo W, Andayani B (2013b) Community hazard map for landslide risk reduction. ICL Landslide Teaching Tools. International Consortium on Landslides, pp 259–266

Lacasse S, Nadim F (2009) Landslide risk assessment and mitigation strategy. In: Sassa K, Canuti P (eds) Landslides-disaster risk reduction. Springer, Berlin, Heidelberg, pp 31–61

UN-ISDR (2006) Developing an early warning system: a checklist. The Third International Conference on Early Warning (EWC III). Bonn, Germany

Model Policy Frameworks, Standards and Guidelines on Landslide Disaster Reduction (WCoE 2014–2017)

A.A. Virajh Dias, Nimesha Katuwala, H.M.J.M.K. Herath, P.V.I.P. Perera, K.L.S. Sahabandu, and N. Rupasinghe

Abstract

The Central Engineering Consultancy Bureau (CECB) has been approved and designated as one of the "World Centres of Excellence on Landslide Disaster Reduction 2014–2017" under the theme of "Model Policy Frameworks, Standards, and Guidelines on Landslide Disaster Reduction" by the Global Promotion Committee of the International Consortium on Landslides (ICL) at the award ceremony of the World Landslide Forum 3 in Beijing, China. The above theme is divided into three thematic areas: first, Developing Conceptual Policy Frameworks to Understand the Causes, Effects and Mitigatory Measures of Landslide Occurrences, secondly to Implement Applicable Guidelines/Teaching Tools to Establish Essential Synergies in Landslide Disaster Phenomena, and thirdly to Originate Pertinent Standards for Humanitarian Activities in support of Effective Risk Reduction and Mitigations on Landslide Occurrences. The amalgamation of these three areas will originate a successive approach to developing a master plan for disaster risk reduction as a cost-effective investment in preventing future losses. The proposal for WCoE submitted by CECB was mainly focused on continuing IPL research activities, building up global partnerships and regional networks and conducting national projects and awareness programmes on Landslide Risk Reduction. This paper illustrates the above activities in a more comprehensive and descriptive manner.

Keywords

Conceptual policy frameworks • Pertinent standards • Essential synergies • WCoE

A.A.V. Dias (✉)
Natural Resource Management & Laboratory Services (NRM & LS), Central Engineering Consultancy Bureau, No. 415, Bauddhaloka MW, Colombo 7, Sri Lanka
e-mail: aavirajhd@yahoo.com

N. Katuwala
Environmental Chemist, Centre for Research & Development, NRM & LS, Central Engineering Consultancy Bureau, No. 415, Bauddhaloka MW, Colombo 7, Sri Lanka
e-mail: nkatuwala@gmail.com

H.M.J.M.K. Herath
Engineering Geologist, Center for Research & Development, NRM & LS, Central Engineering Consultancy Bureau, No. 415, Bauddhaloka MW, Colombo 7, Sri Lanka
e-mail: jmkherath@yahoo.com

P.V.I.P. Perera
Environmental Scientist, Centre for Research & Development, NRM & LS, Central Engineering Consultancy Bureau, No. 415, Bauddhaloka MW, Colombo 7, Sri Lanka
e-mail: ishastha@gmail.com

K.L.S. Sahabandu
Central Engineering Consultancy Bureau, No. 415, Bauddhaloka MW, Colombo 7, Sri Lanka
e-mail: sahabandukls@gmail.com

N. Rupasinghe
Ministry of Megapolis and Western Development, Colombo 7, Sri Lanka
e-mail: secretarymmwd@gmail.com

K. Sassa et al. (eds.), *Advancing Culture of Living with Landslides*,
DOI 10.1007/978-3-319-59469-9_31

Introduction

The Central Engineering Consultancy Bureau (CECB) is one of the foremost consultancy and construction organizations in Sri Lanka today: multi-disciplinary in function and futuristic in approach towards innovative research and technology. CECB joined ICL in year 2008 with the primary objective of strengthening its research capacities and skills in Landslides and Related Earth System Disasters for Global Risk Preparedness. CECB has been an active member of the ICL/IPL-GPC since year 2008 and obtained the membership representing Sri Lanka in the ICL Capacity Development Network, which is a thematic and regional network established to support the regional activities of ICL and IPL. CECB has also been admitted as a Member of the Board Representatives under the status and bylaws of ICL in all matters dealing with landslide initiatives in year 2012. In year 2014, CECB became one of the 15 World Centres of Excellence (WCoE) in Landslide Disaster Reduction for the three consecutive years; 2014–2017 under the title of "Developing Model Policy Frameworks, Standards and Guidelines on Landslide Disaster Reduction".

This paper systematically describes the completed and on-going IPL research components and WCoE activities conducted by CECB for capacity building and knowledge dissemination purposes on Landslide Disaster Reduction.

ICL/IPL Activities of CECB

- The institution has been a member of the ICL/IPL-GPC since 2008 and is a member of the ICL Capacity Development Network.
- CECB has also been admitted as a Member of the Board of Representatives of ICL in all matters dealing with landslide initiatives since 2008.
- CECB participated in the First World Landslide Forum (WLF1) which was held in Japan, 2008 and also represented Sri Lanka from CECB in the WLF2 held in Rome, Italy in 2011 and WLF3 held in Beijing, China in 2014.
- Participated and presented papers at the IPL Symposiums/chairing ICL-GPC sessions.
- Published five peer-reviewed papers based on diversified experiences obtained in the field of landslide disaster phenomena for the WLF3 held in China, 2014.
- Elected as a World Centre of Excellence in Landslide Disaster Reduction for the period 2014–2017 at the WLF3 held in Beijing, China.
- Conducted the International E-conference under the theme of Developing Model Policy Frameworks, Standards and Guidelines in Landslide Disaster Risk Reduction as an initiative for the WCoE activities steered by CECB.

National Projects

As a partner of the IPL-GPC, our institution was strengthened to support many of the national level programmes based on landslide disaster management. We contributed our active participation as a member of the Disaster Impact Assessment checklist systems for the Road Development Sector of the Disaster Management Capacity Enhancement Project (DiMPEP), organized by JICA and DiMPEP. Our expertise was utilized towards the national-level Landslide Disaster Mitigation Action Plan, "Operational Professional Combine" during landslides and floods that occurred in Sri Lanka in 2003.

ICL/IPL Research Projects

During the recent past, CECB has conducted ICL-driven research projects with a problem/solution focus, strengthened by the research team of the Centre for Research & Development Unit of CECB. The project details can be elaborated as follows.

IPL Certified Project M143—Year 2009

Evaluation of Sensitivity of the Combined Hydrological Model (Dynamic) for Landslide Susceptibility Risk Mapping in Sri Lanka.; Team Leader—Eng. A. A. Virajh Dias

This study includes the quantification of the landslide hazard in terms of spatial and temporal occurrences of slope stability in a sub catchment, based on a combined ground water and slope stability dynamic model for the selected area. A comparative analysis of landslide susceptibility was done using the WAA and SINMAP static models (see map in Fig. 1) and using the PC Raster Combined Hydrological Dynamic Model (see map in Fig. 2). The landslide susceptibility maps created from both the models were compared and used for a better understanding of landslide susceptibility for sensitive landslide disaster events, their origins and prioritization of efforts for the reduction and mitigation of future landslide hazards. The sensitivity of both the approaches was fine-tuned using soil strength parameters, geomorphological evidence and field verification techniques.

The above findings indicate the variability of conceptual models and their relative importance with respect to the input databases. Detailed and comprehensive geographical information databases are required to obtain the statistical interpretation and calibration of the model. Similarly, soil saturation conditions and the validity of wetness indices within a watershed also create another avenue to calibrate the analytical model with the inputs of other soil parameters (Dias et al. 2014a).

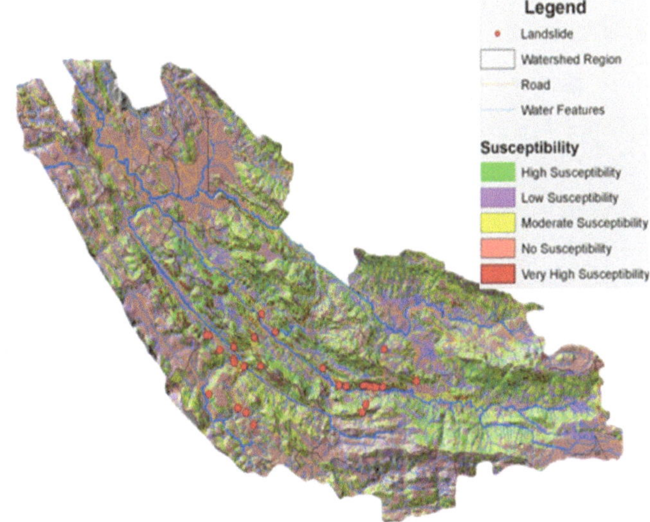

Fig. 1 Detailed landslide susceptibility map of Kalawana division of the Ratnapura District, developed using the SINMAP model, including stream hydrology networks, roads and watershed boundaries

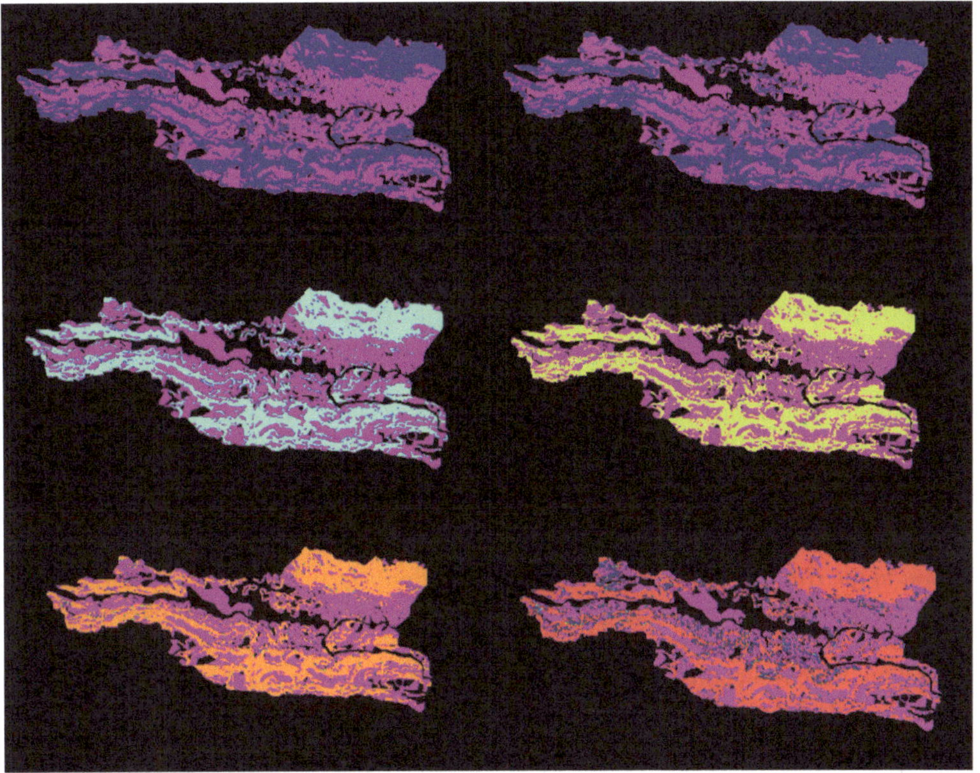

Fig. 2 Landslide susceptibility of a selected area interpreted through combined hydrological dynamic model (PC Raster). *Blue* to *Red* indicates increases in hazard potential due to continuous rainfall and saturation within a period of 6 days

IPL Certified Project 155—Year 2012

Determination of Soil Parameters of Subsurface to be used in Slope Stability Analysis in two Different Precipitations Zones of Sri Lanka; Team Leader—Eng. A. A. Virajh Dias

The characteristics of slopes, saturation and shear strength of soils are the main parameters associated with rainfall-induced slope failures, and these parameters are directly affected by differences in precipitation over prolong periods of time. In most instances a landslide is triggered due to extensive soil saturation and is a function of soil integrity, hydraulic conductivity, density, void content, shear strength and boundary conditions (Dias et al. 2014b). The study on evaluation of E50 (Secant modules) is an experiment set up to understand the behavior of residual soils under changing

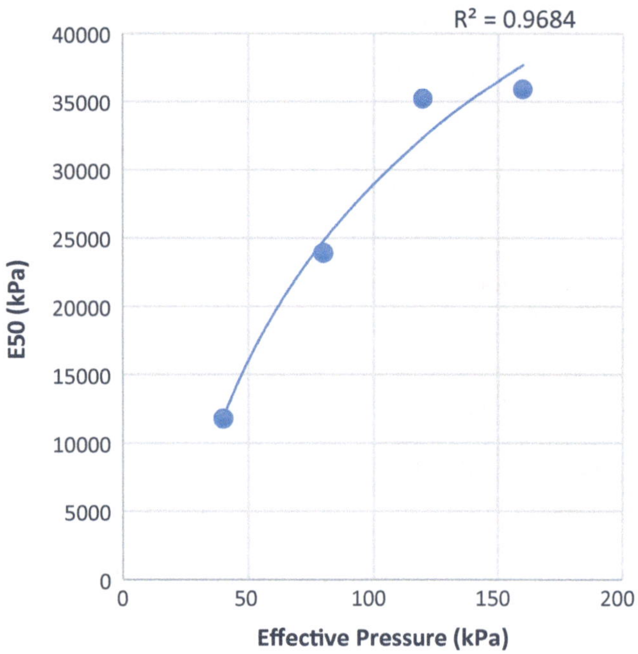

Fig. 3 Relationship of effective confining pressure and soil moldule E_{50}

Fig. 4 The large single root (taproot) and the lateral roots connected to anchor the soil, prevent soil erosion

stress conditions at a site due to various reasons, such as prolong periods of rainfall, movement of soils, unloading effects and the re-loading effect caused by deposition. From the results we conclude a reasonable interdependence of effective (confining) pressure and E50. One of the difficulties faced during the experimental study was to apply the evaluation of degree of landslide density considering a special boundary (area wise) to the selected sample, when comparing prior to evaluation of parametric results which could have influenced the final results, as shown in Fig. 3 (Mallawarachchi et al. 2014).

IPL Certified Project 199—Year 2015

The Effect of Root Systems in Natural Slope Erosion Protection in the Hill Country of Sri Lanka.– on going project; Team Leader—Mr. Ishastha Perera

This study reports the observed details and patterns of vegetation which support slope protection and the roles played by different species in such scenarios. It is understood that not one, but a collection of species contribute to this end through the setting and functions of each type of vegetation and their positioning.

Thus results of the study can be directly used for practical application to critical slopes. For example, the large single root usually which grows straight down, anchors the plant in the ground and the connected lateral roots anchor the soil, preventing soil erosion. The buttress root system, which is distributed on all sides of a shallowly rooted tree, does not penetrate to deeper surface layers (see Fig. 4). It prevents the tree from falling over, while also gathering more nutrients.

IPL Certified Project 200—Year 2015

An assessment of the rock fall susceptibility based on cut slopes adjacent to highways and railways—on going project; Team Leader—Ms. H M Janaki M K Herath

The main target of this research is to carry out appropriate improvements for rock fall hazard assessment by introducing an appropriate Rockfall Hazard Rating System (RHRS). This method indicates various judgment matrices, but it does not clearly define a method of assigning individual weights by prioritizing their significance. Roadside overhanging rock slope failures are commonly due to removal of a passive loading wedge, dipping, continuous joints with adverse orientation, direct exposure to heating or freezing, undulating joint sets, differential erosion, layered strata, disparate rock block sizes, geo-structural discontinuities and the slope.

Approach to Rating Using Geological Evidence

Determining the rating scores for pair-wise geological characters is more convenient than individual geological significance of RHRS. The rating and scoring of evidence still depend on value judgments, but avoids the minimum dependences of other factorial significance and pairwise combined parameters. The study discusses some prominent

Fig. 5 This relatively old cut slope indicates progressive erosion, with unsupported or over-steepened slope conditions and degradation over time. The detachment of larger blocks or volumes of falling rock produces greater impact forces than smaller events. This type of rockfall is commonly caused by erosion that leads to a loss of support, either locally or throughout the slope

characteristics in geological settings, which include time-dependent parameters of rock fall history obtained from the field evidence, similar to that shown in Fig. 5.

The descriptive information related to the study is given below.

Pair 1: Dipping and Slope Height (DSH)

The instability of a cut slope is directly interrelated with the height of the earth cutting. A consideration of the combined parameters of dipping and slope height will demarcate the potential instability more precisely than independent evaluation. Therefore, DSH can be considered as the first of the combined parameters for the evaluation of RHRS.

Pair 2: Block Size and Rock Character (BRC)

It is noticable that different rocks disintegrate to form various sizes of blocks, depending on climatic conditions, weathering, etc. Therefore, the formation of blocks largely depends on the rock character or lithology. Crystalline gneissic rocks with large blocks, well foliated gneissic rocks and bedded/fractured quartzite are considered to be examples in that category.

Pair 3: Persistence and Joint Density (PJD)

Continuous planar joints, discontinuous joints and random joints were common in most of the cases. Random joints and fractures create irregular shaped blocks. Therefore, the persistence and joint density affect detachment of blocks from a cutting. All comparisons were to be made in the vertical or horizontal axis. Planar joints and foliation joints are prominent observations and joint density is essential to defining the potential instability.

Pair 4: Weathering Discontinuity and Aperture Sizes (DAS)

In general, formation of rock blocks is a function of the discontinuities of the intact rock. The potential for rockfall by movement along discontinuities is controlled by the characteristic of joints or the aperture size. The condition of the joints is described in terms of micro and macro roughness. The aperture size always indicates the possibility of absorbing water and saturation potential. In addition, water ingress through discontinuities always reduces the strength of an interface and increase weathering. Therefore the nature of the discontinuity (fracture, joint, fault) of the mass of rock and observations of aperture size is an important combined parameter for the assessment.

Pair 5: Rock Friction and Weathering State (RFWS)

The angle of internal friction of intact rock is usually high in crystalline formations. However, the interface friction of the facture or joint surface will indicate the state of stability. Quartzite and charnockitic formations tend to remain unweathered. Rock friction is a function along a joint, bedding plane, or other discontinuity which is governed by the macro and micro roughness of the surfaces. Macro roughness is the degree of undulation of the joint relative to the direction of possible movement in a rock mass. Micro roughness is the texture of the surface, which is mostly based on the weathering of the slope.

Pair 6: Cutting Plane and Block Movement Direction (CPBMD)

Cutting plane is the exposed face of the cutting which may have an adverse effect on the orientation of a detachment of blocks. The study of cutting faces with respect to the block orientation is used for slopes where, for differential blocks, over-steepening and under-steeping conditions may control or aggravate the possible detachment of overhanging rocks.

Pair 7: Overhang and Fall History (ODH)
Historical rockfalls and the detachment of adjoining blocks at a rock cut is an indicator of future instability events. Typically, the frequency and magnitude of past events is an excellent indicator of the type of events to expect. In addition, overhanging rocks are mainly due to the fall history and therefore this information is an important check on the potential for future failure. The second stage of a rockfall event is when the large overhang becomes unstable and a large volume of rock falls off at one time. Erosion features include over-steepened slopes, unsupported rock units (overhangs), or exposed resistant rocks on a slope, which may eventually lead to a rockfall event. Therefore overhang of rocks with fall history is a useful combined parameter for the revised RHRS (Herath et al. 2014, 2016).

WCoE Activities Conducted by CECB

Developing "Model Policy Frameworks, Standards and Guidelines on Landslide Disaster Reduction"

The Central Engineering Consultancy Bureau was designated as a "World Centre of Excellence on Landslide Disaster Reduction 2014–2017" under the theme of "Model Policy Frameworks, Standards, and Guidelines on Landslide Disaster Reduction" by the Global Promotion Committee of the International Consortium of Landslides (ICL) at the World Landslide Forum 3 held in Beijing, China.

A conceptual model policy framework is a tool with numerous variations and contexts which are used to make conceptual distinctions and organize ideas towards the focus area. The development of a Policy Framework can be done for diverse goals, including the development of a Policy Framework to promote and enrich landslide risk awareness, as well as to promote a multi-disciplinary culture of landslide studies and management, developing policies to breed and nurse scientific culture and outlook, etc. Once the Policy Framework is clearly defined for a particular task, the applicable standards and guidelines can be implemented accordingly.

The International E-Conference 2015 on Landslide Disaster Risk Reduction

CECB has taken the initiative in organizing an Online Global Conference on Landslide Disaster Reduction, finding it an ideal way of providing ample room for respective specialists to assemble the world-wide knowledge and expertise in disaster risk management to better serve and care for the needs of society.

The primary objective of conducting this E-conference is to develop a Master Plan for Landslide Risk Mitigation in the world by identifying the critical factors that may directly affect it and develop a set of Teaching Tools/Guidelines which are more practical and suitable to the level of understanding of the actual victims. The conference preceded under the theme of Developing Model Policy Frameworks, Standards & Guidelines on Landslide Disaster Reduction, which was derived from the theme for work as a WCoE by CECB. It was divided into three subordinate topics; the opening topic was Developing Conceptual Policy Frameworks to Understand the Causes, Effects and Mitigatory Measures of Landslide Occurrences, secondly was the Implementation of Applicable Guidelines/Teaching Tools to Establish Essential Synergies in Landslide Disaster Phenomena, and thirdly to Originate Pertinent Standards for Humanitarian Activities in support of Effective Risk Reduction and Mitigations on Landslide Occurrences. Several important factors discussed throughout the conference are described below (Bhandari 2015).

1. Are landslide hazards natural?

Hazards due to landslides, earthquakes, volcanoes and tsunamis are generally regarded as nature's safety valves which operate to maintain Earth's dynamic equilibrium. It is a hard fact of life that mountains of the world, which are treasure-houses of landslides, are meant by Nature to decay and without their degradation; slope failures and mass movements, there would be no fertile plains for agriculture. Therefore, Nature maintains its dynamic equilibrium by getting rid of human interferences with the mountain system, such as urbanization, illegal constructions, inappropriate land use patterns in cultivation, etc. The sharp increase in tragic landslide events is mostly because of this reason. It is certainly unfair to regard landslides as natural hazards when such massive dominating non-natural causative factors have been added, changing the very character of hazard The importance of an accepted policy framework in land-use planning comes into picture at this stage.

2. How sound are the approaches to landslide risk reduction and mitigation?

Projection of landslide scenarios and estimation of the associated risk are pre-requisites for formulating strategies to reduce or eliminate landslide risk. Risk is defined as "the combination of the probability of an event and its negative consequences". Hazard is defined as "a dangerous phenomenon, human activity or condition that may cause loss of life, injury or other health impacts, property damage, loss of livelihoods and services, social and economic disruption, or environmental damage". It can be calculated risk as the product of hazard and vulnerability.

Within the landslide risk management framework, vulnerability is regarded as potential for damage and loss caused by a hazard and would depend on a random mix of physical, social, economic, and environmental factors, which interact in space and time.

3. Development of a Policy Framework to promote and enrich the landslide risk awareness

The landslide hazard map is another important entity in developing a policy framework to enrich landslide risk awareness. It is essential that the policy context be field validated, reliability certified and user-friendly. It should be effectively utilized, ensuring timely mitigation and expedited response. A special policy context should be outlined for the use of architects, planners, engineers and contractors to certify the safety of structures they plan, design and build. It will be useful to provide these hazard maps and policies to foreign visitors to educate them about the hazards they might face. If an area has been identified as vulnerable to landslides and categorized under the landslide-prone areas, awareness programmes should be immediately conducted for the local community. This should be handled with utmost care, since the local community has been established in these areas making use of the local resources and involved in regional occupations. Thus it is important to build an amicable and lasting relationship with them. The state authorities and intervening organizations should consult these communities on proposing alternative job opportunities and highlighting the benefit of resettling in a safer environment to minimize the impact in catastrophic situations. It is of vital importance to build a strong rapport with these citizens and thereby counteract or avoid misconceptions and negative attitudes towards the policies and action plans proposed by the government (Rupasinghe et al. 2014).

4. Development of a Policy Framework to promote a multi-disciplinary culture of landslide studies and management

The major factor that governs promoting the multi-disciplinary culture of landslide studies is none other than the landslide professionals. It is essential to create a gradational structure, spreading from a large well branched network which acts as a tremendous supportive system of sharing information and services among individuals and groups until it reaches the grass-roots level. This network can be utilized in conducting effective awareness programmes, educating each and every individual in the network on how to avoid, minimize and mitigate the impacts of catastrophic phenomena. Thus every individual should build up their own moral strength. when acting in areas of landslide occurrence, with a sensible approach of leaving such localities and settling down in safer surroundings. This moral strength should be built up bit by bit, until they identify with avoiding vulnerability, including cultivation in slope areas, establishing domestic environs in mountainous regions and occupying landslide-prone regions for employment. Such attempts can only succeed through a greater interaction between professionals pursuing landslide studies and those considering other types of hazards, in pursuit of the common cause of advancement of knowledge and human safety.

5. Do the landslide risk reduction policies breed and nurse a scientific culture and outlook?

Once the landslide risk reduction policy framework has been structured it is worthwhile to analyze whether these policies breed and nurse a scientific culture and outlook. In most developing countries, there is a near absence of policies promoting scientific outlooks in landslide studies. This may be due to the shortfall of resources and inadequate capacities available in third-world countries to manage landslides. Yet there is an urgent requirement to set up policies that drive the culture of scientific scrutiny before allowing investments in landslide disaster risk reduction. All aspects in mitigation strategies should be taken into account to develop deeper insights initially. It is more supportive if this hydra-headed social problem is disassembled into smaller fragments, first considering one at a time and then ultimately as a cluster. For instance, when considering a scientific investigation of a post-landslide occurrence, the most preliminary fact to establish is whether the investigation team is multi-disciplinary in expertise and skilled in exercising engineering judgment to do justice to the landslide study at the moment.

6. How focused are we on what we need to ensure landslide risk reduction?

Classifications based on diagnostics are required by the landslide specialist, whereas decision makers and disaster managers would prefer risk-based classification. Risk is related to damage or losses caused by the occurrence of hazards, both in the short term and in the long term. While in the grip of a disaster, the primary concerns are anticipated losses due to deaths and injuries. Further it could be aimed at persons who are untraceable, damage to buildings, infrastructure, and goods and services. Concerns such as post-disaster lootings, social unrest, environmental degradation, loss of agriculture will also emerge. However the consequences of the phenomena are not easy to quantify. A few of the intangible losses include halting of the process of development caused by disruption and dislocation.

7. Early warning against landslides

Landslides are among those geohazards which are predictable. Yet it generally costs a lot to utilize possible early warning systems against landslides. Thus a third-world country cannot solely depend on its government to establish such facilities. It is essential that the communities are empowered to promote simple and economical instrumentation and trained to foresee the danger of landslides for themselves. People are impressed by the magical power of modern technology, but for the developing world, it would be inappropriate to over-look the advantages of ease, economy, accessibility and time-tested benefits of traditional technologies. It is also time to think in terms of community-centric early warning systems.

8. Viewing landslide studies in the larger context of disaster risk reduction

Landslide risk can no longer be considered in isolation. Other hazardous events such as floods and earthquakes are also entangled with landslides to a certain extent. It can be further introduced as landslide-triggered tsunamis, earthquake-induced landslides, and flood-induced landslides. Since a landslide-prone area could also be flood- and earthquake-prone, it is only by the production of multi-hazard maps that the scientists can connect the universe of landslides with the galaxy of other disasters, and encourage a universal approach to hazard investigation and disaster risk reduction.

9. What do we intend to leave for posterity by way of documentation on diverse and kaleidoscopic aspects of landslides?

The vast rage of research papers and diversified study reports naturally reflect more of the perceptions of the writers, built up on a pile of observations and past experience, and well known for loyalty to the widely accepted trends and theories.

The research teams should be encouraged to re-visit scientifically controversial, but educationally important case-records, not just with a view to either verify or falsify whatever has been concluded earlier, but to let the truth surface naturally through the true spirit of scientific enquiry. Modern technology allows us to repeatedly test the lessons learned and use those very lessons to perform differently. It is depressing to observe the growing disconnect between landslide investigation and landslide remediation, and the gap seems to be widening because of the commercially-driven recourse to a brutal use of technology to fix problematic slopes and landslides.

Finally, Conceptual Model Policy Frameworks will be considered successful if landslide risk reduction becomes a way of life in the national development agenda and breeds the culture of safety in day-to-day lives.

Summary of Statistics of the International E-Conference 2015

The E-conference was an amalgamation of 300 participants representing almost all the continents in the world and 20 keynote experts/facilitators, supporting an effective communication platform throughout the project (see Figs. 6, 7 and 8). The E-conference was carried out through an official website (http://e-conference.crdcecbsl.lk/) specifically created as an open source medium to enable all viewers to connect simultaneously to share their ideas and comments related to the topic.

A Local Forum was also run in parallel with the International E-conference in order to confer in depth on the pressing matters due to landslides of national significance. A consultation workshop was conducted to identify the thematic areas to come up with feasible solutions for the issues faced by vulnerable communities from the grass-roots level. A Question Session was also instigated in order to clarify all the issues that have emerged throughout the conference and utilized the expertise knowledge in finding the best solution for these issues.

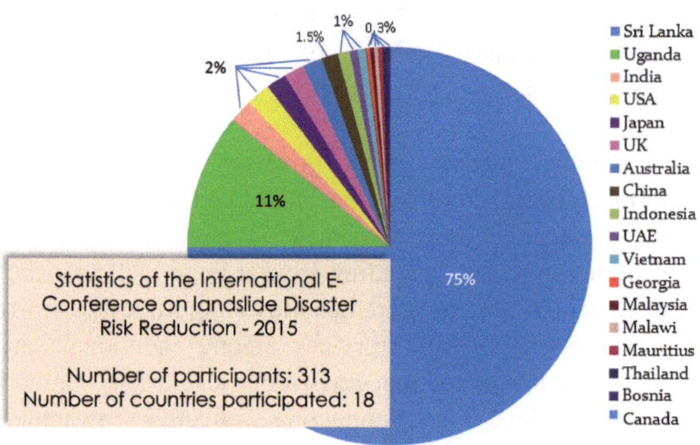

Fig. 6 Percentage of participants by country

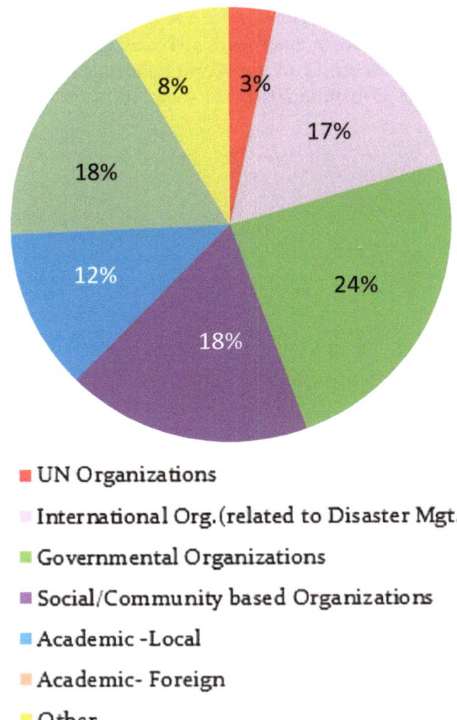

Fig. 7 Percentage of participants by affiliation

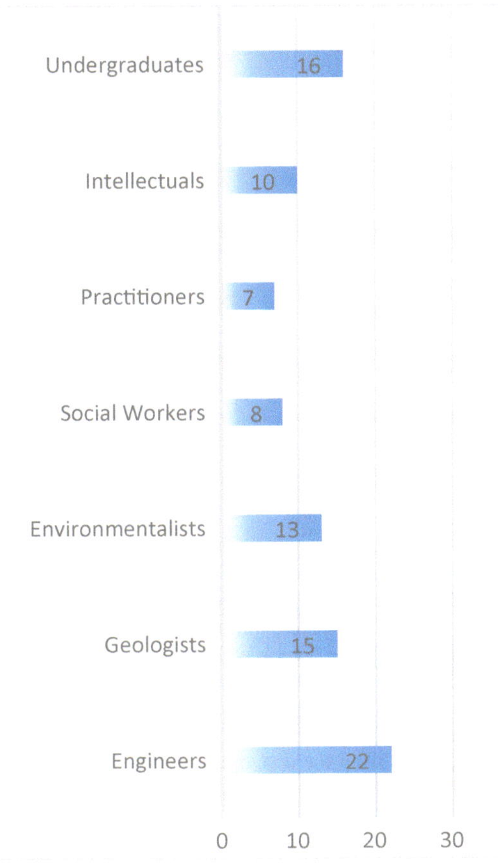

Fig. 8 Percentage of participants by career

The experts reviewed the globally available necessary information, identified research needs and gaps in existing recommendations directly relevant to the respective problems. The summary of the key areas discussed will be published in the conference proceedings.

The main objective of publishing the proceedings is to filter the most significant facts discussed throughout the conference in order to increase the readability and obtaining the maximum usage of it by the society. The institutional target is focused more towards the social aspects, such as enriching landslide risk awareness among citizens (including school students, workmen and laymen) and ensuring the effective transmission of this message to the public to enhance their adaptability and familiarity with the mechanisms involving landslide mitigation and slope protection measures discussed throughout the conference. It is our pleasure to work in partnership with UNDP towards this approach. It was certainly a fruitful global discussion in support of bringing a sustainable humanitarian future to fruition.

Acknowledgements This paper deals with integral parts of the research projects conducted under the International Programme on Landslides: IPL-143, IPL-155, IPL-199 and IPL-200, which have been implemented by the Centre for Research & Development, Natural Resources Management & Laboratory Services, and the Central Engineering Consultancy Bureau (CECB). The CECB is the premier consultancy organisation in Sri Lanka, with a wide range of expertise in different engineering disciplines of the Ministry of Mahaweli Development & Environment. It is published with their permission. The views expressed in the paper are however those of the authors only. Our grateful thanks are due to the Eng. G. D.A. Piyathilaka, Chairman, and Eng. K.L.S. Sahabandu, General Manager, CECB for permission and encouragement.

References

Bhandari RK (2015) Developing conceptual model policy frameworks to understand causes, effects and mitigations of landslide occurrences. In: International E-conference on landslide disaster reduction—2015, Sri Lanka 24th March–24th May 2015

Dias AAV, Gunathilaka AAJK (2014a) Evaluation of sensitivity of the WAA and SINMAP models for landslide susceptibility risk mapping in Sri Lanka. In: Proceeding of the World Landslide Forum3 (WLF3), Beijing, China, 2–6 June 2014, vol 2, Landslide Science for a Safer Geoenvironment, pp 167–173

Dias AAV, Abayakoon SBS, Bhandari RK (2014b) Discrete boundary shear strength of a landslide at high rainfall precipitation zone in Sri Lanka. In: Proceeding of the World Landslide Forum3 (WLF3), Beijing, China, 2–6 June 2014, vol 1, Landslide Science for a Safer Geoenvironment, pp 101–106

Herath HM, Janaki MK, Kodagoda SSI, Dias AAV (2014) Shallow modes of slope failure in road earth cuttings in Sri Lanka. In: Proceeding of the World Landslide Forum3 (WLF3), Beijing, China, 2–6 June 2014, vol 2, Landslide Science for a Safer Geoenvironment, pp 51–58

Herath HM, Janaki MK, Jayasooriya JADNA, Dias AAV (2016) Geological stability of overhanging rock slopes. In: Proceedings of 2016 IPL Symposium, UNESCO, Paris, November 17–18, 2016

Mallawarachchi MASN, Ekanayake EMTM, Kodagoda SSI, Dias AAV (2014) Comparison of soil modulus E50 of residual soil slope failures in two different rainfall zones. In: Proceeding of the World Landslide Forum3 (WLF3), Beijing, China, 2–6 June 2014, vol 1, Landslide Science for a Safer Geoenvironment, pp 135–141

Rupasinghe N, Dias AAV, Hennayake SK (2014) Role of intervening agencies and officials in emergency risk management of landslides, Sri Lanka. In: Proceeding of the World Landslide Forum 3 (WLF3), Beijing, China, 2–6 June 2014, vol 4, pp 700–705

Landslide Hazard and Risk Management (WCoE 2014–2017)

Josef Stemberk, Vít Vilímek, Jan Klimeš, Jan Blahůt, Filip Hartvich, and Jan Balek

Abstract

The World Centre of Excellence (WCoE) on Landslide Risk Reduction entitled "Landslide risk assessment and development guidelines for effective risk reduction" (2014–2017) was designed to contribute to the risk reduction effort formulated in the Sendai Partnership initiative. Several research activities were developed and their results were presented to a broad public through a series of articles, informative web pages and documentary movies. The research focused on improving landslide hazard assessment in a variety of natural environments, including deep-seated as well as shallow landslides. Landslide hazard assessment was applied practically through development projects in Ethiopia and Peru. Within the scope of the WCoE we proposed and conducted two projects of the International Program on Landslides (IPL). One of them is dedicated to compilation and analysis of glacial lake outburst floods (Database of glacial lake outburst floods (GLOFs)–project No. 179) at the global level. This potentially highly damaging natural phenomenon combines characteristics of floods and debris flows and often also involves landslides in the initiation process. The other IPL project focuses on the main challenges of landslide risk reduction in the Czech Republic (Challenges for landslide hazard and risk management in "low risk" regions, Czech Republic, IPL project No. 197), which is a country with abundant landslide-related knowledge and rather low annual occurrence frequencies. Despite that, landslides cause considerable damage and financial losses, which often could be prevented if the available hazard information were to be used.

Keywords

Risk reduction • Landslide hazard • Development projects • Peru • Ethiopia • Czech republic

J. Stemberk · J. Klimeš (✉) · J. Blahůt · F. Hartvich · J. Balek
Czech Academy of Sciences, Institute of Rock Structure & Mechanics, Prague, 182 09, Czech Republic
e-mail: klimes@irsm.cas.cz

J. Stemberk
e-mail: stemberk@irsm.cas.cz

J. Blahůt
e-mail: blahut@irsm.cas.cz

F. Hartvich
e-mail: hartvich@irsm.cas.cz

J. Balek
e-mail: balek@irsm.cas.cz

V. Vilímek
Department of Physical Geography and Geoecology, Charles University, Prague, 128 43, Czech Republic
e-mail: vilimek@natur.cuni.cz

K. Sassa et al. (eds.), *Advancing Culture of Living with Landslides*,
DOI 10.1007/978-3-319-59469-9_32

Landslide Research for Better Hazard and Risk Assessment

Deep-Seated Landslide Monitoring for Improved Hazard Assessment

Deep-seated landslides often act as precursors to episodic and potentially catastrophic movements (Pánek and Klimeš 2016). Because of the poor strength and unfavourable hydrological characteristics of mobilized material, they may represent significant obstacles for infrastructure development (Pánek et al. 2014). Monitoring of this phenomenon requires very precise and robust instruments which allow for precise and long-term movement recordings. For this purpose our research team developed (Košťák 1969) and improved (Klimeš et al. 2012; Rowberry et al. 2016) a very precise, optical-mechanical crack gauge, which records displacement in three dimensions along pre-defined surfaces of ruptures. Study sites equiped with this instrument are located mainly within the Czech and Slovak Republic: (https://www.irsm.cas.cz/ext/tecnet/index_en.php?page=google_mapa_sesuv_en).

Long term monitoring of deep-seated rockslides in a flysh rock complex (Stemberk et al. in press) revealed coupling between landslide creep and tectonic activities. These deformations contribute to a weakening of rock strength along the planes of rupture and thus may contribute to possible future episodic activity of deep-seated landslides.

A fully automated deformation monitoring system with remote access was established on El Hierro Island, Spain (Fig. 1). Its goal is to record possible movement activity of incipient deep-seated landslides, where we so far have recorded creep movement of the order of 1 mm yr^{-1} (Klimeš et al. 2016a, b). Recently, a "big data" approach has been used to improve interpretation of the detected movements, by considering a number of environmental factors which may affect the detected movements (Blahůt et al. submitted).

Possible gravitational or tectonic deformations at the scale of alpine ridges are monitored within the Large Scale Monitoring (LASMO) project at the Grimsel Test Site (Switzerland), dedicated to the long-term management of underground radioactive waste disposal. In this case, precise detection of very slow rock deformation is required and is provided by the TM71 crack gauge.

Fig. 1 One of the monitoring sites on the El Hierro Island, Canary Archipelago, Spain. The automatic crack gauge (in front of the person) records movements along a side scarp of the San Andrés deep-seated landslide

Landslide Hazard in High Mountains and Arctic Regions

Landslides in high mountains may initiate potentially dangerous outburst floods from glacial lakes. Their ability to do so largely depends on their size (volume) and the location of impact point into the lake with respect to the lake outflow. We investigated in depth landslides from side moraines as possible triggers of outburst flood at the Palcacocha Lake, Cordillera Blanca, Peru. The results of slope stability calculations, and models of impact wave propagation, as well as landslide movement detection using satellite radar interferometric analyses, allowed a reliable assessment of hazards related to the outburst floods at this specific site under current conditions (Klimeš et al. online first). The effects of different types of slope movements on glacial lake outburst flood (GLOF) initiation are also included in the susceptibility assessment methods tailored for the Cordillera Blanca Mts. (Emmer and Vilímek 2014). Research on the world-wide patterns of GLOF occurrences and characteristics is further supported by maintaining and improving the GLOF world wide database within the scope of the IPL project entitled "Database of glacial lake outburst floods (GLOFs)—project No. 179" (Emmer et al. 2016).

Rock slope development in the High Arctic (SW Spitsbergen, Norway) was described using several dating and geomorphological methods, while recent rock deformations are being monitored with dilatometers. The monitoring results are applied to the assessment of future possible hazards related to arctic rock slopes affected by climate and environmental change expressed by glacier retreat.

Precipitations as a Landslide Trigger

Case studies in the Czech Republic were used to investigate the relationship between precipitation and debris flow initiation or landslide movement activity. For debris flows that developed in 2010 in granitic mountains, the research concluded that high amounts of antecendent precipitation, as well as high-intensity short-duration precipitation, were required to initiate the flows (Smolíková et al. 2016). The importance of coupling of high ground water levels and short-term high-intensity precipitation for shallow landslide movement acceleration is illustrated by long-term monitoring of near-surface movements of a shallow landslide in Cretaceous sedimentary rocks (W. Bohemia, Czech Republic). The landslide underwent several major movement reactivations, destroying a railroad and local road. The monitoring includes two extensometric profiles measured with manual tape, with extensometer and several piesometers and boreholes where the ground water level is monitored. Results since 1994 show that the major movement accelerations occured during periods with extremly high ground water levels.

Improving Public Landslide Risk Awareness, Czech Republic

Czech Republic has a well developed and up-to date national landslide inventory, as well as a methodology for landslide hazard assessment. Regions highly prone to landslide occurence that have been subjected to several catastrophic landslide events during the last 20 years are also well described. Despite that, landslides repeatedly cause significant damage to infrastructure and incur considerable costs. We asume that one reason is that the general public and responsible authorities do not pay corresponding attention to this potentially damaging phenomenon. To increase awareness about landsliding especially among authorities, we summarized the costs of landslide-related construction and mitigation works claimed in public budgets. These expenses represent the true costs related to landslide occurrence and reactivations and are based on publically available databases which cover the whole territory of the Czech Republic. Therefore this approach represents an objective way to obtain comparable data about expenses related to landslide mitigation works. These data have not been evaluated yet. Resulting annual landslide-related costs do not include all of the expenses but we are convinced that the most costly mitigation works are included. Analysis of the data confirm the high overall costs related to landslides and their uneven distribution within the territory of the Czech Republic.

Application and Dissemination of Landslide Hazard and Risk Assessment Research

Development Cooperation Projects in Ethiopia and Peru

Dissemination of knowledge about landslide hazards and risk assessment on an international level was carried out through projects funded by the Czech Development Cooperation agency. The first one is cooperation with the University of Arba Minch in Ethiopia. The project is focused on teaching (Fig. 2) and field training of undergraduate students in engineering geology, geotechnics and hydrogeology, with special attention to landslide mapping, monitoring and hazard assessment. The final goal of the project is to establish and help to maintain a research center dedicated to hazardous geological processes at the Arba Minch University. This development project takes advantage of the number of research works previously conducted in the landslide-prone regions of Ethiopia (Vařilová et al. 2015). This work points out the importance of the combined negative effects of annual intensive precipitation during rainy seasons and human activities on landslide reactivation and occurence. It also suggests possible measures to reduce the landslide hazard and risk that are applicable to the local technical and social conditions.

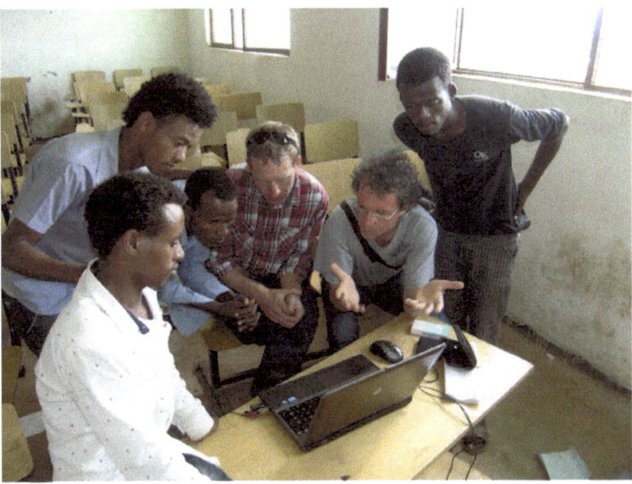

Fig. 2 Close interaction between students and teachers during preparation of thesis works, Arba Minch University, Ethiopia

Another development project was conducted in Peru to implement measures for landslide risk reduction in the village of Rampac Grande (Carhuaz, Ancash). This village has been endangered since 2009 by an active landslide with a high potential for damage. The project follows previously conducted research (Vilímek et al. 2016; Klimeš and Vilímek 2011) and was conducted by Peruvian research institute INAIGEM (Instituto Nacional de Investigación en Glaciares y Ecosistemas de las Montaña), with the collaboration of Czech researchers. The one-year project involved a series of meetings with local authorities and inhabitants during all of its stages. Thanks to the open and repeated interactions with the local people, the project was accepted by the local inhabitants who previously were highly suspicious of any research activity around their village (Klimeš and Vilímek 2011).

The project results include a landslide hazard map and posting of signs showing evacuation routs and construction of two monitoring extensometric profiles. All the project results were presented to the local population (Fig. 3). The purpose and technique of tape extensometric monitoring was explained in the field (Fig. 4). The close communication of all project steps to the local population is a crucial requirement for its success and long-term sustainability.

Fig. 3 Project results were presented at the local school to the village authorities as well as school students and teachers

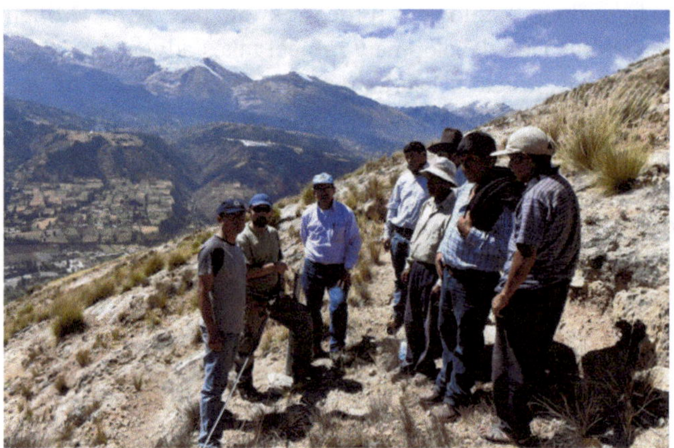

Fig. 4 Explanations of the technique of tape extensometric measurements to the Rampac Grande authorities were conducted in the field with the participation of Peruvian and Czech experts

Knowledge Dissemination in the Czech Republic

We conducted a variety of activities to bring landslide knowledge to the broader public. One of them is online—a publically available archive of historical photographs of Prof. Quido Záruba (https://www.irsm.cas.cz/index_en.php?page=Archiv_qz_en). The archive recently contains 8000 photographs, with some dating back to the 1920s during the 20th century, showing unique images of different geological features and engineering structures, including landslides and landslide-related mitigation works.

We organized several worshops and an international conference dedicated to bringing recent scientific knowledge to the Cezch experts and local authorities responsible for landslide mitigation. We also participated in the preparation of a documentary movie dedicated to landslide processes, which is aimed at explaining the basic information about landslide hazards and risk in the Czech Republic to a wide audience.

Acknowledgements This work was carried out thanks to the support of the long-term conceptual development research organization RVO: 67985891 and financial support of the INGO II project of the Ministry of Education No. LG15007.

References

Blahůt J, Rowberry M, Balek J, Klimeš J, Baroň I, Martí X (submitted) Monitoring giant landslide detachment planes in the era of big data analytics. In: Proccedings of the 4thWLF, 29 May–2 June 2017. Ljubljana, Slovenia

Emmer A, Vilímek V (2014) New method for assessing the susceptibility of glacial lakes to outburst floods in the Cordillera Blanca. Peru Hydrol Earth Sys Sci 18:3461–3479

Emmer A, Vilímek V, Huggel C, Klimeš J, Schaub Y (2016) Limits and challenges of compiling and developing a database of glacial lake outburst floods (IPL project No. 179). Landslides. doi:10.1007/s10346-016-0686-6

Klimeš J, Vilímek V (2011) A catastrophic landslide near Rampac Grande in the Cordillera Negra, northern Peru. Landslides 8:309–320

Klimeš J, Rowberry MD, Blahůt J, Briestenský M, Hartvich F, Košťák B, Rybář J, Stemberk J, Štěpančíkova P (2012) The monitoring of slow-moving landslides and assessment of stabilisation measures using an optical-mechanical crack gauge. Landslides 9:407–415. doi:10.1007/s10346-011-0306-4

Klimeš J, Yepes J, Becerril L, Kusák M, Galindo I, Blahut J (2016a) Development and recent activity of the San Andrés landslide on El Hierro, Canary Islands, Spain. Geomorphol 261:119–131

Klimeš J, Novotný J, Novotná I, Jordán de Urries B, Vilímek V, Emmer A, Strozzi T, Kusák M, Rapre A C, Hartvich F, Frey H (2016b) Landslides in moraines as triggers of glacial lake outburst floods: example of the Palcacocha Lake (Cordillera Blanca, Peru). Landslides. doi:10.1007/s10346-016-0724-4

Košťák B (1969) A new device for in situ movement detection and measurement. Exp Mech 9:374–379

Pánek T, Klimeš J (2016) Temporal behavior of deep-seated gravitational slope deformations: A review. Earth Sci Rev 156:14–38

Pánek T, Hartvich F, Jankovská V, Klimeš J, Tábořík P, Bubík M, Smolková V, Hradecký J (2014) Large Late Pleistocene landslides from the marginal slope of the Flysch Carpathians. Landslides 11:981–992. doi:10.1007/s10346-013-0463-8

Rowberry MD, Kriegner D, Holy V, Frontera C, Llull M, Olejnik K, Marti X (2016) The instrumental resolution of a moire extensometer in light of its recent automatisation. Measurement 91:258–265

Smolíková J, Blahůt J, Vilímek V (2016) Analysis of rainfall preceding debris flows on the Smědavská hora Mt., Jizerské hory Mts., Czech Republic. Landslides 13:683–696. doi:10.1007/s10346-015-0601-6

Stemberk J, Hartvich F, Blahůt J, Rybář J, Krejčí O (in press) Tectonic strain changes affecting the development of deep seated gravitational slope deformations in the Bohemian Massif and Outer Western Carpathians. Geomorphol

Vařilová Z, Kropáček J, Zvelebil J, Štastný M, Vilímek V (2015) Reactivation of mass movements in Dessie graben, the example of an active landslide area in the Ethiopian Highlands. Landslides 12:985–996

Vilímek V, Klimeš J, Torres MZ (2016) Reassessment of the development and hazard of the Rampac Grande landslide, Cordillera Negra. Peru Geoenviron Disasters 3:5. doi:10.1186/s40677-016-0039-8

Mitigation of Landslide Hazards in Ukraine Under the Guidance of ICL: 2009–2016 (IPL-153, IPL-191)

Oleksander Trofymchuk, Iurii Kaliukh, Silchenko Konstantin, Viktoriia Berchun, Taras Kaliukh, and Iaroslav Berchun

Abstract

More than 90% of the territory of Ukraine has complex soil conditions. The number of landslides has increased by a factor of 1.3 in the last 15 years, and by a factor of about 3 over the last 30 years. Ukraine became a member of ICL only in 2009. The main task of Ukrainian division of ICL (UDICL) from 2009 to 2016 was and still is the implementation of the National Plan (the State Programme) on lanslide hazards mitigation. Because of a lack of governmental or any other support during the above years, UDICL has managed to carry out only two projects and is working on one more on a voluntary base. Objectives of the first IPL project were to determine the slopes with a landslide hazard in the Kharkiv region of Ukraine; to develop a database containing the engineering-geological information relevant to descriptors of landslide sites; and to develop targeted GIS on landslides in the Kharkiv region. All the goals of the project were achieved. In 2012–2014 IPL 153 project was implemented: information about landslide protection structures and measures was collected and structured, prospects of their development in the Autonomous Republic of the Crimea of Ukraine (ARCU) were studied, and the target database was created. Since 2015 "Landslide hazard zonation using GIS", the IPL 191 project, is being realized. The main goal of the project was to develop an instrument for landslide hazard forecasting to minimize the impact of landslide activation on people and tangible objects for the Carpathian region of Ukraine. Two Ukrainian standards of construction objects monitoring and building in the landslide sensitive areas will be completed and put into effect in 2017–2018. UDICL plans a training programme concerning these building standards for more than 1000 designers from all the regions of Ukraine.

Keywords

Landslide hazards • Experiment • Mitigation • Monitoring • Heritage • Retaining wall

O. Trofymchuk · V. Berchun · I. Berchun
ITIGS NASU, 13 Chokolivsky Blvd., Kiev, 03186, Ukraine
e-mail: itelua@kv.ukrtel.net

V. Berchun
e-mail: berchun2003@yahoo.com

I. Berchun
e-mail: berchun93@gmail.com

I. Kaliukh (✉) · S. Konstantin
Research Institute of Building Constructions, Preobrajenskaya Str.5/2, Kiev, 03037, Ukraine
e-mail: kalyukh2002@yahoo.com

S. Konstantin
e-mail: kalyukh2002@yahoo.com

T. Kaliukh
Scientific Research Institute of Oil and Gas Industry of National Joint-Stock Company "Naftogaz of Ukraine", Kiev, Ukraine
e-mail: tarasklh@gmail.com

K. Sassa et al. (eds.), *Advancing Culture of Living with Landslides*,
DOI 10.1007/978-3-319-59469-9_33

Landslide Work in Ukraine: A Background

Dr Iurii Kaliukh became aware the ICL after presenting "Theoretical and methodological issues of monitoring of Livadia landslide system and Livadia palace in Crimea, Ukraine" in August 2003 at the National Center of U.S. Geological Survey in Reston (USA). After that he offered to carry out a joint study of landslide hazards in Ukraine, and the Livadia landslide system in particular. He turned to ICL for information. Unfortunately, we became members of ICL only in 2009. From 2003 to 2009 we were trying to find governmental or any other financial support for the ICL entrance fee and further annual fees. However, our attempts had no result. That is why the entrance fee and all further annual fees from 2009 to 2014 were paid at private expense by Dr Olexander Trofymchuk, the director of Institute of Telecommunications and Global Information Space of National Academy of Sciences of Ukraine (NASU). Otherwise Ukraine would not have become a member of ICL

The period from October 2014 to 2016 has been the most difficult for Ukraine since its independence (collapse of the USSR in 1991). There are several reasons: the military aggression of Russian Federation since March 2014, which developed into annexation of the Crimean Peninsula and war in the east of Ukraine (Donbas) a decline in the Ukrainian economy (separation of the Crimea and part of the Donbas, drop in production and investments) a sharp decrease in investments into science and a fall in the living standard of Ukrainian people. In 2015 scientists salaries decreased 3 times in dollar equivalent and Dr Olexander Trofymchuk couldn't pay the annual fee to ICL. In 2016 it was paid from the personal money of Dr Olexander Trofymchuk and Dr Iurii Kaliukh. We are also planning to pay further annual fees together. However, we couldn't participate in many meetings, conferences and congresses of ICL during 2009–2016 because of lack of governmental or any other type of support. Undoubtedly, it has adversely affected the quality and scientific level of our studies of landslide hazards in Ukraine. Because of the aforementioned reasons, in 2009–2016 the Ukrainian division of ICL has managed to realize only two projects and now is working on one more project on a voluntary basis.

Introduction

More than 90% of the territory of Ukraine has complex soil conditions and about 120,000 sq. km of the Ukrainian territory are located in a seismically active area that has earthquakes with magnitudes varying from 6 to 9. Therefore, unpredictable changes of natural geological state and man-made factors that determine the ground conditions may cause dangerous deformation processes at heritage sites in Ukraine. This requires the introduction of additional protection measures, such as the creation of monitoring and early warning systems. Landslide processes in Ukraine occupy first place in the ranks of hazard damage. In general, over 23.1 thousand landslides by 01 December 2011 and 17.4 thousand landslides by 01 December 1997 were detected in Ukraine. Thus, the number of landslides has increased by 1.3 times in the last 15 years, and by 3 times over the last 30 years. Massive landslides took place in Kiev in April 2014:

> The landslides have been activated again in Kiev. Already 131 land areas are moving in different parts of the city (last year there were 125). Experts say that the large-scale falls of ground can be a threat to the city. The moving land areas can damage the roads, buildings and water, heat and gas pipelines.

Periodic exogenic geological processes activation and complex engineering-geological situations within the South Coast of the Crimea result in landscape transformation and deformation of engineering and architecture structures. Growth in the number of landslides is mainly caused by man-made and combined natural and man-made reasons, and less frequently by natural ones alone.

Man-made reasons have caused about 600 human-induced landslides in the territory of the South Coast over the recent 100 years that were registered within the South Coast of the Crimea Landslides Cadaster (the total number of landslides in the Crimea by the end of 2009 was 1576). Thus, about 38% of the South Coast landslides are caused by human activity. The most substantial damage is caused by the landslides threatening the safety of unique architectural, historic and cultural monuments. Over the history of South Coast of the Crimea engineering development, these landslides have caused more damage than earthquakes. At the present moment there is a threatening situation in the vicinity of the Livadia Palace and Park Complex. Intensification of seismic activity in the Vrancea area, the Black Sea region and around the world has led to an increased level of seismic hazard in Ukraine, triggering more landslides. The whole territory of Ukraine (not only just the Carpathian and Crimean regions, as was believed before) is now an area with a high potential risk of future huge earthquakes. The deputy director of the Institute of Geophysics of the NASU, O. Kendzera, says (Newspaper "Segodniya" 2014) "7–8-magnitude effects can be observed in areas with weakened soils (frequent flooding, landslides, etc.)".

"Before the Japanese earthquake (03.11.2011) it was considered that during a 9-magnitude earthquake acceleration of ground can reach 0.4 g. But the Japanese catastrophe showed as much as 2.7 g. It led to much more damage than it was expected. It means that now we must correct all parameters of earthquakes" noted the director of the

Armenian Institute of Geophysics and Engineering Seismology S. Ohasyan. A. Kendzera adds: "…We must correct initial data that will be used by designers in building construction". UDICL is working on these challenges.

2009–2011

In 2010–2011 the IPL project "Landslide hazard zonation in Kharkiv region of Ukraine using GIS" was realized. The main goal of the project was to develop an instrument for landslide hazard forecasting to minimize the impact of landslide activation on people and tangible objects, including construction, transportation services, pipelines etc. Its objectives were to determine the slopes subject to landslide hazards slopes over the Kharkiv region of Ukraine; to develop a database containing the engineering-geological information relevant to descriptors (passports) of landslide sites; and to develop targeted GIS on landslides in the Kharkiv region. All goals of the project were achieved. The results of the project were reported at the 2011 WLF2 conference in Rome (Trofymchuk, Kaliukh et al. (2013a, b, c) and published Trofymchuk et al. (2013d) (Fig. 1).

Ph.D. A.S. Glebchuk has defended a dissertation on the topic "Mathematical modeling of landslide's dangers in condition of flooding and seismic influences" (under the supervision of Dr Olexander Trofymchuk) relevant to the IPL project. The dissertation is devoted to system analysis of landslide dangers using mathematical modeling and GIS-technologies for the creation of models that display landslides processes. Landslides slopes were determined in the territory of the Kharkiv region. Databases, including data descriptors of landslide sites and precipitation in the Kharkiv region, were developed. Developed GIS includes information on relief, steepness of slopes, hydrography, roads and landslide sites of the Kharkov area.

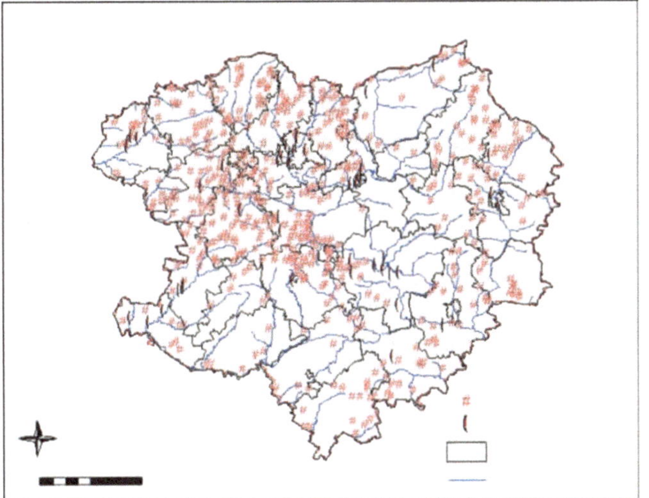

Fig. 1 Map showing the distribution of landslides in the Kharkiv region

Ukraine has been a member of the "Landslides and Cultural & Natural Heritage" (LACUNHEN) thematic Network of the ICL (head of the LACUNHEN is Claudio Margottini, Vice President of the International Consortium on Landslides) since 2012. The scope of the LACUNHEN Network has broadened from studying a single heritage site in isolation to a multidimensional, multiregional and inter-disciplinary approach. Within this broader view, landslides, and more generally slope instabilities, are an important factor endangering cultural heritage sites and causing their degradation (Fig. 1). We used an inter-disciplinary approach based on system analysis principles to create the monitoring and early warning system (ZSUV) of the Livadia Palace, placed on the active Central Livadia Landslide system (CLLS), with the aid of the experimental and analytical studies of natural seismic and rainstorm impacts on landslide movement on the deformation of the buildings of the Livadia Palace. The ZSUV system was designed, programmed, and experimentally worked out by multi-day tests using special precision equipment at the National Technical University of Ukraine "KPI" and installed on the CLLS and Livadia Palace, where it operated in real-time mode. Experimental data were obtained on the impact of natural seismic factors and rainstorms on the Central Livadia landslide system and on the Palace itself. And on the basis of this example, the Ukrainian division of ICL represented the TXT-tool "System approach for monitoring of World Heritage sites placed on the active landslides: the monitoring and early warning system of the Livadia Palace building constructions placed on the active CLLS, Ukraine" to ICL.

For a real-time diagnosis of the lithodynamic conditions, an algorithm was developed, and a computer system for monitoring the CLLS was realized, together with real-time maintenance of the data base of lithogenic and other parameters of the CLLS, and manual updating of data using high-level programming language. The real-time program allows monitoring of the level of groundwater and of the deformation angles of landslide bodies; recording of seismic activity, statistical processing of the obtained data, archiving of the obtained data and statistical processing results. The correlation dependence of the seismic activity of CLLS on solar activity is analyzed by maintaining a data base of the solar and lithogenic parameters. In this case the data are entered into the computer manually with the help of the *ZSUV* because study of these processes does not require any automated system for gathering and accumulation of initial information (the solar cycle of activity increase and decline has a period of 11 years). The solar parameters cover solar activity, changes of temperature and humidity conditions, the character and intensity of fallouts, wind, etc. During the

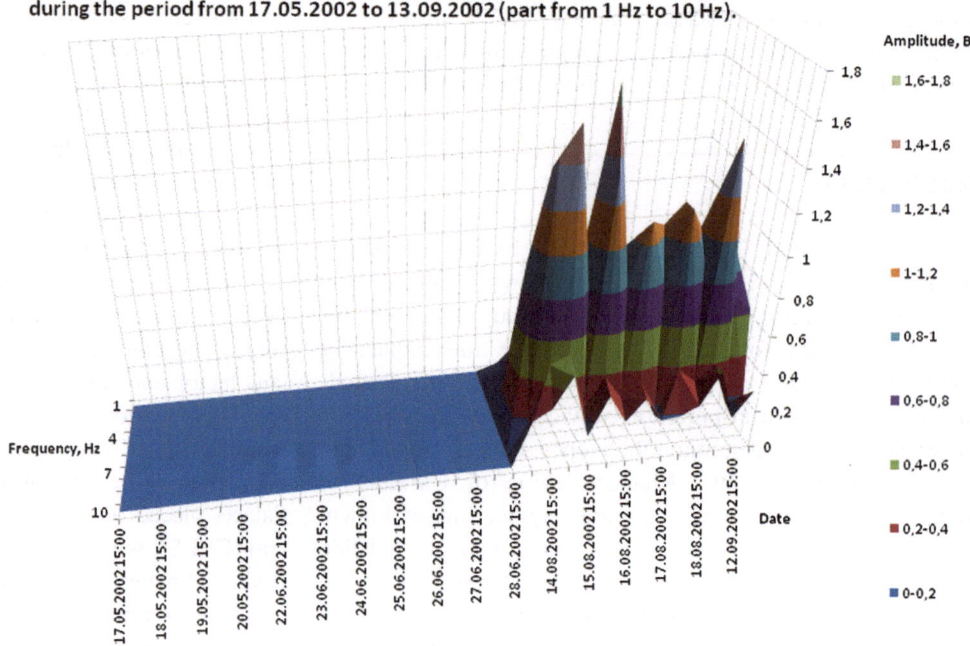

Fig. 2 A fragment of the instant information obtained from one of the numerous channels of high-precision sensors in 2002 during monitoring of the structural elements of the Livadia Palace

period from mid-December 2001 to December 2002 vibrations of the Livadia Palace (Fig. 2.) caused by seismic activity were measured using high-precision sensors. The spatial position was determined by measuring the inclination angle of the building façade.

2012–2014

In 2012–2014 project IPL 153 "Landslide protection structures and their development in the Autonomous Republic of the Crimea, Ukraine (ARCU)" was realized. Within the framework of IPL Project №. 153 "Landslide protection structures and their development in the ARCU" (term: January 2012–December 2014) information about landslide protection structures and measures was collected and structured, prospects of their development in the ARCU of Ukraine were studied, and the target database was created. Some results of the project were presented in the report "Experimental and analytical studies of landslides in the south of Ukraine under the action of natural seismic impacts at the International Symposium on Earthquake-Induced Landslides, Kiryu, Japan, 2012, other conferences and congresses (Trofymchuk et al. 2013a, b, c), and published in the periodical press (Kaliukh et al. 2014; Kaliukh et al. 2015). Ph.D. K.V. Silchenko has defended a dissertation on "Work of landslide retaining structure that consists of short piles" (under the supervision of Dr Iurii Kaliukh) relevant to the IPL project. The dissertation work is devoted to an experimental and theoretical study of the interaction of short piles with sliding soil; the development and implementation of this type of landslide retaining structure to construction practices in the Crimea, Ukraine. For the first time the experimental distribution diagram of landslide pressure on the frontal surface of short piles with varying lengths was experimentally established. The applied calculation method was developed for short piles on creeping and punching in landslide depth and in an area of short piles installed in stable soils under the action of horizontal loading. Mathematical modeling of stress-strain state for landslide slopes with use of retaining structures as short piles under the action of seismic loads and other factors triggering landsliding (it corresponds to regional features of complex soil conditions in ARCU) were further developed. A typical design section is shown in Fig. 3 of a slope for the case of a maximum difference in heights due to cutting for installation of the object under construction. The stage of a construction pit, when arranging the 2nd tier of beams for anchor fixing, is shown in Fig. 4.

2015–2016

Since 2015 "Landslide hazard zonation of Carpathian region of Ukraine using GIS" IPL 191 project has been implemented. The main goal of the project was to develop an instrument for landslide hazard forecasting to minimize the impacts of landslide activation on people and tangible objects, including construction, transportation services, pipelines etc. The objective is to determine the slopes prone

Fig. 3 Typical design section of slope

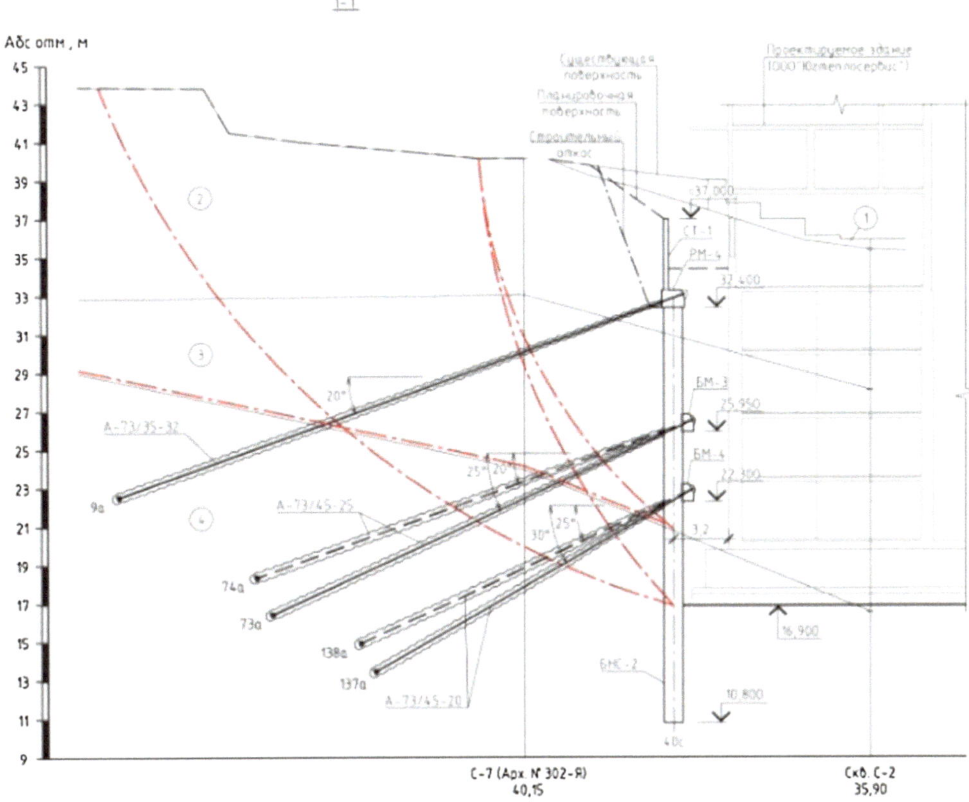

Fig. 4 Stages of construction: pit when arranging the 2nd tier of beams for anchor fixing

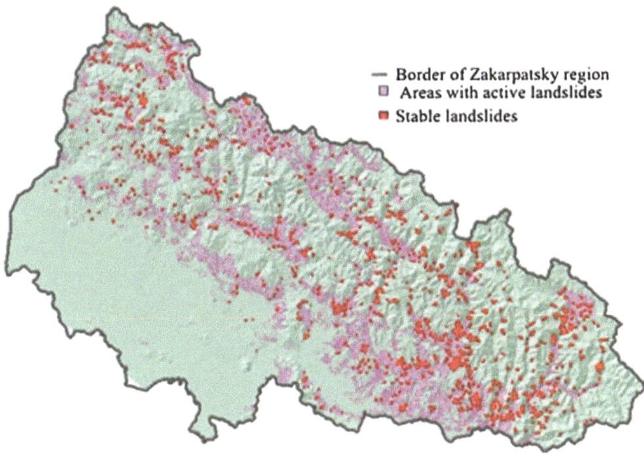

Fig. 5 Remote sensing data on the distribution of landslides sites within the Zakarpatsky region

to landslide hazards over the Carpathian region of Ukraine; to develop a database containing the engineering-geological information relevant to descriptors of landslide sites; and to develop targeted GIS on landslides in the Carpathian region of Ukraine. Preliminary results of the project were presented in the report "Hazardous Activation of Landslides within Western Carpathian Region (Ukraine)" at the WLF3 conference (Kaliukh et al. 2013a), Trofymchuk et al. (2014b) (Fig. 5) and published in the periodical press (Khavkin and Kaliukh 2014). Ph.D. K.A. Khavkin has defended a dissertation on the topic "Landslide hazard and the stress-strain state of retaining walls in seismic regions of Ukraine (on the example of Bukovina)" (supervisor is Dr Iurii Kaliukh) on the subject of the project.

The dissertation work is devoted to experimental analytical research on the stress state of retaining walls and landslide arrays in the difficult ground conditions of Bukovina under dynamic loadings. For the first time the technical and stress state of retaining walls under landslide pressure in Chernovtsy region were investigated using nondestructive methods: integrated vibration diagnostics and ultrasound. Recommendations for repair and reconstruction of retaining walls in the future during a meeting with public servants from the Chernovtsy local municipality are proposed. The elastic and elastic-plastic deformation of ground and variation of its physical and mechanical properties were considered, together with dynamic certification of the current technical state of retaining walls by non-destructive methods (Kaliukh et al. 2016), see Figs. 6 and 7).

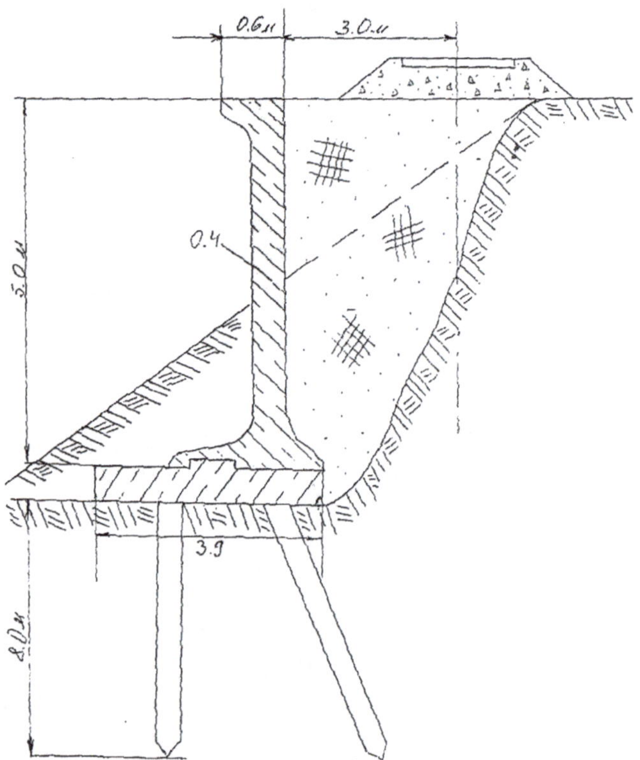

Fig. 6 Scheme of the retaining wall at the railroad haul of Zavalie-Nepolokivtsy

Training and Conferences

Normative document of the Ministry of Regional Development, Construction, and Communal Living of Ukraine. of Ukraine (Trofymchuk et al. 2013a, b, c) ("Construction in seismic regions of Ukraine: DBN V.1.1-12:2013") was introduced into Ukrainian building practice in 2014. More than 500 designers throughout Ukraine have participated in training programmes according to the new Construction Norms DBN V.1.1-12:2013. The type of accreditation is a certificate.

Capacity Development Programme Concerning DRR and Climate Change

Seven conferences were held under the supervision of the Institute of Telecommunication and Global Information Space from 2009 to 2015. One more will be held in October 2016. (Link to 2015: http://itgip.org/14-mizn-nauk-pr-konf/?lang=en [Last accessed: 14 June 2016], Fig. 8).

The journal "Environmental Safety and Natural Resources" is being published by the Institute of Telecommunication and Global Information Space four times per year. Twelve issues that cover problems of DRR and climate change have been published from October, from the beginning of 2000 to October 2016: http://itgip.org/collections_of_scientific_papers/?lang=en. [Last accessed: 14 June 2016]

Planned Future Activities

Two Ukrainian standards of construction for scientific and technical monitoring of construction objects and construction for building in the areas sensitive to landslides will be completed and put into effect in Ukrainian building practice in 2017–2018. The Ukrainian division of ICL is planning training programmes in 2017–2020 concerning these issues, based on the new Construction Norms for more than 1000 designers from all the regions of Ukraine. This will allow

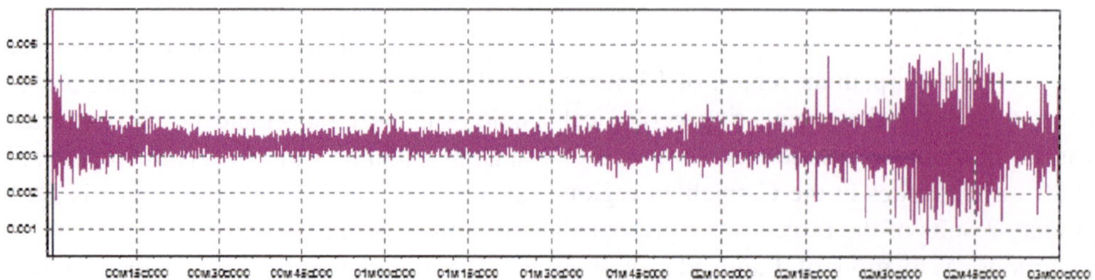

Fig. 7 Graphic of horizontal vibration accelerations of the retaining wall along the X axis at point 2 at microseismic vibrations and train passage. Vertical axis is signal amplitude; horizontal axis is time (m, c)

Fig. 8 Participants at the Conference in 2015

them to correct design monitoring systems of construction objects in landslide-prone territory, in particular, to correctly design landslide-protected buildings in areas sensitive to landslides. At the moment Dr Olexander Trofymchuk is the supervisor of Ph.D. student Iaroslav Berchun, whose Ph.D. is devoted to the study of landslide initiation mechanisms in Neogene clay, and Dr Yurii Kaliukh is the supervisor of Ph. D. student Alexander Ischenko, whose Ph.D. is devoted to the study of the stress-deformed state of retaining walls in towns.

As a member of the "Landslides and Cultural & Natural Heritage" thematic Network of the ICL, we plan the implementation of the new project in 2018–2020. It will be devoted to the study of a single heritage site placed on an active landslide system, in isolation, to one that values a multidimensional, multiregional and inter-disciplinary approach.

Beneficiaries from UDICL

The main beneficiaries from UDICL are Kharkov (IPL project "Landslide hazard zonation in Kharkiv region of Ukraine using GIS"), the Autonomous Republic of the Crimea, Ukraine (IPL project 153 "Landslide protection structures and their development in the Autonomous Republic of the Crimea, Ukraine") and Chernovtsy (IPL project 191 "Landslide hazard zonation Carpathian region of Ukraine using GIS"); local authorities including Kharkov, ARCU and Chernovtsy Regional State Administration; Kharkov and Chernovtsy City State Administration; Kharkov, ARCU and Chernovtsy District State Administrations. Ministry of Environmental Protection, its Kharkov, ARCU and Chernovtsy Regional Branches; ITGIS of NASU and NDIB; Environmental NGOs and. finally, ICL/IPL-GPC.

Acknowledgements Results presented herein have been obtained with the financial support from private consulting firm "Center of Science-Engineering Service", Yalta, Crimea, Ukraine (owner Michail Rijii). National Academy of Science of Ukraine, the Institute of Telecommunications and Global Information Space has been financing UDICL in years 2009–2016. These supports are much appreciated. We would like to express our appreciation to the Ukrainian scientist from National Technical University "KPI" for consultations during the planning and development of the monitoring systems in the Livadia Landslide System and building constructions of the Livadia Palace. Our special thanks are extended to our colleagues Dr. Iurii Gukovskii, Mr. Ruslan Litvinenko and Dr. Viktor Kochin for their support in the site and office work. We wish to thank numerous citizens and the City of Yalta for their contribution to this project, by enabling use of private and City's land for measurement stations establishment.

References

Khavkin K, Kaliukh I (2014) Theoretical and applied issues of dynamic certification of the retaining wall in earthquake-prone regions of Ukraine. J Mod Ind Civ Constr 10(1):5–14

Kaliukh I, Senatorov V, et al (2013a) Experimentally-analytical researches of the technical state of reinforce-concrete constructions for defense from landslide's pressure in seismic regions of Ukraine. In: Proceedings of the Fib Symposium. 22–24 Apr 2013, Tel-Aviv, Israel. pp 625–628

Kaliukh I, Trofymchuk O, et al (2013b) Geotechnical problems of diagnosis, monitoring, calculation and engineering protection of hazards landslides and retaining walls in earthquake-prone regions of Ukraine. Experience of the state enterprise state research institute of building structures for the last 5 years. J Svit Geotech 4(40):25–35

Kaliukh I, Silchenko K, et al (2014) Trench strengthening in the restrained conditions of urban development with allowance for the magnitude 8 seismic loads. In: Proceedings of the XV Danube-European Conference on Geotechnical Engineering, 9–11 Sept, Vienna, Austria, pp 535–540

Kaliukh I, Trofymchuk O, et al (2015) Arrangement of deep foundation pit in restricted conditions of city build-up in landslide territory with considering of seismic loads of 8 points. In: Proceedings XVI ECSMGE, 13th–17th Sept 2015, Edinburgh, Great Britain, pp 535–540

Kaliukh I, Farenyuk G, et al (2016) Experimental and theoretical assessment of structural health of existing reinforced concrete retaining walls under low frequency dynamic loading. In: Proceedings of the Fib Symposium, 21–23 Nov 2016, Cape Town, South Africa (in publication)

Newspaper "Segodniya" (2014) Historic Sites are threatened large-scale landslides in Kiev. url:http://kiev.segodnya.ua/kommunalka/istoricheskim-mestam-kieva-ugrozhayut-masshtabnye-opolzni-512593.html. Last Accessed: 7 Jun 2016

Trofymchuk O, Kaliukh I (2012) The numerical-statistical approach for hazard prediction of landslides and its application in Ukraine. In: Proceedings of the European Geosciences Union General Assembly, poster in session NH3.2. 22–27 Apr 2012. Vienna, Austria. Abstract № EGU 2012-3387

Trofymchuk O, Kaliukh I (2013) Activation of landslides in the south of Ukraine under the action of natural seismic impacts (experimental and analytical studies). J Environ Sci Eng 2(2):68–76

Trofymchuk O, Kaliukh I et al (2013a) Construction in seismic regions of Ukraine: DBN V.1.1-12:2013 (introduced into building practice 2014-10-01). Normative document of the Minregionbud of Ukraine, 118 p

Trofymchuk O, Kaliukh I, et al (2013b) Experimental and analytical studies of landslides in the south of Ukraine under the action of natural seismic impacts. In: Proceedings of the international symposium on earthquake-induced landslides, Kiryu, Japan. Springer, Berlin, pp 883–890

Trofymchuk O, Kaliukh I, et al (2013c) Mathematical and GIS-modelling of landslides in Kharkov region of Ukraine. In: Proceedings of WLF2. Landslide science and practice. Volume 3. Spatial analysis and modelling. Springer, Berlin, pp 347–352

Trofymchuk O, Kaliukh I, Glebchuc A, et al (2013d) Modelling of landslide hazards in Kharkov Region of Ukraine using GIS. Landslides: global risk preparedness. Springer, Berlin, pp 273–283

Trofymchuk O, Kaliukh I, et al (2014a) Use accelerogram of real earthquakes in the evaluation of the stress-strain state of landslide slopes in seismically active regions of Ukraine. In: Proceedings of the Engineering Geology for Society and Territory—Volume 2. Springer, Berlin, pp 1343–1346

Trofymchuk O, Yakovlev E, et al (2014b) Hazardous activation of landslides within Western Carpathian Region (Ukraine). In: Proceedings of the WLF3. Landslide science for a safer geoenvironment. Volume 2. Methods of landslide studies. Springer, Berlin, pp. 533–536

Trofymchuk O, Kaliukh I, et al (2017) Landslide stabilization in building practice: methodology and case study from Autonomic Republic of Crimea. In: Proceedings of the WLF4. Springer, Berlin (in publication)

Development of a Hazard Evaluation Technique for Earthquake-Induced Landslides Based on an Analytic Hierarchy Process (AHP) (IPL-154)

Daisuke Higaki, Eisaku Hamasaki, and Kazunori Hayashi

Abstract

In this study, we developed a hazard evaluation technique for earthquake-induced landslides that is based on topographical and geological factors extracted by an analytic hierarchy process (AHP). Several past earthquake cases that have caused multiple landslides in Japan were analyzed. With this method, through buffer movement analysis, we were able to obtain factor data on the respective sizes of terrain impacted by landslides and the magnitude of the landslides in the target area. In addition, we incorporated a method to provide predictive values for the evaluation through blunder probability analysis. The area distribution of the coherent landslides following the Mid-Niigata Prefecture Earthquake in 2004 corresponded well with the high-scoring areas derived by our evaluation model. This paper presents the results of the IPL project (IPL-154) titled "Development of a methodology for risk assessment of the earthquake-induced landslides".

Keywords

Earthquake-induced landslide • AHP (analytic hierarchy process) • Hazard • Landslide susceptibility map

Background and Objectives

Landslides are a type of natural disaster that can be triggered by earthquakes, and these slides occur along destabilized slope areas. In recent years, researchers have analyzed the topography and geology, as well as the conditions related to seismic motion, in several areas where earthquake-induced landslides have occured (e.g., Keefer 1984; Rodríguez et al. 1999; Yagi et al. 2009). Earthquakes are frequent events in Japan, and The Japan Landslide Society has developed a method for earthquake-induced landslide hazard zoning based on local topographic and geological conditions (Higaki et al. 2015).

This study was aimed at the development of a method for hazard zoning roughly at a scale of a 1:50,000 topographic map, rather than seismic risk evaluations at the scale of individual slopes. An evaluation at such a scale can be utilized as the basis for forecasting seismic damage in a given area, or for creating disaster mitigation plans. In this paper, we provide an overview of the proposed method, as well as the results yielded from the application of this method to landslides caused by the Mid Niigata Prefecture Earthquake in 2004.

D. Higaki (✉)
Hirosaki University, Faculty of Agriculture and Life Science, Bunkyo-Cho 3, Hirosaki, 036-8561, Japan
e-mail: dhigaki@hirosaki-u.ac.jp

E. Hamasaki
Advantechnology Co., Ltd. Aoba-Ku Kakyoin 1-4-8, Sendai, 980-0013, Japan
e-mail: hamasaki@advantechnology.co.jp

K. Hayashi
Okuyama Boring Co., Ltd. Aoba-Ku Futsukamachi 13-18 , Sendai, 980-0802, Japan
e-mail: k.hayashi@okuyama.co.jp

Methodology of Hazard Zonation

In recent years, much research has been conducted on evaluation methods for identifying landslide-prone hazardous areas. This work typically involves performing

K. Sassa et al. (eds.), *Advancing Culture of Living with Landslides*,
DOI 10.1007/978-3-319-59469-9_34

statistical analyses on the geographic characteristics of landslide-prone areas and displaying the data with a geographic information system (GIS). The analytic hierarchy process (AHP) has been frequently used for landslide hazard mapping, as it can incorporate both quantitative and qualitative factors as part of the evaluation; then, expert opinions are used to score the relative importance of those factors (e.g., Kamp et al. 2008; Yalcin et al. 2011; Miyagi et al. 2014).

The Japan Landslide Society has summarized the characteristics of many landslides caused by past earthquakes in Japan (The Japan Landslide Society 2012), and the distribution of landslides caused by the 2011 Great East Japan Earthquake were recently added to this database (The Project team for Collaborative Research and Development of River and Erosion Control, Japan Landslide Society 2013); by referencing prior studies for relevant characteristics and upon brainstorming among experts. This work has revealed the topographical and geological conditions that are most prone to disruptions that will initiate landslides following earthquakes (Figs. 1 and 2). The key findings can be summarized as follows: shallow disrupted landslides are particularly likely to occur during earthquakes; coherent landslides occur more often in geological terrain from or after the Neogene period; large-scale disrupted landslides occur more often in geological terrain from or before the Paleogene period and mainly in accretionary prisms. In addition, large-scale disrupted slides and shallow disrupted landslides have occurred frequently in the hills covered in volcanic ash deposits from the Quaternary period (Sugimoto et al. 2012).

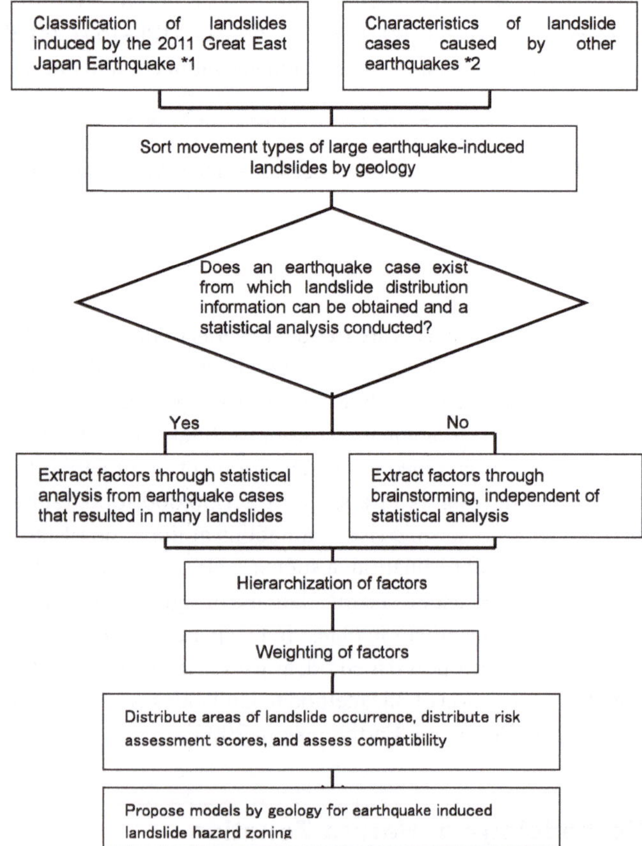

Fig. 1 Research flow chart (Higaki et al. 2015). *1 The Project team for Collaborative Research and Development of River and Erosion Control, Japan Landslide Society (2013), *2 The Japan Landslide Society (2012)

In this study, we conducted GIS statistical analyses with respect to topographical and geological factors on the following earthquakes that caused multiple landslides: the Tokachi-Oki Earthquake in 1968, Izu-Ohshima Kinkai Earthquake in 1978, Western Nagano Prefecture Earthquake in 1984, the Mid Niigata Prefecture Earthquake in 2004, and the Iwate–Miyagi Nairiku Earthquake in 2008. On the basis of these results, we constructed a hazard zoning model by using the AHP (Fig. 3). Next, we evaluated the factors contributing to the spatial spread based on a scale that matched the area of landslide occurrence or the degree of the slopes.

We also incorporated buffer movement analysis (Hamasaki et al. 2015) in order to create GIS data that takes scale into account. In addition, in reality, many landslides occur in places with high evaluation scores; thus, to increase the predictive values of the evaluation, scores should be allocated by factors to create a wider gap in evaluation points between slopes where landslides do and do not occur. To achieve this, we incorporated blunder probability analysis (Hamasaki et al. 2015; Hayashi et al. 2015).

Factor Data Creation and Weighting of Factors—a Sample Study of the Deep-Seated Coherent Landslides Caused by the Mid Niigata Prefecture Earthquake in 2004

Distribution of Landslides and Obtaining Factor Data

In this section, we discuss the case study of the Mid Niigata Prefecture Earthquake in 2004. Following this earthquake, the distributions of shallow disrupted landslides and coherent landslides were characterized (Yagi et al. 2009). By referencing past studies, we identified the following seven factor categories with regard to coherent landslides: (1) slide potential, (2) ease of shaking during an earthquake, (3) erosion potential at the toe of slopes, (4) water collectability, (5) geology, (6) reduction in ground strength, and (7) geological structure (Hayashi et al. 2015). To calculate the topographical sizes, 10 m mesh elevation data were used

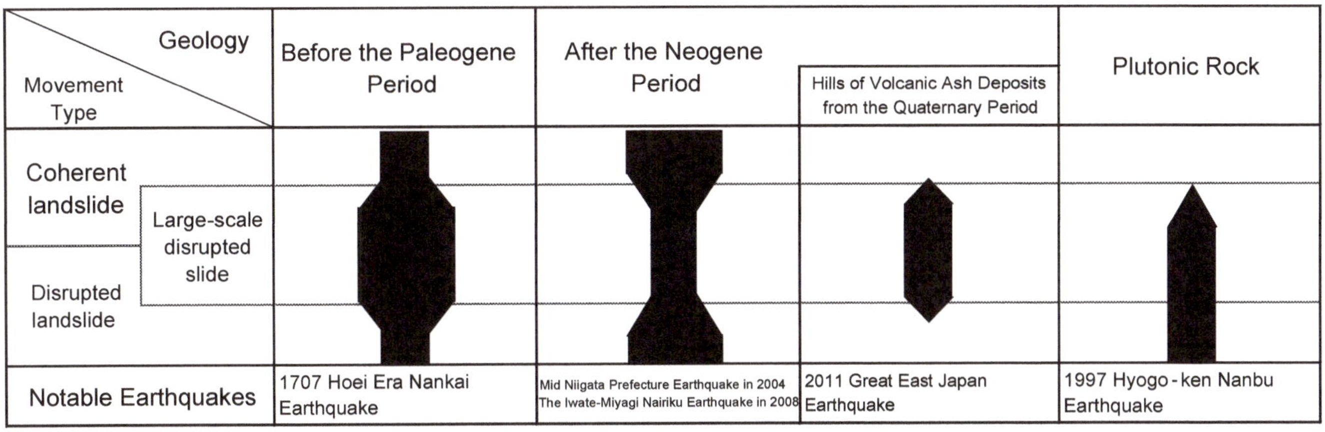

Movement Type \ Geology	Before the Paleogene Period	After the Neogene Period	Hills of Volcanic Ash Deposits from the Quaternary Period	Plutonic Rock
Coherent landslide				
Large-scale disrupted slide				
Disrupted landslide				
Notable Earthquakes	1707 Hoei Era Nankai Earthquake	Mid Niigata Prefecture Earthquake in 2004 The Iwate-Miyagi Nairiku Earthquake in 2008	2011 Great East Japan Earthquake	1997 Hyogo-ken Nanbu Earthquake

Fig. 2 Types of landslides that have occurred following past earthquakes in Japan, and geological regions prone to each (Higaki et al. 2015)

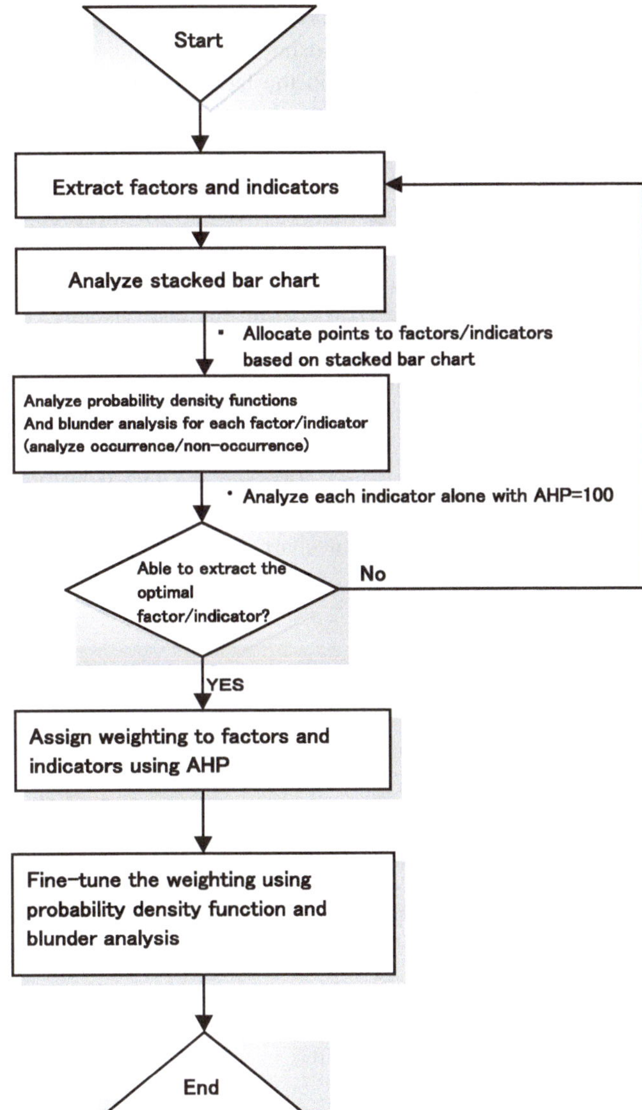

Fig. 3 Flow chart for developing the methodology for earthquake-induced landslide hazard zoning using AHP

(Map of Hokkaido, GIS MAP Terrain; hereinafter referred to as DEM). We used slope in factor (1). In factor (2), since convex slopes are prone to amplify seismic motion, we used the convex–concave index. Here, the convex–concave index (Hamasaki et al. 2015) refers to the value obtained by dividing by distance d, the absolute value of the differences between the central altitude and the eight mean altitudes located a specified distance “d” in eight directions from the target mesh. This value indicates the degree of relief in the periphery of the target mesh. In factor (3), we used overground-openness as an indicator, since sloped surfaces become unstable if the slope edges are susceptible to erosion. Overground-openness (Yokoyama et al. 1999; Fig. 4) was calculated as the average of the values taken from each of the eight directions as follows: at a given point located on the terrain cross section within a specified distance (L) of the target mesh, subtract the maximum value of the angle of elevation from the vertical 90° upward angle. The value becomes smaller as it moves to deep valleys with advanced erosion.

Hydrological conditions during earthquakes are important factors to consider when evaluating landslide occurrence

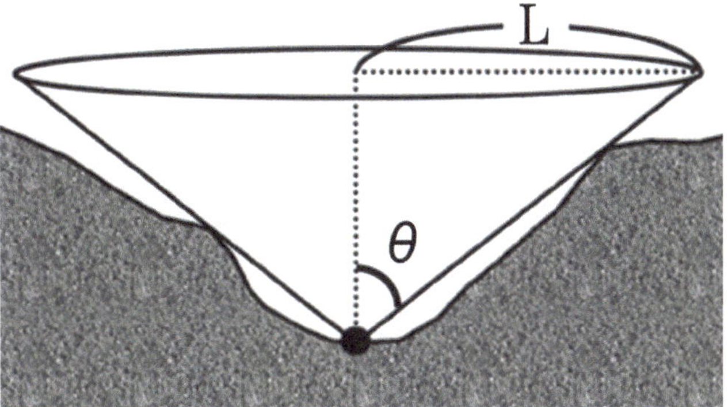

Fig. 4 The measurement of overground-openness in a given location (Yokoyama et al. 1999). Set L and solve for θ as shown in the figure

(Chigira et al. 2012). Thus, in factor (4), we calculated the Topographic Wetness Index, which reflects the topographical water collectability (Beven and Kirkby 1979).

Slope material is also an important factor for landslide occurrence. In factor (5), we used seven lithological classifications to characterize the slope material; this scheme roughly categorizes the geology of Japan by the sliding resistance force (Hamasaki et al. 2015). We re-classified the geological portion of the 1:50,000 numerical geologic map (Takeuchi et al. 2004) and converted it to data for the 10 m mesh. As many coherent landslides from the Mid Niigata Prefecture Earthquake in 2004 occurred within landslide bodies that had less strength compared to bedrock (Has et al. 2012), we divided the target mesh into those points located in the relevant landslide body and those that were not, and this information was used as the indicator for (6). In (7), due to the fact that sedimentary rocks developed in the bedding planes resulted in numerous large-scale coherent landslides along slopes with a dip slope structure (Hayashi et al. 2015), and that numerous shallow disrupted slides occurred in anti-dip slopes (Yagi et al. 2005), we calculated the difference between the tilt directions of the slope and stratum (β) and the visual tilt angle of the strata in the tilt direction of the slope (γ). These were divided and used in the 10 m mesh.

The data for the GIS statistical analysis were collected by deriving the landslide area ratio as the ratio of mesh including landslides within each buffer to the in-buffer mesh, while moving the circular buffer of a fixed search radius (R) (circle shown in Fig. 5) sequentially in one direction. At the same time, we obtained the above-mentioned topographical and geological factor data in each buffer. Then, while keeping the buffer at a constant distance and moving from each in one direction so as to not create gaps, we gathered data in the target area. The above-described method is referred to as buffer movement analysis (Hamasaki et al. 2015). At this point, for the search radius (R), we set the buffer size so that it contained the majority of the area where coherent landslides or shallow disrupted landslides have occurred. In the case of coherent landslides, we set R = 250 m.

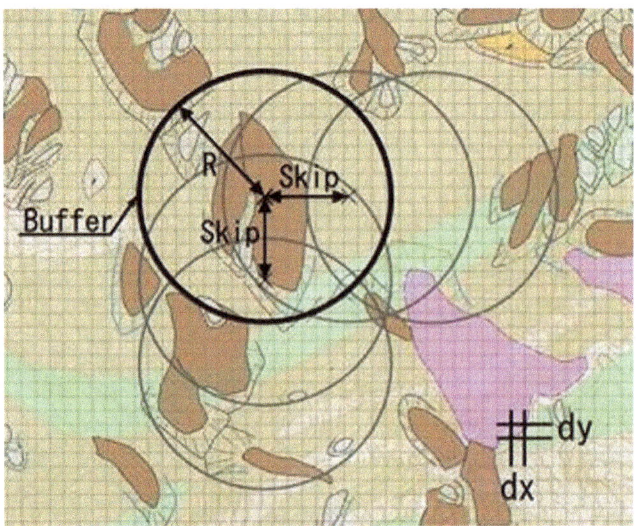

Fig. 5 Methodology for buffer movement analysis (Hamasaki et al. 2015). The brown colored area represents the landslide occurrence area and the background is the distribution of the numerical topographical/geological data

Weighting the Factors

Next, we sorted the data gathered in the buffer movement analysis into occurrence data, which included both landslide occurrence data and nonoccurrence data. Taking as an example the factors that can be expressed by rank, it can be said that a particular rank is closely connected to the landslide occurrence if a larger landslide area ratio increases the ratio of that certain rank within the factor. Thus, we determined the total buffer count for each landslide area ratio for a given factor. If the ratio of a given rank within the buffer count increased as the landslide area ratio increased, we gave a higher weighting to that rank. Conversely, if the ratio of the rank decreased with a larger collapse area ratio, the rank was considered to contribute less to the collapse and was assigned a smaller score. To perform this task visually, we created a stacked bar chart as displayed in Fig. 6. Using Fig. 6, we will walk through the example of the slopes where landslides occurred. The landslide area ratio was divided into six levels, five of which were broken up into 10% intervals between 0 and 40% and the last level consisted of 40%+ data. If we view the ratio of the slope rank for each 5° in the stacked bar chart, we can determine that

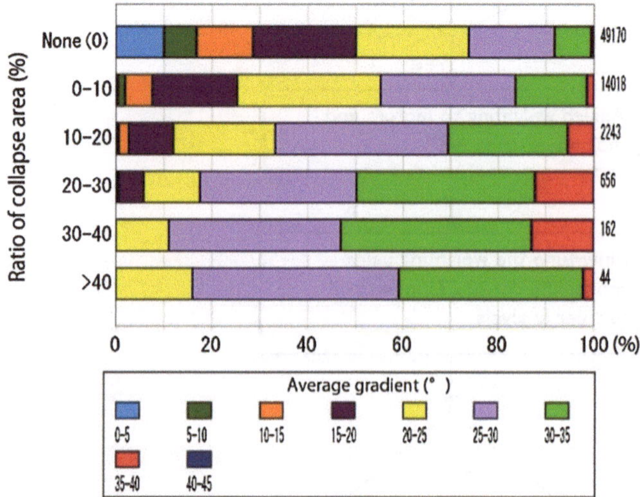

Fig. 6 Stacked bar chart showing changes in occupancy rate by slope rank and landslide area ratio (the numbers to the right of the bar chart are the total number of buffers by landslide area ratio)

Mid Niigata Prefecture Earthquake in 2004— Coherent landslide

Hierarchy level II – Predisposition 3 items

	Geology	Topography	Groundwater	Geometric Mean	Weight a
Geology	1.0	1.0	7.0	1.913	0.47
Topography	1.0	1.0	7.0	1.913	0.47
Groundwater	0.14	0.14	1.0	0.273	0.07
			sum=	4.099	1.00

Hierarchy level III – Geology 2 items

	Geology (Lithological)	Landslide Clod	Geometric Mean	Weight b	Weight a×b	Initial Model
Geology (Lithological)	1.0	2.0	1.414	0.67	0.31	31.1
Landslide Clod	0.5	1.0	0.707	0.33	0.16	15.6
		sum=	2.121	1.00		

Hierarchy level III – Topography 2 items

	Convex-Concave Index	Overground-Openness	Geometric Mean	Weight b	Weight a×b	Initial Model
Convex-Concave Index	1.0	2.0	1.414	0.67	0.31	31.1
Overground-Openness	0.5	1.0	0.707	0.33	0.16	15.6
		sum=	2.121	1.00		

Hierarchy level III – Groundwater 2 items

	Stream Order	Geometric Mean	Weight b	Weight a×b	Initial Model
Stream Order	1.0	1.000	1.00	0.07	6.7
	sum=	1.000	1.00		
			sum=	1.00	100.00

Fig. 7 AHP hierarchy and weighting of topographical, geological, and groundwater factors in a paired comparison

collapses are more likely to occur in slopes of 30° or more. Between 25° and 30°, we observed a slight upward trend in the buffer count, but conversely between 15° and 25°, there was a downward trend. Below 10°, the buffer count decreased as the landslide area ratio increased, so the score allocation was small. Thus, we allocated the following weights to the data: 1.0 for 30° or more, 0.6 to 25°–30°, 0.3 to 15°–25°, and 0.1 to 15° or less.

The factor data described above are values that can be numerically rank classified. Qualitative factors such as the geological classification were represented by the mode of the mesh counts in the buffer. Additionally, we binarized whether the buffers include or not include the body of old coherent landslides.

On the other hand, the relative weighting between each factor was determined by a paired comparison and through brainstorming work (Fig. 7). To determine the first-level factors most responsible for causing coherent landslides induced by earthquakes, we selected all the geological, topographical, and groundwater factors and assigned weightings for each. Since the coherent landslides were caused by earthquakes rather than rainfall, the groundwater weighting was assigned at 1/7 of the other two factors. In addition, for geological factors, it was shown that at one rank lower in the factor hierarchy, lithological factors (rock type) were the most important, while the presence of a body of the coherent landslide contributed to secondary landslides. For this reason, it was assigned 1/2 in the weighting. Among the topographical factors were the convex–concave index and overground-openness. The groundwater was assumed to be related to the stream order for a given location.

Through the hierarchical structure and process of factor weighting described above, AHP evaluation scores were calculated for any location where relevant data could be obtained.

Study of the Optimal Model Based on Blunder Probability

In order to achieve the optimal model, we evaluated the degree to which the factors were relevant to coherent landslides or shallow disrupted slides using blunder probability analysis (Hamasaki et al. 2015). First, we sorted the frequency distribution of the evaluation scores within the target range into landslide occurrence and nonoccurrence data and compared the two after approximating their frequency distribution to the normal distribution (Fig. 8). After determining the mean score, we set the mean score of each as the threshold. We then determined the proportion of

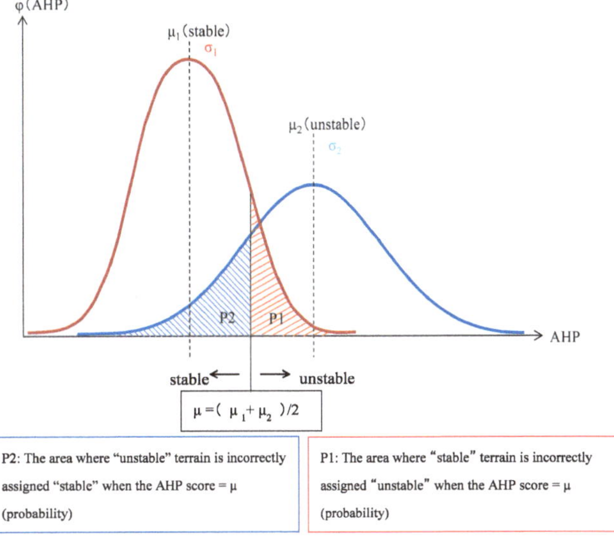

Fig. 8 Approximation using the normal probability density function and compatibility assessment using the blunder probability (P) (Hamasaki et al. 2015)

Fig. 9 AHP score distribution in the target area during Mid Niigata Prefecture Earthquake in 2004 and score distribution in occurrence/nonoccurrence areas for deep-seated coherent slides

Item	Geology (Lithological)	Landslide Body	Convex–Concave Index	Overground -Openness	Stream Order	
Weight	30	20	25	15	10	
Nonoccurrence			Occurrence			Blunder probability (P)
μ1	σ1	p1	μ2	σ2	p2	(p1+p2)/2
60.9	25.5	0.378	76.7	22.4	0.362	0.370

μ: Mean σ: SD p: probability

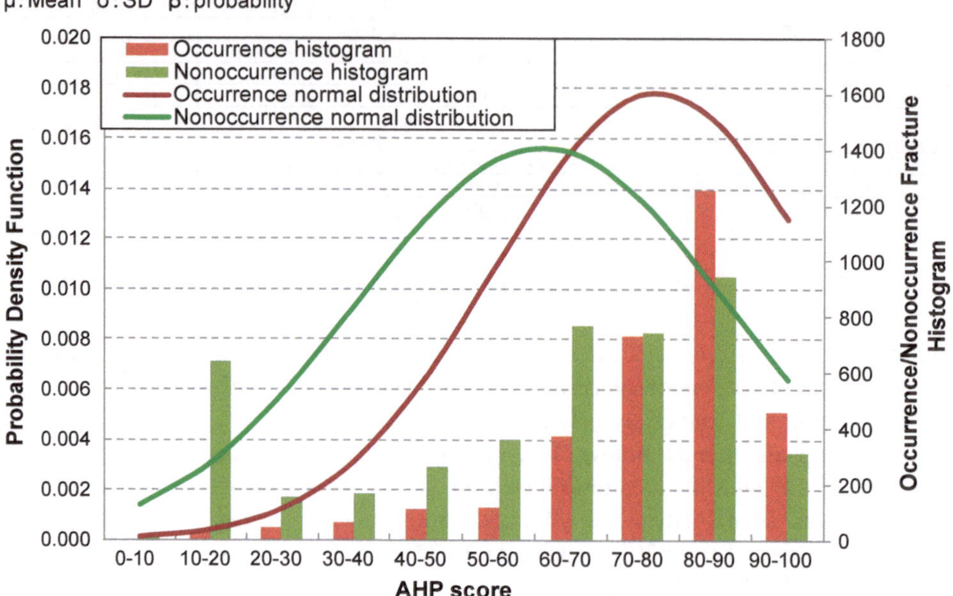

nonoccurrence data that fell higher than that score (P1) and the proportion of occurrence data that fell lower (P2). The blunder probability P was defined as the average of P1 and P2. It can be said that a smaller P signifies a good model that can better separate occurrence and non-occurrence data.

We evaluated the coherent landslides caused by the 2004 Niigata–Chuetsu Earthquake by adjusting the allotted points to the five factors shown in Fig. 9. We were able to slightly reduce P in comparison to the initial model by allocating scores as shown in the table in Fig. 9.

Applicability of AHP Risk Evaluation Results

In the final evaluation model described above, we viewed the distribution of AHP scores and occurrences of deep-seated slides during the Mid Niigata Prefecture Earthquake in 2004 (Fig. 10). We were able to see that the Imo River basin area, where large-scale coherent landslides had been re-activated and landslide dams were formed, had a concentration of high scores that were 80 points or more (Fig. 10).

Conclusions

On the basis of data from past studies that investigated multiple landslide events following large earthquakes in Japan, and through the application of GIS statistical analyses, we have proposed a method for earthquake-induced landslide hazard zoning that employs an AHP. In this study, we used the 2004 Mid Niigata Prefecture Earthquake in 2004 as an example to evaluate the proposed method. The distribution data for coherent landslides were used to conduct a factor analysis of the key features related to landslide occurrence, and data

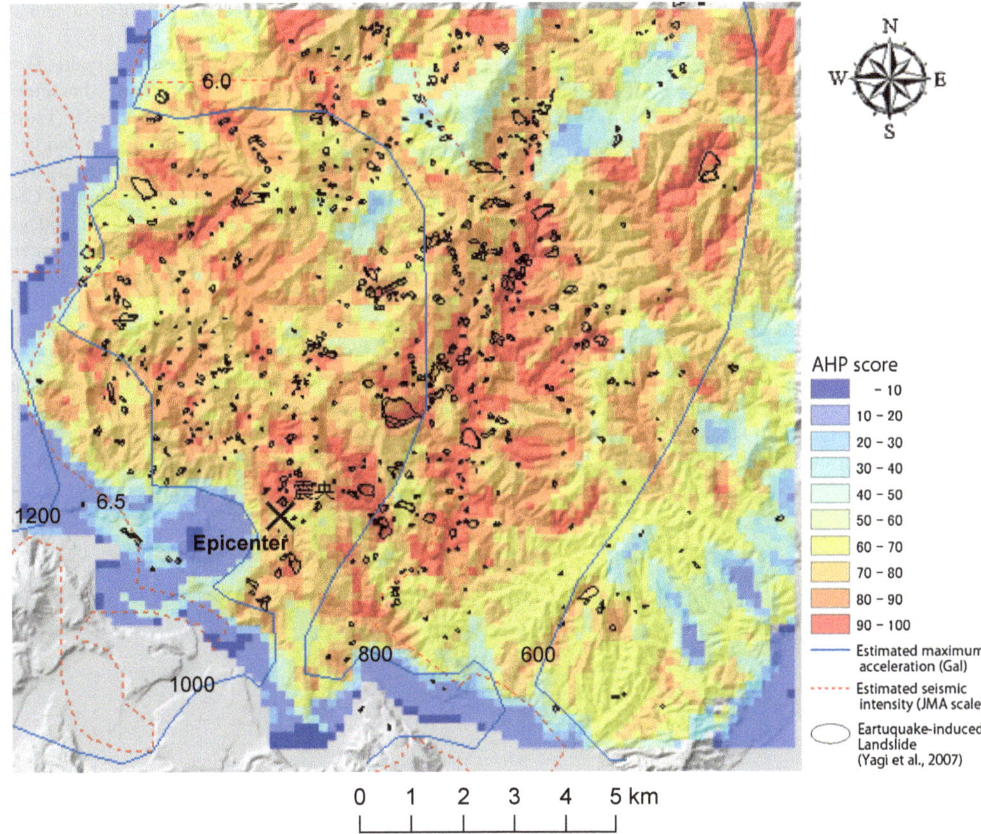

Fig. 10 Distribution of AHP scores and landslides in and around the Imo River basin following Mid Niigata Prefecture Earthquake in 2004

acquisition was achieved through buffer movement analysis. In addition, we improved the evaluation model through blunder probability analysis. As a result of the verification, the methodology described above was shown to be effective as a topographical and geological approach to identify earthquake-induced landslide hazard areas.

Acknowledgements This research was part of the following projects: "Research and Development Efforts for River and Erosion Control" at the Ministry of Land, Infrastructure, and Transport; "Create a Methodology to Evaluate Earthquake-Induced Landslide Hazard Areas based on Typology," which was conducted during the 2011–2013 fiscal year. These initiatives were commissioned by the National Institute for Land and Infrastructure Management (NILIM) through the Japan Landslide Society. We express deep gratitude to the supporters of the relevant organizations.

References

Beven KJ, Kirkby MJ (1979) A physically based, variable contributing area model of basin hydrology. Hydrol Sci Bull 24(1):43–69

Chigira M, Nakasuji A, Fujiwara S, Sakagami M (2012) Catastrophic landslides of pyroclastics induced by the 2011 Off the Pacific Coast of Tohoku Earthquake. Earthquake-Induced Landslides: Proceedings of the International Symposium on Earthquake-Induced Landslides. Kiryu, Japan. pp 139–147

Hamasaki E, Higaki D, Hayashi K (2015) Buffer movement analysis and blunder probability analysis for GIS-based landslide susceptibility mapping–A case study of the 2008 Iwate-Miyagi Nairiku Earthquake, Japan. J Japan Landslide Soc 52(2):51–59 (in Japanese with English abstract)

Has B, Maruyama K, Noro T, Nakamura A (2012) An approach of susceptibility analysis for deep-seated landslide induced by earthquake within pre-existing landslide topography using logistic regression. J Japan Landslide Soc 49(1):12–21 (in Japanese with English abstract)

Hayashi K, Hamasaki E, Yagi H, Higaki D (2015) Development of landslide susceptibility mapping model for earthquake-induced landslides based on buffer movement analysis and blunder probability analysis. J Japan Landslide Soc 52(2):60–66 (in Japanese with English abstract)

Higaki D, Hayashi K, Hamasaki E, The Project team of Collaborative Research for River and Erosion Control, Japan Landslide Society, Kanbara J (2015) A report on the collaborative research for river and erosion control by the Japan landslide society and ministry of land, infrastructure, transportation and tourism–Susceptibility mapping of earthquake-induced landslides. J Japan Landslide Soc 52 (2):85–92 (in Japanese)

Kamp U, Growley BJ, Khattak GA, Owen LA (2008) GIS-based landslide susceptibility mapping for the 2005 Kashmir earthquake region. Geomorphol 101(4):631–642

Keefer DK (1984) Landslides caused by earthquakes. Geol Soc Am Bull 95:406–421

Miyagi T, Prasad G B, Tanavud C, Potichan A, Hamasaki E (2014) Landslide risk evaluation and mapping-Manual of aerial photo interpretation for landslide topography and risk management. Reports of the National Research Institute for Earth Science and Disaster Prevention, No 66, pp 75–136

Rodríguez CE, Bommer JJ, Chandler RJ (1999) Earthquake-induced landslides: 1980–1987. Soil Dynamics and Earthquake Engineering 18(5):325–346

Sugimoto H, Takeshi T, Uto T, Takeshi T, Honma H (2012) Geomorphologic and geologic features of landslides induced by the 2011 Off the Pacific Coast of Tohoku Earthquake, in Shirakawa Hills, Fukushima Prefecture. Earthquake-Induced Landslides: Proceedings of the International Symposium on Earthquake-Induced Landslides. Kiryu, Japan. pp 189–201

Takeuchi K, Yanagisawa Y, Miyazaki J, Ozaki M (2004) 1:50,000 Digital Geological Map of the Uonuma region, Niigata Prefecture (Ver. 1), GSJ Open-file Report, No. 412. https://www.gsj.jp/data/openfile/no0412/index.html

The Japan Landslide Society (2012) Earthquake-induced landslides, p 302 (in Japanese)

The Project team for Collaborative Research and Development of River and Erosion Control, Japan Landslide Society (2013) Characteristics and their classification of landslides induced by the 2011 Great East Japan Earthquake. J Japan Landslide Soc 50(2):25–30 (in Japanese)

Yagi H, Sato G, Higaki D, Yamamoto M, Yamasaki T (2009) Distribution and characteristics of landslides induced by the Iwate-Miyagi Nairiku Earthquake in 2008 in Tohoku District, Northeast Japan. Landslides 6:335–344

Yagi H, Yamasaki T, Moriiwa T (2005) Characteristics of landslides triggered by the Niigata Chuetsu earthquake in 2004 with special reference to geology and geomorphology. Proceed Intern'l Symp Landslide Hazard in Orogenic Zone drom the Himalaya to Island Arc in Asia, pp. 407–416, Japan landslide Society

Yalcin A, Reis S, Aydinoglu AC, Yomralioglu T (2011) A GIS-based comparative study of frequency ratio, analytical hierarchy process, bivariate statistics and logistics regression methods for landslide susceptibility mapping in Trabzon. NE Turkey Catena 85(3):274–287

Yokoyama R, Shirasawa M, Kikuchi Y (1999) Representation of topographical features by opennesses. J Japan Soc Photogramm Remote Sens 38(4):26–34 (in Japanese with English abstract)

The Croatian-Japanese SATREPS Joint Research Project on Landslides (IPL-161)

Željko Arbanas, Snježana Mihalić Arbanas, Kyoji Sassa, Hideaki Marui, Hiroshi Fukuoka, Martin Krkač, Martina Vivoda Prodan, Sanja Bernat Gazibara, and Petra Đomlija

Abstract

The Croatian-Japanese joint research Science and Technology Research Partnership for Sustainable Development (SATREPS) project 'Risk Identification and Land-Use Planning for Disaster Mitigation of Landslides and Floods in Croatia' was performed from 2009 to 2014. Key objectives of the project were landslides and floods hazard analysis and the development of guidelines for use in urban planning. This project is also designated as on-going IPL project 161. The aims of the working groups dealing with landslides were to establish a methodology of comprehensive real time monitoring at two most important landslides in Croatia based on the results of previous investigations and new in situ and laboratory testing and behavior analysis; laboratory soil testing and numerical modelling of static and dynamic landslide behavior; development of landslide inventories using direct sensing and remote sensing techniques followed by the development of methodologies of landslide hazard analysis and zonation in three pilot areas in Croatia. In this paper we will present the most important achievements of working groups related to landslide studies at the project pilot areas: two in Primorsko-Goranska County (the Rječina River Basin and the Dubračina River Basin) and one in the City of Zagreb (a hilly area of Medvednica Mt.). The identification and mapping of existing landslides in the hilly area of Medvednica Mt., Dubračina River Basin and Rječina River Basin so as establishment and results of the monitoring systems installed on the Grohovo Landslide and the Kostanjek Landslide will be described.

Keywords

Landslide • Monitoring • Early warning system • Hazard zonation • Risk mitigation

Ž. Arbanas (✉) · M. Vivoda Prodan · P. Đomlija
Faculty of Civil Engineering, University of Rijeka, Radmile Matejčić 3, 51000 Rijeka, Croatia
e-mail: zeljko.arbanas@gradri.uniri.hr

M. Vivoda Prodan
e-mail: martina.vivoda@gradri.uniri.hr

P. Đomlija
e-mail: petra.domlija@gradri.uniri.hr

S. Mihalić Arbanas · M. Krkač · S. Bernat Gazibara
Faculty of Mining, Geology and Petroleum Engineering, University of Zagreb, Pierottijeva 6, 10000 Zagreb, Croatia
e-mail: snjezana.mihalic@rgn.hr

M. Krkač
e-mail: martin.krkac@rgn.hr

S. Bernat Gazibara
e-mail: sanja.bernat@rgn.hr

K. Sassa
International Consortium on Landslides, Tanaka Asukai-Cho 138-1, Kyoto, 606-8226, Japan
e-mail: sassa@iclhq.org

H. Marui · H. Fukuoka
Niigata University, Research Institute for Natural Hazards and Disaster Recovery, 8050Igarashi, Nino-Cho, Niigata 950-2181, Japan
e-mail: maruihi@cc.niigata-u.ac.jp

H. Fukuoka
e-mail: fukuoka@cc.niigata-u.ac.jp

K. Sassa et al. (eds.), *Advancing Culture of Living with Landslides*,
DOI 10.1007/978-3-319-59469-9_35

Introduction

The Croatian-Japanese Joint Research Project entitled "Risk Identification and Land-Use Planning for Disaster Mitigation of Landslides and Floods in Croatia" was launched in 2008 when it was selected as one of the initial projects of the Science and Technology Research Partnership for Sustainable Development (SATREPS) programme. The project was approved as an ongoing IPL project (IPL-161) by the IPL Global Promotion Committee in 2009 (Han et al. 2017). SATREPS is a Japanese government program that promotes international joint research. The program is structured as a collaboration between the Japan Science and Technology Agency (JST), which provides competitive research funds for science and technology projects, and the Japan International Cooperation Agency (JICA). The SATREPS program enables joint research of Japanese and Croatian researchers by financing travel of the project members between project institutions and donating equipment for implementation of project activities. The main Japanese partner institutions in the Project were Research Institute for Natural Hazards and Disaster Recovery at Niigata University, the International Consortium on Landslides (ICL), the Disaster Prevention Research Institute of Kyoto University (DPRI). Three Croatian universities, the University of Rijeka (Faculty of Civil Engineering), the University of Zagreb (Faculty of Mining, Geology and Petroleum Engineering and Faculty of Agriculture) and the University of Split (Faculty of Civil Engineering, Architecture and Geodesy) as well as the Croatian Geological Survey were Croatian partner institutions in the project. The five-year project, started in 2009 and completed in 2014, involved about 15 researchers from Japan and included collaborative research conducted in Japan and Croatia to evaluate and mitigate landslide and flood hazards and risks in Croatia. Key objectives of the project were landslides and floods hazard analysis and the development of guidelines for use in urban planning. The project aimed to contribute to sustainable development through appropriate land use in Croatia. The Project researches covered the pilot areas around the cities where the three partner universities are located, namely Zagreb, the capital of Croatia, Rijeka, a port city, and Split, whose historic center is a UNESCO World Heritage Site. Researchers from Japan and Croatia carry out the following investigations and analyses (Mihalić and Arbanas 2013): (1) identification and mapping of landslides, (2) comprehensive monitoring of landslides, (3) continuous monitoring of sediment flows, (4) testing of the physical and mechanical properties of soils and rocks, (5) modeling of landslide dynamics, (6) modeling of flood and debris flow propagation; (7) landslide susceptibility and hazard zonation, (8) establishment of early warning systems and (9) development of risk mitigation measures through the urban planning system.

The project activities are organized into three working groups: Working Group on Landslides (WG1), Working Group on Flash Floods and Debris Flows (WG2) and Working Group on Landslide Mapping (WG3). The aim of Working Group 1 on Landslides was to establish a methodology of comprehensive real time monitoring at two most important landslides in Croatia: the Grohovo Landslide in Primorsko-Goranska County and the Kostanjek Landslide in the City of Zagreb based on the results of previous investigations and new in situ and laboratory testing and behavior analysis so as a laboratory soil testing and numerical modelling of static and dynamic landslide behavior. The activities of Working Group 3 on Landslide Mapping (WG3) were aimed at the development of landslide inventories using direct sensing and remote sensing techniques followed by the development of methodologies of landslide hazard analysis and zonation in three pilot areas in Croatia: two in Primorsko-Goranska County and one in the City of Zagreb. In this paper we will present the most important achievements of working groups related to landslide studies.

Landslide Monitoring

One of the most important issues of the Working Group on Landslides (WG1) was establishing the monitoring systems ate two most important active landslides in Croatia: the Grohovo Landslide in Primorsko-Goranska County and the Kostanjek Landslide in the City of Zagreb.

By definition, landslides are characterized by movement. Knowledge of the movement magnitude and velocity distribution along the slope, are the most important data for all landslide analysis (Mihalić Arbanas and Arbanas 2014). Monitoring is required to observe the changing conditions that may lead to total failure of the slope where slope movement is occurring, where safety factors against sliding are low, or where high risk is present from a possible slope failure. Landslide movement monitoring expressed via ground surface displacements and deformation of structures (including the landslide body) related to landslides can be accomplished using different types of monitoring systems and techniques that are classified according to Savvaidis (2003) as follows: satellite and remote sensing techniques, photogrammetric techniques, geodetic or observational techniques, and geotechnical or instrumentation or physical techniques.

Usually, different types of monitoring techniques and instrumentation are used in different combinations and connected in a unique comprehensive landslide monitoring system. Following the previously described statements and principles in establishing of landslide monitoring systems, two different concepts of a system were applied to the Grohovo Landslide and Kostanjek Landslide based on previous cognition of landslide characteristics, magnitudes of motion, possibility of equipment installation, visibility of landslide area and availability of powers supply for monitoring equipment.

The Grohovo Landslide Monitoring System

The Grohovo Landslide, the largest active landslide along the Croatian part of the Adriatic coast, is located on the north-eastern slope of the Rječina Valley out of the City of Rijeka, Croatia. The last complex retrogressive landslide was reactivated in December 1996, after long dormant period and about $1.0x10^6$ m^3 moved down the slope and buried the Rječina river-bed. After initial movements, the landslide retrogressively developed up the slope, Fig. 1. Slip surfaces are considered to be on the contact of superficial deposits and flysch bedrock. After removing displaced mass from the landslide foot, further significant movements stopped, and monitoring indicated that only small displacements developed in the upper part of the slope (Benac et al. 2005; Arbanas et al. 2012b). Comprehensive landslide remediation projects were never conducted, but further movements of the landslide were reduced (Benac et al. 2011).

The Grohovo Landslide monitoring system was designed to consist of geodetic and geotechnical monitoring. Geodetic monitoring will include geodetic surveys with a robotic total station and displacement measurements of GPS points.

The geotechnical monitoring system includes two vertical inclinometers, and four wire extensometers; 13 long- and three short-span extensometers, four pore-pressure gauges, a weather station and a rain gauge. Pore-pressure gauges, inclinometers and vertical extensometers are installed at two locations inside the central part of the landslide body at the same locations as the GPS rovers and prisms. Long-span extensometers (NetLG 501E Osasi, 13 pcs) are installed in a continuous line from the Rječina riverbed to the limestone mega-blocks at the top of the slope, while short-span extensometers (NetLG 501E Osasi, 3 pcs) are installed over the open cracks at the top of the landslide (Arbanas et al. 2014a), Fig. 2. All monitoring equipment is connected in one system with continuous monitoring and export of the data to a central computer unit located at the Faculty of Civil Engineering, University of Rijeka (Arbanas et al. 2012b).

Geodetic monitoring was established with an automatic total station measuring 25 geodetic benchmarks (prisms) and

Fig. 1 Aerial view of the Grohovo landslide reactivated in 1996; the Rječina River channel is situated in the bottom

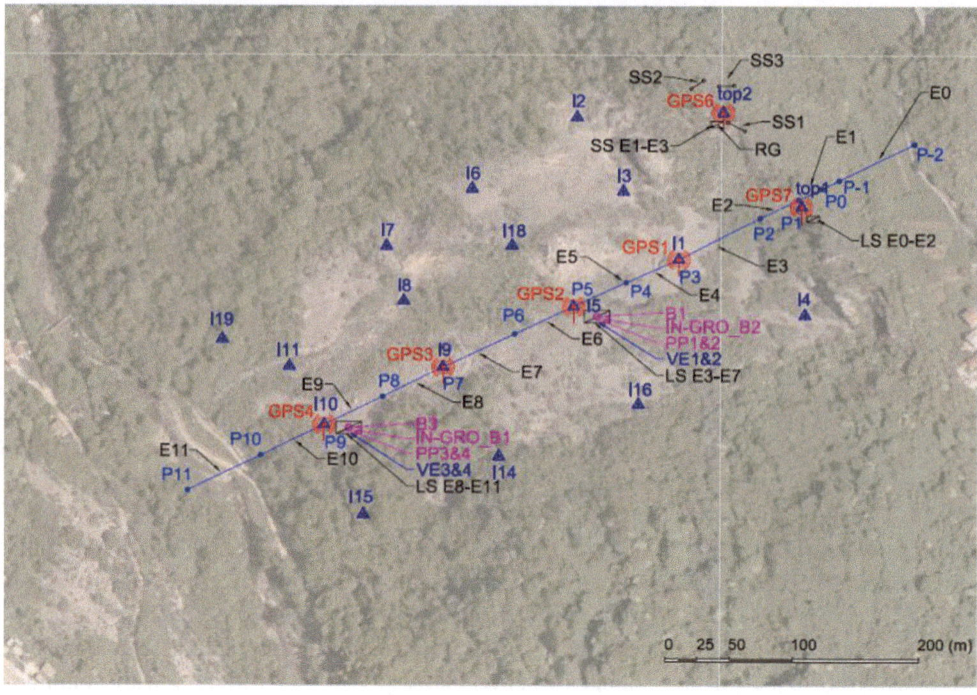

Fig. 2 Installed sensors at the Grohovo Landslide: *GPS* GPS rover; *I* prism; *E* long span extensometer wire, *P* extensometer pole; *SS* short span extensometer; *B* position of borehole; *IN* inclinometer casing in borehole; *VE* vertical extensometer in borehole; *PP* pore pressure gauge in borehole; *LS* long span extensometer data logger; *RG* rain gauge

GPS master unit with 9 GPS receivers (rovers). The robotic total station and the GPS master unit are located in a relatively stable area, on the top of the opposite slope. The master unit consists of the robotic total station *TM30 Leica*, single frequency GPS master unit *GMX901 Leica* (combined receiver and antenna), meteorological sensor, and web cam. The robotic station measures 25 benchmarks (prisms) located on the landslide body, on the top of the main scarp and, as reference points, around the landslide every 30 min (Fig. 2.). The GPS network is composed of the GPS master unit which is a reference station for 4 single frequency GPS rovers located in the landslide body, 3 single frequency GPS rovers located on top of the limestone scarp above the landslide, 1 single frequency GPS rover located on dam near landslide, and 1 reference single frequency GPS rover located on Faculty building (Arbanas et al. 2012a, b, 2014a, b). All prisms, for robotic total station measurement, were installed on the same poles with GPS receivers. Data are transmitted by Wi-Fi system from each rover to the GPS master unit connected to industrial master unit PC in which the GNSS Spider software creates measurement data files. The master unit PC is reachable remotely by an UMTS module from the main working station at the University. Instruments in the master unit and GPS rovers are powered from solar panel installations (Arbanas et al. 2012b).

Installation of the monitoring equipment started in May 2011 and establishment of the monitoring was mostly completed at the end of 2011 when the data collection started. All measurement's techniques showed on higher landslide activity in the upper part of the slope while the compression of material and heaving of the GPS rovers and prisms are outlined in the lower part of the landslide body and in the landslide foot.

A relationship between rainfall, ground water level, pore pressures and landslide movement is still very hard to establish because of numerous interruptions of in monitoring system working caused by low energy production of mini solar power plant that supply the main unit (robotic total station, main GPS unit, and site PC unit) during the winter period from end of October to early March when the landslide activity is the most expressed. The second reason is very slow landslide activity (<5 mm/year) that requires longer uninterrupted monitoring period.

The Kostanjek Landslide Monitoring System

The Kostanjek landslide is the largest active landslide in the Republic of Croatia. It is located in the western residential area of the City of Zagreb at the base of the southwestern slope of Medvednica Mt. Since its activation in 1963, this landslide has caused substantial damage to buildings and infrastructure in the residential zone, as well as to factories and commercial buildings. The landslide was mainly caused by anthropogenic factors, including mining and excavation in a marl quarry for cement production at the foot of the hills (Mihalić and Arbanas 2013).

The Kostanjek Landslide is a reactivated, translational type landslide without a clearly defined main scarp or landslide borders. According to Ortolan and Pleško (1992)

and Ortolan (1996), the Kostanjek Landslide extends over an area of approximately 1.20 km^2 with a total volume of displaced mass of 32.6×10^6 m^3. Ortolan and Pleško (1992) correlated movements on the landslide surface with soil layers in the geological profile and deduced three different slip surfaces, the deepest at 90 m and two subparallel slip surfaces at depths of 65 and 50 m. The positions of the slip surfaces were defined on the basis of unfavorably oriented bedding planes in sediments and sedimentary rocks of Sarmatian and Pannonian age.

Landslide velocities have been changing over the last 50 years, from landslide activation until today, in a range from extremely slow to very slow. According to the photo interpretation of aerial stereo pairs from 1963 to 1988, horizontal displacements of the ground surface in the period 1963–1988 were detected in a range 3–6 m (average 12–24 cm per year) (Mihalić Arbanas et al. 2014). Monitoring results of recent movements from the period 2010–2012 at the 35 stable geodetic points shows very similar movement directions to historical data (Fig. 3).

The Kostanjek Landslide monitoring system encompasses approximately 40 sensors for the monitoring of landslide movement and landslide causal factors. Figure 4 provides the layout of the sensor network installed at the Kostanjek Landslide (Krkač et al. 2014a, b, c). The monitoring system consists of various sensors for the measurement of: (1) external triggers (1 rain gauge and 7 accelerometers); (2) displacement/deformation/activity (15 GNSS sensors, 7 extensometers, 4 borehole extensometers and 1 inclinometer); (3) hydrological properties (3 pore pressure gauges and 3 water level sensors in boreholes and domestic wells, 2 water level sensors at outflow weirs).

A Trimble GNSS monitoring system consists of fifteen double-frequency NetR9 TI-2 GNSS reference stations with Zephyr Geodetic 2 GNSS antennas installed on each of GNSS reference points. GNSS receivers are fixed to 4 m

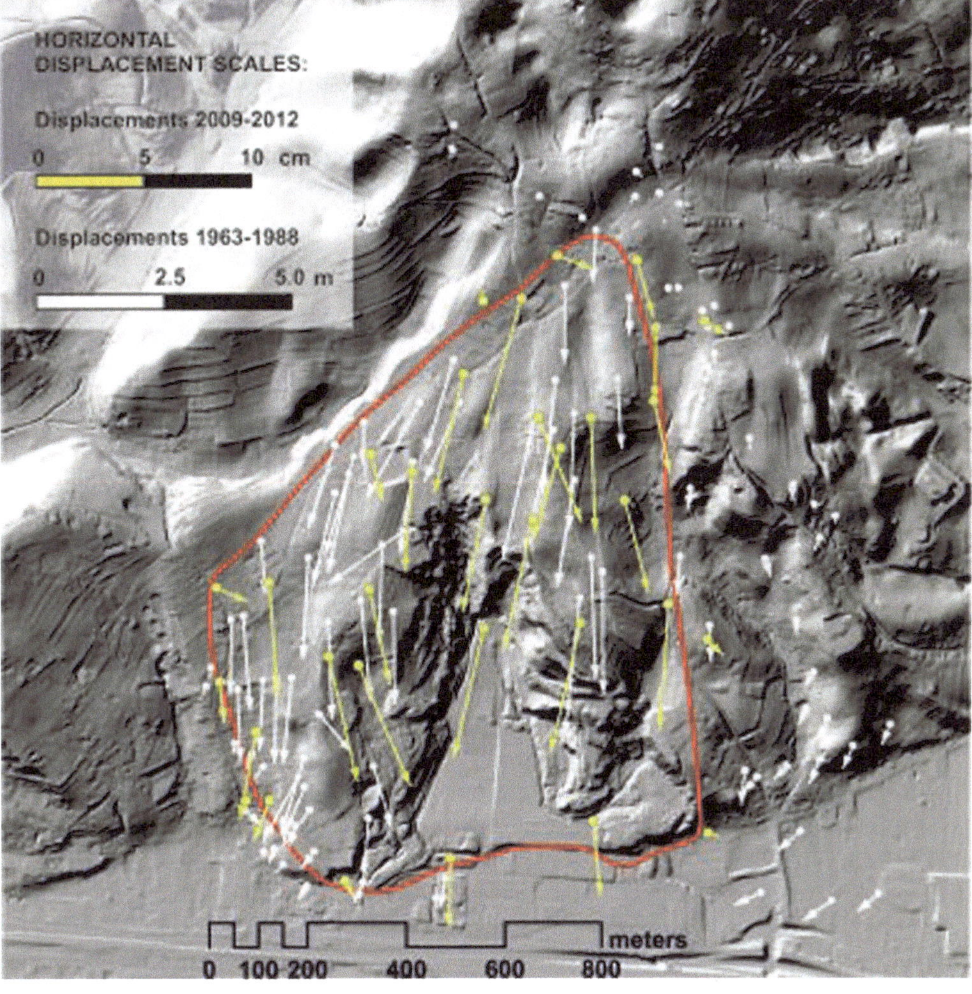

Fig. 3 Recent horizontal displacements (*yellow arrows*) at the Kostanjek landslide area in the period 2010–2012 (Županović et al. 2012) compared with historical horizontal displacement (*white arrows*) for the period 1963–1988 (Ortolan and Pleško 1992). Red line depicts Kostanjek landslide outline according to Ortolan (1996)

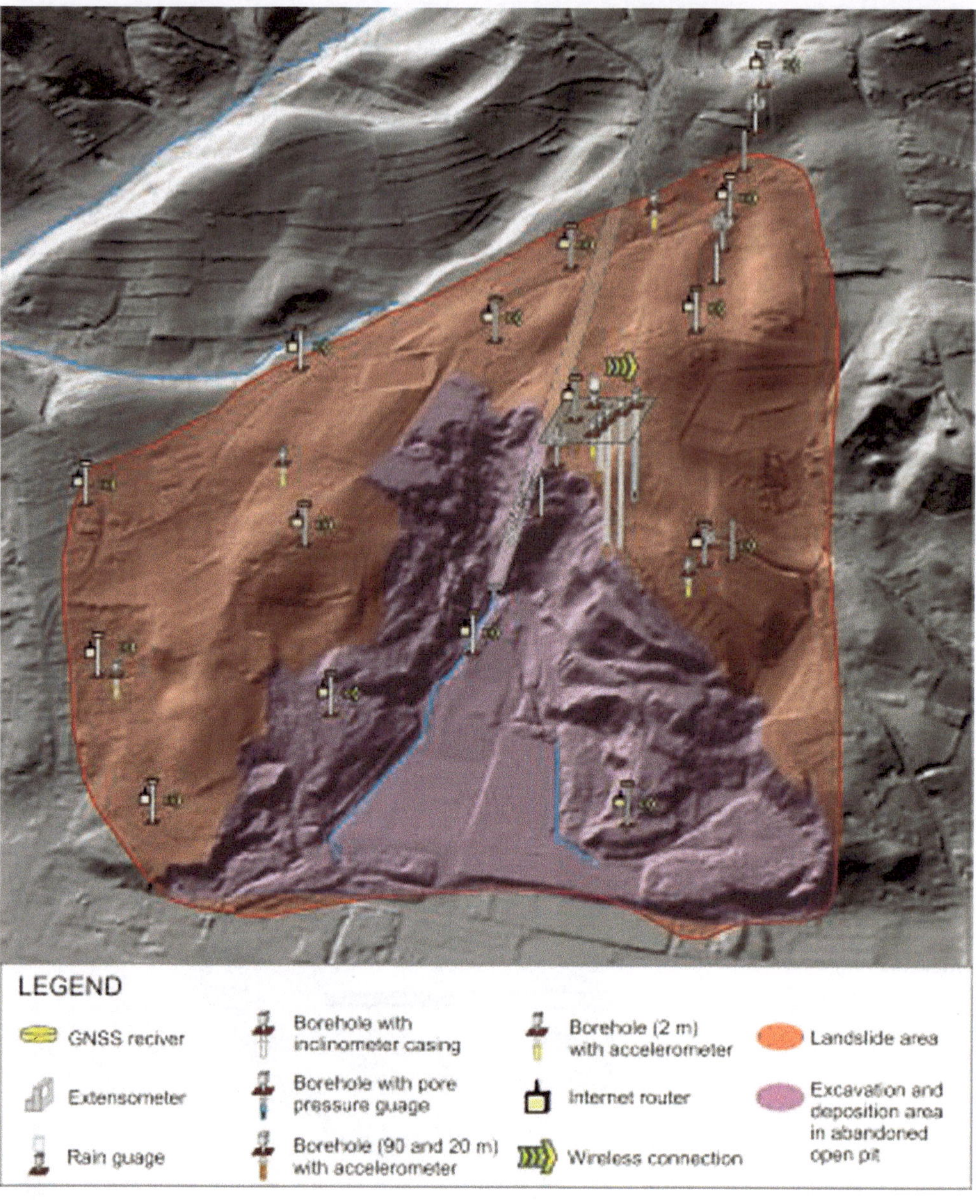

Fig. 4 Sensor network at the Kostanjek landslide area established in the framework of Japanese-Croatian SATREPS FY2008 scientific joint research project (Krkač et al. 2014a)

high poles with 1 m deep reinforced foundations. All monitoring stations are supplied with electricity from public network. Recievers collect GNSS raw data and deliver this data in real-time, over communication lines (using routers), to Trimble 4D Control software (T4DC) installed on an application/data server in a data center at the Faculty of Mininig, Geology and Petroleum Engineering UNIZG. GNSS receivers provide data on absolute positions of surficial points with precisions in the range of cm to mm (Krkač et al. 2014a). To be able to calculate very precise coordinates of 15 GNSS reference stations at the landslide area, the system needs at least one GNSS reference stations outside of the landslide zone located in Gornji Stupnik, which is 7 km away from Kostanjek (Krkač et al. 2014a).

The long- and short-span wire extensometers, type NetLG-501E Osasi, provide data on absolute deformation with submillimetre-level precision. Five long-span extensometers are placed from the top of the most stable point above the landslide; one short-span extensometer is placed perpendicular to the left landslide flank; two long-span extensometers are installed to cross the crown of an artificial steep slope where the highest magnitude of displacement is expected; and one short-span extensometer is installed in the underground, in a tunnel which crosses the sliding surface.

One inclinometer casing is installed in a 100 m deep vertical borehole in the middle of the landslide for measurements of the inclination of the pipe by a high-precision probe. The depth of the present-day major active shear surface is at 62.5 m (determined on the basis of two measurements, in May 2012 and February 2013). There is also some evidence of shallower sliding in the same borehole, at an approximate depth of 30 m (Krkač et al. 2014a).

The geophysical measurement sensor network encompasses seven accelerometers installed inside the landslide area for the purpose of (i) monitoring local micro-earthquake activity in the landslide area; (ii) monitoring regional earthquake activity, including strong motion; and (iii) monitoring of any ground tremors associated with the landslide, including possible ground inclination. This is a low cost and hi-fidelity broad-band monitoring system consisting of three-component MEMS accelerometer and three-channel autonomous broadband digital recorders with GPS to keep accurate synchronization between each other. A Seismic Source DAQ3-3 3CH high-fidelity digital logger with accurate GPS clocking enables continuous recording, with data harvesting by the attached USB memory every three weeks.

Three accelerometers (Colibrys SF1500S.A) are installed in three boreholes at the central monitoring station in the middle part of the landslide: one is in 90 m deep borehole below the sliding surface; one is in a 20 m deep borehole inside the landslide body; and one is in a shallow borehole near the surface. Four accelerometers (Colibrys SF3000L) are installed near the surface, in shallow boreholes at depths of approximately 1.5 m (Krkač et al. 2014a, b).

Monitoring sensors started recording landslide movements and one of the most active period was landslide reactivation due to external triggers in the winter period of 2012/2013. During the period from September 2012 to March 2013 the total cumulative precipitation was 793.7 mm and horizontal displacements were in the range of 4–20 cm (Mihalić Arbanas et al. 2014). The installed monitoring sensor network proved to provide reliable data for the establishment of relations between landslide causal factors and landslide displacement rates aimed at establishing threshold values for early warning system. Monitoring results from the winter period of 2012/2013 has been described in Krkač et al. (2014b, c).

Results of 2-year (2013–2014) monitoring at the central station, consisting of movement observations gathered by GNSS station (GNSS 08) and GWL depth observations gathered by water level sensor (WLS 01), together with precipitation data from Zagreb Grič meteorological station, located 9 km east of Kostanjek, were used for the establishment of the phenomenological model for the prediction of landslide movements (Krkač 2015).

The established phenomenological method enabled the prediction of slow landslide movements by means of landslide velocity and displacements from the precipitation data. The described method encompasses the modeling of groundwater level change rate from the daily and historical precipitation data followed by modeling of landslide velocity from the predicted daily groundwater level depths (calculated from groundwater level change rates). The modeling was performed using random forests, which enabled the evaluation of model parameter importance and development of the prediction model (Krkač 2015, 2016).

The model is established on 2-year monitoring data from the Kostanjek landslide, enabling the prediction of cumulative landslide displacements for periods up to 30 days (Fig. 5). The rainfall–groundwater level depth model was developed using 75 variables placed in six main groups: (1) antecedent precipitation, (2) effective precipitation, (3) precipitation events, (4) modified precipitation, (5) time periods since a certain precipitation event, and (6) climatological indexes. The groundwater level depth–landslide movement model was developed using 10 variables belonging to four main groups: (1) groundwater level depth,

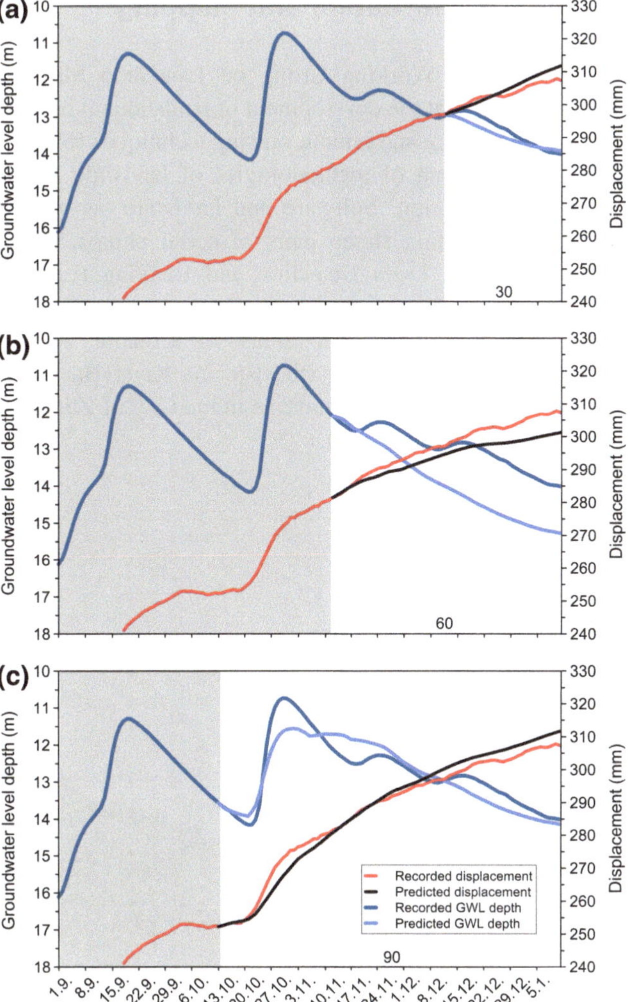

Fig. 5 Prediction of landslide displacements from predicted groundwater levels for periods of **a** 30 days, **b** 60 days, and **c** 90 days with displacement record (GNSS 08) (Krkač et al. 2016)

(2) groundwater level change rate, (3) groundwater change acceleration, and (4) climatological indexes. The most important predictors for movement prediction are groundwater level depth, 7-day groundwater level change rate, and groundwater level change rate on the day after the day for which velocity is predicted.

Random forests prove to teach different patterns of landslide behavior under various conditions, which can be described by a large number of variables. Considering the described reliability of the prediction results, simplicity, and robustness, a random forests is proposed for application in the prediction of landslide movements. Automated predictions updated on a daily basis would produce improved everyday predictions, which are an important source of reliable information for decisions related to crisis management in the case of risky landslide movements (Krkač et al. 2016).

Landslide Identification and Mapping

The activities of Working Group on Landslide Mapping (WG3) are aimed at the development of landslide inventories using direct sensing and remote sensing techniques followed by the development of methodologies of landslide hazard analysis and zonation. Software and hardware, as well as input data (including stereo pairs of aerial photos, digital elevation models, Light Detection and Ranging (LiDAR) data and satellite images) have been donated by the Japanese government for three pilot areas in Croatia: two in Primorsko-Goranska County (the Rječina River Basin and the Dubračina River Basin) and one in the City of Zagreb (a hilly area of Medvednica Mt.).

Hilly Area of the Medvednica Mt. (City of Zagreb)

The hilly area of the southern foothills of Medvednica Mt. of 180 km^2 is mostly urbanized and densely populated. The elevations in this area range from 115 to 612 m a.s.l., the prevailing slope angles (59%) range from 6° to 24° and 84% slopes have slope angles >3° which are potentially prone to sliding. The dominant types are small and shallow landslides that mostly endanger residential structures and roads. The area is composed of Upper Miocene and Quaternary sediments (Fig. 6). The Upper Miocene deposits are stratified sands, silts and marls, with moderately to slightly-inclined bedding (bedding slope angle in range of 10°–20°). The top parts of Miocene deposits are fine-grained soils, mostly silts. The Quaternary deposits are heterogeneous mixtures of unfoliated, mostly impermeable clayey-silty soils. The geologic contact between the Miocene sandy-silty soils and the Miocene or Quaternary clayey-silty soils is highly susceptible to sliding. The most frequent triggering factors are rainfall and man-made activities (Mihalić and Arbanas 2013).

The objective of identification and mapping of landslides at the area of hills of Medvednica Mt. was landslide inventory mapping, based on remote sensing. One of the task was to find technology appropriate for determination of landslide boundaries at the ground surface of very small to moderately large landslides (primarily < 1000 m^2) mostly covered by deciduous vegetation and partially masked by urbanization. Part of project activities was also identification and mapping of landslides using conventional visual interpretation of stereoscopic aerial photographs (Podolszki 2014). Innovative method applied at the same pilot area was analysis of surface morphology with very-high-resolution

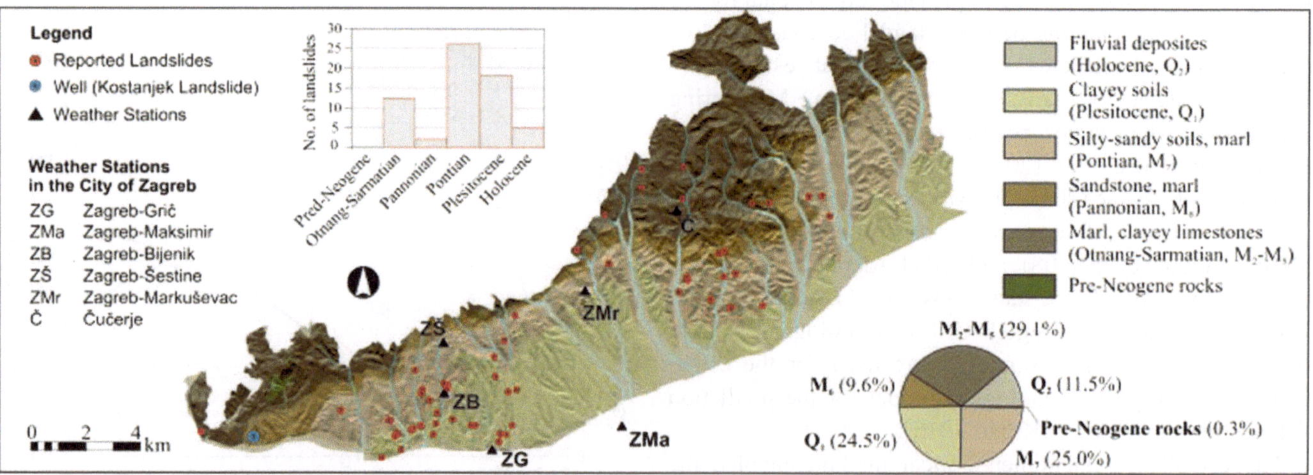

Fig. 6 Precipitation triggered landslide inventory map for the period from 1st January to 7th April 2013 (Bernat et al. 2014a). Landslides are depicted by dots on generalized geological map together with weather stations and one well where continuous water level measurement is available (Krkač et al. 2014a). *Pie chart* shows the relative distribution of main stratigraphic units; histogram shows the number of (re) activated landslides per stratigraphic unit

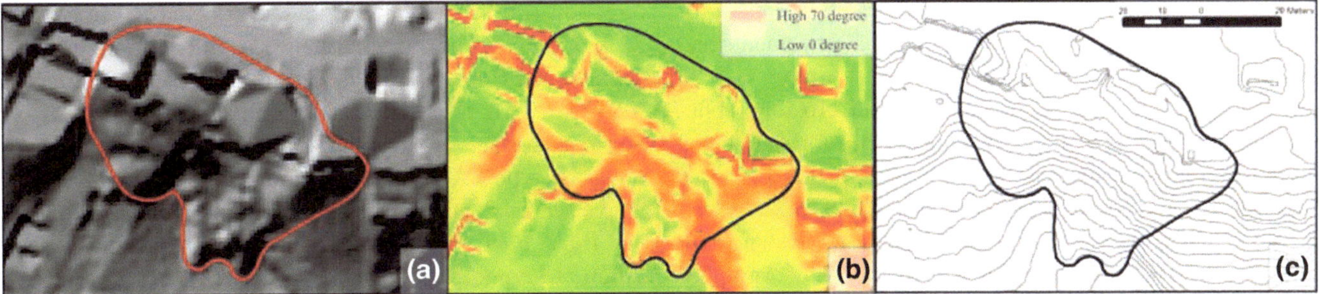

Fig. 7 Composite displays of three different topographic derivative maps of the Črešnjevec landslide. The estimated extent of displaced mass has a *red* and *black* contours: **a** Hillshade map generated and draped over a bare earth DEM. **b** Slope map showing areas of high slope angle in warmer colors (*red*, *orange*, *yellow*) and areas of low slope angle in cooler colors (*green*). **c** Contour map generated with a 1 m contour spacing (Mihalić Arbanas et al. 2014)

digital elevation models (DEMs) captured by airborne LiDAR (Light Detection and Ranging) (Mihalić and Arbanas 2013).

For the LiDAR data that are used in the project, a test flight (covering 24 km^2) took place in April 2011 and a flight over the entire study area (180 km^2) was undertaken in December 2013, which corresponds to periods after leaf fall in Croatia. LiDAR ground-surface measurements were acquired at an average density of five points per square meter. The raw data were post-processed and a 1-m resolution bare earth DEM was interpolated. Slope maps, contour line maps and hillshade maps were created from the DEM by using standard tools in the ArcGIS. A contour line map was created with 1-m intervals. Landslide identification was performed by visual analysis and interpretation of the representation of the topographic surface (Mihalić Arbanas et al. 2016).

The visual identification of the landslides was based on the recognition of landslide features on three topographic derivative maps: a contour map with a 1-m contour span, a hillshade map and a slope map. The mapped landslides were characterized by visible main landslides features such as main scarps, landslide boundaries and toes, zones of depletion and zones of accumulation.

Figure 7 presents one of the typical identification results obtained for the previously known Črešnjevec landslide, which is characterized by multiple reactivations of the upper part of the slope after unsuccessful remedial measures in 2004. The red and black lines depict the contours of an active slow-moving retrogressive landslide. Combining the hillshade, slope and contour maps from the airborne LiDAR data easily distinguishes only the lower part of the displaced mass, while the main scarp, which has a vertical displacement of approximately 10–30 cm, is not clearly expressed. The identification of the landslide contours in the built-up area of the Črešnjevec landslide with airborne LiDAR data was followed by field checking and mapping (Mihalić Arbanas et al. 2014, 2016).

Dubračina River Basin (Primorsko-Goranska County)

The Dubračina River Basin represents a lower part of the elongated unique morphostructural unit Rječina Valley—Bakar Bay—Vinodol Valley. The central part of the Vinodol Valley, which belongs to the Dubračina river catchment area, is about 13 km long and 1.5–5 km wide and stretches parallel to the Adriatic coast in the northwest-southeastern direction. The Vinodol valley has an asymmetrical cross section in this part with a prominently longer northeastern and shorter southwestern slope. The Dubračina riverbed is mostly situated on the littoral ridge (Benac et al. 2010).

The elevations in this area range from 5 to 920 m a.s.l., and the prevailing slope angles (58%) between 5° and 20° are present in the middle part of the basin. Along the northeast and southeast border, the valley is surrounded by steep cliffs whose peaks reach 922 m. The basin has an asymmetrical cross section (in the NE-SW direction) with a prominently longer northeast slope and a shorter southwest slope. The ridge peaks on the southwest side reach 357 m, and the valley bottom is 30 m a.s.l. Karstified carbonate rocks (mostly Upper Cretaceous and Paleogene limestones) are visible at the top of the slopes and cover 55% of the basin area. The clearly marked rocky scarps represent the rim of the karstic plateau at the top of the NE slopes (Benac et al. 2010). Siliciclastic rocks of Eocene age (siltstones and sandstones interlayered with breccia, conglomerates and limestone) are situated in the lower slopes and in the bottom of the valley, but they are mostly covered by slope deposits (Đomlija et al. 2014).

Figure 8 presents seven geomorphological units distinguished in the Dubračina River Basin: alluvial plain; proluvium; terrace slope; hills; mountain slope; limestone wall; and karst plateau. Basic characteristics of the outlined landform units are given with regard to geological settings, geomorphological features and hydrological conditions in Bernat et al. (2014b). The main geological hazards in the

Dubračina River Basin are active geomorphological processes in the form of different types of slope movements and erosion processes.

Several types of slope movements and erosional phenomena are identified that are characteristic for delineated landforms of the Dubračina River Basin. In this phase of investigation, the aim was only to identify all types of slope movement processes that are characteristic for each landform unit. Types of identified active and historical geomorphological processes and phenomena are the following: active and historical slides, creeping phenomena, active gully erosions, screes, rockfalls, topplings, and traces of historical debris flows (Bernat et al. 2014b).

Rječina River Basin (Primorsko-Goranska County)

The Rječina River basin is located in continental part of the Primorsko-Goranska County, and it stretches in NW-SE direction (Fig. 8). Total area of the Rječina River Basin is 22 km^2 and it is mostly unurbanized, with approximately 12 settlements. Current land-use at the area of the Basin is as follows: 19.5 km^2 is forest and semi natural areas, 2 km^2 are artificial surface (roads and buildings), and $\sim$1 km^2 are other use types. Altitude of this area ranges from 0 to 600 m above sea level, and the slope angles mostly range from 0° to 30° (88% area, Fig. 9). Along the main watercourse, three different geomorphological zone exist: first from the river source at the foot of Gorski Kotar Mountains to the Lukeži Village; second zone from the Lukeži Village to the canyon entrance near the Pašac Bridge; and third from the canyon area to the alluvial plain on the river mouth in the centre of the City of Rijeka (Benac et al. 2011).

The most interesting part of the basin is the central part of the Rječina River Valley regarding active geomorphological processes and historical landslide phenomena. This part of the valley is 3 km long and 0.8–1.5 km wide. Cross section of the valley is symmetrical with bottom at 150–200 m

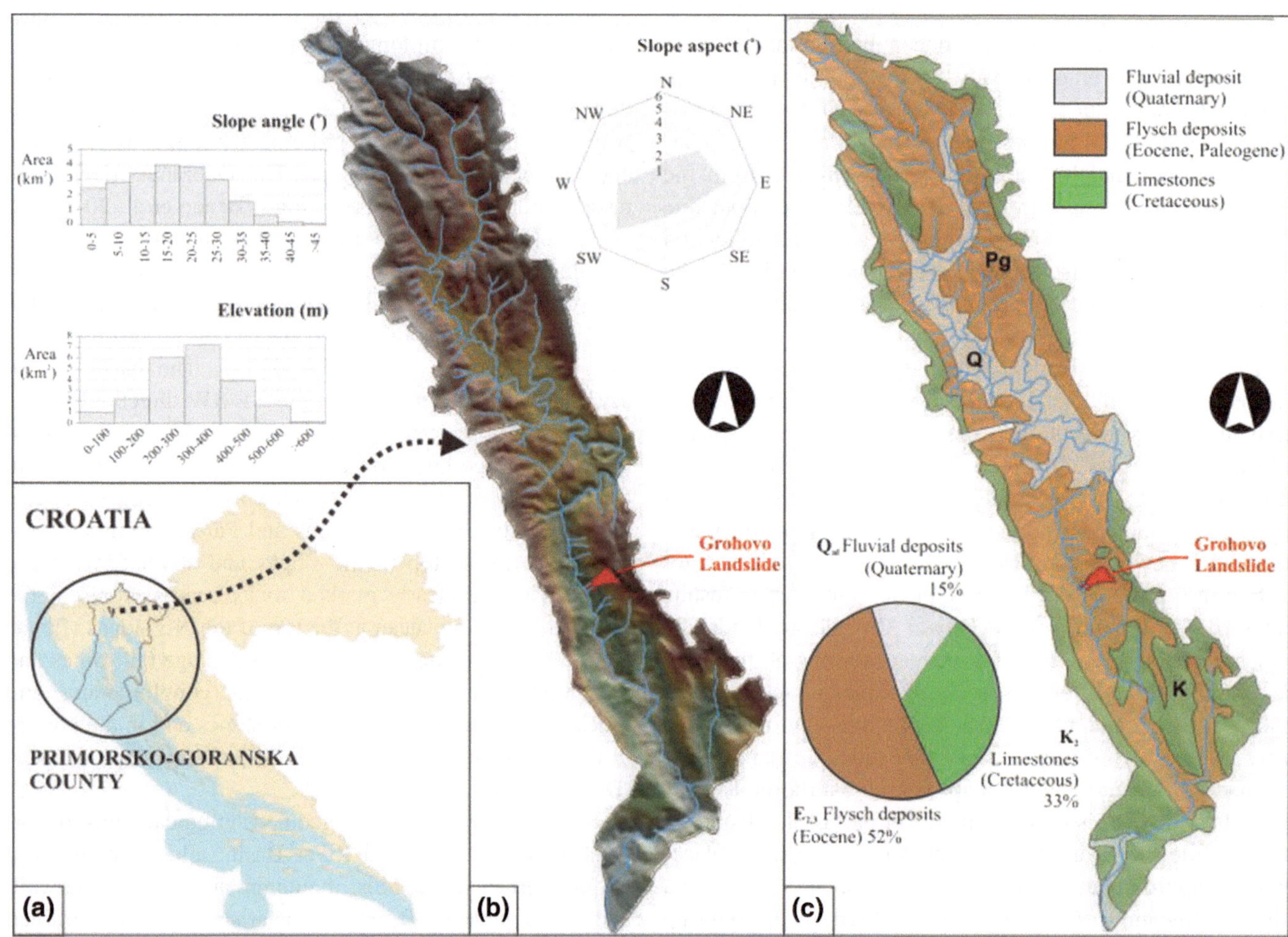

Fig. 8 Thematic data available for the Rječina River Basin. **a** Geographic location; **b** elevation map from the 25 m × 25 m DEM. Histograms show the distribution of elevation and slope angle computed from the DEM. Rose diagram shows distribution of slope aspect (in km^2). **c** geological map showing the main stratigraphic units (Mihalić and Arbanas 2013)

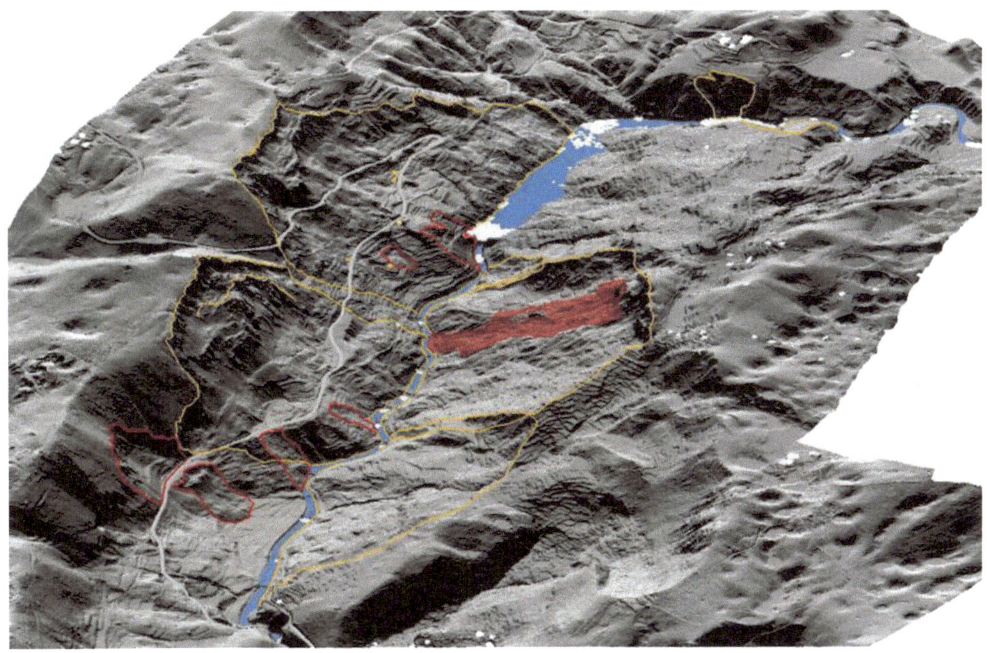

Fig. 9 Oblique aerial view created from the LIDAR-derived bare-earth DEM of the Rječina River Valley. Landslide features of an unnamed landslide can be clearly identified. Yellow borders contour historical landslides from 1885 and 1750 on the *left side* and 1893 on the *right side*. *Red coloured zone* is area of reactivated landslide from 1996. *Red borders* contour younger landslides (Arbanas et al. 2014a)

above sea level, and the peak height at 432 m in the SW and 412 m in the NE. Karstified limestone rock masses are visible on the top of the slope in the central part of the valley, while the flysch rock mass are placed in the lower slopes and at the bottom of the valley. The clearly marked rocky scarps represent the rim of the karstic plateau above the SW slopes and partially the NW slopes which could be seen at simplified geological map (Fig. 8). Both slopes are covered with potentially unstable superficial deposits. This area is geodynamically very active, and landslides are the main geological hazard. Different types of slope movements can be distinguished: relict and dormant deep seated landslides, reactivated rockfalls from limestone cliffs at the top part of the slope, reactivated sliding of coluviall deposits (predominantly coarse fragments and limestone blocks) over the flysch bedrock (Benac et al. 2005).

Numerous historical descriptions, figures, photographs and maps describing landslides were found in the Croatian State Archive in Rijeka which provides evidence of the occurrence of landslides in the Rječina Valley near the Grohovo Village. Sliding was first documented in 1767, when numerous landslides and rockfalls in the Rječina River Valley were caused by the 1750 earthquake. A large landslide occurred on the southwestern slope in 1870, and after reactivation of the slide in 1885 part of the Grohovo Village was buried by a rock avalanche. Next large landslide occurred in 1893 on the northeastern slope of the Rječina River Valley at the location of the recent landslide, and the Rječina River channel was shifted to the south by approximately 50 m. Numerous landslides occurred during the first half of the 20th century but did not cause significant damage to structures on the river banks. New problems with landslides in the Rječina valley occurred during construction of the Valići Dam in 1960, when landslides appeared on the northeast slope near the dam. The large landslide, so called the Grohovo Landslide, occurred in December 1996 at the location of the landslide from 19th century on the northeast slope described in the chapter The Grohovo Landslide monitoring system.

With the knowledge about numerous landslide occurrences in the Rječina River Valley in the history and without landslide inventory which would offer insight in real landslide distribution in the area, the first step for necessary analyses was identification of existing landslides in the research area and assessment of their features (Mihalić Arbanas and Arbanas 2014). As an appropriate technique for landslide identification in the Rječina River Valley the analyses of very-high resolution DEMs obtained by airborne laser scanning in combination with field mapping was chosen. The visual identification of landslides is based on the recognition of landslide features on the following types of topographic derivative maps: contour map, hillshade map, slope map, curvature map and topographic roughness map. The main landslide features (main scarp, flanks etc.) are checked by field mapping.

This procedure was repeated for identification of all known historical landslides in the valley on the basis of historical descriptions and all these landslides are located and determined. During very-high resolution DEMs analyses a notable number of unknown landslides was identified and most of them are completely hidden covered by vegetation. Time of their occurrence is also unknown but scars and expression of their features indicate on relatively young phenomena. Historical landslides in the Rječina River

Valley are major instabilities and their volumes vary from 6.1 to 25 million m^3, while younger landslides are significantly smaller with volumes from 10 thousand to 3 million m^3, and mostly occurred as reactivated parts of the older landslides. When all these landslides are presented on same DEM it is clearly visible that the both slopes of the valley are intensively affected by sliding, Fig. 9 (Arbanas et al. 2014a).

Soil Testing and Landslide Simulation

Soil Testing

Detailed laboratory soil testing as a base for prediction of landslides behaviour are performed in a ring-shear apparatus that is designed for testing under static and dynamic conditions for deep seated large landslides in Croatia. The ring-shear apparatus was designed initially to investigate the residual shear resistance under the drained condition along the sliding surface at large shear displacements in landslides because it allows unlimited deformation of the specimen. Professor Sassa with his team has developed the undrained high speed ring shear apparatus to reproduce a rapid landslide motion after failure and to measure the generated pore pressure and the shear resistance mobilized on the sliding surface during motion (Sassa et al. 2003, 2004; Okada et al. 2004; Fukuoka et al. 2006). The new developed apparatus is (Fig. 10), compared to previous ones, much smaller in dimensions and has higher performances. It can keep undrained condition up to 1 MPa of pore water pressure, up to 3 times more than in previous versions of apparatus and load normal stress up to 1 MPa. This makes it suitable for testing of soil samples in stress condition as on surfaces of rupture in deep seated landslides.

The soil samples from the Grohovo Landslide were taken from the flysch outcrop in the central part of the landslide body. Speed control test was conducted under constant shear speed of 0.002 cm/s in undrained conditions. Sample was sheared until the shear displacement reached 1.0 m and the steady state conditions were obtained. As a results of this test, the basic parameters values (peak, mobilized and apparent friction angle, so as cohesion) as well as steady state normal and shear stress of soil sample were obtained (Oštrić et al. 2012; Vivoda et al. 2014).

The Integrated Model of Landslide Simulation

The LS-Rapid software is the first landslide simulation model possible to integrate the whole process of stable state, failure, post-failure strength reduction, motion and deposit of sliding mass (Sassa et al. 2010, 2014). In the simulation, the friction angle and cohesion will be reduced from their peak values to the normal motion time values within the source area in the determined distribution of the unstable mass. The strength reduction will be started in the moment when the travel length will become equal to shear displacement at the start of strength reduction. The strength reduction will be completed and the normal motion simulation will be started when the travel length will reach the value of shear displacement at the end of strength reduction.

The topography of the Rječina River Valley (Fig. 11) was determined using original DEM data. The limestone rock mass is situated at the top of the slopes, while the siliciclastic

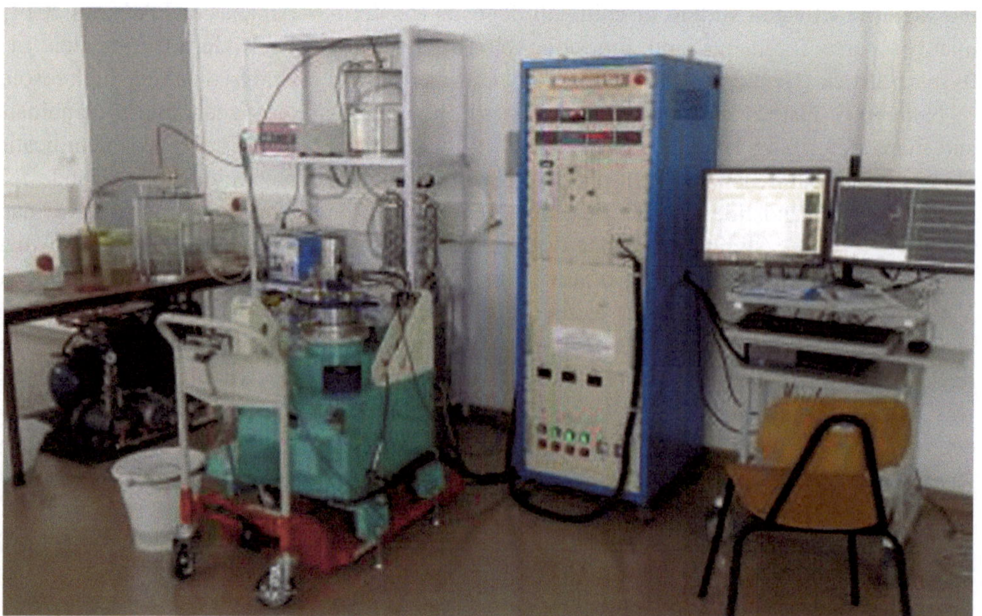

Fig. 10 New developed portable ring-shear apparatus designed for testing under static and dynamic conditions for deep-seated large landslides in Croatia

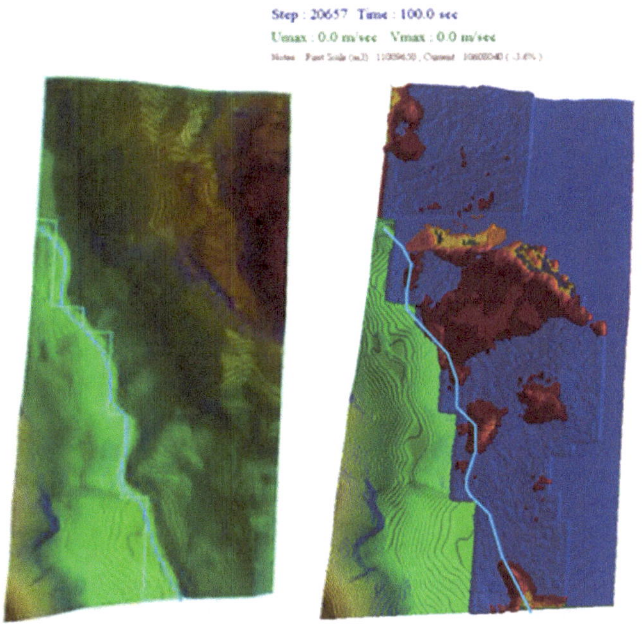

Fig. 11 Numerical simulation of the Grohovo landslide using LS-Rapid software (Vivoda et al. 2014). Digital elevation model on the *left side* and displaced mass on the *right side*

rocks and flysch are situated on the lower slopes and the bottom of the valley. Depth of the sliding mass varies from 3 to 10 m over the flysch bedrock and from 0.0 to 0.5 m over the limestone rock mass. This assumption is based on knowledge that the existing slip surface is positioned at the contact between superficial slope deposits and flysch bedrock (Benac et al. 2005). The long-term rainfalls and consequent ground water level rising were the main triggering factor for the existing landslide occurrences in the Rječina River Valley. This ground water level rising in the model was expressed by excess of the pore pressure ratio until the value of $r_u = 0.60$, which is correspondent to the ground water level equal to terrain surface (Vivoda et al. 2014).

The simulation results are shown on Fig. 11. The blue colored zones represent the stable areas or areas with movement velocity less than 0.1 m/s. The orange and red colored zones represent areas where the sliding occurred. The results of conducted simulation very clearly suggest that the new slides, caused by future unfavorable hydrogeological conditions, would be occurred in the area of the existing Grohovo Landslide and this fact confirm the correct selection of the Grohovo Landslide for monitoring and early warning system establishment.

Gradiški et al. (2014) analyzed behavior of the Kostanjek landslide (Fig. 12) using the LS-RAPID software. Existing landslide model from Ortolan (1996) was modified on the basis of monitoring results by creation of ellipsoidal sliding surface with maximum depth of 65 m in the central part of the landslide body (according to maximal displacements). For more reliable interpretation of the sliding surface depths, additional subsurface investigations and monitoring are necessary. Parameters used for these analyses were determined from drained test of samples in ring shear apparatus.

According to the results of the analyses, the most unstable part of the landslide is the central part of the landslide body, i.e. the slopes of the abandoned open marl pit. In the analyses the movements started in the central part of the landslide body, and the failure area will expand around the initial failure zone. At the end of simulation the area of the whole landslide mass corresponds to the landslide contour from historical landslide model according to Ortolan (1996). This is also in accordance to the new surface deformations (cracks, bulging, and subsidence) developed by very recent landslide movement in 2013 (Fig. 12) (Mihalić Arbanas et al. 2014).

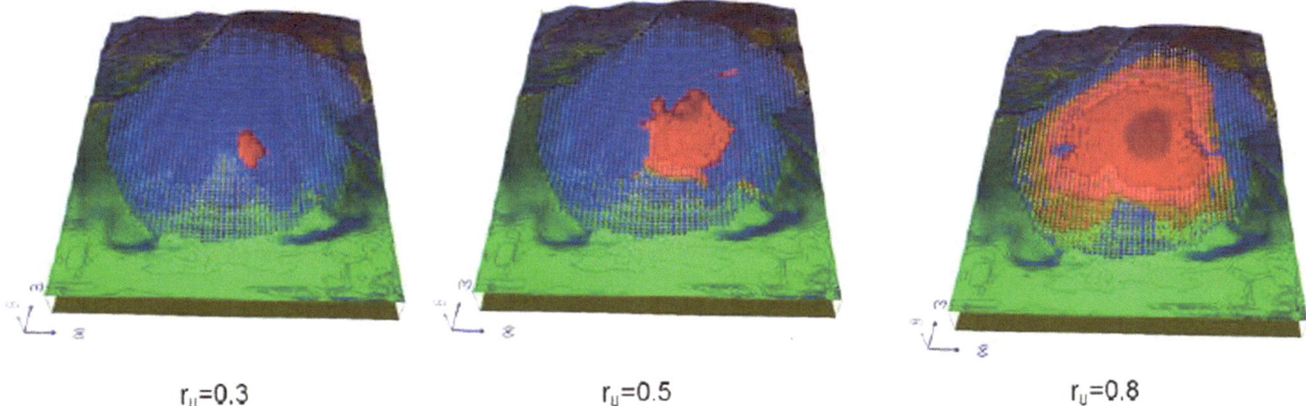

Fig. 12 The Kostanjek landslide simulation at $r_u = 0.3$, $r_u = 0.5$, $r_u = 0.8$ (Gradiški et al. 2014)

Conclusions

The Croatian-Japanese joint research Science and Technology Research Partnership for Sustainable Development (SATREPS) project 'Risk Identification and Land-Use Planning for Disaster Mitigation of Landslides and Floods in Croatia' was performed from 2009 to 2014. Key objectives of the project were landslides and floods hazard analysis and the development of guidelines for use in urban planning. The main Japanese partner institutions in the Project were Research Center for Natural Hazards and Disaster Recovery at Niigata University, Disaster Prevention Research Institute of Kyoto University (DPRI) and the International Consortium on Landslides (ICL). Three Croatian universities, the University of Rijeka (Faculty of Civil Engineering), the University of Zagreb (Faculty of Mining, Geology and Petroleum Engineering and Faculty of Agriculture) and the University of Split (Faculty of Civil Engineering, Architecture and Geodesy) as well as the Croatian Geological Survey were Croatian partner institutions in the project. The five-year project, started in 2009 and completed in 2014, involved about 15 researchers from Japan and included collaborative research conducted in Japan and Croatia to evaluate and mitigate landslide and flood hazards and risks in Croatia.

The project activities are organized into three working groups: Working Group on Landslides (WG1), Working Group on Flash Floods and Debris Flows (WG2) and Working Group on Landslide Mapping (WG3). The aim of Working Group 1 on Landslides was to establish a methodology of comprehensive real time monitoring at two most important landslides in Croatia: the Grohovo Landslide in Primorsko-Goranska County and the Kostanjek Landslide in the City of Zagreb based on the results of previous investigations and new in situ and laboratory testing and behavior analysis so as a laboratory soil testing and numerical modelling of static and dynamic landslide behavior. The activities of Working Group 3 on Landslide Mapping (WG3) were aimed at the development of landslide inventories using direct sensing and remote sensing techniques followed by the development of methodologies of landslide hazard analysis and zonation in three pilot areas in Croatia: two in Primorsko-Goranska County and one in the City of Zagreb.

The results of the project were published at the four project workshops held in Dubrovnik (2010), Rijeka (2011), Zagreb (2013) and Split (2014). The project also resulted with regional cooperation and establishment of Adriatic Balkan Network of ICL and the third project workshop held in Zagreb (2013) was organized as the 1st Regional Symposium on Landslides in Adriatic Balkan Region.

Acknowledgements The results presented herein have been obtained with financial support from the Japan Agency for Science and Technology (JST) and the Japan Agency for International Cooperation (JICA) as a FY2008 SATREPS (Science and Technology Research Partnership for Sustainable Development) programme for 2008–2014. Co-financing was ensured from the Ministry of Science, Education and Sports of the Republic of Croatia, the City Office for Physical Planning, Construction of the City, Utility Services and Transport, City of Zagreb and the Emergency Management Office of the City of Zagreb and City of Rijeka. These supports are gratefully acknowledged.

References

Arbanas Ž, Jagodnik V, Ljutić K, Dugonjić S, Vivoda M (2012a) Establishment of the Grohovo Landslide monitoring system. Proceedings of the 2nd workshop of the project risk identification and land-use planning for disaster mitigation of landslides and floods in Croatia, 15–17 December 2011. Rijeka, Croatia, pp 29–32

Arbanas Ž, Sassa K, Marui H, Mihalić S (2012b) Comprehensive monitoring system on the Grohovo Landslide, Croatia. Proceedings of the 11th international and 2nd North American symposium on landslides: landslides and engineered slopes: protecting society through improved understanding, June 2–8, 2012. Banff, Canada, pp 1441–1447

Arbanas Ž, Mihalić Arbanas S, Vivoda M, Peranić J, Dugonjić Jovančević S, Jagodnik V (2014a) Identification, monitoring and simulation of landslides in the Rječina River Valley, Croatia. Proceedings of the SATREPS Workshop on Landslides in Vietnam, Hanoi, Vietnam, pp 200–213

Arbanas Ž, Sassa K, Nagai O, Jagodnik V, Vivoda M, Dugonjić Jovančević S, Peranić J, Ljutić K (2014b) A landslide monitoring and early warning system using integration of GPS, TPS and conventional geotechnical monitoring methods. Proceeding of In proceeding of world landslide forum 3, landslide science for a safer geoenvironment, vol: 2, methods of landslide studies, 13–16 June 2014. Beijing, China, pp 631–636

Benac Č, Arbanas Ž, Jurak V, Oštrić M, Ožanić N (2005) Complex landslide in the Rječina River valley (Croatia): origin and sliding mechanism. Bull Eng Geol Env 64(4):361–371

Benac Č, Mihalić S, Vivoda M (2010) Geological and geomorphological conditions in the area of Rječina river and Dubračina river catchments (Primorsko-goranska County, Croatia). Proceedings of 1st workshop of the Japanese-Croatian SATREPS FY2008 project, 22–24 November 2010. Zagreb, Croatia. pp 39–39

Benac Č, Dugonjić S, Vivoda M, Oštrić M, Arbanas Ž (2011) Complex landslide in the Rječina Valley: results of monitoring 1998–2010. Geologia Croatica 64(3):239–249

Bernat S, Mihalić Arbanas S, Krkač M (2014a) Inventory of precipitation triggered landslides in the winter of 2013 in Zagreb (Croatia, Europe). Proceedings of the 3rd world landslide forum, vol. 2, 2–6 June 2014. Beijing, China. pp 829–836

Bernat S, Đomlija P, Mihalić Arbanas S (2014b) Slope movements and erosion phenomena in the Dubračina River Basin: a geomorphological approach. Proceedings of the 1st Regional symposium on landslides in the Adriatic-Balkan region 'Landslide and flood hazard assessment', 6–9 Mar 2013. Zagreb, Croatia, pp 79–84

Đomlija P, Bernat S, Arbanas Mihalić S, Benac Č (2014) Landslide inventory in the area of Dubračina river basin (Croatia). Proceedings of the 3rd world landslide forum, vol. 2, 2–6 June 2014. Beijing, China. pp 837–842

Fukuoka H, Sassa K, Wang G, Sasaki R (2006) Observation of shear zone development in ring-shear apparatus with a transparent shear box. Landslides 3:239–251

Gradiški K, Sassa, K, He B, Krkač M, Mihalić Arbanas S, Arbanas Ž, Oštrić M, Kvasnička P (2014) Application of integrated landslide simulation model using LS-rapid software to the Kostanjek Landslide, Zagreb, Croatia. Proceedings of the 1st regional symposium on landslides in the Adriatic-Balkan Region 'Landslide and flood hazard assessment', 6–9 Mar 2013. Zagreb, Croatia, pp 11–16

Han Q, Sassa K, Kan FM, Margottini C (2017) International programme on landslides (IPL): objectives, history and list of IPL projects. Advancing culture for living with landslides, vol 1 ISDR-ICL Sendai partnerships 2015–2025 (this volume)

Krkač M (2015) A phenomenological model of the Kostanjek landslide movement based on the landslide monitoring parameters. PhD thesis, University of Zagreb (in Croatian)

Krkač M, Mihalić Arbanas S, Nagai O, Arbanas Ž, Špehar K (2014a) The Kostanjek landslide—monitoring system development and sensor network. Proceedings of the 1st regional symposium on landslides in the Adriatic-Balkan Region 'Landslide and flood hazard assessment', 6–9 Mar 2013. Zagreb, Croatia, pp 27–32

Krkač M., Mihalić Arbanas S, Arbanas Ž, Bernat S, Špehar K, Watanabe N, Osamu N, Sassa K, Marui H, Furuya G, Wang C, Rubinić J, Matsunami K (2014b) Review of monitoring parameters of the Kostanjek landslide (Zagreb, Croatia). Proceedings of the 3rd world landslide forum, vol 2, 2–6 June 2014. Beijing, China, pp 637–645

Krkač M, Mihalić Arbanas S, Arbanas Ž, Bernat S, Špehar K (2014c) The Kostanjek landslide in the City of Zagreb: forecasting and protective monitoring. Proceedings of the XII IAEG congress, 15–19 Sept 2014. Torino, Italy, pp 715–719

Krkač M, Špoljarić D, Bernat S, Mihalić Arbanas S (2016) Method for prediction of landslide movements based on random forests. Landslides (published online). doi:10.1007/s10346-016-0761-z

Mihalić S, Arbanas Ž (2013) The Croatian–Japanese joint research project on landslides: activities and public benefits. In: Sassa K, Rouhban B, Briceño S, McSaveney M, He B (eds) Landslides: Global Risk Preparedness. Springer, Heidelberg, pp 333–349

Mihalić Arbanas S, Arbanas Ž (2014) Landslide mapping and monitoring: review of conventional and advanced techniques. Proceedings of the 4th symposium of Macedonian association for geotechnics, 25–28 June 2014. Struga, Macedonia, pp 57–72

Mihalić Arbanas S, Krkač M, Bernat S, Arbanas Ž (2014) Landslide mapping and monitoring in the City of Zagreb (Croatia, Europe). Proceedings of the SATREPS workshop on landslides in Vietnam, Hanoi, Vietnam, pp 214–226

Mihalić Arbanas S, Krkač M, Bernat S (2016) Application of advanced technologies in landslide research in the area of the City of Zagreb (Croatia, Europe). Geologia Croatica 69(2):231–243

Okada Y, Sassa K, Fukuoka H (2004) Excess pore pressure and grain crushing of sands by means of undrained and naturally drained ring-shear tests. Eng Geol 75:325–343

Ortolan Ž, Pleško J (1992) Repeated photogrammetric measurements at shaping geotechnical models of multi-layer landslides. Rudarsko-geološko-naftni zbornik 4:51–58

Ortolan Ž (1996) Development of 3D engineering geological model of deep landslide with multiple sliding surffaces (Example of the Kostanjek Landslide). PhD thesis. Faculty of Mining, Geology and Petroleum Engineering, University of Zagreb, Zagreb (in Croatian)

Oštrić, M., Ljutić K., Krkač M., Setiawan H., He, B., Sassa, K. (2012) Undrained ring shear tests performed on samples from Kostanjek and Grohovo landslide. In: Sassa K, Takara K, He B (eds) Proceedings of the IPL symposium. Kyoto, Japan, pp 47–52

Podolszki L (2014) Stereoscopic analysis of landslides and landslide susceptibility on the southern slopes of Medvednica Mt. PhD thesis, Faculty of Mining, Geology and Petroleum Engineering, University of Zagreb, Zagreb, Croatia (In Croatian)

Sassa K, Wang G, Fukuoka H (2003) Performing undrained shear tests on saturated sands in a new intelligent type of ring shear apparatus. ASTM Geotech Test J 26(3):257–265

Sassa K, Fukuoka H, Wang G, Ishikawa N (2004) Undrained dynamic-loading ring-shear apparatus and its application to landslide dynamics. Landslides 1:7–19

Sassa K, Nagai O, Solidum R, Yamazaki Y, Ohta H (2010) An integrated model simulating the initiation and motion of earthquake and rain induced rapid landslides and its application to the 2006 Leyte landslide. Landslides 7(3):219–236

Sassa K, He B, Dang K, Nagai O, Takara K (2014) Progress in landslide dynamics. Proceeding of world landslide forum 3, landslide science for a safer geoenvironment, vol: 1, methods of landslide studies, 13–16 June 2014. Beijing, China, pp 37–67

Savvaidis PD (2003) Existing landslide monitoring systems and techniques. Proceedings of the conference from stars to earth and culture, In honor of the memory of Professor Alexandros Tsioumis. Thessaloniki, Greece, pp 242–258

Vivoda M, Dugonjić Jovančević S, Arbanas Ž (2014) Landslide occurrence prediction in the Rječina River valley as a base for an early warning system. Proceedings of 1st regional symposium on landslides in the Adriatic-Balkan Region "Landslide and flood hazard assessment", 6–9 Mar 2013, Zagreb, Croatia, pp 85–90

Županović L, Opatić K, Bernat S (2012) Monitoring of movements of the Kostanjek landslide using relative static method (GNSS technology). Ekscentar 15:46–53 (In Croatian)

Results of a Technical Cooperation Project to Develop Landslide Risk Assessment Technology along Transport Arteries in Vietnam (IPL-175)

Dinh Van Tien, Nguyen Xuan Khang, Kyoji Sassa, Toyohiko Miyagi, Hirotaka Ochiai, Huynh Dang Vinh, Lam Huu Quang, Khang Dang, and Shiho Asano

Abstract

Like other South-East Asia countries, Vietnam is a country with mountainous terrain, complicated geological structure and high rainfall, and as a result, landslides occur regularly, with serious consequences for the mountain road networks in the rainy season. Due to economic difficulties and a lack of deep knowledge of the phenomena, activities to prevent and mitigate landslides are not effective. The SATREPS project of research cooperation between Japanese and Vietnamese researchers in the years 2011–2016 has not only helped Vietnam in the development of human resources, research equipment and development of a standard system of landslide investigation, monitoring, forecast and early warning, but has also contributed to disaster prevention and reduction in Vietnam in the future. This project is considered as a success for a new landslide-training tool, in cooperation with Asia members of the International Consortium on Landslides (ICL), especially South-East Asia countries, for the mitigation of natural disasters.

Keywords

Technical cooperation • Landslide • Risk assessment • Transport • Vietnam

D. Van Tien (✉) · N.X. Khang · H.D. Vinh · L.H. Quang
Institute of Transport Science and Technology, Hanoi, Vietnam
e-mail: dvtien.gbn@gamil.com

N.X. Khang
e-mail: khangluc@itst.gov.vn

H.D. Vinh
e-mail: huynhdangvinh@gmail.com

L.H. Quang
e-mail: lhqlinh@yahoo.com

K. Sassa
International Consortium on Landslides, Tanaka Asukai-cho 138-1, Kyoto, 606-8226, Japan
e-mail: sassa@iclhq.org

T. Miyagi
Graduate School of Human Informatics, Tohoku-Gakuin University, Sendai, Japan
e-mail: miyagi@mail.tohoku-gakuin.ac.jp

H. Ochiai
Forestry and Forest Products Research Institute, Tsukuba, Japan
e-mail: hirotakaochiai@gmail.com

K. Dang
International Consortium on Landslides, Kyoto, Japan
e-mail: khangdq@gmail.com

K. Dang
VNU University of Science, Hanoi, Vietnam

S. Asano
Department of soil and water conservation, Forestry and Forest Products Research Institute, Tsukuba, Japan
e-mail: shiho03@ffpri.affrc.go.jp

K. Sassa et al. (eds.), *Advancing Culture of Living with Landslides*,
DOI 10.1007/978-3-319-59469-9_36

Landslides Along Transport Arteries in Vietnam

Landslides have been a persistent problem in mountainous regions, and especially those distributed along transportation corridors. Landslides not only cause damage to properties (houses, buildings, vehicles, etc.) and large numbers of casualties, but also disrupt utility services and economic activities. Located on the Eastern Indochina Peninsula, Vietnam has mountainous areas that cover up to 3/4 of the area with steeply sloping terrain, due to the powerful tectonics of the earth's crust. Moreover, it also has a complex geological structure and a tropical monsoon climate, with an average annual rainfall of as much as 3000–4500 mm/year in some regions. Consequently, Vietnam is a typical tropical country, and has the most serious landslide disasters in Southeast Asia and the Mekong sub region.

Vulnerability to landslide hazards is the probability of movement of slopes by landsliding. There are many causes of landslide vulnerability, such as conditions of topography, geomorphology, geology, climate and artificial activities. Landslide vulnerability assessment is a major aspect, especially for risk assessment of reactivated landslides and landslide susceptibility (Tien et al. 2016a, b). To mitigate the effects of this phenomenon on human life, landslide risk assessment is a requirement.

Landslides Risk Assessment Project

For more insight into the phenomenon of landslides in general, as well as to control and mitigate the losses from this natural phenomenon for the traffic system, as well as for new projects in mountainous areas, a Technical Cooperation Project named "Development of Landslides Risk Assessment Technology along Transport Arteries in Vietnam" was proposed by the Vietnam Institute of Transportation Science and Technology (ITST) and ICL. It was one of SATREPS projects, which was established in 2008 as a part of the new "Science and Technology Diplomacy" implemented jointly by the Ministry of Foreign Affairs (MOFA) though (JICA) and the Ministry of Education, Culture, Sports, Science and Technology (MEXT through JST). The project was approved as an ongoing IPL project (IPL-175) by the IPL Global Promotion Committee in 2011 (Han et al. 2017). It is the second SATREPS project proposed and implemented by ICL members after IPL-161 Croatia Project 2008–2014 (Arbanas et al. 2017).

The overall objective of the project is to socially implement the developed landslide risk assessment technology and early warning system, which will contribute to ensuring the safety of transportation arteries and residents in mountainous communities in Vietnam.

The project started in 2011 and ended in 2016, with a full implementation time of 5 years (Fig. 1).

Overview of Results of the Technical Cooperation Project

All results of the project were divided among four work groups (WG) as follows:

- Preparation of integrated guidelines for the application of developed landslide risk assessment technology and capacity development by a WG1 Joint Team of all groups;
- Wide-area landslide mapping and identification of landslide risk area by the WG2 Mapping Group;
- Development of landslide risk assessment technology based on soil testing and computer simulation by the WG3 Testing Group;
- Risk evaluation and development of an early warning system based on landslide monitoring by the WG4 Monitoring Group.

Development of Landslide Risk Assessment Technology and Education

Based on technological transfer from Japan to Vietnam, Vietnamese researchers have drafted intergrade guidelines for landslide risk assessment in the following six parts, with 33 guidelines, which cover (1) Mapping and Site Prediction, (2) Material Tests, (3) Monitoring, (4) Landslide flume experiments and (5) Software application. Those guidelines will be the first step for a strategy of national standard development for landslide risk assessment in Vietnam (Project report—Sassa et al. 2016).

In terms of education and human resources development, three doctor and five master certifications have been obtained at the end 2016 through the training courses for Master (for 2 years) or Doctor (for 3 years) or short training programs at Kyoto University, Tohoku Gakuin University, University of Shimane Prefecture, Shizuoka University and Gunma University. Five other doctoral candidates are currently studying.

Wide-Area Landslide Mapping and Landslide Risk Identification

Based on landslides that have occurred in the study area, a general method to prevent and mitigate landslide activity along roads in humid tropical regions also was proposed. The

Fig. 1 Some pictures of landslides that have taken place and their location on a map of the Vietnamese transport system

core of the study is a new strategy to reduce the effects of vulnerability to landslides, by using a combination of landslide risk assessment maps, which are developed from risk assessments of landslides that have occurred, and landslide susceptibility maps, which are developed from evaluation of the sensitivity to sliding of natural slopes from landslide causative factors. A flowchart of WG2 is presented in Fig. 2.

After five years, six sheets of landslide inventory maps for the Ho Chi Minh route and detailed scale landslide distribution map (1:12,000) for a 60 km long section were established. Landslide risk assessment maps for the mentioned region and susceptibility maps for 150 km along the Ho Chi Minh route have been developed using an analytic hierarchy process (AHP) approach (Le et al. 2014; Tien et al. 2016a). For air-photo interpretation for the landslide inventory, a section of 25 km of National Road No.7 from Muong Xan to Tam Quang was the target for mapping (Ngo 2016).

The technology for identification of the precursor stage of landslides through a pattern analysis of a digital surface model (DSM) of the forest cover and others was developed. Technical analysis based on tree crown deformation using the comparative study of UAV and aerial photo data for the landslide survey was developed at a mangrove forest in Iriomote island, Okinawa, then applied to the Halong landslide trial site. (The application to identify the initial stages of landslide deformation is being used for the Aratozawa landslide, Kurikoma, Japan (Myiagi project report, 2016).

Those study results can contribute to forecasting, preventing and mitigating the negative impacts of landslides for planning, land use, construction of infrastructure, ensuring the safety of existing traffic roads and mountainous residential areas in Vietnam. It can be applied to areas with similar conditions to the study area.

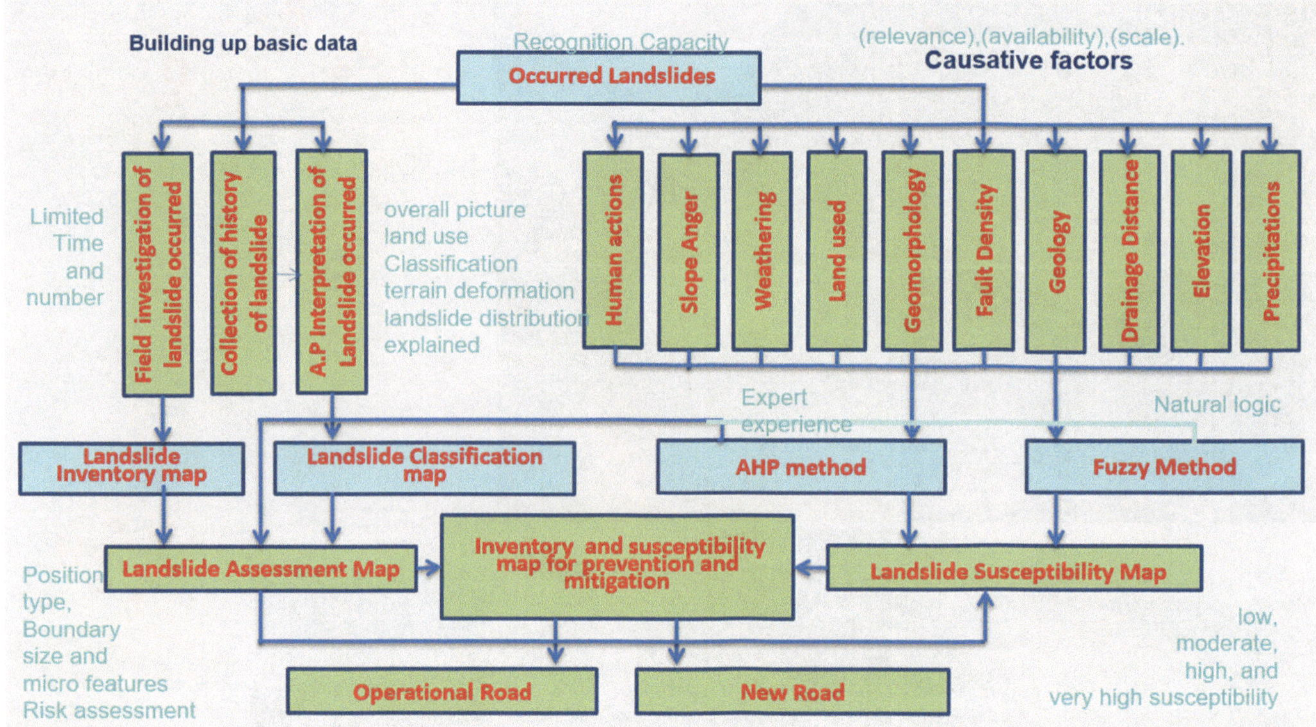

Fig. 2 Flow chart of research for landslide mapping

Soil Testing—Computer Simulation of Landslide Initiation and Motion

A testing group developed a high-stress undrained dynamic loading ring shear apparatus (ICL-2) which can be applied to deep landslides more than 100 m thick. The composition of new undrained dynamic loading ring shear apparatus is presented in Fig. 3. Its development and its first application to the Unzen Mayuyama Landslide, which killed 15,000 people in Japan, was reported in Landslides, Vol. 11, No. 5 in 2014.

The developed ring shear apparatus was revised in 2014–2015, based on the experiences of testing by Vietnamese short-term trainees as well as long-term trainees. A major revision was to add new two safety systems to protect the apparatus from mishandling by testing persons. The revised apparatus was installed in ITST in June 2015 and now it is available for testing (Lam, project report 2016).

Samples were taken from the ground and the drilled cores at various depths in the Hai van Landslide were tested using ICL-2, and a computer simulation was conducted based on the measured parameters by Vietnamese researchers. Output of the WG3 "Development of landslide risk assessment technology based on soil testing and computer simulation" was completed. A paper has already been written.

Adding the function that simulates tsunami generated by landslides was one of the targets of JST research that was completed to integrate the tsunami simulation code developed by the Intergovernmental Oceanographic Commission (IOC) and the landslide simulation code (LS-RAPID). The paper was accepted in February 2016 by the journal Landslides. This function was applied to assess tsunami levels possibly triggered by a large-scale rapid landslide from the Hai van slope.

Landslide Monitoring and Development of an Early Warning System

Haivan landslide is a deep-seated landslide, and a national railway runs over its body. It was selected as a target area for monitoring and early warning. For installation of monitoring equipment, both topographic and geology surveys had been carried out. Three boreholes were drilled. Figure 4 presents the 80 m depth bore-hole log at Haivan Landslide.

An integrated monitoring system, including rain gauges, extensometers, inclinometers, total station, and GNSS, was developed here. Most monitoring equipment was installed in Hai Van Station landslide until March 2015. However rainfall and slope deformation monitoring was started in May 2013 and number of slope deformation records during heavy rainfalls have been monitored from September 2013 until the present.

Installation of the data transferring and displaying system was finished in Haivan and the project office in ITST, Hanoi

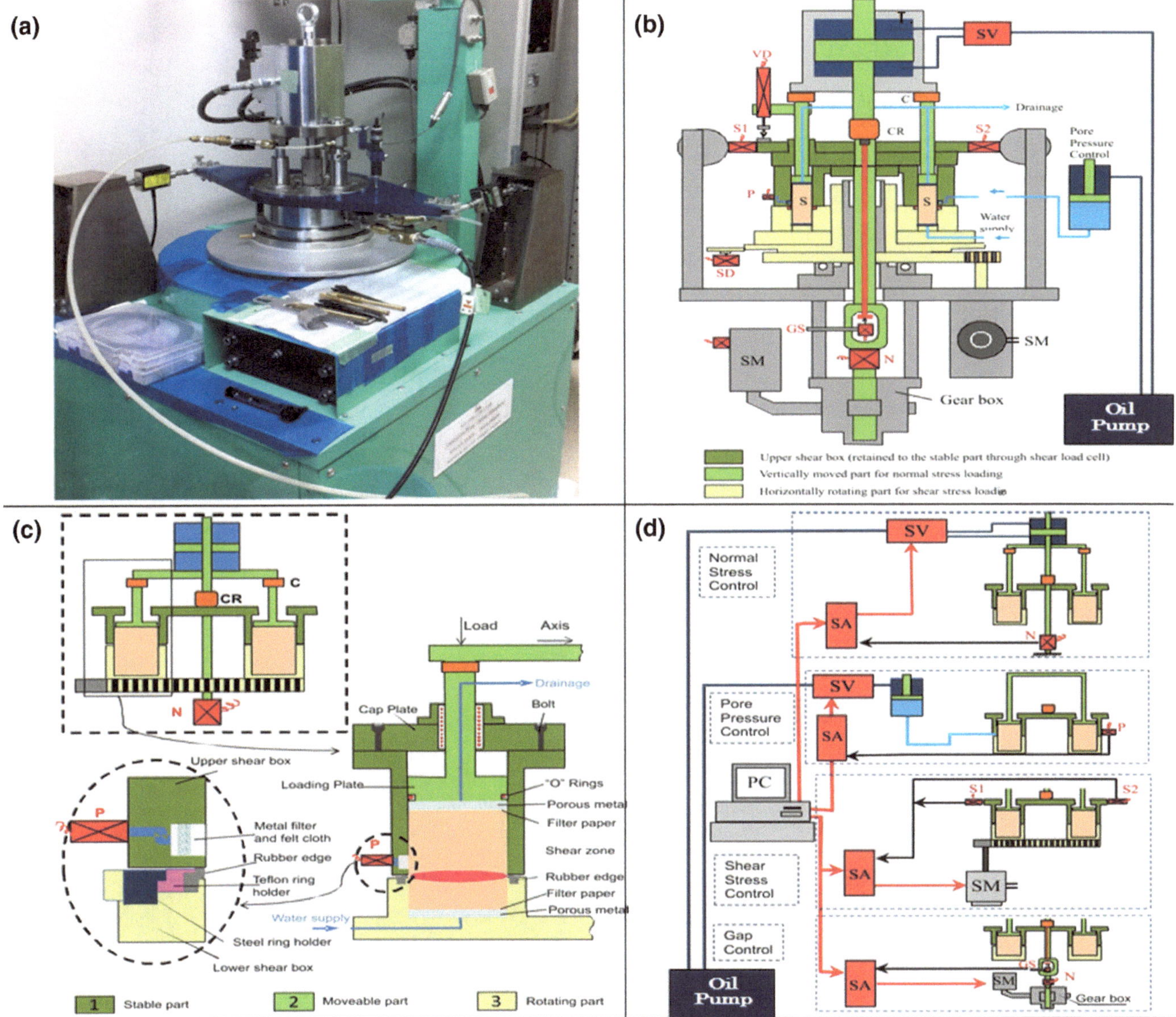

Fig. 3 Diagrams illustrating the new undrained dynamic loading ring shear apparatus

in March 2016. This monitoring system would allow monitoring in real time.

The landslide experimental facilities, including a landslide flume and data logging system and pore water pressure sensors, were provided and adjusted for ITST. The first landslide experiment using river sand was conducted in November 2015. The displacement of first landslide experiment using river sand is shown in Fig. 5.

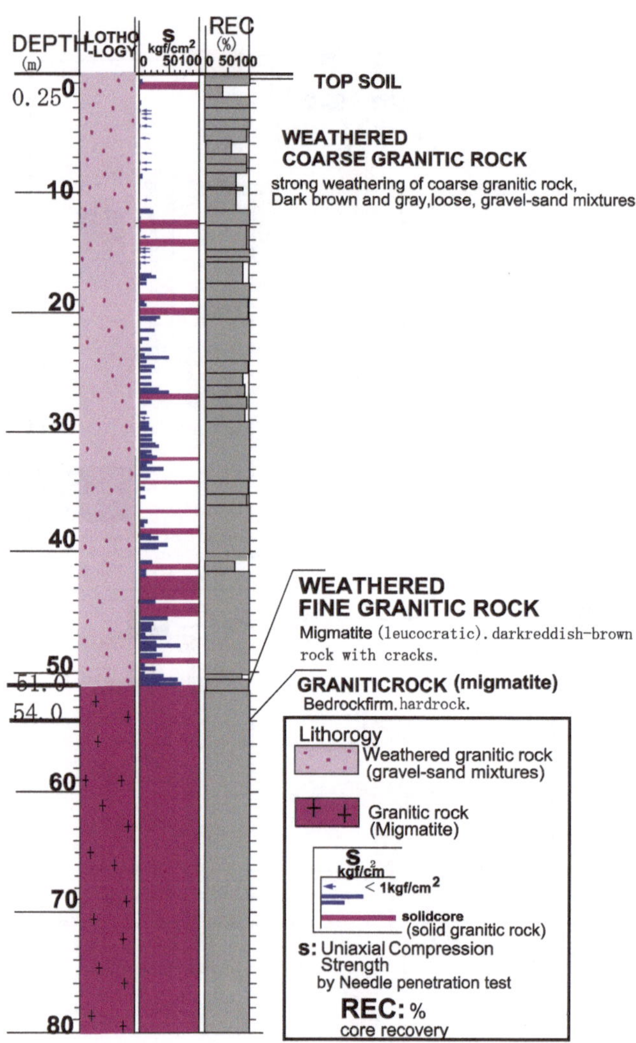

Fig. 4 80 m depth borehole log at Haivan landslide

Fig. 5 The displacement of the first landslide experiment using river sand

Reproduction of a landslide test using the granitic soils taken from the Hai van slope was a success. And new multi-depth wireless tensiometers were developed in Japan and utilized in landslide experiments by ITST.

Discussion

The new technology of interpretation of paired photos taken from a UAV for landslide identification of micro-features was developed as a trial. The scope of the application should be taken into consideration. The first trial was successful using a topographic landslide map with a scale of 1:500–1:1000.

For the Haivan landslide, the landslide initiation mechanism of the deep-seated landslide, which took place in the past, could be understood and explained using a ring shear apparatus. However, over time a transition in the landslide from occurrence to termination occurred. Minor landslides appeared on the body of the large landslide, together with erosion. So the sensitivity to mass sliding should be considered based on a multi-slip surface at the depth for early waning.

The principle for landslide early waning on Haivan is based on a prediction method using the inverse number of Velocity (Fukuzono 1985). However, as we have just gathered measurements on displacement over time for such a short period, the trend of landslide inverse velocity is not clear. It should be determined after a longer monitoring period.

Conclusion

Vietnam is a country on the coastal area of the Pacific Ocean. Due to its mountainous terrain, complicated geological structure and high rainfall, landslides occur regularly and cause serious damage to the mountain road network in the rainy season.

For the project, a wide-area landslide map and identification of landslide risk areas had been studied and developed by the WG2 Mapping Group. Development of landslide risk assessment technology based on soil testing and computer simulation had been carried out by the WG3-Testing Group. The relationship between landslide displacement and other causative factors such as precipitation, and pore water pressure by depth was studied at the landslide experimental facilities, and included landslide flume and data logging systems and pore water pressure sensors. The early warning system based on real-time landslide monitoring was installed by the WG4 Monitoring Group. The results of the project are not only very important developments in the scientific field of risk assessment but also in developing the research capacity of ITST in part and to Vietnam in general. Landslide risk assessment is a very important issue in the strategy to "proactively prevent natural disasters" of the Vietnamese government. This success from the research results could be applied in other tropical countries with similar conditions.

References

Arbanas Z, Mihalic-Arbanas S, Sassa K, Fukuoka H, Krkač M, Prodan MV, Gazibara SB, Domlija P (2017) The Croatian-Japanese SATREPS joint research project on landslides (IPL-161). Advancing culture for living with landslides, vol 1 ISDR-ICL Sendai partnerships 2015–2025 (this volume)

Han Q, Sassa K, Kan FM, Margottini C (2017) International programme on landslides (IPL): Objectives, history and list of IPL Projects. Advancing culture for living with landslides, vol 1 ISDR-ICL Sendai partnerships 2015–2025 (this volume)

Le HL, Miyagi T, Abe S, Hamasaki E, Tien DV (2014) Detection of active landslide zone from aerial photograph interpretation and field survey in central provinces of Vietnam—Proceedings of world landslide forum 3, 2–6 June 2014, Beijing

Ngo DD, Miyagi T, Luong LH, Hamasaki E, Hayashi K, Tien DV, Daimaru H, Abe S (2016) Trial of landslide Topography mapping using ALOS W3D data – Case study along the national road No. 7 in center of Vietnam. – Transactions, Japanese Geomorphological Union, vol 37(1); January, 2016; (ISSN 0389-1755)

Sassa K, Miyagi T, Ochiai H (2016) Project report for final evaluation. Development of landslide risk assessment technology along transport arteries in Vietnam. Proceedings of the final SATREPS workshop on landslides in Hanoi (331 p). ITST and ICL (ISBN:978-4-9903382-3-7)

Tien DV, Miyagi T, Abe S, Hamasaki E, Yoshimatsu H (2016a) Landslide susceptibility mapping along the HCMR in central Vietnam—an application of an AHP approach to humid tropical area. Transactions. Japanese Geomorphological Union, vol 37–1

Tien DV, Abe S, Ngo DD, Do NH, Miyagi T(2016b) Outline, typology and the causes of landslides in Vietnam. Transactions, Japanese Geomorphological Union, vol 37-1

Study of Slow Moving Landslide Umka Near Belgrade, Serbia (IPL-181)

Biljana Abolmasov, Miloš Marjanović, Svetozar Milenković, Uroš Đurić, Branko Jelisavac, and Marko Pejić

Abstract

The IPL project No 181 titled "Study of slow moving landslide Umka near Belgrade" started in November 2012. The study area is located on the right bank of Sava River, 25 km south west of Belgrade, Serbia. The basic objective of the Project was to enable the analysis, correlation and synthesis of data obtained from various phases of investigation of Umka landslide after 35 years of research. Apart from this, the analysis of data from monitoring conducted during certain phases of research was compared with data from automated GNSS monitoring over the last six years, although during numerous investigations various research methods were used for research and monitoring. The project was focused on: analysis of previous detail site investigations and field instrumentation from 1990–2005, analysis of aerial photos and orthophoto images from 1957–2010, analysis of automated GNSS monitoring results from 2010 to end of the Project and analysis of precipitation and levels of the Sava River. Project beneficiaries are local community and local and regional authorities. In this paper we will present results of the proposed project targets performed by Project participants.

Keywords

Landslide • Slow moving • Active • Deep seated

Introduction

The Republic of Serbia is located on the Balkan Peninsula in south-east Europe, covers an area of 88,361 km^2 and has a population of 7,181,505 (http://stat.gov.rs) (Fig. 1). Because of its complex geological history and terrain composition, and morphological and climate characteristics, 15.08% of Serbian territory is affected by landslides (active, suspended and dormant) (Dragićević et al. 2011). The greatest numbers of landslides are in Tertiary and Quaternary sediments. Tertiary sediments in Serbia are prevalent in the Pannonia basin and its northerly rim, as well as in remains of isolated lake basins in the central region south of the Sava and Danube rivers. Around 18% of Serbian territory is underlain by Neogene

B. Abolmasov (✉) · M. Marjanović
Faculty of Mining and Geology, University of Belgrade, Đušina 7, 11000 Belgrade, Serbia
e-mail: biljana.abolmasov@rgf.bg.ac.rs

M. Marjanović
e-mail: milos.marjanovic@rgf.bg.ac.rs

S. Milenković · B. Jelisavac
The Highway Institute, Kumodraška 257, 11000 Belgrade, Serbia
e-mail: svetozar.milenkovic@highway.rs

B. Jelisavac
e-mail: branko.jelisavac@highway.rs

U. Đurić · M. Pejić
Faculty of Civil Engineering, University of Belgrade, Bul Kralja Aleksandra 84, 11000 Belgrade, Serbia
e-mail: udjuric@grf.bg.ac.rs

M. Pejić
e-mail: mpejic@grf.bg.ac.rs

K. Sassa et al. (eds.), *Advancing Culture of Living with Landslides*,
DOI 10.1007/978-3-319-59469-9_37

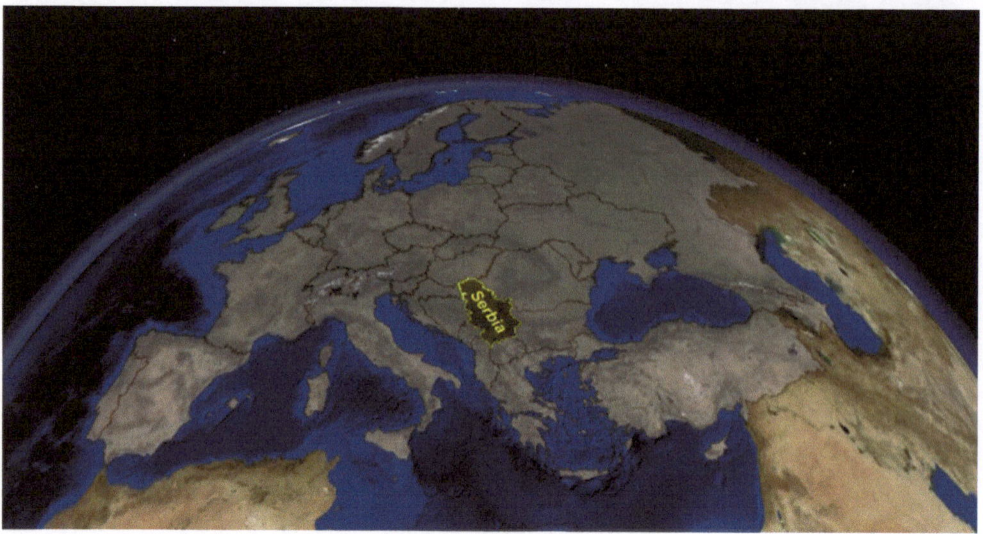

Fig. 1 Geographical position of the Republic of Serbia within Europe and the Balkan Peninsula

sediment complexes, with clays, marls, soft limestones, sands and gravels in a variety of spatial ratios. In terrains underlain by the Tertiary complexes, more than 25% of the territory is affected by landslides (Abolmasov et al. 2015).

Especially interesting are the areas of outer Belgrade and slopes of the right banks of the Sava and Danube rivers, which are known for their instability. Luković (1951), Vujanić et al. (1981, 1984), Lokin et al. (1988), Rokić et al. (1998, 2002), Ćorić et al. (1994, 1996) have all written about this phenomenon. The basic cause of the instability is the complex geological and morphological evolution of the terrain, further influenced by intensive anthropogenic activity, so many dormant landslides have reactivated and many new ones became active as well. According to the latest landslide inventory from 2010, which includes the inner area of the General Plan of Belgrade and covers an area of 437 km^2 (1/3 of the total area of the city), over 30% of the territory is composed of active and suspended landslides (Lokin et al. 2010).

The initiative to collaborate with the International Consortium on Landslides was started in September 2009. The faculty of Mining and Geology of the University of Belgrade became a member of ICL in 2011, and a member of the ICL Adria-Balkan Network in 2012 (Mihalić Arbanas et al. 2013). In March 2012, the Faculty of Mining and Geology and The Highway Institute applied for an IPL project and during the 7th Session of IPL-GPC in Paris in 2012, a joint project number 181 was approved. It was entitled "Study of Slow Moving Landslide Umka near Belgrade Serbia". This paper will show results obtained during four years of Project conduct, as described in the project plan and program.

Project Description

The basic objective of Project 181: "Study of slow moving landslide Umka near Belgrade, Serbia" was to analyse, correlate and synthesise data obtained from various phases of investigation after 35 years of research. Apart from this, the analysis of data received from geotechnical monitoring conducted during certain phases of research would be compared with data from automated GNSS monitoring over the last six years. Synthesis of research results would help define the mechanism and dynamics of movement of this large active slow landslide, with the objective of proposing adequate remedial measures. Project results would also aid a better understanding of other landslides on the right banks of the Sava and Danube.

The Project is organized by the University of Belgrade, Faculty of Mining and Geology and Faculty of Civil Engineering and The Highway Institute Belgrade. University, with Institute staff providing all necessary documentation for Project finalization. Maintenance of equipment will be organized by both institutions. Project Leader is Associate Professor Biljana Abolmasov from University of Belgrade,

Faculty of Mining and Geology. Core members of the Project are: Svetozar Milenković, MSc, Branko Jelisavac, MSc, Uroš Djurić, Ph.D. student, and Assistant Professor Miloš Marjanović.

Results

Study Area

Umka landslide is located in the territory of Belgrade, Municipality of Čukarica, in the right meander of the Sava River (Fig. 2). The Umka landslide has been investigated in detail for a number of years, leading to extensive geotechnical documentation, as well as the publication of a great number of papers in the last 35 years. The greatest amount of research was conducted to prepare technical documentation for various phases of design and planning of the E-763 motorways, whose route would cross the landslides of Umka and Duboko. To a lesser extent, research was carried out for urban development in the settlement of Umka. The last phase of research for the level of the preliminary design for the E-763 motorway was completed in 2005, and no further research has been conducted since then, as no remedial measures were taken and the motorway was not built. Automatic GNSS monitoring was introduced in March 2010 in the body of the Umka landslide as part of the TR36009 Project supported by the Ministry of Education, Science and Technological Development of the Republic of Serbia (Abolmasov et al. 2012b). The rates of displacement, oscillation of the levels of the Sava River, as well as hydro-meteorological parameters (type and intensity of precipitation) have been followed daily since then.

General Landslides Features

Umka landslide is in the shape of a triangular fan, around 900 m long and 1450 m wide at the toe, with a total surface area of around 1.8 km^2, an average depth of 14 m and with a total volume of approximately 14,000,000 m^3. The average inclination of the slope is 9°, apart from the main scarp zone and minor scarps where it can reach up to 25°. The main scarp of the landslide rises up 25 m. The left flank has developed a distinct side scarp, while the right flank of the landslide is concealed along the zone of the local road. The landslide is traversed by numerous secondary scarps of 1–10 m in height. The toe of the landslide is beneath the water level of the Sava River. Comparative morphological analysis of the orthophoto images from 1957 to 2010

Fig. 2 Orthophoto image of Umka landslide

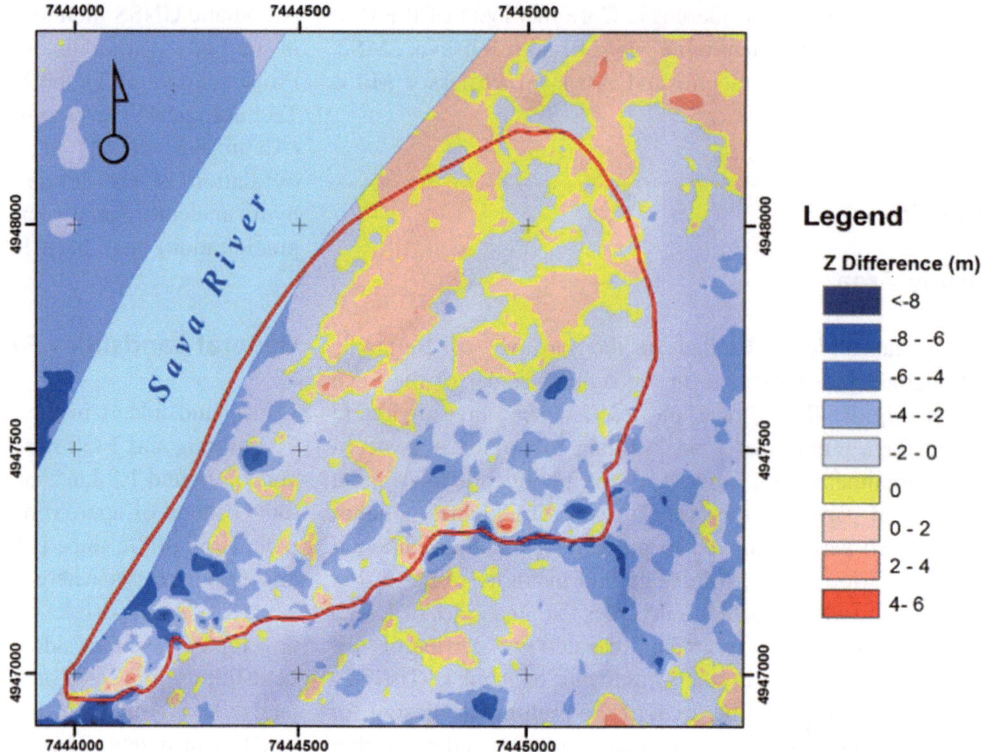

Fig. 3 Results of comparative morphological analysis of DEMs from 1957 to 2010

indicate a clear mass deficit in the zone of the frontal scarp and secondary scarps, i.e., enlargement of the landslide body in the toe (Abolmasov et al. 2012a, 2015) (Fig. 3).

The geological setting of the terrain is composed of Neogene and Quaternary sediments. Neogene sediments are represented by silty-clay and massive Pannonian marls (M_3^2L) over 200 m thick. Quaternary sediments are represented by loess (Q_2l) and diluvial clays (Q_2dl), which form a relatively thin cover (2–15 m) over the Pannonian marls.

The upper 25 m of the Pannonian marls is altered to clayey marls (M_3^2GL) and represents the weathered crust of the bedrock. The contact of these two units, i.e., the contact of marly clays (M_3^2GL) and fresh gray marls (M_3^2L) is where the principal (deepest) sliding surface was developed. As determined by engineering-geological mapping and confirmed by inclinometer measurements, it is not a uniform surface, but instead can be segregated into three parts that define three separate blocks of the landslide body (blocks A, B, C) (Fig. 4). The deepest displacements have been registered in block A, with a maximum depth of 26 m. The type of movement is translational, with a total shift during the period March–May 2005 recorded by inclinometers measurements of over 6 cm. The landslide depth varies from 4 to 26 m. Numerous secondary scarps and shallow slip planes that additionally separate certain blocks, and which are visible on the terrain surface, have also been mapped along the Sava River banks (Fig. 5).

As a consequence, groundwater distribution is complexly structured. It is usually scattered in hydraulically isolated accumulations, i.e., isolated free-level aquifers formed in the depletion zone, from the ground surface to a depth of 6–8 m, which was confirmed by piezometer observations during 1991–1993 (Vujanić et al. 1995). Very high levels of groundwater were recorded in boreholes and observation wells during January–August 2005 (Jelisavac et al. 2006). In the middle and foot sections of the landslide, the water was almost at the surface and in many locations was flowing diffusely. There is no other data of continuously monitoring of groundwater levels between these periods (Vujanic et al. 1995; Jelisavac et al. 2006).

Material Properties

To characterize the colluvium and bedrock sediments in terms of its implications for slope stability, an extensive number of laboratory tests were carried out on more than 200 samples, both disturbed and undisturbed. The focus of the testing was on the main physical and mechanical characteristics (grain size distribution, natural water content, unit weight, degree of saturation, Atterberg limits and strength parameters). All the values of available geotechnical parameters from the laboratory tests are plotted against depth, on the triangle diagram and on the A-line plasticity

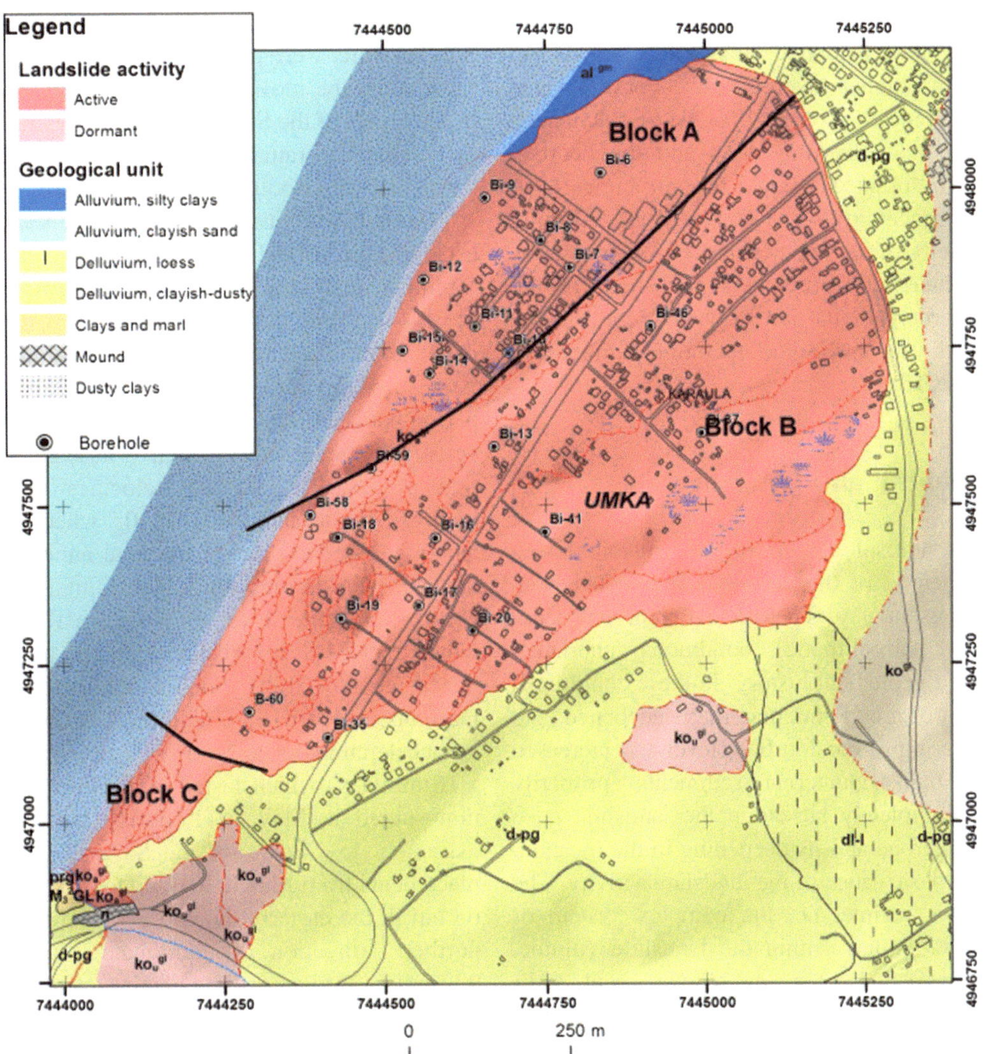

Fig. 4 Engineering geological map of Umka landslide

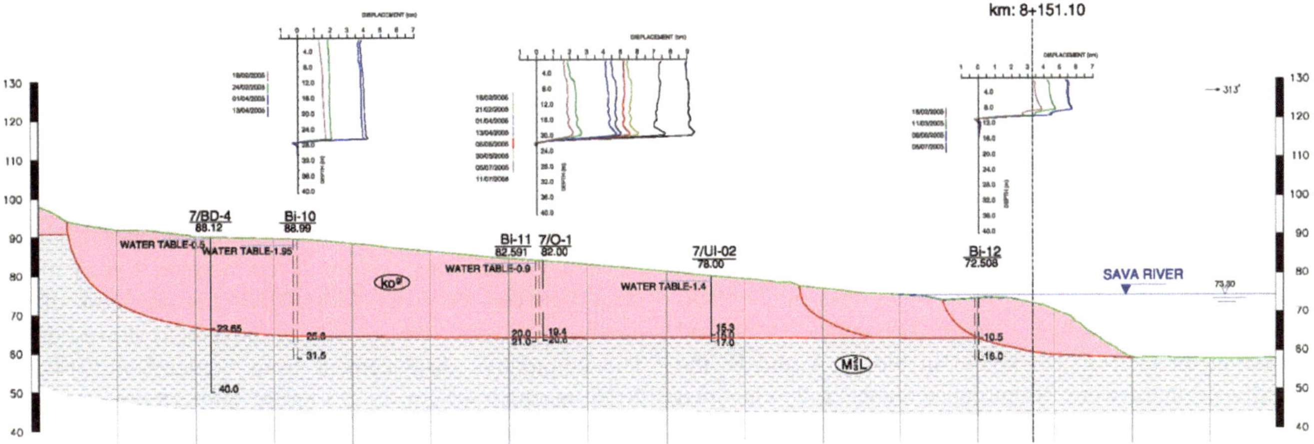

Fig. 5 Engineering geological cross section of Umka landslide toe—Block A

chart (Abolmasov et al. 2015). Because of the large area and different depths of the landslide, the geotechnical soil parameters show substantial heterogeneity. For the same reason, sharp differences in many physical-mechanical properties between the colluvium mass and the bedrock are not apparent. The drained residual friction angles measured in shear tests ranges from 4 to 21.5 degrees, reflecting the large heterogeneity of the material properties too. These results were compared with previously published correlations based on Atterberg limits and residual friction angle, and residual values are inside the literature correlations (Abolmasov et al. 2015).

Landslide Mechanism and Dynamics

Umka landslide is a result of prolonged erosion during migration and deepening of the Sava River riverbed from Holocene to the present day. Therein, different paleoecological (primarily paleoclimatic) conditions alternated throughout its recent geological history. In these conditions, the weathering crust of grey Pannonian marls deepened, and because of the lowering of erosion base level and increased erosion of the slope banks, Quaternary sediments—primarily loess, are almost completely absent in the landslide area. Narrowing of the riverbed and its deepening in the meander curve remained the basic cause of the landslide activity. The specific groundwater regime, i.e. the complex system of isolated groundwater bodies within the landslide volume, contributed to the present mechanisms of movement, additionally compounded by numerous secondary movements within certain blocks. Due to the deep sliding surface, intensification of displacement and secondary movements occur during long-term precipitation or sudden snow thaw. These are also the periods of sudden surges and subsequent drops in the water level of the Sava River, which is described as a dominant hydraulic triggering factor.

The dynamics of the landslide are not easy to determine, as there has been no complete or continuous monitoring over a longer period, as would be required for such a complex and large landslide. Based on the analysis of inclinometer readings during 2005, which was also recorded as one of the years when the landslide activity was more intensive, the following conclusions were inferred. A total of 24 inclinometers were installed during February 2005, and depending on their spatial layout in blocks A and B, the majority of them were discontinued in the following two months, while some remained functional until July 2005. The greatest displacements were recorded in block B, where the inclinometers were discontinued after only two month of observation (Jelisavac et al. 2006; Abolmasov et al. 2015). Also in block A, a sliding plane was registered at 26 m, and greatest displacement also occurred in March–April 2005. Similar shifts were measured in all 24 inclinometers. At the same time, the movements are synchronised with increases of the levels of the Sava River and an increase in the average daily air temperature, which caused thawing of 30 cm of snow in 24 h (Abolmasov et al. 2014).

Automated continuous real-time GNSS monitoring was established in March 2010 (Abolmasov et al. 2012b, 2013). Simultaneously, the levels of the Sava River were observed in near-real time, i.e., on daily basis, as well as the average daily temperature and type and amount of precipitation. After the monitoring system was established, during the period between March 2010 and October 2013, the landslide velocity varied, with distinctive acceleration and deceleration phases. The highest velocities of movement were recorded in March–July 2010, when the total rainfall amounted to one third of the total annual rainfall. Simultaneously with the unusually high rainfall, a rise and then an abrupt fall in the level of the Sava River were recorded, which caused 2 cm displacement in a single day (Abolmasov et al. 2012b). Similar behaviour was registered in December 2010 when a high level of the Sava River caused an acceleration in landslide velocity from 0.54 mm/day to 1.16 mm/day, which lasted until February 2011. Deceleration started in May 2011, during the drought period and extremely low levels of the Sava River. Cumulative displacements from March 2010 up to December 2013 equalled 65 cm in the easterly direction (y coordinate), 42 cm in the northerly direction (x coordinate), and a subsidence of 22 cm (z coordinate). The displacement vector of the monitoring point clearly shows a downhill trend towards the Sava River, with a relatively uniform constant linear movement (Fig. 6).

During extreme floods in Serbia in May 2014, the GNSS monitoring system was broken and it was installed again in September 2014. The structure of the automated monitoring system (Abolmasov et al. 2012b) was kept, but the monitoring point-station on Block B was moved 10 m to the SW to another position.

Analysing the nature of the 3-D movements during September 2014–July 2016, it is clear that trend is close to linear, which confirms that there were no sudden slides in the 22-month monitoring period, except acceleration of movement between February–May 2015. For the rest of the monitoring period, the velocity of movement slowed down. The cumulative displacement from September 2014–July 2016 reached 40 cm E, 25 cm N and 15 cm in height (Fig. 7). One can also conclude that Umka landslide has had a relatively constant velocity of movement towards the Sava River during the new GNSS point position monitoring period, as well as during previous research (Abolmasov et al. 2015).

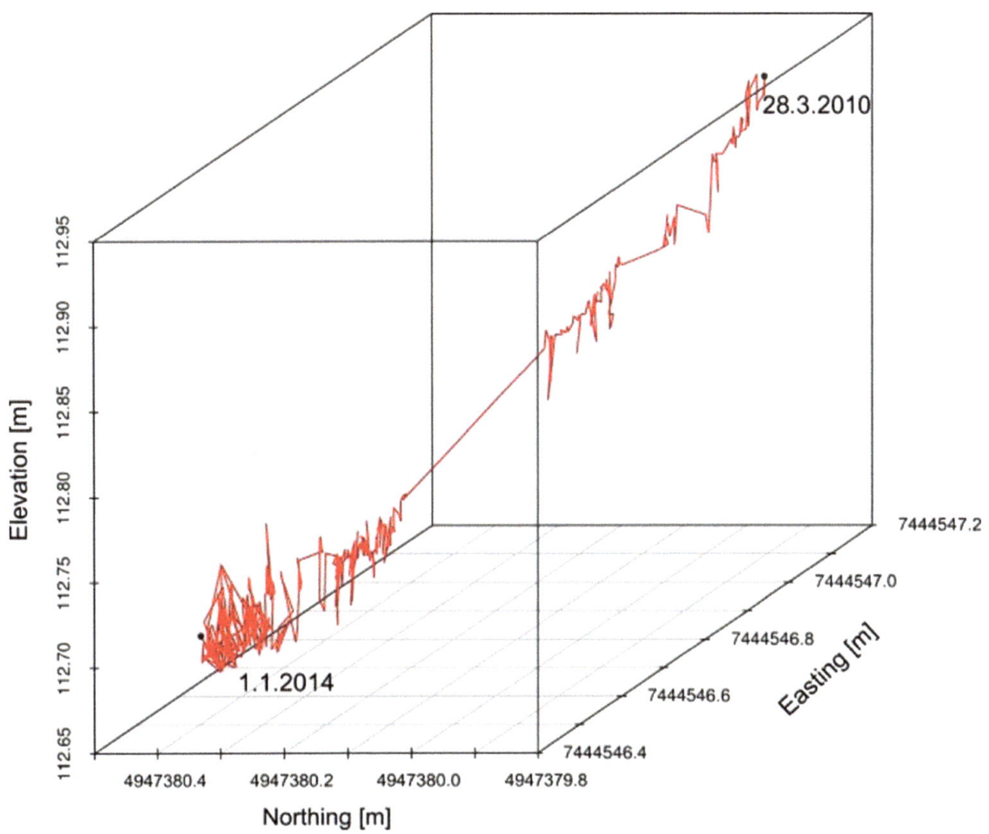

Fig. 6 Three-dimensional displacement vector of a GNSS monitoring station from 2010 to 2014

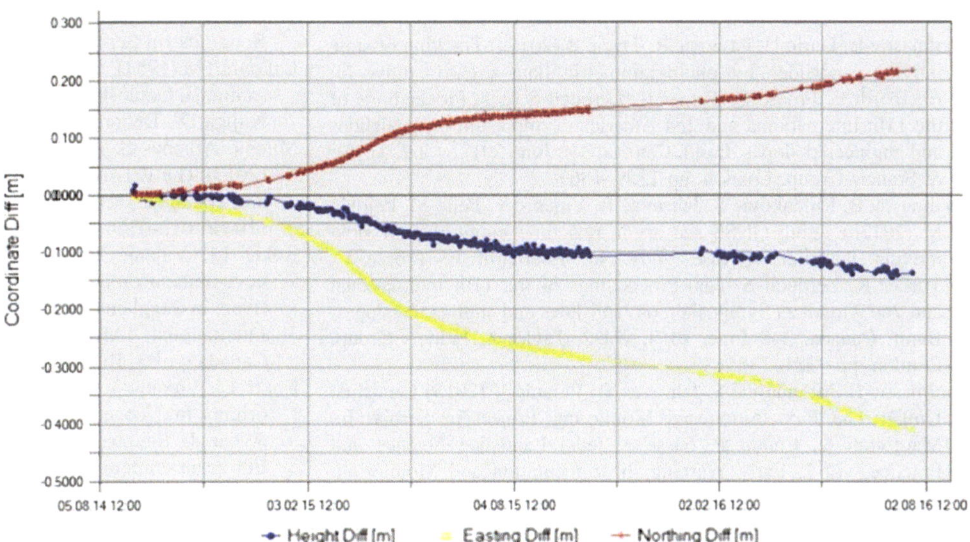

Fig. 7 Three-dimensional displacement plot from September 2014 to July 2016

Conclusion

The analysis of field investigations, laboratory results and monitoring data confirmed that Umka can be characterized as a compound landslide in stiff fissured clayey marls in active stage, with alternating phases of intensive and slow movements, while the velocity can be characterized as slow to very slow (Abolmasov et al. 2015).

Further research during the IPL 181 Project will focus on details of A, B and C block movement mechanisms and their synchronisation with data on the triggers (the Sava River level and rainfall) and correlation with actual GNSS monitoring results. Also, coupling current surface real-time monitoring GNSS system with a near-real time inclinometer measurements is also planned in the nearer future. This would support the geotechnical model and reveal the connection between ground displacement and actual displacement on the slip surface level. Finally, the continuation of the current monitoring campaign will be further support for the geotechnical model development and evaluation of additional numerical analyses.

Acknowledgements The IPL Project 181 would have not been possible without support from VekomGeo d.o.o and Republic Hydrometeorological Service of Serbia. The research was supported by the Ministry of Education, Science and Technological Development of the Republic of Serbia Project No TR 36009.

References

Abolmasov B, Đurić U, Pavlović R, Trivić B (2012a) Tracking of slow moving landslides by photogrammetric data–a case study. In: Eberhardt E, Froese C, Turner K, Leroueil S (eds) Proceedings of the 11th International and 2nd American symposium on landslides and engineered slopes, Banff, Canada, 3–8 June 2012, vol 2, Taylor & Francis Group, London, pp 1359–1363

Abolmasov B, Milenković S, Jelisavac B, Vujanić V, Pejić M, Pejović M (2012b) Using GNSS sensors in real time monitoring of slow moving landslides–a case study. In: Eberhardt E, Froese C, Turner K, Leroueil S (eds) Proceedings of the 11th International and 2nd American Symposium on landslides and engineered slopes, Banff, Canada, 3–8 June, 2012, vol 2, Taylor & Francis Group, London, pp 1381–1385

Abolmasov B, Milenković S, Jelisavac B, Vujanić V (2013) Landslide Umka: The First Automated Monitoring Project in Serbia. In: Margottini C, Canuti P, Sassa K (eds) Landslide Science and Practice, vol 2: Early Warning, Instrumentation and Monitoring, Springer, Berlin, pp 339–346

Abolmasov B, Pejić M, Šušić V (2014) The analysis of landslide dynamics based on automated GNSS monitoring. In: Sassa K, Mihalić Arbanas S, Arbanas Ž (eds) Proceedings of the 1st Regional Symposium on landslides in the Adriatic-Balkan Region—1st ReSyLAB 2013, Zagreb 6–9 March 2013, University of Zagreb, Faculty of Mining, Geology and Petroleum Engineering and University of Rijeka, Faculty for Civil Engineering, Zagreb, Croatia, pp 187–191

Abolmasov B, Milenković S, Marjanović M, Đurić U, Jelisavac B (2015) A geotechnical model of the Umka landslide with reference to landslides in weathered Neogene marls in Serbia. Landslides 12 (4):689–702. doi:10.1007/s10346-014-0499-4

Ćorić S, Božinović D, Vujanić V, Jotić M, Jelisavac B (1994) Slope instability analysis in Neogene clays and marls. In: Lisboa Portugal, Oliveira R, Rodrigues LF, Coelho AG, Cunha AP (eds) Proceedings of 7th International IAEG Congress, 5–9 Sept 1994, vol 3, Balkema, Rotterdam, pp 1759–1770

Ćorić S, Božinović D, Vujanić V, Jotić M, Jelisavac B (1996) Geotechnical characteristics of old landslides in Belgrade area. In: Senneset K (ed) Proceedings of the 7th International symposium on landslides, 17–21 June 1996, Trondheim, Norway, Vol 2, Balkema, Rotterdam, pp 689–694

Dragićević S, Filipović D, Kostadinov S, Ristić R, Novković I, Živković N, Anđelković G, Abolmasov B, Šećerov V, Đurđić S (2011) Natural hazard assessment for land-use planning in Serbia. Int J Environ Res 5(2):371–380

Jelisavac B, Milenković S, Vujanić V, Mitrović P (2006) Geotechnical investigations and repair of the landslide Umka–Duboko on the route of motorway E-763 Belgrade-South Adriatic. International Workshop-Prague-Geotechnical days, Prague

Lokin P, Sunarić D, Cvetković T (1988) Landslides in Neogene sediments on the right Danube bank, Yugoslavia. In: Ch Bonnard (ed) Proceedings of the 5th international symposium on landslides, 10–15 July 1988, Lausanne, vol 1, Balkema, Rotterdam, pp 213–217

Lokin P, Pavlović R, Trivić B (2010) Projekat istraživanja terena za izradu katastra klizišta područja Generalnog urbanističkog plana područja Beograda. J.P. Direkcija za građevinsko zemljište grada Beograda (in Serbian). Unpublished material

Luković TM (1951). Važniji tipovi naših klizišta i mogućnosti njihovog saniranja. Geološki Vesnik Savezne uprave za geološka istraživanja, Knjiga IX, Beograd, pp 275–310 (in Serbian)

Mihalić Arbanas S, Arbanas Ž, Abolmasov B, Mikoš M, Komac M (2013) The ICL Adriatic-Balkan network: analysis of current state and planned activities. Landslides 10:103–109. Doi:10.1007/s10346-012-0364-2

Rokić Lj, Vujanić V, Jotić M (1998). Forecast of the landslide development processes based on the study of erosion processes of rivers in the plains. In: Moore D, Hungr O (eds) Proceedings of 8th International IAEG Congress, 21–25 Sept 1998, Vancouver, Canada, vol 3, Balkema Rotterdam, pp 1485–1491

Rokić Lj, Vujanić V (2002) A contribution on the study of landslide origins in Neogene sediments of Danube river coastal area. In: Rybar J, Stemberk J, Wagner P (eds) Proceedings of the 1st European conference on landslides, Prague, Czech Republic, June 24–26 2002, Balkema Publishers, pp 291–298

Vujanić V, Jotić M, Jelisavac B, Božinović D, Ćorić S (1995) Sinteza rezultata gotehničkih istraživanja klizišta na Savi: Umka i Duboko. Zbornik radova Drugog simpozijuma Istraživanje i sanacija klizišta, 6–9 juni 1995, Donji Milanovac, pp 335–351 (in Serbian)
Vujanić V, Livada N, Božinović D (1984) On an old landslide in Neogene Clays on the right bank of the Sava near Belgrade. Proceedings of 4th international symposium on landslides, Toronto, Canada, 1984, vol 2, pp 227–233
Vujanić V, Livada N, Jotić M, Gojković S, Ivković J, Božinović D, Sunarić D, Šutić J (1981) Klizište "Duboko" na Savi kod Beograda. Zbornik radova Simpozijuma istraživanje I sanacija klizišta, Bled 1981. Knjiga 1:119–134 (in Serbian)

Influence of Post-Earthquake Rainfall on the Stability of Clay Slopes (IPL-192)

Binod Tiwari, Beena Ajmera, and Duc Tran

Abstract

Rainfall and earthquakes are considered two of the major causes of landslides worldwide. These landslides cause billions of dollars in property damage and revenue losses, as well as the deaths of thousands of people each year. While researchers have been examining the effect of either rainfall or earthquakes on the deformation and stability of slopes, the combined effect of rainfall and earthquakes on deformation and slope stability has not been evaluated systematically. In this study, a series of model slopes were constructed in a Plexiglas container placed on top of a shake table. The model slopes were prepared to have different initial void ratios of 0.89, 1.0 and 1.2 and various slope inclinations of 30°, 40°, and 45°. These slopes were instrumented with accelerometers, tensiometers and inclinometers and subjected to a number of sinusoidal seismic motions with different seismic accelerations from 0.1 to 0.3 g, with several frequencies ranging from 1 to 3 Hz for various durations ranging from 10 cycles to 50 cycles of loading. Following the earthquake event, a rain simulator system was used to induce rainfall at intensities of either 18, 30 or 60 mm/h. The seepage velocity, spatial variation of suction and the deformation of the slopes were determined. The results obtained were compared to those obtained from similar slopes subjected to rainfall without an earthquake event. The study showed that the seismic shaking resulted in a reduction in the seepage velocity in the slope, which led to an increase in the factor of safety of the slope with time.

Keywords

Earthquake-induced landslides • Rainfall-induced landslides • Partially saturated soils • Shake table • Seepage velocity

Introduction

Globally, rainfall and earthquakes are two of the most prominent causes of landslides, which result in thousands of injuries and deaths, and billions of dollars of property damage and revenue losses each year. In recent history, the 2014 Oso Landslide in Washington, which was said to be triggered by heavy rainfall, is among the worst landslide disasters in the United States (Keaton et al. 2014; Iverson et al. 2015). Researchers, including Tiwari et al. (2016), Tran (2016), Collins and Znidarcic (2004), Kim et al. (2012), Orense et al. (2004), Ling and Ling (2012), Lu et al.

B. Tiwari
Department of Civil and Environmental Engineering, California State University Fullerton, 800°N. State College Blvd.E-419, Fullerton, CA 92831, USA
e-mail: btiwari@fullerton.edu

B. Ajmera (✉)
California State University Fullerton, 800°N. State College Blvd. E-318, Fullerton, CA 92831, USA
e-mail: bajmera@fullerton.edu

D. Tran
Department of Civil and Environmental Engineering, Graduate Student California State University Fullerton, 800°N. State College Blvd, Fullerton, CA 92831, USA
e-mail: ductran@csu.fullerton.edu

K. Sassa et al. (eds.), *Advancing Culture of Living with Landslides*,
DOI 10.1007/978-3-319-59469-9_38

(2012), and Tiwari and Caballero (2015), have studied the effect of various parameters such as slope geometry, soil density, and intensity of rainfall on the stability of rainfall-induced slope failures. Furthermore, databases containing over 3400 landslides after the 2011 Tohoku Earthquake in Japan (Wartman et al. 2013), over 14,000 landslides after the 2015 Gorkha Earthquake in Nepal (Tiwari et al. 2016), over 56,000 landslides after the 2008 Wenchuan Earthquake in China (Dai et al. 2011) and over 10,000 landslides after the 1999 Chi-Chi Earthquake in Taiwan (Khazai and Sitar 2004) clearly demonstrated the disastrous consequences of earthquake-induced landslides. Following strong ground shaking, a major concern for residents and the local officials in the region is the potential damage and consequences that could result if additional landslides are triggered during the future rainstorms and monsoon seasons. Despite this, there has been little research geared at understanding the influence of post-earthquake rainfall on the stability of slopes. This study aims to fill that gap by examining the effect of rainfall and seismic shaking on model soil slopes and numerical simulations using the GeoStudio software suite.

The soil used in the model soil slopes was collected from a housing development project in Mission Viejo, California. At this project site, slope movements were reported to have occurred on January 20, 2005 near the residences on Encorvardo Lane on the eastern side of Ferrocarril. Heavy rainfall from December 28, 2004 through January 11, 2005 resulted in almost 254 mm of precipitation. It was following this storm that initial movements were recorded on a 70 ft tall graded slope constructed in 1967. Model soil slopes tested in this study were prepared from compacted fill slope material obtained from this housing development project. Several standard laboratory tests were also conducted to determine the material properties. This paper will detail the influence of post-earthquake rainfall on the stability of slopes by presenting the results of the model soil slopes and the numerical simulations conducted.

Materials and Methods

The soil used in this study was collected from the housing development project in Mission Viejo, California. It was used to measure the grain size distribution (ASTM D422), the index properties (ASTM D4318), maximum dry density and the optimum moisture content (ASTM D1557), the saturated coefficients of permeability (ASTM D5084), the soil water characteristic curve and the shear strength parameters (ASTM D3080). Table 1 outlines the material properties for the soil properties.

The slope was constructed in a 1.2 m × 1.2 m × 1.2 m Plexiglas container mounted on a shake table. At the bottom of the container, a drainage layer consisting of a 5-cm-thick gravel layer and perforated plastic pipe drainage network were installed. To prevent the soil from clogging the drainage layer and the perforated pipe, a geo-textile was placed on top of the gravel drainage layer. The soil used to construct the model slopes was first sieved through a #40 sieve using a mechanical shaker and mixed with sufficient water to have an initial moisture content of 12%, corresponding to the field moisture content. The soil was placed in 5 cm thick lifts compacted to a void ratio of 1.2 to create a slope inclined at 40°. The completed model slope is pictured in Fig. 1.

Tensiometers and accelerometers were installed at various locations and depths (Figs. 2 and 3) within the slope to instrument the slope in order to measure the pore water pressure during the seismic loading and the post-earthquake rainfall event. Decagon T5 tensiometers with an active surface area of 0.5 cm^2 and a 5 mm diameter ceramic tip were used in this study. They are capable of recording pore water pressures ranging from +100 to −85 kPa. The holes made to install the tensiometers were backfilled with the slope material and the surface was covered with a bentonite slurry to prevent preferential movement of water through those locations.

The slope was then subjected to different ground motions using the shake table at California State University, Fullerton. The 1.5 m × 2 m shake table uses a 25 kN actuator and has a ±6.5 cm horizontal capacity. The model slope was subjected to sinusoidal motions with accelerations ranging in amplitude from 0.1 g to 0.3 g with frequencies between 1 and 3 Hz. Each motion was applied for 10 to 50 cycles of loading. The ground motion recorded at Station 90095 during the 1994 Northridge Earthquake was also applied to the slope model. The ground motions applied are shown in Fig. 4. Following the seismic loading, a rain simulator system was placed on top of the slope. The system was used to apply rainfall at an intensity of 3.6 cm/h until the slope became completely saturated, as determined from the location of the wetting front. A photograph of the rain simulator system applying rainfall to the model slope is presented in Fig. 5.

A second model was prepared following the procedure described above and instrumented as illustrated in Fig. 2. In this model, the slope was subjected to rainfall without any seismic excitation. The results obtained served as baseline in order to compare the effect of the seismic loading on the observed responses.

The slope geometry and material properties were used to perform numerical simulations using the SEEP/W, SIGMA/W, and QUAKE/W modules in the GeoStudio software suite. SEEP/W was used obtain the pore-water pressure distribution. The seismic loading was simulated using QUAKE/W, in which the input motion was the same as that used in the experimental models, as shown in Fig. 4.

Table 1 Soil properties

Property	Value
Gravel (%)	0
Sand (%)	15
Silt (%)	65
Clay (%)	20
Liquid Limit (%)	57
Plasticity Index (%)	30
USCS Classification	CH
Maximum Dry Density (pcf)	110
Optimum Moisture Content (%)	15
Saturated permeability at (m/s) Void ratio of 0.89 Void ratio of 1.00 Void ratio of 1.20	4×10^{-9} 5×10^{-9} 9×10^{-9}
Friction angle (°) and cohesion (kPa) at void ratio of 1.2 at Degree of Saturation of 33% Degree of Saturation of 100%	27°, 34 kPa 26°, 0 kPa
Rate of increase in shear strength relative to matric suction	27.2°

Fig. 1 Slope model used in this study

The seepage patterns as a result of the rainfall were simulated using SIGMA/W.

Results and Discussion

The movement of the wetting fronts with time for the two models prepared in the laboratory that were subjected to post-earthquake rainfall and to rainfall without earthquake shaking are compared and presented in Fig. 6. The locations of the tensiometers, denoted by T, are also shown in this figure. Unfortunately, all except one of the tensiometers malfunctioned in the slope model that was subjected to rainfall without any shaking (Fig. 6b). Figure 6 also contains the values of the seepage velocity computed, based on the time required for the wetting front to reach Point A. It is clear from Fig. 6 that more time was required for the complete saturation of the model that was subjected to

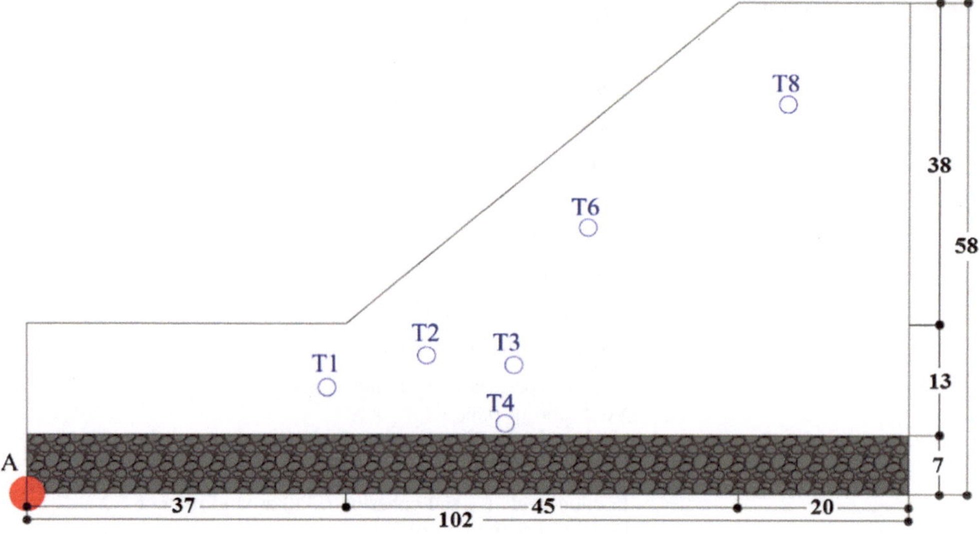

Fig. 2 Locations of tensiometers

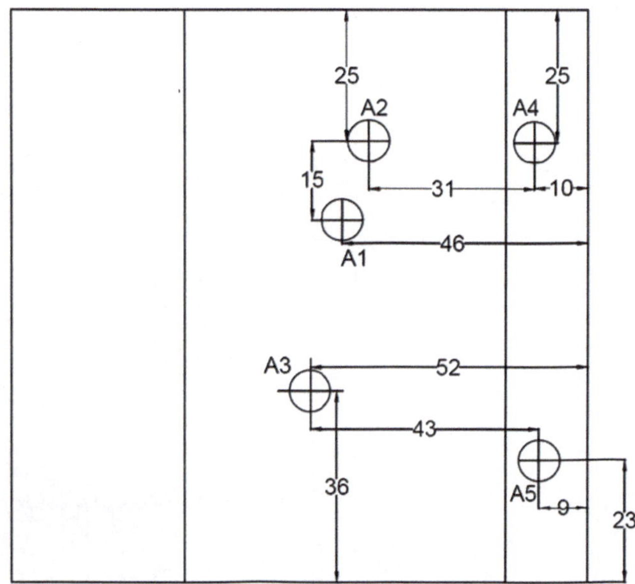

Fig. 3 Plan view of slope showing locations of the accelerometers

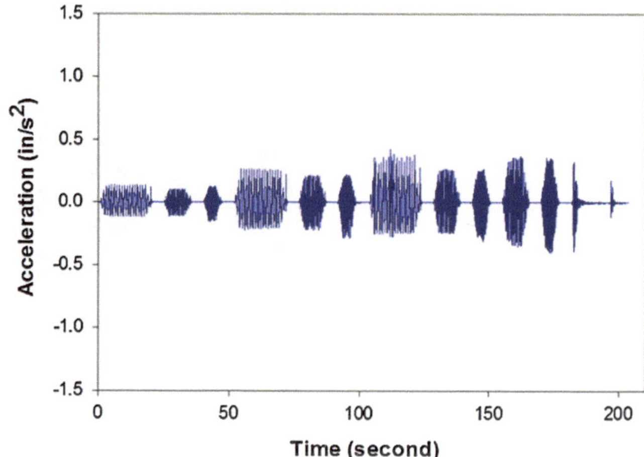

Fig. 4 Applied cyclic loading

earthquake loading than the time required for complete saturation of the model that was not subjected to any shaking. Specifically, the seepage velocity in the slope subjected to post-earthquake rainfall was 14.6 cm/h in comparison to the seepage velocity of 25.7 cm/h in the slope subjected to rainfall without earthquake loading. A comparison of the results from the tensiometers (Fig. 7) also agrees with the observations made of the wetting fronts. Specifically, the results show that approximately 2.7 h were required for the suction to reduce to a value of zero at the location of T1 in the slope subjected to rainfall without earthquake loading, while approximately 3.2 h were required in the slope subjected to post-earthquake rainfall.

Results from Numerical Simulations

Figure 8 shows the movement of the wetting front with time obtained from the numerical simulations conducted as part of this study. The results in Fig. 8 were used to calculate the seepage velocity to point A and compared with the experiment results. As it can be observed from Fig. 6b and Fig. 8, the seepage velocity from the numerical simulations are similar to those obtained from the experimental models.

SLOPE/W was used to calculate the factor of safety of the slope subjected to post-earthquake rainfall. The variation in the factor of safety during the earthquake loading is presented in Fig. 9, while the results for the factor of safety during the post-earthquake rainfall are presented in Fig. 10. Figure 10 also contains the results for the factor of safety for the slope subjected to rainfall without any earthquake loading. The figure shows that the factor of safety for the

Fig. 5 Rain simulator system on slope model

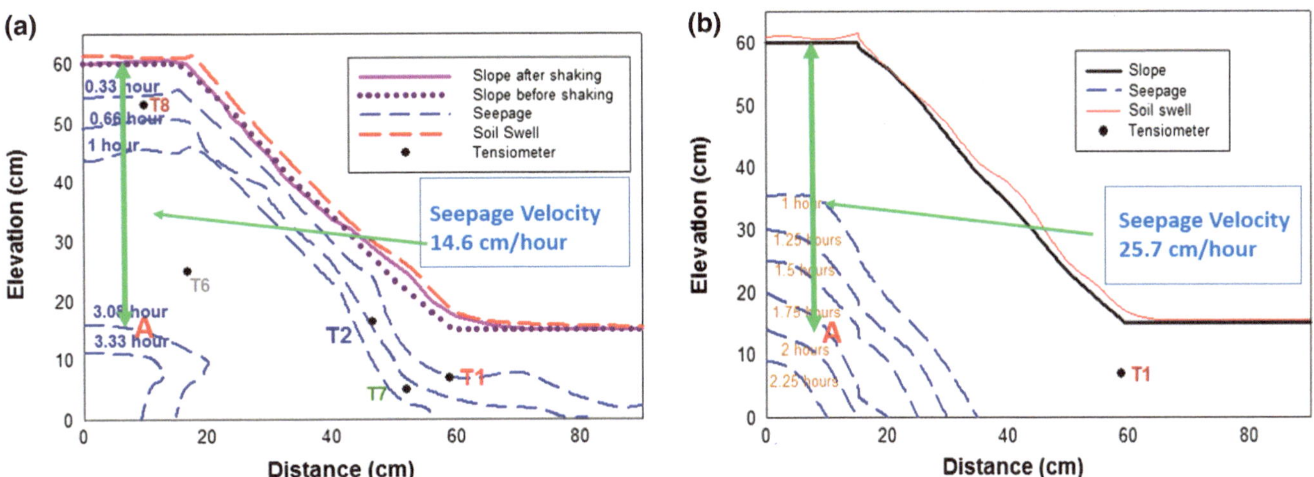

Fig. 6 Comparison of the wetting fronts and seepage velocities for slopes subjected to **a** post-earthquake rainfall and **b** rainfall without earthquake loading

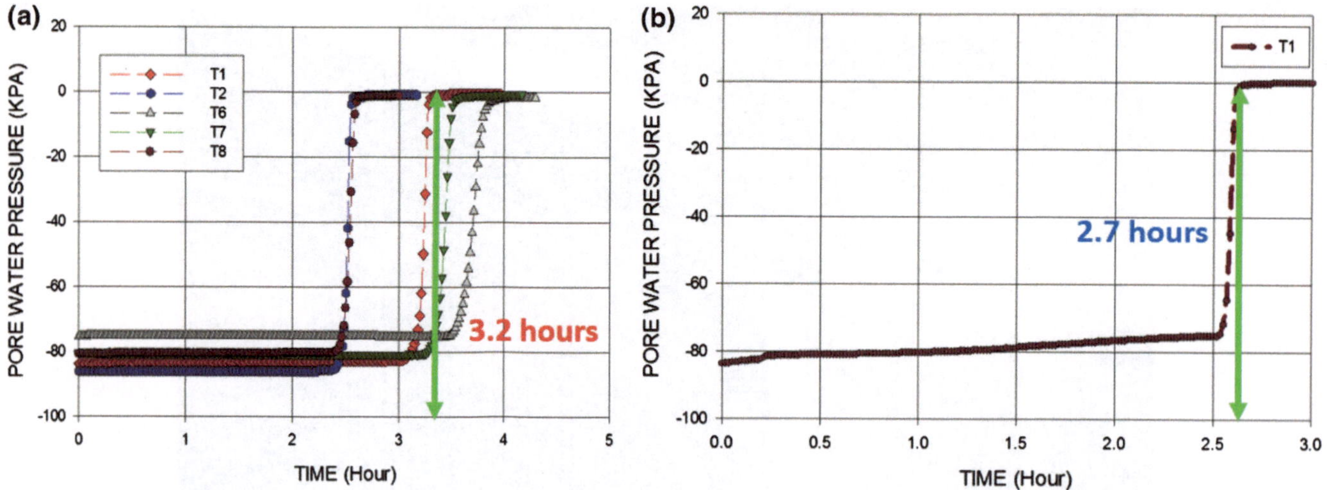

Fig. 7 Comparison of suction recorded at tensiometer 1 (T1) for slopes subjected to **a** post-earthquake rainfall and **b** rainfall without earthquake loading

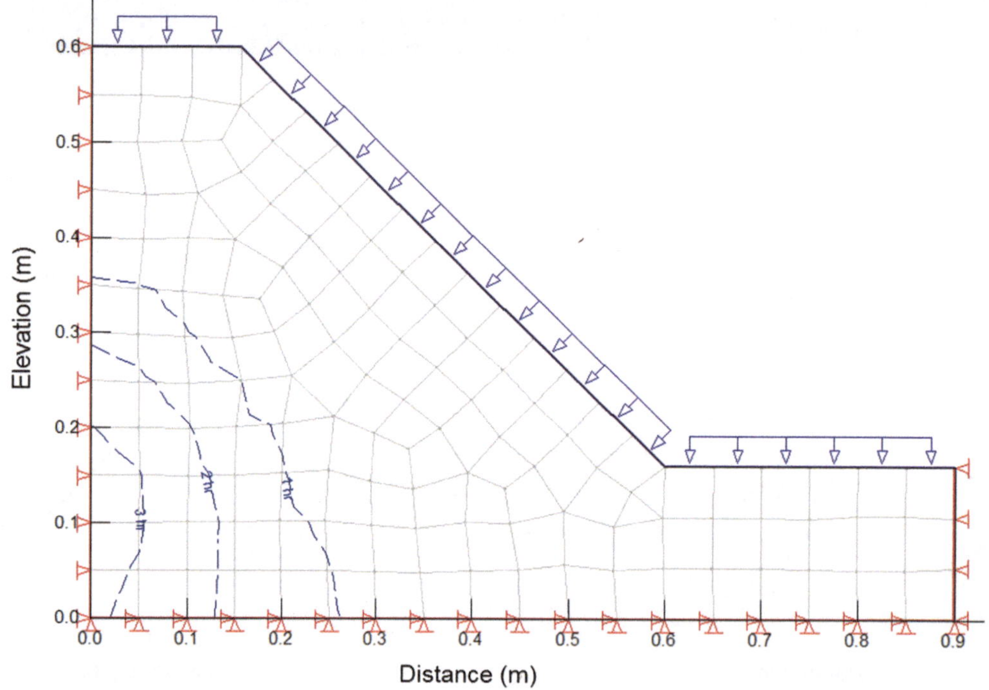

Fig. 8 Movement of the wetting front with time from numerical simulations

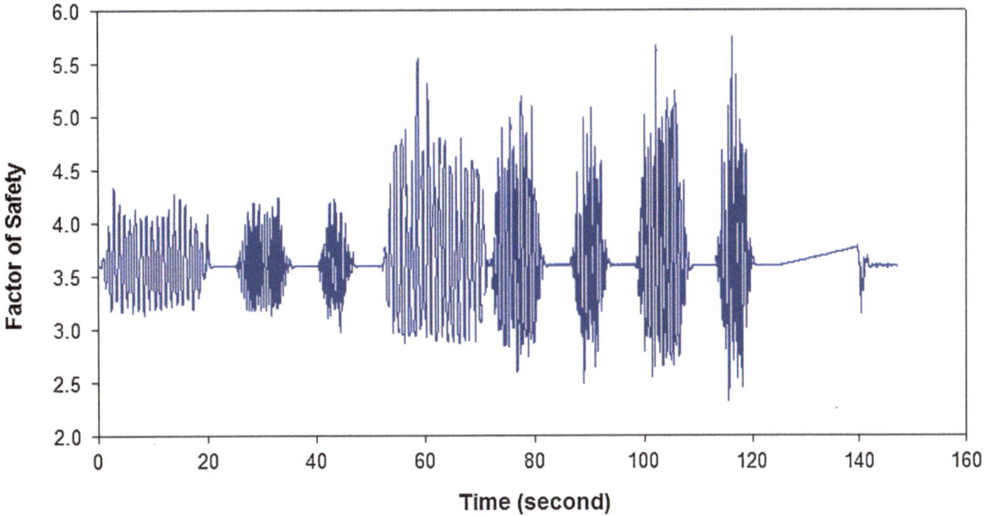

Fig. 9 Factor of safety computed from SLOPE/W during earthquake loading

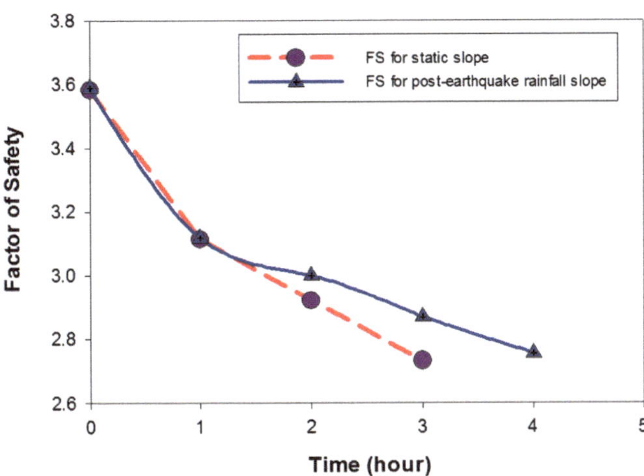

Fig. 10 Factor of safety computed from SLOPE/W in a slope subjected to rainfall following earthquake shaking and a slope subjected to rainfall without any earthquake loads

slope subjected to rainfall following earthquake shaking was higher than the slope subjected rainfall without any earthquake shaking. This can be attributed to the fact that the seepage velocity in the slope subjected to post-earthquake rainfall was lower than in the slope subjected to rainfall without earthquake loading.

Conclusions

In order to determine potential causes for the computed factors of safety, soil was collected from a landslide site and used to prepare the laboratory slope models in a Plexiglas container. The model slopes, instrumented with tensiometers and copper wires, were subjected to rainfall using a rain simulator system. A second model prepared to the same slope geometry and density was constructed on top of a shake table and subjected to a series of sinusoidal earthquake loading functions, as well as to the motion recorded from the 1995 Northridge Earthquake, before being subjected to rainfall at the same intensity as the previous model. The results showed that the slope subjected to post-earthquake rainfall required more time to become completely saturated and had a lower seepage velocity than the slope subjected to rainfall without earthquake shaking. The results suggest that the earthquake shaking was responsible for increasing the density of the slope material and thereby effectively reducing the permeability of the material, resulting in a more stable slope.

Acknowledgements The authors would like to thank the California State University, Fullerton Intramural Grant 3361, the Engage in STEM grant and the STEM^2 grant for supporting students working on this project through stipends, as well as for funds required to purchase the materials and supplies necessary for this study.

References

ASTM D1557 (2012) Standard test method for laboratory compaction characteristics of soil using modified effort. Am Soc Test Mater

ASTM D3080 (2011) Standard test methods for direct shear test of soils under consolidated drained conditions. Am Soc Test Mater

ASTM D422 (2002) Standard test methods for particle-size analysis of soils. Am Soc Test Mater

ASTM D4318 (2010) Standard test methods for liquid limit, plastic limit, and plasticity index of soils. Am Soc Test Mater

ASTM D5084 (2016) Standard test methods for measurement of hydraulic conductivity of saturated porous materials using a flexible wall permeameter. Am Soc Test Mater

Collins BD, Znidarcic D (2004) Stability analyses of rainfall induced landslide. J Geotech Geoenvironmental Eng 130(4):362–372

Dai FC, Xu C, Yao X, Xu L, Tu XB, Gong QM (2011) Spatial distribution of landslides triggered by the 2008 Ms 8.0 Wenchuan Earthquake, China. J Asian Earth Sci 40(4):883–895

Iverson RM, George DL, Allstadt K, Reid ME, Collins BD, Vallance JW, Schilling SP, Godt JW, Cannon CM, Magirl CS, Baum RL (2015) Landslide mobility and hazards: implications of the 2014 Oso disaster. Earth and Planet Sci Lett 412:197–208

Keaton JR, Wartman J, Anderson S, Benoît J, deLaChapelle J, Gilbert R, Montgomery DR (2014) The 22 March 2014 Oso landslide, Snohomish County, Washington, National science foundation geotechnical extreme events reconnaissance report

Khazai B, Sitar N (2004) Evaluation of factors controlling earthquake-induced landslides caused by Chi-Chi Earthquake and comparison with the Northridge and Loma Prieta events. Eng Geol 71(1–2):79–95

Kim J, Jeong S, Regueiro R (2012) Instability of partially saturated soil slopes due to alteration of rainfall pattern. Eng Geol 147:28–36

Ling H, Ling HI (2012) Centrifuge model simulations of rainfall-induced slope instability. J Geotechn Geoenvironmental Eng 138:1151–1157

Lu N, Wayllace A, Oh S (2012) Infiltrating-induced seasonally reactivated instability of a highway embankment near the Eisenhower tunnel, Colorado, USA. Eng Geol 162:22–32

Orense R, Shimona S, Maeda K, Towhata I (2004) Instrumented model slope failure due to water seepage. J Nat Disaster Sci 26 (1):15–26

Tiwari B, Ajmera B, Dhital S (2016) Photographic database and geospatial analyses of co-seismic landslides triggered by the main shock and aftershocks of the 2015 Gorkha earthquake. Report Submitted to the Department of Civil and Environmental Engineering, California State University, Fullerton

Tiwari B, Caballero S (2015) Experimental model of rainfall induced slope failure in compacted clays. Geotechn Spec Publ 256:1217–1226

Tiwari B, Tran D, Ajmera B, Woli H, Stapleton J (2016) Effect of pre and post earthquake rainfall events on deformation and stability of slopes. In: Proceedings of geotechnical and structural engineering congress

Tran D (2016) Effect of rainfall and seismic activities on compacted clay slopes having different void ratios and inclinations. Masters Thesis, Submitted to the Department of civil and environmental engineering, California State University, Fullerton

Wartman J, Dunham L, Tiwari B, Pradel D (2013) Landslides in Eastern Honshu induced by the 2011 off the pacific coast of Tohoku earthquake. Bull Seismol Soc Am 103(2b):1503–1521

Public Awareness and Education Programme for Landslide Management and Evaluation Using a Social Research Approach to Determining "Acceptable Risk" and "Tolerable Risk" in Landslide Risk Areas in Malaysia (IPL-194, IPL-207)

Ab Rashid Ahmad, Zainal Arsad Md Amin, Che Hassandi Abdullah, and Siti Zarina Ngajam

Abstract

Although early records of landslides in Malaysia have existed since the beginning of the last century, national attention on landslides increased in earnest in the wake of the 1993 Highland Towers landslides. In 2003, an economically devastating rockslide in Bukit Lanjan led to the establishment of the Slope Engineering Branch (Cawangan Kejuruteraan Cerun—CKC). One of CKC's first achievements upon formation was to carry out a National Slope Master Plan study to reduce risks and losses from landslides. One of the studies explores and devises methods for assessing risk that combine traditional and risk-based approaches. It introduces a risk assessment-based approach that looks beyond the fulfilment of Factors of Safety; it evaluates a slope based on its risk or probability of slope failure occurrence and assesses the consequence or damage caused by the failure. Most significantly, it compares the derived risk assessment results with the acceptable risk level of the public and residents. In essence, it becomes a decision-making tool for slope planners and developers to determine whether to proceed with the construction of a new slope or how much mitigation work should be put into an existing failing slope. One of the study components, Public Awareness and Education, launched a national awareness and education campaign to get create awareness of landslide risks and mobilize various stakeholders in the public, private, civil society and community levels into taking proactive measures for mitigation and prevention. It culminated in a programing conveying four main key messages, which are "Learn, Monitor, Maintain and Report".

Keywords

Public awareness • Slope safety • Learn • Monitor • Maintain • Tolerable risk • Acceptable risk • Risk criteria • ALARP • Quantitative risk analysis • Societal risk • Economic losses

A. Rashid Ahmad (✉) · Z.A.M. Amin
C.H. Abdullah · S.Z. Ngajam
Public Works Department (Jabatan Kerja Raya Malaysia), Slope Engineering Branch Ministry of Works, Jalan Sultan Salahuddin, 50582 Kuala Lumpur, Malaysia
e-mail: ABRashidAhmad.jkr@1govuc.gov.my

Z.A.M. Amin
e-mail: ZArsad.jkr@1govuc.gov.my

C.H. Abdullah
e-mail: Hassandi.jkr@1govuc.gov.my

S.Z. Ngajam
e-mail: SitiZarina.jkr@1govuc.gov.my

K. Sassa et al. (eds.), *Advancing Culture of Living with Landslides*,
DOI 10.1007/978-3-319-59469-9_39

Introduction

This paper will discuss two of the important elements in Malaysia's National Slope Master Plan (NSMP)—the Public Awareness Education Program and the Loss Assessment Measures.

Records of landslides in Malaysia have existed since the beginning of the last century, but national attention to landslide hazards increased in the wake of the 1993 Highland Towers landslides. In 2003, an economically devastating rockslide in Bukit Lanjan led to the establishment of the Slope Engineering Branch (Cawangan Kejuruteraan Cerun—CKC) of the Public Works Department Malaysia (National Slope Master Plan 2009–2023 (2009)).

The Slope Engineering Branch ran a public awareness and education programme to create awareness among the public to minimize the effects of landslides. The objective was to create a society that emphasized risk reduction and resiliency to landslide disasters. Through outreach activities, the programme identified and disseminated information on actions and measures that can be taken by community members, as well as by government and private owners of slopes. As it is human nature to focus on safety only after a disaster happens, the public awareness program aims to get people thinking about the slope safety before any landslides occur.

On other hand, when it comes to urban development on slopes, there are many requirements and guidelines stipulated by local, state, and federal authorities. However, these stipulations are technical and engineering-driven, such as factors of safety, buffer zones, and maximum berm height.

However, stakeholders in slope management should extend beyond land-use planners and technical personnel. Members of the public and communities in hillside areas that experience the consequences of failing slopes, whether it be loss of property, depreciation in real estate value, or even loss of lives, have to be included. Thus, slope planning approaches that incorporate the requirements of both the technical engineer and hillside residents are needed.

This study explores and formulates methods for assessing risk that combine conventional and risk-based approaches. It introduces a risk assessment-based approach that looks beyond the fulfilment of factors of safety; it evaluates a slope based on its risk or probability of slope failure occurrence and assesses the consequence or damage caused by the failure.

Most significantly, it compares the derived risk assessment results with the risk level considered acceptable by the public and residents. In essence, it becomes a decision-making tool for slope planners and developers to determine whether to proceed with the construction on a new slope or how much mitigation work should be put into an existing failing slope.

This study is an interdisciplinary endeavour that encompasses geotechnical engineering, risk management, mathematical modelling, psychology and social sciences, and economic assessment. The objective of the study is the application of the risk assessment methodology for two practical purposes:

(1) Guiding the design and approval of new developments on slopes, and
(2) Prioritizing treatment and monitoring efforts for existing developments.

Public Awareness and Education Program

Although the main target groups of the program were the communities-at-risk and the general public, there were other target groups consisting of the state and local governments, private slope owners, media, universities and schools (JKR 2010).

The objective of the awareness program was to convey two key messages to the public. The first was to let the public know that there is a body of useful information that is available to the public on the phenomenon of landslides and tips on monitoring and maintenance. The second is that there is a government agency dedicated to safeguarding the interest of public safety.

These messages were encapsulated in the campaign theme of **"Learn, Maintain, Monitor and Report"** and all activities of the awareness program were centred around this theme. The motif that tied all these activities together was the slogan "Safe Slopes Save Lives", courtesy of the Geotechnical Engineering Office in Hong Kong.

Learn

Before any community action can be taken, residents and the general public needed some knowledge of the concepts of slope safety. Through seminars and public talks, the public was given a briefing on slopes and landslides. Residents were taught about what is a slope, types of landslides, factor of landslides, triggers of landslides and key concepts on slope water control, slope cover, retaining walls and geological aspects. This basic knowledge enabled residents to better understand the reasons behind the measures prescribed by CKC and ensure the sustainability of the programme.

Monitor

Although landslides sometime occur without warning, there are usually signs which residents in hillside communities overlook, as they do not recognize the signs of landslides. With urban development encroaching onto hillsides, monitoring for signs of slope failures is vital to slope safety. Man-made slopes in particular are becoming more and more prevalent and they need to be routinely checked. For this purpose, CKC compiled a collection of images showing signs of landslides as they appear on and around the slopes, inside and outside houses and buildings. Armed with this knowledge, residents are able to carry out visual slope inspections on slopes within their community.

Maintain

Even well-designed slopes will fail if they are not maintained. Maintenance is simple to do, yet often neglected. Proper maintenance can make the difference between safety and disaster. Maintenance starts with an inspection of a slope. This can be done by residents, local authorities, and private slope owners with some basic knowledge of maintenance guidelines.

One of the key messages for this activity is that there must be a regular schedule of maintenance to be done by laymen residents and by professional engineers. Ideally, there are three different kinds of maintenance to be undertaken: simple maintenance such as clearing drains and returfing slope vegetation, which can be done by residents, and the other is a more technical slope inspection by qualified engineers who inspect slope features such as rock anchors and retaining walls. The third kind is inspection of water pipes for leakages and seepage.

While the onus of maintenance falls on the slope owners, community members are encouraged to engage in maintenance as it affects their livelihood and safety.

Report

The key message of monitoring for signs of landslides would not be effective if the observations were not reported to the relevant authorities for follow-up repair or mitigation action. Thus, residents are told to report any signs that are deemed to require action, such as broken drains that allow water to seep into the slope. Usually the authority to be contacted is the engineering department of the local authorities, as they deal with infrastructural matters on public property. If the slope in question is on private property, the local authority would contact the land owner for action. In some cases, where there is no response from the land owner, a Notice to Take Action, would be issued. As a slope agency, CKC receives its share of reports from the public, and in these cases CKC will direct the case to the appropriate agency or land owner.

Methodology of the Program

To provide maximum exposure, the methodology of the programme entailed presenting its messages and material through various mediums.

Logo and Slogan

There is the adage that the messenger is just as important as the message. If people distrust the agency or are not aware that it exists, then the messages will not be received by the target audience. During a pre-programme survey, many respondents in target areas revealed that they were not aware of the existence of CKC. Others said that the Public Works Department (JKR) as a whole does not disseminate information to the public. To create the image that CKC's role is to engage and teach residents on slope safety, a friendly face was put onto the agency. After several candidate cartoon depictions of a mascot, a youthful, energetic, pan-Asian JKR engineer was created as a logo. He is seen leaning forward to meet residents halfway and to convey an attitude of attentiveness. Mr. JKR Logo is shown in Fig. 1.

"Mr. JKR" is surrounded by the slogan 'Safe Slopes Save Lives", courtesy of GEO Hong Kong, which encapsulates the same objective and goal of saving lives. This underscores CKC's philosophy that slope safety is not an engineering issue, but a people and quality of living issue.

Printed Material

A series of brochures written in Bahasa Malaysia and English was created to convey information on the four key messages of the campaign which is "Learn, Monitor, Maintain and Report". Designed with bright colours and replete with illustrations and photos, the brochures provide hands-on, action-oriented tips and guides to readers. Figure 2 shows the brochures written in Bahasa Malaysia and English.

Another set of publications created were posters written in Malay Language to post up in public places ('Learn, Monitor, Maintain, And Report). The posters are shown in Fig. 3.

Advertorials

The mass-media plays an important role in raising the awareness of the public. During the peak months of the monsoon season, when landslides are likely to happen, CKC publishes a series of advertorials in newspapers in four main languages,

Fig. 1 Mr. JKR Logo was created to give friendly face to the message, as well as to the image of JKR

Fig. 2 A series of brochures used during the public awareness campaign

which are English, Malay language, Mandarin and Tamil languages to renew awareness and provide useful information to readers nationwide in upgrading their surveillance of slopes and signs of landslides. Advertorials are shown in Fig. 4.

TV Commercials

In addition to the print media, CKC has used broadcast media to run several seasons of TV commercials to raise awareness of slope safety. The commercials are produced to create awareness among people about the importance of being vigilant and keeping an eye on the environment, especially those who live on risk areas such as slopes. The commercials insert scenes of learning and teaching to have a deep impact on the audience watching the commercials. It is hoped that by watching the commercials, people will acquire the knowledge to detect signs before the occurrence of landslides, know what should be done if there are signs in their area, and what actions should be carried out in the event of a landslide near their resident area.

As with the advertorials, the infomercials are run at the end of the year during the monsoon season.

Awareness Billboards

To make slope monitoring, maintenance and reporting a regular practice in landslide-prone areas identified in the National Slope Master Plan, billboards near at-risk communities have been set up to remind residents to keep an eye out for the signs of landslides and report them to the authorities. This message is conveyed to urban audiences. For the rural communities, a different message is provided. Slope construction in rural and agricultural areas is not as regulated as in the urban areas, and there have been incidents of slopes that have been constructed not following proper guidelines. For this audience, a sterner message has been crafted, saying that improper construction of slopes on unauthorized land can lead to landslides.

Billboards are merely reminders of prescriptive actions, and of themselves will not change social behaviour unless

Fig. 3 A series of posters used during the public awareness campaign

Fig. 4 A series of advertorials published in newspaper

accompanied by grassroots and outreach programs. However, the billboards have garnered attention by media whenever a landslide or mudflow incident occurs in the adjacent area. In Bukit Antarabangsa, where the billboard greets all residents and visitors coming up a particularly rocky slope which is the entrance to the housing developments, it serves as reminders to always keep a vigilant eye. Figure 5 shows the example of billboard displayed during the campaign.

Evaluation Determining "Acceptable Risk" and "Tolerable Risk" in Landslide Risk Areas in Malaysia

Methodology

As the practice of utilizing risk assessment using F-N charts is new in Malaysia, a suitable and proven method-

Fig. 5 Billboard illustrations for the public awareness campaign

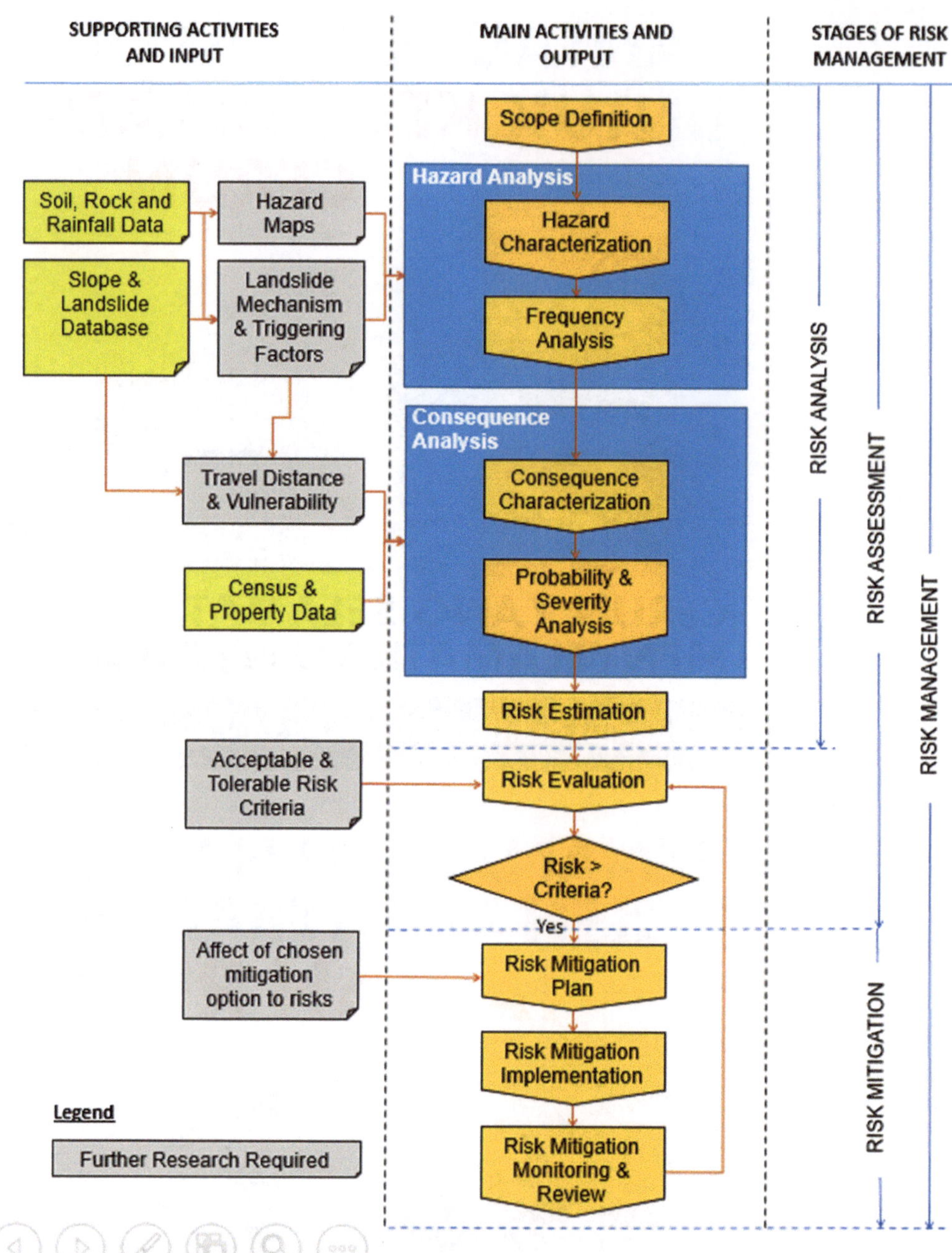

Fig. 6 Risk management flowchart

ology had to be researched, evaluated, and tested. An extensive literature review led to the development of a methodology that incorporates quantitative risk assessment (QRA), historical risk assessment, psychometrics, risk perception and risk criteria theories, loss assessment quantification, and cost-benefit analysis (JKR 2015).

The methodology selected was based on the proceedings of the International Conference on Landslide Risk Management, which is a collection of risk management papers edited and written by seasoned geotechnical risk assessment (Finlay et al. 1999). Figure 6 shows the localized framework based on their model.

The main activities of the study are to:

(i) Calculate the risk assessment line representing the probability of slope failure and its consequences (called the projected "F-N curve").
(ii) Identify the acceptable and tolerability risk thresholds of stakeholders (called the "risk criteria curve").
(iii) Juxtapose both curves on the F-N chart and determine whether the risk level of the slope project falls within the tolerable, acceptable, or unacceptable range as defined by the resident stakeholders.

(iv) Determine which option of action to take, depending on the outcome of the F-N chart.
(v) Recommend whether the proposed slope project is economically feasible if further re-designs or mitigation works are required.

These activities are carried in the following key deliverables and actions.

Determining Mortality Rates

One of the objectives of this study is to determine the actual frequency or the probability of failures and fatalities. Plotted on an accumulated frequency–fatality (F-N) chart on a log-log scale, the risk and likelihood of failure can be clearly projected. This is done by using historical data on landslide events with fatalities. Figure 7 indicates the FN Chart for Malaysia as compared with other countries.

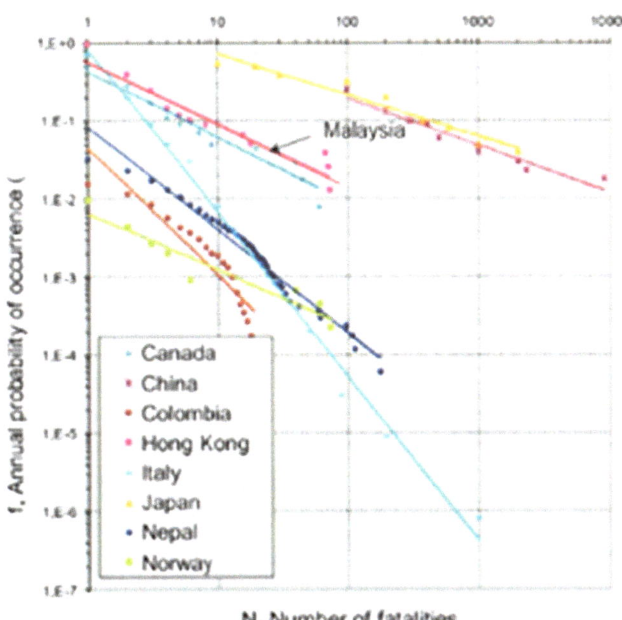

Fig. 7 F-N chart for Malaysia compared with other countries

Determining Rate of Economic Losses

It has been observed that some of the most devastating landslides in Malaysia have been those incurring major economic losses, not high fatalities. Thus, the process for determining the rate of economic losses have been discussed in this study. It follows the same methodology as for F-N curves, as a reference model has not been addressed in worldwide studies and warrants further research in this area Fig. 7.

Identifying Social Survey Methodology and Design

The F-N curve alone merely shows the frequency and consequences of fatal landslide events. To see whether these are acceptable by the general public or by specific communities, they need to be compared with risk criteria, which are curves on the F-N chart that determine the thresholds for acceptable, tolerable, and unacceptable risks. Interpretation of the curves on the F-N chart is fairly straightforward: any part of the F-N curve that falls under the acceptable risk criteria is fine; no preventive or mitigation actions are needed.

Parts of the F-N curve that fall into the tolerable range needs to be worked on; they are only tolerable as long as there is some mitigation or preventive action being taken. And finally, anything that falls above the unacceptable curve is out of the question by the stakeholders; action to relocate or engage in major mitigation work is required. These risks, experienced by many in a population, is called societal risk.

Coming up with the risk thresholds on the other hand is not so straightforward. To identify the social research approach, various theories and methods of social research were reviewed. After reviewing numerous revealed and expressed preference approaches, a psychometric approach modelled after a study by University of New South Wales Ph.D. recipient P. J. Finlay and Professor Emeritus Dr. Robin Fell was selected. The study, which encompasses a survey questionnaire applied to nine communities in Australia and Hong Kong, was adopted for application in Malaysia.

Deriving Risk Criteria Thresholds for Malaysia

Risk criteria are locale-specific and the same risk criteria cannot be applied from country to country, or even from region to region, as the thresholds depend on feedback from the population being surveyed.

The risk criteria curve used in Hong Kong for example cannot be applied in Malaysia. For the risk criteria in the study area, which consists of communities in Ulu Klang (representing urban communities) and Ulu Langat (for rural communities), a combination of two derivation approaches were used.

Derivation entails taking the results from the survey questionnaires on acceptable frequency and number of fatalities acceptable to people living in these areas, and converting them into a chart. One approach is the Risk Matrix method used in hazard-sensitive industries such as the aviation industry, and the other is the Risk Criteria method used by Finlay and Fell.

Both methods produced the same results and validated the resultant risk criteria curves. They shall be called the Interim Malaysian Risk Criteria, as they represent the acceptable and tolerable levels for the study areas of Ulu Klang and Ulu Langat.

Determining Actions to Be Taken by Authorities

The results of the F-N chart will be reviewed by land-use planners and approving urban development agencies, as well as by technical agencies responsible for undertaking repair and rehabilitation works. These activities will be largely affected by financial considerations and whether the cost of repair or mitigation will be commensurate with the benefits derived (i.e., avoiding the losses). For this, a cost-benefit analysis needs to be carried out.

Financing the Actions

All preventive, mitigation, or rehabilitation options require financing. Thus, a study was conducted to determine the willingness to pay by the residents living in the affected hillside communities. Using the contingent valuation method, the willingness to pay was derived using survey questionnaires on how much residents were willing to pay for different solutions that correspond with varying risk levels and resultant 'savings in lives'.

Framework for Risk-based Slope Decision Making in Malaysia

Finally, a risk-based approach to slope management cannot materialize without a formal framework for which stakeholders, from developers to local authorities to communities, can follow. Thus, a framework is proposed which shows the steps from the preparation of the Development Order by the developer, to the review and approval process by the hillside and highland review committee at the local authority level, to final review by the One Stop Centre at the state and local authority levels.

Findings

Findings show that the tolerable and acceptable risk criteria for the study areas in Malaysia are higher than the interim Hong Kong criteria. The risk criteria derived from this area shall be called 'Interim Malaysia Risk Criteria'. The tolerable frequency for Hong Kong is one landslide with fatalities every 1000 years, while the tolerable threshold for the Malaysian study areas is once every 50 years. Figure 8 shows The Interim Risk Criteria for Malaysia.

Using the loss assessment data from the National Slope Master Plan, the analysis suggests that economic losses amounting to RM 1 billion per annum is possible once in a hundred years.

In determining the mortality rates the study indicates that the landslide mortality rate for Malaysia is the same as Hong Kong. When plotted on a cumulative frequency-fatality (F-N) chart, the equation for the curve could be written as $F = 1.99\ N-1.15$. This means that on average in Malaysia 2 fatal landslides occur annually, and a landslide claiming more than 100 lives would occur once in a hundred years.

Urban residents were willing to pay between RM151 to RM305 annually for maintenance or repair works, while rural residents could commit RM22 to RM112 annually. This shows that Malaysians are willing to pay for safer slopes.

Conclusion

CKC's awareness and education programme has had an impact in ways that extend beyond outreach to the communities. It has created awareness among the local authorities, some of which have instituted their own slope units and have started requesting further training to enhance their existing knowledge and know-how in slope management. CKC's ultimate goal is to mainstream slope safety so that it impacts all relevant aspects of land-use planning and governance. However, this is an ongoing process and as such, it is important that it continues its programme to sustain the momentum that it has started. All programs on awareness and education are focusing on the element of "Learn, Maintain, Monitor and Report" which demand the participation of general public to avoid any untoward incident due to landslides.

A risk-based approach is a decision-making instrument that can provide decision makers with more information about a slope. Following the adage that even with a high Factors of Safety (FOS), the probability of failure can be high:

i. For a nationwide F-N curve, historical data is used; for site-specific F-N curves, a Quantitative Risk Assessment (QRA) approach is used
ii. FOS will still be used in risk assessment of slopes, but we can improve on risk reduction by including the F-N chart method
iii. The Interim Malaysian F-N curve can replace or eliminate the buffer zone requirement, which is a point of contention between developers and approving authorities
iv. The Interim Malaysian risk criteria is higher (more accepting of risks) than the Hong Kong risk criteria
v. A deterministic approach is used for QRA, as opposed to the static probability approach. ALARP (As Low As

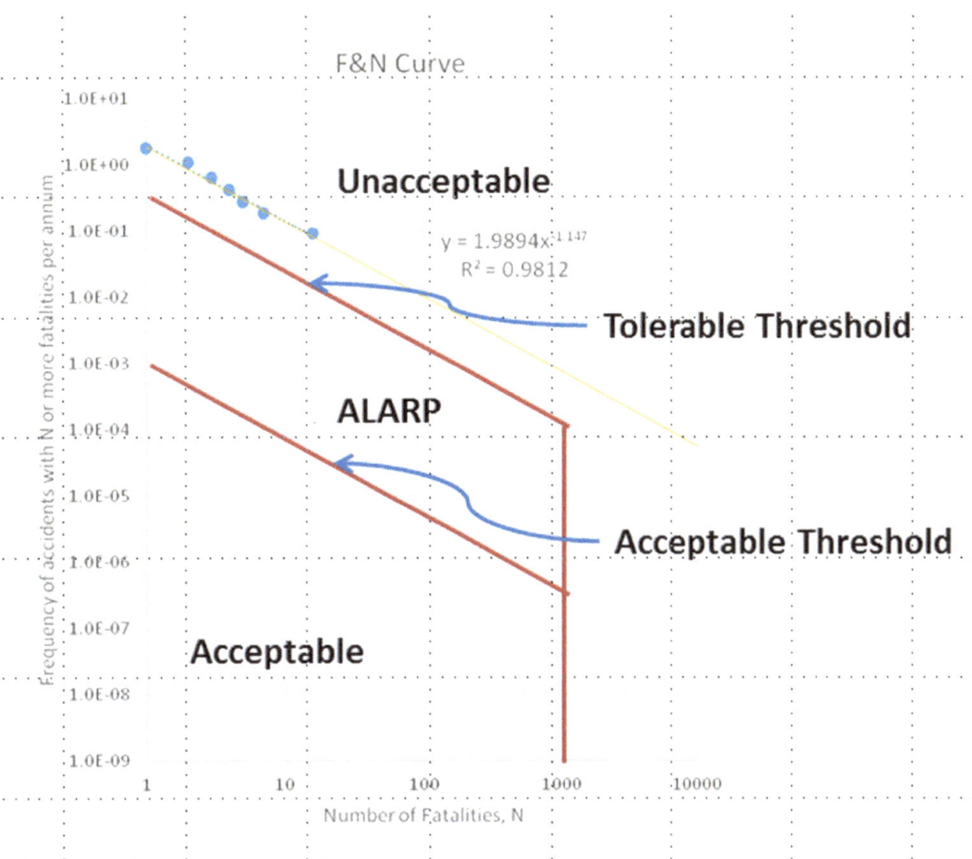

Fig. 8 Malaysian F-N curve with risk criteria from study area

Reasonably Possible) is the same as tolerable risk, except that it takes into consideration that the costs of reducing risk should never exceed the benefits

vi. More surveys in other parts of the country should be carried out to develop national risk criteria.

References

Finlay PJ, Mostyn GR, Fell R (1999) Landslide: prediction of travel distance and guidelines for vulnerability of persons. In: Proceeding of 8th Australia/New Zealand conference on geomechanics, Hobart. Australian geomechanics, vol 1. pp 105–113

JKR (2010) Final report public awareness and education program on landslide and slope safety. Public Works Department Malaysia, Kuala Lumpur

JKR (2015) A study on determining 'Acceptable Risk" and "Tolerable Risk" criteria. Public Works Department Malaysia, Kuala Lumpur

National Slope Master Plan 2009–2023 (2009), Public Works Department Malaysia, Kuala Lumpur

Geotechnical Site Characterization of a Mud Eruption Disaster Area Using CPTu for Risk Assessment and Mitigation (IPL-195)

Paulus P. Rahardjo, Adityaputera Wirawan, and Andy Sugianto

Abstract

A mud eruption in East Java that occured on May 29, 2006 is well known worldwide. The mechanisms of the causes of the eruptions are still in debate, whether it was triggered by gas well drilling or by pressurised fluid reactivated by the Jogjakarta earthquake of May 26, 2006. This debate is not the main issue in this report. Instead, this paper discusses mainly the results of CPTu tests recently conducted and mitigation and risk reduction. The volume of the mud discharge is estimated at 5000 m^3. Dykes were constructed to contain the mud, which covered areas reaching 650 ha (Sofyan 2015). The location of the disaster is in the middle of the town of Porong in the district of Sidoardjo, near Surabaya International airport, and mud has blocked the major arterial roads from north to south of East Java. The soil condition of the site is deep soft clays which causes instability of the dykes. Some dyke failures occurred, endangering residential areas due to the flow of the mud (Rahardjo 2015). This paper describes the characterics of the soil conditions from a number of drillings and CPTu tests conducted by the authors for designing the replacement of the arterial road and for dyke reinforcement, and also in the middle of the mud. The paper discuss the geotechnical problems of land subsidence over large areas and differential settlement that cause damage to infrastructure, including roads, gas pipes, railways, bridges and buildings, and is of particular importance to the safety of the dykes. The mud has been discharged through the Porong River, and sedimentation is part of the problem.

Keywords

Mud eruption • Soft soils • Geotechnical failures • Land subsidence • CPTu • Mitigation and risk reduction

P.P. Rahardjo (✉) · A. Wirawan · A. Sugianto
Universitas Katolik Parahyangan, Jl. Ciumbuleuit No. 94, Bandung, 40142, Indonesia
e-mail: rahardjo.paulus@gmail.com

A. Wirawan
e-mail: teknik@gec.co.id

K. Sassa et al. (eds.), *Advancing Culture of Living with Landslides*,
DOI 10.1007/978-3-319-59469-9_40

Introduction

The mud disaster that occurred in 2006 in east Java is a special event that is very rare. A mud eruption occured on May 29, 2006 and is the biggest of its kind in the world. It was preceded by drilling for gas exploration in this area. However, until this day, there is still some speculation about whether the eruption was triggered by the drilling or by the Jogjakarta earthquake of May 26, 2006. It is still debated, but in recent discussions (Rahardjo 2015) Massini et al. (2010) claimed that there is a trend toward the opinion that the cause of the eruption is largely natural.

The disaster caused 14 fatalities, displaced 30,000 people (13,000 families), closed 30 factories and hundreds of small business, and flooded more than 10,000 homes with mud in more than a dozen villages (BPLS 2007). The team of Universitas Katolik Parahyangan (Unpar) has conducted a number of cone penetration tests using a piezocone with pore pressure sensors (CPTu) during the construction of the relocated arterial road to the west side and also in the area of the mud eruption and a nearby control dyke. This paper mainly deals with the geotechnical site characterization of the area using CPTu and the discussion of mitigation and risk reduction, which is the concern of the government. The location of this mud eruption is at Porong, the southern district of Sidoardjo regency, about 12 km south of the town of Sidoardjo. Figure 1 shows the location of the mud eruption.

The site is a residential area and surrounded by an industrial area in East Java. The Surabaya Gempol toll

Fig. 1 Location of the mud eruption in East Java, Indonesia

Fig. 2 The vent of the mud eruption in East Java, Indonesia, surrounded by protection dykes (Boston.com 2008)

highway Surabaya-Malang–Pasuruan Banyuwangi, which is an important economic route in East Java, is directly influenced by this mud eruption.

Despite of many efforts carried out to reduce the eruption, the scale of the disaster is beyond the capability of human beings. As a result, the last effort is merely defence to save the lives of people by containing the mud in cells and to limit the widening effect to the surrounding by pumping the mud into the Porong River.

The disaster has become the world's most destructive mud volcano (Wayman 2011). But at present, after almost 9 years, the discharge has dropped in volume, and this gives hope that the eruption is almost over. Some geologists however predicted that the flow of mud will end 30 years from the start of the disaster, so it seems that it is not certain and only a rather hasty conclusion.

The mud ejected through this volcano is hot, and for samples collected nearby, temperatures are in the range of 80–90 °C (Figs. 2, 3 and 4).

The Cause of the Mud Eruption

Prior to the mud eruption, there was drilling for gas exploration by a company in Indonesia, known as Banjar Panji 1 (Fig. 5), which had reached about 3000 m depth. Richard

Fig. 3 The center of the mud eruption (BPLS 2007)

Davis, a geologist at Durham University in England believes that the gas released by the drilling, created failure in the surrounding rock, with materials being washed out to the surface. Figure 6 shows the mechanism, as illustrated in National Geographic magazine (2006). Michael Manga, a geologist at the University of California Berkeley, explained the overlying sediments compress the lower layers and pressure builds as the upper layers get thicker and heavier, and the squeezed water has no where to go. If a path to the surface opens, the highly pressurized water will shoot up (article rewritten by Wayman 2011).

However some other opinions, such as that of Adriano Mazzini of the University of Oslo, have suggested the occurrence was reactivation of the fault nearby due to the Yogyakarta shallow earthquake of magnitude M = 6.3 which was also believed to have reactivated the Opak Fault from the south of Yogjakarta to Central Java near Prambanan Temple. Stephen Miller, a geophysicist at the University of Bonn who led research on seismic energy from the quake, suggested it was reflected and focused by the surrounding rocks and became sufficiently concentrated to liquefy the mud source. According to the study, the mud fluid then injected itself into the adjacent Watukosek Fault (Fig. 7) and caused it to slip, thus linking it to a hydrothermal volcanic system deeper down, so that the previously trapped system completely rearranged itself and, fatally, became connected to the surface. Then the mud came.

Since 8 April 2010, there have been dramatic changes. The old Lusi crater (active since May 29, 2006) has died. The current speed of the flow at minimum conditions is

Fig. 4 The spreading of the mud eruption in the district of Sidoardjo and the cells with protection dykes (CRISP)

generally <5000 m^3 a day, compared to the peak rate at the beginning of 180,0000 m^3/day decreasing to around 100,000 m^3 in 2008/2009. Currently the model of Lusi Volcano is very much similar to Bledug Kuwu (in Central Java), with a kick of mud without flow or waves (Hady Prasetyo 2010).

Characteristics of the Mud and Mud Removal

The mud is very soft, consisting of about 70% water, with the material being clayey silts and very fine sands with gravel. This mud is difficult to handle because of its high liquidity and it cannot be easily removed except by mud pump (Rahardjo 2015). Figures 8 and 9 show the mud characteristics in wet and in drying conditions. After it is sufficiently mixed with water, the mud is pumped to the river, which in turn may cause other problems with sedimentation (see Fig. 10). This has caused environmental problems along the river and at the estuary, where a lot of sediment can not flow freely.

Damage Due to the Mud Eruption

The main damage to infrastructure is mainly due to mud flooding thousands of buildings, including houses, schools, mosques, and 30 factories, and damage to the toll roads of

Fig. 5 Photograph of Banjar Panji 1 well (Didiek Djarwadi 2015)

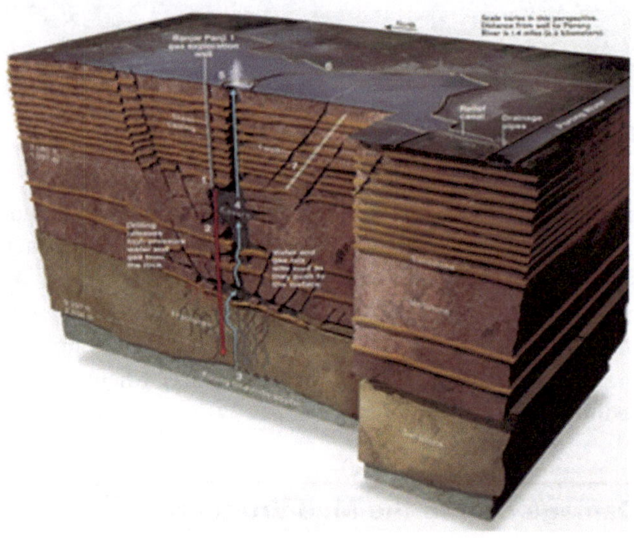

Fig. 6 The mechanism of the start of mud eruption (National Geographic 2006)

Surabaya-Gempol, Arterial Road Surabaya-Malang, and to railways, gas pipes, drainage canals and others structures, causing economic losses in the order of billions of US dollars. Both economic and social activities are significantly direct or indirectly affected.

Figure 11 shows the range of spreading of the mud in 2007, when the highway of Surabaya-Gempol was still in use. The author reviewed the condition of the dyke, and water had started seeping at the toe of the slope, which meant that the dyke was not safe. In the same year, this highway was flooded by the mud and is no longer used (Rahardjo 2015).

Geotechnical Site Characterization Using CPTu

Soil Conditions at the Site

Based on laboratory tests in this area, the soils are highly plastic materials, with a natural water content ranging from 40–100%. Generally the upper part is slightly stronger, showing slight overconsolidation. However the void ratio could be as high as 1.5–3.0. Laboratory tests also show that the soft soils are still consolidating. Compressibility of the soils, measured by its compression index is very high, with a range of 0.5–1.5, so the settlement is large (Soleman 2012) (Fig. 12).

Based on in situ testing (CPTu and SPT), the soil upper layers are very soft, with a thickness of 15–25 m, dominated by clays to silts and silty sands. The silty sands are mixed with clay. This soil condition has a very low bearing capacity and may cause very substantial settlement upon loading. The possibility of squeezing of lower soil layers are among the problems that need to be considered.

CPTu Tests at the Mud Site

The use of CPTu (Cone Penetration Test with pore pressure measurement) has been popular in Indonesia since 1990, especially for soft soils. The increasing use of the CPTu is due to a number of factors:

- It is handy, fast and accurate for soil profiling and not dependent on the operators.

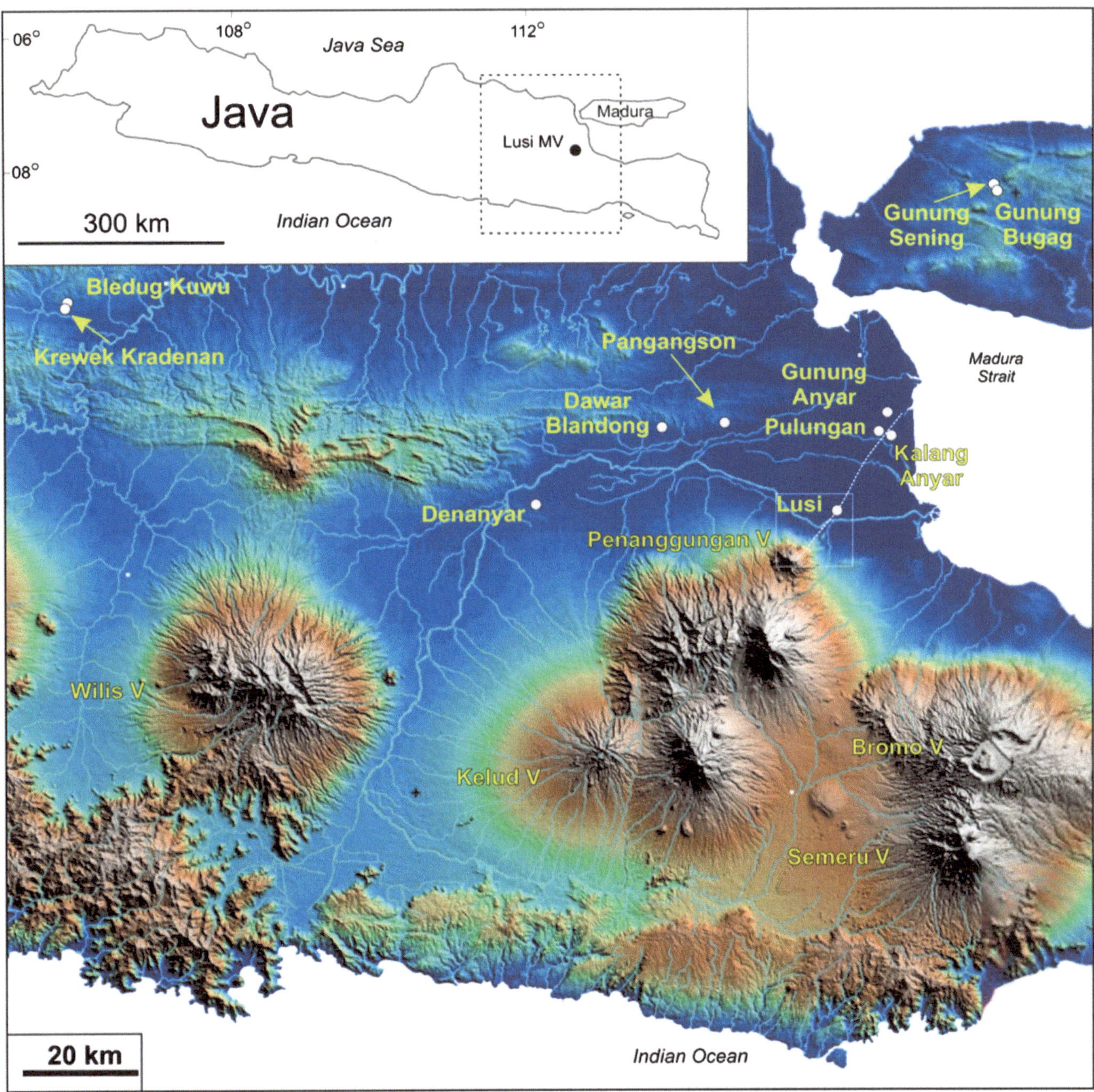

Fig. 7 Location of lusi along a fault (Mazzini et al. 2009)

Fig. 8 The visual appearance of the wet mud (BPLS 2007)

- It can distinguish both the soil resistance and the pore pressure, hence the effective reaction of the soils can be measured and it can recognize the drained and undrained response of the soils.
- The interpretation of soil properties, although heavily reliant on empirical correlation, are accurate due to the many available measurements for comparison and justification.
- Dissipation tests can be conducted to measure the permeability and consolidation characteristics of the soils, which provides more reliable methods compared to laboratory tests.

The authors have gained a lot of experience in many projects throughout the northern coast of Jakarta and also in many places where soft soil deposits create instability during construction and the problems of long-term settlement.

The equipment consists of an electronic cone with cone tip to measure tip resistance, a friction sleeve to measure friction resistance of the soils during the course of

Fig. 9 Condition of the mud upon drying (hufingtonpost.com, July 22, 2013)

penetration and a pore pressure sensor to measure pore pressure—both the hydrostatic pore pressure as well as the excess pore pressure due to penetration of the cone into the ground. The measurements are conducted through cable wired to the cone and the ADU (analog digital unit) connected to the computer for data aquisition. Figure 13 shows the cone tip and the accessories for CPTu.

For reinforcement of the dyke, a series of CPTu test were conducted. Drilling soils on sites are impossible due to the existence of shallow gas and the psychological condition of the people who have suffered trauma from the drilling. However the soil investigations used for describing the condition of the soils on site may be done by using old soil tests as well as from the new soil investigation for the construction of the dyke. Figure 14 shows the position of the CPTu tests on mud eruption site. Eight CPTus are next to the dyke and two CPTus are in the position of the mud (Fig. 15).

Figure 16 shows a typical condition of the soil next to the dyke on a mud site from CPTu tests. The soils are not only soft but there is indication that the soft foundation soils are still consolidating, judging from the value of the pore pressure ratio $Bq > 0.7$. The upper part is clayey silts and a water

Fig. 10 Removal of the mud via the Porong River (BPLS 2007)

table is detected at about 1.0 m below ground surface. The variation of qc magnitude, friction ratio and pore pressure ratio show the differences in the soil behavior and soil type.

Figure 17 is the results of CPTu-09 located in the mud site. It clearly shows the thickness of the mud of 14 m, then a thin layer of sand of about 1 m, then the foundation soils. A test in the center of the excavation is shown in Fig. 18, where down to 28 m the condition is dominated fully by the mud. Further penetration was impossible due to an uplifted anchor.

It is interesting that the initial elevation of the site is about + 5.0 m above sea level, and the elevation of the mud at this site is about + 14.0 m. Hence, if there is no settlement, the mud should be about 9 m thick. However the test shown that the mud thickness is more than 28 m, which means that this point has settled more than 19 m. The mud is still consolidating as shown by the high value of the Bq.

The elevation of the mud has been surveyed from 2014 until today. Figure 19 shows the results of elevation measurements in July 2015 (Sofyan 2015).

Fig. 11 Damage to area infrastructure (BPLS 2007)

Stability of Dykes Used to Contain the Mud

Due to the unknown duration of the mud eruption, the only possibility of saving human beings is by containing the mud in cells, in which the cells boundaries were dykes. Figure 20 shows the dyke boundaries in 2007. But then the boundary was kept as it is due to the possibility that the mud could be removed through the Porong River.

The dykes were constructed using gravelly sands and most were not well designed. The reason is simply because there was not sufficient time for investigation nor for compaction. Figure 21 shows emergency construction of dykes as the mud spread toward the arterial road. The compaction was conducted mainly by bulldozers and quality control was not possible. Another problem related to the dykes' safety is due to the fact that the foundation soils are mostly soft, with thicknesses from 5 to 30 m depth (Fig. 22).

Due to the soft foundation soils, the dykes frequently failed and move laterally, which could endanger the people in the surrounding area. Figure 23 shows maintenance of

Fig. 12 Damage to houses in the muddy flood (BPLS 2007)

dykes in the center, which now has reached 25 m high due to the continuous settlement. As the dykes became higher, stability has become particular concern. To reduce the amount of fill, large sandbags have been used in the fill embankment.

A study was conducted by Augustawijaya et al. (2012) to analyse the stability of the dykes. The particular location for his study was the dyke at number 10D. The results show that the stability is marginal.

In August 22, 2010, a particular accident accompanied a sudden flood landslide (like a cold lahar) in the south east, a phenomena that has occured in debris flows and sediment gravity flows (Hardy Prasetyo 2010). In a matter of minutes, three dredgers that were operating there were driven as far as 200 m from their original position (Figs. 24 and 25).

Land Subsidence and Settlement Problems

At the center of the vent, the land sank because so much water and mud from beneath the ground was erupted. To maintain the height of the dykes, large sandbags were used.

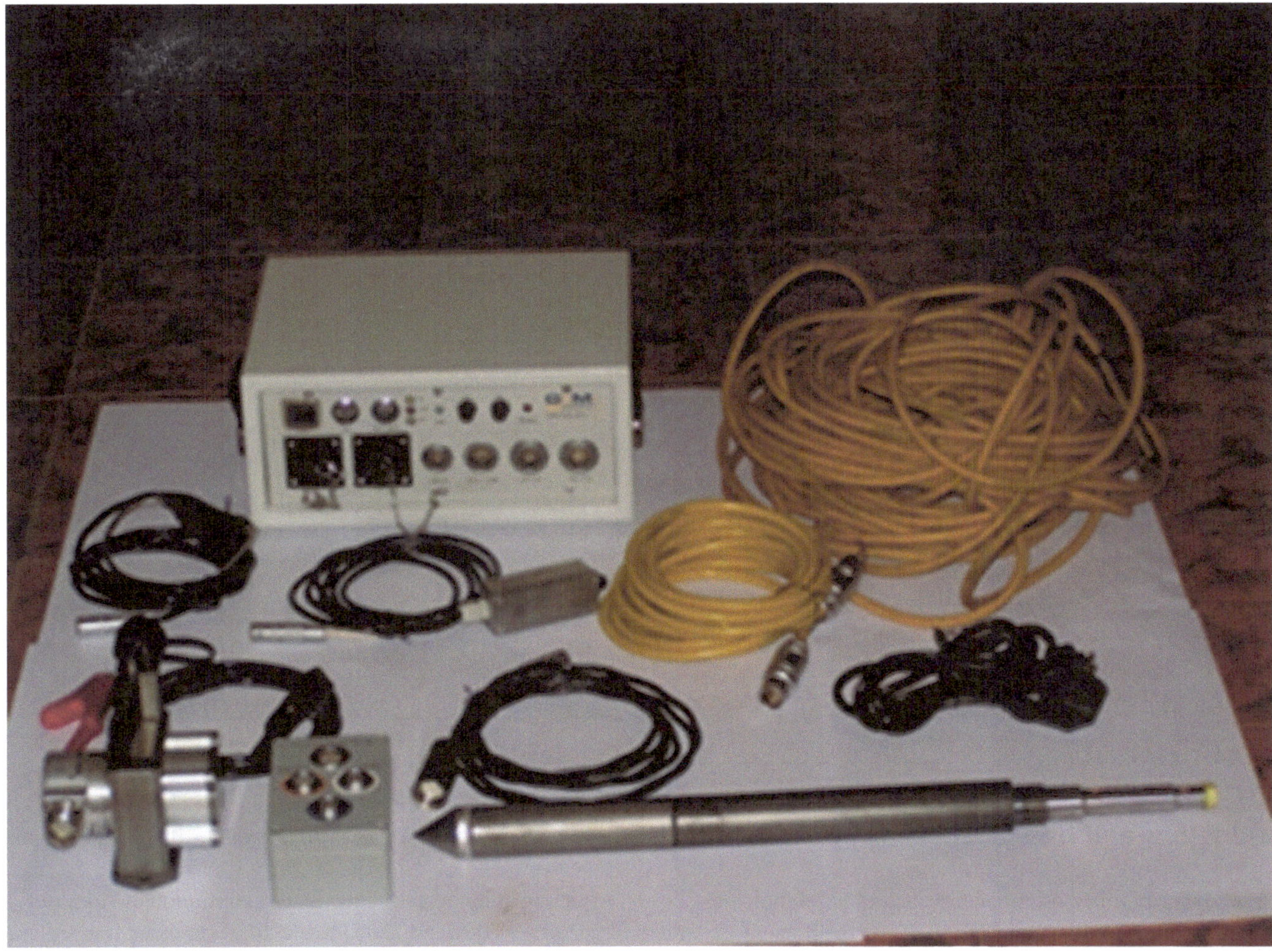

Fig. 13 CPTu equipment for the research of lusi

But also the surrounding land shows signs of settlement, and based on measurements, the affected area has a radius of 1.5 km. Pressure in the pond has caused settlement of the surrounding area, including serious settlement along the dykes. Feasibility Study of Settlement (TKKP) from ITS since November 2009 recorded settlement of the order of 10 cm per month. The average drop of settlement within a period of six month may reach 150 cm. On the other hand, land in the western and Jatirejo Siring Village has been raised by about 1.0–3.0 m. This could be due to deep sliding.

Fig. 14 Location of CPTu tests

Mitigation and Implementation

There are a number of geotechnical problems that must be solved, including the safety of the dykes, settlement or land subsidence, and risk of overflow. For dyke safety, it should be monitored visually as well as using instruments such as settlement points, inclinometers and pizometers. Those instruments will serve to measure the performance of the dyke, as well as act as a warning system if the dyke is not stable.

For subsidence it is necessary continuously monitor a large area. The land subsidence may damage houses and hence the boundary of the area that is influenced shall be relocated. For transportation toward Malang, a new arterial road has been completed. The relocation is to west side, assuming that the east side has lower elevation.

Fig. 15 Location of the CPTus tests on mud sites

Conclusions

Based on the study, some conclusions are made:

- The soil conditions underneath the mud eruption and the area of mud containment are generally very soft. This condition causes serious problems with stability and subsidence. In most cases, the soils are still consolidating and are sensitive to disturbance. Additional loading will be very dangerous and may trigger dyke failures.
- The first year of the study has been basically to identify the geotechnical problems of the mud eruption disaster. We need more detailed studies of awareness and preparedness.
- CPTu is the best tool to study the mud conditions, as well as the foundation soils.
- There is a chance for improvement due to the significant decrease in mud discharge. Reclamation may be proposed. At least in part it will be possible to compress the mud into a stronger layer along the periphery of the dyke.
- The mud discharge to Porong River may cause long-term problems due to sedimentation along the river and in the estuary area.

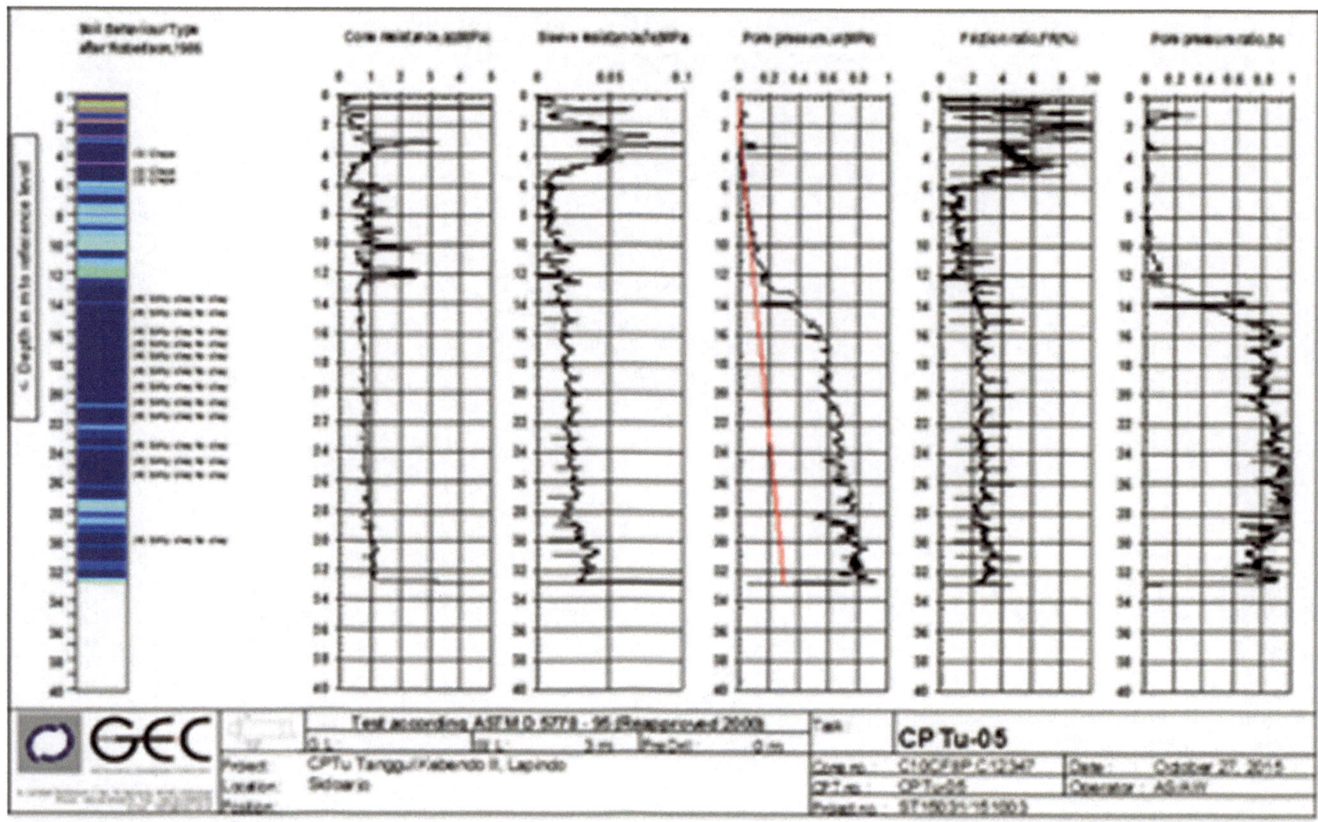

Fig. 16 CPTu tests on the outer side of the dyke

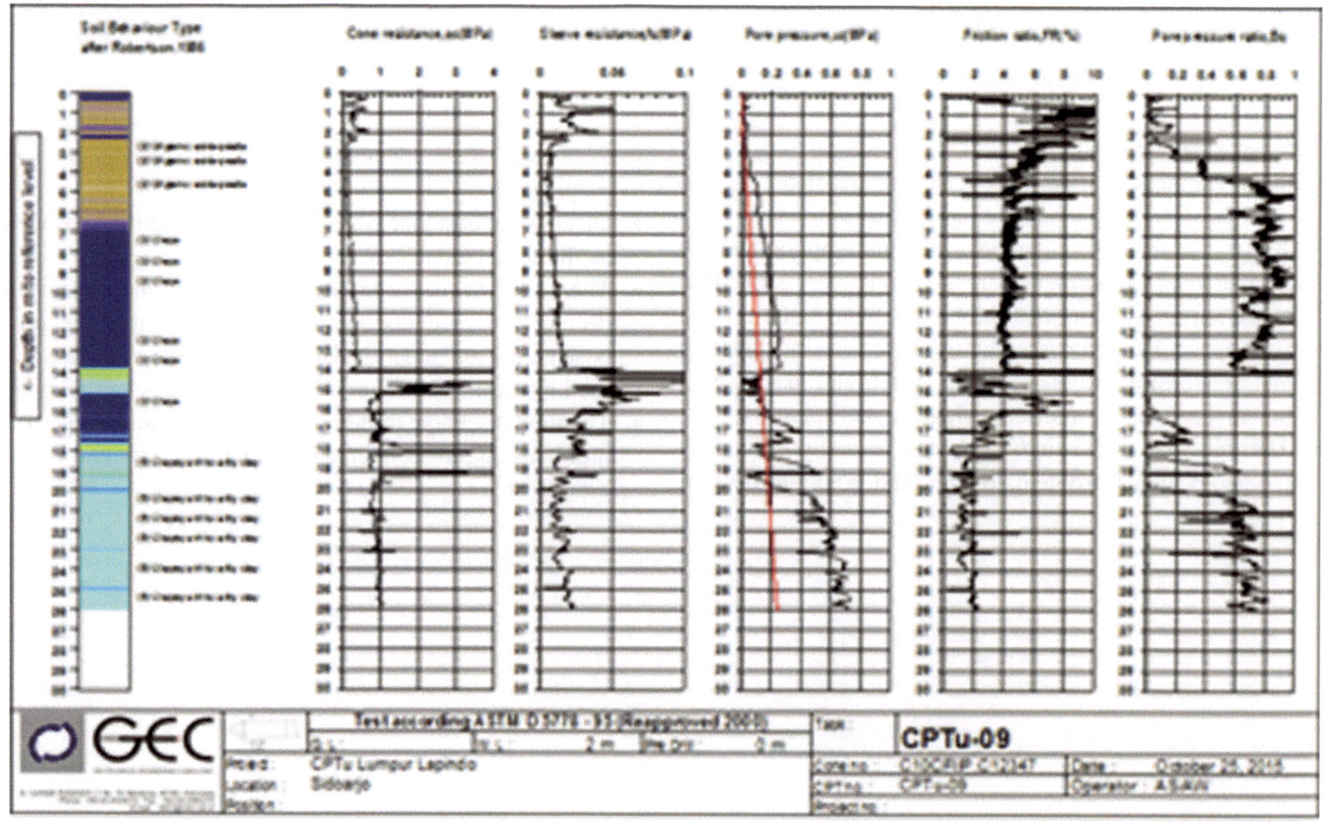

Fig. 17 CPTu-09 located in the mud eruption site

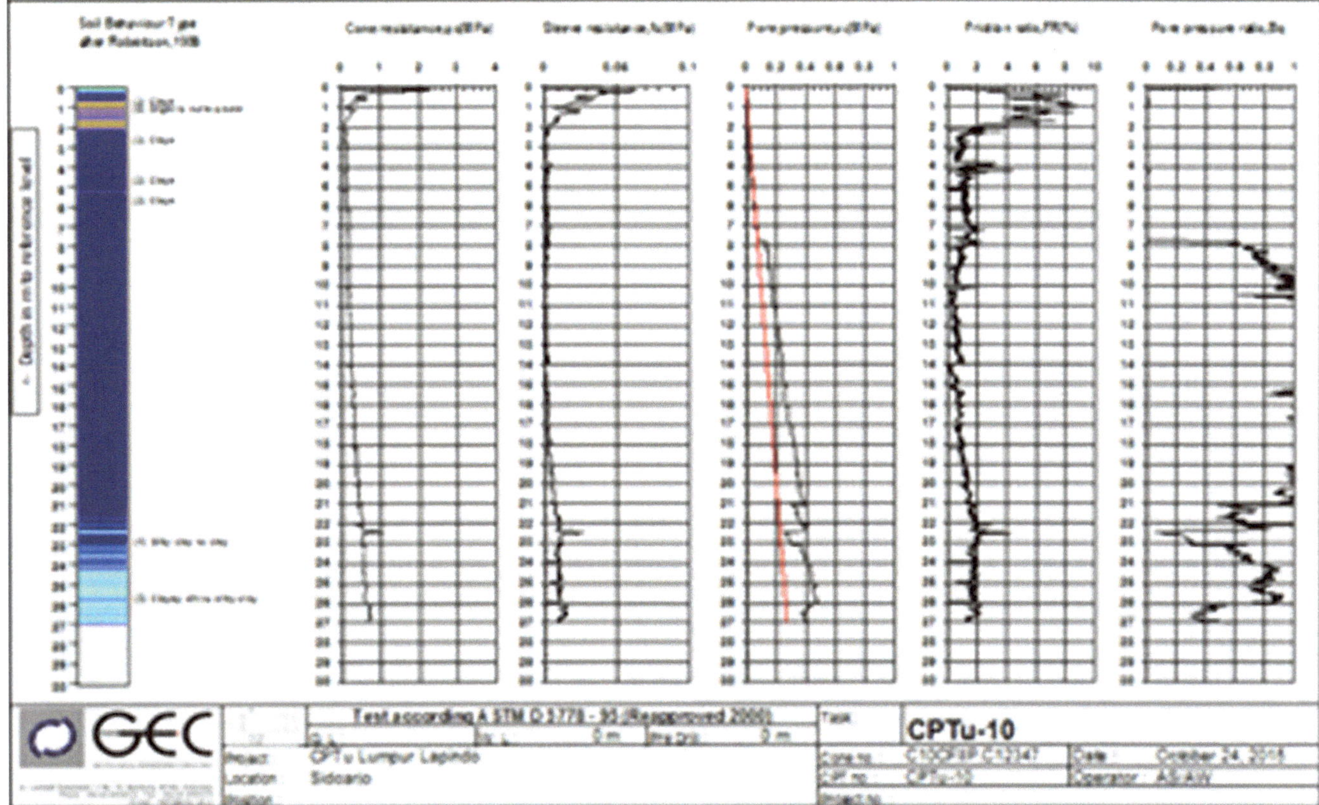

Fig. 18 CPTu at the center of the mud eruption

It is suggested to install geotechnical monitoring system which will provide information on the performance of the dykes.

Acknowledgements The authors appreciate the assistance of Lembaga Penelitian dan Pengabdian Masyarakat (LPPM) of Parahyangan Catholic University for providing funds to do work on this research and to BPLS and engineers of PT GEC for providing information and data.

Fig. 19 Survey elevation of the mud (Sofyan 2015)

Fig. 20 Boundaries of cells to contain the mud (Tim Nasional 2007)

Fig. 21 Construction of permanent dykes (BPLS 2007)

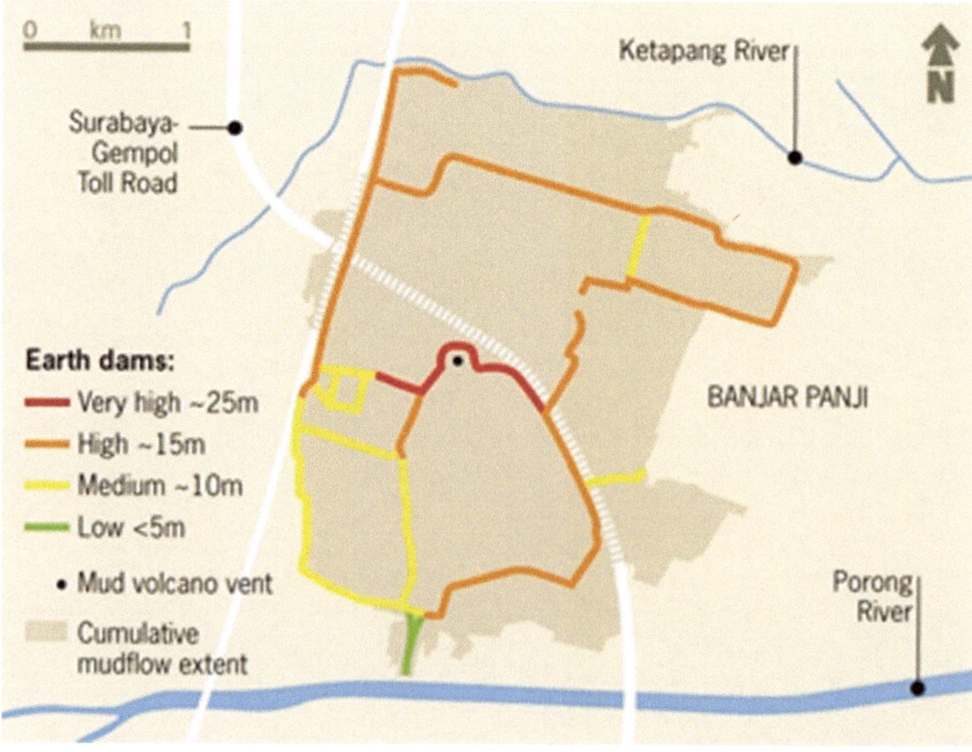

Fig. 22 The heights of dykes used as boundaries of cells

Fig. 23 Maintenance of dykes

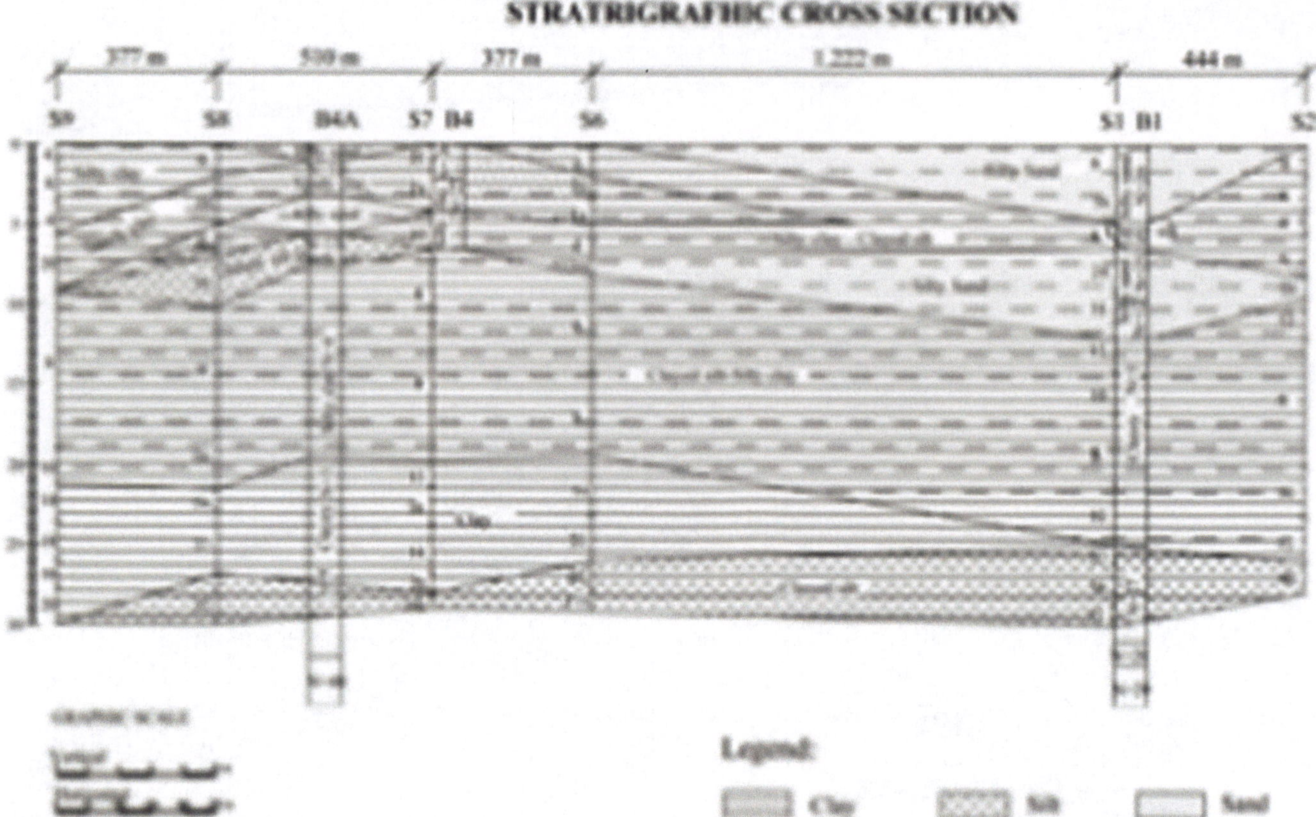

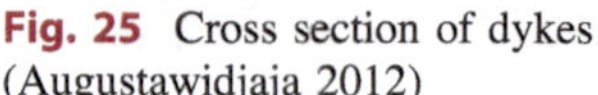

Fig. 24 Geotechnical cross section of foundation soils (Augustawijaya et al. 2012)

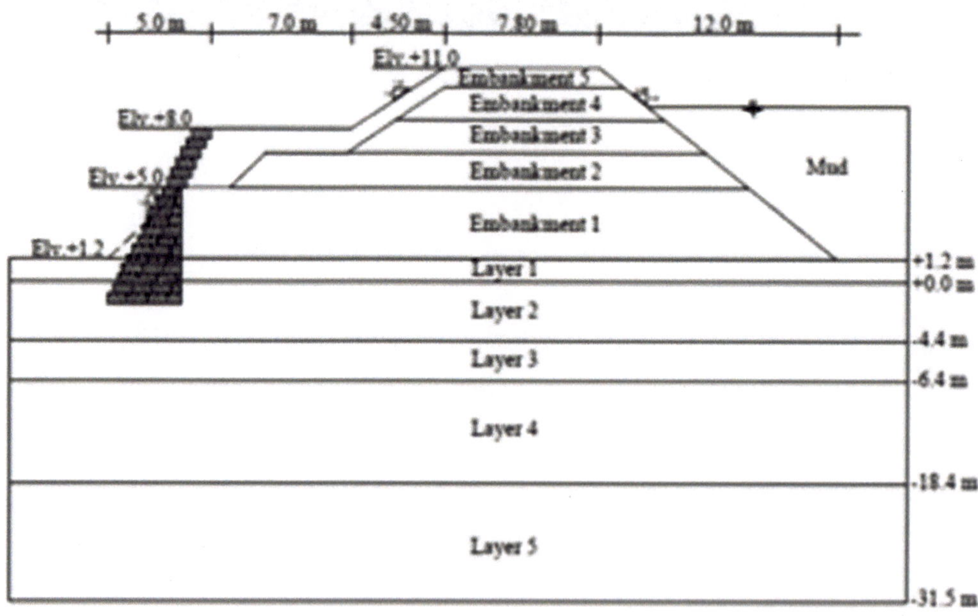

Fig. 25 Cross section of dykes (Augustawidjaja 2012)

References

Agustawijaya DS, Sukandi (2012) The stability analysis of the Lusi mud volcano embankment dams using FEM with special reference to the dam point P10D. Civ Eng Dimension 14(2):100–109

BPLS (Badan Penangulangan Lumpur Sidoardjo) (2007) Personal communication with Mr. Sofyan

Boston.com (2008) Sidoardjo's man-made mud volcano

GEC PT (2009) Soil investigation report using CPTU for the alternative by pass road at Porong (Report)

Guntoro A (2011) Understanding the origin of Sidoardjo mud volcano in relation to longevity estimation based on regional study and seismic interpretation. Lectures at Geological Department Trisakti University, Jakarta

Kompas (2010) Lapindo mud resulted in building fracture, Surabaya, 22 June 2010

Rahardjo PP (2011) Geotechnical aspect of Sidoardjo mud eruption (in Indonesian), report, Center for Infrastructures and Urban Development, Parahyangan Catholic University

Rahardjo PP (2015) Geotechnical aspect of mud eruption disaster in East Java. In: Proceeding 19th Asian regional conference on soil mechanics and foundation engineering, Fukuoka, Japan

Mazzini A, Nermoen A, Krotkiewski M, Podladchikov S, Planke S, Svensen H (2009) Strike-slip faulting as a trigger mechanism for overpressure release through piercement structures: implications for the lusi mud volcano, Indonesia. Mar Petrol Geol 26(9):1751–1765

Sofyan (2015) Report on the latest situation of Lumpur Sidoardjo. In: Proceedings international conference on landslides and slope stability, Bali

Soleman AR (2012) Karakterisasi Tanah Lempung Lunak di Porong Sidoardjo berdasarkan Uji CPTu dan Uji Laboratorium, tesis Magister. Universitas Katolik Parahyangan, Bandung

Wayman E (2011) The World's Muddiest disaster, Smithsonian.com, 2 Dec 2011

Massive Landsliding in Serbia Following Cyclone Tamara in May 2014 (IPL-210)

Biljana Abolmasov, Miloš Marjanović, Uroš Đurić, Jelka Krušić, and Katarina Andrejev

Abstract

The IPL project No 210, titled "Massive landsliding in Serbia following Cyclone Tamara in May 2014", started in March 2016. The study area is located in the Western and Central part of the Republic of Serbia territory affected by Cyclone Tamara in May 2014. The project aims to summarize and analyse all collected relevant data, including historic and current rainfall, landslide records, aftermath reports, and environmental features datasets from the May 2014 sequence. Objectives of the proposed project include: collecting all available and acquired landslide data, analysing the trigger/landslide relation in a feasible time span and in the May 2014 event, relating the landslide mechanisms and magnitudes versus the trigger, identifying spatial patterns and relationships between landslides and geological and environmental controls, proposing an overview susceptibility map of the event and numerical modelling of the site-specific location and landslide mechanisms. The Project will be organized by University of Belgrade, Faculty of Mining and Geology and Faculty of Civil Engineering. Project beneficiaries are local community and local and regional authorities. In this paper we will present preliminary results of the proposed project targets performed by project participants.

Keywords

Landslides • Extreme precipitation • Flooding • Republic of Serbia

Introduction

The Republic of Serbia is located on the Balkan Peninsula in south-east Europe, covers an area of 88,361 km^2 and has a population of 7,181,505 (http://stat.gov.rs) (Fig. 1).

Serbia's climate varies between a continental climate in the North, with cold winters, and hot, humid summers with well distributed rainfall patterns, and a more Adriatic climate in the South, with hot, dry summers and autumns and relatively cold winters with heavy inland snowfall. Differences in elevation and large river basins, as well as exposure to the winds account for climate differences, especially for annual

B. Abolmasov (✉) · M. Marjanović · J. Krušić · K. Andrejev
Faculty of Mining and Geology, University of Belgrade,
Đušina 7, 11000 Belgrade, Serbia
e-mail: biljana.abolmasov@rgf.bg.ac.rs

M. Marjanović
e-mail: milos.marjanovic@rgf.bg.ac.rs

J. Krušić
e-mail: jelka.krusic@rgf.rs

K. Andrejev
e-mail: katarina.andrejev@rgf.rs

U. Đurić
Faculty of Civil Engineering, University of Belgrade,
Bul Kralja Aleksandra 84, 11000 Belgrade, Serbia
e-mail: udjuric@grf.bg.ac.rs

K. Sassa et al. (eds.), *Advancing Culture of Living with Landslides*,
DOI 10.1007/978-3-319-59469-9_41

Fig. 1 Geographical position of the Republic of Serbia

precipitation sums. Annual precipitation increases with altitude. In lower regions, annual precipitation ranges from 540 to 820 mm. Areas with altitudes over 1000 m have on average 700–1000 mm of precipitation, and some of the mountainous summits in the South Western part of Serbia have heavier precipitation of up to 1500 mm. June is the rainiest month, with an average of 12–13% of the total annual rainfall. Because of complex geological history and terrain composition, and morphological and climate characteristics, 15.08% of Serbian territory is affected by landslides (Dragićević et al. 2011).

In the third week of May 2014, Serbia and Bosnia and Herzegovina experienced its severest floods in the last 120 years caused by Cyclone Tamara (Fig. 2). Huge amounts of rainfall of 250–400 mm for three days caused sudden and extreme flooding of several rivers—in particular the Sava River, but also the Drina, Bosna, Una, Sana, Vrbas, Kolubara, and Morava, as well as their tributaries. In Western and Central Serbia for instance, daily precipitation on May 15 exceeded the expected average of the entire month. Urban, industrial and rural areas were completely submerged under water, cut off without electricity or communications, and roads and transport facilities were damaged.

As a result, 1.6 million persons (one fifth of the population) were directly or indirectly affected in Serbia. The floods and landslides caused 51 casualties and around 32,000 people were evacuated. The Serbian Recovery Needs Assessment (RNA) revealed that the total effects of the disaster in the 24 affected municipalities cost up to EUR 1.525 billion (equal to 3% of the Serbian Gross Domestic Product).

The initiative to collaborate with the International Consortium on Landslides was started in September 2009. University of Belgrade, Faculty of Mining and Geology

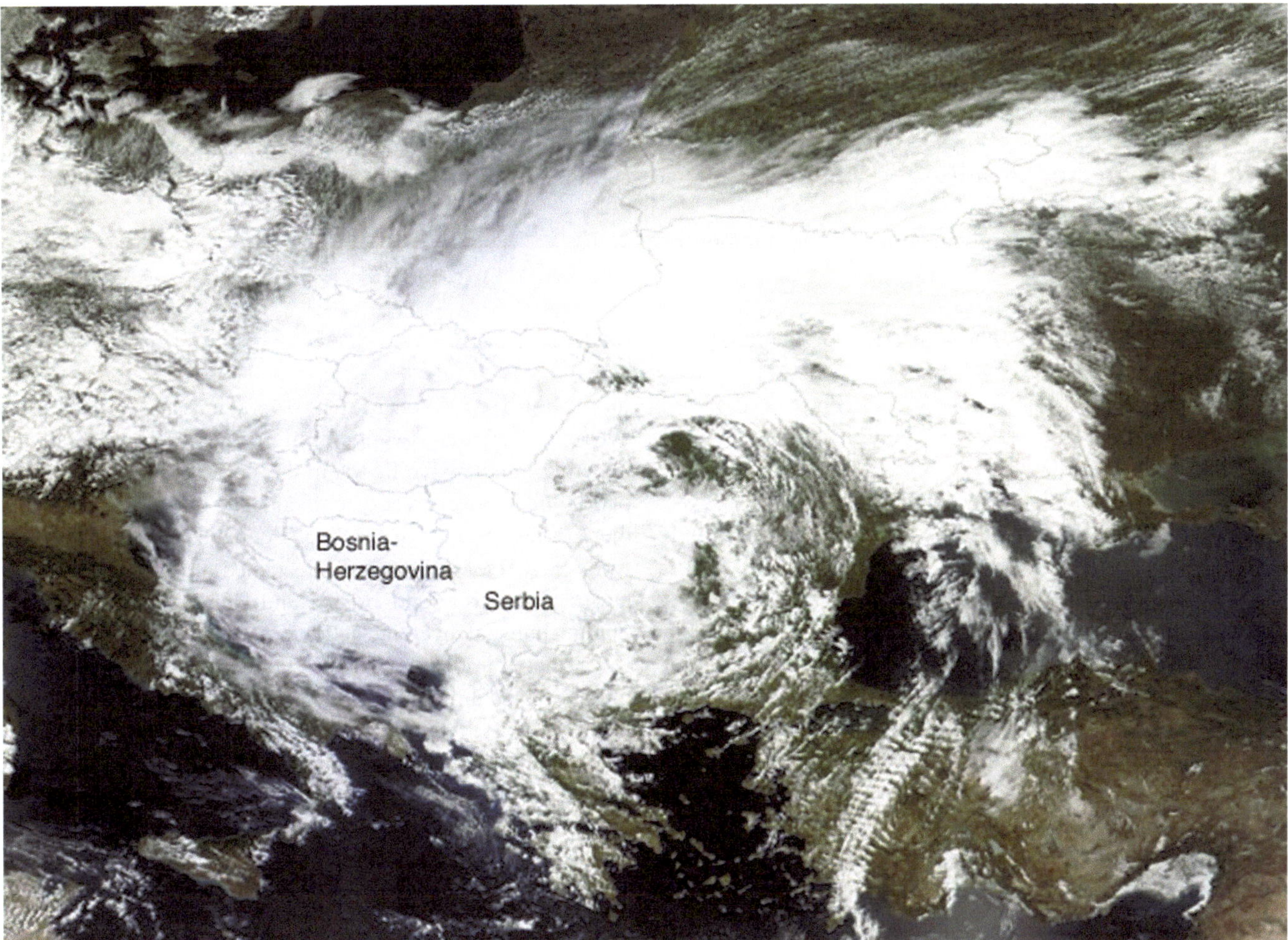

Fig. 2 MODIS satellite image of extratropical storm Yvette (Tamara) taken on May 15, 2014. (*Credit* LANCE Rapid Response/MODIS/NASA)

became a member of ICL in 2011, and a member of the ICL Adria-Balkan Network in 2012 (Mihalić Arbanas et al. 2013). In March 2016, the Faculty of Mining and Geology applied for the IPL project and during the 11th Session of IPL-GPC in Kyoto in 2016, a joint project number 210 was approved. It was entitled “Massive landsliding in Serbia following Cyclone Tamara in May 2014”.

This paper will show partial results obtained during less than a year of conducting the project, as described in the project plan and program.

Project Description

Objectives

Landslides are amongst the most dangerous natural threats to human lives and property, especially in times of dramatic climate change effects on one hand, and urban sprawl and land consumption on the other.

The project attempts to determine if the May 2014 extreme landsliding event was preconditioned by soil saturation, caused by a high precipitation yield, within several weeks before the event. All relevant data, including historic and current rainfall, landslide records, aftermath reports, and environmental features datasets, have to be analyzed for characterizing the extreme nature of the event and identifying key environmental controls of landslide occurrences.

In this respect, it was essential to produce unified large-scale inventories of the May 2014 event and use them for the state-of-the-art hazard analysis. Thus, the project aimed at summarizing and analyzing collected landslide information from the May 2014 sequence. Following this, the objectives of the proposed project include: (1) collecting all available (existing) and acquiring new landslide data, (2) analyzing the trigger/landslide relations for a feasible time span (past 15 years) and in the May 2014 event, (3) relating the landslide mechanisms and magnitudes versus the trigger and its aftermath, (4) locating spatial patterns and relationships between landslides and geological and environmental controls, (5) proposing an overview susceptibility map of the event and (6) numerical modeling of site-specific locations and landslide mechanisms.

Work Plan-expected Results

The following activities are planned during the duration of the project:

- Collection, review and harmonization of the landslide data (Phase 1)
- Analysis of trigger and landslide data (Phase 2)
- Analysis of landslides versus geological and environmental controls (Phase 3)
- Proposing a landslide susceptibility map (Phase 4)
- Numerical modeling of site-specific locations and landslide mechanisms (Phase 5)
- Compilation and analysis of all results (Phase 6).

After certain activities, it was planned to prepare partial reports, and to prepare a comprehensive report at the end. Preparation of papers for the Landslide journal was also foreseen. Deliverables and time frames are as follow:

- Report 1. Compilation of results of Phase 1 and Phase 2 (end of 1st year)
- Report 2. Compilation of results Phase 3 (end of 18th month)
- Report 3. Proposing landslide susceptibility map Phase 4 (end of 24th month)
- Report 4. Numerical modeling on site specific locations/landslide mechanism Phase 5 (end of 30th month)
- Report 5. Final report-Phase 6 (end of 3rd year).

Personel—Beneficiaries

The Project will be organized by University of Belgrade, Faculty of Mining and Geology and Faculty of Civil Engineering. The University and staff will provide all necessary documentation for Project finalization. The Project Leader is Associate Professor Biljana Abolmasov from University of Belgrade, Faculty of Mining and Geology. Core members of the Project are: Assistant Professor Miloš Marjanović from University of Belgrade Faculty of Mining and Geology, Uroš Djurić, Ph.D. student from University of Belgrade Faculty for Civil Engineering, Jelka Krušić, Ph.D. student from University of Belgrade Faculty of Mining and Geology and Katarina Andrejev, Ph.D. student from University of Belgrade Faculty of Mining and Geology.

Direct beneficiaries will be local communities and municipalities affected by landslide occurrences during the May 2014 event. Other beneficiaries include local and regional authorities—the housing sector, infrastructure authorities, civil protection units and land-use sectors within the affected area.

Preliminary Results

Rainfall Event

In the third week of May 2014, a massive low-pressure cyclone, Tamara, swept through the Western Balkans, resulting in extensive floods in the Sava River system and in part in the Morava river catchment. The cyclone moved from the Adriatic Sea to the Balkan Peninsula very slowly, and from 14 to 16 May pressure deepened at all altitudes in the territory of Serbia and Bosnia and Herzegovina. The result of that unusual cyclone activity was extreme precipitation for a short period that caused floods, torrential floods and massive landsliding in the Western and Central part of Serbia.

The analysis of precipitation data included available monthly and daily precipitation from the Hydrometeorological Service of the Republic of Serbia for April and May 2014. The rainfall with highest statistical significance for a 48 h duration was registered at the Loznica Main Meteorological Station (MMS), where precipitation of 160 mm corresponded to a 1000 year return period (Fig. 3), while the MMS in Valjevo (Fig. 4) and Belgrade (Fig. 5) recorded precipitation of a 400 year return period for the same duration (Prohaska et al. 2014). The highest precipitation for a 72 h duration was recorded at Loznica (213 mm), Valjevo (190 mm) and Belgrade (174 mm) MMS. The flood event (14–15 May 2014) and landslides occurrences (15–18 May 2014) were caused simultaneously by extreme Cyclone Tamara activity, but the massive landsliding was additionally initiated by the antecedent rainfall from April 15 to May 14 (Alleoti 2004). The main triggering factor for all landslide activities was extreme cumulative precipitation from April 15 up to May 18, in which the precipitation amount exceeded one half of the yearly average precipitation for one month in Western and Central part of Serbia (Marjanović and Abolmasov 2015). The analysis of monthly precipitation for April and May 2014 is shown in Figs. 6 and 7.

Study Area

The study area covered 11,840 km^2, i.e. 23 of 27 municipalities affected by different type of landslides in the Western

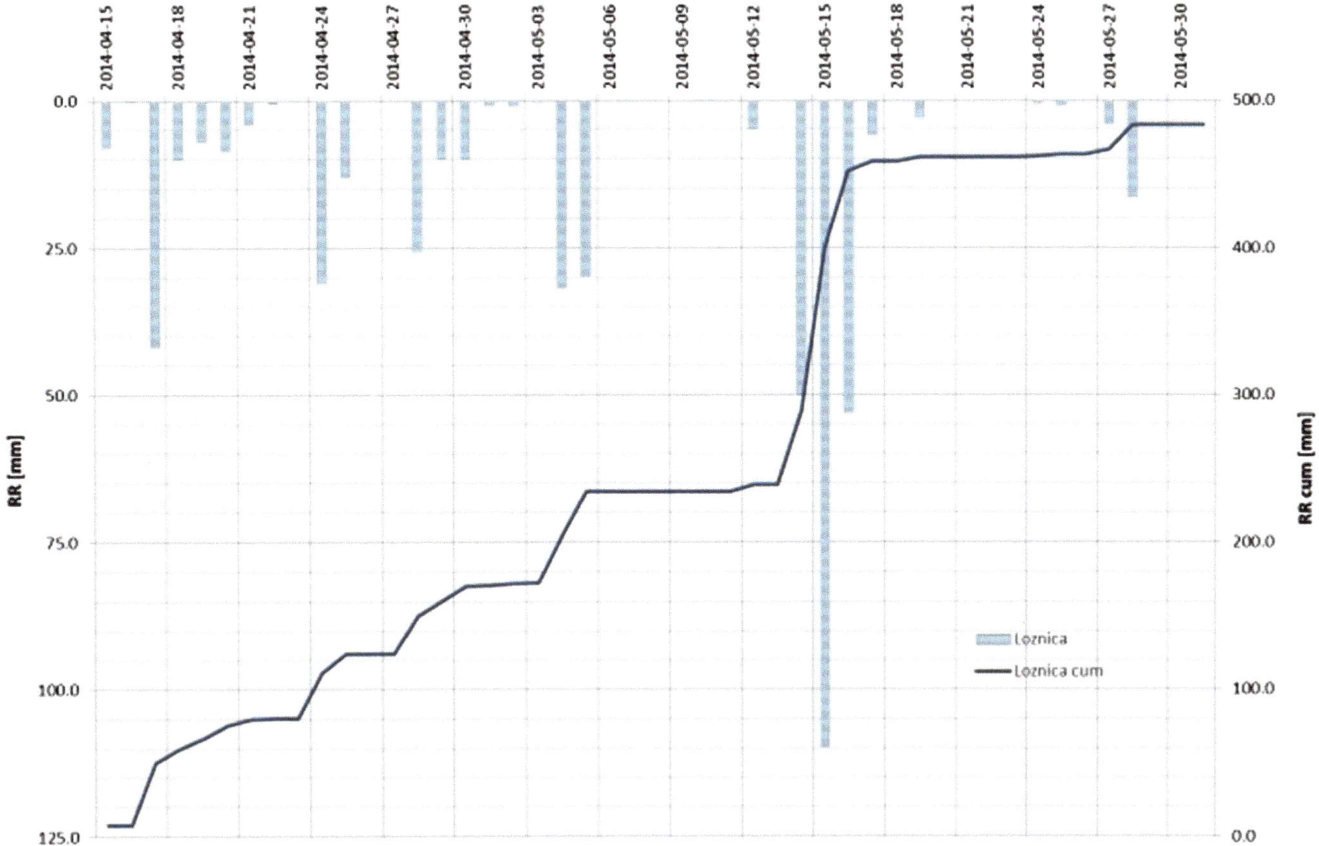

Fig. 3 Cumulative precipitation for MMS Loznica (Western Serbia) from April 15 to May 30 2014

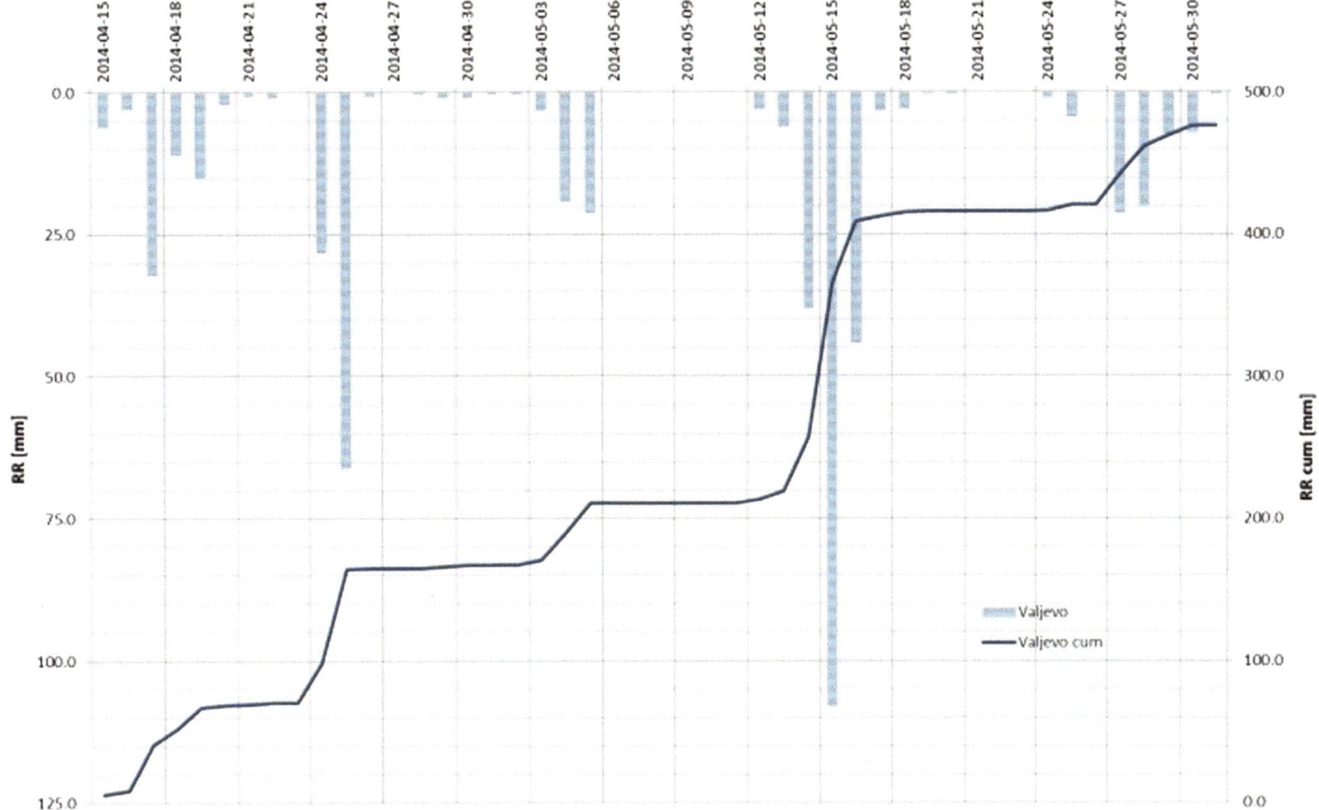

Fig. 4 Cumulative precipitation for MMS Valjevo (Western Serbia) from April 15 to May 30 2014

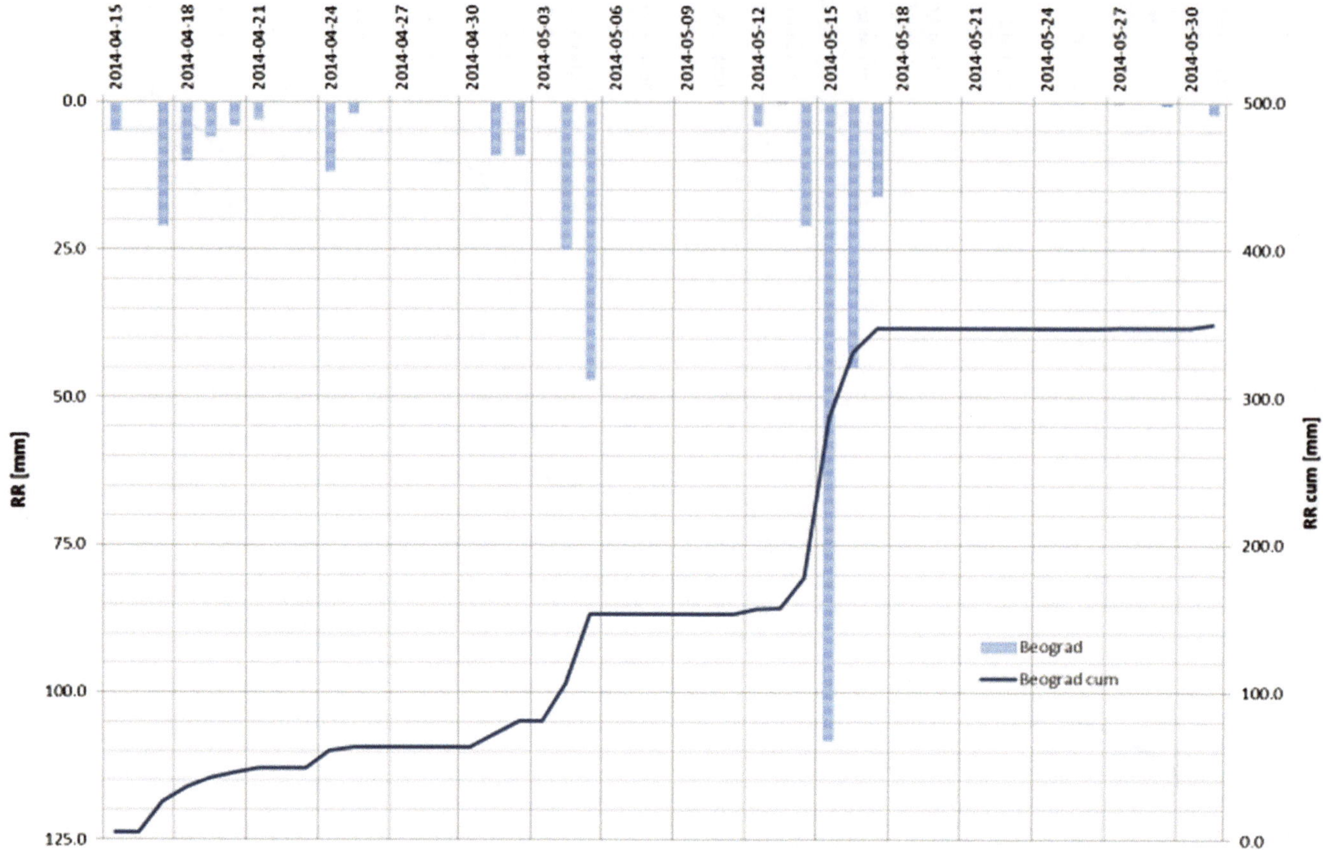

Fig. 5 Cumulative precipitation for MMS Belgrade (Central Serbia) from April 15 to May 30 2014

and Central part of the Republic of Serbia (Fig. 8). These municipalities were recognized as the most vulnerable to floods, torrential floods and landslides by the UNDP Office in Serbia during the post-disaster phase after the May 2014 event. Four municipalities were excluded from IPL 210 Project activities because no landslides occurred during the May 2014 rainfall episode; there was only flood damage. The geological and geomorphological settings are very complex, as well as other environmental conditions in such a wide area. The type of movement and type of material involved (Cruden and VanDine 2013) were dependent on lithological type, local geomorphological characteristics, engineering geological properties, degree and depth of weathering substratum, as well as the amount of precipitation received during the May 2014 event.

Landslide Data

The most common landslide triggers are floods and high-yield rainfall, which was the case in the catastrophic cyclone Tamara episode that struck Serbia and surrounding countries in May 2014. At the time, the effects of the disasters were closely followed by the media and public and handled by responsible state services, such as Civil Protection offices, and volunteers, but little has been done after the waters retreated and the landslides settled, especially with regard to landslide analysis and mitigation. Landslide reports (in analogue form) greatly understated the realistic number of landslides (concentrating more on urgent and acute cases), while report quality standard and consistency was uneven (because they were collected by different institutions, depending on the acute needs), so the resulting inventories remain incomplete and far from standardized. In this respect, it was essential to produce unified large-scale inventories of the May 2014 event and beyond, and use them for further analysis.

Based on the classification of Cruden and VanDine (2013), a harmonized landslide data report was created (Fig. 9). The total number of 2203 landslides are mapped as open data file reports, according to the BEWARE Project deliverables. Different type of movement and type of material involved were registered during an extensive field campaign (Fig. 10). A total number of 1888 different type of movement were certified by supervisors (1539 slides, 78

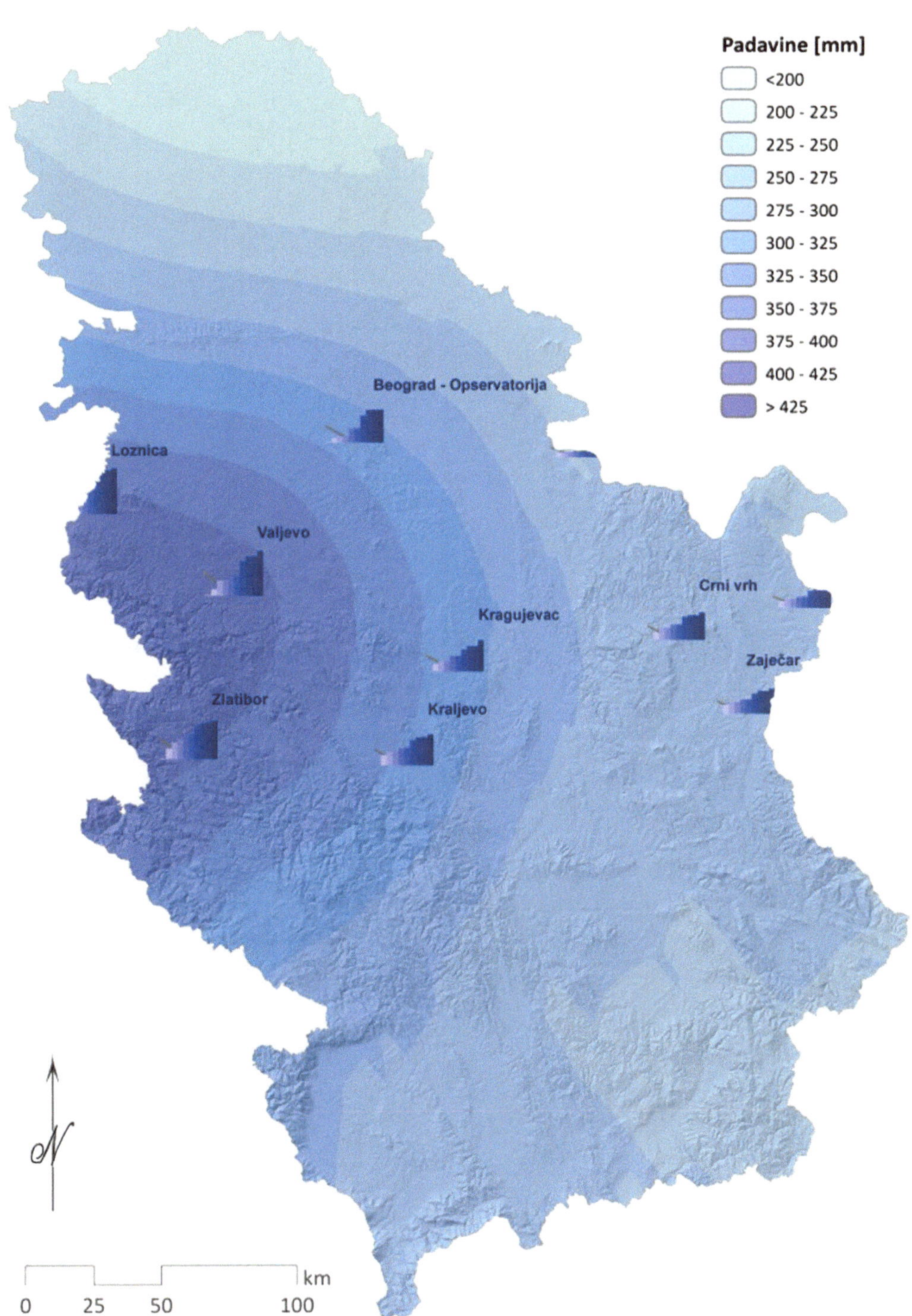

Fig. 6 Distribution of sum of precipitation for April 2014

flows, 48 falls, 1 topple, 23 complex, 138 flows and slides, 55 falls and slides and 6 falls and flows). Based on the material involved, 925 type of movement were formed from debris, 894 from earth, 20 from rock, 33 from mixed and 16 from artificial material. The simple analysis performed based on landslide distribution by municipalities shows that the highest number of landslide occurrences were recorded in the Western part of Serbia (Fig. 11).

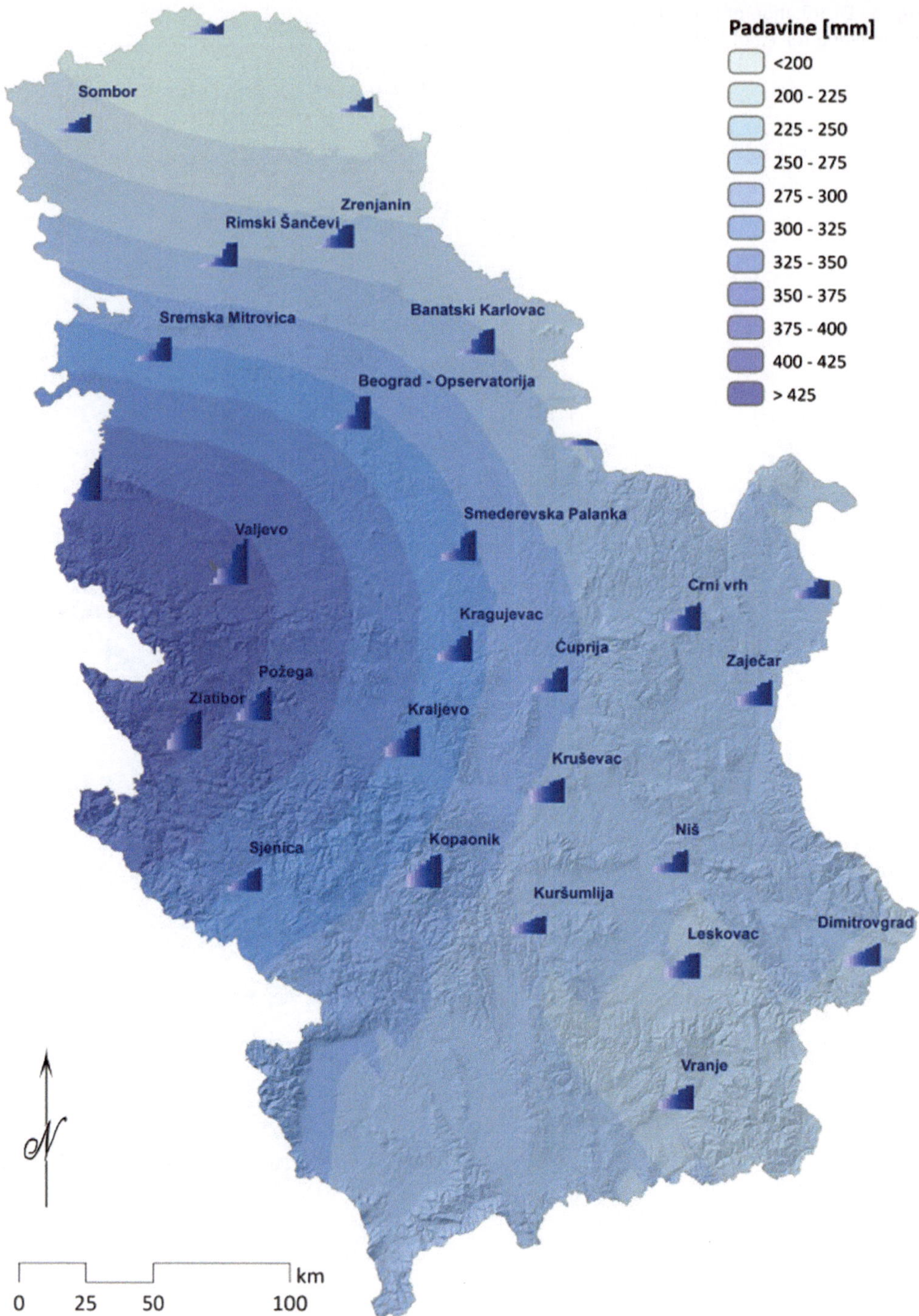

Fig. 7 Distribution of sum of precipitation for May 2014

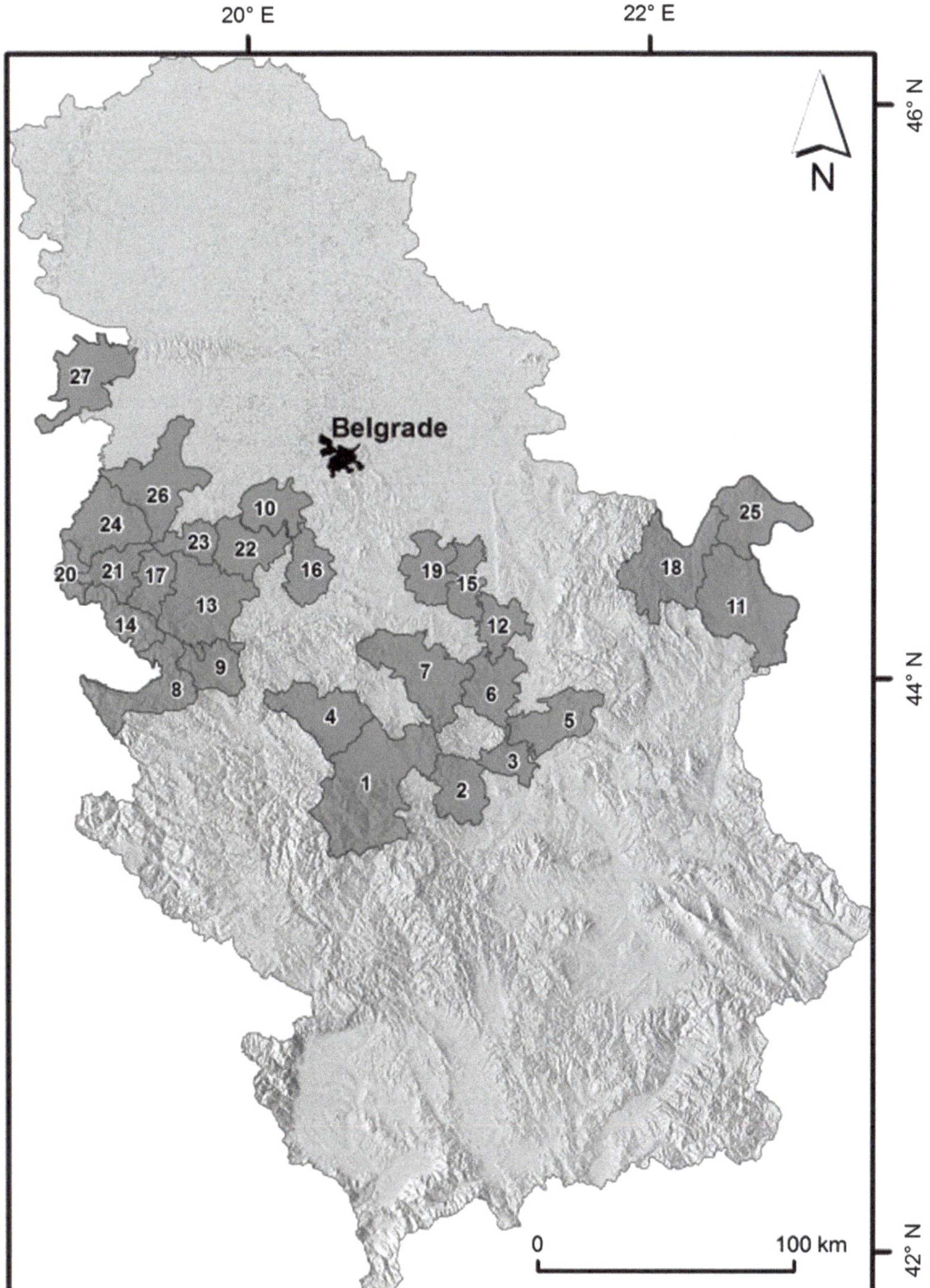

Fig. 8 Geographical position of 23 municipalities included in the IPL 210 Project analysis (four municipalities were excluded–No 11, 18, 25 and 27)

BEWARE KATASTARSKI LIST EVIDENTIRANOG KLIZIŠTA – POJAVE NESTABILNOSTI

ID BROJ*

1. OPŠTI PODACI	
Opština*	
Lokalnost*	
Koordinate*	
Datum registr.*	
Datum aktiviranja	
Datum reaktiviranja	
Odgovorni istraživač:	

2.1 OPŠTI PODACI O PROCESU			
Tip pojave*		Vrsta pokrenutog materijala*	
Odronjavanje	☐	Stena	☐
Prevrtanje	☐	Drobina	☐
Klizenje	☐	Organski materijal	☐
Bočno širenje	☐	Antropogeni materijal	☐
Tečenje	☐	Heterogeni materijal	☐
Složeno	☐	Tlo: Glina ☐ Pesak ☐	
BEWARE		Tlo: Prašina ☐ Šljunak ☐	

2.2 OPŠTI PODACI O PROCESU					
Sadržaj vode*		Brzina kretanja*		Aktivnost*	
Suvo	☐	Ekstremno sporo	☐	Uslovno stabilna padina	☐
Vlažno	☐	Vrlo sporo	☐	Aktivan	☐
Vlažno na granici tečenja	☐	Sporo	☐	Trenutno umiren	☐
U tečnom stanju	☐	Umereno	☐	Reaktiviran	☐
		Brzo	☐	Neaktivan	
		Veoma brzo	☐	Privremeno umiren ☐	Saniran-stabilizovano ☐
		Ekstremno brzo	☐	Umiren ☐	Fosilan ☐

2.3 OPŠTI PODACI O PROCESU				Skica klizišta u preseku
Trend kretanja*		Način kretanja*		
Progresivno uz padinu	☐	Pojedinačno	☐	
Progresivno niz padinu	☐	Sukcesivno	☐	
Progresivno bočno	☐	Višestruko	☐	
Progresivno u dva pravca	☐	Mešovito	☐	
Smirivanje	☐	Kompleksno	☐	

3 OPŠTI PODACI O TERENU				
Genetski tip reljefa		Morfološki oblik		Geološka građa
Fluvijalni tip reljefa	☐	Čelenka	☐	Vrsta osnovne stene - litološki sastav:
Padinski reljef	☐	Kosa	☐	
Jezerski	☐	Brdo	☐	Starost:
Tektonski	☐	O padini		
		Visina	m	Struktura:
		Nagib	°	

Hidrologija				Stepen raspadnutosti stenske mase
Vodotoci:				Zemljasta raspadina ☐
Stalni	☐	Povremeni	☐	Sitna drobina ☐
Ostalo:				Drobina ☐
				Blokovi ☐

Hidrogeološka funkcija
H.G. funkcija:
Tip izdani (nivoi, hranjenje, pražnjenje):
Pojave:
Oka ☐ Izvori ☐ Pištevine ☐ Difuzno pražnjenje ☐

MOLIMO VAS DA OBRAZAC POPUNITE ČITKO - ŠTAMPANIM SLOVIMA - HEMIJSKOM OLOVKOM

1

* OBAVEZNA POLJA ZA UNOS PODATAKA

Fig. 9 Unified landslide data report from BEWARE Project (in Serbian)

Fig. 10 The Valjevo municipality earth flow

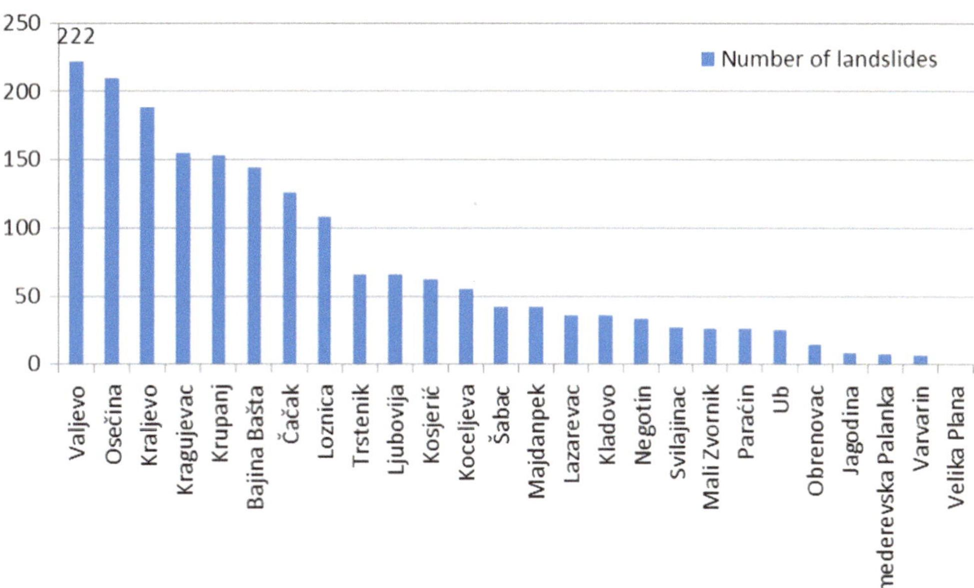

Fig. 11 Landslide distribution by municipality http://geoliss.mre.gov.rs/beware/

Conclusion

First research results from the IPL 210 Project after six months of project conduct are presented in the paper. The analysis, correlation and synthesis of large volumes of data are currently being performed. Following the Project activities, the next steps will be focused on analyzing: (1) the trigger/landslide relation in a feasible time span (past 15 years) and in the May 2014 event and (2) relating the landslide mechanisms and magnitudes to the trigger and its aftermath.

Acknowledgements IPL Project 210 would not be not possible without Project BEWARE (BEyond landslide aWAREness) funded by People of Japan and UNDP Office in Serbia (grant No. 00094641). The project was implemented by the State Geological Survey of Serbia, and the University of Belgrade Faculty of Mining and Geology. All activities are supported by Ministry for Energy and Mining and Ministry for Education, Science and Technological Development of the Republic of Serbia Project No. TR36009, too.

References

Aleotti P (2004) A warning system for rainfall-induced shallow failures. Eng Geol 73(3–4):247–265. doi:10.1016/j.enggeo.2004.01.007

Cruden D, VanDine DF (2013) Classification, description, causes and indirect effects—Canadian technical guidelines and best practices related to landslides: a national initiative for loss reduction. Geol Surv Can Open File 7359:2013

Dragićević S, Filipović D, Kostadinov S, Ristić R, Novković I, Živković N, Anđelković G, Abolmasov B, Šećerov V, Đurđić S (2011) Natural hazard assessment for land-use planning in Serbia. Int J Environ Res 5(2):371–380

Marjanović M, Abolmasov B (2015) Evidencija i prostorna analiza klizišta zabeleženih u maju 2014. Časopis Izgradnja 69(5–6):129–134 (in Serbian)

Mihalić Arbanas S, Arbanas Ž, Abolmasov B, Mikoš M, Komac M (2013) The ICL Adriatic-Balkan Network: analysis of current state and planned activities. Landslides 10:103–109. doi:10.1007/s10346-012-0364-2

Prohaska S, Đukić D, Bartoš Divac V, Božović N (2014) Statistical significance of the rainfall intensity that caused the may 2014 flood in Serbia. Water Res Manage 4(3):3–10

Part III
Landslides and Society

Landslides and Society—A Foreword

Irasema Alcántara-Ayala

Abstract

This chapter provides a general account of the contributions of the ICL community to the implementation of the Sendai Framework for Disaster Risk Reduction. By addressing the particular thematic issue of "Landslides and Society", the ICL group is committed to face the big challenge of developing linkages between landslide science and decision making and practice. The chapter introduces all articles contained within this special issue. Some of the contributions of this suite of papers have hinted at additional challenges that need to be further addressed through the establishment of linkages between landslide science and policy making and practice.

Keywords

Landslides • Society • Science • Disaster risk reduction • Decision making • Practice

On the Nature of Landslide Hazards

As landslides involve the movement of materials down slope due to gravity, they are controlled by a series of natural factors and physical processes, and therefore can be considered as natural hazards. However, in recent decades the influence of human transformations on the slopes is so vast that nowadays it seems to be rather impossible to find landslides 100% of natural origin. Anthropic activities induce a series of constant disturbances on the slopes producing a lack of equilibrium and a higher susceptibility to landsliding that affect all kind of societies (Alcántara-Ayala 2016). Consequently, landslides can be regarded as socio-natural hazards, that is to say, resulted from the interaction between social practices and the environment (Lavell 2003).

I. Alcántara-Ayala (✉)
National Autonomous University of Mexico (UNAM), Institute of Geography, Circuito Exterior, Ciudad Universitaria, Coyoacán, 04510 Mexico City, Mexico
e-mail: irasema@igg.unam.mx

Deficient or non-existing territorial planning, urbanization, degradation of the environment, exploitation of resources, deforestation and population growth are some of the main landslide disaster risk drivers that deserve attention and improved strategies for policy making and practice (Alcántara-Ayala et al. 2017).

This chapter brings together a series of manuscripts focused on different international initiatives directed towards the reduction of disaster risk associated with landslides.

The nine papers in this issue reveal different technical and practical perspectives to assess different aspects of landslide disaster risk under varies spatio-temporal scenarios. Mapping approaches and results reflect a series of problems related to data availability at various scales, including local, national, regional and global levels, while social analysis are highly complex as landslide disaster risk perception is shaped through experience and is strongly influenced by context. Combining instrumentation, monitoring and awareness has been also depicted as a major strategy for the preservation of World Heritage Sites, although this approach can be with no doubt, applied to Geoparks, Geomorphosites and other relevant sites that require protection.

K. Sassa et al. (eds.), *Advancing Culture of Living with Landslides*,
DOI 10.1007/978-3-319-59469-9_42

Landslides: Space and Time

Understanding landslide disaster risk requires the comprehension of its occurrence through time and space. Under such approach, da Silva Pereira et al. (2017) prepared a large database of landslides that occurred in Portugal from 1865 to 2015. These records were used to analyze the spatio-temporal occurrence of damaging landslides; the frequency and patterns of fatal landslides, in order to further explore the distribution of landslide fatalities in space and time. Likewise, they also identified the most deadly landslide types, the gender tendencies in landslide mortality, and evaluated the individual and societal risk. By analyzing the data, the authors came to the conclusion that in Portugal, the spatial patterns of landslide mortality can be associated with the heterogeneous distribution of landslide controlling factors, changes in the land use and exposure, as well as social vulnerability to landslide hazards.

Landslide mapping is fundamental for hazard and risk assessment. It has to be undertaken at different scales being the national level quite significant for decision making and policy development. Landslide inventory mapping in the fourteen Northern provinces of Vietnam was carried out by Hung and collaborators (2017). Five difficulties were identified in terms of interpretation of aerial photographs and field surveys: (1) lack of availability of multi-date air-photos; (2) the need of human resources with high expertise on interpretation of aerial photographs; (3) field validation in remote sites (4) lack of historical records of small or medium size events; (5) Unavailable means to update field surveys. Recognition of such constrains leads to improvement and also offers the possibility to identify the actual contributions to map landslides in time and space. The most important achievement was the development of the landslide inventory, which includes 10,149 landslides: 50% are of small volume, 33% are moderate, 16% are large, and 0.30% is of very large volume. Based on the inventory, different maps were printed out as well as published on a landslide national WebGIS. Information regarding historical and spatial extent of landslides has been provided to local communities and authorities. The inventory will be used as a basic input for the development of susceptibility and hazard mapping.

Safe and sustainable transport is a major concern for the Federal Ministry of Transport and Digital Infrastructure of Germany. Under such framework, Klose and collaborators (2017) focused on the assessment of the landslide hazard potential for the federal transport system under the influence of climate change. On a Geographical Information System Platform, they combined a landslide susceptibility map with regional climate change projections. The general strategy is based on landslide identification and inventory mapping; landslide hazard mapping; landslide impact and vulnerability assessment; and hazard communication. Specifically, they investigated three landslide sites situated in west and northwest Germany. Linkages between rainfall, temperature, and landslide activity were analyzed in order to be combined with the available climate projections. Owing to future climate scenarios, results suggest a potential increase in the occurrence of shallow landslides and debris flow activity in the summer, whereas in winter, a raise in landslide and rockfall activity is expected.

Preserving Nature and Culture

Cesaro et al. (2017) have provided a comprehensive analysis to portray the essential role that risk identification, monitoring and awareness can play to address the challenges faced by The Petra Archaeological Park, a World Heritage Site established since 1985. In addition to the to the reconstruction of geomorphological dynamics and monitoring of active slope processes, particular attention is paid to the potential contribution of stakeholders and local communities in terms of managing a geo-archaeological site, as a significant endeavor for disaster risk reduction. Specific goals included the implementation of landslide risk mitigation measures by promoting capacity building for national authorities and the delineation and accomplishment of a strategy on hazard awareness. Accordingly, public awareness and communication on natural hazards can be regarded as a very useful non-structural technique for the management and mitigation of landslide risk in World Heritage Sites, such as Petra.

Landslides, Knowledge, Awareness and Management

From an historical perspective, Wohlers and co-workers (2017) examined the occurrence of landslides in the highway network of the Lower Saxon Uplands, NW Germany. Their investigation aimed at understanding the interactions between landslide risk, and public and private landslide risk awareness. Landslide events were categorized and classified in terms of landslide types, processes, damages, and executed mitigation measures. Results indicated that the increasing frequency and magnitude of landslides have lead to the amplification of public landslide risk awareness in last 20 years, which has been expressed through the implementation of structural mitigation measures as local decision makers have invested in expensive long-term stabilization projects, including soil anchoring, rock nailing, and steel-reinforced concrete walls.

Chiu and Eidsvig (2017) measured the perception of landslide risk management by adapting and applying the Risk Management Index (RMI) developed by Cardona et al.

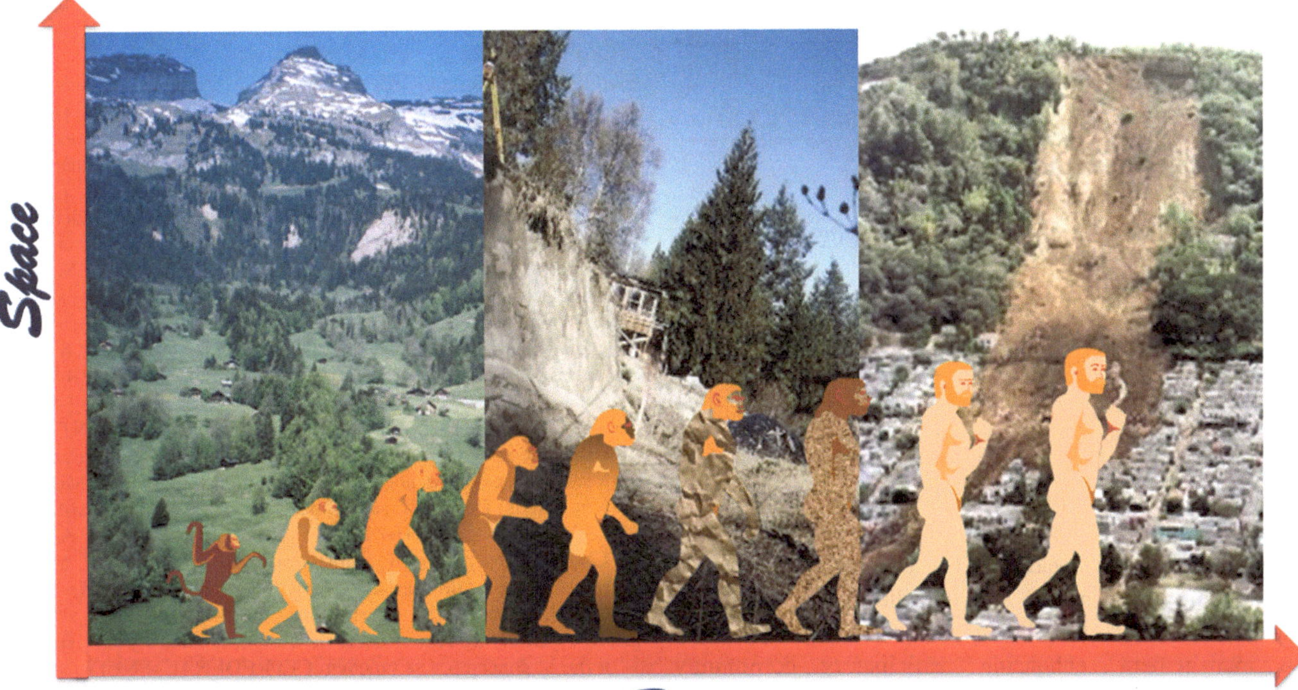

Fig. 1 Human agency as a main contributor for landslide occurrence

(2004) to local practitioners in landslide risk reduction in Norway. Two time scenarios, 2015 and 2050, and three administrative levels, national, county, and municipality, were included. Results of the investigation suggested the need of improving varies aspects of landslide risk management, including the prioritization of landslide hazard evaluation and mapping, improvement and reconstruction of damaged assets, and more importantly, enhancing inter-institutional organization, allocation and use of financial resources for landslide management.

Undoubtedly, landslide susceptibility maps are a cardinal tool for land use planning and civil protection emergency management. However, quite frequently they are neglected as decision making and practice are not always based on scientific contributions. Bad and good experience regarding this issue must serve as a reference for future strategies. In Portugal, for instance, concerns about land use planning and emergency management have induced a series of regulations regarding practice, prevention and risk management, which have been lately promoted through diverse legal instruments. Following this criteria, Oliveira and colleagues (2017) developed a landslide susceptibility map for the Loures municipality, on which, exposed elements and their associated potential risk were also identified and included. Science and decision making linkages have proved to be good enough so that these documents have been included in the Loures Municipal Master Plan, and are currently used by the Municipal Civil Protection Service to produce estimations of people exposed to landslides.

Landslides occur under different environments and societal conditions. Nonetheless, people's awareness plays a key role in all types of societies as hazard understanding can be regarded as the first step towards prevention and management. To this regard, Piangiamore and Musacchio (2017) addressed the significance of education to merge knowledge and understanding as a requirement for disaster risk reduction. By using a participatory approach—a non structural measure—they presented two study cases from Italy, to document effective dissemination and communication on hydrometeorological hazards by using educational tools and engagement initiatives to raise public awareness and hazard understanding. They developed this strategy particularly by means of engaging students and teachers on a disaster risk reduction attitude for a resilient community.

Last, but undoubtedly not least, Sapač and colleagues (2017) advocate for "More Room for Landslides", a campaign claiming for more space for control of potential landslides induced by anthropic activities; more space for torrents; more space for water and sediment storages and a reduction on the impact on the stability of slopes, which also means a higher security for societies. By depicting three study cases from Slovenia, the authors illustrate the way improperly designed or maintained forest roads and skid trails have involved a greater landslide exposure and damage. This situation has been exacerbated particularly in the last 20–30 years, as the number of skid trails and forest roads has increased considerably nationwide.

Concluding Remarks

The human influence on the landscape during the last decades has provided enough evidence to understand that to a great extent, human agency can be regarded as a main contributor for landslide occurrence (Fig. 1). This suite of papers presented in this chapter. "Landslides and Society" highlights the contribution of the International Consortium on Landslides to the Sendai Framework for Disaster Risk Reduction. Some of the contributions have hinted at additional challenges that need to be further addressed through the establishment of linkages between landslide science and policy making and practice. These challenges include development of integrated and transdisciplinary landslide disaster risk science, landslide capacity building for disaster risk management, enhancing participatory community approaches to guarantee the co-production of risk knowledge, and promoting territorial management, and the establishment of associated legal frameworks.

Acknowledgements Special thanks are due to all contributors of this chapter Landslides and Society.

References

Alcántara-Ayala I (2016) On the multi-dimensions of integrated research on landslide disaster risk. In: Aversa S, Cascini L, Picarelli L, Scavia C (eds) Landslides and engineered slopes, experience, theory and practice. CRC Press, Taylor & Francis Group, Balkema, UK, pp 155–168

Alcántara-Ayala I, Murray V, Daniels P, McBean G (2017) On the future challenges for the integration of science into international policy development for landslide disaster risk reduction. In: Sassa K, Mikos M, Yin Y (eds) ISDR-ICL sendai partnerships 2015–2025, vol 1

Cardona OD, Hurtado JE, Duque G, Moreno A, Chardon AC, Velásquez LS, Prieto SD (2004) Disaster risk and risk management benchmarking: a methodology based on indicators at national level. IDB/IDEA program on indicators for disaster risk management. Universidad Nacional de Colombia, Manizales, p 101p

Cesaro G, Delmonaco G, Khrisat B, Salis S (2017) Geological conservation through risk mitigation and public awareness at the Siq of Petra, Jordan. In: Sassa K, Mikos M, Yin Y (eds) ISDR-ICL sendai partnerships 2015–2025, vol 1

Chiu J, Eidsvig U (2017) Surveying perception of landslide risk management performance, a case study in Norway. In: Sassa K, Mikos M, Yin Y (eds) ISDR-ICL sendai partnerships 2015–2025, vol 1

Hung LQ, Van NTH, Son PV, Ninh NH, Tam N, Huyen NT (2017) Landslide inventory mapping in the fourteen Northern provinces of Vietnam: achievements and difficulties. In: Sassa K, Mikos M, Yin Y (eds) ISDR-ICL sendai partnerships 2015–2025, vol 1

Klose M, Auerbach M, Herrmann C, Kumerics C, Gratzki A (2017) Landslide hazards and climate change adaptation of transport infrastructures in Germany. In: Sassa K, Mikos M, Yin Y (eds) ISDR-ICL sendai partnerships 2015–2025, vol 1

Lavell A (2003) La gestión local del riesgo, nociones y precisiones en torno al concepto y la práctica. CEPREDENAC–PNUD

Oliveira SC, Zêzere JL, Guillard-Gonçalves C, Garcia RAC, Pereira S (2017) Integration of landslide susceptibility maps for land use planning and civil protection emergency management. In: Sassa K, Mikos M, Yin Y (eds) ISDR-ICL sendai partnerships 2015–2025, vol 1

Pereira S, Zêzere JL, Quaresma I (2017) Landslide societal risk in Portugal in the last 155 years. In: Sassa K, Mikos M, Yin Y (eds) ISDR-ICL sendai partnerships 2015–2025, vol 1

Piangiamore GL, Musacchio G (2017) Participatory approach to natural hazard education for hydrological risk reduction. In: Sassa K, Mikos M, Yin Y (eds) ISDR-ICL sendai partnerships 2015–2025, vol 1

Sapač K, Humar N, Brilly M, Kryžanowski A (2017) More room for landslides. In: Sassa K, Mikos M, Yin Y (eds) ISDR-ICL sendai partnerships 2015–2025, vol 1

Wohlers A, Kreuzer T, Damm B (2017) Case histories for the investigation of landslide repair and mitigation measures in NW Germany. In: Sassa K, Mikos M, Yin Y (eds) ISDR-ICL sendai partnerships 2015–2025, vol 1

Landslide Societal Risk in Portugal in the Period 1865–2015

Susana Pereira, José Luís Zêzere, and Ivânia Quaresma

Abstract

In Portugal, social impacts caused by landslides occurred in the period 1865–2015 are gathered in the DISASTER database. This database includes social consequences (fatalities, injuries, missing people, evacuated people and homeless people) caused by landslides documented in newspapers. The DISASTER database contains 291 damaging landslides that caused 238 fatalities. In this work we aim to: (i) analyse the spatio-temporal analysis of damaging landslides occurred in the last 150 years; (ii) analyse the frequency and the temporal evolution of fatal landslides; (iii) analyse the spatio-temporal distribution of landslide fatalities; (iv) identify the most deadly landside types; (v) verify gender tendencies in landslide mortality; and (vi) evaluate the individual and societal risk. Individual risk is evaluated computing mortality rates for landslides, which are calculated based on the annual average population and the annual average of fatalities. The societal risk is evaluated by plotting the annual frequency of landslide cases that generated fatalities. The results demonstrate the absence of any exponential growth in time of both landslide cases and landslide mortality in Portugal. The highest number of landslide cases and related mortalities occurred in the period of 1935–1969 in relation to very wet years. Most of landslide fatalities mainly occurred in the north of the Tagus valley where the geologic and geomorphologic conditions are more prone to landslides. The Lisbon area registered a mortality hotspot, which is explained by natural conditions combined with the high exposure of population to landslide risk. Falls and flows were responsible for the highest number of fatalities associated with landslides. Males were found to have the highest frequency of fatalities. In conclusion, the spatial patterns of landslide mortality can be related to the unequal distribution of predisposing conditions to landslides, changes in the land use and exposure and social vulnerability to landslide hazards.

Keywords

Landslides • Disaster database • Societal risk

Introduction

Landslides activity usually causes direct human impacts (fatalities, injuries), but the quantitative assessment of landslide societal risk has been developed in a limited number of studies performed at global (Petley 2012), European (Haque et al. 2016) and national scales (Guzzetti 2000; Salvati et al. 2010; Pereira et al. 2016).

S. Pereira (✉) · J.L. Zêzere · I. Quaresma
Institute of Geography and Spatial Planning, Edifício IGOT, Universidade de Lisboa, Rua Branca Edmée Marques, Cidade Universitária, 1600-276 Lisbon, Portugal
e-mail: susana-pereira@campus.ul.pt

J.L. Zêzere
e-mail: zezere@campus.ul.pt

K. Sassa et al. (eds.), *Advancing Culture of Living with Landslides*,
DOI 10.1007/978-3-319-59469-9_43

The existence of reliable databases on disasters is crucial to study mortality due to landslides in terms of temporal trends, spatial distribution and epidemiological topics (Pereira et al. 2016). Features on fatalities caused by different types of hazards can be found for instance in natural hazard databases (e.g. EM-DAT, DISASTER) based on documental sources, demographic statistics, death certificates and civil protection authorities.

There are some constraints in the inclusion criteria of mortality data in disaster databases. For instance, the EM-DAT only record disasters that have caused at least 10 fatalities, while the DISASTER database includes every landslide that caused fatalities regardless of their number. EM-DAT underestimates the numbers of fatalities and the fatalities are usually associated to the triggering event, such as earthquake, storm and floods (Haque et al. 2016).

Another constraint to study the mortality caused by landslides is the relatively short time span of the existing databases and the lack of field validation (Petley 2012).

In Portugal, social impacts caused by landslides and floods occurred in the period 1865–2015 are gathered in the DISASTER database. This database includes social consequences (fatalities, injuries, missing people, evacuated people and homeless people) caused by landslides documented in newspapers (Zêzere et al. 2014).

The DISASTER database contains 291 damaging landslides that caused 238 fatalities. In the present work we explore the mortality patterns resulting from damaging landslides occurred in Portugal since 1865.

In this work we aim to: (i) analyse the spatio-temporal analysis of damaging landslides; (ii) analyse the frequency and the temporal evolution of fatal landslides; (iii) analyse the spatio-temporal distribution of fatalities generated by landslides; (iv) identify the most deadly landside types; (v) verify gender tendencies in mortality resulting from landslides; and (vi) evaluate the individual and societal risk.

Data and Methods

Landslide mortality data for the period 1865–2015 was gathered in the DISASTER database (Zêzere et al. 2014). A set of national and regional newspapers were used to collect data from which landslide DISASTER cases and DISASTER events were identified. A DISASTER case is a unique landslide occurrence, which fulfils the DISASTER database criteria (i.e., any landslide that, independently of the number of affected people, caused casualties, injuries, or missing, evacuated, or homeless people), and is related to a unique geographic location and a specific period of time (i.e., the specific place and time where the harmful consequences of the landslide occurred) (Zêzere et al. 2014). A DISASTER event is a set of DISASTER cases sharing the same trigger, which may have a widespread spatial extension and a certain magnitude (Zêzere et al. 2014).

A content analysis of the newspapers reports of DISASTER cases was made in order to organize the information in a standardized format. Each DISASTER case includes details on the disaster characteristics and damages (Zêzere et al. 2014). The first includes data on landslide subtype, date of occurrence, location, and triggering factor. The second includes structural damages (damage to buildings and damages to rail and road networks) and social consequences (human damage, gender of fatalities, and circumstances surrounding the fatalities). The total number of fatalities resulting from landslide disasters is certainly underestimated because deaths that did not occur immediately after the landslide would not have been reported, and then it is not possible to asses these mortalities using newspapers as the single data source.

Details on the circumstances related to fatalities associated with landslides were also obtained from newspaper reports. These circumstances were divided in three classes: inside a building, outdoors and inside a vehicle (motor or train). For cases where fatalities occurred due to debris accumulation of a landslide, or building collapse or a failed attempt to escape from inside a building, the circumstance of the fatality was classified as "inside a building". Individual risk was evaluated using mortality rates for landslides, which were calculated based on years with population census in Portugal (usually every 10 years) as proposed by Pereira et al. (2016). The annual mortality rates were computed for each decade using the annual average of fatalities, which were then divided by the annual average population. The result was multiplied per 100,000 to scale it according to the size of population per unit time (Pereira et al. 2016). Societal risk was evaluated by plotting F-N curves representing the annual frequency of landslide cases that generated fatalities.

Results

Frequency and Temporal Trends of Mortality

Figure 1 shows the number of fatalities in each landslide case against its rank in the Portuguese regions (1865–2015), from the highest to the lowest, represented using a log–log scale. South Region includes Alentejo and Algarve NUTs II.

The maximum rank achieved by the landslide fatalities distribution in Portugal and in the North region was 103 and 54, respectively. The relationships are almost linear on the

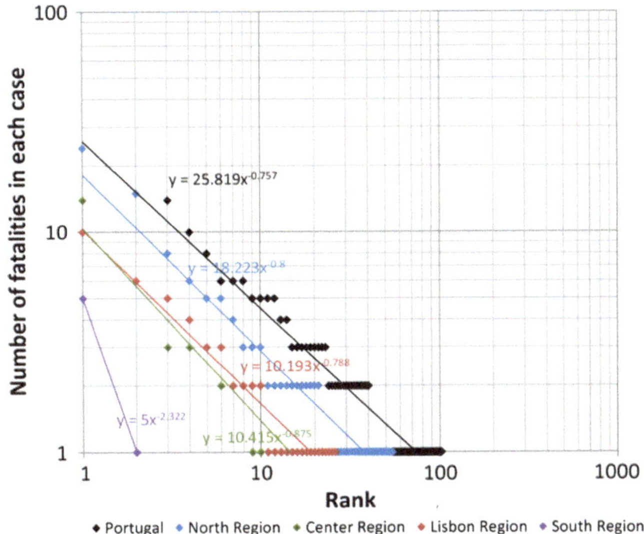

Fig. 1 Intensity of the landslide fatalities cases versus their rank in the Portuguese regions (1865–2015)

plot, indicating a power law distribution. Fatalities are higher for landslides occurred in the North region for all ranks.

Individual landslide cases that occurred in the North region generated more deaths when compared to other regions.

The evolution of the annual distribution of landslide DISASTER cases and the annual number of fatalities for the period 1865–2015 are illustrated in Fig. 2. For the 151 years three distinct time periods were identified by Zêzere et al. (2014): (i) 1865–1934; (ii) 1935–1969; and (iii) 1970–2015. Table 1 summarizes the statistics of landslide cases that generated fatalities in Portugal.

The first period (1865–1934) was characterized by a below-average number of cases (0.7 landslides per year) (Table 1). However, average mortality per year (1.1 fatalities/year) is similar to average values for the complete time series (1.6 fatalities/year). This time period accounts for 33.6% of the total number of fatalities associated with landslides (Table 1). During this period, the number of fatalities was not regularly distributed, with two peaks of annual fatalities occurred in 1904 (24 fatalities) and 1912 (14 fatalities) for single landslide cases (Fig. 1).

The second period (1935–1969) was shorter in duration but registered the highest mean annual number of landslide cases (3.8 landslides per year) (Table 1). The mean annual mortality associated was highest during this period (2.6 per year) and also the percentage of fatalities (38.7%) (Table 1).

The third period (1970–2015) accounts on average 2.4 landslides per year. Fatalities associated with landslides correspond to 27.7% and the annual average is 1.4 fatalities (Table 1). The year 1981 stands out with the second highest annual landslide mortality (15 fatalities) for the complete period (Fig. 2).

There are clear differences among the Portuguese regions concerning the number and density of landslide disaster cases and landslide fatalities (1865–2015) as it can be seen in Table 2.

The first period registered the lowest percentage of landslide disaster cases in all the regions, but the highest percentage of landslide fatalities was registered in the Center region. North region on the contrary registered approximately 1/3 of landslide fatalities in each period, despite the number of landslide disaster cases reached 44% in the period of 1935–1969 and 39.7% in the most recent period. The highest percentage of landslide disaster cases and fatalities registered in the Lisbon region occurred in the period 1935–1969, 58.8 and 63.8%, respectively. South region only registered landslide fatalities in the last period (1970–2015).

These results suggest that population exposure to damaging landslides have changed over time.

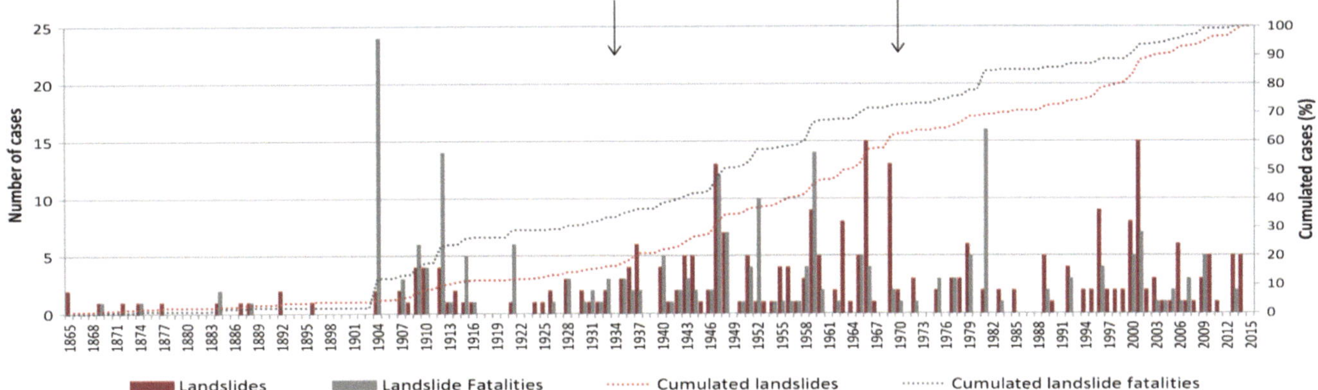

Fig. 2 Annual distribution of landslide disaster cases (*red*) and fatalities (*grey*) occurred in Portugal during the period of 1865–2015. Note *arrows* delimit the time periods referred in the text and in Tables 1 and 2

Table 1 Landslide disaster cases and fatalities per time period

	1865–1934	1935–1069	1970–2015	1865–2015
Number of cases	47	133	111	291
% of total cases	16.2	45.7	38.1	100.0
Landslide cases/year	0.7	3.8	2.4	1.9
Number of fatalities	80	92	66	238
% of total fatalities	33.6	38.7	27.7	100.0
Fatalities/year	1.1	2.6	1.4	1.6

Table 2 Number and density of landslide disaster cases and fatalities per regions and time period (1865–2015)

	Area (km^2)	1865–1934				1935–1969				1970–2015			
		% DC	DCD #10^3 km^2	% F	FD #10^3 km^2	% DC	DCD #10^3 km^2	% F	FD #10^3 km^2	% DC	DCD #10^3 km^2	% F	FD #10^3 km^2
North	21,286	16.4	0.9	33.8	2.1	44.0	2.4	33.1	2.0	39.7	2.2	33.1	2.0
Center	28,199	16.4	0.4	55.3	0.9	36.1	0.8	29.8	0.5	47.5	1.0	14.9	0.2
Lisbon	3002	16.5	5.3	18.2	3.3	58.8	19.0	63.6	11.7	24.7	8.0	18.2	3.3
South	36,602	0.0	0.0	0.0	0.0	20.0	0.1	0.0	0.0	80.0	0.3	100.0	0.2
Portugal	89,089	15.6	0.5	33.6	0.9	46.0	1.5	38.7	1.0	38.4	1.2	27.7	0.7

Note DC disaster cases; *DCD* disaster cases density; *F* fatalities; *FD* fatalities density

Figure 3 shows the monthly distribution of disastrous landslides cases and related fatalities occurred in Portugal during the 1865–2015 period. Landslide disaster cases and landslide fatalities distribution along the climatological year evidences a clear concentration during the autumn and the winter seasons from December to March (74.8% of landslide disaster cases and 72.6% of landslide fatalities).

Landslide occurrence depends also on the local topography, geology, and hydrologic processes acting on the slope (Zêzere et al. 2015). The monthly regional distribution of landslides that caused fatalities presents a small variation. In the North region landslide fatalities were found to be more frequent in December and February (57.7% of landslide fatalities) and were typically associated with debris flows. In the Center region, landslide fatalities were dominant in February (36.2% of landslide fatalities) and in the Lisbon region in the later winter months (36.4% of landslide fatalities occur in March). In the South region all landslide fatalities occurred in August.

Spatial Distribution

The location of disastrous landslides and the number of fatalities generated by landslides during the period of 1865–2015 is shown in Fig. 4a, b. Landslide fatalities are almost

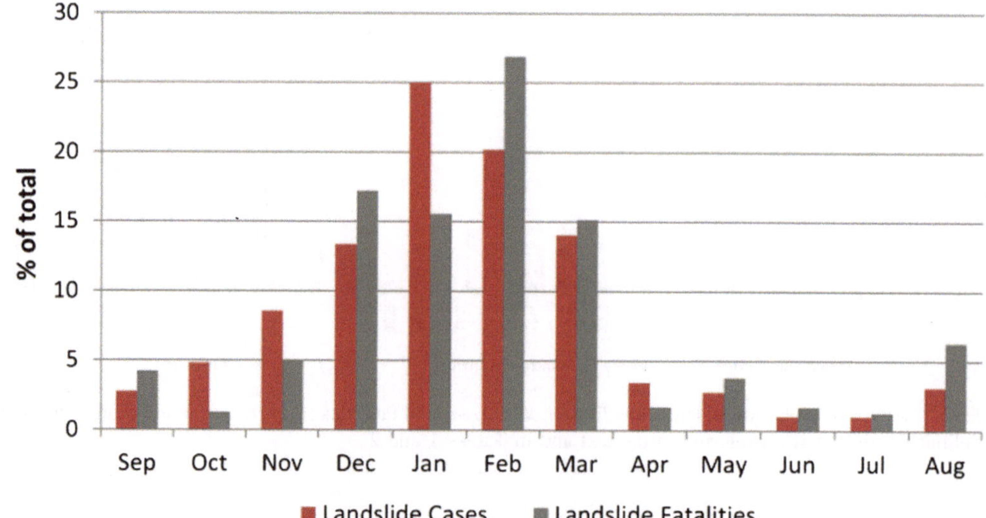

Fig. 3 Monthly distribution of landslide disaster cases and landslide fatalities (1865–2015)

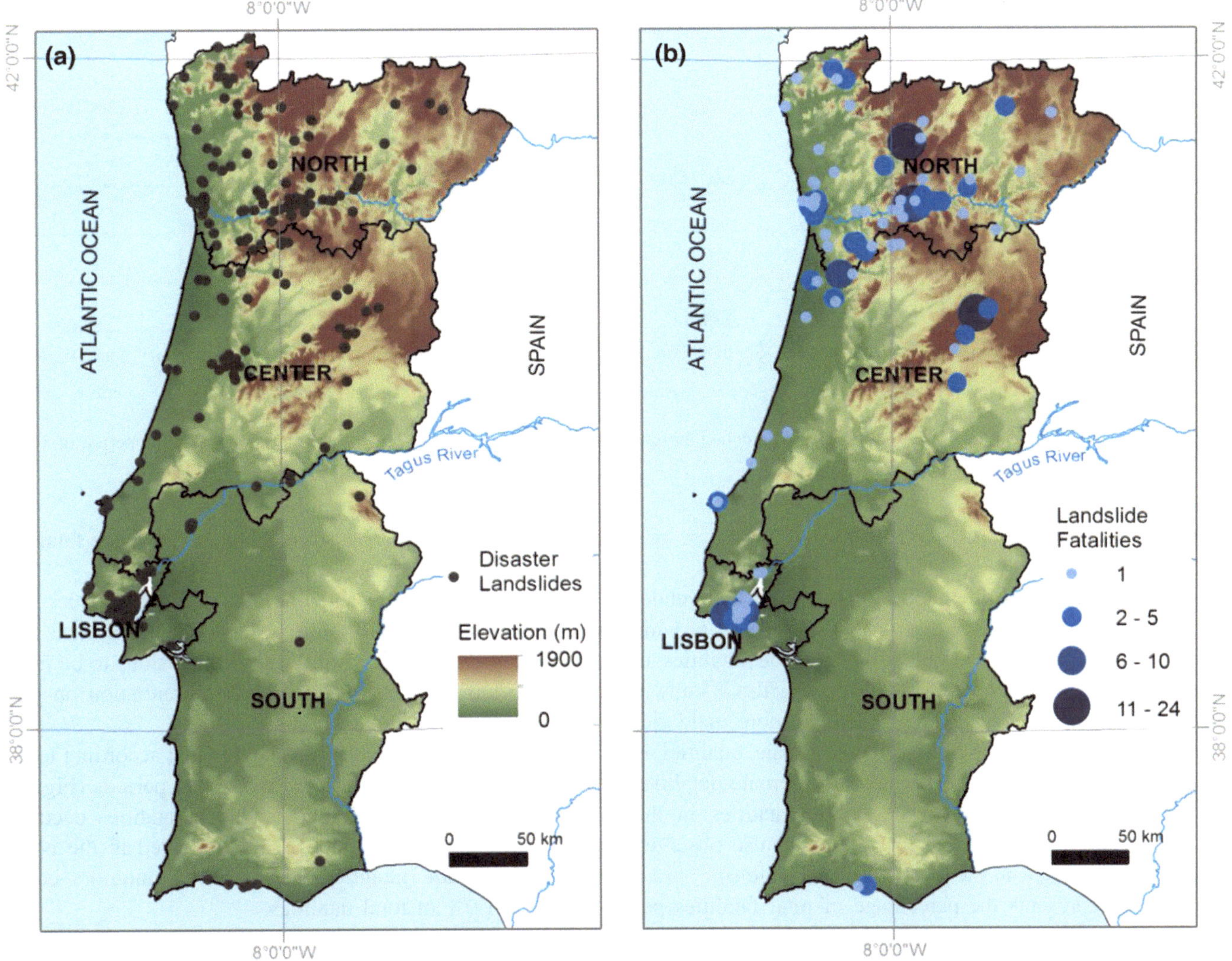

Fig. 4 Number of disaster landslides (**a**) and number of landslide fatalities (**b**) in the period 1865–2015

completely constrained northwards of the Tagus valley where geologic and geomorphologic conditions are more landslide prone than in the southern part of the country (Zêzere et al. 2014; Pereira et al. 2016). The main mortality hotspot observed in the Lisbon region is explained both by natural conditions and also by the high population numbers that increase exposure to landslide risk.

Landslide fatalities in the south region have been constrained to coastal cliffs, and have grown in number in recent years, reflecting the increasing exposure associated with careless intensive use of the coastal areas for tourism and leisure.

In Portugal, 60% of the landslide Disaster cases were related to falls, which were also associated with the highest percentage of fatalities (54%). These were followed by flows (36%). Figure 5 represents the frequency of landslide disaster cases (a) and landslide mortality (b) by landslide type for each Portuguese region.

The North region contains 41.2% of disaster landslides and more than a half of the landslide fatalities (58.4%) where the dominant fatal landslide types are falls and flows (Fig. 5). This is a mountainous region with metamorphic and granitic rocks where fatal flows and falls frequently occur (Pereira et al. 2014).

The Lisbon region is similar to the North region with respect to the number of landslide cases (33.3% of the country's total). However, the landslide mortality is lower (23.2%), reflecting the dominance of slow-moving landslides affecting clay rich sedimentary rocks and soils. The Center region registers 21% of total landslide cases, mostly of fall type, and 19.7% of landslide fatalities were generated by falls and flows that occurred in the Central Massif. The South region registered the lowest number of both landslide disaster cases and landslide fatalities which is explained by the region's low landforms.

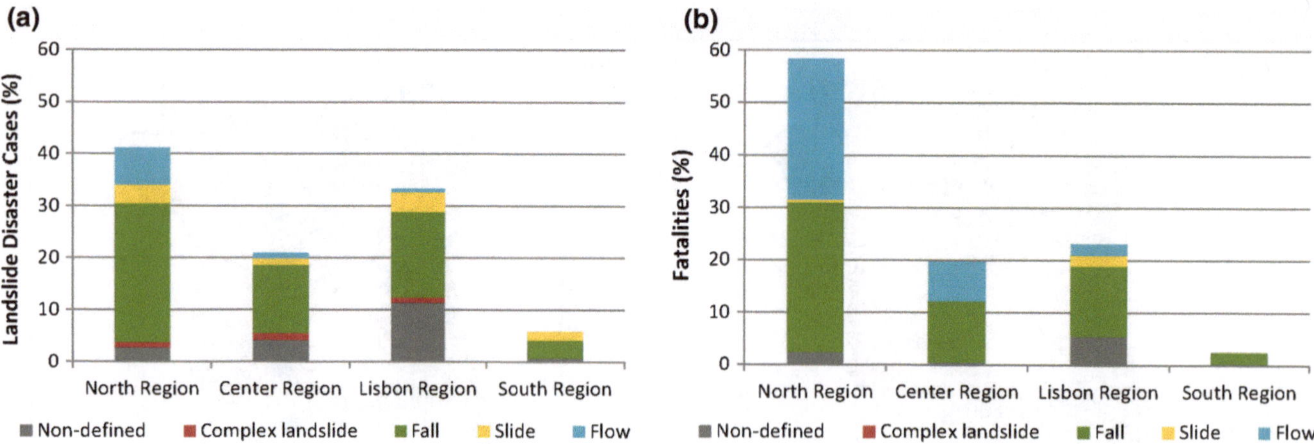

Fig. 5 Frequency of landslide disaster cases (**a**) and landslide mortality (**b**) by landslide type for each Portuguese region in the period of 1865–2015

Gender Tendencies in Mortality

The percentage of landslide fatalities per regions and gender is summarized in Table 3. Newspapers did not provide information on gender for 19.3% of landslide fatalities in Portugal. For cases where the gender of the victim is known, it is clear that disaster landslides generated more male victims. Indeed, the number of documented male fatalities is more than double of the equivalent for female fatalities (67.7% for males and 32.2% of female fatalities in the country). The dominance of male fatalities is also observed in all regions, except in the South region (Table 3).

Figure 6 represents the percentage of total fatalities per region and gender and periods in order to verify if there are different gender trends among regions and periods.

Excluding the South region where female fatalities were dominant in the most recent analyzed period (1970–2015), in the other regions male fatalities are predominantly dominants in spatial and temporal terms. The period of 1865–1934 concentrates the highest percentage of landslide fatalities with gender not reported due to the limited number of newspapers available and reports about the gender of the victims.

Male fatalities were more frequent in the period of 1935–1969 in the North, Center and Lisbon regions. In the last period (1970–2015) the prevalence of male fatalities is still very important in the North region. The other regions registered a decrease in the percentage of landslide fatalities but male fatalities are still dominant.

Also frequency of landslide fatalities per gender and period does not evidence any particular trend for each Portuguese region, not allowing for any conclusions to be made regarding territorial inequalities and differentiation on gender mortality in Portugal.

Landslide fatalities were also aggregated according to the circumstances surrounding the fatality and periods (Fig. 7).

In Portugal 49.2% of total landslide fatalities occurred inside buildings, from which 32.8% occurred in the North region. Landslide fatalities that occurred outdoors correspond to 31.9% of total fatalities.

In the period of 1865–1934 fatalities that occurred inside buildings were the most frequent in the North and Center regions (14.3 and 8.8%, of total fatalities, respectively). In the period 1935–1969 mortality occurred outdoors was more frequent in the North and Lisbon regions while in the Center region mortality inside vehicles was more frequent.

In the last period (1970–2015) North region still stands out with the highest frequency of total fatalities inside buildings and inside vehicles (10.9 and 5.9%, respectively). In the North region fatalities occurring inside vehicles have been increasing in time while fatalities occurring outdoors registered a slightly decrease and inside buildings remain stable.

Table 3 Percentage of landslide fatalities per regions and gender

	% M	% F	% NR
North region	54.6	20.8	24.6
Center region	44.7	27.7	27.7
Lisbon region	67.3	30.9	1.8
South region	16.7	83.3	0.0
Portugal	54.6	26.1	19.3

Note M male; *F* female; *NR* gender not reported

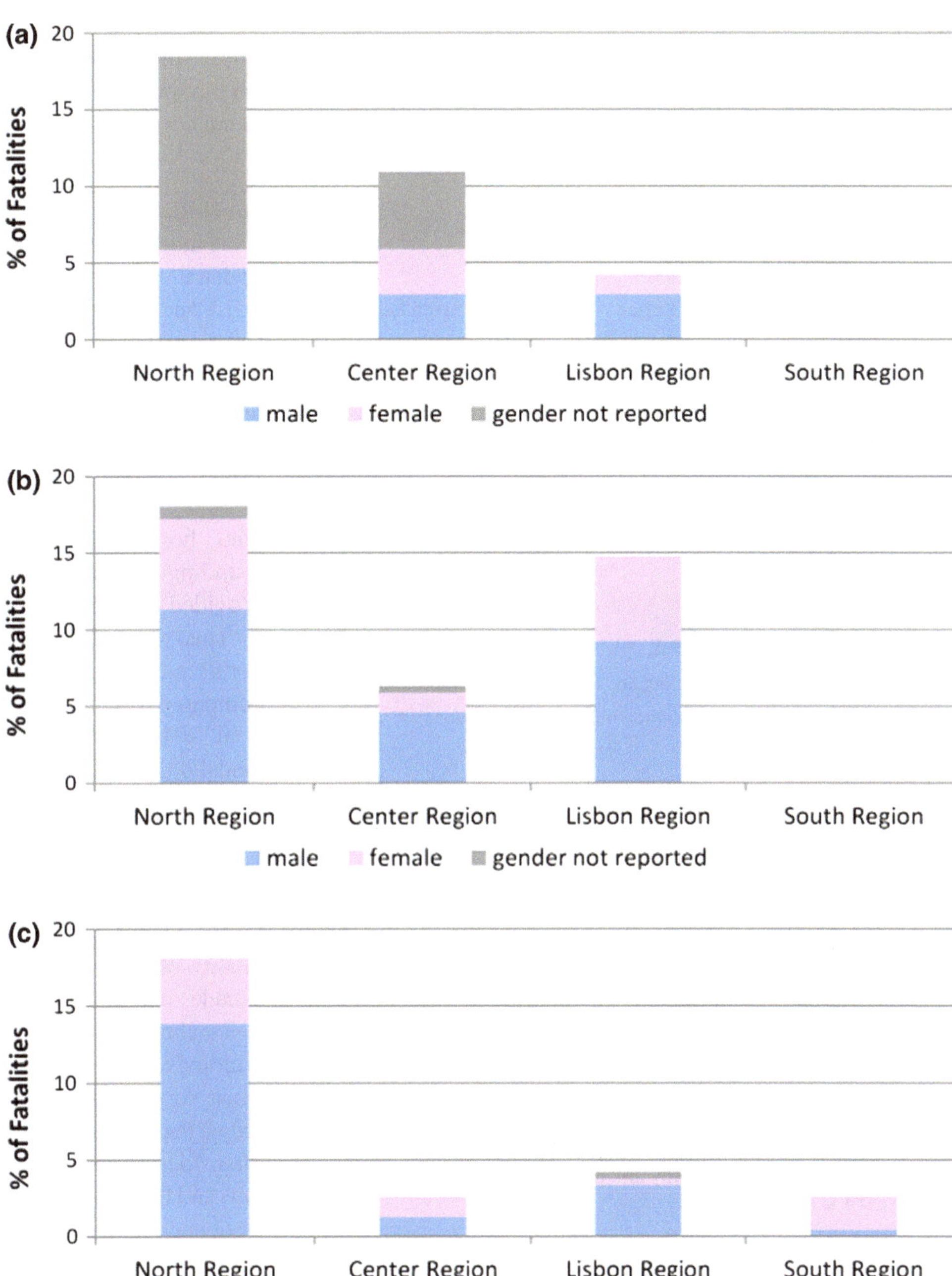

Fig. 6 Percentage of total fatalities per region and gender and periods **a** 1865–1934; **b** 1935–1969; **c** 1970–2015

Individual and Societal Risk

The mean annual average mortality rate for landslides per regions was computed for time periods defined by the dates of population census (Fig. 8a). Landslide mortality rates reached three maximums, one in the decade of 1901–1911 that includes a debris flow that occurred at Peso da Régua and caused 24 fatalities in 1904 (Pereira et al. 2016). Moreover, the mortality rate peak ($0.150/10^5$ inhabitants) is also a consequence of the low number of exposed population at that time (17.3×10^5 inhabitants in the North region).

The second peak of mortality occurred in the decade of 1921–1930 in the Lisbon region ($0.079/10^5$ inhabitants). Other landslide mortality peak ($0.078/10^5$ inhabitants) was registered in 1981 again in the North region, as a consequence of a debris flow that caused 15 fatalities at Cabeceiras de Basto (Pereira et al. 2016).

Landslide mortality rates in the North region registered an important decrease since the 1982–1991 decade while in the

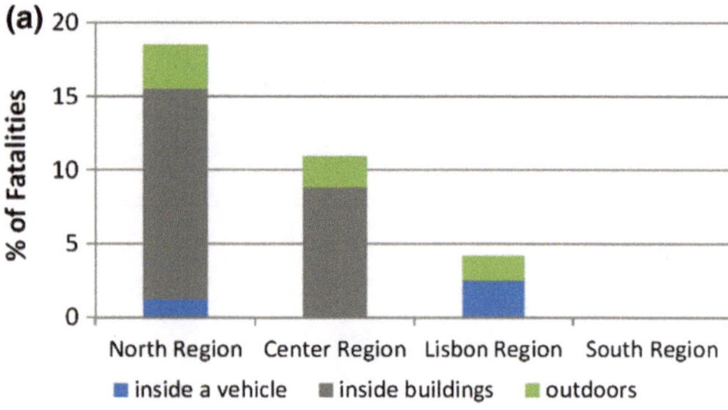

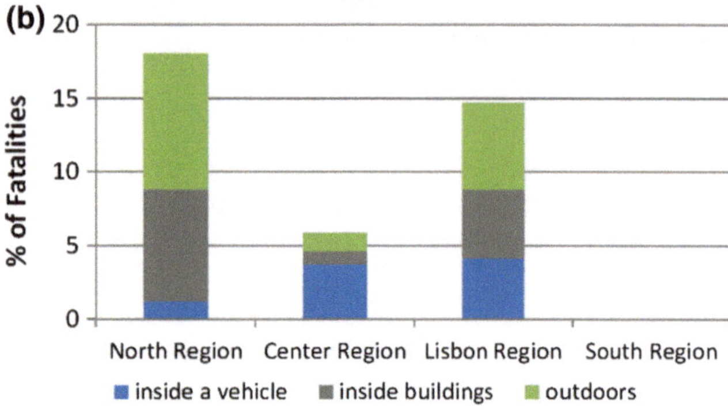

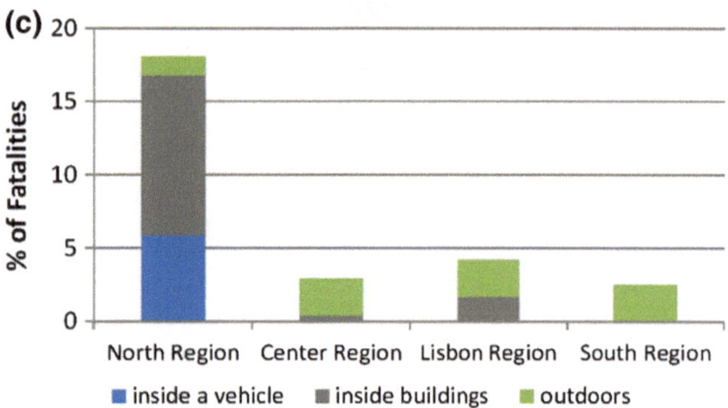

Fig. 7 Percentage of total fatalities per region according to the circumstances related to the fatality and periods **a** 1865–1934; **b** 1935–1969; **c** 1970–2015

Lisbon region this decrease started two decades earlier (1961–1971).

According to Pereira et al. (2016) the landslide mortality rate shows a slight upward trend in the Portuguese regions, except in the Center region where a slight decrease is observed.

Figure 8b shows the curves of frequency against consequences for landslides that caused deaths in Portugal (1865–2015). F-N curves for landslides that caused deaths in Portugal were described by Zêzere et al. (2014) and compared with similar curves obtained for other countries.

In Portugal the annual frequency of fatal landslides is highest in the North region (Fig. 8b) and lowest in the south region where the maximum annual number of fatalities is below 10. Taking into account these results and according to the most commonly used risk acceptable criteria (Fell et al. 2005) the societal risk in Portugal is considered unacceptable for landslides.

Conclusions

Landslide societal risk in Portugal was based on an updated version of the Disaster database until the year 2015 and updates the work developed by Pereira et al. (2016) about landslide fatalities. The Disaster database was built under the assumption that social consequences (specially fatalities) of landslides are relevant enough to be reported by newspapers, which were used as a data source (Pereira et al. 2016). Also information about the mortality, i.e., time, place, loss description (how many, who, how, and what) and the description of the surrounding environments in which the fatality occurred was provided by the newspaper reports.

This work shows a more detailed vision of the temporal and spatial distribution of landslide fatalities when compared with the results of other studies (e.g. Haque et al. 2016) that did not used detailed data on damaging landslides for such a long time as the Disaster database.

The temporal trends of disaster landslides and corresponding fatalities demonstrate the absence of any exponential growth with time (151 years) in Portugal.

The temporal trend of landslide mortality rate in Portugal did not show a decrease. Despite the increasing quality of the building environment observed in the last decades due to the adoption and use of building construction techniques and codes most of fatalities occurred inside buildings. These buildings improvements are not enough to resist very rapid landslides such as falls and flows. Even today these incidents are frequent because buildings are often located on hazardous slopes. In the last 151 years, falls and flows were responsible for the majority of fatalities associated with landslides in Portugal.

Spatial distribution of landslide mortality can be related to factors other than climate, such as the unequal distribution of predisposing conditions to landslide occurrence (geomorphologic and hydrologic), changes on land use and population exposure to landslide hazard and the evolution of social vulnerability (Pereira et al. 2016).

Gender distribution of landslide fatalities showed that landslide victims are mostly male, which are more than double the number of female victims. It is apparent that males are more exposed to disastrous landslides in Portugal, which can be explained by cultural reasons, related to the social role of the breadwinner that exposes men to hazardous occupations and men often assume risk behaviors outdoors and act with a false sense of security when are driving vehicles.

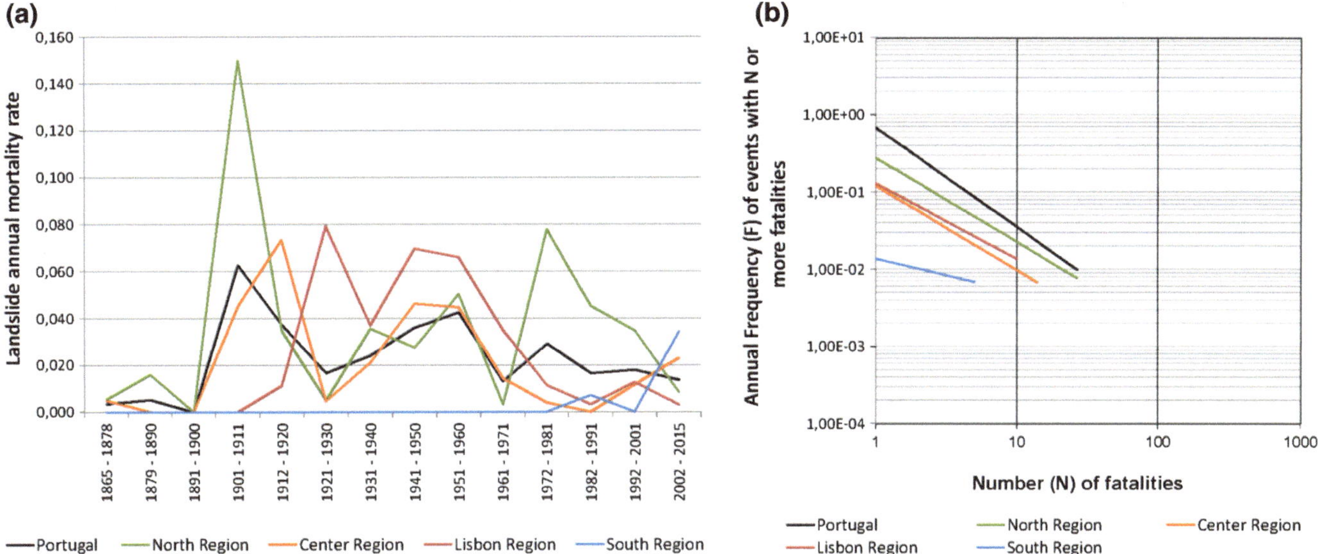

Fig. 8 Mean annual mortality rates per time period adjusted to population census (**a**) and frequency versus consequences (**b**) for landslide cases that caused fatalities in Portugal per region

Acknowledgements S. Pereira is a Post-Doc fellow of the Fundação para a Ciência e a Tecnologia, Portugal (FCT) (SFRH/BPD/69002/2010). This work was also financed by national funds through FCT—Portuguese Foundation for Science and Technology, I.P., under the framework of the project FORLAND—Hydro-geomorphologic risk in Portugal: driving forces and application for land use planning (PTDC/ATPGEO/1660/2014).

References

Fell R, Ho K, Lacasse S (2005) A framework for landslide risk assessment and management. In: Hungr O, Fell R, Couture R, Eberhardt E (eds) Landslide risk management. Taylor & Francis Group, London, pp 3–25

Guzzetti F (2000) Landslide fatalities and the evaluation of landslide risk in Italy. Eng Geol 58:89–107. doi:10.1016/S0013-7952(00) 00047-8

Haque U, Blum P, da Silva PF, et al (2016) Fatal landslides in Europe. Landslides, pp 1–10. doi:10.1007/s10346-016-0689-3

Pereira S, Zêzere JL, Quaresma ID, Bateira C (2014) Landslide incidence in the North of Portugal: Analysis of a historical landslide database based on press releases and technical reports. Geomorphology 214:514–525. doi:10.1016/j.geomorph.2014.02.032

Pereira S, Zêzere JL, Quaresma I, et al (2016) Mortality Patterns of hydro-geomorphologic disasters. Risk Anal 36:1188–1210. doi:10.1111/risa.12516

Petley D (2012) Global patterns of loss of life from landslides. Geology 40:927–930. doi:10.1130/G33217.1

Salvati P, Bianchi C, Rossi M, Guzzetti F (2010) Societal landslide and flood risk in Italy. Nat Hazards Earth Syst Sci 10:465–483

Zêzere JL, Pereira S, Tavares AO et al (2014) DISASTER: a GIS database on hydro-geomorphologic disasters in Portugal. Nat Hazards 72:503–532. doi:10.1007/s11069-013-1018-y

Zêzere JL, Vaz T, Pereira S et al (2015) Rainfall thresholds for landslide activity in Portugal: a state of the art. Environ Earth Sci 73:2917–2936. doi:10.1007/s12665-014-3672-0

Landslide Inventory Mapping in the Fourteen Northern Provinces of Vietnam: Achievements and Difficulties

Le Quoc Hung, Nguyen Thi Hai Van, Pham Van Son, Nguyen Hoang Ninh, Nguyen Tam, and Nguyen Thi Huyen

Abstract

The State-Funded Landslide Project (SFLP) is a national program to systematically assess landslide susceptibility, hazard and risk for all of prone areas in Vietnam. Under this SFLP, in the first phase of SFLP (2012–2014), activities of landslide inventory mapping were implemented over the fourteen Northern mountainous provinces. As the achievements, 10,149 historic landslides were mapped by field surveys and 9405 locations with landslide signs were interpreted from air-photos and analysis of 3D relief. Approximately 83% of the surveyed landslides locate in accessible areas, with small and medium dimensions and partly as a result of the slope cuts. About 76% of the interpreted landslides locate in inaccessible areas, and only 24% of the interpreted locate in accessible areas, of which 65% were found active landslides at the time of surveying, naturally occurred with large dimensions. However, the inventory exposes some major drawbacks: (1) The unavailability of multi-date air-photos; (2) The lack of human resources with enough experiences in image interpretation; (3) The difficulties of verifying the interpreted landslides, especially for the inaccessible sites; (4) Few or no sources of historic information due to the isolated sites or little memory of small or medium size events; (5) No updates developed by the surveyors after they finished their tasks. Those drawbacks can lead to the insufficiency of adequate data on the types, sizes and characteristics of the slope failures, especially the exact dates of occurrences. Despite of those difficulties, the achieved inventory database have been updated and then used as basic input for the susceptibility and hazard mapping as well as preliminary results of SFLP to inform the local authorities and communities about real situations of landslides in their areas.

Keywords

Landslide • Inventory mapping • Air-photo interpretation • Field survey • Vietnam

Introduction

Vietnam is located in the tropical monsoon area, one of the five storm-prone areas with high rainfall in the Asia Pacific region. Three quarters of the mainland (or 37/63 provinces and cities) are hilly and mountainous regions, which have been recently playing a critical role in national socio-economic development. Over the past two decades, many areas in the mountains have attracted intensive flows of immigration to explore the natural resources. However, those regions are often threatened by a number of hazards, particularly landslides, causing losses and damages to people, properties, economics and environment (Hung et al. 2015). The problems of landsliding have been increasing under the impact of global climate change and expressed through many extreme rainfall events. They are also induced

L.Q. Hung (✉) · N.T.H. Van · P.V. Son ·
N.H. Ninh · N. Tam · N.T. Huyen
Vietnam Institute of Geosciences and Mineral Resources,
67 Chien Thang Street, Ha Dong District, Hanoi City, Vietnam
e-mail: le.quoc.hung@vigmr.vn

K. Sassa et al. (eds.), *Advancing Culture of Living with Landslides*,
DOI 10.1007/978-3-319-59469-9_44

by human activities, such as deforestation, mining, slope-cutting for constructions of houses and roads. The incorporation of landslide mapping into regional and local land-use planning is an important measure to reduce the impact of landslides (Fell et al. 2008).

In Vietnam, landslide-related studies were carried out for a long time ago by means of traditional geological surveys, mainly by transect-walks. The investigations on landslide hazard and risk have evolved since the last two decades thank to the advancement of GIS and remote sensing technology (Hung et al. 2016). However, like in many other developing countries, as summarized by van Westen et al. (2006), there are many kinds of obstacles that make the execution of hazard risk zonation difficult in Vietnam. Most of projects on landslide mapping in Vietnam have been conducted at small scales (less than 1:100,000) in large regions (more than 1000 km^2) (Hung et al. 2016). Several advanced-worldwide technologies for landslide zonation and warning at larger scales (>1:50,000) have been applied to several pilot areas, but not yet systematically adapted to the Vietnamese context (Hung et al. 2015). Those projects are often carried out by several organizations with different specialties for a number of purposes, and different criteria. Therefore, the results of previous works have not yet been integrated into one unified national database. Furthermore, the outcomes of those projects have not been transferred to the end-users, especially to the local communities and the national organizations for natural disaster prevention and control (Hung et al. 2015).

To overcome the above-mentioned limitations, the country needs a national program to systematically conduct landslide inventory mapping, assess landslide susceptibility, hazard and risk for all of prone areas. The programs should apply the advanced techniques for zoning and warning of landslides in mountainous areas in Vietnam and it is urgently needed in order to serve the operational effectiveness for prevention and mitigation of landslides in the context of climate change.

The State-Funded Landslide Project (SFLP)

The State-Funded Landslide Project (SFLP), namely "Investigation, assessment and warning zonation for landslides in the mountainous regions of Vietnam", was ratified by the Prime Minister of the Socialist Republic of Vietnam on 27 March 2012 under the Decision number 351/QĐ-TTg. The total budget for the whole Project, which sources from the Government of Vietnam, was approved by the Minister of Natural Resources and Environment on 29 August 2012 under the Decision number 1409/QD-BTNMT. This SFLP has been carried out by 10 organizations of Ministry of Natural Resources and Environment (MONRE) and 5 organizations of other ministries, in which the Vietnam Institute of Geoscience and Mineral Resources (VIGMR) is the coordinating organization. The project has been planned to be conducted in ten years from 2012 to 2020, and is divided into two phases: the first phase (2012–2015) and the second phase (2016–2020). The studied areas of this project consist of 37 mountainous provinces in the whole territory of Vietnam (Fig. 1).

The objectives of SFLP are: (1) to establish a standard national database on landslides and generation of landslide hazard maps at 1:50,000 scale for all mountainous provinces in Vietnam, and at 1:10,000 scale for specific areas in 17 Northern mountainous provinces; and (2) to design of an Early Warning System for landslides, and implementing that in a number of test areas. To obtain those two targets, the following main are proposed to execute in a period of 10 years:

1. Collecting, analyzing and compiling all data and documentation related to landslide hazard and risk;
2. Landslide inventory mapping, assessment of landslide causes and consequences, zoning of landslide susceptibility, hazard and risk for all mountainous regions of Vietnam;
3. Establishing a set of atlas of landslide warning zonation maps at scales of 1:50,000 for 37 mountainous provinces and at scales of 1:25,000–1:10,000 for high landslide hazard risk areas;
4. Establishing a scientific relevance for the setup of a landslide monitoring and early warning network in mountainous areas;
5. Pilot instrumentation for landslide monitoring at some landslide hotspots;
6. Developing a set of guidelines for the use of landslide databases, landslide hazard risk zonation maps, and landslide monitoring network;
7. Transferring of the Project results to the end-users to improve their capacity of landslides mitigation and management.

Landslide Inventory Mapping Under the SFLP

To fully carry out a landslide risk assessment, especially for analysis of landslide hazard, it is required to include information on historic landslides and involved losses. Therefore, landslide inventory mapping is the fundamental phase to evaluating landslide hazard and risk. Some countries in the world have been successful in establishing complete landslide inventories such as Hong Kong, Italy, Turkey, etc. (Guzzetti et al. 1994; Chau et al. 2004; Duman et al. 2005). Depending on the purpose and the available resources, landslide inventory maps are compiled at different scales using various techniques.

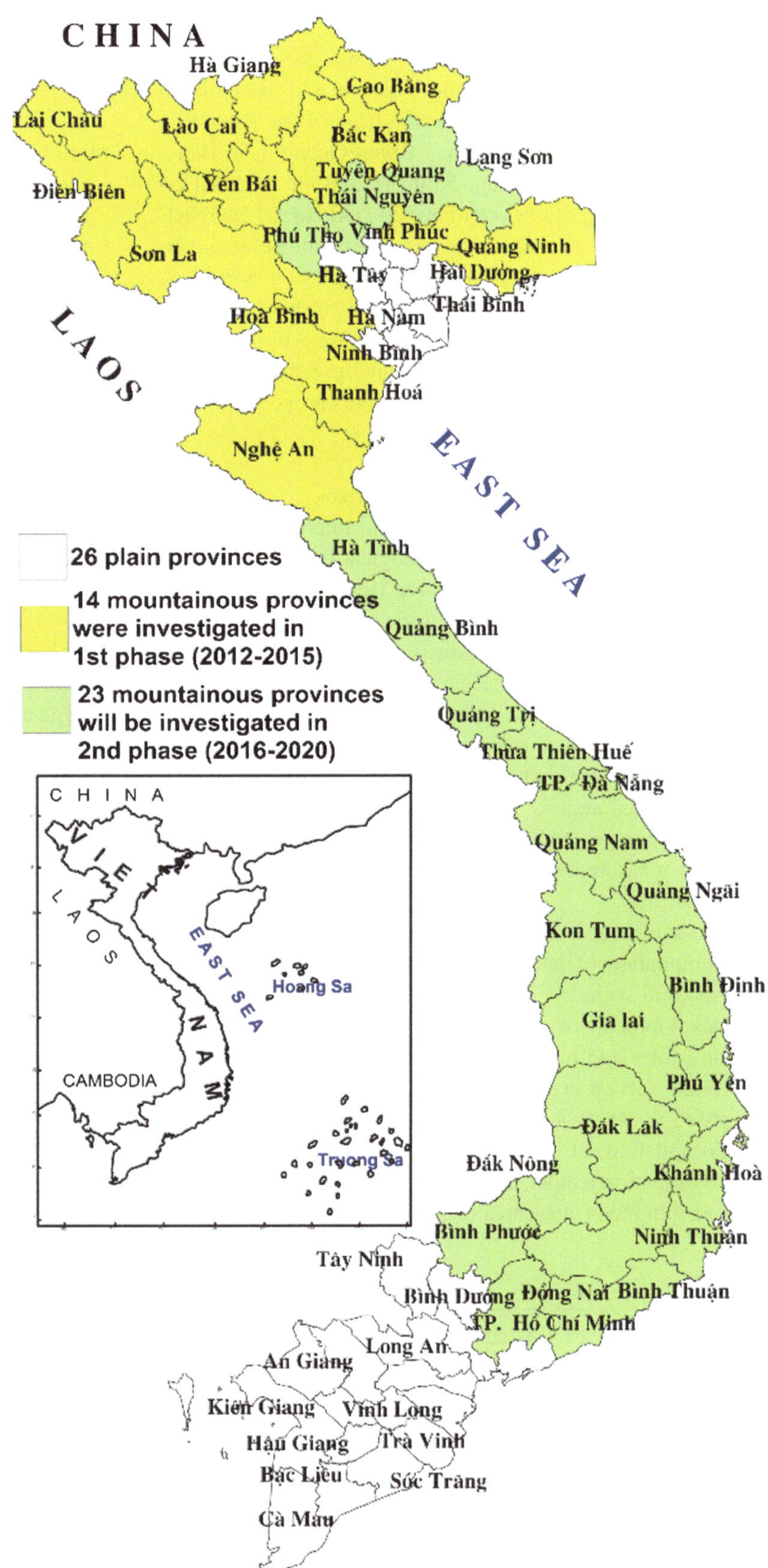

Fig. 1 Map showing the administrative boundaries of 63 provinces in the mainland of Vietnam, in which 37 mountainous provinces are studied areas of the State-Funded Landslide Project (SFLP)

Under the SFLP, activities of landslide inventory mapping at 1:50,000 scale have been carried out since September 2012. In the first phase, mapping took place in the fourteen Northern mountainous provinces: Lai Chau, Dien Bien, Son La, Lao Cai, Yen Bai, Ha Giang, Tuyen Quang, Cao Bang, Bac Kan, Bac Giang, Quang Ninh, Hoa Binh, Thanh Hoa and Nghe An (Fig. 1). More than 500 staffs of 10 MONRE institutions were involved. Due to the difficulties in searching public media (such as journals, articles, newspapers, technical and scientific reports…) to collect historic information on past landslide events, the mapping activities of the SFLP mainly relied on the air-photo interpretation and field surveys (Fig. 2). The following tasks were carried out during the inventory mapping for each province:

1. Collection and review of journals, newspapers, technical/scientific reports/books and internal documents on landslides, other related geological hazards and controlling factors in the study area (topography, geology, meteorology, hydrology, historic landslides and rock landslide other geological disasters, remote sensing…);
2. Preparation for the field surveys: analysis of 3D relief based on 1:10,000 scale topographical maps; air-photo interpretation and analysis of other types of remote sensing images; and planning survey profiles over each investigated area.
3. Field surveys are based on landslide inventory forms: field measurements, interviews of local people and authorities.
4. Soil sampling and laboratory tests.
5. Compilation of landslide inventory maps per district at 1:50,000 scale.
6. Development of a national database on inventoried landslides and related geohazards.
7. Detail survey at some landslide hotspots that are proposed to set up a monitoring and early warning system.
8. Compilation of a provincial report, synthesis of the investigation results, and recommendations of prevention and mitigation measures (Fig. 3).

Main Results of Landslide Inventory Mapping in the First Phase of the SFLP

The Landslide Inventory Database

The most important result of the first phase of the SFLP is the national landslide inventory database that visually displays in a set of 142 landslide inventory maps at 1:50,000 scale for each districts of the fourteen investigated provinces. One example layout of those inventory maps is shown in Fig. 4. The resulting products are as follows:

- 9405 locations with landslide signs were delineated by interpretation of air-photos (captured in 2000) and analysis of 3D relief (interpolated from 1:10,000 topographic maps, which were constructed in 2000). 2218 locations (~24% of the interpreted) were accessible in the field, of which 1446 locations (65% of the accessible interpreted) were found active landslides at the time of surveying (Table 1).
- 10,149 historic landslides were mapped by field surveys; 5076 landslides are of small volume (less than 200 m^3); 3346 landslides are of moderate volume (200–1000 m^3); 1614 landslides are of large volume (1000–10,000 m^3); 83 landslides are of very large volume (10,000–100,000 m^3), and 30 landslides are of extreme large volume (more than 100,000 m^3) (Table 2).

The results show that approximately 83% of the surveyed landslides locate in accessible areas (along roads and inside/nearby residential areas) with the following characteristics: (a) having small and medium dimensions (from small to moderate volumes); (b) partly occurring as a result of the slope cuts; and (c) causing high losses in terms of economic values. On the contrary, about 76% of the interpreted landslides locate in inaccessible areas (remote areas/high mountains); only 24% of the interpreted locate in accessible areas, of which 65% locations are active landslides with the following characteristics: (a) having large dimensions (from moderate to very large volumes); (b) naturally occurring; and (c) causing losses in terms of both economic values and human lives. The achieved inventory maps have been handed over to the above-mentioned provinces and involved organizations in order to inform the local authorities and communities about real situations, and improve the effectiveness of disaster prevention and mitigation in the investigated areas.

In addition, a part of the database has been published on the website as a WebGIS on landslides (http://www.canhbaotruotlo.vn/). This online spatial database includes maps of landslide inventory and controlling factors at 1:50,000 scale for all the investigated provinces (Fig. 3). This system has been used as a simple tool for updating historic events as well as broadcasting urgent information through the Web browsers. It is also considered as one of national interactive interfaces among scientists, managers and local communities for landslide early warning; especially for fulfilling the missing information of historic

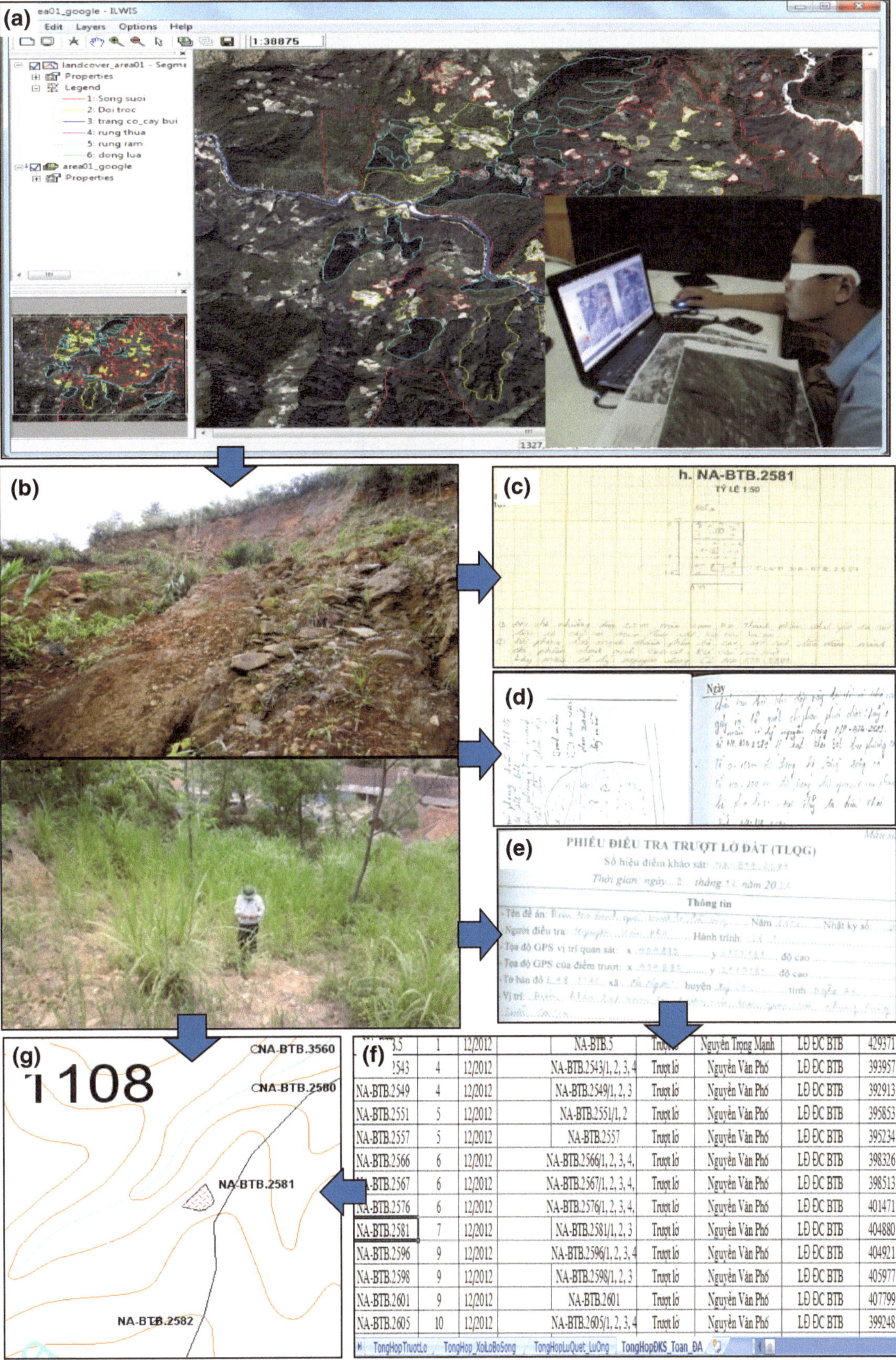

Fig. 2 Main techniques for landslide inventory mapping in Vietnam under the State-Funded Landslide Project include: **a** air-photo interpretation and 3D relief analysis; **b** field survey; **c** soil description sheet; **d** field note; **e** inventory form; **f** landslide database; and **g** GIS mapping

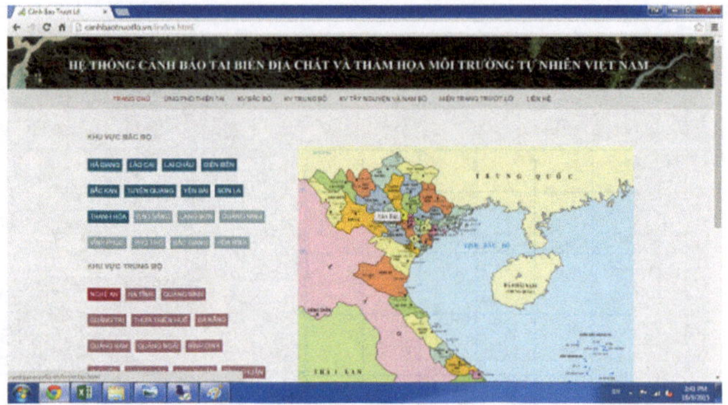

Fig. 3 Main interface of the WebGIS on landslides as one of main achievement of the SFLP in the first phase

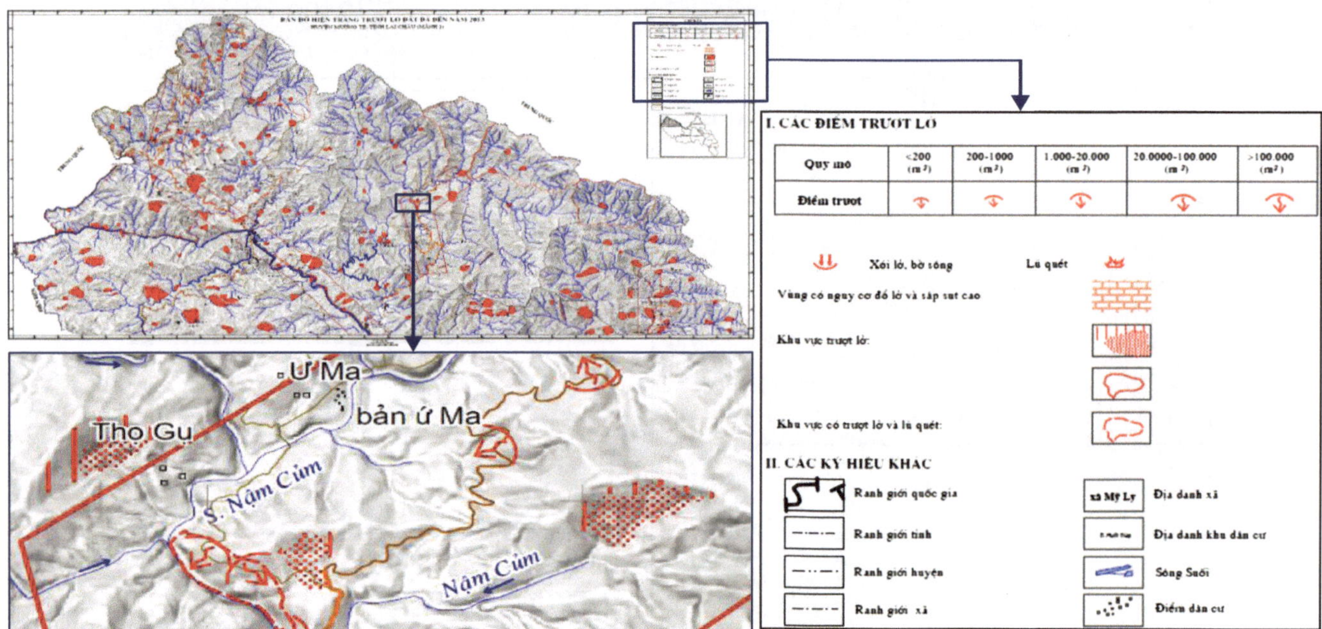

Fig. 4 An example layout of a landslide inventory map constructed at 1:50,000 scale for each district and then handed over to the end-users in mountainous provinces in Vietnam

landslides and promptly describe the new landslides or provide warning. All provided information will be used to inform the local authorities and communities about real situations, and improve the capacity of scientific research as well as the effectiveness of disaster prevention and mitigation in the mountainous areas of Vietnam.

The Promulgation of a Field Guidelines for Landslide Inventory Mapping at 1:50,000 Scale in Vietnam

Another achievement of the landslide inventory mapping in the first phase of the SFLP was the establishment of field

Table 1 Summary of air-photo interpretation results for the landslide inventory mapping in the 14 mountainous provinces of Vietnam

Provinces	Number of interpreted landslides	Number of accessible interpreted landslides	Number of active landslides at the accessible interpreted	% verified landslides
Lai Chau	905	372	229	62
Dien Bien	748	172	92	53
Son La	1791	255	181	71
Lao Cai	551	152	98	64
Yen Bai	708	251	183	73
Ha Giang	1161	112	68	61
Tuyen Quang	266	84	35	42
Cao Bang	268	124	65	52
Bac Kan	317	150	90	60
Bac Giang	35	4	1	25
Quang Ninh	46	30	8	27
Hoa Binh	68	45	20	44
Thanh Hoa	1194	87	67	77
Nghe An	1347	380	309	81
Total	9405	2218	1446	
%		24		65

Table 2 Summary of field-surveyed results for the landslide inventory mapping in the 14 mountainous provinces of Vietnam

Nr.	Provinces	Sum of surveyed landslides	Small volume (<200 m^3)	Moderate volume (200–1000 m^3)	Large volume (1000–10,000 m^3)	Very large volume (10,000–100,000 m^3)	Extremely large volume (>100,000 m^3)
1	Lai Châu	970	337	325	280	18	10
2	Điện Biên	673	335	181	139	12	6
3	Sơn La	1694	795	622	266	11	
4	Lào Cai	534	316	162	53	3	
5	Yên Bái	1165	580	385	187	9	4
6	Hà Giang	963	519	289	150	2	3
7	Tuyên Quang	246	151	94	1		
8	Cao Bằng	88	21	42	25		
9	Bắc Kạn	720	305	282	123	9	1
10	Bắc Giang	302	192	94	16		
11	Quảng Ninh	374	162	141	67	4	
12	Hòa Bình	184	69	81	34		
13	Thanh Hóa	938	630	223	78	7	
14	Nghệ An	1298	664	425	195	8	6
	Total	10,149	5076	3346	1614	83	30
	Percentage (%)	100	50.01	32.97	15.90	0.82	0.30

guidelines for landslide inventory mapping at 1:50,000 scale. It was developed by SFLP's key staffs and other international experts, then promulgated by MONRE on the 20th of November, 2013 under the Decision number 2321/QĐ-BTNMT (MONRE 2013). These guidelines comprise of 4 chapters, 22 main clauses and 11 annexes. The most important contribution of these guidelines is the development of a landslide inventory form for field surveys. All members of the staff from different organizations and background followed these guidelines.

Major Difficulties in Landslide Inventory Mapping During the First Phase

Lack of Multi-date Data

There are insufficiency of multi-date air-photos as well as base maps for developing the inventory. In the beginning of the SFLP, several base maps (e.g., topography, land cover, forest, road systems, geomorphology, ect. at 1:200,000–1:10,000 scales) were collected from other institutions. However, most of them were problematic due to the overlap of digitization, incorrect attributes, etc. In addition, some types of satellite images of low resolution (e.g., Landsat TM/ETM, SPOT) were available at VIGMR, but they required experienced experts on image processing to detect old landslides. Processing those images would require more time, budget and human resources. Therefore, only topographical maps at 1:10,000 scale and air-photos captured in 2000 were used in the project.

Lack of Human Resources with Experiences

The staffs involved in developing the landslide inventory mapping mainly came from geological institutions and had little skills on landslide investigation; particularly, they were not familiar with air-photo processing. At the beginning of the first phase, only few members of the staff could identified historic landslides on the aerial photographs; many others needed training and practice for image interpretation. At the end of the inventory phase, there were incomplete landslide records due to the lack of skills and experience of the field surveyors.

Insufficient Measures for Field-Checks

A lot of landslides were found along the road cuts, in which landslide deposits are usually removed very soon after failure. Furthermore, many slides were not visible or were missed in the interpretation of air-photos because of their small sizes, the coverage of tree plantation, removal of soil masses or new build-ups after the events. Especially in the areas with fast development of constructions, the trace of a failure can be replaced by a new or larger cut or new buildings in a short time right after the event. Information on historic landslides in the isolated places as well as in the crowded areas are always difficult to recover from the local people. Therefore, it is difficult to validate the interpreted landslides in the field.

Little Sources of Archived Information on Historic Landslides

In Vietnam, there is a top-down structured system for the disaster management, namely the "Department of Natural Disaster Prevention and Control". Those organizations are in charge of collecting historic/urgent information before/during and after disastrous events. All the collected information are timely/periodical reported to the higher level of the management system as well as broadcasted by public media. However, the collected information is not yet sufficient because:

- The content of their reports mostly contains the summary of the events that are always related to the typhoons, heavy rains, and floods.
- Landslide is often mentioned as a concatenated event or typhoon/rain-induced disaster; or as a single event if it did not happen during a typhoon/heavy rain or caused serious impacts to a particular province or region or the whole country.
- The reports are kept in a short time without archiving all of important data in a good database.

Conclusions

Landslide inventory mapping is a crucial step for landslide research. During the first phase of the project (2012–2015), more than 9400 locations with landslide signs were mapped by means of air-photo interpretation and 3D-relief analysis. About 24% of those interpreted landslides were accessible in the field, of which 65% were found active landslides and 35% were not able to trace slope failure. The most important achievement is the development of the database comprising the historic landslide inventory, which includes 10,149 landslides. Out of the total, 50% are of small volume, 33% are moderate, 16% are large, and 0.30% are of extreme large volume. The produced data has been printed on maps as well as published on a national WebGIS on landslides.

Along with those above-mentioned achievements in the first phase of the project, the landslide inventory mapping has exposed the following major drawbacks: (1) The unavailability of multi-date air photos; (2) The lack of human resources with enough experiences in image interpretation; (3) The difficulties of verifying the interpreted landslides, especially for the inaccessible sites; (4) Few or no sources of historic information due to the isolated sites or little memory of small or medium size events; (5) No updates developed by the surveyors after they finished their tasks. Those drawbacks can lead to the insufficiency of adequate data on the types, sizes and characteristics of the slope failures, especially the exact dates of occurrences. Therefore, it would be difficult to correlate the landslide with a triggering event as different landslides have different meteorological triggers.

To overcome the above-mentioned obstacles in the second phase, some of the recommendations included: (1) Making the most advantage of available Google Earth, satellite and radar images; (2) Providing and/or participating in the training courses for image processing and interpretation as well as for improving their capacity of inventory mapping; (3) Involving the local communities in field surveys; (4) Upgrading and maintaining the online spatial database - WebGIS on landslides, and training the local authorities and people to use the WebGIS for validating inventoried landslides, as well as for updating the historic and recent landslides.

Despite of many obstacles, a set of landslide inventory maps at 1:50,000 scale for the fourteen mountainous provinces have been handed over to the local authorities and officers of the Department of Natural Disaster Prevention and Control. The published maps have been used by the local people and officers as a preliminary early warning on landslide sites. The landslide database and digital maps and in Argus, MapInfo and Web-GIS have been used by both local and central authorities and officers as a tool for promptly updating and warning about the occurrence of new landslides reported and included in the database by the local people.

Acknowledgements The authors would like to thank all the members of the staff of VIGMR and to the other involved institutions under the MONRE, for their contributions to the SFLP. The authors also credit all local people and authorities for their support and international landslide experts for their guidance to complete the mapping activities.

References

Chau KT, Sze YL et al (2004) Landslide hazard analysis for Hong Kong using landslide inventory and GIS. Comput Geosci 30 (4):429–443

Duman TY, Can T et al (2005) Landslide inventory of northwestern Anatolia, Turkey. Eng Geol 77(1–2):99–114

Fell R, Corominas J et al (2008) Guidelines for landslide susceptibility, hazard and risk zoning for land use planning. Eng Geol 102(3–4):85–98

Guzzetti F, Cardinali M, Reichenbach P (1994) The AVI Project: a bibliographical and archive inventory of landslides and floods in Italy. Environ Manage 18(4):623–633

Hung LQ, Van NTH, Tam N et al (2015) Overview of landslide inventory mapping at 1:50,000 scale in the mountainous areas of Vietnam (2012–2014). J Geol Min Resour 11:154–160 (in Vietnamese)

Hung LQ, Van NTH, Duc DM, Ha LTC, Son PV, Khanh NH, Binh LT (2016) Landslide susceptibility mapping by combining the analytical hierarchy process and weighted linear combination methods: a case study in the upper Lo River catchment (Vietnam). Landslides 13(5):1285–1301

Khien NX, Hung LQ et al (2012) Report of project "Assessment of status geohazards in four mountainous provinces in North of Viet Nam: Ha Giang, Bac Kan, Cao Bang, Tuyen Quang. Identification of sources, prediction and propose preventive measures and consequences mitigation" sponsored by Ministry of Natural Resources and Environment. Center for Information and Archives of Geology, Hanoi, 318 p (in Vietnamese)

MONRE (2013) Decision No.2321/QĐ-BTNMT issued on 20 November, 2013 on "Technical guidelines for landslide inventory mapping at 1:50,000 scale in the mountainous areas of Vietnam". Vietnam Ministry of Natural Resources and Environment, Hanoi, 68 p (in Vietnamese)

Thinh DV et al (2004) Report of project "Geohazard surveys in the Northwestern mountainous regions of Vietnam" sponsored by Ministry of Industry. Center for Information and Archives of Geology, Hanoi 305 p (in Vietnamese)

van Westen CJ, van Asch TWJ et al (2006) Landslide hazard and risk zonation: why is it still so difficult? Bull Eng Geol Environ: Off J Int Assoc Eng Geol Environ: IAEG 65(2):67–184

Varnes DJ (1978) Slope movement types and processes. In: Schuster RL, Krizek, RJ (eds) Landslides: analysis and control. Transportation Research Board, National Academy Press, Special report 176, National Research Council, Washington, D.C., pp 11–33

Geological Conservation Through Risk Mitigation and Public Awareness at the Siq of Petra, Jordan

Giorgia Cesaro, Giuseppe Delmonaco, Bilal Khrisat, and Sabrina Salis

Abstract

The Petra Archaeological Park, a UNESCO World Heritage Site since 1985, characterized by a spectacular geo-archaeological landscape, is also a fragile site facing a wide diversity of natural phenomena (landslides, flash floods, earthquakes) that pose a major threat to the heritage as well as to the visitors. The UNESCO Office in Amman, in partnership with the Department of Antiquities of Jordan and the Petra Archaeological Park has engaged in a long term strategy aimed at the prevention and mitigation of natural hazards at the site. Specific attention has been devoted to the case of the Petra Siq, a 1.2 km naturally formed gorge in the sandstone mountains serving as the only tourist entrance to the site, which is particularly at risk due to its narrow pathway, limited access points and recent active slope processes. Drawing on this approach, the UNESCO "Siq Stability" project has been developed to design a strategy towards prevention and mitigation of instability phenomena at the Siq of Petra. After an initial phase devoted to the reconstruction of geomorphological dynamics and monitoring of active slope processes, the current phase of the project focuses on the implementation of landslide risk mitigation measures, the capacity development of the national authorities and the development and implementation of an awareness strategy on natural hazards. Main focus has been placed on project activities undertaken, results achieved and suggestions for steps ahead, aiming to present a useful case study on the management of natural hazards applied to heritage sites leading to the conservation of a unique World Heritage property.

Keywords

Landslide risk • Community awareness • Petra archaeological park • UNESCO world heritage

G. Cesaro (✉) · S. Salis
Culture Sector, UNESCO Amman Office, Yacoub Ammari Street, Amman, 11181, Jordan
e-mail: g.cesaro@unesco.org

S. Salis
e-mail: s.salis@unesco.org

G. Delmonaco
ISPRA, Italian National Institute for Environmental Protection and Research, Via V. Brancati 48, 00144 Rome, Italy
e-mail: giuseppe.delmonaco@isprambiente.it

B. Khrisat
Department of Conservation Sciences, Queen Rania Institute of Tourism and Heritage, Hashemite University, Zarqa, Jordan
e-mail: bilal@hu.edu.jo

K. Sassa et al. (eds.), *Advancing Culture of Living with Landslides*,
DOI 10.1007/978-3-319-59469-9_45

Introduction

In the last decades, the number of disasters and associated losses has been progressively increasing (IFRC 2015; Munich Re 2016).

In landslide risk management, non-structural measures, as community preparedness, public awareness and communication strategies, can be successfully applied to mitigate landslide risk especially in developing countries (Anderson 2013). Such strategies, along with the implementation of active measures for reducing landslide hazard, demonstrate the importance of involving the affected population and other stakeholders in the decision-making process for risk reduction (Nadim 2014).

According to the Hyogo Framework for Action 2005–2015, the main UN-wide policy on the subject of Disaster Reduction, increasing awareness on the importance of disaster reduction policies is a key objective to ensure the substantial reduction of disaster losses and enhance the resilience of communities to respond to disasters (UNISDR 2005).

Building on the Five Priorities for Action defined by the *Hyogo Framework*, the UNESCO World Heritage Committee approved at its 31st session in 2007 the Strategy for Risk Reduction at World Heritage properties having per objective to strengthen the protection of World Heritage sites and contribute to sustainable development by integrating concern for heritage into national disaster reduction policies and within management plans for World Heritage properties in their territories (UNESCO 2006).

The strategy was prepared by the World Heritage Centre, in co-operation with the States Parties, Advisory Bodies, and other international agencies and non-governmental organizations concerned by emergency interventions.

This same strategy is also reported on the Resource Manual on "Managing Disaster Risks at World Heritage Properties" produced by the World Heritage Centre in cooperation with ICCROM, ICOMOS and IUCN (UNESCO 2010).

In line with the above, in the last decade the UNESCO Office in Amman has engaged in the implementation of activities geared towards ensuring that preventive measures are in place in the main touristic areas of the site, and, specifically, in the Siq, so to evaluate how the site can best be protected and preserved against natural risks and ensure that it is safe for the thousands of tourists who visit each year

Scope of Work

The archaeological site of Petra (Fig. 1) lies in a large valley surrounded by mountain ranges. Its geology is dominated by Palaeozoic sandstone rocks that form most of the hand-carved Nabataean rock monuments of Petra. The Siq is a 1.2 km naturally formed gorge in the sandstone rocks that represents the main entrance to the archaeological site. Because of the religious niches and water management features, the Siq, in its entirety, is considered as a monument of religious and historic significance, considerably contributing to the Outstanding Universal Value of Petra. The width of the Siq ranges from 3 m to 15.70 m. It is formed by very steep slopes with variable height from the ground level, from few meters at the entrance to several tens of meters in some areas of the path.

Petra is also a very fragile site facing a wide diversity of risks, ranging from those posed by environmental factors, such as natural and geological hazards, as well as those attributed to tourism and the lack of adequate site management and emergency measures for tourist and monument safety. In recent years, natural phenomena, such as earthquakes, floods and landslides were registered as increasingly impacting the site, and most specifically the Siq posing a major threat to cultural heritage and visitors.

During the rainy season, water flows into the Siq from the surrounding *wadis*. In 1963, 24 tourists died as a result of a sudden flash-flood in the Siq. Water management and the hydraulic system created by the Nabataeans, protected the monuments and the people from life threatening flash-floods, however those systems are now deteriorated and no longer protecting the site or visitors. A survey of the Nabataean hydraulic network in the Siq and the areas with direct impact on the Siq was conducted from 1996 to 2002 (PNT 2003). As a result of this project, the velocity of water flow during flash floods was reduced by restoring the existing floor of the Siq to its original pavement and grade.

Despite this intervention, the risk posed by landslides is still present in the Siq due to its specific geomorphology. In the last decade, several landslide events, mostly rock falls and rock slides, with different magnitude (volumes from <1 to >10 m^3) have occurred in the Siq (2009, 2015) and in the core area of the site (2009, 2010, 2016).

These recent events have prompted UNESCO Amman Office, in cooperation with the local authorities, to initiate a process for analysis, monitoring and urgent and long-term mitigation of landslide risk. Awareness and communication activities on natural hazards have been among the non-structural mitigation strategies implemented.

UNESCO Petra Risk Assessment and Mitigation Strategies

Preserving Petra's Outstanding Universal Value for which the site has been inscribed in the World Heritage List (UNESCO 2016), is one of the corporate UNESCO priorities for culture actions in Jordan, in line with the UNESCO Strategy for Risk Reduction at World Heritage properties (UNESCO 2006).

Since 2009, the UNESCO Office in Amman has supported the Petra Archaeological Park and the Department of Antiquities in assessing, managing and mitigating natural hazards in Petra World Heritage site.

Within the framework of the project "Risk Mapping at the Petra Archaeological Park" (2011–2012), a strategic partnership was established with the government and several partner organizations, including national and international

Fig. 1 Location of Petra (*left*) and satellite image of the Siq (*red circle*) and Petra archaeological area

universities, to map and document the natural and human-made risks in the core area of the property. A proposal for risk management at the Petra Archaeological Park to identify and prioritize continuous threats with cumulative and slow effects (not disaster risks) was elaborated and handed over to the government in 2012 (Paolini et al. 2012).

From 2009 to 2015, UNESCO engagements focused on addressing for the first time the impact of landslides phenomena in the Siq. In 2009 a technical expertise in engineering geology was provided to the national authorities to support the consolidation of a fractured block in the Siq.

Through the implementation of the projects "Rapid Risk Assessment" (2011) and "Siq Stability", Phase I (2009–2015) actions have been focusing on the analysis of the stability conditions of the Siq slopes based on a comprehensive documentation of the site, the installation of an integrated monitoring system for the detection and control of deformation processes and the definition of mitigation measures against rock instability (Delmonaco et al. 2013a, b, c, 2014, 2015).

The Petra "Siq Stability" project Phase II (2015–2016) aims to operationally implement the mitigation of landslide risk in the Siq through the (a) application of priority and urgent landslide mitigation interventions in the upper Siq plateau and on the Siq slopes to address immediate slope hazards in the short term; (b) capacity development of the national authorities to address the management of landslide risk at the site and implement mitigation measures in coordination with international experts; (c) awareness raising among different levels of stakeholders on landslide and other natural hazards occurring within the Petra Archaeological Park and specifically in the Siq.

Public awareness and communication among a broad set of stakeholders, ranging from decision makers to the local community, fall into the third project component as non-structural mitigation strategies against natural hazards and shall be later incorporated into a wider management strategy for the site.

Risk Awareness Methodology and Implementation in Petra

The local community and the tourists are generally unaware of how the geological and hydraulic processes that shaped spectacular landscapes can be hazardous to people. Informed visitors can instead assume a certain degree of risk and responsibility for their own safety when visiting natural, cultural or recreational environments. In view of this, park public safety programs shall involve the communication of site-specific hazards to visitors, education and information programs that encourage self-reliance, cooperation with other departments, non-governmental organizations, tourism operators, concessionaires, and service providers (NPS 2006).

Data gathered showed that awareness of natural risks preparedness and mitigation, mostly at the community level, can be the foundation for risk prevention in Petra. Stakeholders and local communities can play a key role in the management of a geo-archaeological site, in particular in relation to disaster risk reduction.

The approach adopted in the "Siq Stability" project aims at supporting the Petra Archaeological Park in raising awareness on heritage management and conservation, focusing on natural risks preparedness and mitigation. Different typologies of stakeholders were identified (decision makers, governmental institutions, NGOs and UN agencies, professionals/researchers, site business beneficiaries, tour guides, children, local community) and a set of targeted activities was selected for each of them (Table 1).

Overall, the strategy aims at: (a) achieving best practices on preservation and management of the site supported and endorsed by the national authorities; (b) making local communities, site beneficiaries, and other stakeholders engaged in the site with different capacities, aware of the activities undertaken in the Siq for the prevention of natural hazards; (c) ensuring that best practices are adopted by tourists when visiting the Siq in regards to the impact of natural hazards that might occur on site; (d) making the international community and the national authorities aware of the work that UNESCO is conducting in the Siq.

In parallel to the priority landslide mitigation interventions carried out in the upper Siq plateau and on the Siq slopes from March to July 2016, as part of the project "Siq Stability", Phase II, a number of communication and public awareness activities on geological and geo-hydrological hazards were implemented, according to the strategy developed.

While some of the stakeholders' categories could be addressed through convening meetings or field visits (decision makers, NGOs, UN agencies), some others required specific outreach methodologies because of their primary involvement during the implementation of the landslide risk mitigation works on site, as in the case of tourists, tour guides, site business beneficiaries and local communities. The type of activities implemented varied in relation to the target group addressed and the timing of implementation (before or during the field works).

In order to ensure effective and efficient implementation, a comprehensive coordination and management system was set in place in cooperation with the Petra Archaeological Park in advance of each field mission and according to a preliminarily agreed check-list of actions.

Before the implementation of the field activities, community awareness workshops involving site business beneficiaries and tour guides from the local community were carried out in coordination with the project experts and the local authorities. The primary aim of the workshops was to raise awareness on the geomorphological and hydraulic hazards characterizing the Siq, promote a more responsible behavior conducive to risk prevention and, thus, ensure their cooperation during the upcoming field activities. This measure would apply to business beneficiaries as horse-driven carriage riders transporting tourists unrelentingly from the beginning to the end of the Siq, often at high speed and with limited interest in the surrounding environment: having them informed on the specificity of the site and possible natural hazards can enhance their sense of responsibility and produce a transfer of information to the visitors.

As part of the actions taken before the implementation of works, communication materials as project brochures and

Table 1 Overview of target groups and related activities

Target groups	Type of activities					
	Field visits Technical meetings Workshops	Informal meetings and site visits	Informative material (ENG and ARB)	Presentations Lectures	Workshop High level conference	Interactive sessions
1. Decision makers					X	
2. Other Gov. institutions	X					
3. NGOs and UN agencies	X					
4. Professionals researchers			X	X		
5. Site beneficiaries		X				
6. Tour guides		X				
7. Local community		X	X			
8. Children			X			X
9. Tourists			X			

informative flyers were disseminated to the hotel management units within the surrounding village of Wadi Musa and a warning on the upcoming activities was posted on the Petra Archaeological Park web site.

During the landslide risk mitigation works (Fig. 2), awareness raising activities were performed on site and focused on public awareness with tourists and site business beneficiaries. Ad hoc awareness materials on the project were prepared for distribution at the Petra Visitor Centre (project flyers, Fig. 3).

The same materials were distributed to tourists in the Siq, at the beginning and at the end of the work site. Access to the work area (normally about 30 m in length) was temporarily blocked for tourists by using white and red striped tape and placing project banners (Fig. 4).

Dedicated UNESCO and Petra Archaeological Park staff stood at the beginning and at the end of the work site to share project flyers and provide general information on the activities being implemented (Fig. 5). The project team could also rely on the substantial support provided by the Jordan Civil Defense, the Petra Archaeological Park rangers

Fig. 2 Landslide risk mitigation works

Fig. 3 Flyer of the risk mitigation works 2016

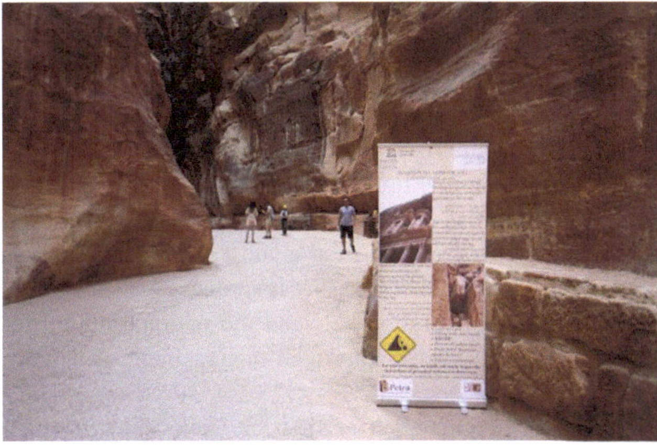

Fig. 4 Banner of the risk mitigation works 2016

and the Tourism Police, whom engaged in public awareness activities with the tourists and the local community alongside their more institutional duties.

The whole project team, from the project experts to the workmen (all belonging to the local community) were involved in communicating with the general public on the risk stability phenomena present in the Siq.

Fig. 5 Communication session on landslide mitigation works to visitors

Future actions (September–December 2016) will be targeting those categories that have not been yet recipient of the awareness programme as children and school students among the local communities surrounding the Petra Archaeological Park.

It is foreseen that the final achievements of this work will be compiled and integrated in the management plan for the whole site, whose preparation is currently being developed by UNESCO and the local authorities.

Conclusions

This study demonstrates the essential role that public awareness and communication on natural hazards can play as non-structural technique in the management and mitigation of landslide risk in the Siq of Petra and further support the Petra Archaeological Park in the management of environmental risks at the site.

Despite the time limitations, project related experience reported and the challenges faced in the management of the tourist flow at one of the most visited sites in the world, this initiative can be regarded as a successful example of cooperation between the national authorities and UNESCO for the improved management of a World Heritage Site.

Similar actions shall be then integrated in a management plan for the protection of the site for which is currently being elaborated by UNESCO Amman in cooperation with the national authorities.

Acknowledgements This study has been conducted in the framework of the UNESCO, Italian funded, project "Siq Stability—Mitigation of Immediate Hazards in the Siq of Petra" (Phase II). A special acknowledgement is, thus, addressed to the Italian Development Cooperation for having entrusted the UNESCO Amman Office with the implementation of such a strategic project at the Petra World Heritage Site.

The authors wish also to acknowledge the substantial support provided by the Government of Jordan represented by the Petra Archaeological Park and the Department of Antiquities for their continuously cooperation in the implementation of this pioneering project.

References

Anderson MG (2013) Landslide risk reduction in developing countries: perceptions, successes and future risks for capacity building. In: Canuti P, Margottini C, Sassa K (eds) Landslide science and practice. Social and economic impact and policies, vol 7. Springer, Heidelberg, pp 247–256. ISBN 978-3-632-31312-7

Delmonaco G, Margottini C, Spizzichino D (2013a) Rock-fall hazard assessment in the Siq of Petra, Jordan. In: Canuti P, Margottini C, Sassa K (eds) Landslide science and practice. Risk assessment, management and mitigation, vol 6. Springer, Heidelberg, pp 441–449. ISBN 978-3-632-31312-6

Delmonaco G, Margottini C, Spizzichino D (2013b) Slope dynamics, monitoring and geological conservation of the Siq of Petra (Jordan). In: Bilotta E, Flora A, Lirer S, Viggiani C (eds) Geotechnical engineering for the preservation of monuments and historic sites. Taylor & Francis Group, London, pp 325–334. ISBN 978-1-138-00055-1

Delmonaco G, Margottini C, Spizzichino D (2013c) Slope dynamics, monitoring and geological conservation of the Siq of Petra (Jordan). In: Canuti P, Margottini C, Sassa K (eds) Landslide science and practice. Risk assessment, management and mitigation, vol 6. Springer, Heidelberg, pp 441–449. ISBN 978-3-632-31312-6

Delmonaco G, Margottini C, Spizzichino D, Khrisat B (2014) Rock slope potential failures in the Siq of Petra (Jordan). In: Sassa K, Canuti P, Yueping Y (eds) Landslide science for a safer geoenvironment. Targeted landslides, vol 3. Springer, pp 341–347. ISBN 978-3-319-04995-3

Delmonaco G, Leoni G, Margottini C, Spizzichino D (2015) Implementation of advanced monitoring system network in the Siq of Petra (Jordan). In: Lollino et al G (eds) Engineering geology for society and territory, vol 8. Springer, Cham, pp 299–303. ISBN 978-3-319-09407-6

International Federation of Red Cross and Red Crescent Societies (2015) World disaster report. In: Hanza M (ed) ISBN 978-92-9139-226-1, p 270

Munich Re (2016) Topics Geo—Annual review of natural catastrophes 2015. Analyses, assessments, positions. https://www.munichre.com/site/corporate/get/documents_E1273659874/mr/assetpool.shared/Documents/5_Touch/_Publications/302-08875_en.pdf. Last accessed 30 Aug 2016

Nadim F (2014) Introduction: risk reduction strategy. In: Sassa K, Canuti P, Yueping Y (eds) Landslide science for a safer geoenvironment. Methods of landslide studies, vol 2. Springer, pp 743–744. ISBN 978-3-319-05049-2

Paolini A, Vafadari A, Cesaro G, Santana Quintero M, Van Balen K, Vileikis O, Fakhoury L (2012) Risk management at heritage sites: a case study of the Petra world heritage site. UNESCO, Amman. ISBN 978-92-3-001073-7

PNT (2003) The Petra Siq: nabataean hydrology uncovered. Petra National Trust, Amman. ISBN deposit number 2003-6-1223

UNESCO (2006) Issues related to the state of conservation of world heritage properties: strategy for reducing risks from disasters at world heritage properties, adopted by the Committee at in 30th session. WHC-06/30, COM/7.2, Vilnius, p 1. http://whc.unesco.org/en/disaster-risk-reduction/#strategy. Last accessed 30 Aug 2016

UNESCO (2010) Managing disaster risks for world heritage. http://whc.unesco.org/en/managing-disaster-risks/. Last accessed 30 Aug 2016

UNESCO (2016) World Heritage Centre, Petra world heritage site. http://whc.unesco.org/en/list/326. Last accessed 30 Aug 2016

UNISDR (2005) The hyogo framework for action 2005–2015: building the resilience of nations and communities to disasters at the world conference on disaster reduction. http://www.unisdr.org/2005/wcdr/intergover/official-doc/L-docs/Hyogo-framework-for-action-english.pdf. Last accessed 30 Aug 2016

Case Histories for the Investigation of Landslide Repair and Mitigation Measures in NW Germany

Annika Wohlers, Thomas Kreuzer, and Bodo Damm

Abstract

To understand the complex interactions between landslide risk, public and private risk awareness, including land use practices and repair and mitigation measures in a complete manner, case histories were developed and analyzed using the example of the highway network of the Lower Saxon Uplands, NW Germany. The case histories utilize datasets extracted from the German landslide database that includes information of historical and current landslide impacts, elements at risk as well as land use practices and provide an overview of spatio-temporal changes in the exposure and vulnerability to landslide hazards over the past 250 years. For the developed case histories the recorded landslide events were categorized and classified at representative sites, according to landslide types, processes, and damages as well as applied repair and mitigation measures. In a further step, data of recent landslides are compared with historical and modern mitigation measures and are correlated with concepts of risk management. As a result, it is possible to identify some complex interactions between landslide hazard, hazard awareness and damage impact. The case histories show that especially since the last 20 years public risk awareness rose due to an apparent increase in landslide frequency and magnitude at some sites. Before the 1990s landslide mitigation measures were mainly low cost prevention measures such as the removal of loose rock and vegetation, rock blasting, catch barriers, and temporal or perpetual traffic lane closure. Recently there is a shift towards the implementation of expensive mitigation measures in order to minimize landslide occurrence. Local decision makers increasingly invest in expensive long-term stabilization projects like soil anchoring, rock nailing, and steel-reinforced concrete walls.

Keywords

Structural landslide mitigation • Central Europe • Case histories

A. Wohlers (✉) · T. Kreuzer · B. Damm
Applied Physical Geography, University of Vechta,
Driverstraße 22, 49377 Vechta, Germany
e-mail: Annika.wohlers@uni-vechta.de

T. Kreuzer
e-mail: Thomas.kreuzer@uni-vechta.de

B. Damm
e-mail: Bodo.damm@uni-vechta.de

K. Sassa et al. (eds.), *Advancing Culture of Living with Landslides*,
DOI 10.1007/978-3-319-59469-9_46

Introduction

Landslides occurring in the German Uplands, NW Germany are usually restricted to small areas with small to medium volumes involved. Nonetheless, the consequences of landslides impose a high risk to the public by disrupting essential infrastructure. Little is known about social and economic impacts of landslides regarding infrastructures due to the complexity of the resulting economic losses (e.g., Klose et al. 2016; Zêzere et al. 2007).

Risk management describes four phases after a landslide event: response, recovery, mitigation and preparedness (Alexander 2002; Godschalk 1991). Mitigation is defined as application of measures in order to prevent an event or reduce its damage potential. Mitigation is related to structural (i.e., construction) and non-structural (i.e., planning, hazard mapping) measures and realization can impose a high financial burden for the involved communities, transportation offices and regional/federal construction authorities. As a consequence, preventive measures are realized in a sporadic way since public decision makers consider landslide risks acceptable, even though cost-benefit studies confirm the financial advantage of mitigation over post-disaster measures (Altay et al. 2013).

The presented study gives first insights into the development of repair and mitigation measures in Southern Lower Saxony, NW Germany. From historical data sets case histories at representative sites were created. Mitigation measures are used as indication of public risk awareness.

Study Area

The presented case histories are located in the Upper Weser area in Lower Saxony, NW Germany (Fig. 1). The region is characterized by a moderate population density of ~150 persons per km^2 (Destatis 2016). The transportation infrastructure is represented by a well-established federal highway system ("*Bundesstraße*"), which connects the regional urban centers (Kassel, Göttingen, Hannover) with the national highway A7 ("*Autobahn*"), one of the major transportation routes in Germany (Fig. 1).

The area is part of the Solling anticline between the Rhenish massif in the West and the Harz Mountains in the East. The bedrock formation of the Middle Lower Triassic consists of interbedded fine to medium-grained sandstone and fragile silt—and claystone with a thickness of 600–700 m. The bedrock is covered by a Quaternary layer of slope debris, which is susceptible to landslides in the case of high soil moisture content (Damm 2005). Due to river incisions up to 350 m of the Fulda, Werra and Weser rivers, a moderate relief was formed with elevations ranging between 50 and 500 m a.s.l.

The geologic predisposition to landslides is well known from previous studies in the area (e.g., Damm 2005; Klose et al. 2012). Hazard mapping indicates that 7% of federal highway length is exposed to landslides (Klose et al. 2016). Mitigation is documented since modern road construction and ranges from low-cost, (i.e., removal of material from susceptible outcrop areas, rock blasting and road closure) to expensive measures (i.e., catch fences, mesh wiring and concrete injections).

Methods and Data

Landslide Database

The basis of the case histories developed in this study is a landslide database for Germany (Damm and Klose 2015). Most information within the database is gathered by means of archive studies from inventories of emergency agencies, state, press and web archives, company and department records as well as scientific and geotechnical literature. Furthermore, the database contains in addition to geoscientific, data from various sources, including field surveys, climatic records, and satellite imagery. The database stores information related to landslide characteristics, dimensions and dynamics as well as data concerning the relationship of landslides to soil and lithologic properties, geomorphometry and climatic conditions. Beyond that, it includes information about land use effects, damage impacts and economic losses. For the here studied area the database covers a time period of about 150–250 years (Damm and Klose 2015).

Case Histories to Study Mitigation Measures

In hazard assessment a key concept is to study historical data sets in order to analyze impacts and costs of landslides (Varnes and IAEG 1984; Glade 2001). Case histories are used to understand the interaction of landslide activity and land use practices including mitigation measures (Calcaterra et al. 2003; Klose et al. 2016). Although case histories are limited to a local area by design, they can be used in the area of the Upper Weser to differentiate landslide mitigation over time in order to identify local construction patterns. The main purpose is to understand the correlation of landslide activity, resulting vulnerability, concepts of mitigation as measures of public risk perception, and the resilience of infrastructures.

Case Histories

Site 1: Landslide Mitigation at Urban Infrastructures

The site is located at a federal highway within an urban area. The slope of 300 m length and 50–70 m height was formed as a river cut bank. The natural slope gradient varies between 30° and 50°. During road construction the slope was cut and

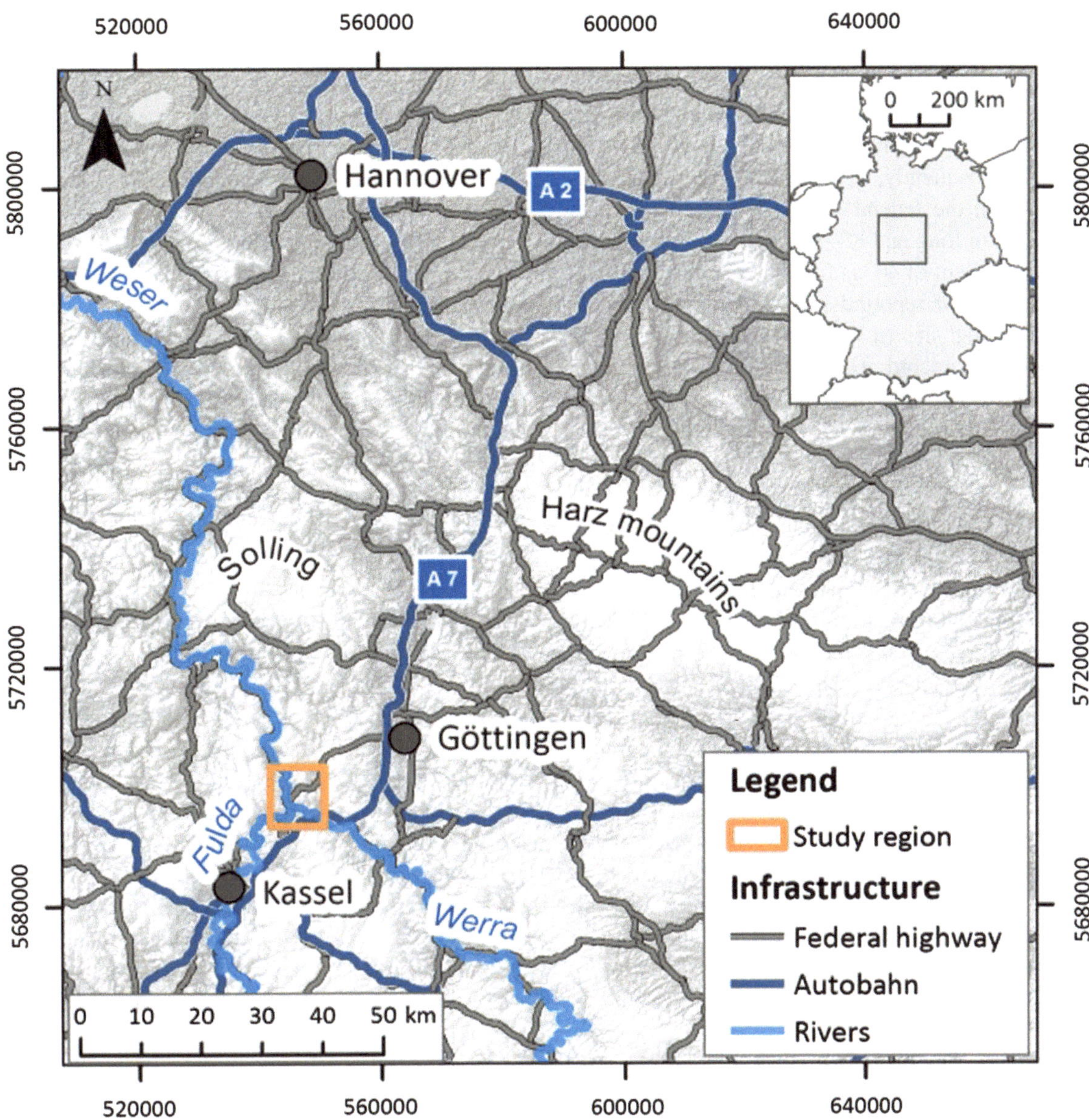

Fig. 1 Overview of the highway network in the Upper Weser area, NW Germany (*sources* Aster, OSM)

the material was used to create an embankment fill for the latter road (Damm 2005).

The first small to medium sized (50–150 m^3) landslides were registered during a phase of road construction in 1880–1882. First mitigation measures were realized (masonry gravity walls). Contemporary high financial relevance is linked with increased construction costs and mitigation during that time according to expert opinion (Klose et al. 2016). Subsequent landslides ($n \approx 30$) can be clustered in events in the 1920s (n = 2), 1936/1937 (n = 5), 1960s (n = 3), 1970–1975 (n = 6), 1999–2001 (n = 7). In these small to medium sized events (50–200 m^3) upslope material collapsed due to sliding or falling (Damm 2005).

For long time mitigation measures were very simple and aimed at the removal of loose rock and vegetation from susceptible outcrop areas within the slope (1924, 1936, 1961/1962 and 1994). In 1937, 1994 and 2000 a wooden barrier was constructed to protect road traffic against rock-falls. In addition, in 1994 wire meshes were installed. In 2001 a comprehensive reconstruction of the slope was undertaken including rock blasting, concrete injections and a rear anchored concrete wall (Damm 2005).

Site 2: Rockfall Mitigation at a Rural Highway

The site is located at a federal highway between two small cities. The slope is situated in the federal state of Hesse and therefore in the responsibility of a different state office for transportation. Consequently, it is possible to compare mitigation measures in the federal states of Hesse and Lower Saxony. The 2.2 km long and 80–120 m high slope with an average natural gradient of 40° is formed as a river cut bank.

The total number of reported landslides is considerably less than at the first site (n = 9), probably linked to an incomplete landslide record. Under the consideration that most data are extracted from newspaper articles, it can be assumed that newspaper coverage focused on events in urban areas. Clusters can be found in the 1890s/1900s (n = 3), 1960s (n = 1), 1980s (n = 3) and 2010s (n = 2). In all events small to medium sized volumes were involved in landslides and rockfalls (Landslide Database of Germany).

The mitigation measures are similar to those of site 1. Apart from drywall construction in 1907 to stabilize the slope, mitigation focused on removal of loose material. In the 1960s a gravity wall was constructed for slope stabilization. In 1983/1984 the slope was comprehensively reconstructed implementing the use of rock nailing, new gravity walls, wooden barriers, catch fences and wire meshes. A cascade system was installed to prevent gully erosion. In 2015/2016 the mitigation was modernized. While the gravity walls and wooden barriers are still intact, catch fences with a system of dynamic brakes and anchored wire meshes are installed. Gullies are reshaped to control erosion (Landslide Database of Germany).

Site 3: Landslide/Rockfall Mitigation at an Important Infrastructural Junction

The third site represents a 550 m long and 70 m high slope formed as a river cut bank which is situated at an important infrastructural junction. The importance of the site for the city is emphasized by occurrence of a federal highway in the vicinity of a river port and a railway line. Even though the port is no longer in use, and the railway only seldom for heavy load transportation, the site is still of some infrastructural importance for the Upper Weser area since it represents the alternative route to the national highway in case of full closure.

The importance is reflected by the early road construction in the 1770s with first documented landslides during that time (n = 4). The number of subsequent landslides (in total n ≈ 30) can be clustered in events in the 1880s (n = 2), 1900s (n = 2), 1925–1930s (n = 3), 1950–1960s (n = 8), 1970–1980s (n = 2), 2000s (n = 4) and 2010s (n = 3).

Early mitigation measures focused on slope stabilization on both sides of the road (slope cut and roadbed) by constructing gravity walls (1770s, 1930s). Aside from drywall construction, sliding and falling material in general was removed from the road for a long time (until the 1950s). Some of the small to medium scale (50–150 m^3) events are related to construction works.

With the broadening of the highway from 1962–1975 the slope was reshaped and mitigation measures were installed. To decrease the slope gradient 7000–8000 m^3 material was removed. A concrete barrier with retention space, wire meshes and a drainage system was realized. The concrete barrier was shortly replaced by a wooden barrier. In 1986 the wire meshes were renewed. In 2015/2016 new mitigation measures were installed including concrete injections, anchored wire meshes covered with geofabrics and shotcrete (Landslide Database of Germany).

Development of Mitigation Measures in the Upper Weser Area

The reported case histories show that for the last 150 years clusters of landslide events can be found in many decades, especially since the 1950s (Fig. 2). An apparent increase in the landslide events must be reflected under the consideration of a general worldwide increase of event reports in the last century linked to a more complete documentation in recent years (cf. Raška et al. 2014; Taylor et al. 2015). However, clusters of landslides can be found in the 1770s, 1880s, 1900s, 1930s, 1950s and 1970s. For the last four decades (since the 1980s) the number of landslides is

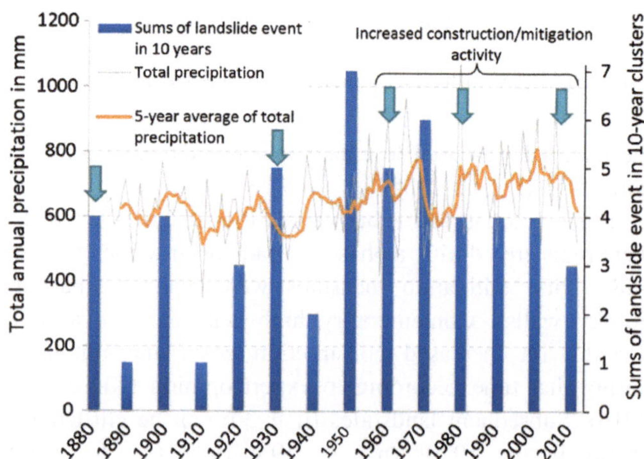

Fig. 2 10-year clusters of the landslide events in the Upper Weser area in combination with total precipitation since 1880. Arrows indicate phases of increased construction/mitigation (*sources* DWD, Landslide Database of Germany)

Fig. 3 Development of mitigation measures in southern Lower Saxony, NW Germany. **a** Rockslide event (02/1937) in the Upper Weser area. The drywall in the front stabilizes slope with no protection against rockfalls/slides. **b** Removal of loose rock and vegetation in the Upper Weser area in 1936. **c** Combination of steel buttress/wooden planks as a barrier against rockfall. Construction in the Harz Mountains, NW Germany. **d** Rockfall mitigation in the Upper Weser area. The barrier wall in the left and wooden barrier in the right foreground are constructed in 1983, rock drapery and catch fence constructed in 2015 (*photos* retrieved from Landslide Database of Germany)

constant. The peaks of landslides events can be explained by construction phases on one side (i.e., 1770s, 1880s and 1930s), and high precipitation phases on the other in the 1960s and 1970s (Fig. 2, Damm and Klose 2015).

Corresponding to the event clusters, construction and mitigation phases can be identified in the 1770s, 1880s (both construction), 1930s, 1960s, 1980s and 2010s (mitigation combined with road extension). Clusters of landslide events are followed by an intensified period of mitigation (Fig. 2).

The mitigation phases in the Upper Weser area can be differentiated by distinct measures. In the road construction phase in the 1770s and 1880s mitigation focused on slope stabilization by constructing dry walls (Fig. 3a). Protection against rockfall was not achieved, falling material had to be removed during road closure. The construction of higher dry walls was intensified at the beginning of the 20th century and especially in the 1930s. After a period of high landslide activity, engineered mitigation was planned at some sites but not realized. Instead, loose material from cover beds and vegetation was removed by clearing and rock blasting (Fig. 3b).

Engineered mitigation was executed in the 1960s. Slope stabilization was realized by constructing concrete walls and piles. Protection against rockfall was achieved by building wires meshes and barriers in a combination of steel buttress and wooden planks (Fig. 3c).

In the 1980s the use of wire meshes and rock drapery increased. In addition to the 1960s, they were nailed to the ground. Rockfalls were mitigated by installing catch fences. Furthermore, drainage systems were realized to improve slope stability.

Comprehensive mitigation measures have been installed since the turn of the millennium to stabilize sediment layers with concrete injections and anchor them deep in the ground. Dynamic catch fences are not only designed to protect against rockfalls but against falling trees likewise (Fig. 3d). In modern mitigation measures esthetic and ecologic aspects are considered as well (cf. Popescu and Sasahara 2009). Geofrabics and colored shotcrete are used to maintain the impression of a natural slope.

Discussion

The increase in mitigation efforts and resultant costs in the study area (Southern Lower Saxony and Northern Hesse) is linked with increased risk awareness. The risk management in the Upper Weser area follows a cycle of event with a short-term response in terms of material removal from the affected site, recovery, which means the reinstatement of the infrastructure, and mitigation measures. Finally, a period of awareness/preparation should follow (Crozier and Glade 2005; Alexander 2002) to maintain high sensibility to the risk. That includes investments into existing mitigation measures. Instead, public awareness to landslide vulnerability decreases and subsequently mitigation measures are neglected which may result in a decrease of structural integrity. As a consequence, landslide events cannot be prevented and instead dysfunctional mitigation measures become more damaging than before (i.e., loose wire meshes and wooden planks from barriers).

With a new phase of landslide events the risk management cycle starts from the beginning. Public risk awareness raises and new mitigation measures are installed. An increase in mitigation effort and costs is intensified by technical standards. Introduced in the 1960 s technical standards are created to define slope stability and calculate modes of failure (cf. Eurocode 2014). With the technical development the standards for slope stability increased and consequently mitigation measures become outdated and must be reconstructed.

The trend of installing cost intensive and maybe oversized mitigation measures is enhanced by producers of relevant systems. Technical warranties of 50 years and more are common, even though construction engineers argue that these numbers are hardly realistic. Decision makers in transportation offices and involved communities might be under the impression that measures with extended warranties are highly efficient long-term solutions to landslide hazards. Mitigation concepts from the 1980s show, that this argument is only valid, if regular maintenance is applied.

Conclusions

The paper presents preliminary results of the study of repair and mitigation measures in the Upper Weser area, NW Germany. Several phases of increased landslide activity are followed with a small delay by a period of intense mitigation construction. Linked to risk awareness, after a series of landslide events, awareness raises together with public need for mitigation. In a phase of functional mitigation, only a few and small volume landslides occur. As a result, risk awareness decreases and aged mitigation measures lose their integrity. In a next phase of high landslide activity mitigation is inefficient and infrastructure is vulnerable again.

Acknowledgements The research work is funded by the Ministry of Science and Culture of Lower Saxony (MWK Niedersachsen 11.2 76202-10-1/07) as well as the German Research Foundation (DFG DA 452/6-1). The authors appreciate funding from these institutions. Thanks to the Department of Transportation in Lower Saxony (NLStBV) and Hessen (hessen.mobil) and the municipal office of Hann. Münden for their friendly collaboration.

References

Alexander D (2002) Principles of emergency planning and management. Oxford University Press, 340 p. ISBN 978-0195218381

Altay N, Prasad S, Tata J (2013) A dynamic model for costing disaster mitigation policies. Disasters 37(3):357–373

Calcaterra D, Parise M, Palma B (2003) Combing historical and geological data for the assessment of the landslide hazard: a case study from Campania, Italy. Nat Hazards Earth Syst Sci 3:3–16

Crozier MJ, Glade T (2005) Landslide hazard and risk: issues, concept and approach. In: Glade T, Anderson MG, Crozier MJ (eds) Landslide hazard and risk.Wiley, Chichester, pp 1–40. ISBN 978-0471486633

Damm B (2005) Gravitative Massenbewegungen in Südniedersachsen. Die Altmündener Wand – Analyse und Bewertung eines Rutschungsstandorts, Zeitschrift für Geomorphologie NF, Suppl.-Bd. 138:189–209

Damm B, Klose M (2015) The landslide database for Germany: closing the gap at national level. Geomorphology 249:82–93

Destatis (2016) Bevölkerungsstand Kreise am 31.12.2014. URL: https://www-genesis.destatis.de/genesis/online/. Last accessed 31 Aug 2016

Eurocode 7 (2014) Geotechnical design—Part 1: General rules; German version EN 1997-1:2004 + AC:2009 + A1:2013

Glade T (2001) Landslide hazard assessment and historical landslide data—an inseparable couple? In: Glade T, Frances F, Albini P (eds) The use of historical data in natural hazard assessment, Springer, Dordrecht, pp 153–168. ISBN 978-0792371540

Godschalk DR (1991) Disaster mitigation and hazard management. In: Thomas ED, Hoetmer GJ (eds) Emergency management: principles and practice for local government. International City Management Association, Washington, DC, pp 131–160. ISBN 978-0873260824

Klose M, Damm B, Gerold G (2012) Analysis of landslide activity and soil moisture in hillslope sediments using a landslide database and a soil water balance model. GEOÖKO 33(3–4):204–231

Klose M, Maurischat P, Damm B (2016) Landslide impacts in Germany: a historical and socioeconomic perspective. Landslides 13:183–199

Popescu MH, Sasahara K (2009) Engineering measures for landslide disaster mitigation, In: Sassa K, Canuti P (eds) Landslides—disaster risk reduction. Springer, Berlin, pp 609–631. ISBN 978-3-540-69966-8

Raška P, Zábranský V, Dubišar J, Kadlec A, Hrbáčová A, Strnad T (2014) Documentary proxies and interdisciplinary research on historic geomorphologic hazards: a discussion of the current state from a central European perspective. Nat Hazards 70:705–732

Taylor FE, Malamud BD, Freeborough K, Demeritt D (2015) Enriching Great Britain's National Landslide Database by searching newspaper archives. Geomorphology 249:52–68

Varnes DJ, IAEG (1984) Landslide hazard zonation: a review of principles and practice. UNESCO, Paris, 60 p. ISBN 92-3-101895-7

Zêzere JL, Oliveira SC, Garcia RAC, Reis E (2007) Landslide risk analysis in the area North of Lisbon (Portugal): evaluation of direct and indirect costs resulting from a motorway disruption by slope movements. Landslides 4:123–126

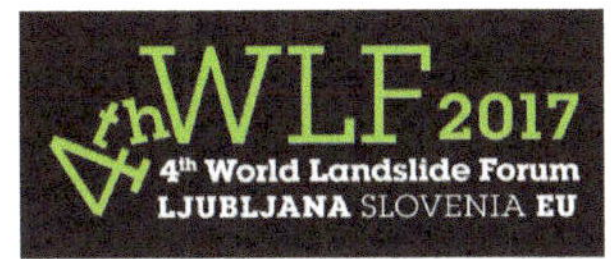

Surveying Perception of Landslide Risk Management Performance, a Case Study in Norway

Jessica Chiu and Unni Eidsvig

Abstract

The effectiveness of landslide risk management should be assessed to optimize the implementation of landslide risk mitigation measures. The Risk Management Index (RMI) of Cardona et al. (Disaster risk and risk management benchmarking: a methodology based on indicators at national level. IDB/IDEA Program on Indicators for Disaster Risk Management, Universidad Nacional de Colombia, Manizales, p 101, 2004) provides useful procedures to holistically measure perceptions of risk management for natural hazards from selected actors. This paper uses Norway as a case study to present a modified RMI for surveying perceptions of landslide risk management at two time scenarios: 2015 (present) and 2050 (future), and for various administrative levels: national, county, and municipality. All survey respondents are practitioners in landslide risk management in Norway. The survey results are able to reflect some viewpoints of these experts on landslide risk management in Norway. Factors considered for assessing the future performance of landslide risk management by respondents are also studied to understand how respondents project their expectations. This paper also demonstrates how areas of improvement in landslide risk management in Norway can be identified based on the survey results. Due to limited responses, limited knowledge of respondents and the subjective nature of perceptions, the survey results are associated with uncertainties and should thus be used with care. Upon simplification of technical terms, the survey can be applied to survey public perceptions. The survey can also be regarded as a starting point for developing a common language/terminology for landslide risk management in Norway. This research activity has been funded by the Norwegian Centre of Innovation Klima 2050 (www.klima2050.no).

Keywords

Landslide risk management • Opinion survey • Landslide risk management index

Introduction

Effective management of landslide risk ensures that mitigation measures are optimally implemented. Therefore, it is important to assess the effectiveness of landslide risk management in order to identify areas for improvement and to prioritise future work plans for landslide risk reduction.

One of the ways to assess the performance of landslide risk management is to collect opinions and perceptions. The Risk Management Index (RMI) of Cardona et al. (2004) is a well-established method to holistically measure perceptions of risk management for natural hazards from selected actors. The RMI method involves an opinion survey to practitioners, such as technical staff, decision-makers, and stakeholders, in any stages of landslide risk management. It considers both structural and non-structural landslide risk

J. Chiu (✉) · U. Eidsvig
Norwegian Geotechnical Institute, Sognsveien 72, 0855 Oslo, Norway
e-mail: kyc@ngi.no

U. Eidsvig
e-mail: uke@ngi.no

K. Sassa et al. (eds.), *Advancing Culture of Living with Landslides*,
DOI 10.1007/978-3-319-59469-9_47

mitigation strategies, the latter including monitoring and warning systems, disaster management, and risk transfer.

Based on the RMI method of Cardona et al. (2004), a survey to local practitioners in landslide risk reduction in Norway was conducted between mid-September and late October 2015. The survey was modified to fit with Norwegian conditions. It aimed at measuring the performance of landslide risk management in Norway perceived for 2015 (present) and for 2050 (future) at national, county, and municipality levels. Measurement based on different administrative levels enables comparisons among different spatial scales, whereas measurement based on present and future time scenarios allows one to project future expectations.

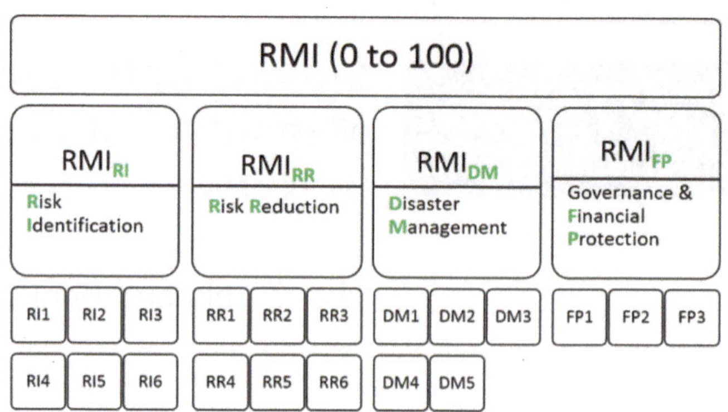

Fig. 1 Structure of the RMI (adopted from Chiu and Eidsvig 2016). Title of indicators is provided in Fig. 3

Background

The RMI method was first proposed by Cardona et al. (2004). It was later modified and applied periodically to measure performance of risk management of natural disasters in countries in Latin America and the Caribbean (Cardona et al. 2005; Carreño et al. 2007; Cãrdona and Carreño 2011).

Chiu (2015) implemented the RMI method to study the perceptions on landslide risk management at county and national levels in Norway for 2004, 2014 and 2024. A total of nine responses were received from both public and private Norwegian institutes which are involved in landslide risk management in the country. The present survey simplify the questionnaires by Chiu (2015) to increase survey responses and reliability of the survey results. Useful survey data from Chiu (2015) (data for 2014) were also collaborated to the new data from the present survey.

Methodology

The RMI

The RMI measured in the present survey consists of four public policies, which are represented by policy indices, namely Risk Identification index (RMI_{RI}), Risk Reduction index (RMI_{RR}), Disaster Management index (RMI_{DM}), and Governance and Financial Protection index (RMI_{FP}). Each policy index comprises of three to six indicators. Figure 1 shows the structure of the RMI and Table 1 gives a brief description of each public policy.

Procedures of how the RMI is obtained are schematically presented in Fig. 2. Each policy index is quantified by an aggregation of the weighed values of its indicators. The weighed values are calculated from performance levels and relative weights, which are collected by separated questionnaires. The title of each indicator is provided in the vertical axis of Fig. 3.

Five management performance levels are designated, corresponding to linguistic expressions from 'low', to 'optimal' or numerically from 1 to 5, respectively. Each performance level has its benchmark criteria. The criteria were modified from Cardona et al. (2005) and Lahidji (2009) to become representative for landslide hazards in Norway. Based on a brief scope of indicators and detailed benchmarking criteria, survey participants have to assign a performance level for each indicator, but they can also select 'not relevant' or 'not able to answer' if appropriate. Each management performance level represents a membership function of a fuzzy set (Fig. 2a). A fuzzy set gives flexibility to modelling of linguistic or qualitative expressions.

Relative importance of each indicator within a public policy is allocated using the Analytic Hierarchy Process (AHP). AHP decomposes a decision-making problem into hierarchy to enable evaluation of a problem based on both qualitative and quantitative aspects. Via AHP, indicators in a public policy are compared pairwise by two steps: (1) 'Which is more important?' and (2) 'In which degree? Rate from 1 to 9.' A degree of 1 represents equal importance between two indicators, whereas a degree of 9 represents that one is 9 times more important than the other one. Results of each comparison can be tabulated into a comparison matrix (see example in Fig. 2), from which relative weights are calculated using an eigenvector technique, with consistency checked against the consistency ratio after Saaty (1987) and Carreño et al. (2007). The standardised relative weights determined by AHP for each indicator give the height to the membership function of each fuzzy set (Fig. 2b). The policy index is then calculated as the centroid of area of the fuzzy set (Fig. 2b).

Table 1 Public policies in RMI (adopted from Cardona et al. 2005; Carreño et al. 2007; Chiu and Eidsvig 2016)

Public policy	Policy index	Description
Risk identification (RI)	RMI_{RI}	Individual and social risk awareness of landslide hazards; methodological approaches in landslide hazard assessment
Risk reduction (RR)	RMI_{RR}	Prevention and mitigation measures against landslides
Disaster management (DM)	RMI_{DM}	Response and recovery following a disaster
Governance and financial protection (loss transfer) (FP)	RMI_{FP}	Allocation and use of financial resources for dealing with disaster

The RMI is defined as the mean of the four public policy indices. Each policy index, as well as the RMI, can vary from 0 to 100. The lowest and highest values correspond to the poorest and best performance respectively.

The Survey

Target participants were invited from organisations that are involved in any stages/disciplines of landslide risk management in Norway, including government agencies, local authorities, consultants, contractors, research institutes as well as academic bodies.

The performance levels are assessed in terms of two time scenarios (2015 and 2050) and three administrative levels (national, county, and municipality). All participants were invited to rate the performance level but only those whose work or expertise is related to the public policy were invited to assign relative weights for the indicators in the policy.

Information about participants' technical background was also surveyed. In addition, participants were asked to explain the predominant factor(s) that influence(s) their performance rating for 2050. These factors include anthropogenic activities, climate, demography, knowledge and technology, socio-economy, risk perception, as well as other possible factors provided by the participants.

Results

Background of Respondents

Twenty eight (61% of all invitees) responses were received. Seventy percent of the respondents work in the public sector, among which the majority works in municipalities and counties. Around 20% works in research institute and universities. In addition, around 10% works in the private sector, including manufacturer, consultant, financial and insurance companies. In addition, more respondents work in landslide risk assessment and public information and community participation (17–30%) than other disciplines such as emergency response, physical mitigation measures, research, insurance, legislation, and finance ($\leq 10\%$).

RMI Results

Figure 3 presents the medians of performance levels and average AHP weights of different indicators. These numbers are used to calculate the policy indices and RMI and the results are presented in Fig. 4.

Policy indices at any administrative level in 2015 and 2050 range from 17 to 45 and 37 to 77 respectively (Fig. 4). The most popular factors considered for projecting the 2050 performance include advanced knowledge and technology, climate change that leads to increased landslide hazards, increased risk awareness, expanded development, and adequate mitigation measures and planning.

Policy indices as well as RMIs are higher for national level than sub-national levels. RMI_{RI} at national level is the highest among all policy indices in both years, especially in 2050. RMI_{RR} and RMI_{DM} are similar at any administrative level. RMI_{FP} is the lowest among all policy indices in both years, but this public policy is perceived to improve the most (more than 100% increment in RMI_{FP} from 2015 to 2050).

Discussion

The higher RMIs at national than sub-national level, as obtained in this study, may be related to individuals' assumption that other municipalities/countries have better performance so the country as a whole should be better. This may also reflect that respondents have higher expectations for local levels than the national level.

High national RMI_{RI} for 2050 may indicate a much higher expectation on risk awareness and landslide hazard assessment overall in Norway in the future.

Within RR and DM public policies, indicators related to upgrading, retrofitting, and reconstruction of assets (RR4,

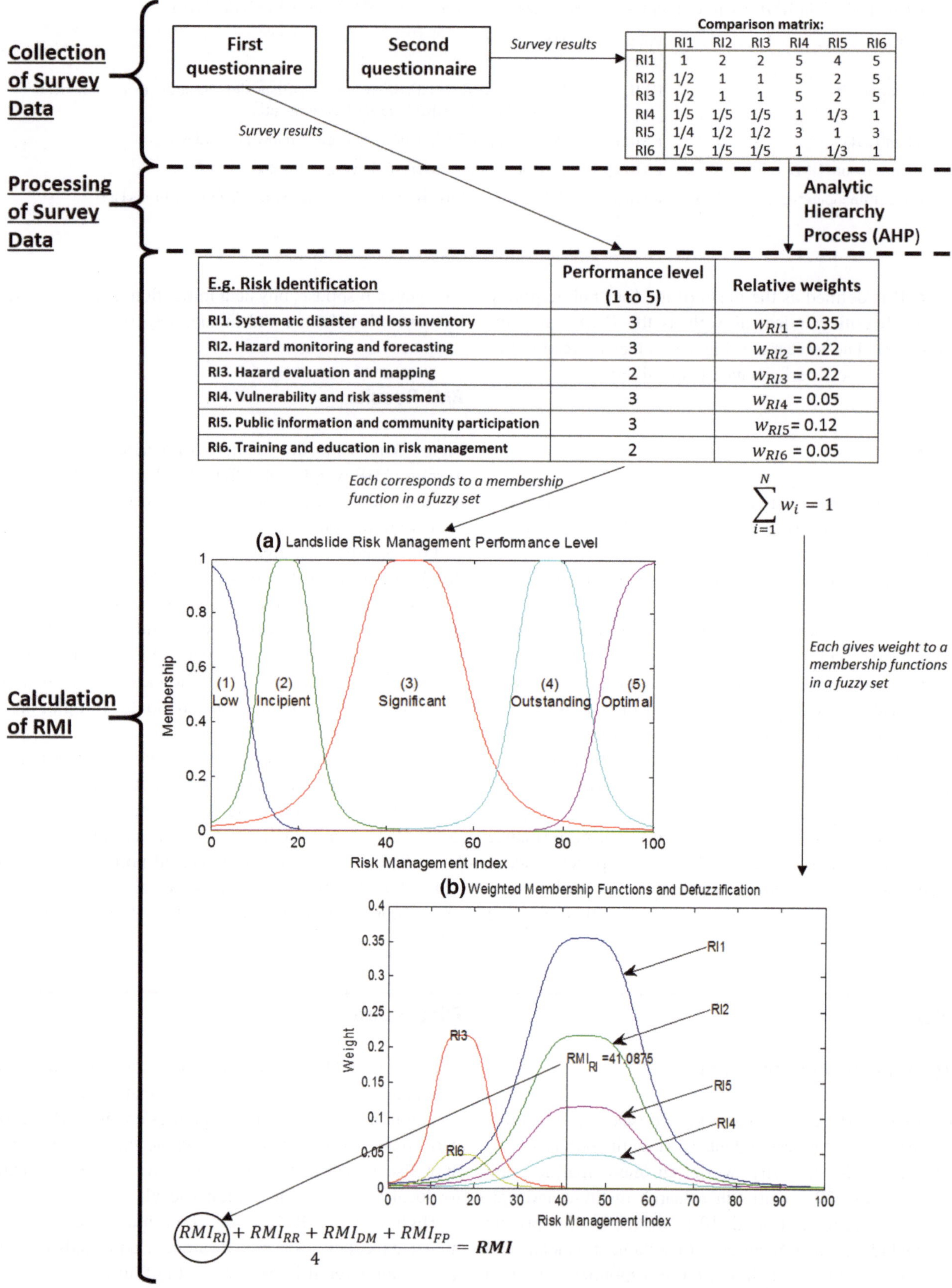

Fig. 2 Procedures for obtaining a policy index (e.g. RMI_{RI}) and RMI (adopted from Chiu and Eidsvig 2016; modified from Chiu 2015)

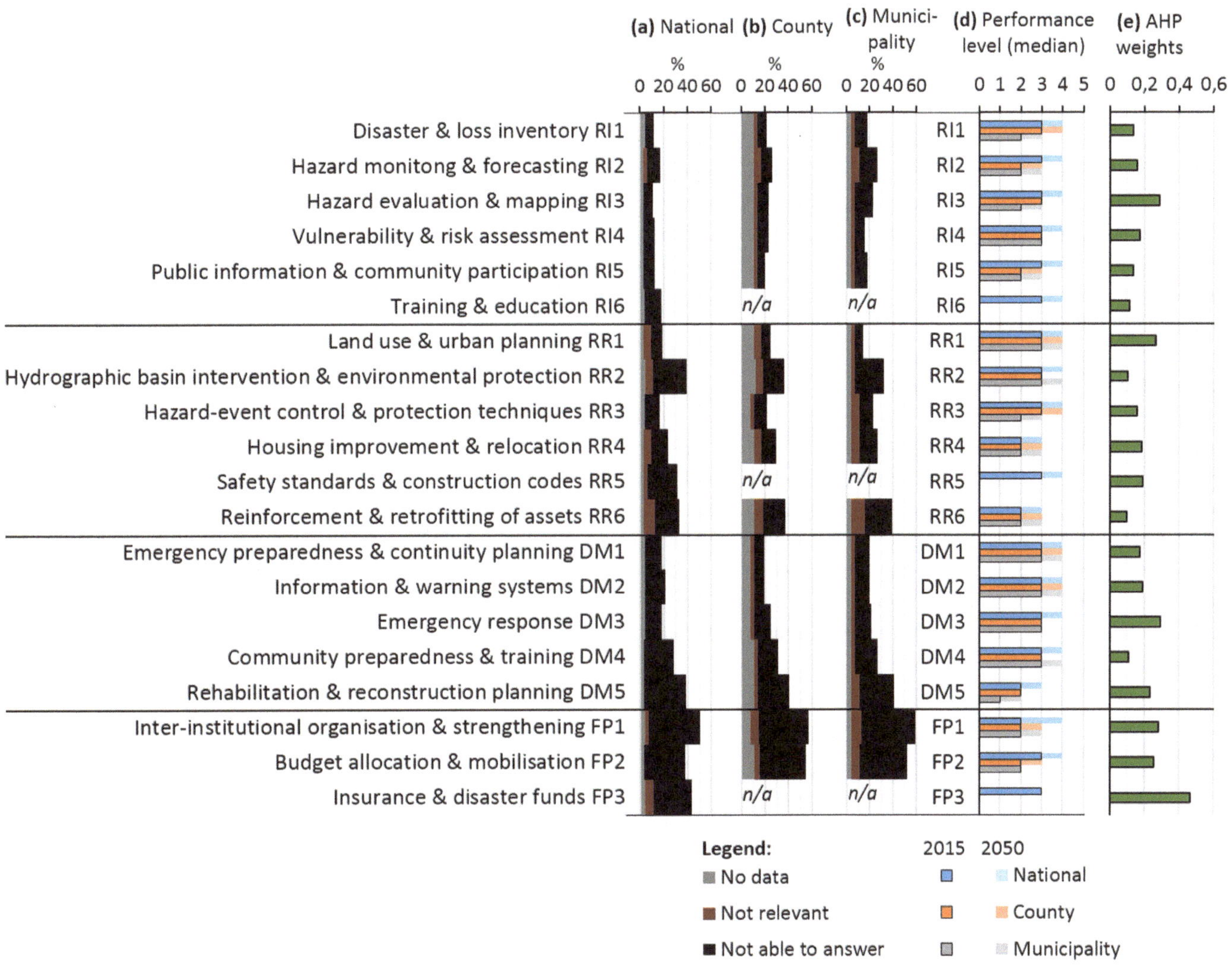

Fig. 3 Survey results—performance levels and relative weights (adopted from Chiu and Eidsvig 2016)

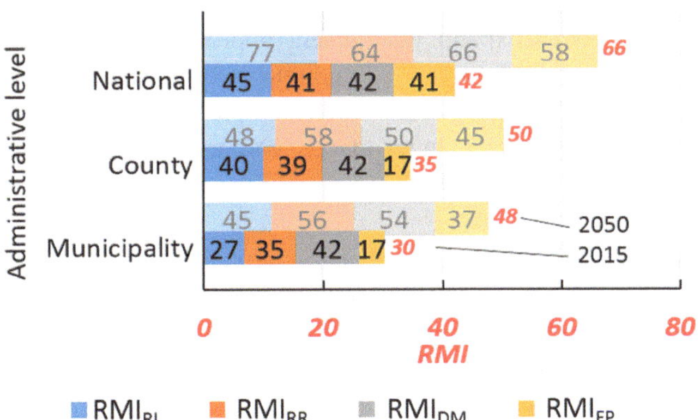

Fig. 4 Results of policy indices and RMI (adopted from Chiu and Eidsvig 2016)

RR6, and DM5) have the poorest performance in both years but also most answers for 'not relevant' or 'not able to answer'. This may reveal that these indicators are unfamiliar to many respondents or irrelevant/inapplicable in Norway.

The large increment from 2015 to 2050 obtained for RMI_{FP} may indicate the urge for improvement in today's inter-institutional organisation and allocation of financial resources for tackling landslide risk. However, results for RMI_{FP} have relatively large uncertainty since many respondents chose 'not relevant' or 'not able to answer' for FP indicators (Fig. 3a–c). The large number of non-performance level answers is probably due to limited knowledge in finance/insurance by most respondents.

The survey results also enable identification/prioritisation of direction of improvement by studying both the performance levels and relative weights. Indicators with poor performance but high relative weights may imply the need for more focus on improvement of these indicators. As shown in Fig. 3d, e, three such indicators can be identified: RI3. Hazard evaluation and mapping (relatively poor performance at municipality level), RR4. Housing improvement and relocation from prone-areas, and DM5. Rehabilitation and reconstruction training.

Perception for Future Scenario

Despite most respondents agree that climate change and urban development will likely increase landslide hazards in Norway in the future, a trend of improvement in landslide risk management is perceived. This may imply that respondents also believe that other factors such as knowledge and technology, risk perception and mitigation measures are the key to the adaptation of increasing landslide risk in the future.

Reliability of Survey Results

Degree of understanding of individual indicators, subjective perceptions, as well as the limited number of survey responses limit the reliability of the survey results. Therefore, extra care has to be taken while using the results. Nevertheless, results obtained from the present survey are still able to give examples of viewpoint from local practitioners on landslide risk management in Norway.

Applications

As demonstrated above, the method presented is useful for prioritisation of areas of improvement and decision-making in landslide risk management.

Rød (2013) mentioned that public's perceptions on the usefulness of risk information and risk communication may influence people's willingness to follow evacuation instructions. Since the public is the beneficiary of landslide risk management, their perceptions should also be considered. The current approach can thus be applied to survey public perceptions on landslide risk management in Norway. Perceptions from experts and the public can then be compared to assess if landslide risk communication between experts and the public is effective.

In addition, the survey can be regarded as a starting point to develop a common terminology/language in landslide risk management in Norway to facilitate mutual understandings among people with different backgrounds. In order to achieve this, a survey with simpler terms should be used so that the questions are easier to understand. Practitioners in local authorities, such as municipalities, should be consulted for appropriate terminology.

Conclusion

A method for surveying perceptions of landslide risk management in Norway at various administrative levels and time scenarios is presented. Respondents of the survey are practitioners in any stage of landslide risk reduction in Norway. The survey results reflect some viewpoints by these experts on landslide risk management in Norway in different spatial scales and time scenarios under a holistic framework. The investigation of factors respondents considered for assessing the future performance of landslide risk management also helps to understand how respondents project their expectations.

The survey results indicate that several aspects of landslide risk management in Norway can be improved. For example, landslide hazard evaluation and mapping should be prioritised in Norway. Upgrading and reconstruction of assets may also be included in the landslide risk reduction strategies in Norway. In addition, there should be more focus on inter-institutional organisation as well as allocation and use of financial resources for dealing with landslides. However, the survey results are associated with uncertainties and should be used carefully.

This paper also demonstrates how one can identify areas of improvement in landslide risk management in Norway based on the survey results. It is further recommended to expand the present approach by surveying public perceptions so as to compare perceptions between experts and the public. Such comparisons can help to assess the effectiveness of risk communication between experts and the public.

Acknowledgements We thank all the survey participants for their valuable time and opinions. This research activity has been funded by the Norwegian Centre of Innovation Klima 2050 (www.klima2050.no). The support is gratefully acknowledged.

References

Cardona OD, Carreño ML (2011) Updating the indicators of disaster risk and risk management for the Americas. J Integr Disaster Risk Manag 1(1):27–47

Cardona OD, Hurtado JE, Duque G, Moreno A, Chardon AC, Velásquez LS, Prieto SD (2004) Disaster risk and risk management

benchmarking: a methodology based on indicators at national level. IDB/IDEA Program on Indicators for Disaster Risk Management, Universidad Nacional de Colombia, Manizales, p 101
Cardona OD, Hurtado JE, Duque G, Moreno A, Chardon AC, Velásquez LS, Prieto SD (2005) System of indicators for disaster risk management: program for Latin America and the Caribbean: main technical report. IDB/IDEA Program on Indicators for Disaster Risk Management, Universidad Nacional de Colombia, Manizales, p 216
Carreño ML, Cardona OD, Barbat AH (2007) A disaster risk management performance index. Nat Haz 41:1–20
Chiu JKY (2015) Landslide risk management perceptions in territories—comparative case studies of Hong Kong and Norway. MSc thesis, University of Oslo, Oslo, Norway
Chiu JKY, Eidsvig U (2016) Klima 2050 report no. 2 surveying perception of landslide risk management. SINTEF Building and Infrastructure, Trondheim. (ISBN 978-82-536-1509-7), p 75
Lahidji R (2009) Appendix G—measuring the capacity to cope with natural disasters. NGI report 20071600-1 risk assessment and mitigation measures for natural- and conflict-related hazards in Asia-Pacific. Norwegian Geotechnical Institute, Oslo, pp 139–151
Rød SK (2013) Risk communication in relation to an imminent rockslide and tsunami. PhD thesis, Norwegian University of Science and Technology, Trondheim, Norway
Saaty TL (1987) The analytic hierarchy process—what it is and how it is used. Math Model 9:161–176

Landslide Hazards and Climate Change Adaptation of Transport Infrastructures in Germany

Martin Klose, Markus Auerbach, Carina Herrmann, Christine Kumerics, and Annegret Gratzki

Abstract

This paper provides insights into a new landslide hazards project which is part of a national research program on safe and sustainable transport in Germany funded by the Federal Ministry of Transport and Digital Infrastructure (BMVI). Here we report on a work in progress and present selected results of a pilot study conducted prior to the launch of the research program in 2016. The main goal of the landslide hazards project is to assess the future landslide hazard potential for the federal transport system under the influence of climate change. A federal road-related pilot study with focus on developing an approach to this type of hazard assessment was a first step in this direction. The developed approach is based upon a Geographic Information System (GIS) as mapping tool to combine a landslide susceptibility map with spatial datasets of regional climate change projections. Here we present the basic framework of this approach only, and provide information on landslide activity and climate change. This information refers to findings from three example landslide sites in Germany. The purpose of this paper is to introduce these landslide projects of German transport research against the backdrop of the existing national strategy of climate change adaptation.

Keywords

Landslide hazards • Transport infrastructure • Climate change adaptation • Germany

Introduction

The transport system of the Federal Republic of Germany is one of the densest and most developed transport systems of the world (Fig. 1). Being an essential part of the Trans-European Transport Network (TEN-T), it serves as a backbone of Germany's export economy and the European Union (EU) single market (e.g., European Commission, 2011). The resilience of transport systems in Germany and Europe depends to a large extent on the success of climate change adaptation (e.g., Nemry and Demirel 2012; Conference of European Directors of Roads 2013; European Environment Agency 2014). In order to lay the foundation for safe and sustainable transport in the 21st century, thousands of kilometers of road, waterway, and railroad infrastructure throughout the country need to undergo a scientific

M. Klose (✉) · M. Auerbach
Federal Highway Research Institute (BASt), Brüderstraße 53, 51427 Bergisch Gladbach, Germany
e-mail: Klose@bast.de

M. Auerbach
e-mail: AuerbachM@bast.de

C. Herrmann
Federal Railway Authority (EBA), Heinemannstraße 6, 53175 Bonn, Germany
e-mail: HerrmannC@eba.bund.de

C. Kumerics
Landslide Research Centre at the Johannes Gutenberg-University Mainz, Mombacher Straße 49-53, 55122 Mainz, Germany
e-mail: fsr@geo-international.info

A. Gratzki
German Meteorological Service (DWD), Frankfurter Straße 135, 63067 Offenbach, Germany
e-mail: Annegret.Gratzki@dwd.de

K. Sassa et al. (eds.), *Advancing Culture of Living with Landslides*,
DOI 10.1007/978-3-319-59469-9_48

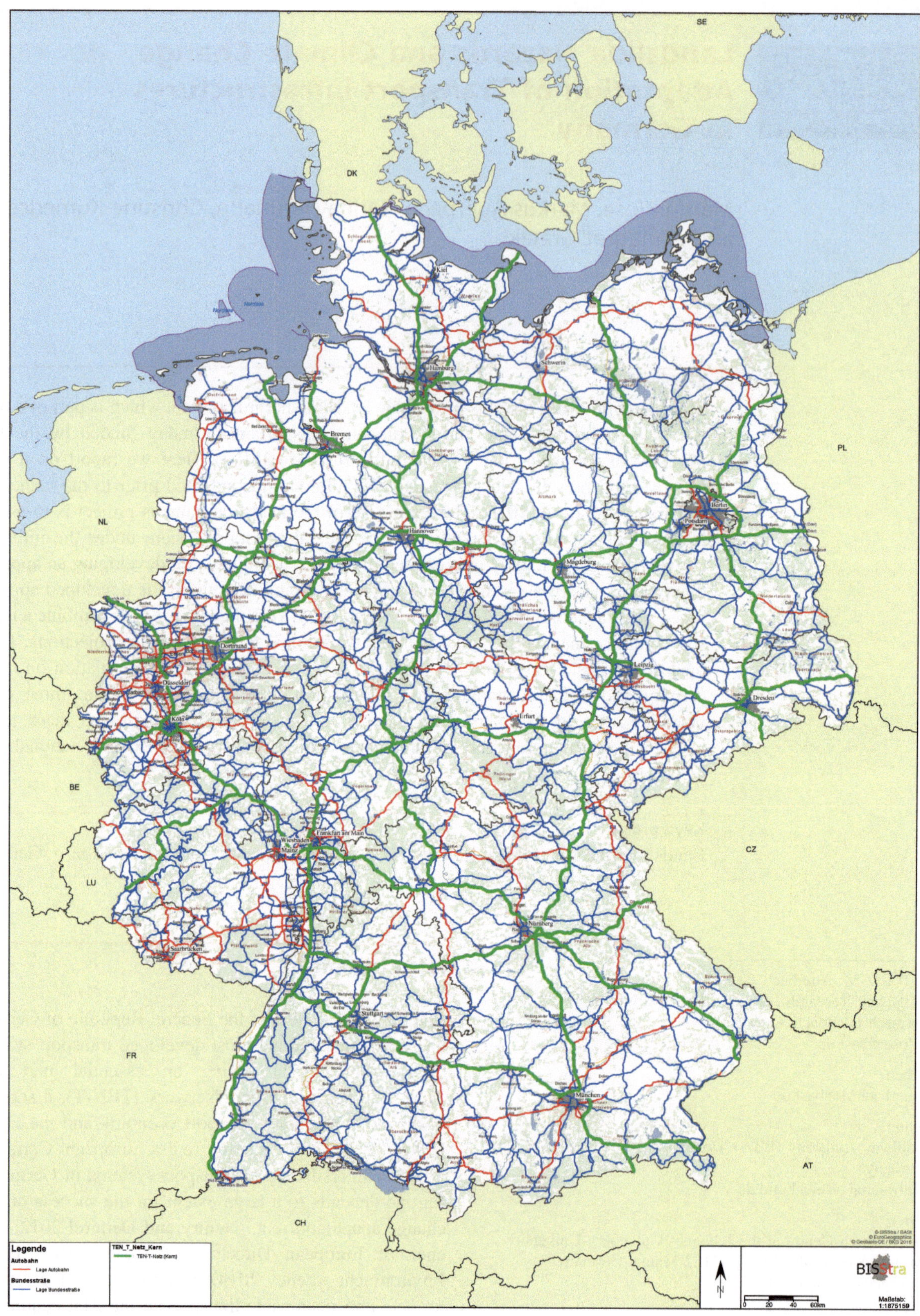

Fig. 1 The federal road network of Germany as one of the densest and most developed road networks of the world (*red*: *Bundesautobahnen*; *blue*: *Bundesstraßen*). This map also illustrates the core network (*green*) of the Trans-European Transport Network (TEN-T). *Source* BISStra 2016, Federal Highway Research Institute (BASt), ©BISStra/BASt, ©EuroGeographics, ©Geobasis-DE/BKG 2016

assessment of potential climate change impacts, both at the system and object level. The criterion of climate change policy in Germany is the German Strategy for Adaptation to Climate Change (The Federal Government 2008; Federal Ministry for the Environment, Nature Conservation and Nuclear Safety 2012; Die Bundesregierung 2015). This strategy applies to the whole public sector, including transport and infrastructure, and constitutes a legal basis for putting science and strategy into practice.

Climate change adaptation requires assessing climate change impacts (e.g., Noble et al. 2014). It is due to their dependency on rainfall extremes and anomalies that landslides serve as an example for climate change impacts on roads worldwide (e.g., World Road Association 2015). The German transport system, and especially road transport, is recognized to be affected by landslide damages, both direct and indirect (e.g., Keller and Atzl 2014; Klose et al. 2015; Fig. 2). Information on the exposure and vulnerability to and the potential impact of these damages provide a first basis for scientific approaches to safe and effective road operations under the influence of climate change. As elsewhere in Europe (e.g., Spizzichino et al. 2010), landslides affecting roads in Germany are usually local, but widespread phenomena, interacting with rainfall extremes and anomalies over time (e.g., Krauter et al. 2012). Climate change adaptation in the sense of landslide risk reduction is a challenging task as being confronted with these complex interactions and their direct implications on risk assessment and management.

This paper reports on research activities in the field of climate change adaptation by using the example of landslide hazards and transport infrastructures in Germany. Here we present a new landslide hazards project which contributes as work in progress to the current strategic and scientific efforts for enabling effective climate change adaptation, either at national or international level. This is followed by a summary of selected results of a pilot study conducted before the beginning of the new landslide hazards project in 2016.

Research Program

Overview

The topic of landslide hazards represents the central theme of a project within a national research program of the Federal Ministry of Transport and Digital Infrastructure (BMVI). This new research program currently funds three research fields consisting of several individual projects each (http://www.bmvi-expertennetzwerk.de). One of the main goals across these research fields is to address the challenges related to climate change and safe and sustainable transport through collaborative research following an interdisciplinary approach bringing together scientists and practitioners from research fields such as climatology, civil engineering, and geosciences. The participating research institutes and agencies include the Federal Highway Research Institute (BASt), the German Meteorological Service (DWD), the Federal Railway Authority (EBA), the German Federal Institute of Hydrology (BfG), the Federal Waterways Engineering and Research Institute (BAW), the Federal Maritime and Hydrographic Agency (BSH), and the Federal Office for Goods Transport (BAG). This network of research institutes and agencies operates at the interface between science and practice, with the goal of providing effective knowledge transfer and advice for decision makers in government and industry. Research and development aims at implementing scientific solutions and best practices for climate change adaptation and sustainable development of the federal transport system of Germany.

Landslide Hazards

The landslide hazards project contributes to a research field addressing the topic of adaptation to climate change and extreme weather events. High priority within this research field is given to hazard and risk assessments for different modes of transport (road, rail, water) and various types of climatic extremes or secondary geologic hazards. The main goal of the landslide hazards project is to evaluate both the current exposure of federal transport routes to landslides and the future landslide hazard potential under the influence of climate change. In order to reach this goal and to make data products available for national hazard and risk assessments in Germany, it is intended to apply and combine existing and new methods, including landslide inventory, hazard mapping, as well as impact and vulnerability assessment. The research and innovation approach of this project covers the risk assessment cycle and is characterized by four main steps:

(1) *Landslide identification and inventory*. Information about historical and current landslide events and the resulting damages is critical for hazard and risk assessment. The goal therefore is to develop and maintain a landslide inventory for the federal transport system of Germany.
(2) *Landslide hazard mapping*. A key to best practices in planning and maintenance are landslide hazard maps that take account of the potential influence of climate change. It is intended to optimize methods enabling to combine landslide susceptibility maps with regional climate change projections.
(3) *Landslide impact and vulnerability assessment*. Landslide risk is generally understood as a function of exposure and vulnerability. In order to improve resilience through effective risk assessment, potential impacts and vulnerability will be investigated, both at the object and network level.

Fig. 2 The Upper Middle Rhine Valley, southwest Germany—a transport corridor of trans-European economic importance. As a natural transit route for road, rail, and water transport passing through the Rhenish Slate Mountains, this deeply incised river valley is characterized by a spatial concentration of critical transport infrastructures, with some of which being located in areas prone to landslides and related types of hazards (*Photo* M. Klose, BASt)

(4) *Hazard communication*. Research and development to the benefit of climate change adaptation requires policy guidance and knowledge transfer to practitioners. This should be realized by making use of innovative geospatial and web mapping technologies.

Pilot Study

Background

The following sections present a pilot study conducted prior to the landslide hazards project of the national research program. Here it is referred to a specific road-related landslide study now serving as an ideal starting point for this new project. The goal of the pilot study was to develop a method to assess the future landslide hazard potential along the German federal road network. Two research projects in the context of the AdSVIS program (Auerbach et al. 2014) were therefore commissioned by the Federal Highway Research Institute (BASt) in the recent past. Further information on the specific goals and tasks of one of these projects is given in the report prepared by Krauter et al. (2012). In the following two sections, we report on some of the main outcomes of the pilot study, with special emphasis on data sources and the methodological approach.

Method

The key element of the method developed in this pilot study is a Geographic Information System (GIS). It serves as a mapping tool that provides GIS functionality to organize, analyze, and illustrate geospatial and climatic datasets. ESRI ArcGIS 10.1 was used as software to implement the method and its different components. The approach to hazard mapping consists of two specific models that in combination enable a GIS-based assessment of the future landslide hazard potential along federal road corridors throughout the country (Fig. 3). A heuristic landslide susceptibility model resting upon expert opinion in factor selection and classification specifies the current exposure to landslides on the basis of experiences about predisposing factors of historical and current landslides in Germany. Alternatively, two types of regional climate models (RCMs) provide projections for climate parameters acting as potential triggering factors of landslides on soil and rock slopes within a 500 m buffer at each side of the road corridors. Both parts of this method are embedded into a GIS tool that integrates and processes the input data by means of qualitative map combination as well as spatial data overlay and intersection (Fig. 3).

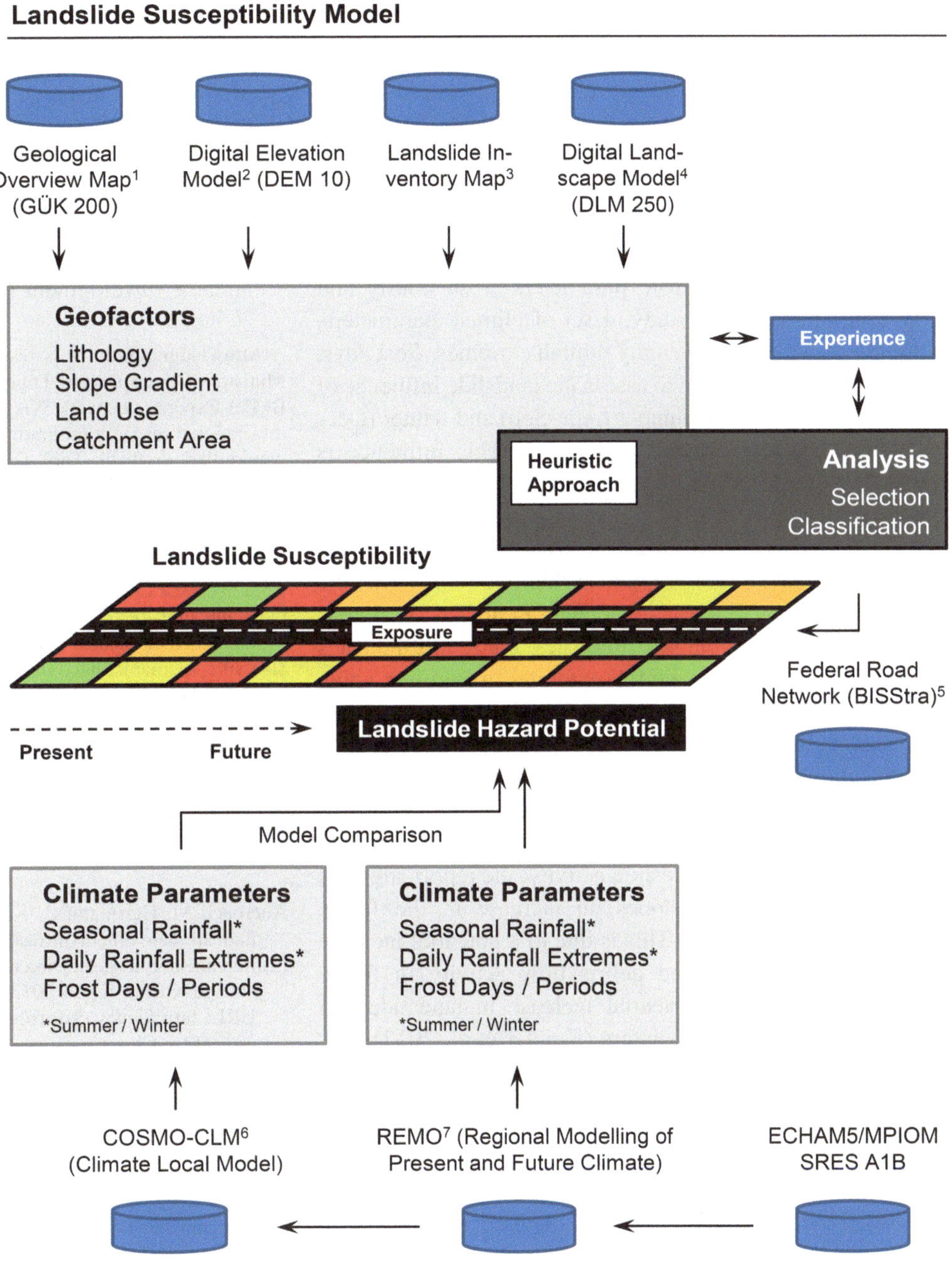

Fig. 3 Methodological approach to the assessment of the future landslide hazard potential along federal roads in Germany. The diagram also illustrates the data material necessary for the application of the different methods of this approach. *Data material* [1]Federal Institute for Geosciences and Natural Resources (*BGR*); [2]Federal Agency for Cartography and Geodesy (*BKG*); [3]For model validation only, multiple sources; [4]Federal Agency for Cartography and Geodesy (*BKG*); [5]Federal Highway Research Institute (*BASt*); [6]International scientific consortium (http://www.clm-community.eu/; http://www2.cosmo-model.org/), German Meteorological Service (*DWD*); [7]Max Planck Institute for Meteorology (MPI-M), German Meteorological Service (*DWD*)

Results

The results of the pilot study and subsequent research activities are in part of preliminary character. This mainly applies to the hazard maps developed using the methodological approach illustrated in Fig. 3. Here it is therefore referred to the results discussed in Krauter et al. (2012) only. Using the example of three landslide sites in west and northwest Germany (see also Damm 2005), the historical relationship between rainfall, temperature, and landslide activity was reviewed and analyzed, with the goal of linking this analysis with datasets of climate projections in a qualitative way.

These datasets for different climatic parameters were derived from the regional climate model REMO (Regional Modelling of Present and Future Climate). REMO is a dynamical RCM based on the global climate model (GCM) ECHAM5/MPIOM. With a maximum spatial grid cell resolution of approximately 10 km for Germany, it provides datasets for climate parameters at an hourly and daily basis. In this pilot study, a set of climate parameters, including seasonal rainfall, daily rainfall extremes, frost days, and frost periods was used to assess the landslide influence of climate change for both summer (Apr–Sep) and winter (Oct–Mar) on a 30-year average until 2100. This influence is defined by the future parameter changes compared to the reference period 1971–2000, with seven future time intervals ranging from 2011–2040 to 2071–2100. The emission scenario (SRES) taken into account in this pilot study was the moderate SRES A1B scenario (Krauter et al. 2012).

For the three example landslide sites, the report indicates a future increase in daily rainfall extremes in summer, while the total summer rainfall is expected to decrease. In winter, by contrast, both total rainfall and daily rainfall extremes will potentially increase, which is accompanied by a decrease in the number of frost days and frost periods, as indicated by this report. Given the datasets on the historical influence of climate on landslide activity, the report suggests for both soil and rock slopes an increase in the future landslide hazard potential. This is due to a potential increase in shallow landsliding and debris flow activity in future summers as well as a potential increase in landslide and rockfall activity in future winters (Krauter et al. 2012).

Conclusions

The pilot study was a first attempt to assess the future landslide hazard potential along the German federal road network under the influence of climate change. Although some of its results are still of preliminary character, it now serves as an ideal starting point for further research and development in this context. The results will be incorporated into the ongoing landslide hazards project that started its work after the launch of the national research program at the beginning of 2016. This work in progress is focused, among others, on expanding the methods available, updating the climate modeling components, and integrating other modes of transports into the method as well. The knowledge base for this specific field of transport planning has significantly improved in recent times. Landslide inventories, hazard maps, and risk assessments along with innovative mapping or monitoring technologies are now playing an increasingly important role in landslide research globally (e.g., Guzzetti et al. 2012; Van Den Eeckhaut and Hervás 2012; Corominas et al. 2014). In order to translate this scientific progress into research and development at national level, the project aims at making use of the existing global knowledge base to meet the requirements of transport planning and policy within the country. The project contributes to the national research program in ways that enable reaching the overarching goal of safe and sustainable development of transport systems in Germany.

Acknowledgements This research work was funded by the Federal Ministry of Transport and Digital Infrastructure (BMVI). It is part of the BMVI-Expertennetzwerk Wissen—Können—Handeln. The funding of this national research program is gratefully acknowledged. The authors would like to thank Dipl.-Geol. Michael Bürger, Federal Highway Research Institute (BASt), for his personal advice and the supervision of previous research projects. The pilot study discussed in the paper refers to the projects FE 89.238/2009/AP and FE 05.0170/2011/MRB commissioned by the BASt. We very much appreciate the support received from all parties involved in the two projects. Special thanks are due to the Landslide Research Centre at the Johannes Gutenberg-University Mainz which conducted both projects on behalf of the BASt. We would also like to thank the different research institutes and agencies that provided data material for these projects. A detailed list of these research institutes and agencies is given in Fig. 3. The authors are grateful to Dipl.-Ing. Jens Kirsten, BASt, who supported the preparation of Fig. 1.

References

Auerbach M, Herrmann C, Krieger B, Mayer S (2014) Klimawandel und Straßenverkehrsinfrastruktur. Straße und Autobahn 7:531–539

Conference of European Directors of Roads (2013) Adaptation to climate change. CEDR report 2013/07. CEDR's Secretariat General. p 168. URL: http://www.cedr.fr/home/fileadmin/user_upload/Publications/2013/T16_Climate_change.pdf. Last accessed 25 Aug 2016

Corominas J, van Westen C, Frattini P, Cascini L, Malet J-P, Fotopoulou S, Catani F, Van Den Eeckhaut M, Mavrouli O, Agliardi F, Pitilakis K, Winter MG, Pastor M, Ferlisi S, Tofani V, Hervás J, Smith JT (2014) Recommendations for the quantitative analysis of landslide risk. Bull Eng Geol Env 73(2):209–263

Damm B (2005) Gravitative Massenbewegungen in Südniedersachsen. Die Altmündener Wand—Analyse und Bewertung eines Rutschungsstandortes. Zeitschrift für Geomorphologie NF, Suppl.-Bd. 138:189–209

Die Bundesregierung (2015) Fortschrittsbericht zur Deutschen Anpassungsstrategie an den Klimawandel. Stand: 16, November 2015. URL: http://www.bmub.bund.de/fileadmin/Daten_BMU/Download_PDF/Klimaschutz/klimawandel_das_fortschrittsbericht_bf.pdf. Last accessed 25 Aug 2016

European Commission (2011) White paper. Roadmap to a single European transport area—towards a competitive and resource efficient transport system. URL: http://ec.europa.eu/transport/themes/strategies/doc/2011_white_paper/white_paper_com%282011%29_144_en.pdf. Last accessed 25 Aug 2016

European Environment Agency (2014) Adaptation of transport to climate change in Europe—challenges and options across transport modes and stakeholders. EEA Report 08/2014. Publications Office of the European Union, Luxembourg, p 60 (ISBN 978-92-9213-500-3)

Federal Ministry for the Environment, Nature Conservation and Nuclear Safety (2012) Adaptation Action Plan for the German Strategy for Adaptation to Climate Change. Federal Ministry for the Environment, Nature Conservation and Nuclear Safety (BMU), Public Relations Division, Berlin, Germany, p 79

Guzzetti F, Mondini AC, Cardinali M, Fiorucci F, Santangelo M, Chang K-T (2012) Landslide inventory maps: new tools for an old problem. Earth Sci Rev 112(1–2):42–66

Keller S, Atzl A (2014) Mapping natural hazard impacts on road infrastructure—the extreme precipitation in Baden-Württemberg, Germany, June 2013. Int J Disaster Risk Sci 5(3):227–241

Klose M, Damm B, Terhorst B (2015) Landslide cost modeling for transportation infrastructures: a methodological approach. Landslides 12(2):321–334

Krauter E, Kumerics C, Feuerbach J, Lauterbach M (2012) Abschätzung der Risiken von Hang- und Böschungsrutschungen durch die Zunahme von Extremwetterereignissen. Berichte der Bundesanstalt für Straßenwesen. Straßenbau, Heft S75. Wirtschaftsverlag NW, Bremerhaven, p 61

Nemry F, Demirel H (2012) Impacts of climate change on transport: a focus on road and rail transport infrastructures. JRC Scientific and Policy Reports. Report EUR 25553 EN. European Commission, Joint Research Centre, Institute for Prospective Technological Studies. Publications Office of the European Union, Luxembourg, p 89 (ISBN 978-92-79-27037-6)

Noble IR, Huq S, Anokhin YA, Carmin J, Goudou D, Lansigan FP, Osman-Elasha B, Villamizar A (2014) Adaptation needs and options. In: Field CB, Barros VR, Dokken DJ, Mach KJ, Mastrandrea MD, Bilir TE, Chatterjee M, Ebi KL, Estrada YO, Genova RC, Girma B, Kissel ES, Levy AN, MacCracken S, Mastrandrea PR, White LL (eds) Climate change 2014: impacts, adaptation, and vulnerability. Part A: global and sectoral aspects. Contribution of Working Group II to the Fifth Assessment Report of the Intergovernmental Panel on Climate Change. Cambridge University Press, Cambridge, United Kingdom and New York, NY, USA, pp 833–868

Spizzichino D, Margottini C, Trigila A, Iadanza C, Linser S (2010) Chapter 9: landslides. In: European Environment Agency (ed) Mapping the impacts of natural hazards and technological accidents in Europe: an overview of the last decade. EEA Technical report 13/2010. European Environment Agency. Publications Office of the European Union, Luxembourg, pp 81–93 (ISBN 978-92-9213-168-5)

The Federal Government (2008) German Strategy for Adaptation to Climate Change. URL: http://www.bmub.bund.de/fileadmin/bmu-import/files/english/pdf/application/pdf/das_gesamt_en_bf.pdf. Last accessed 25 Aug 2016

Van Den Eeckhaut M, Hervás J (2012) State of the art of national landslide databases in Europe and their potential for assessing landslide susceptibility, hazard and risk. Geomorphology 139–140:545–558

World Road Association (2015) International climate change adaptation framework for road infrastructure. PIARC Report 2015R03EN. Authors: Toplis C, Kidnie M, Marchese A, Maruntu C, Murphy H, Sébille R, Thomson S. World Road Association (PIARC), La Défense cedex, France, p 88 (ISBN 978-2-84060-362-7)

Integration of Landslide Susceptibility Maps for Land Use Planning and Civil Protection Emergency Management

Sérgio C. Oliveira, José Luís Zêzere, Clémence Guillard-Gonçalves, Ricardo A.C. Garcia, and Susana Pereira

Abstract

Landslides are one of the most relevant geomorphological hazards in Portugal, by the high levels of people affected, destruction of assets and disruption of economic and social activities. Regarding the Portuguese territorial land use planning and emergency management, regulation, practice, prevention and risk management have been promoted in different ways. In Portugal, the areas susceptible to landslides are included in the 'National Ecological Reserve', which is a public utility restriction legal figure that rules the land use planning at the municipal level. In addition, the Municipal Emergency Plans include landslide susceptibility maps that are combined with the map of the exposed elements, allowing the assessment of exposure to landslides. This study is applied to the Loures municipality located to the north of Lisbon. In this municipality 621 landslides registered in a landslide inventory (rotational slides, deep-seated translational slides and shallow translational slides) that affected 1,469,577 m^2 (0.87%) of the Loures territory. The final landslide susceptibility map shows that in Loures municipality 1347 ha are associated to a Very high landslide susceptibility and 2372 ha to High landslide susceptibility, which corresponds both to 22.1% of the entire municipality, and constitutes the larger fraction of the National Ecological Reserve, related to landslides. These areas do not present geomorphological and geotechnical suitability for building structures or infrastructures. From the civil protection and emergency management point of views 34 classes of exposed elements were identified in the municipality, with point, linear and polygonal representations. The elements at risk located in the Very High or High landslide susceptibility classes were summarized and correspond to: high voltage poles; wind turbines; transmission/reception antennas; industrial areas; water tanks; silo; gas station/tank; service area; buildings of educational institutions; worship buildings; buildings of electricity facilities; regular buildings; gas pipeline; motorways; national roads; and municipal roads.

Keywords

Landslides • Susceptibility • Land use planning • Emergency management

S.C. Oliveira (✉) · J.L. Zêzere · C. Guillard-Gonçalves
R.A.C. Garcia · S. Pereira
Institute of Geography and Spatial Planning, Universidade de Lisboa, Edifício IGOT, Rua Branca Edmée Marques, Cidade Universitária, 1600-276 Lisbon, Portugal
e-mail: cruzdeoliveira@campus.ul.pt

J.L. Zêzere
e-mail: zezere@campus.ul.pt

C. Guillard-Gonçalves
e-mail: cguillard@campus.ul.pt

R.A.C. Garcia
e-mail: rgarcia@campus.ul.pt

S. Pereira
e-mail: susana-pereira@campus.ul.pt

K. Sassa et al. (eds.), *Advancing Culture of Living with Landslides*,
DOI 10.1007/978-3-319-59469-9_49

Introduction

Landslides are one of the most relevant geomorphological hazards in Portugal, due the high levels of people affected, destruction of assets and disruption of economic and social activities. Regarding the Portuguese territorial land use planning and emergency management, regulation, practice, prevention and risk management have been recently promoted through different legal instruments.

In terms of the territorial planning since 2007 the National Program on Politics for the Territorial Planning assumes that preventive risk management is a priority vector for the spatial planning politics and a major conditioning factor within the Portuguese territorial model at all planning levels. Nevertheless, the land use regulation for both public institutions and private owners is only effective at the municipal scale. For landslide hazard the restrictions of land use have been safeguarded by the inclusion of areas susceptible to landslide occurrence in the public utility restriction legal figure named 'National Ecological Reserve' (NER), which includes the typology "Slope instability prone areas" (Decree-Law 166/2008 of August 22nd, changed and republished by the Decree-Law 239/2012, of November 2nd).

The NER integrates areas of ecological and sensitivity value and areas that require special protection due to excessive exposure and susceptibility to natural hazards. The NER establishes a set of constraints to the land occupation, use and transformation. In particular, along those areas classified as NER the following actions are forbidden: (a) housing development; (b) urbanization construction or expansion; (c) road construction; (d) excavation and embankment; and (e) destruction of vegetation cover, not including the necessary actions to the regular development of cultural operations of agricultural use and the exploitation of forest areas.

In Portugal the main objectives of the civil protection are the prevention and mitigation of collective risks and restoration of people's life routines following the occurrence of a severe accident or catastrophe. To accomplish these goals, rescue and assistance of people and other living beings are assumed as the priority action field of civil protection as well as the protection of endangered assets, cultural values and the environment. The operational response level, typically reactive, is clearly expressed in the Civil Protection emergency plans. These, with respect to risk framework, should incorporate the characterization of the risk reference framework, the chronology of past occurrences and the definition of the sensible and strategic infrastructures for the civil protection operations. For the Portuguese Municipal Emergency Plans, landslide susceptibility maps are overlaid with the map of the exposed elements, which allows the assessment of exposure to landslides, although the risk is not quantified.

Following the previous approach present in Guillard and Zêzere (2012) the main objectives of this work are: (i) to assess the landslide susceptibility at the municipal scale; (ii) to integrate landslide susceptibility maps on land use planning by setting National Ecological Reserve "Slope instability prone areas"; and (iii) to integrate landslide susceptibility maps on civil protection and emergency management actions accounting exposed elements. The study area is the Loures municipality (168 km^2), located north of Lisbon in Portugal (Fig. 1).

Data

Landslide Inventory

In this work we consider only landslides of the slide type, following the classification of Cruden and Varnes (1996). Falls and flows were excluded from the municipal landslide inventory because they are residual in number and typically of small size in the study area.

The Loures municipality landslide inventory was based on (1) stereoscopic interpretation of aerial photographs obtained in 1982 and 1983 at the 1:15,000 scale, (2) interpretation of digital orthophotomaps (pixel = 0.5 m) dated from 2004, and (3) detailed geomorphological field mapping made in the study area since 1980 with regular updates.

The landslide inventory contains 621 slides, including rotational slides (RS), shallow translational slides (STS) with slip surface depth ≤ 1.5 m and deep-seated translational slides (D-STS) with slip surface depth >1.5 m (Fig. 1). The total affected area is approximately 1.5 km^2, which represents 0.87% of the total study area and the landslide density is 3.7 landslides/km^2. The main morphometric characteristics and landslide typologies of the municipal landslide inventory are summarized in Table 1.

Landslide Predisposing Factors

A dataset of seven predisposing factors were used to assess landslide susceptibility for each type of landslide referred in Table 1: slope, aspect, plan slope curvature, slope over specific catchment area ratio, geology, soil type and land

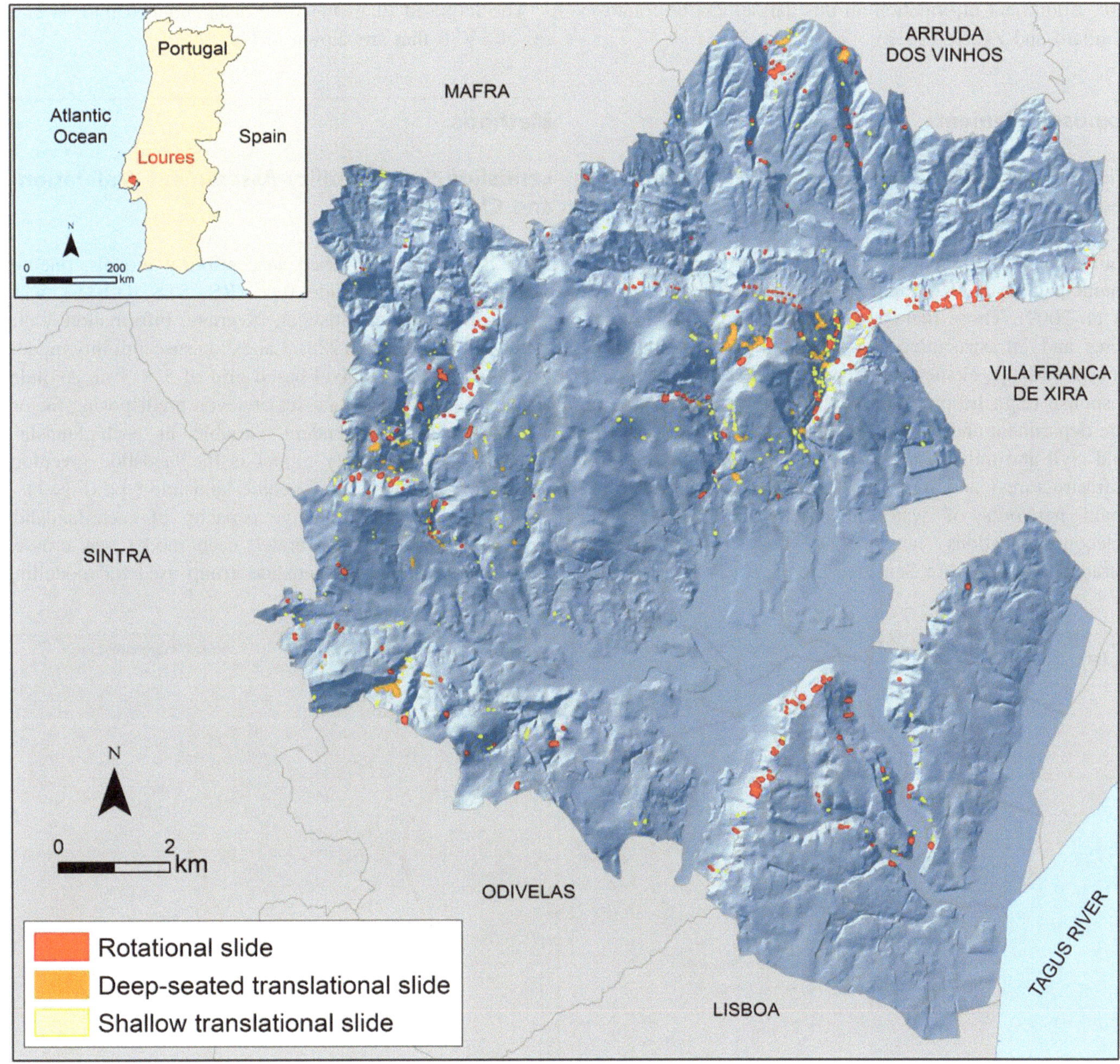

Fig. 1 Landslide inventory of Loures municipality and Loures municipality location (*top-left figure*)

Table 1 Main characteristics of the landslide inventory of the Loures municipality

	Landslide type			
	RS	D-STS	STS	Total
Number	246	62	311	621
Min. area (m^2)	193	237	65	65
Max. area (m^2)	31,245	48,623	7899	48,623
Mean area (m^2)	3278	5761	976	2366
Std. Dev. (m^2)	3749	8109	1094	3897
Total area (ha)	80.6	35.7	30.3	147
Density (#/km^2)	1.5	0.4	1.8	3.7
Affected area (%)	0.48	0.21	0.18	0.87

use. Additional information on this dataset can be found in Guillard and Zêzere (2012).

Exposed Elements

The relevant exposed elements considered for the Loures municipality were identified and classified according to the Portuguese Methodological Guide for the production of municipal risk mapping and for the development of municipal-based Geographic Information Systems (Julião et al. 2009). Those include the urban and industrial structures and infrastructures (e.g. regular buildings, industrial areas) and the elements that are required for emergency response (e.g., hospital and health system; school system; fire department and facilities of other civil protection agents and civil and military authorities; road/railway/air transport infrastructures) and for basic support to population (e.g., main networks of water, electricity and fuel supply; telecommunications networks), which are considered Strategic, Vital and/or Sensitive Elements (SVSE).

The Exposed Elements were disaggregated into 34 classes of SVSE that are shown in Fig. 2a, b.

Methods

Landslide Susceptibility Assessment, Validation and Classification

Three landslide type-based susceptibility models, one for each considered landslide type (RS, STS, D-STS), were produced using the statistical, bivariate, Information Value method (Yin and Yan 1988). Landslide susceptibility models were elaborated at a pixel-based grid of 5 × 5 m. As independent variables the data-set of seven predisposing factors were used. The dependent variable in each landslide type-based susceptibility model is the landslide inventory subset corresponding to a specific landslide type (Fig. 1).

To evaluate the predictive capacity of each landslide type-based susceptibility model, each model was crossed with the corresponding landslide group used for modelling

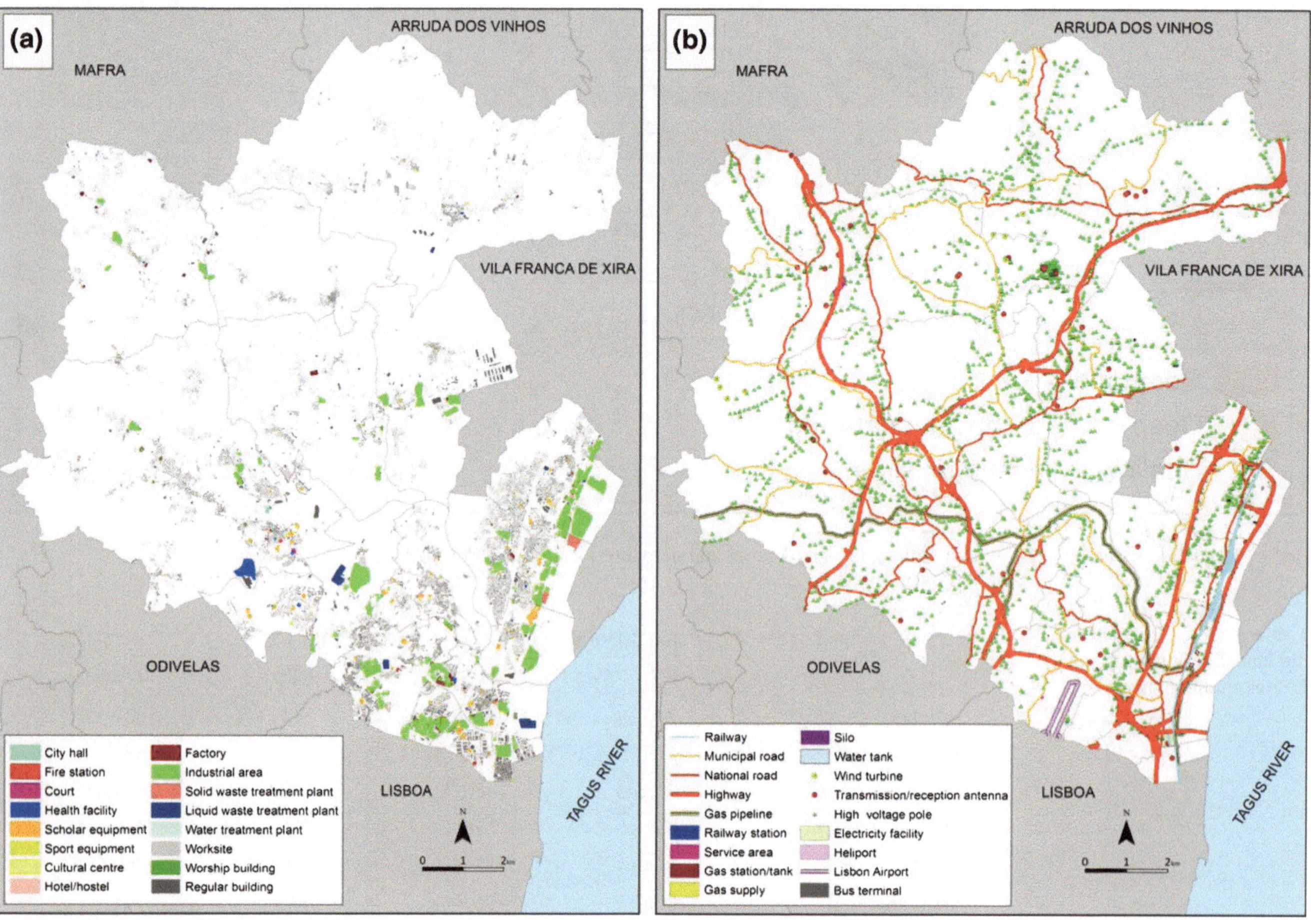

Fig. 2 Exposed elements of Loures municipality considered as Strategic, Vital and/or Sensitive Elements

(model success). Validation results were graphically expressed by success-rate curves and the models global quality was quantified calculating the Area Under the Curve (AUC).

Each landslide type-based susceptibility map was classified in five classes: Very high, High, Moderate, Low and Very low. The upper limit of each class defined by the proportion of study area necessary to accommodate the following cumulative landslide area of the landslide validation group: 50; 70; 90; 95; and 100%.

The final landslide susceptibility map was obtained by combining the three landslide type-based susceptibility maps produced at a first moment individually for each landslide type. The three landslide type-based susceptibility maps were overlapped and the final susceptibility value attributed to each pixel of 25 m^2 within the study area was the highest susceptibility class found, considering the three landslide type-based susceptibility maps in that terrain unit.

Integration of Landslide Susceptibility Maps in Land Use Planning

The criteria for the integration of landslide areas and landslide prone areas in NER are defined for Portugal at the municipal level through a set of strategic guidelines. Accordingly, the following areas should be included in the NER: (i) scarps with slope over 100% and the protection buffers to the scarp upper limit and lower limit that should be at least equal, to the scarp height; (ii) the complete landslide inventory and a surrounding protection area of 10 m for each individual landslide; and (iii) the landslide susceptible area obtained with the Information Value statistical method that ensures the validation (inclusion) of a fraction not less than 70% of total area of the municipal landslide inventory.

Integration of Landslide Susceptibility Maps in Civil Protection Emergency Planning

The assessment of the exposure to landslides is one of the main objectives of the integration of spatial information related to landslide susceptibility to civil protection actions. To evaluate the exposure to landslides the 34 categories of SVSE were overlapped to the final landslide susceptibility map, and the exposed elements that fall into the Very high and High landslide susceptibility classes are considered to be potentially at risk.

Results and Discussion

Landslide Susceptibility Assessment, Validation and Classification

The classified landslide type-based susceptibility maps resulting from the application of the Information Value method to the set of seven independent predisposing factors and to each landslide modelling group (RS, STS and D-STS) are presented in Fig. 3a, b, c, respectively. The final landslide susceptibility map for the Loures municipality resulting from the integration of the above-mentioned landslide susceptibility maps is shown in Fig. 3d.

Table 2 summarizes the area corresponding to each landslide susceptibility class of the three landslide type-based susceptibility maps and of the final landslide susceptibility map. The areas classified as Very high and High susceptible to slide occurrence validate the 70% of landslide occurrences within the municipality and are the ones that should integrate the NER. Considering only landslide type-based susceptibility maps, these classes correspond to 11.2, 10.3, and 16.5% of the Loures territory, respectively. In all models the Low and Very low susceptibility classes are dominant ranging from 66 to 78.6% of the total area.

The final municipal landslide susceptibility map is similar to the landslide type-based susceptibility maps. In fact more than half of the municipality (51.8%) is associated to Low or Very low landslide susceptibility (Table 2), not justifying special concerns, on the assumption that future interventions in the territory will not cause drastic changes in the current topography. The Moderate susceptibility class is larger in area (26% of total area) when compared with the landslide type-based susceptibility maps. Interventions in these areas are possible, but should be avoided both, slope cuts and embankments not supported by an engineering geology project. The High and Very high susceptibility is observed in 22.1% of the territory. Due to its geomorphological and geotechnical characteristics, these areas are not suitable for building structures or infrastructures implementation.

The overall landslide susceptibility models fit, expressed by the corresponding success-rate curves obtained in the validation process are shown in Fig. 4.

The level of adjustment of each landslide type-based susceptibility model to the respective landslide inventory is also observed by the AUC values calculated for each success-rate curve. The landslide susceptibility model that presents the highest adjustment is the one constructed for the

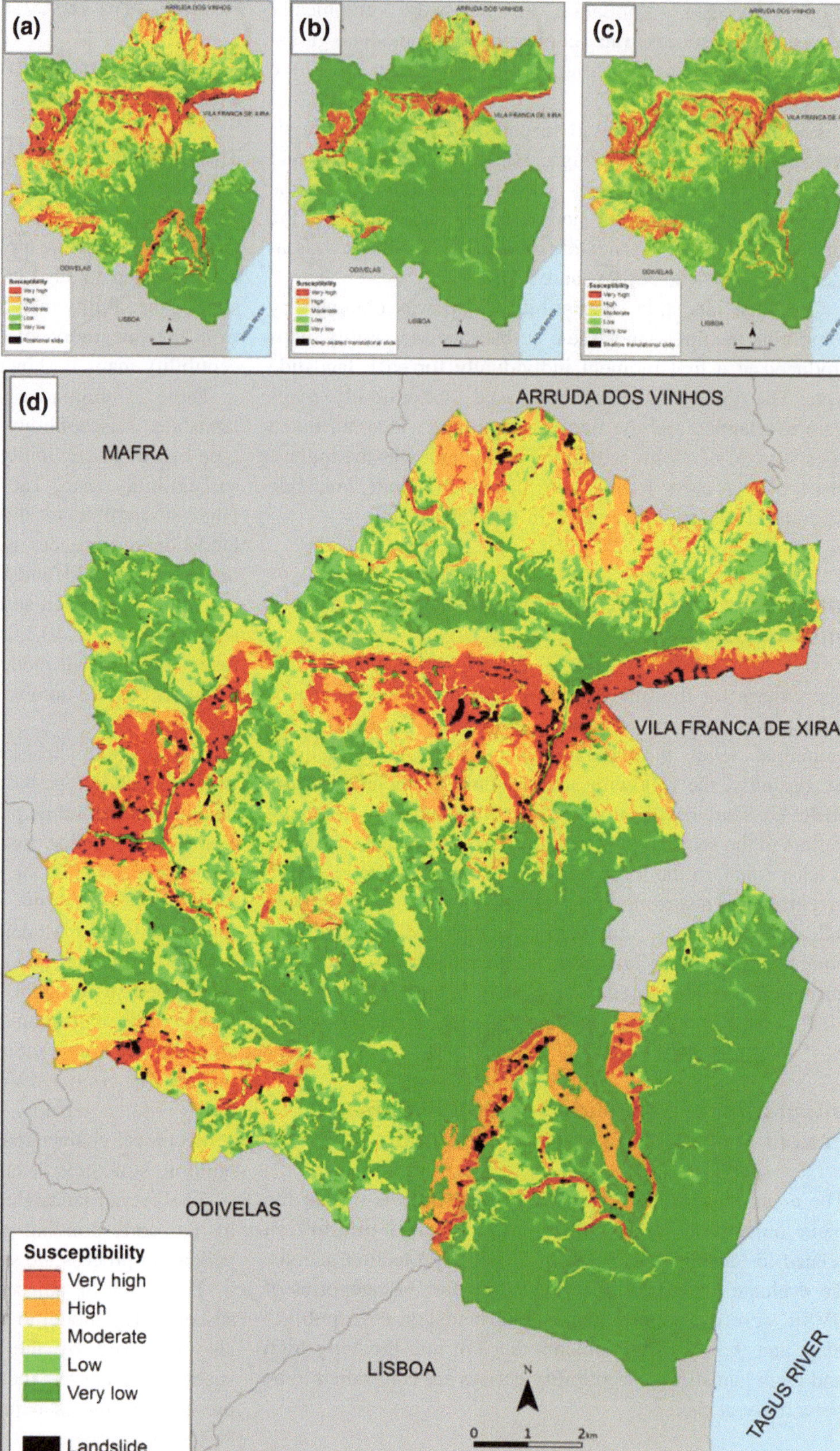

Fig. 3 Landslide susceptibility models of Loures municipality. **a** rotational slides; **b** deep-seated translational slides; **c** shallow translational slides; **d** Final landslide susceptibility map

Table 2 Percentage of Loures municipality area in each class of each landslide susceptibility map

Susceptibility class (% cumulative predicted landslide area*)	Landslide susceptibility maps			
	RS	D-STS	STS	Final
Very high (50%*)	5.1	4.4	7.0	8.0
High (70%*)	6.1	5.7	9.5	14.1
Moderate (90%*)	19.0	11.3	17.5	26.1
Low (95%*)	16.7	10.1	11.9	7.7
Very low (100%*)	53.1	68.5	54.1	44.1

*Not valid for the final landslide susceptibility map

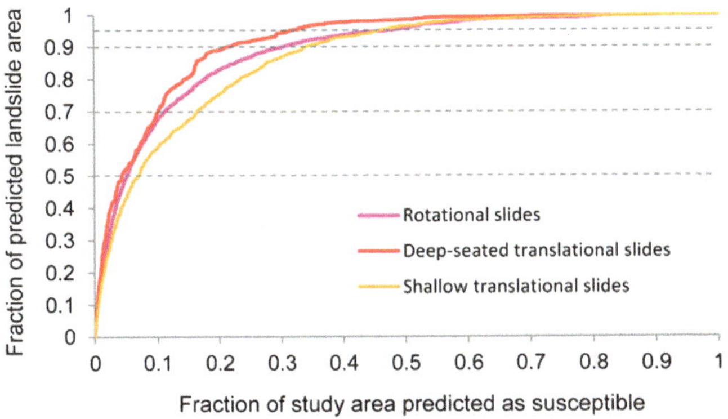

Fig. 4 Success-rate curves produced for each landslide type-based susceptibility model

Deep-seated translational slides (AUC = 0.913). This model is followed by the models constructed for the rotational slides (AUC = 0.889) and for the shallow translational slides (AUC = 0.868).

The performance associated to each curve, for each class threshold of the landslide susceptibility maps can be inferred from Table 2. The deep-seated translational slides model has the best performance, expressed by the lowest percentage of the study area present in the highest susceptibility classes (Very high and High = Validation of 70% of the landslide area constrained to only 10.3% of the study area) In the case of the final landslide susceptibility map, 97% of the total unstable area identified in the Loures municipality are included in the Very High and High susceptibility classes, which cover 22.1% of the study area.

Integration of Landslide Susceptibility Maps in Land Use Planning

The final NER map is shown in Fig. 5. This map includes: (i) the intersected area covered by the two highest landslide susceptibility classes that ensure the validation of 70% of each of the three considered landslide subsets (RS, STS and D-STS) corresponding to 3441.3 ha. This area constitutes the largest contribution from the perspective of landslides to NER; (ii) the landslides of the landslide inventory and an additional protection buffer zone of 10 m to each individual landslide (269.8 ha); and (iii) the scarps with slopes over 100% (45°) (8.9 ha). As all scarps are included within the Very high or High landslide susceptibility classes, it was not necessary to define the protection buffers to the scarp upper limit and lower limit.

There is some overlapping between the three components of the NER and the final area covered by the NER in the study area is 3494 ha (20.8% of the study area).

Integration of Landslide Susceptibility Maps in Civil Protection Emergency Planning

The spatial distribution of the 34 classes of exposed elements considered in the municipality of Loures, superimposed to the final landslide susceptibility map are showed in Figs. 6a, b. The potential risk associated to the exposed elements location results from the intersection of areas classified with Very high or High landslide susceptibility and the exposed elements distribution are shown in Fig. 2a, b. The elements that fall in these classes are more likely to be damaged in a non-specified time span due to the occurrence of landslides that are typically triggered in the study area by rainfall (intense rainfall episodes or long lasting rainfall periods) and eventually by earthquakes (Zêzere et al. 2015; Vaz and Zêzere 2015).

Tables 3 and 4 summarize the elements exposed to landslide hazard in the study area. It should be highlighted that 18 types of exposed elements never occur in High or Very high susceptibility classes, which let them in a more favorable condition with respect to landslide hazard. These types of exposed elements are: city hall; court; fire station; solid waste treatment plant; liquid waste treatment plant; water treatment plant; bus terminal; health facilities; sport equipment; cultural center; railway station; heliport; Lisbon airport; gas supply; factory; worksite; hotel/hostels; and railway.

Among the elements exposed to landslide hazard (i.e., located in areas classified with Very high or High susceptibility to landslide occurrence, it was possible to identify (Tables 3 and 4): 393 high voltage poles; 5 wind turbines; 5

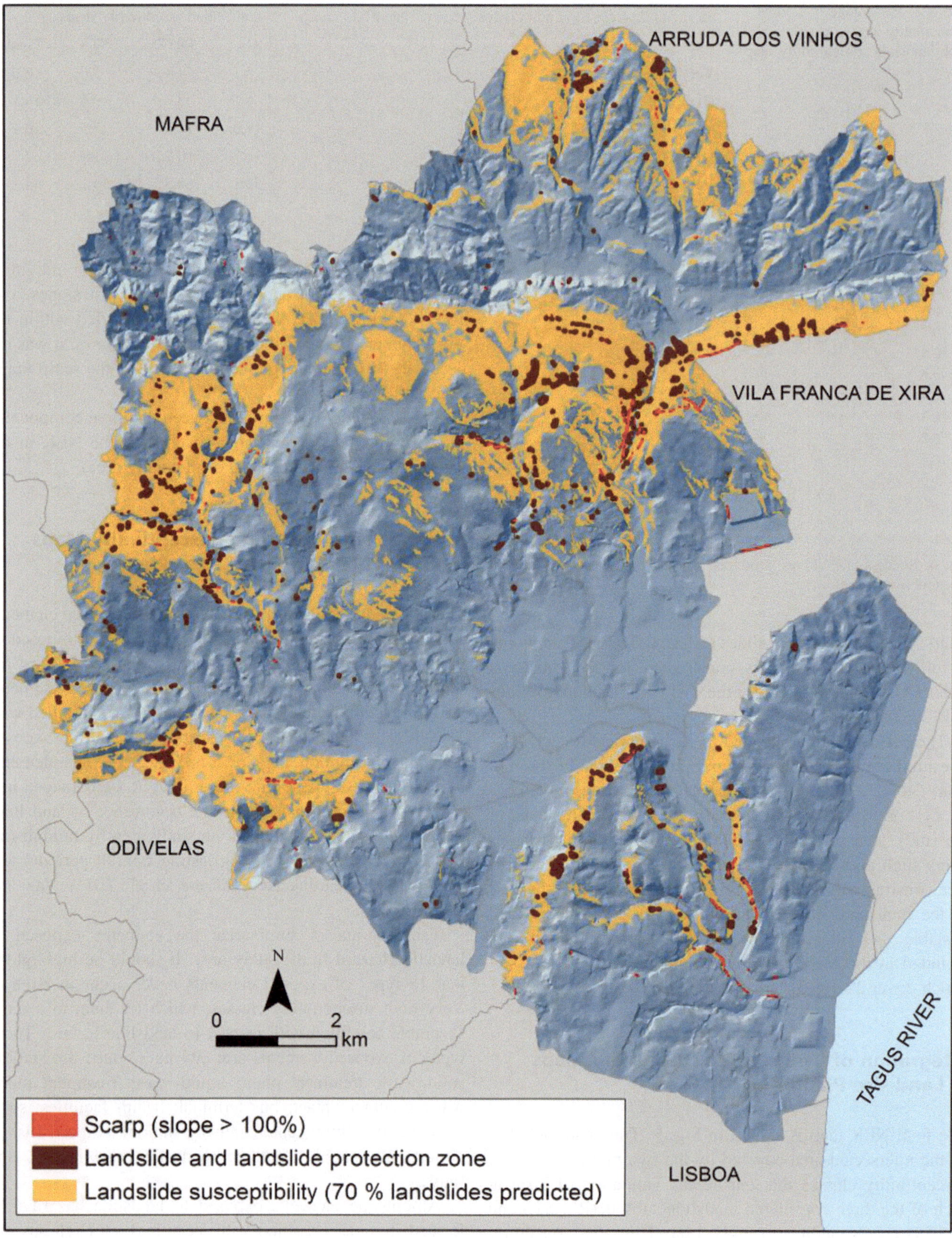

Fig. 5 Distribution of each landslide susceptibility component to the Loures municipal NER definition

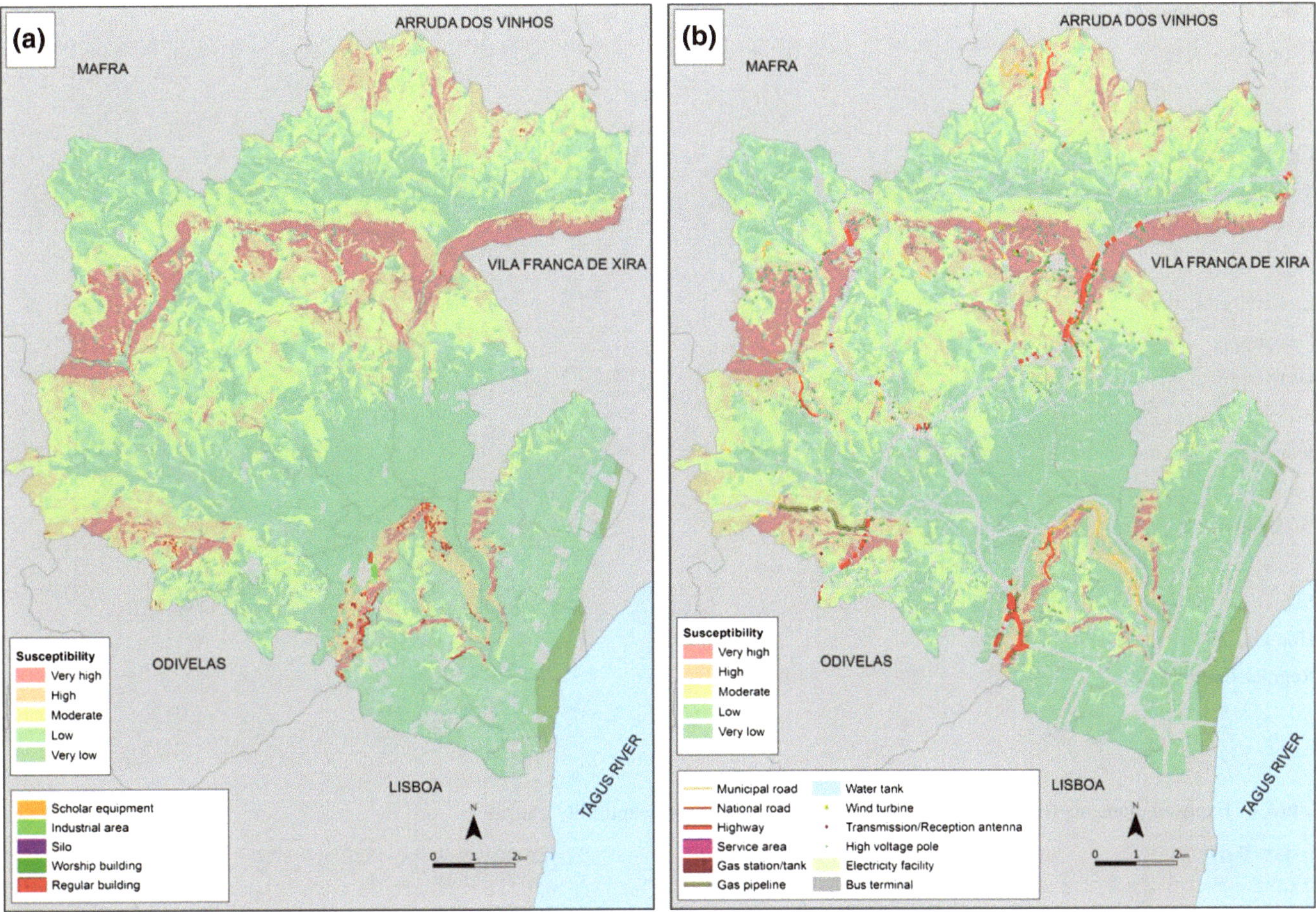

Fig. 6 Exposed elements overlay with the landslide susceptibility in Loures municipality

Table 3 Exposed elements (points and polygons) in Very high and High landslide susceptibility classes

Exposed elements	Total # buildings	Buildings in very high and high susceptibility classes	
		(#)	(%)
City hall	2	0	0
Court	2	0	0
Fire station	9	0	0
Solid waste treatment plant	2	0	0
Liquid waste treatment plant	7	0	0
Water treatment plant	4	0	0
Bus terminal	3	0	0
Health facility	29	0	0
Scholar equipment	244	2	0.8
Sport equipment	66	0	0
Cultural Centre	1	0	0
Worship building	117	3	2.6
Railway station	8	0	0
Heliport	5	0	0

(continued)

Table 3 (continued)

Exposed elements	Total # buildings	Buildings in very high and high susceptibility classes	
		(#)	(%)
Lisbon Airport	1	0	0
Transmission/Reception antenna	59	5	8.5
Wind turbine	8	5	62.5
High voltage pole	2696	393	14.6
Electricity facility	440	28	6.4
Gas supply	2	0	0
Gas station/tank	54	1	1.9
Service area	51	1	2.0
Industrial area	150	8	5.3
Factory	1327	0	0
Water tank	422	39	9.2
Worksite	46	0	0
Silo	51	1	2.0
Hotel/hostel	9	0	0
Regular building	92,813	4689	5.1

Table 4 Exposed elements (linear) in Very high and High landslide susceptibility classes

Exposed elements	Total (km)	Extension in very high and high susceptibility classes	
		(km)	(%)
Highway	88.8	9.7	10.9
National road	89.6	9.7	10.9
Municipal road	69.2	11.8	17.1
Railway	32.9	0	0
Gas pipeline	23.3	3.5	15.0

transmission/reception antennas; 8 industrial areas; 39 water tanks; 1 silo; 1 gas station/tank; 1 service area; 2 buildings of educational institutions; 3 worship buildings; 28 buildings with electricity facilities; 4689 regular buildings; 3.5 km of gas pipeline; 9.7 km of motorways; 9.7 km of national roads; and 11.8 km of municipal roads.

Conclusions

Landslide susceptibility was evaluated considering each landslide typology and further combined in order to produce a final landslide susceptibility map, more conservative, for the entire municipality territory. This final landslide susceptibility map allows evaluate that in Loures municipality, 1347 ha are associated to a Very high landslide susceptibility, and 2372 ha to High landslide susceptibility, which corresponds both to 22.1% of the entire municipal territory. These areas do not present geomorphological and geotechnical suitability for building structures or infrastructures and are integrated into the National Ecological Reserve.

From the civil protection and emergency management point of views the elements at risk located in the Very high or High landslide susceptibility classes were identified and may suffer damage resulting from the occurrence of landslides. Among these: critical power, water, gas distribution and storage facilities, industrial areas, national road networks, industrial areas and 2 buildings of educational institutions.

The final landslide susceptibility map of the Loures municipality is now included in the new version of the Municipal Master Plan. In addition, the map combining the landslide susceptibility with the Exposed elements is being used by the Municipal Civil Protection Service that cross this information with the Population Census data in order to estimate people exposed to landslides in the study area.

Acknowledgements This work was supported by the project FOR-LAND—Hydrogeomorphologic risk in Portugal: driving forces and application for land use planning [PTDC/ATPGEO/1660/2014] funded by the Portuguese Foundation for Science and Technology (FCT). S.C. Oliveira and S. Pereira are Post-Doc fellows of the Portuguese Foundation for Science and Technology (FCT) [grant number SFRH/BPD/85827/2012 and SFRH/BPD/69002/2010], respectively.

References

Cruden DM, Varnes DJ (1996) Landslide types and processes. In Turner AK, Schuster RL (eds) Landslides investigation and mitigation. Transportation Research Board. National Academic Press, Washington DC Special Report 247, pp 36–75

Guillard C, Zêzere JL (2012) Landslide susceptibility assessment and validation in the framework of municipal planning in Portugal: The case of Loures Municipality. Environ Manage 50(4):721–735. doi:10.1007/s00267-012-9921-7

Julião RP, Nery F, Ribeiro JL, Branco MC, Zêzere JL (2009) Guia metodológico para a produção de cartografia municipal de risco e para a criação de sistemas de informação geográfica (SIG) de base municipal. Autoridade Nacional de Protecção Civil, Direcção-Geral do Ordenamento do Território e Desenvolvimento Urbano, Instituto Geográfico Português

Vaz T, Zêzere JL (2015) Landslides and other geomorphologic and hydrologic effects induced by earthquakes in Portugal. Nat Hazards. doi:10.1007/s11069-015-2071-5

Yin K, Yan T (1988) Statistical prediction models for slope instability of metamorphosed rocks. In: Bonnard C (ed) Landslides. Proceedings of the fifth international symposium on landslides, Balkema, Rotterdam, pp 1269–1272

Zêzere JL, Vaz T, Pereira S, Oliveira SC, Marques R, Garcia RAC (2015) Rainfall thresholds for landslide activity in Portugal: a state of the art. Environ Earth Sci 73(6):2917–2936. doi:10.1007/s12665-014-3672-0

Participatory Approach to Natural Hazard Education for Hydrological Risk Reduction

Giovanna Lucia Piangiamore and Gemma Musacchio

Abstract

Modern Society needs interactive public discussion to provide an effective way of focusing on hydrological hazards and their consequences. Embracing a holistic Earth system Science approach, we experiment since 2004 different stimulating educational/communicative model which emotionally involves the participants to raise awareness on the social dimension of the disaster hydrogeological risk reduction, pointing out that human behavior is the crucial factor in the degree of vulnerability and the likelihood of disasters taking place. The implementation of strategies for risk mitigation must include educational aspects, as well as economical and societal ones. Education is the bridge between knowledge and understanding and the key to raise risk perception. Children's involvement might trigger a chain reaction that reinforce and spread the culture of risk. No matter how heavy was the rain that hit our land in the past and recent seasons, we still are not prepared. If on one hand we need to fight against worsening Global Warming that trigger extreme meteorological events, we should also work on sustainable land use and promote landscape preservation. Since science can work on improving knowledge of phenomena, technology can provide modern tool to reduce the impact of disasters, children and adults education is the flywheel to provide the change. We present here two cases selected among the wide range of educational activities that we have tested and to which more than 2,000 students and adults have participated within a period of 12 years. They include learn-by-playing, hands-on, emotional-learning activities, open questions seminars, learning paths, curiosity-driven approaches, special venues and science outreach.

Keywords

Natural hazard • Hydrogeological risk • Prevention • Territory • Participatory approach • Awareness raising • Resilience

Introduction

The public understanding of science has a strong impact on the social debate upon natural hazards, environment, resources and sustainability. However although Eurobarometer 2011 data reveals that 75% of EU citizens are positive about science, since 2005 the share of Europeans experiencing trust in science has declined from 78 to 66%. The

G.L. Piangiamore (✉)
Istituto Nazionale di Geofisica e Vulcanologia,
Sez. Roma 2—Sede di Portovenere, via Pezzino Basso, 2,
19025 Fezzano di Portovenere (SP), Italy
e-mail: giovanna.piangiamore@ingv.it

G. Musacchio
Istituto Nazionale di Geofisica e Vulcanologia, Amministrazione Centrale, via di Vigna Murata 605, 00143 Rome, Italy
e-mail: gemma.musacchio@ingv.it

K. Sassa et al. (eds.), *Advancing Culture of Living with Landslides*,
DOI 10.1007/978-3-319-59469-9_50

largest decline in trust has taken place in Germany, Italy and Poland.

The improvement of science understanding has on one end supported citizens towards independent opinions and participation to crucial decisions; on the other end it seems to build skepticism towards the institutional settings where knowledge generation takes place. Due to the complexity, and yet the uncertainty, of the nowadays process of building up knowledge in science, new discoveries and claims can be contested, leave ample room for different interpretations, and implant suspicion in non-experts.

Good communication is certainly a necessary condition for improving trustworthiness, but engagement of citizens in the process of building, spreading and responsible use of science is thought to have a high potential for success (European Commission 2013).

Public engagement with science strengthens citizenship skills and empowerment; it increases awareness of the cultural relevance of science, and recognition of the importance of multiple perspectives and domains of knowledge to scientific endeavors (Annual Report AAAS 2015).

In this frame a participatory approach to Natural Hazard education has a high value. Here we describe two cases study that focuses on student engagement in hydrogeologic risk reduction. We start from local memory of catastrophic events that occur in the Italian region involved in our program, which is tested in Liguria. La Spezia and Genova Provinces has been repeatedly affected by severely damaging floods throughout its history. The intervals between floods are too often very short and people are afraid when heavy rain strikes.

Land, Hazard and Risk

Mountains and hills correspond to the 77% of Italian land, most of which has steep slopes or clay-based composition (Fig. 1). Recent mapping has shown for the northern part of the country and in Emilia, Liguria and Tuscany regions a high level of hazard for flood and landslides. After World War II Italy underwent heavy urbanization without taking into account the areas with high level of hydrogeologic hazards. Moreover population moved from mountains to cities and abandoned land.

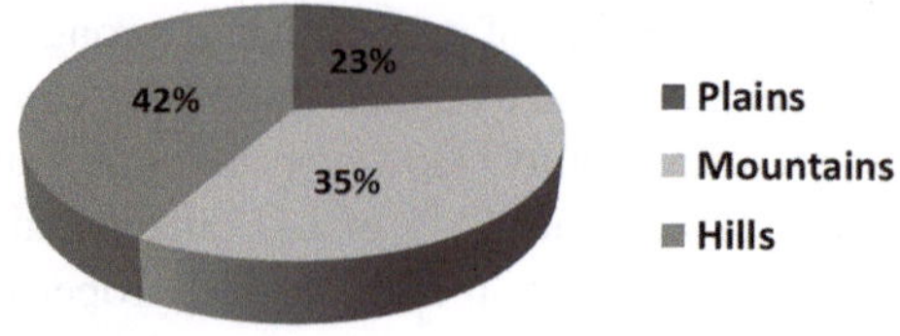

Fig. 1 Land distribution in Italy

Terraces, traditionally sustained by dry stonewalls, occupy about thirty percent of the territory of Liguria. If constantly maintained, they effectively contribute to slow down the natural slope erosion. When no longer managed, terraces may increase geomorphological risk along the slopes and, consequently, at the bottom of the valley (Brandolini et al. 2012).

Bad weather such as flash rain worsen the background hazard causing severe damages and devastation across Italy and weight over the economy of a country that has to faces a large variety of disasters. Italy faces emergency but efforts on prevention are not enough to deal with the problem. Emergency often announced with "state of alert", and results in schools closure to try minimize the number of people on the roads during heavy rain. In addition to damage to infrastructure and housing, bad weather also affects agriculture. According to Coldiretti (Italy's largest agricultural association) heavy rain, overflown rivers and landslides will cost millions of euros to the agricultural sector due to land and crop damage. Farmland overflown with water, vegetables production and plants cultivation located near rivers registered the worst images.

The terraced landscape is very common in the Ligurian region, covering about of the 30% of the territory. Agricultural terraces are sustained by dry stonewalls and reduce the slow down the natural slope erosion, but if no regularly maintained, can worst the risk of landslides and "detritus flow".

Since 1970 recurring floods affect Genoa where the Bisagno Stream catchment flows across the eastern part of city center. The most recent and tragic was in October 2014 (Faccini et al. 2016).

The 5 Terre peculiarity of little towns cling to sheer hills along a narrow, rugged strip of land between the Maritime Alps, the Apennine Mountains and the Ligurian Sea is results into a heavy geo-hydrological risk. The fragile land conditions turn into instability during unusual meteorological conditions that causes heavy flooding. This happened in October 2011 when the river rushed violently down dislocating and destroying several towns, redefining the natural architecture of the affected the Vara and Magra Valleys near La Spezia (Piangiamore et al. 2015) (Figs. 2 and 3).

Materials and Methods

Floods and landslides may strongly affect lives of many people. Emotional-learning activities can activate a *life long learning* process developing skills that might end up being fundamental for the own safety (Piangiamore et al. 2012).

To engage students into an active learning path and a flipped-up learning strategy we have to listen them and exchange ideas and experiences.

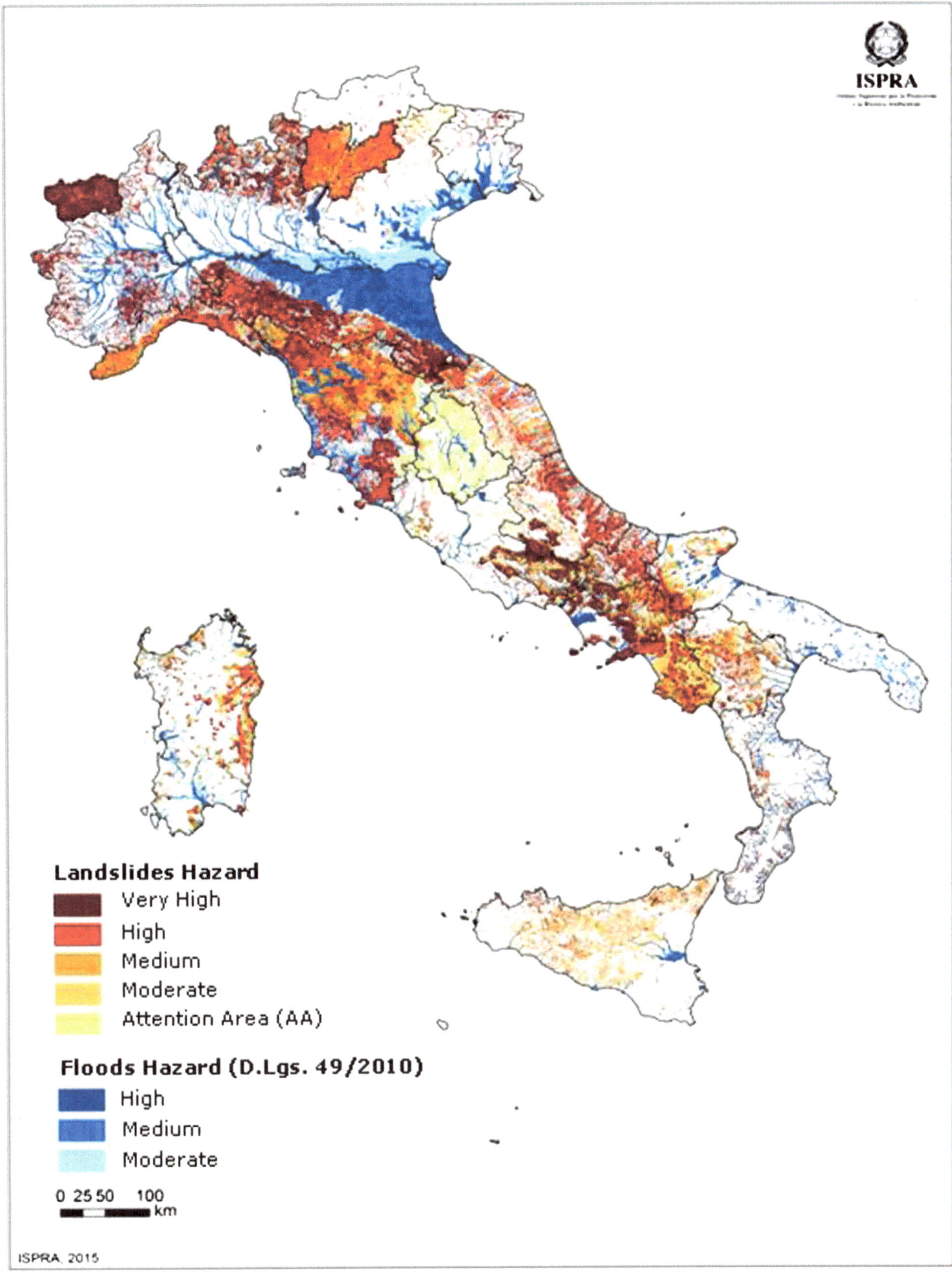

Fig. 2 Landslides and floods hazard map (Trigila et al. 2015)

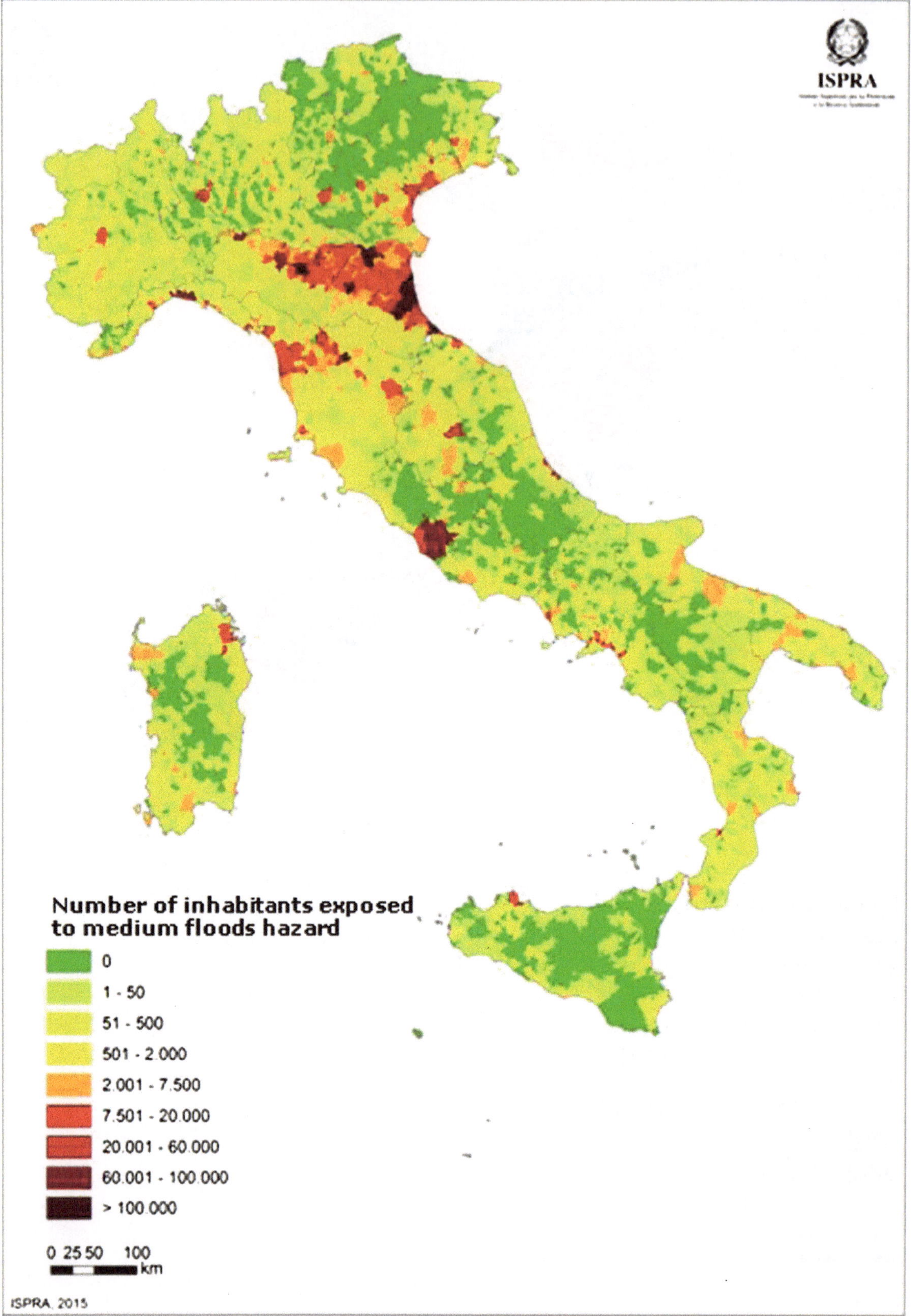

Fig. 3 Inhabitants exposed to medium floods hazard (D.Lgs. 49/2010)

More than lack of knowledge is a question of understanding their needs and interests.

In approaching risk communication and learning, we test different ways and methods from 2004 to nowadays. Here we describe two cases study we develop in the last four years in which students and experts have collaborated in preparing (1) a participated exhibition on the 2011 flood that severely hit the land of La Spezia and surrounding, and (2) a flipped-up learning tool.

The participated exhibition is *Piovono Idee! (Cloudy with a Chance of ideas!)* and it is the result of experts' interaction directly in school, focus groups and hands on activities. Students involved had experience the disaster and could put their emotion into their products. Exhibits of the interactive path tell us their own history.

Primary and secondary schools in areas strongly affected by the flood of the 25th October 2011—Borghetto Vara, Brugnato, Monterosso, Vernazza, Riomaggiore, Pignone, Riccò del Golfo, Aulla and Ragnaia (Aulla)—and secondary schools from the city of La Spezia were directly involved in the participatory action. Planning, creation and exhibition were the three main phases of the action to improve their specific scientific knowledge with hands-on activities, while exploring feelings and emotions triggered by the experience of a flood. The learn-by playing approach is also used to instil appropriate behaviours and the awareness that every actions we take derives from a choice: we can chose whether or not be respectful towards nature. Even little actions can turn into negative or positive consequences as it is sudden evident in *Piovono idee!*. Indeed the learning path starts with the question: '*Is nature scary?*' shows a colored and smiling Nature full of gifts, which can became angry simply poking on it. The discovery of hydrogeological risk continues focusing on the importance to respect territory and on the correct land use with different totems, exhibits, experiments, scientific games. It emotionally involves the participants to favor reflective learning (Piangiamore et al. 2014, 2015). The latest activities are two role games to enforce eco-friendly behaviors for hydrological risk mitigation.

A few years later, *Piovono idee*! was updated and the exhibition got fully portable since 2015 and ready to travel the schools of the country to involve as many citizens as possible. The second edition of *Piovono idee*! (Fig. 4) is also an example of peer-education activity as students of the High School of La Spezia and of the Secondary school of Lerici, that were not directly and emotionally involved in the disaster, have played the role of the guides for general public (Bernhardsdóttir et al. 2012; Piangiamore et al. 2016).

The flipped-up learning tool is the EAS *Floods: what to do?, an* application of the Episodes of Situated Learning (EAS) method to natural hazard education (Rivoltella 2014; Piangiamore et al. 2015). The EAS method acts on problem solving abilities that the classroom shared among peer and with the mentor and ends with a reflective learning approach where concepts are reworked and restructured. We applied this interactive approach to natural hazards at school to stimulate best practice for the good of all. Researchers and teachers worked side by side designed *Floods: what to do*? on hydrogeological hazard active learning to promote knowledge and safety.

This is a way for education of the new generations intriguing students to arouse their interest towards environmental problems, generally underestimated by formal education (Muttarak and Lutz 2014). *Floods: what to do*? asks students to prepare a tool capable of effective communication to their peers and to the public, involving them in three phase: (1) the *preparation phase* based on problem-solving; (2) the *activity phase* based on learning by doing; and (3) *the debriefing phase* based on reflective learning.

In (1) the teacher gives inputs to make them curious on hydrogeological matter using Internet and web browsing. Students watch short-shoots of videos, presentations or a text inputs surfing on selected websites at home prior lesson.

In (2) the teacher's conceptual framework on hydrogeological, geomorphological, water drainage risks and climate change. Then students do a cooperative work gathered in small groups by means of specific application to build comics (e.g., Comic Life, Comic Maker, Bistrips, Pixton Comics) freely available on the Internet. The results are comics to communicate the safe behaviors and the basics of self-protection in case of natural extraordinary events.

In (3) the teacher makes assessments, discusses misconception and define concepts. Students analyze their classmates' products, discusses with them and reflects on crucial aspects of behaviors that can save lives discovering together the territory they live in.

Students are familiar with comics, their products are set in a familiar environment such as at home, at school, in a narrow road, in an open space, next to a river bank, in a woods, in the fields, etc. This is an amusing way to raise awareness on disaster reduction and risk education enforcing the ability to foster hazards before the occur of extreme events.

Results

The two case studies were useful to document educational tools and engagement initiatives to ensure effective dissemination and communication and raise public awareness and understanding in countries exposed to hydro and geomorphological hazard. Global promotion of understanding and reducing hydrogeological disaster risk is one of the ambitious mandate for the next 15 years at the 3rd United Nations World Conference in Japan. The principle of 'risk-informed decision-making' involve application of risk

Fig. 4 The second edition of *Piovono idee*! with peer-education activity. Students of the high school and of the secondary school are the guides of the learning path. Hands-on activities are completely interactive and made of simple materials (mainly cardboard) to improve scientific knowledge while exploring emotions triggered by the experience of a flood

information at all levels (UNISDR 2005, 2015). We promote responsibility for enhancing international science and technology cooperation by engaging students and teachers on disaster risk reduction attitude for a resilient community.

Resilience is the capacity to withstand and recover quickly from extreme dissemination strategies to instill a culture of safety will exploit some new aspects for the promotion of best practice.

Italy bad weather emergency reveals the country's lack of readiness for hydrogeological risk prevention plans. Flood and landslide risk education can contribute to advancing culture of living with natural hazards and to reduce landslide and flood disaster risk. The necessity to dialogue about this theme has to be a priority for the risk governance and management to establish an effective hydrogeological reduction. The experiences here designated has reached students and adults with excellent feedbacks, supported by questionnaires and press releases. These example of good cooperation incorporating many relevant stakeholders into a risk dialogue, enforcing the common aim to build a sustainable system of prevention for prepared future citizens able to respect natural environment and Nature.

Acknowledgements About *Piovono idee!* we are grateful to D. Modenesi and F. Brasini (ConUnGioco *Onlus)*, R. Camassi (INGV Bologna), M. Miconi (INGV Roma), M. Bocchia, G. Mancini, C. Cassinoni, S. Lualdi, M. Lombardo and G. Savoldi, M. Biso (La Spezia Provincial Civil Protection Service); M. Casarino, P. Milano and O. Zocco (La Spezia Provincial Europe Service, Coordination of Municipalities), G. Forlani, M. S. Ariodante, A. M. Bimbi (La Spezia Prefect's Office's Civil Protection Department), MARIDIPART, MARICOMMI, the voluntary Civil Protection associations and the GEV of the Province of La Spezia and many thanks to all the Province of La Spezia primary and lower secondary schools contributing to our initiative (Borghetto Vara, Brugnato, Monterosso, Vernazza, Riomaggiore, Pignone, Riccò del Golfo, Aulla e Ragnaia (Aulla) and "V. Alfieri", "U. Mazzini" and "J. Piaget" of La Spezia. About *Piovono idee! second edition* we acknowledge D. Modenesi and F. Brasini (ConUnGioco *Onlus*), R. Camassi (INGV Bologna), S. Merlino

(CNR-ISMAR of La Spezia), the secondary schools of "ISA 10" (Lerici) and the High School "Pacinotti" (La Spezia). About *EAS* (Episodes of Situated Learning) for natural risk reduction at school we would like to thank Monica De Vecchi (Comprehensive Institute "Voltri 1" of Genova) and Alessandra Carenzio (Professor at the University "Cattolica del Sacro Cuore" of Milan). *Piovono idee*! was funded by Province of Spezia within the framework of the 'Laboratories of shared and participatory citizenship- IV edition' project promoted by the Directorate-General for the Third Sector and Social Formations- Italian Ministry of Labour. *Piovono idee! second edition* was funded by MIUR (Italian Ministry of Education, University and Research) inside the Dissemination of Scientific and Technological Culture call for the year 2014 within the framework on '*MATER – Pianeta Terra-Mare (*Planet Earth and Sea)'.

References

Bernhardsdóttir AE, Thorvaldsdóttir S, Sigbjörnsso R, Musacchio G, Nave R, Falsaperla S, D'Adda S, Sansivero F, Zonno G, Sousa ML, Carvalho A, Raposo S, Ferreira MA, Nunes JC, Jimenez MJ (2012) Disaster prevention strategies based on an education information system. In: Proceedings of the 15th WCEE, 2012. Lisbona, Portugal. http://hdl.handle.net/2122/8148

Brandolini P, Cevasco A, Firpo M, Robbiano A, Sacchini A (2012) Geo-hydrological risk management for civil protection purposes in the urban area of Genoa (Liguria, NW Italy). Nat Hazards Earth Syst Sci 12:943–959. www.nathazards-earth-syst-sci.net/12/943/2012/. doi:10.5194/nhess-12-943-2012

Commission European (2013) Science for an informed, sustainable and inclusive knowledge society, Policy paper by President Barroso's Science and Technology Advisory Council, 29 August 2013. Brussels, Belgium

Faccini F, Giostrella P, Paliaga G, Piana P, Sacchini A (2016) The role of historical agricultural terraces in geo-hydrological risk reduction: a case study from the Bisagno Stream Catchment (Genoa, Italy). Geophysical Research Abstracts, vol 18. EGU2016. Vienna, Austria, p 7441

Muttarak R, Lutz W (2014) Is education a key to reducing vulnerability to natural disasters and hence unavoidable climate change? Ecol Soc 19(1):42. doi:10.5751/ES-06476-190142

Piangiamore GL, Fanelli E, Furia S, Garau D, Merlino S, Musacchio G, Centineo MC (2016) MATER—Pianeta Terra-Mare: an interactive and multidisciplinary approach to Geosphere sciences. Geophysical Research Abstracts, vol 18. EGU 2016. Vienna, Austria, pp 10298–10303

Piangiamore GL, Musacchio G, Devecchi M (2015) Situated learning episodes: natural hazards active learning in a smart school. In: Interactive learning: strategies, technologies and effectiveness. NOVA, New York, USA. https://www.novapublishers.com/catalog/productinfo.php?products_id=56608

Piangiamore GL, Musacchio G, Pino NA (2015) Natural hazards revealed to children: the other side of prevention. In: Peppoloni S, Di Capua G (eds) Geoethics. The Role and Responsibility of Geoscientists, Geological Society Special Publications, London, p 419. http://dx.doi.org/ 10.1144/SP419.12

Piangiamore GL, Musacchio G, Bocchia M (2014) Piovono idee! (cloudy with a chance of ideas!): an interactive learning experience on hydrogeological risk and climate change. In: Lollino G, Arattano M, Giardino M, Oliveira R, Peppoloni S (eds) Engineering geology for society and territory, vol 7. Education, Professional Ethics and Public Recognition of Engineering Geology, Turin, Italy, pp 121–124

Piangiamore GL, Musacchio G, Bocchia M (2013). Challenging risk reduction through education and preparedness. In: Proceedings of the 32th GNGTS, 2013, Trieste, Italy, vol 2, pp 446–452

Piangiamore GL, Pezzani A, Bocchia M (2012) ERiNat project (training on natural risks): from informed children to knowledgeable adults. In: Proceedings of the 7th EUREGEO—European congress on regional geoscientific cartography and information systems, 2012, vol 1, Bologna, Italy, pp 321–322

Rivoltella PC (2014) Episodes of situated learning. A new way to teaching and learning. In: Research on education and media, VI, N. 2, pp 79–87. http://ojs.pensamultimedia.it/index.php/rem_en/article/view/1070

The American Association for the Advancement of Science (AAAS) (2015) Annual report. Innovation, information and imaging. http://annualreport.aaas.org/?utm_source=aaasorg&utm_medium=aboutpg&utm_campaign=2015AnnualReport

Trigila A, Iadanza C, Bussettini M, Lastoria B, Barbano A (2015) Dissesto idrogeologico in Italia: pericolosità e indicatori di rischio. Annual report. ISPRA, 233/2015

UNISDR (2005) Hyogo framework for action 2005–2015: building the resilience of nations and communities to disasters, international strategy for disaster reduction. http://www.unisdr.org/2005/wcdr/intergover/official-doc/L-docs/Hyogo-framework-foractionenglish.pdf

UNISDR (2015) Sendai framework for disaster risk reduction 2015–2030. http://www.preventionweb.net/files/43291_sendaiframework fordrren.pdf

More Room for Landslides

Klaudija Sapač, Nina Humar, Mitja Brilly, and Andrej Kryžanowski

Abstract

Since ancient times, and more intensively from the mid-19th century, land in the mountain region is developing and the space belonging to land and water has been reduced. On the surface that previously belonged to the river have developed agriculture, transport routes and settlements. At the end of the 20th century, development spread over hazardous areas, many torrent flowed in highly confined channels, and ground water recharges drop down, instability of land surface and security for inhabitants decreased. This resulted in the changes reducing water resources of appropriate quality, reducing space for sediment deposit, increasing erosion, affecting natural habitats, causing major flood damage, decreasing groundwater stock, and deteriorating water quality. The water regime integrates all events across space from landslides, debris flow and is manifested in river regime in low lands. This problem is partially covered by many United Nations (UN) and UNESCO documents and reports. Proper actions are also suggested in the Ministerial Declaration from the 7th World Water Forum, where the first mentioned action is the significance of appropriate land management in relation to sustainable water management and planning. More room for landslides control means more space for potential landslide control out man made impacts that cause land slope instability, more space for torrents, more space for water and sediment storages, less impact on the slope stability and higher security for the peoples. We should change paradigm of space planning and development, especially in countries in development under intensive urbanization. The aim of this paper is to present particularly bad examples from Slovenia in order to support this proposal.

Keywords

Landslide • Flood • Spatial planning • Damage • Hazard • Vulnerability • Risk

K. Sapač · M. Brilly (✉) · A. Kryžanowski
Faculty of Civil and Geodetic Engineering, University of Ljubljana, Jamova Cesta 2, Ljubljana, Slovenia
e-mail: mitja.brilly@fgg.uni-lj.si

K. Sapač
e-mail: klaudija.sapac@fgg.uni-lj.si

A. Kryžanowski
e-mail: andrej.kryzanowski@fgg.uni-lj.si

N. Humar
Institute for Water of the Republic of Slovenia, Dunajska Cesta 156, Ljubljana, Slovenia
e-mail: humar.nina@gmail.com

Introduction

The campaign *More Room for Landslides* is designed on the basis of successfully implemented campaign *More Room for Water*. The Campaign More Room for Water was started by the Slovenian Committee of UNESCO IHP in the year 2013. The idea of this action is based on an important European project from the Netherlands *Room for the River* (Klijn et al. 2013), of which the main objective is to give back to the river at least some of the space that it once possessed. With the growing population, industrialization and urbanization,

K. Sassa et al. (eds.), *Advancing Culture of Living with Landslides*,
DOI 10.1007/978-3-319-59469-9_51

the inundated areas and wetlands have been consumed and, through river engineering, watercourses have been regulated such that the space belonging to water has been reduced (Brilly 2015).

Since ancient times, and more intensively from the mid-19th century, the development of human settlements, infrastructure and economic activities is expanding more and more in the mountain region and the space belonging to land and water has been reduced. On the surface that previously belonged to the river have developed agriculture, transport routes and settlements. At the end of the 20th century, development spread over hazardous areas, many torrent flowed in highly confined channels, and ground water recharges dropped down, instability of land surface and security for inhabitants decreased. This resulted in reduction of water resources of appropriate quality, reduction of space for sediment deposit, increase of erosion, impacts on natural habitats, causing major flood damage, decrease of groundwater stock, and deterioration of water quality. The water regime integrates all events across space from landslides, debris flow and is manifested in river regime in low lands (Brilly 2015).

This problem is partially covered by many United Nations (UN) and UNESCO documents and reports. Proper actions are also suggested in the Ministerial Declaration from the 7th World Water Forum (World Water Council 2015), where the first mentioned action is the significance of appropriate land management in relation to sustainable water management and planning.

More Room for Landslides actually means more space for control of potential landslides caused by human activities that cause slope instability; more space for torrents; more space for water and sediment storages and less impact on the slope stability; consequentially higher security for the people.

The occurrence of landslides in Slovenia can be expected at approx. 1/3 of the total surface area (Ribičič et al. 2005). However, more and more intense and concentrated use of space is increasing risk of landslide occurrence. Furthermore, there are also spatial interventions at unstable areas by which conditionally stable areas change into unstable. Landslides in Slovenia most often occur as consequence of heavy rainfall, which means that landslides are inseparably linked to water.

In our paper we present three cases of landslides that have occurred in the past on the territory of the Republic of Slovenia, and discuss how could measures in the field of spatial planning reduce risk and consequential damage in these cases. We have focused on cases where analysis has shown that due to improperly designed or maintained forest roads and skid trails, the damage was greater: Stože (2000), Laze Dolsko and Slapnica (2009), Upper Savinja Valley (1990).

Network of roads in the forest land have multiple functions: easy access to forest resources for extraction, regeneration, protection, and recreation activities (Akay et al. 2008; Khalipoor et al. 2008). Skid trails, highlighted in this paper are used in ground skidding systems and are recognized as the source of erosion (Jusoff and Majid 1996), in many cases, they also serve as a secondary channel for the transport of the material from the hinterland directly to the settlements.

The construction of forest roads is an economic activity which includes the design, construction and maintenance of engineering structures. The Construction Act (2002) gives legislative framework in this field in the Republic of Slovenia. 5064 km of forest roads were recorded in the social and private forests in Slovenia in the year 1970 (Remic 1971). After this year the length of the network of forest roads started to increase rapidly. The extent of the construction of forest roads per year is shown in Fig. 1 (Robek et al. 2006).

The most intensive period of construction was between years 1982 and 1984, when more than 300 km of paved roads were built per year. In the year 2004 the length of all recorded forest roads (paved roads, unpaved roads, skid trails) in Slovenia was 12,683 km (Slovenia forest service 2004). This means that the density of the forest roads on the entire surface of the Slovenian is approx. 0.63 km/km^2 (Fig. 2).

On average 1 m of newly built forest road requires an excavation of 3 m^3 of soil, 1 m of newly built forest trail requires excavation of 1 m^3 and 1 m of reconstruction requires 0.5 m^3 excavation of soil. Based on these data, it was estimated that half a million m^3 of soil was excavated from hillslopes in the Republic of Slovenia between 2000 and 2005 (Robek et al. 2006). It is true that this is still four times less than in the period of most intensive construction, but for every 3 m^3 of harvested timber at least 1 m^3 of forest soil is transported. Most of these excavations is on the skid trails, which are more than roads exposed to erosion processes (Robek et al. 2006).

Overview of Landslide Cases in Slovenia

In the next three cases we are highlighting the issue of the impact of inadequate interference into watercourse and scarce maintenance of forest roads on occurrence of landslides and extent of the damage caused by them.

One of the most notable characteristic of the torrential flows is that watercourses have most of the time extremely low flow rates or are even dried up, but the rates increase sharply almost immediately after heavy rains and increase greatly torrent's transport capacity. Since the channels are not able to conduct such large quantities of water often

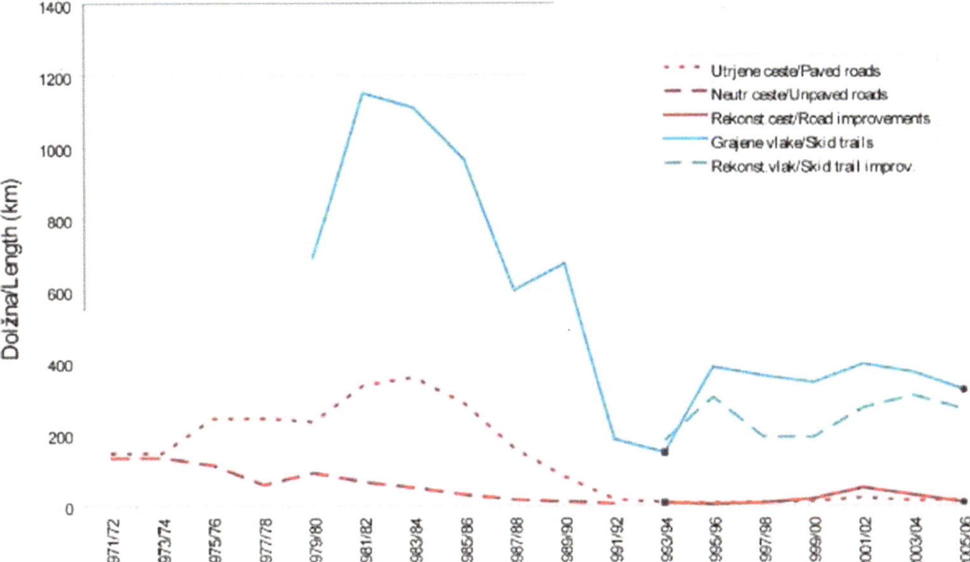

Fig. 1 Average annual extent of constructions and reconstructions in forestry in Slovenia after 1970 (*black square* (■) and *black circle* (●) are missing data)

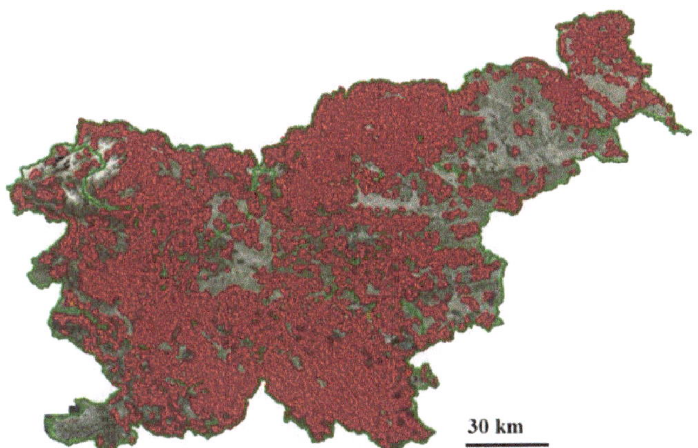

Fig. 2 Density of the forest roads in the Republic of Slovenia (EGC 2016)

mixed also with substantial quantities of gravel or mud the water finds the way and spills over on the nearby terrain.

Very often there are roads near the watercourses and the water uses the road as a secondary channel that offers the easiest way down to the valley. In Slovenia it has been observed that in many cases especially if they were not planned adequately (without adequate drainage) and in the absence of maintenance, the forest roads form a preferential route and redirect the torrential flow in unfavorable directions and thus can cause disastrous consequences.

Laze, Dolsko and Slapnica, 2009

One of the events of this kind happened on 10th July 2009 in municipality of Dol pri Ljubljani. In the summer of 2009 the area of the Middle Sava River suffered many rain storms and subsequently large amounts of rainfall. Heavy rain transformed the normally small creeks into powerful torrents, giving them a devastating power and great transport abilities. This, combined with weathering and rot prone flysch basis resulted in enormous quantity of debris that got transported by the torrent from hinterland to the flatter part of the gorges. The debris filled the culverts in the lower part of the stream course, but with the increase of the flow the water started to exit the channel already before it reached the critical part. With no way to go in the original riverbed the torrent flooded the embankments on the transition to the flatter part, where the valleys also slightly expand. The streams took the easier path and started to flow on the nearby forest roads using them as a new water way. Carrying large quantities of debris and mud the water found its way between the houses, depositing the material in flatter parts or near the obstacles (houses and other objects/buildings) and finally on the road and railroad before flowing into the Sava River. The water mixed with mud and debris caused enormous damage on 8 houses and on the outbuildings as well as on main railroad Ljubljana—Zidani most which has been closed for several days (Kambič 2009) (Fig. 3).

Laze pri Dolskem is a settlement located on the right bank of the Sava River, which occupies the greater part of the relatively narrow band between the river and the footstep of the hills of Lipavčev grič and Janče, which steeply rise from Sava River plains. Between village and Sava river runs one of the most important railway connections—the railway line Ljubljana—Zidani most as well as the local road connection between Laze and Jevnica.

The area is composed predominantly by unstable flysch basis, which are extremely sensitive to athmospheric influences (Kambič 2009). The water catchment area is mostly

Fig. 3 Consequences of the storm in the Dol pri Ljubljani, 2009 (Zurnal24 2009)

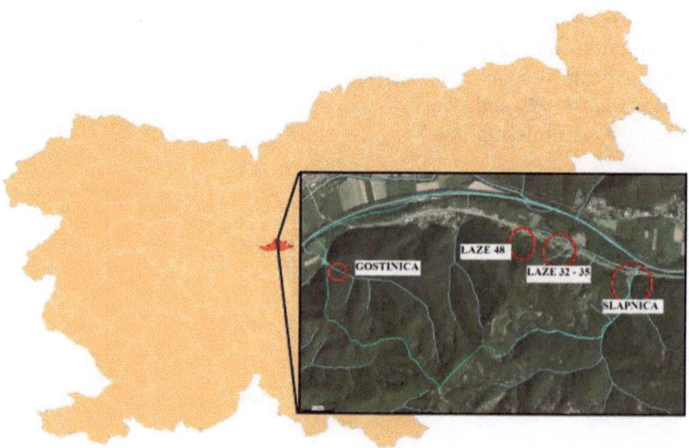

Fig. 4 The most affected areas in Laze pri Dolskem (Atlas okolja 2016)

covered with forest, which is in terms of water retention favorable; however inappropriate interference in such space (inadequate felling, unmaintained forest roads and skid trails) caused devastating damage on a very small area between Gostinica and Jevnica. Due to frequent rainfall during whole summer the soil was relatively soggy and it was not able to absorb large quantities of water anymore. Heavy rain on the night between 9th and 10th July washed large quantities of debris from the steep slopes of Lipavčev grič into watercourses, sediment the material in flatter or less transitional zones of the watercourses and changed the direction of the riverbed. Due to the already very wet ground precipitation has triggered some landslides, which overwhelmed local road, three residential houses and a railway line (Fig. 1).

Less than a month later (on 4th August 2009), abundant rainfall once again covered the area Laze, Dolsko and Jevnica, and additionally soaked hinterland creeks, causing again erosion and triggering slides of material in the headwaters (Kambič 2009). Subsequent analyses of the event showed that, due to large quantities of the debris many torrents in the area burst their banks and the water has paved the way for unmaintained forest roads and skid trails, shifting their way from the regular stream or redirecting water and material into torrent flowing nearby.

The area near houses Laze 32, Laze 33, Laze 34 and Laze 35 was the most affected (Fig. 4). The damage was caused by two parallel streams (one of them almost inactive before) flowing less than 150 m apart. Large damage was caused also by the torrent Slapnica, and the torrent near house Laze 48 and the one near the fire station. Another critical area was the area affected by Gostinica stream.

One of the affected areas was also the area next to the house at Laze 48. The stream with catchment area of 0.06 km^2 runs normally as a creek, but precipitation before 10th July 2009 filled the creek abundantly and transformed into a powerful torrent. A riverbed was considerably deepened and the water carried a large quantity of debris, depositing it on the backyard and garden of the house Laze 48. Since the torrent cuts into two the land plot and divide the house from the garden and the forest and the access to the building is possible only via the channel of the stream the owner lined the fairly under-dimensioned pipes into the torrent to have an access to the house and further on to the forest. During the event the pipes got clogged immediately and the water took its way through the access road.

The worst damage was caused by stream passing the houses Laze 32, Laze 33, and Laze 34 (Fig. 4). The stream catchment area is covered mostly with tame forest and extends on 2.1 km^2. A forest road runs along the stream, which crosses the watercourse two times and has been used to access forest in the hinterland. In the last period before the storm the road has been used rarely and the lack of maintenance and cleaning of the forest could have been observed.

Less than 1 km to the south-east there is another stream that diverted from the original riverbed and found its way through the forest road causing a devastating damage. The Slapnica stream is overgrown with forest, however the headwater is crossed with forest access roads (Fig. 5). The forest as well as the forest roads are mostly unmaintained and in bad conditions (Fig. 6).

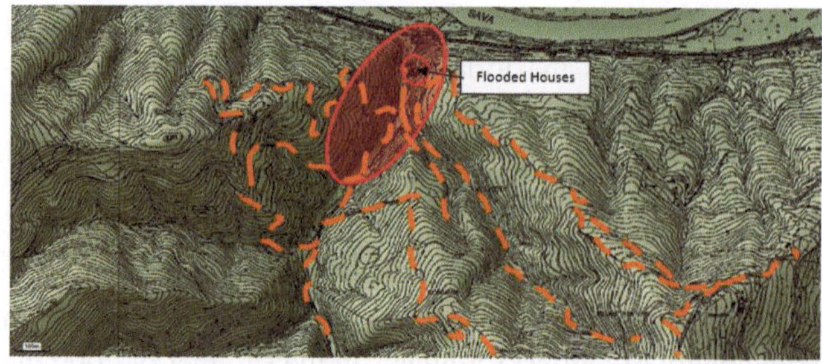

Fig. 5 The devastated area of Slapnica. *Red colour* marking the devastated area and *orange lines* the forest roads

Fig. 6 Debris deposited outside the narrow "canyon" near Laze 35. The water used the access forest road (*red arrow*) as its riverbed (Kambič 2009)

Fig. 7 The consequences of flooding it's clearly visible; the torrent diverted from original channel and took the forest road as its riverbed (Kambič 2009)

Apart from the debris the water brought down to the flatter land branches and pieces of wood. On the watercourse there is a debris retention dam with a fairly big accumulation. However, the transported material exceeded the volume of the storage available and the water mixed with the debris started to flow over the access roads and latter over the meadow and field and flooded the houses and deposited the transported material (Fig. 7).

Upper Savinja Valley, 1990

The case of Savinja valley highlights even better the influence of inadequately planned service roads. The 1990 floods were the most severe in the Savinja valley in the previous century. They were caused by excessive rainfall; in the period of 26–31 October 100 mm of precipitation fell, causing two small flood waves. The October precipitation was followed by 220 mm of precipitation on 1–2 November, which was catastrophic (Pristav 1991). Three major landslides and many small landslides were triggered. The many landslides surprised the services in charge of emergency procedures during the event. The phenomenon was also affected by the construction of new forest roads, because in the 1970 there were only 16.55 km of roads while in 1990 there were already 79.7 km of roads in the area of Podvolovljek. Most of the forest roads were built on slopes, and as much as 87.7% on slopes with more than a 50% gradient (Kladnik 1991).

A major landslide was triggered at 10 p.m. on the right bank of the Lučnica River, the right tributary of the Savinja in the Upper Savinja valley. The river runs in a narrow valley with steep slopes consisting of andesitic tuffs. The houses in Podvolovljek, Podveža, and Luče at the confluence with the Savinja River are situated along the river channel. The landslide was triggered 2 km from the confluence with the Savinja and dammed the Lučnica River channel. A 20 m high and 200 m wide dam was formed. Behind the dam a lake was formed which quickly filled up and flooded the lower part of the Podvolovljek village, including 4 houses. A 97-year old woman drowned because she was unable to escape from the quickly increasing water (Meze 1991). Civil protection services ordered evacuation of the threatened buildings above and below the dam where there was dam failure risk; flood wave design was also required. In the early morning hours a lake in a length of 1.5 km was formed, with approx. 2 million m^3 of water. On the 2nd November at 5:20 a.m. the water overtopped the weir and formed a flood wave of a height of 2 m and in a duration of 3–5 min, which demolished two houses downstream, in Luče. The flood wave caused by dam failure carried away two thirds of the landslide and thus substantially reduced the lake. The wave was then registered at the water station Nazarje on the Savinja at 10 a.m, with a height 80 cm (Natek 1991; Planki 1991). Except for Luče, where all inhabitants of the houses at risk were evacuated in time, the wave in the Savinja did not cause major damage because the water level was lower than the peak of the previous flood wave. A great problem for the functioning of the protection and rescue services was disconnected road connections and demolished bridges at the Lučnica and Savinja Rivers.

Stože 2000

Due to the tragic consequences and scope, the Stože landslide was well studied and documented in papers in professional and research journals. The landslide was caused by excessive precipitation at the end of the October 2000 and lasted throughout the following month (Mikoš et al. 2002). It rained 2–3 days each week. The event occurred in two parts. The first landslide was triggered on 15th November 2000 in the Stože area, at between 1400 and 1600 m above sea level. The Mangart stream channel was filled with material in a height of 10 m and a length of 1450 m (Majes, 2001). The landslide area is rather steep with a gradient of 30 degrees. The surface of the moraine consisted of clay and scree, while below it there was cracked and permeable dolomite lying on impermeable bedrock. As the Mangart stream was dammed it almost entirely infiltrated the landslide, which filled the stream channel but did not form a special reservoir space for water due to the large gradient of the channel. Because the stream stopped flowing, the Civil Defense Service raised the alarm and moved people out of the houses at risk in the village Log pod Mangartom. The stream started to flow again on 17th November, without causing any special land mass movements or increasing its erosion. This also led to the conclusion that landslide risk no longer existed and that inhabitants could return to their homes, even though the risk level was not called off (Ušeničnik 2001). The inhabitants returned to their homes. The heavy rains from 16th to 17th November triggered another landslide and debris flow a few minutes after midnight reached Log pod Mangartom and caused seven fatalities. The debris flow descended along the steep slope and in the valley filled the channel of the Koritnica with scree. The debris flow reached the village of Log pod Mangartom in 4–5 min and, according to the travel distance of 2200 m, the mass travelled with the speed of 30 km per hour, i.e. 8 m per second, so the population had no chance of a timely withdrawal (Fig. 8). Clay fractions of the moraine were carried away by the river, turbidity on the Soča River at Kobarid was measured at 8112 per m^3 (Komac 2001). The surface of the landslide comprised of 200,000 m^2, a mass of 1,500,000 m^3 was moved, and of which debris flow carried away 1,000,000 m^3 so that 1,500,000 m^3 of unstable material was left at the site (Majes 2001). In Log pod Mangartom 700,000 m^3 of material was deposited at a surface area of 150,000 m^2, and at least 100,000 m^3 was transported downstream as turbidity.

Fig. 8 Debris flow Stože has reached the village Log pod Mangartom (Ribičič 2001)

Laboratory investigations revealed very interesting characteristics of the moraine which composed the landslide mass and later the debris flow (Majes et al. 2002). The results of measurement of shear stress and moisture or suspension concentration show that the material with a relatively high shear stress in dry condition (up to a moisture of 20%) loses its moisture relatively quickly by the addition of water and liquefies already at 25% moisture. 5% of moisture will practically completely change material properties.

Discussion and Conclusions

People excessively and especially thoughtless affect riparian areas (e.g. by building homes or other important infrastructure too close to potential hotspots, and in the desire to find life easier violates or does not respect natural laws) and areas where there may be collection of water (ravines, hollows etc.).

In all three presented cases there are forest roads or skid trails that "served" as a secondary riverbed in the hinterland of the affected zones. In all three cases they run next to the stream and crossed it several times. The forest roads and skid trails were unmaintained and were made without proper reflection on its spatial placement and the construction has been carried out regardless on basic principles of drainage and headwaters arrangements. Being on unflattering and irrationally chosen routs and with lack of proper maintenance the diversion form the original channel was just a matter of time.

In the last 20–30 year in Slovenia the number of skid trails and forest roads has increased considerably. Unfortunately, the skid trails are poorly maintained. These skid trails now represent potential paths for water, which, of course, select the easiest way (places with lower potential energy). If the skid trails are not maintained properly, then the drainage does not work (if it was even arranged) because of the buried and blocked channels. Furthermore, if the trails are full of deep ruts, which are result of driving on the muddy road, then the water collects in ruts and direct it on the trails.

In many cases skid trails are located on potentially dangerous or unstable areas. Since the trails are leading from economic or residential buildings into the forest the water can bring material directly to infrastructure or bury transport infrastructure. And if that happens, then the damage is much greater than if it were to happen in an area where there are no human settlements and infrastructure. However, since this can not be completely avoided, we want with action *More Room for the Landslides* promote measures to reduce the risk against landslides, more specifically to reduce or even better avoid economic damage and loss of life.

References

Akay AE, Erdas O, Reis M, Yuksel A (2008) Estimating sediment yield from a forest road network by using a sediment prediction model and GIS techniques. Build Environ 43:687–695

Atlas okolja (2016) Environmental Atlas. http://gis.arso.gov.si/atlasokolja/profile.aspx?id=Atlas_Okolja_AXL@Arso. Last accessed: 1 July 2016

Brilly M (2015) More room for water. Hydrological sciences and water security: past, present and future. In: Proceedings of the 11th Kovacs Colloquium, Paris, France, June 2014. IAHS Publ. 366

EGC (2016) Evidenca gozdnih cest = Register of forest Roads. Slovenia forest service. URL: http://zgs.gisportal.si/javno/profile.aspx?id=EGC@ZGS [Last accessed: 1 July 2016]

Jusoff K, Majid NM (1996) The impacts of skid trails on the physical properties of Hill forest soils. Pertanika 9:311–321

Kambič A (2009) Poplave na pritokih Save v občini Dol pri Ljubljani – pregled škode in analiza vzrokov s predlogi. Zbornik 20. Mišičevega vodarskega dne, 26 November 2009. Maribor, Vodnogospodarski biro, pp 14–21

Khalipoor H, Hosseini SA, Lotfalian M, Kooch Y (2008) The assessment of sediment production yield from forest road using sediment production model. J Appl Sci 8:1944–1949

Kladnik D (1991) Ujma 1990 v Podvolovljeku. Ujma 5:51–53

Klijn F (2013) Design quality of room-for-the-river measures in the Netherlands: role and assessment of the quality team (Q-team). Int J River Basin Manag 11(3):287–299

Komac B (2001) Geografski vidiki nesreče = Geographical Aspects of the Disaster in Log pod Mangartom. Ujma 14–15:60–66

Majes B (2001) Analiza plazu in možnosti njegove sanacije = Analysis of landslide and its rehabilitation. Ujma 14–15:80–91

Majes B, Petkovšek A, Logar J (2002) Primerjava materialnih lastnosti drobirskih tokov iz plazov Stože, Slano blato in Strug. Geologija 45 (2):457–463

Meze D (1991) Ujma 1990 v Gornji Savinjski dolini, med Lučami in Mozirsko kotlinico = Effect of Flooding in Upper Savinja Valley between Luče and the Mozirje Basin. Ujma 5:39–50

Mikoš M, Vidmar A, Šraj M, Kobold M, Sušnik M, Uhan J, Pezdič J, Brilly M (2002) Hidrološke analize na plazu Stože pod Mangartom = Hydrologic analayses of the Stože landslide. Ujma 16:326–335

Natek K (1991) Plazovi v Gornji Savinjski dolini = Landslides in the Upper Savinja Valley. Ujma 5:62–65

Planki J (1991) Delovanje civilne zaščite ob poplavah v Lučah = Operation o. Ujma 58:133–134

Pristav J (1991) Razpored padavin in njihov vpliv na poplave = Distribution of Rainfall and its influence on Flood Waves). Ujma 5:10–15

Remic C (1971) Stanje mehanizacije v izkoriščanju gozdov SR Slovenije koncem leta 1970 = Condition of mechanization in forest exploitation in Socialist Republic of Slovenia at the end of 1970-Biotechnical faculty. Institute for forestry and wood economy of Slovenia, Business Association of forest management organizations, Ljubljana 26 pp

Ribičič M (2001) Značilnosti drobirskega toka Stože pod Mangartom = debris flow at Log pod Mangartom. Ujma 14–15:102–108

Ribičič M, Komac M, Mikoš M, Fajfar D, Ravnik D, Gvozdranovič T, Komel P, Miklavčič L, Fras M (2005) Final report of the project "Updating and upgrading of the landslide's information system and inclusion into the database GIS_Ujma" No. 145-KSH/d-87. University of Ljubljana, Faculty of Civil and Geodetic Engineering, pp 16

Robek R, Klun J, Vončina R (2006) Dosežki in izzivi pri graditvi gozdnih prometnic v Sloveniji = Achievements and challenges in forest traffic way construction in Slovenia. Gozdarski vestnik = Slovenian Forestry J 64(10):509–525

Slovenia forest service (2004) Data from the regional plans 2001 −2010. Slovenia forest service, CD

The Construction act (2002) Official Gazette of the Republic of Slovenia No. 102/04 with changes

Ušeničnik B (2001) Posledice in ukrepanje ob nesreči = Consequences of and responses to the Disaster. Ujma 14–15:67–80

World Water Council (2015) Ministerial declaration of 7th world water forum. Republic of Korea, Gyeongju

Zurnal24 (2009) Vojska na terenu. URL: http://www.zurnal24.si/vojska-na-terenu-clanek-2525. Last accessed: 1 July 2016

Landslide Technology and Engineering in Support of Landslide Science

Kyoji Sassa
Executive Director of the International Consortium on Landslides

The World Landslide Forum (WLF) is the triennial conference of the International Consortium on Landslides (ICL) and the International Programme on Landslides (IPL). The IPL is a programme of the International Consortium on Landslides, managed by ICL and its supporting organizations: UNESCO, WMO, FAO, UNISDR, UNU, ICSU, WFEO and IUGS.

IPL and WLF contribute to the United Nations International Strategy for Disaster Reduction. The World Landslide Forum provides an information and academic-exchange platform for landslide researchers and practitioners. It creates a triennial opportunity to promote worldwide cooperation and share new theories, technologies and methods in the fields of landslide survey/investigation, monitoring, early warning, prevention, and emergency management. The forum's purpose is to present achievements of landslide-risk reduction in promoting the sustainable development of society.

Advancements in landslide science and disaster-risk reduction are supported by developments in landslide technology and engineering. Here we invited ICL supporters who support the publication of the international full-color journal "Landslides: Journal of the International Consortium on Landslides", the companies advertising in the seven volumes of "Landslide Science and Practice: Proceedings of the Second World Landslide Forum 2011" and the companies exhibiting at the Third World Landslide Forum 2014 to introduce their landslide technology and engineering. Eight companies applied to exhibit in this book, their names, addresses, contact information and a brief introduction are given below:

1. MARUI & Co. Ltd.

1-9-17 Goryo, Daito City, Osaka 574-0064, Japan

URL: http://marui-group.co.jp/en/index.html

Contact: hp-mail@marui-group.co.jp

MARUI & Co. Ltd is the leading manufacturing and sales company in Japan since 1920 of material testing machines for soil, rock, concrete, cement and asphalt. Marui engineers built and assisted in development of the series of stress and speed control ring-shear apparatus by DPRI and ICL to study landslides since 1982. The latest versions of the undrained dynamic-loading ring shear apparatus are ICL-1 and ICL-2.

2. Okuyama Boring Co., Ltd.

10-39 Shimei-cho, Yokote City, Akita 013-0046, Japan

URL: http://www.okuyama.co.jp/

Contact: info@okuyama.co.jp

The Okuyama Boring Company Ltd specializes in landslide investigation, analysis of landslide mechanisms, and design of landslide remedial measures. The company uses its own software to analyze the initiation and motion of landslides, including the tsunami generated by landslides into reservoirs.

3. GODAI KAIHATSU Corporation

1-35 Kuroda, KANAZAWA-City, ISHIKAWA Pref. 921-8051, Japan

URL: http://www.godai.co.jp

Contact: pp-sale@godai.co.jp

Godai Kaihatsu Cooperation has developed a variety of software related to mitigate landslide disasters and social infrastructure, analysis, simulation, and monitoring. The company assisted in development of the landslide simulation software and the landslide monitoring and data transfer system by ICL.

4. Japan Conservation Engineers & Co., Ltd.

3-18-5 Toranomon, Minato-ku, Tokyo 105-0001, Japan

URL: http://www.jce.co.jp

Contact: hasegawa@jce.co.jp

Japan Conservation Engineers & Co, Ltd develops landslide-simulation software and shear-testing apparatus, including slip-surface direct-shear apparatus and ring-shear apparatus to measure the shear strength mobilized on the sliding surface of landslides. Japan Conservation Engineers is a consulting company for landslide investigation, reliable monitoring, data analysis and the design of landslide-risk reduction works.

K. Sassa et al. (eds.), *Advancing Culture of Living with Landslides*,
DOI 10.1007/978-3-319-59469-9

5. KOKUSAI KOGYO Co., Ltd.

2 Rokuban-cho, Chiyoda-ku, Tokyo 102-0085, Japan

URL: http://www.kk-grp.jp/english/

Contact: overseas@kk-grp.jp

Kokusai Kogyo has undertaken aerial surveys, infrastructure development projects for road and harbor facilities, and landslide-disaster prevention and mitigation works since its foundation in 1947. The company has recently developed remote-sensing technology using the laser profiler, satellite synthetic aperture radar, and a new monitoring system called <Shamen-net> integrating GPS and other monitoring devices, all of which contribute to landslide-disaster prevention and mitigation.

6. OSASI Technos, Inc.

65-3 Hongu-cho, Kochi City, Kochi 780-0945, Japan

URL: http://www.osasi.co.jp/en/

Contact: info-tokyo@osasi.co.jp

OSASI Technos, Inc. develops and markets the slope disaster monitoring system called OSASI Network System (OSNET). The monitoring devices use a built-in lithium battery and operate without external electricity supply in mountainous areas. The system enables a network of up to 64 units with up to 1 km distance between units. OSNET is suitable for quickly establishing monitoring systems on landslides in emergencies.

7. OYO Corporation

7 Kanda-Mitoshiro-cho, Chiyoda-ku, Tokyo 101-8486 Japan

URL: https://www.oyo.co.jp/english/

Contact: Seihin@oyo.jp, https://www.oyo.co.jp/english/contacts/

OYO Corporation was established in 1957, located in Tokyo, Japan. The Corporation provides engineering works regarding investigation, analyses, projection, diagnoses, evaluation for landslides, slope failure, earthquake disasters, and storm and flood damage. The company has developed and sales remote monitoring system. The company assists the stability assessment, risk management and the planning of countermeasures.

8. PROTEC ENGINEERING, INC.

5322-26, Oaza Hasugata, Seiro-machi, Kitakanbara-gun, Niigata 957-0106, Japan

URL: http://www.proteng.co.jp/en/

Contact: info@proteng.co.jp

PROTEC ENGINEERING have been developing several kind of products which protect human life and properties against natural disasters, mainly focusing on rock fall, slope failure and snow avalanche. The products include GEO ROCK WALL, ARC FENCE, SLOPE GUARD FENCE, and debris flow barriers.

Full-color presentations from these eight exhibitors focusing on their landslide technology are shown in the following pages. The progress of landslide science is supported by advances in landslide technology. The success of landslide risk-reduction measures needs effective landslide engineering. The International Consortium on Landslides seeks expressions of interest in contributing to "Landslide Technology and Engineering to Support Landslide Science" at the Fifth World Landslide Forum to be held on 2–6 November 2020, in Kyoto, Japan. We may call for presentations on landslide technology and engineering in the proceedings, as well as through exhibitions at the site. Those interested in this initiative are requested to contact the Secretariat of the International Consortium on Landslides <secretariat@iclhq.org>. We will send invitations to interested applicants when further details become available.

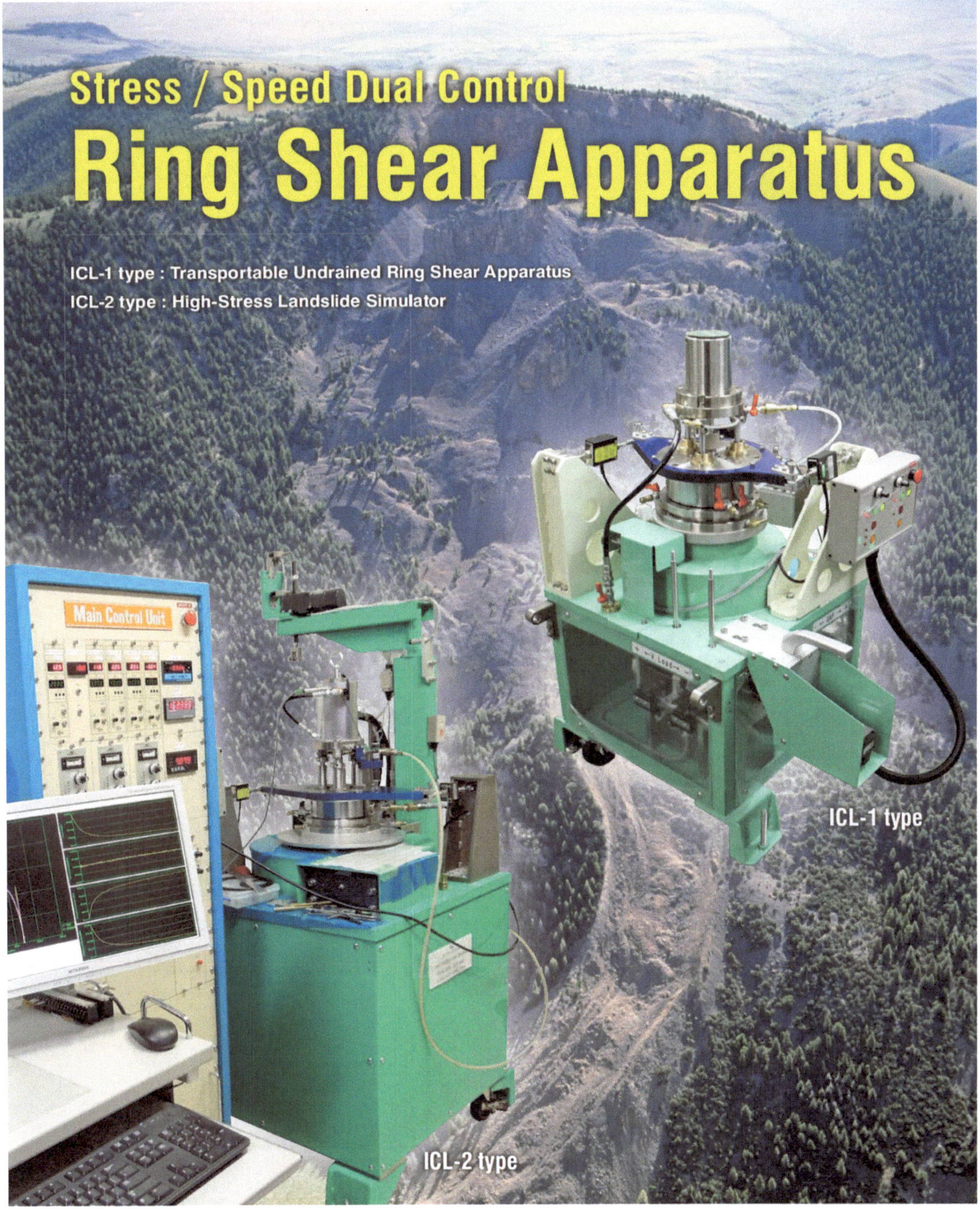

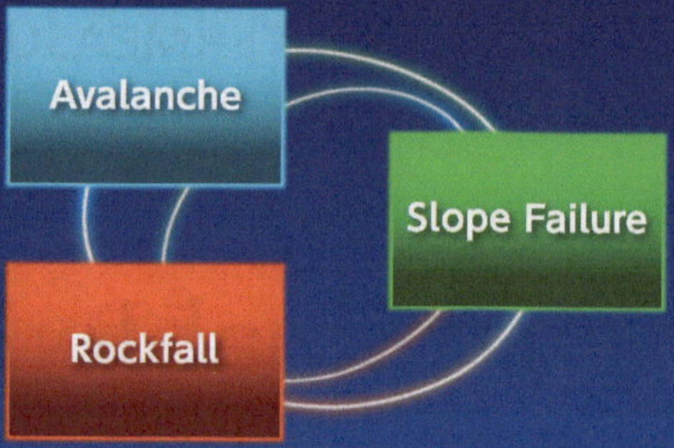

International Consortium on Landslides

An international non-government and non-profit scientific organization promoting landslide research and capacity building for the benefit of society and the environment

President: Yueping Yin (China Geological Survey)

Vice Presidents: Irasema Alcantara-Ayara (UNAM), Mexico, Matjaz Mikos (University of Ljubljana), Slovenia Dwikorita Karnawati (Gadjah Mada University, Indonesia)

Executive Director: Kyoji Sassa (Prof. Emeritus, Kyoto University, Japan), Treasurer: Kaoru Takara (Kyoto University, Japan)

ICL Supporting Organizations:

The United Nations Educational, Scientific and Cultural Organization (UNESCO) / The World Meteorological Organization (WMO) / The Food and Agriculture Organization of the United Nations (FAO) / The United Nations International Strategy for Disaster Reduction Secretariat (UNISDR) / The United Nations University (UNU) / International Council for Science (ICSU) / World Federation of Engineering Organizations (WFEO) / International Union of Geological Sciences (IUGS) / International Union of Geodesy and Geophysics (IUGG) / Government of Japan

ICL Members:

Albania Geological Survey / The Geotechnical Society of Bosnia and Herzegovina / Geological Survey of Canada / Chinese Academy of Sciences, Institute of Mountain Hazards and Environment / Northeast Forestry University, Institute of Cold Regions Science and Engineering, China / China Geological Survey / Nanjing Institute of Geography and Limnology, Chinese Academy of Sciences / Tongji University, College of Surveying and Geo-Informatics, China / Universidad Nacional de Columbia, Columbia / City of Zagreb, Emergency Management Office, Croatia / Croatian Landslide Group (Faculty of Civil Engineering, University of Rijeka and Faculty of Mining, Geology and Petroleum Engineering, University of Zagreb) / Charles University, Faculty of Science, Czech Republic / Institute of Rock Structure and Mechanics, Department of Engineering Geology, Czech Republic / Cairo University, Egypt / Joint Research Centre (JRC), European Commission / Technische Universitat Darmstadt, Institute and Laboratory of Geotechnics, Germany / National Environmental Agency, Department of Geology, Georgia / Universidad Nacional Autónoma de Honduras, UNAH, Honduras / Amrita Vishwa Vidyapeetham, Amrita University / National Institute of Disaster Management, India / University of Gadjah Mada, Indonesia / Parahyangan Catholic University, Indonesia / Research Center for Geotechnology, Indonesian Institute of Sciences, Indonesia / Building & Housing Research Center, Iran / University of Calabria, DIMES, CAMILAB, Italy / University of Firenze, Earth Sciences Department, Italy / Istituto de Ricerca per la Protezione Idrogeologica (IRPI), CNR, Italy / Italian Institute for Environmental Protection and Research (ISPRA) - Dept. Geological Survey, Italy / Forestry and Forest Product Research Institute, Japan / Japan Landslide Society / Kyoto University, Disaster Prevention Research Institute, Japan / Korea Forest Research Institute, Korea / Korea Infrastructure Safety & Technology Corporation, Korea / Korea Institute of Civil Engineering and Building Technology / Korea Institute of Geoscience and Mineral Resources (KIGAM) / Korean Society of Forest Engineering / Slope Engineering Branch, Public Works Department of Malaysia / Institute of Geography, National Autonomous University of Mexico (UNAM) / International Centre for Integrated Mountain Development (ICIMOD), Nepal / University of Nigeria, Department of Geology, Nigeria / International Centre for Geohazards (ICG) in Oslo, Norway /Grudec Ayar, Peru / Moscow State University, Department of Engineering and Ecological Geology, Russia / JSC "Hydroproject Institute", Russia / University of Belgrade, Faculty of Mining and Geology, Serbia / Comenius University, Faculty of Natural Sciences, Department of Engineering Geology, Slovakia / University of Ljubljana, Faculty of Civil and Geodetic Engineering (UL FGG), Slovenia / University of Ljubljana, Faculty of Natural Sciences and Engineering (UL NTF), Slovenia / Geological Survey of Slovenia / Central Engineering Consultancy Bureau (CECB), Sri Lanka / National Building Research National Organization, Sri Lanka /Taiwan University, Department of Civil Engineering, Chinese Taipei / Landslide group in National Central University from Graduate Institute of Applied Geology, Department of Civil Engineering, Center for Environmental Studies, Chinese Taipei/ Asian Disaster Preparedness Center, Thailand / Ministry of Agriculture and Cooperative, Land Development Department, Thailand / Institute of Telecommunication and Global Information Space, Ukraine / California State University, Fullerton & Tribhuvan University, Institute of Engineering, USA & Nepal / Institute of Transport Science and Technology, Vietnam / Vietnam Institute of Geosciences and Mineral Resources (VIGMR), Vietnam

ICL Supporters:

Marui & Co., Ltd., Osaka, Japan / Okuyama Boring Co., Ltd., Yokote, Japan / GODAI Development Corp., Kanazawa, Japan / Japan Conservation Engineers & Co., Ltd, Tokyo / Kokusai Kogyo Co., Ltd., Tokyo, Japan / Ohta Geo-Research Co., Ltd., Nishinomiya, Japan / OSASI Technos Inc., Kochi, Japan / OYO Corporation, Tokyo, Japan / Sabo Technical Center, Tokyo, Japan / Sakata Denki Co., Ltd., Tokyo, Japan

Contact:

International Consortium on Landslides, 138-1 Tanaka Asukai-cho, Sakyo-ku, Kyoto 606-8226, Japan
Web: http://icl.iplhq.org/, E-mail: secretariat@iclhq.org
Tel: +81-774-38-4834, +81-75-723-0640, Fax: +81-774-38-4019, +81-75-950-0910

Author Index

K. Sassa et al. (eds.), *Advancing Culture of Living with Landslides*,
DOI 10.1007/978-3-319-59469-9